Biosensors

Biosensors—Recent Advances and Future Challenges

Editors

Paolo Bollella
Evgeny Katz

MDPI • Basel • Beijing • Wuhan • Barcelona • Belgrade • Manchester • Tokyo • Cluj • Tianjin

Editors
Paolo Bollella
Department of Chemistry &
Biomolecular Science,
Clarkson University
USA

Evgeny Katz
Department of Chemistry and
Biomolecular Science,
Clarkson University
USA

Editorial Office
MDPI
St. Alban-Anlage 66
4052 Basel, Switzerland

This is a reprint of articles from the Special Issue published online in the open access journal *Sensors* (ISSN 1424-8220) (available at: https://www.mdpi.com/journal/sensors/special_issues/Biosensors_Recent_Advances_Future_Challenges).

For citation purposes, cite each article independently as indicated on the article page online and as indicated below:

LastName, A.A.; LastName, B.B.; LastName, C.C. Article Title. *Journal Name* **Year**, *Volume Number*, Page Range.

ISBN 978-3-03943-887-7 (Hbk)
ISBN 978-3-03943-888-4 (PDF)

Cover image courtesy of Wikipedia Public Domain.

© 2020 by the authors. Articles in this book are Open Access and distributed under the Creative Commons Attribution (CC BY) license, which allows users to download, copy and build upon published articles, as long as the author and publisher are properly credited, which ensures maximum dissemination and a wider impact of our publications.
The book as a whole is distributed by MDPI under the terms and conditions of the Creative Commons license CC BY-NC-ND.

Contents

About the Editors vii

Paolo Bollella and Evgeny Katz
Biosensors—Recent Advances and Future Challenges
Reprinted from: *Sensors* **2020**, *20*, 6645, doi:10.3390/s20226645 1

Pankaj Vadgama
Monitoring with In Vivo Electrochemical Sensors: Navigating the Complexities of Blood and Tissue Reactivity
Reprinted from: *Sensors* **2020**, *20*, 3149, doi:10.3390/s20113149 7

Mahshid Padash, Christian Enz and Sandro Carrara
Microfluidics by Additive Manufacturing for Wearable Biosensors: A Review
Reprinted from: *Sensors* **2020**, *20*, 4236, doi:10.3390/s20154236 37

Magnus Falk, Carolin Psotta, Stefan Cirovic and Sergey Shleev
Non-Invasive Electrochemical Biosensors Operating in Human Physiological Fluids
Reprinted from: *Sensors* **2020**, *20*, 6352, doi:10.3390/s20216352 65

Leif K. McGoldrick and Jan Halámek
Recent Advances in Noninvasive Biosensors for Forensics, Biometrics, and Cybersecurity
Reprinted from: *Sensors* **2020**, *20*, 5974, doi:10.3390/s20215974 93

Taiki Adachi, Yuki Kitazumi, Osamu Shirai and Kenji Kano
Development Perspective of Bioelectrocatalysis-Based Biosensors
Reprinted from: *Sensors* **2020**, *20*, 4826, doi:10.3390/s20174826 109

Paolo Bollella and Evgeny Katz
Enzyme-Based Biosensors: Tackling Electron Transfer Issues
Reprinted from: *Sensors* **2020**, *20*, 3517, doi:10.3390/s20123517 131

Julia Alvarez-Malmagro, Gabriel García-Molina and Antonio López De Lacey
Electrochemical Biosensors Based on Membrane-Bound Enzymes in Biomimetic Configurations
Reprinted from: *Sensors* **2020**, *20*, 3393, doi:10.3390/s20123393 163

Melisa del Barrio, Gabriel Luna-López and Marcos Pita
Enhancement of Biosensors by Implementing Photoelectrochemical Processes
Reprinted from: *Sensors* **2020**, *20*, 3281, doi:10.3390/s20113281 181

Bettina Neumann and Ulla Wollenberger
Electrochemical Biosensors Employing Natural and Artificial Heme Peroxidases on Semiconductors
Reprinted from: *Sensors* **2020**, *20*, 3692, doi:10.3390/s20133692 203

Susana Campuzano, María Pedrero, Maria Gamella, Verónica Serafín, Paloma Yáñez-Sedeño and José Manuel Pingarrón
Beyond Sensitive and Selective Electrochemical Biosensors: Towards Continuous, Real-Time, Antibiofouling and Calibration-Free Devices
Reprinted from: *Sensors* **2020**, *20*, 3376, doi:10.3390/s20123376 227

Fernando Otero and Edmond Magner
Biosensors—Recent Advances and Future Challenges in Electrode Materials
Reprinted from: *Sensors* **2020**, *20*, 3561, doi:10.3390/s20123561 . . . . **249**

Chunmei Li, Yihan Wang, Hui Jiang and Xuemei Wang
Biosensors Based on Advanced Sulfur-Containing Nanomaterials
Reprinted from: *Sensors* **2020**, *20*, 3488, doi:10.3390/s20123488 . . . . **267**

Reem Khan and Silvana Andreescu
MXenes-Based Bioanalytical Sensors: Design, Characterization, and Applications
Reprinted from: *Sensors* **2020**, *20*, 5434, doi:10.3390/s20185434 . . . . **295**

Nataliya Stasyuk, Oleh Smutok, Olha Demkiv, Tetiana Prokopiv, Galina Gayda, Marina Nisnevitch and Mykhailo Gonchar
Synthesis, Catalytic Properties and Application in Biosensorics of Nanozymes and Electronanocatalysts: A Review
Reprinted from: *Sensors* **2020**, *20*, 4509, doi:10.3390/s20164509 . . . . **315**

Mohamed Sharafeldin, Karteek Kadimisetty, Ketki S. Bhalerao, Tianqi Chen and James F. Rusling
3D-Printed Immunosensor Arrays for Cancer Diagnostics
Reprinted from: *Sensors* **2020**, *20*, 4514, doi:10.3390/s20164514 . . . . **357**

Manikandan Santhanam, Itay Algov and Lital Alfonta
DNA/RNA Electrochemical Biosensing Devices a Future Replacement of PCR Methods for a Fast Epidemic Containment
Reprinted from: *Sensors* **2020**, *20*, 4648, doi:10.3390/s20164648 . . . . **381**

Maria Smith, Kenneth Smith, Alan Olstein, Andrew Oleinikov and Andrey Ghindilis
Restriction Endonuclease-Based Assays for DNA Detection and Isothermal Exponential Signal Amplification
Reprinted from: *Sensors* **2020**, *20*, 3873, doi:10.3390/s20143873 . . . . **397**

Rimsha Binte Jamal, Stepan Shipovskov and Elena E. Ferapontova
Electrochemical Immuno- and Aptamer-Based Assays for Bacteria: Pros and Cons over Traditional Detection Schemes
Reprinted from: *Sensors* **2020**, *20*, 5561, doi:10.3390/s20195561 . . . . **411**

Arshak Poghossian and Michael J. Schöning
Capacitive Field-Effect EIS Chemical Sensors and Biosensors: A Status Report
Reprinted from: *Sensors* **2020**, *20*, 5639, doi:10.3390/s20195639 . . . . **439**

Mohammed Sedki, Ying Chen and Ashok Mulchandani
Non-Carbon 2D Materials-Based Field-Effect Transistor Biosensors: Recent Advances, Challenges, and Future Perspectives
Reprinted from: *Sensors* **2020**, *20*, 4811, doi:10.3390/s20174811 . . . . **471**

About the Editors

Paolo Bollella is a Research Assistant Professor at Clarkson University in the Department of Chemistry and Biomolecular Science since September 2019. He received his PhD in Pharmaceutical Science under the supervision of Prof. Riccarda Antiochia in the Department of Chemistry and Drug Technologies at Sapienza University of Rome. During the PhD studies, he joined to the bioelectrochemistry laboratory at Lund University (Lund, Sweden) led by Prof. Lo Gorton. After his PhD, he was involved in a collaboration with Prof. Anthony E.G. Cass (Imperial College) about the modification of microneedle electrode array for non-invasive detection of different biomarkers (e.g., glucose, lactate etc.). In August 2018, he joined the Department of Analytical Chemistry at Åbo Akademi in Turku (Finland) with a Johan Gadolin PostDoc fellowship awarded from the board of Johan Gadolin Process Chemistry Center. In October 2018, he joined the group of "Bioelectronics & Bionanotechnology" led by Prof. Evgeny Katz. In March 2019, he was awarded with the Minerva Prize for the Scientific Research – Merit Mention for the achievements obtained during his doctoral studies. He was also recipient of the 2020 MDPI Sensors Travel Award. He is author of 56 papers on peer-reviewed international journals, 3 book chapters, 1 student book, 2 proceedings and almost 60 oral or poster contributions to national and international conferences.

Evgeny Katz received his Ph.D. in Chemistry from Frumkin Institute of Electrochemistry (Moscow), Russian Academy of Sciences, in 1983. He was a senior researcher in the Institute of Photosynthesis (Pushchino), Russian Academy of Sciences, in 1983-1991. In 1992-1993 he performed research at München Technische Universität (Germany) as a Humboldt fellow. Later, in 1993–2006, Dr. Katz was a Research Associate Professor at the Hebrew University of Jerusalem. From 2006 he is Milton Kerker Chaired Professor at the Department of Chemistry and Biomolecular Science, Clarkson University, NY (USA). He has (co)authored over 480 papers in peer-reviewed journals/books with the total citation more than 35,000 (Hirsch-index 90) and holds more than 20 international patents. He edited five books on different topics, including bioelectronics, molecular and biomolecular computing, implantable bioelectronics and forensic science. Two books fully written by him on switchable electrochemical systems and enzyme-based computing were published recently. He was an Editor-in-Chief for IEEE Sensors Journal (2009–2012) and he is a member of editorial boards of many other journals. His scientific interests are in the broad areas of bioelectronics, biosensors, biofuel cells and biomolecular information processing (biocomputing). In 2019 he received international Katsumi Niki Prize for his contribution to bioelectrochemistry.

Editorial

Biosensors—Recent Advances and Future Challenges

Paolo Bollella * and Evgeny Katz *

Department of Chemistry and Biomolecular Science, Clarkson University, Potsdam, NY 13699-5810, USA
* Correspondence: pbollell@clarkson.edu (P.B.); ekatz@clarkson.edu (E.K.)

Received: 11 November 2020; Accepted: 13 November 2020; Published: 20 November 2020

Biosensors are analytical devices that are able to convert a biological response into an electrical signal. The "golden" biosensor must be highly specific, independent of physical parameters (e.g., pH, temperature, etc.), and should be reusable. The research within the biosensing field requires a multidisciplinary approach that involves different branches of science such as chemistry, biology, and engineering. Biosensors can be categorized based on the biorecognition mechanism: with the biocatalytic group comprising enzymes, the bio-affinity group including antibodies and nucleic acids, and the microbe-based group containing microorganisms. The present Special Issue aimed at summarizing the most recent findings and future challenges regarding biosensors.

In the last six decades, several biosensors have been reported as end-user and time-saving analytical methods for the detection of multiple analytes (e.g., food, clinical, and environmental analytes). In 1962, Professor Leland C. Clark published the first example of an enzyme electrochemical biosensor by entrapping glucose oxidase in a dialysis membrane over a Clark-type oxygen electrode [1]. Moreover, Guilbault and Montalvo reported on glass electrodes coupled with urease to measure urea concentration by means of potentiometry [2]. Besides these first examples, electrochemical transducers have been combined with enzymes, antibodies, and DNA as biochemical recognition components. Nowadays, they represent the largest category of biosensors for food, clinical, and environmental sensing.

The increasing number of scientific publications focusing on biosensors indicates growing interest in the broader scientific community (Figure 1). The present collection of the papers is devoted to all aspects of biosensing in a very broad definition, including, but not limited to, biomolecular composition used in biosensors (e.g., biocatalytic enzymes, DNAzymes, abiotic nanospecies with biocatalytic features, bioreceptors, DNA/RNA, aptasensors, etc.), physical signal transduction mechanisms (e.g., electrochemical, optical, magnetic, etc.), engineering of different biosensing platforms, operation of biosensors in vitro and in vivo (implantable or wearable devices), self-powered biosensors, etc. The biosensors can be represented with analogue devices measuring concentrations of analytes and binary devices operating in the YES/NO format, possibly with logical processing of input signals.

In this collection, we combined twenty outstanding contributions focusing on different aspects of the biosensing field, mostly highlighting recent advances and future challenges of DNA detection, immunosensing, in vivo electrochemical biosensors, redox enzyme-modified electrode surfaces, photoelectrochemical processes, field-effect transistor-based biosensors, etc., which can be considered as biosensing sub-topics, as reported in Figure 2. A brief summary of each accepted contribution is provided below to encourage the readers to go through them and "visualize" the state of the art within the field of biosensing.

Among the big question marks in biosensing development, Vadgama certainly addressed one of the main challenges regarding continuous and in vivo monitoring in complex media like blood or human tissues. Based on recent findings, electrochemical sensors offer one of the few routes to obtain continuous read-out and implantable devices information referable to specific tissue locations [3]. In this regard, wearable devices are at the forefront in both academic and industrial research on biosensors. The main advantage for wearable technologies is the remote monitoring of human health by biomarkers detection on the skin (e.g., continuous glucose self-monitoring in diabetic patients). Nowadays, the minimally invasive collection of the sample implies the integration of wearable biosensor platforms with microfluidic systems that allow information from the sample to be transmitted directly from the skin to the electrode surface [4,5]. Moreover, the continuous and minimally invasive monitoring of biomarkers has also become of fundamental importance in forensic, biometric, and cybersecurity fields. McGoldrick et al. [6] reported on the possibility of using different bodily fluids for metabolite analysis. This provides an alternative to the use of DNA in order to avoid the backlog that is currently the main issue with DNA analysis by providing worthwhile information about the originator.

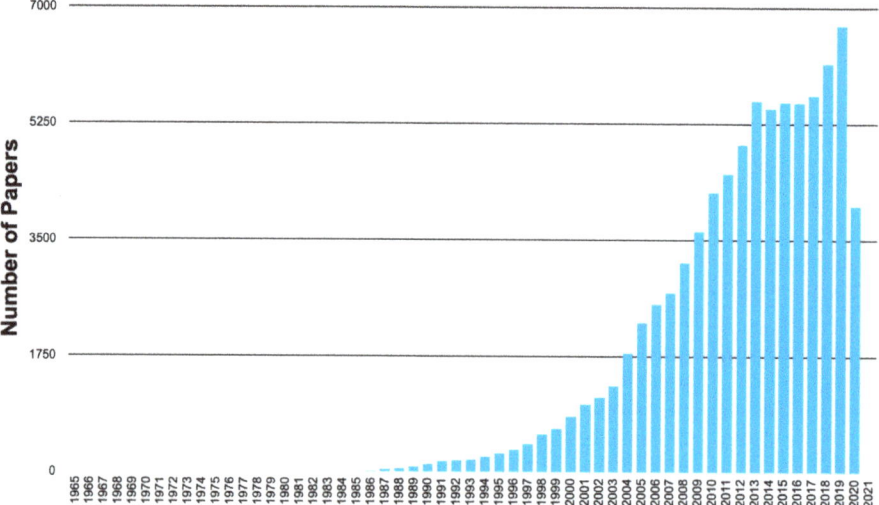

Figure 1. The number of published papers mentioning "biosensors" derived from statistics provided by the Web of Science. The search was performed for the keyword "biosensors" in the topic. (The statistics for 2020 was not complete).

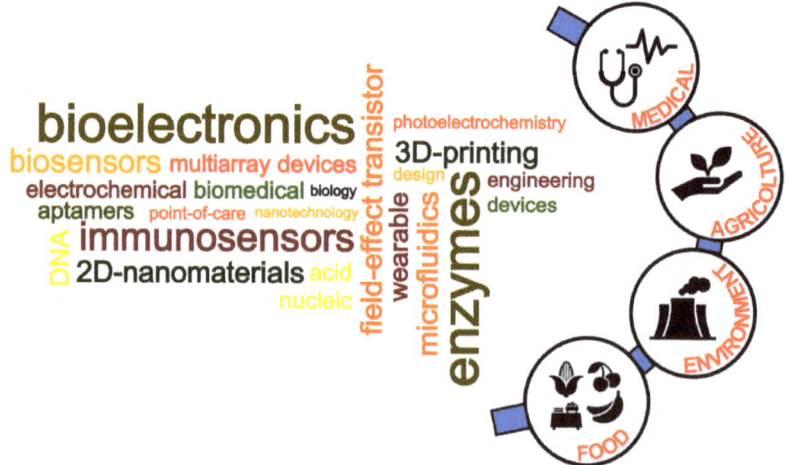

Figure 2. Biosensing topics and sub-topics and possible applications as analytical devices.

Despite the efforts of the scientific community towards the development of minimally invasive and wearable electrochemical biosensors for continuous and in vivo self-monitoring, the fundamental theory behind electrochemical biosensors development still remains a landmark. In particular for enzymes-based biosensors, most bioelectrochemists have focused their attention on possible solutions to tackle direct electron transfer (DET) issues, which are important for enhancing the selectivity and sensitivity of biosensors [7]. Particular attention has been devoted to the case of glucose oxidase (GOx). Despite the huge number of publications on this subject, which unfortunately account for thousands of citations, there is no solid evidence to support DET in GOx, as demonstrated by a stunning statement made by George Wilson: "based on recent experimental results, the observed electrochemical signal corresponds to the FAD cofactor non-covalently bound to the enzyme scaffold that comes out from the redox enzyme upon application of potential, getting adsorbed onto the electrode surface" [8].

Beyond the use of GOx as a redox enzyme, there are several enzymes that are able to transfer electrons according to direct or mediated pathways. In nature, many enzymes are attached or inserted into a cell membrane, having hydrophobic subunits or lipid chains for this purpose. Their reconstitution on electrodes allows them to maintain their natural structural characteristics and enables the optimization of their electrocatalytic properties and stability. In this regard, Alvarez-Malmagro et al. [9] discussed different biomimetic strategies to modify electrode surfaces in order to accommodate membrane-bound enzymes, including the formation of self-assembled monolayers of hydrophobic compounds, lipid bilayers, or liposome deposition.

Besides the "classical" enzymes-based biosensors, in the last two decades, many enzymes have been coupled with semiconductive electrodes containing a light-harvesting material in order to develop photoelectrochemical sensing devices. Del Barrio et al. [10] reported on the integration of nanomaterials, such as quantum dots and titanium oxide (TiO_2) nanoparticles with redox enzymes (e.g., acetylcholinesterase (AChE), glucose oxidase (GOx), etc.), in order to enhance device sensitivity. Considering the successful results in this specific field, future research trends will certainly involve the investigation of different combinations of semiconductor materials and biomolecules and will also consider the possibility of tuning the wavelength to develop a multi-analyte photoelectrochemical biosensor. In particular, Neumann et al. [11] reported on the possibility of combining artificial and natural heme peroxidases with semiconductive electrodes in order to offer new read-out possibilities for hydrogen peroxide and phenolic compounds detection.

Moreover, the continuous and renown efforts toward the development of nanomaterial-modified electrodes represent another aspect that has been deeply disclosed in the present collection. In particular, Campuzano et al. [12] covered the topic of the modification of electrode surfaces with antibiofouling reagents, which will eventually prevent the non-specific adsorption of biological species on the electrode surface. This is an important topic, especially considering the research on multiplexed and point-of-care devices as cost-effective and selective multianalyte detection methods. Among all the strategies currently available to develop antibiofouling surfaces, the modification of electrode substrates with different biomaterials, including monolayers, transient polymeric coatings, or multifunctional peptides, is particularly attractive and promising.

In this collection, the use of structured materials, such as nanoporous metals, graphene, carbon nanotubes, and ordered mesoporous carbon, for biosensing applications has been deeply discussed [13].

Recently, sulfur-containing nanomaterials and their derivatives/composites have been extensively employed for the development of alternative biosensing devices. Li et al. [14] summarized the recent findings and future challenges of employing metallic sulfide nanomaterial-modified electrodes, particularly disclosing their specific properties, namely, nanometric scale, water dispersibility, large specific surface area, excellent catalytic activity, conductivity, biosafety, photoluminescence (PL) quenching abilities, photoactivity, and fascinating optical properties.

Beyond graphene and graphene-like-2D-nanomaterials (e.g., sulfur-containing nanomaterials etc.), Khan et al. [15] reported on the possibility of exploiting MXenes as 2D-layered nanomaterials that provide unique capabilities for bioanalytical applications. These include high metallic conductivity, large surface area, hydrophilicity, high ion transport properties, low diffusion barrier, biocompatibility, and ease of surface functionalization.

Considering special features of nanomaterials, Stasyuk et al. [16] summarized the recent findings about nanozymes. Nanozymes are defined as nanomaterials with enzyme-biomimicking features (e.g., gold nanoparticles that mimic oxidases activity, etc.). This contribution gives an overview of the classification of the nanozymes, their advantages vs. natural enzymes, and their potential practical applications, devoting particular attention to the different synthesis methods developed so far.

Beyond enzyme-based biosensors, immunosensors are also used for the development of point-of-care devices. In particular, Sharafeldin et al. [17] reviewed the most recent findings on 3D-printed immunosensing devices for cancer detection. In the last few years, 3D-printing platforms have been used to produce complex sensor devices with high resolution.

Moreover, aptasensors and DNA-modified electrodes have also been identified as point-of-care devices that are especially useful for quick diagnostics during pandemic emergencies. Santhanam et al. [18] summarized the most recent findings about DNA/RNA-based biosensors, especially considering classical detection method pitfalls, such as for reverse transcription PCR (RT-PCR) and real-time PCR (qPCR), which are considered time-consuming and require specialized professionals and instrumentation.

On this specific topic, researchers are not focused only on the development of new detection platforms, but they are also addressing potential issues about biosensor sensitivity through different signal amplification methods. Smith et al. [19] reported on recent findings and future challenges surrounding DNA detection based on a direct restriction endonuclease (REase) assay. This assay allows for detection at an attomolar level through an exponential signal amplification method based on a cascade of self-perpetuating restriction endonuclease reactions, which induce continuous cleavage of amplification probes, thus leading to exponential signal amplification. The proposed approach provides a cost-, time-, and labor-effective alternative DNA detection method.

Besides the detection of DNA or antigens, immunosensors, DNA/RNA biosensors, and aptasensors are currently considered in microbiology as powerful tools for the detection of bacteria cells at a single cell level. These biosensors allow for the specific detection of bacteria in complex biological matrices, often in the presence of excessive amounts of other bacterial species [20].

The research within the biosensing field is not only focused on electrochemical and optical transduction techniques but is also currently considering different approaches to obtain a direct electronic read-out, like for electrolyte-insulator-semiconductor (EIS) field-effect sensors, which belong to a new generation of electronic chips. Poghossian and Schöning [21] gave an overview on recent advances and current trends in the research and development of chemical sensors and biosensors based on the capacitive field-effect EIS structure—the simplest field-effect device, which represents a biochemically sensitive capacitor. Similarly, Sedki et al. [22] reported the most recent findings on non-carbon 2D-materials-FET biosensors, discussing how transition metal dichalcogenides (TMDCs), hexagonal boron nitride (h-BN), black phosphorus (BP), and metal oxides impacted the development of the FET-based biosensors.

Acknowledgments: We would like to acknowledge all the authors for their valuable contributions in making this collection successful. Further, we are thankful to the *Sensors* editorial team for their continuous cooperation throughout this work. Lastly, we are grateful to the anonymous reviewers for their valuable input, comments, and suggestions for the submitted papers.

Conflicts of Interest: The authors declare no conflict of interest.

References

1. Clark, L.C., Jr.; Lyons, C. Electrode systems for continuous monitoring in cardiovascular surgery. *Ann. N. Y. Acad. Sci.* **1962**, *102*, 29–45. [CrossRef] [PubMed]
2. Guilbault, G.G.; Smith, R.K.; Montalvo, J.G. Use of ion selective electrodes in enzymic analysis. Cation electrodes for deaminase enzyme systems. *Anal. Chem.* **1969**, *41*, 600–605. [CrossRef]

3. Vadgama, P. Monitoring with In Vivo Electrochemical Sensors: Navigating the Complexities of Blood and Tissue Reactivity. *Sensors* **2020**, *20*, 3149. [CrossRef] [PubMed]
4. Padash, M.; Enz, C.; Carrara, S. Microfluidics by Additive Manufacturing for Wearable Biosensors: A Review. *Sensors* **2020**, *20*, 4236. [CrossRef] [PubMed]
5. Falk, M.; Psotta, C.; Cirocovic, S.; Shleev, S. Non-Invasive Electrochemical Biosensors Operating in Human Physiological Fluids. *Sensors* **2020**, *20*, 6352. [CrossRef] [PubMed]
6. McGoldrick, L.K.; Halámek, J. Recent Advances in Noninvasive Biosensors for Forensics, Biometrics, and Cybersecurity. *Sensors* **2020**, *20*, 5974. [CrossRef]
7. Adachi, T.; Kitazumi, Y.; Shirai, O.; Kano, K. Development Perspective of bioelectrocatalysis-based biosensors. *Sensors* **2020**, *20*, 4826. [CrossRef]
8. Bollella, P.; Katz, E. Enzyme-Based Biosensors: Tackling Electron Transfer Issues. *Sensors* **2020**, *20*, 3517. [CrossRef]
9. Alvarez-Malmagro, J.; García-Molina, G.; López De Lacey, A. Electrochemical Biosensors Based on Membrane-Bound Enzymes in Biomimetic Configurations. *Sensors* **2020**, *20*, 3393. [CrossRef]
10. Del Barrio, M.; Luna-López, G.; Pita, M. Enhancement of Biosensors by Implementing Photoelectrochemical Processes. *Sensors* **2020**, *20*, 3281. [CrossRef]
11. Neumann, B.; Wollenberger, U. Electrochemical Biosensors Employing Natural and Artificial Heme Peroxidases on Semiconductors. *Sensors* **2020**, *20*, 3692. [CrossRef] [PubMed]
12. Campuzano, S.; Pedrero, M.; Gamella, M.; Serafín, V.; Yáñez-Sedeño, P.; Pingarrón, J.M. Beyond sensitive and selective electrochemical biosensors: Towards continuous, real-time, antibiofouling and calibration-free devices. *Sensors* **2020**, *20*, 3376. [CrossRef] [PubMed]
13. Otero, F.; Magner, E. Biosensors—Recent Advances and Future Challenges in Electrode Materials. *Sensors* **2020**, *20*, 3561. [CrossRef] [PubMed]
14. Li, C.; Wang, Y.; Jiang, H.; Wang, X. Biosensors Based on Advanced Sulfur-Containing Nanomaterials. *Sensors* **2020**, *20*, 3488. [CrossRef] [PubMed]
15. Khan, R.; Andreescu, S. MXenes-Based Bioanalytical Sensors: Design, Characterization, and Applications. *Sensors* **2020**, *20*, 5434. [CrossRef] [PubMed]
16. Stasyuk, N.; Smutok, O.; Demkiv, O.; Prokopiv, T.; Gayda, G.; Nisnevitch, M.; Gonchar, M. Synthesis, Catalytic Properties and Application in Biosensorics of Nanozymes and Electronanocatalysts: A Review. *Sensors* **2020**, *20*, 4509. [CrossRef] [PubMed]
17. Sharafeldin, M.; Kadimisetty, K.; Bhalerao, K.S.; Chen, T.; Rusling, J.F. 3D-printed Immunosensor arrays for cancer diagnostics. *Sensors* **2020**, *20*, 4514. [CrossRef]
18. Santhanam, M.; Algov, I.; Alfonta, L. DNA/RNA Electrochemical Biosensing Devices a Future Replacement of PCR Methods for a Fast Epidemic Containment. *Sensors* **2020**, *20*, 4648. [CrossRef]
19. Smith, M.; Smith, K.; Olstein, A.; Oleinikov, A.; Ghindilis, A. Restriction Endonuclease-Based Assays for DNA Detection and Isothermal Exponential Signal Amplification. *Sensors* **2020**, *20*, 3873. [CrossRef]
20. Jamal, R.B.; Shipovskov, S.; Ferapontova, E.E. Electrochemical Immuno-and Aptamer-Based Assays for Bacteria: Pros and Cons over Traditional Detection Schemes. *Sensors* **2020**, *20*, 5561. [CrossRef]
21. Poghossian, A.; Schöning, M.J. Capacitive field-effect EIS chemical sensors and biosensors: A status report. *Sensors* **2020**, *20*, 5639. [CrossRef] [PubMed]
22. Sedki, M.; Chen, Y.; Mulchandani, A. Non-Carbon 2D Materials-Based Field-Effect Transistor Biosensors: Recent Advances, Challenges, and Future Perspectives. *Sensors* **2020**, *20*, 4811. [CrossRef] [PubMed]

Publisher's Note: MDPI stays neutral with regard to jurisdictional claims in published maps and institutional affiliations.

© 2020 by the authors. Licensee MDPI, Basel, Switzerland. This article is an open access article distributed under the terms and conditions of the Creative Commons Attribution (CC BY) license (http://creativecommons.org/licenses/by/4.0/).

Review

Monitoring with In Vivo Electrochemical Sensors: Navigating the Complexities of Blood and Tissue Reactivity

Pankaj Vadgama

School of Engineering and Materials Science, Queen Mary University of London, Mile End, London E1 4NS, UK; p.vadgama@qmul.ac.uk

Received: 11 May 2020; Accepted: 31 May 2020; Published: 2 June 2020

Abstract: The disruptive action of an acute or critical illness is frequently manifest through rapid biochemical changes that may require continuous monitoring. Within these changes, resides trend information of predictive value, including responsiveness to therapy. In contrast to physical variables, biochemical parameters monitored on a continuous basis are a largely untapped resource because of the lack of clinically usable monitoring systems. This is despite the huge testing repertoire opening up in recent years in relation to discrete biochemical measurements. Electrochemical sensors offer one of the few routes to obtaining continuous readout and, moreover, as implantable devices information referable to specific tissue locations. This review focuses on new biological insights that have been secured through in vivo electrochemical sensors. In addition, the challenges of operating in a reactive, biological, sample matrix are highlighted. Specific attention is given to the choreographed host rejection response, as evidenced in blood and tissue, and how this limits both sensor life time and reliability of operation. Examples will be based around ion, O_2, glucose, and lactate sensors, because of the fundamental importance of this group to acute health care.

Keywords: metabolite sensors; sensor biocompatibility; ion selective electrodes; foreign body reaction; O_2; glucose; lactate

1. Introduction

Physiological processes operate under highly dynamic conditions that are controlled by a multitude of biofeedback systems operating on both long and short term timescales. These establish homeostatic control within finite, set, limits. It is the essence of any multi-cell and tissue organism that it is able to maintain relative internal stability in the face of unpredictable, and often undesired, environmental change. Within the cell itself, sensitive surveillance mechanisms recognise status deviation and effect timely responses, on an ultrafast basis if deemed necessary. These responses are all the more effective where delivered through specialist tissues and the major internal organs. Complexities at this cell level are only partially reflected in changes in the extracellular space. However, it is only the changes in the latter that we are able to monitor and assess through available sampling and measurement capabilities. For some extracellular parameters, these changes take place on highly compressed time scales and justify frequent, if not continuous, measurement, for both a better fundamental understanding and the better management of disease. These variables can be considered to be highly labile and their dysregulation represents major failure of homeostatic control. Currently, these centre on ions, gases, and small metabolites. Their potential for rapid change is also indicative of the potential value of continuous monitoring to track their trajectory and help titrate therapy. The timing of any therapy is of equal importance to its amount in certain types of critical illness and might influence recovery and survival.

For all practical purposes, only extracellular events are trackable. Moreover, the fractal and multi-organelle architecture of the cell does not readily offer simplified messages that allows for easy conclusions about its status or, indeed, that of the whole individual organism. This remains the case, even if robust microsensors were to become available. A clear differentiation between the normal state and disease is vital clinically and, whatever the biological value of intracellular monitoring, any information thus secured will be a complicated case mix influenced by cell ageing, the cell cycle, the micro-compartment sampled, the reaction of the cell to interrogation, and a myriad of other unknowns that are not easily synthesised into workable diagnostic formulations. The occasional exception is where disease disruption is so overwhelming that it leaves a substantial intracellular signature leading to structural change, such as when an insoluble end product accumulates or a core metabolic process is involved, as in some inherited metabolic diseases.

The earliest efforts with sensors were indeed directed at intracellular monitoring, but for enhancing biological understanding. This involved pioneering work with wire and micropipette electrodes. The latter as ion selective electrodes were particularly prominent, and were used to variously follow intracellular H^+, K^+, and Na^+, e.g., to study cell plasma membrane ion exchanges. A particular interest was in excitable tissues, such as nerve and muscle [1], where ultra-rapid ionic exchanges mediate membrane depolarisation and launch action potentials. Such work initiated our understanding of cell physiology and the cell's ability to function at high speed. Oxygen was the other intracellular target, enabled through more robust wire-based electrochemical sensors [2]. Priority interest in oxygen emanated from, and still remains, its central importance to the energy economy of the cell and the dire consequences of its deficiency.

The cell, tissue, and whole organism hierarchy gives us a dimensional scale across which we can select monitoring options at the supra-cellular level. For clinical purposes, it is at the whole organism end of the range that we can see direct clinical value. This larger scale is fortunately readily accessible to us via the blood circulation, accommodating invasive probes. Whole body physiology targeted this vascular space in order to understand change in the intact organism. The current medicine paradigm remains the use of blood as the ultimate pre-mixed representative of whole-body change. Whilst blood is core to our disease understanding, it is still an approximate messenger, being differentially affected by various tissues and with a compositional change that may also only be a diluted version of events at local level. Our access and assay technologies in the clinical laboratory are still inevitably directed at blood. A typical diagnostic biochemistry repertoire is 200 parameters, a triumph for analytical science [3], but one less attentive to following rapid change other than by more frequent discrete measurement. Yet, we already know there is advantage to continuously monitoring oxygen, certain ions, and metabolites in blood.

Other fluids, such as urine and CSF, might have discrete measurement value, but continuous monitoring is unlikely to be of added benefit. The tissue biopsy for targeted analysis lies at the opposite extreme end of the scale for repeat measurement need [4]. Our developing blood biochemistry repertoire has made it possible to gain a better picture of the timescale for variation. Thus, homeostatic control, far from creating a biochemically static environment, effects a pattern of pre-set cycles and modulations. Chronobiologists have elucidated such patterns for the healthy state, segregated variously into ultradian, circadian, and also longer-term and less well understood forms of biorhythm [5]. In disease, there is not only a deviation from normal set points for a biochemical parameter, but moment to moment fluctuations that might change in dynamic character and could contain added information. This has gone unrecognised because continuous tracking is not available, and the minor fluctuations nominally trivial. Rapid change possibilities were understood in relation to monitored blood gasses: pH, pO_2, and pCO_2 [6,7]. However, from what we now know of glucose and lactate, these and other intermediate metabolites may well also exhibit rapid fluctuations.

In early examples relating to ions, continuous monitoring of blood was undertaken in animals using ion selective electrodes (ISEs) in early physiology studies [8–11]. Striking speeds of change were seen. These studies were remarkable in being well before the advent of the microfabrication toolkit for reproducible sensor miniaturisation. In one example, rapid change to arterial blood pH was shown as a result of increased inspired CO_2 composition (Figure 1) [12]. The conversion of CO_2 to carbonic acid by carbonic anhydrase in red cells, we know, is rapid, but its manifestation as a matching fast outcome seen through blood [H^+] is a useful dynamic indicator. More remarkable, perhaps, was the observation, as in this study, of breath to breath pH oscillations due simply to normal breath to breath pulmonary tidal pCO_2 variation. The oscillations are stable, and remarkable for being observable in the highly buffered entity that is whole blood. Interestingly, the amplitude of these oscillations parallels respiratory CO_2 excursions. The nature of these oscillations in respiratory illness and compromised blood buffering remains an unknown, but could provide new clinically relevant information. Only a fast response ISE with its millisecond response has the capability to unmask these hidden variables. Sensor advances, combined with signal processing, might provide a valuable step forward in extracting and processing such hidden information. Technically, the active membrane components used in ISEs have changed little, a tribute to the early chemists. However, other newer design iterations could be usefully pursued. An ion selective membrane for bicarbonate ion has yet to be made, yet bicarbonate ion has huge importance for the acid-base status of the body in disease, and its dynamic variation is unknown.

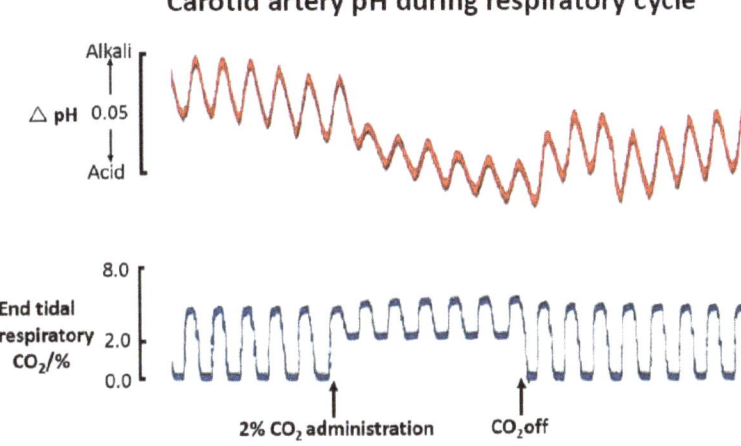

Figure 1. Arterial blood pH monitored extracorporeally at a carotid artery loop in an anaesthetised cat using a glass pH electrode. End tidal CO_2 was monitored by an infrared CO_2 analyser. Administration of 2% CO_2 led to increased end tidal CO_2 and a drop in arterial pH. The pH trace also shows breath to breath arterial pH oscillations. Adapted from [12].

A classical basis for the in vivo sensor is its ability to follow a trend and thereby to pick up deviation early, even whilst a parameter remains within the bounds of normality. With wider deployment of such in vivo sensors, there will be greater identification of early variation and of the hidden dynamic patterns that are linked to disease. Minor oscillatory and other minimal changes will need high resolution sensing. Rapid response will also be crucial, as shown with pH. However, on the speed of response, it is probable that a slower biosensor tracking of metabolic fluctuations will still be sufficient. Oscillatory cycles for basal insulin and glucagon have been observed in blood, and are relatively slow, on timescales of minutes. They also have a synchrony with a glucose oscillation of similar periodicity [13], though the glucose cycling has an amplitude of merely 0.05 mM. Future sensors that are able to track such minor cycling behaviour could allow linkage to be made to metabolite control and its dysregulation.

The dynamics of oxygen change at tissue level were the focus of some early studies. These helped to unravel interrelationships between oxygen delivery centrally and localised tissue oxygen uptake.

The latter is modulated by the local microcirculation, itself in a state of dynamic variation. Silver [14,15] investigated normal and solid tumour tissue pO2 using microelectrodes; Lubbers [16,17] used multi-wire electrodes and tracked both oxygenation variability and microenvironment heterogeneity. The time dependent shunting of O_2 was regularly seen, involving complex cycling topologically connected to the organisation of the vascular bed. The interstitial tissue space was thus demonstrated, for oxygen at least, as the seat of complex gradients, which alsoreflected the balance of cell respiration against tissue level fluctuations in blood supply.

A physiology study may only require a sensor to operate for a limited period, and this also under highly supervised conditions. The progression to medical use poses a more severe challenge where simplified, robust operation is the key. Technical refinements and miniaturisation have moved the field along in analytical chemistry terms, but stable in vivo deployment and local tissue communication has proved to be a more protracted challenge. In the absence of reference methods, monitored output is always a combination of true metabolic readout and an uncertain, artefactual change due to local implant site tissue change, including that leading to fouling on the device sensing surface. The balance might well be towards a meaningful readout and not to artefact, but the latter can prove a remarkable mimic of real change. The communication interface with tissue or blood is thus a weak link. Interface stability is under constant threat from assembled biological reactivity that changes the very environment intended as the sampling window into whole body changes. Our understanding of the complexity of both the reactive process and it structural outcomes is still limited. Innumerable sensor design approaches, including of new transduction routes, have been reported to combat the adverse effects found in the host environment, but success has been partial, at best. For tissue, especially, the sensor is merely another foreign body intrusion, and it is of no consequential difference to the host biology as to whether it has been architectured as an 'intelligent' material or some other standard, unreactive biomaterial. Emphasis is needed now on the biology to find a route to more reliable monitoring and the promise that it holds. Recent work has begun to consider these extra-sensor processes and their key elements. The balance of work, though, still does not reflect the centrality of the biological question. Existing biomaterials research certainly provides a guide, but sensing research needs its own approach now. On this aspect, the review summarises the response basics of blood and tissue, and how these can affect sensor performance. Examples of dynamic monitoring, further, show why the effort is worthwhile. Therefore, the review is really about the bio-interface. Detailed descriptions of transduction chemistries and electrode designs can be found in the many reviews already published on these aspects.

2. Sensors for Continuous Intravascular Monitoring

2.1. Ion Selective Electrodes

The ISE has two intrinsic advantages for monitoring. Firstly, it responds on the basis of surface ion binding and does not require slow diffusive access to deeper structures. This delivers a response within milliseconds. Secondly, response is on the basis of equilibrium ion binding, so a maintained ion flux to the device is not needed. The first allows for unmasking of rapid transients (Figure 1), the second is much less affected by surface fouling, other than possibly by a slower dynamic readout or if a deposited layer itself has charge properties that alter ISE potential. The special feature to recognise in practical monitoring is that ISE emf response is log-linear and, for a monovalent cation M^+, this is approximated in the Nernst equation by:

$$E = E - 2.303\frac{RT}{nF}Log[M^+] \tag{1}$$

where R, T, F, and n have their usual meaning. This means that concentration resolution that is based on the emf will be considerably better at lower concentrations than at high ion levels. Only because of this can we differentiate pH 7.35–7.45, the reference range for blood pH. The resolution of millimolar concentration ions is also readily achievable, but measurement is more challenging for a divalent cation, such as Ca^{2+}. Not only is the Nernstian slope of 61mV/decade (body temperature) halved, but finer emf resolution is demanded than for Na^+/K^+; the reference range for blood Ca^{2+} is a mere 1.05–1.3 mM. A further practical issue for blood use, likely to be made more complicated in a tissue matrix, is changes to the liquid–liquid junction potential at the reference electrode. This is a combined function of sample ionic strength/colloid composition through to cellul sedimentation and streaming potential, viz an encounter with flowing blood. Finally, it should be born in mind that the ISE responds to ion activity and not concentration; to that extent, it is a true thermodynamic measurement. The activity coefficients at biological fluid ionic strength are around 0.65, so not only are matched calibrants critically important to measurement, but any background ion change will affect the measured values simply via activity change, even if true ion concentration is unmoved. Despite these uncertainties, ISEs have seen effective clinical laboratory use through use of meticulous quality control and use of reference samples. Further extrapolation to in vivo monitoring makes this challenging, especially without sample dilution, so it is fortuitous that clinical value here is not conformity to accuracy *per se* as needed for standard clinical decision making, but in picking up trends fast and in their timely management.

ISE biological sample use has been mostly without modification to device membranes. Indeed, a material as unpromising and as bioincompatible as a glass pH membrane is usable in biological fluids in the first place because of its independence from continued ion flux for a stable reading. The more relevant issue for ISEs in vivo, though, is the potential toxicity of incorporated ionophores and plasticisers. Though quantitatively small in amount, they are biologically active and toxicologically risky; the standard plasticiser 2-nitrophenyl octyl ether, for example, is pro-inflammatory and active ionophores will have cell membrane effects.

In the main, in vivo studies have focused on K^+. It is of predominantly cellular origin, and its abnormal release can both reflect and cause instability in excitable neuromuscular tissues. In an early animal study, Hill et al. [18] devised a flexible intravascular catheter sensor for K^+ with a membrane comprising the K^+ ionophore valinomycin in PVC and achieved low drift monitoring with a catheter mechanical compliance suited to intravascular use. A tip diameter of 0.6mm enabled safe small vessel insertion without flow obstruction. Femoral vein catheterisation in humans was achieved with low drift (<3 mV/h) and enabled monitoring of rapid K^+ efflux from muscle during exercise; even K^+ transients of <0.1 mmol/L were picked up [19]. A high pressure vessel poses an added practical catheterisation risk, but arterial catheterisation has also been reported. In an animal study using a carotid artery catheter, rapid K^+ release from carotid body oxygen chemoreceptors was monitored during their hypoxic stimulation [20]. The benefits of precise localisation of a sensor tip were shown using a catheter advanced into the coronary sinus of the heat in human studies. This blood vessel serves as the common venous conduit for blood draining the heart. In patients suffering from myocardial ischaemia, K^+ transients were shown that would have been undetectable within the general circulation [21]. Moreover, K^+ release from myocardium correlated with the severity of ischemia [22].

Rapid blood Ca^{2+} transients have also been shown using indwelling catheters. In one report, a synthetic Ca^{2+} ionophore was used with a plasticised PVC membrane [8]. In a study on dogs, cardiac venous drainage monitoring showed Ca^{2+} perturbations due to injection of an ionised X-Ray contrast agent [23]. With glass being unacceptable for in vivo use, pH catheters have been made using H^+ affinity polymersand ionophores, e.g., octadecyl isonicotinate [24]. When applied to coronary sinus monitoring, blood pH change could be tracked in cardiac ischaemia patients [25].

Much less has been accomplished with the fibreoptic monitoring of ions in medicine. Few chromionophores are available for high selectivity binding to alkali cations, where ion binding is both selective, and leads to high resolution optical change. Additionally, there is the risk of reagent leaching and photobleaching, more likely during extended operation. However, fluorescent weak acid/base dyes are readily available, and some have allowed monitoring. This includes of tissue [26] and intravascular pH monitoring [27] in animal studies. In the latter, immobilisation of the fluorescent dye within a sol gel matrix provided protection from photobleaching and use of a haemocompatible outer 2-Methacryloyloxyethyl phosphorylcholine (MPC)-cellulose membrane offered protection from blood clotting, observed overeight hours. The operational advantage of an optical route is indifference to background electrical noise and avoidance of a reference electrode. Ratiometric measurement of fluorophore fluorescence with pH offers internal self-referencing to compensate for extraneous artefactual optical changes. Miniaturisation is also readily possible without compromise to fibre robustness, flexibility or integrity. Because response depends on equilibrium binding, biofouling, as with ISEs, should be less of a problem, but response times are outer membrane diffusion constrained and, hence, much longer than for ISEs.

2.2. ISE Biocompatibility

The active ionophore of an ISE is typically a high affinity, high binding reversibility agent, and if able to exit the membrane could pose a risk in vivo. The direct consequence would be permeabilization of the cell plasma membrane. The quantities used for small in vivo ISE membranes are clearly insufficient for systemic toxicity, but they could still pose local tissue risk; regulatory approval in any case would necessitate extensive testing compliant with standards. Discovery research for new ISE membranes will be able to extend our analytical repertoire for ions, but should preferably now combine biological with chemical screening. Cánovas et al. [28] undertook such combined evaluation studies of ISE membrane components and tested cytotoxicity in vitro, notably for valinomycin, the most efficient K^+ ISE ionophore to date. Given its potential toxicity, the antibiotic mutacin, a polycyclic peptide, was suggested as a possible alternative ionophore for K^+. Much, of course, depends on of the extent to which an ionophore will leach out, and this, in turn, will be a function of membrane permeabilization via the co-entrapped plasticiser. The plasticisers assessed in this study showed varying degrees of toxicity, and reinforce the desirability of pre-use screening. The polymeric ISE membrane itself should not be a toxicity concern. A polymer is only really toxic in so far as it releases its small molecule constituents. This could possibly arise from polymer biological degradation, as seen with polymeric biomaterials. Indeed the reactive implant site has high degradation potential with its constituent cellular hydrolytic enzymes, lowered pH, and free radical release from activated phagocytic cells. So again, biological screening needs to be part of any new polymer development, other than, possibly, in the case of established PVC or polyurethane.

ISE surface modification and coating for safe retention of diffusible membrane components is unlikely in future, given the parallel need for target ion access. However, a possibility does exist for reducing surface biofouling by a coating. Pharmacologically active agents for stabilising blood platelets at the surface could also help to mitigate fouling, and surface hydrophilic layers, such as of polydopamine [29], and hydrogels, such as poly(2-hydroxyethyl methacrylate) (pHEMA) [30], have been reported. Zwitterionic phosphorylcholine is an integral component of the outer red cell membrane surface, and when used, can confer a high degree of haemocompatibility [31]. Other biologically inspired molecules have also shown effectiveness. Surface immobilised heparin [32] has been used, and a NO adduct in a membrane released thrombus countering free NO [33]. Heparin works through binding antithrombin and thereby concentrating its anticoagulation effect if used at a surface, and the ubiquitous signalling molecule NO provides surface protection through its platelet passivating action. As with any small molecule agent, including mediators used for second generation enzyme biosensors, whenever a new component is contemplated, the risks of agent toxicity also need to be considered; the analytical benefits alone do not confer in vivo usability.

A further factor for any intravascular component is progressive thrombosis at a point other than the sensing surface. The result could be flow blockage or disseminated thrombi to a distant tissue location. Materials for surface biocompatibility are often only tested over limited periods, and usually in vitro, so, whatever promise is shown might not transfer to in vivo deployment. One complicating factor is physical vascular flow; platelets are highly environmentally sensitive, and they become activated even by local flow turbulence, so surface deposits may occur due to smooth flow disruption, regardless of any high material surface haemocompatibility.

Future designs for ISEs, if used intravascularly, will need an integrated approach, whereby not just the sensing surface, but also the flow compatibility of the entire construct in a confined blood vessel will need optimisation. The starting point, though, is a recognition of the scale of this in vivo biological challenge with blood. Testing with blood sub-components in focused studies and under controlled conditions can only serve for initial understanding. There is a limit to the practical value of such a reductionist strategy, with a resultant ever-present risk of overoptimistic commentary about some new material or surface delivering almost completely what is needed. Whole blood is a high alert, rapid response system that features multiple cooperative systems. It is able to harness a combination of cellular effectors just like a tissue, andso is not dependent simply on diffusible humoral signalling agents. It has evolved to recognise, package, and potentially degrade any foreign surface intrusion, both as part of a fail-safe haemostasis and as a means of partitioning any 3D object exposed within the circulation. Later parts of the blood response, in fact, begin to resemble those of tissue more closely. From the first, transient, foreign surface encounter, it generates a coagulation cascade for thrombus formation which accelerates through multiple enzymes and finally reaching an explosive rate in the mass generation of the final fibrin crosslink layer. It also utilises a parallel complement cascade that delivers surface coating protein (C3b) that promotes phagocytosis, amongst other effects, and the promotion of inflammatory change in the clot. An ISE in vitro to in vivo transition is thus challenging and warrants greater balance in blood vs. sensor basic research if the early gains of sensor design are to be translated to routine clinical use. It might seem attractive to consider an intermediate solution with the use of an extracorporeal sensor as part of a controllable external blood circuit. However, this also is not a simple solution; whilst greater control over blood flow dynamics, coagulation, and calibration are achieved, complicated pump flow control and sterility protection are now needed. A cumbersome platform and secure fluidic supply can allowusage only be appropriate in high dependency clinical care environments.

2.3. Oxygen Electrodes

After the adaptation of O_2 polarography to in vitro blood use, via the Clark electrode, the measurement principle has changed little. Electrodes use gas selective membranes that are able to reject other solutes and ions while retaining an electrolyte film for stable cathodic O_2 reduction. Also excluded are cells and colloids from the sample so eliminate working electrode fouling. The electrochemical reaction is substantially more complex than the summarising four electron, −0.65 V (vs. Ag/AgCl), reaction typically cited:

$$O_2 + 4\,e^- + 4\,H^+ \rightarrow 2\,H_2O \tag{2}$$

It is dependent variously on oxygen adsorption, surface reactions and electrode material catalytic properties, and is affected by solution conditions—alkaline vs. acid. Gold is the preferred working electrode for oxygen. The electroreduction process here involves first the adsorption of hydrated O_2, then converted to hydroxyperoxide (OOH), a key intermediatesurface [34]. Subsequent reduction may go by two 2e steps with H_2O_2 intermediary or by a combination of these with the above 4e reaction. These also only represent some of the possible electron transfer reactions. The practical aspect of H^+ utilization is a possible pH induced drift in response due to alkalinisation of the low volume electrolyte film of the Clark electrode. The electrode adsorption-reaction cascade here is also a reminder of the ever-present risk of surface contamination effects. Low molecular weight species in biofluids, especially, have the capacity to adsorb and disturb the catalytic surface, so such adsorption is not limited to just macromolecules. This is also what Leland Clark effectively avoided with his gas permeable outer membrane. This is less easy to avoid with porous membranes and a problem therefore exists for the glucose sensor (vide infra). Bbiocompatibility problems are the consequences of diffusible species warrant study, especially since only in the Clark electrode with its blocking polypropylene or PTFE can internal contamination be totally discounted.

Oxygen sensor miniaturisation for intra-arterial use has been achieved [35], but it is difficult given the need for seamless attachment of a relatively inert hydrophobic membrane barrier material to the body of the device. Equally, surface chemical modification forhaemocompatibility is difficult; functionalisation here needs harsh treatment. Also, any deposited coating might delaminate, andany residual exposed hydrophobic surface, will promote fouling through the extra tendency of hydrophobic surfaces to denature adsorbed protein. Denaturation is more likely to trigger a greater host response than with a non-denatured protein. Surface fouling for an oxygen sensor is important because its response requires a continuous, stable flux of oxygen for a plateau response, in contrast to the ISE.

Vascular catheters as a monitoring route have been reported [36,37] in early studies, and a more structurally refined catheter model has used the catheter wall itself, e.g., silicone, as the gas membrane. NO release at such an electrode offered partial suppression of surface coagulation. In one example, a double lumen silicone cannula was used where NO was electrochemically released from the second lumen containing a nitrite reservoir [38]. Here, stabilised O_2 output was seen during acute monitoring of hypoxia (Figure 2). Surface coagulation, even though not entirely eliminated, could still be made sufficiently low for extended monitoring subject to a design for sustained NO release. At such a blood contacting device there is also the risk, in principle, that blood cells, notably nucleated cells, highly active metabolically, can act as an oxygen sink, depressing measured O_2 values. Platelets are also active in this regard, though not the mitochondria deficient RBCs. Ultimately, the blood interfacing problem could be amenable to resolution through synergistic use of locally delivered and surface immobilized anticoagulant agent, along with refined catheter shape to sustain normal blood flow profile. The extracorporeal answer to this is a multiparameter system available for neonatal use (VIA LVM Blood Gas and Chemistry Monitoring System, VIA Medical) [39]; reliability here is achieved through blood flow alternating with heparinised calibrant solution.

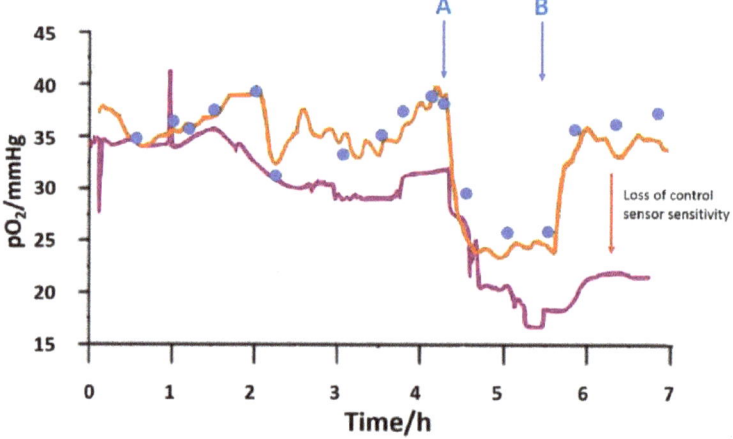

Figure 2. Double lumen intravascularly placed O_2 catheter with haemo-protective NO delivery used in rabbit jugular vein. (-) NO flow protected O_2 sensor, (-) control O_2 sensor. Blue filled circles are intermittently sampled venous blood pO_2 values assayed by in vitro analyser. (**A**) 100% inspired O_2 was switched to 21%. (**B**) Return to 100% inspired O_2 Adapted from [38].

Numerous dye functionalised fibreoptic sensors for intravascular pO_2 monitoring have been reported and for a period available in a commercial clinical intravascular catheter [37], where a triplet of pH, pO_2, and CO_2 was monitored. Oxygen monitoring is universally based on dye fluorescence quenching. Oxygen reversible binding to the dye leads to non-radiative transfer of energy and thus reduced fluorescent emission/lifetime (Stern–Volmer relationship). This approach has a huge theoretical advantage over an electrochemical sensor in that a sustained flux of O_2 is not needed for response and so external transport constraints in vivo are reduced. Nevertheless, problems of surface coagulation in blood and the dangers of thrombus generation are not avoided, and these can stilllead to artefactual output change in measured pO_2. Measurement uncertainty is further compounded by the catheter wall effect, where catheter tip impaction against a vessel wall blocks off sensor surface blood contact. Otherwise, practical performance is similar to that of electrochemical sensors, with dynamic response, for example, being set by the membrane barrier interposed not the internal chemistry. The similarity of reliability problems in blood for the contrasting type of devices attests to the often limited benefit achieved with radical changes in transduction method.

3. Blood as a Reactive Sample Matrix

3.1. Protein Surface Interactions

Blood-surface recognition utilises multiple, complex pathways that are, as of yet, incompletely understood. Both plasma proteins and the formed elements of blood, other than RBCs, have a high tendency to adsorb to surfaces. Protein deposition commences within milliseconds, and is later amplified through complement and coagulation cascades that deliver high mass surface coatings. This is the start of the thrombus and, though it might be structurally indeterminate, it advances through highly organised, controlled pathways. The speed and multi-factorial nature of the process makesexperimental study difficult. Consideration of the idealised situation of a single protein as a monolayer provides a model to understand the initial events. Immediately after the deposition and attachment of a protein to any surface, conformational remodeling is initiated due to non-covalent binding interactions with the surface and desolvation changes. Essentially, this is protein denaturation, which leads to peptde chain unravelling and molecular dimensional expansion. Thermodynamicallly, enthalpy lowering drives these surface attachments, but, since attachment also leads to a reduced entropy, in order to compensate, available free loops unravel to thereby increase entropic freedom. Such a molecule has been considered to have a 'loopy' conformation. It's result is that its surface footprint increases in area. The extent of this unravelling process depends on time and, so for a given mass, the area occupied will increase (Figure 3) [40]. If in a biofouling study, the assessment period is a short one, then in this idealised situation, there will be a higher surface mass per unit area than if the experimental time scale is a long one when the protein molecules have expanded with fewer needed for full coverage, ie jamming. This makes for uncertainty in study comparisons. In the limit, all of the molecules unravel and molecular surface density reaches a finite minimum.

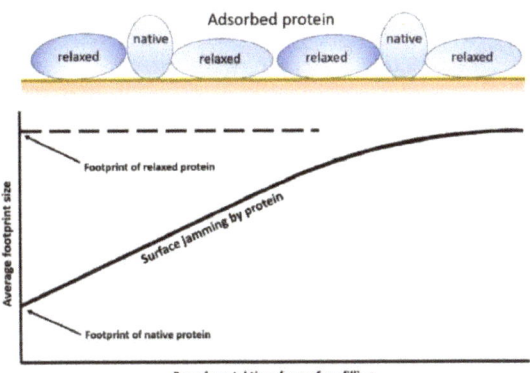

Figure 3. Schematic of progressive relaxation of adsorbed protein layer and increase in surface occupancy per molecule over time. Amount of protein needed for total confluent coverage (jamming) of surface is reduced as time of experimental observation increases. Adapted from [40].

The adsorptive behaviour of proteins from blood is orders of magnitude more complicated, but still driven by the thermodynamics. This complexity is partly summarised by the Vroman effect [41]. This, in any multi-protein system, there is competitive protein surface binding and exchange, not yet fully understood. Early protein adsorbates from high concentration proteins are later displaced in this model by slower arrivals with stronger surface affinity. Typically, here, fibrinogen eventually replaces albumin. This shifting protein interface creates a changing contact surface, and is also the trigger for later biologically mediated cell and fibrin coagulum deposits. Protein denaturation is inevitable at a surface, and this is also a stimulus for blood activation through its presentation of new protein motifs (epitopes). Later, a different type of protein depositionbecomes activated, via the complement cascade; opsonization, andthis is designed to facilitate phagocytosis by polymorphonuclear leucocytes.

Efforts to ultimately achieve zero protein adsorption would seem unrealistic with respect to effectiveness against intact blood biology. Brash suggested an interesting alternative [42]. This envisages that if deposition cannot be avoided, then surface directed selective protein deposition might prove effective. Thus, a surface might be able to selectively invest itself with a defined functional property, such as fibrinolysis (plasminogen adsorption) or anticoagulation (antithrombin adsorption). Reports on haemocompatible sensor surfaces still indicate a continued quest for the single 'magic bullet' solution where none may exist. Nevertheless, general rules may be derived from such studies, such as general rules for hydrophobic/hydrophilic balance for lowered fouling and specific surface chemical motifs that link to complement activation [43]. The surface protein profile together with its later remodeled form [44] presents the real final contact layer for all the subsequent cellular processes organised by blood. So from the start, the original engineered or chemically designed surface ceases to be the direct material. Despite this masking, blood recognition continues and its reactivity remains as long as the device is in place.

3.2. Blood Biological Reactivity

Following the protein interaction stage, the intrinsic coagulation pathway is initiated by Factor XII surface binding. Complement C3, the core driver of the separate complement cascade, causes independent protein coating and opsonisation. Complement C3 is triggered to fragment autocatalytically and produces C3b adsorbate for surface opsonisation; a surface that is now an attractant for inflammatory cells [45]. There is cooperation between the coagulation and complement pathways, and this later leads to the incorporation of inflammatory cells within the developing surface thrombus (Figure 4) [46,47].

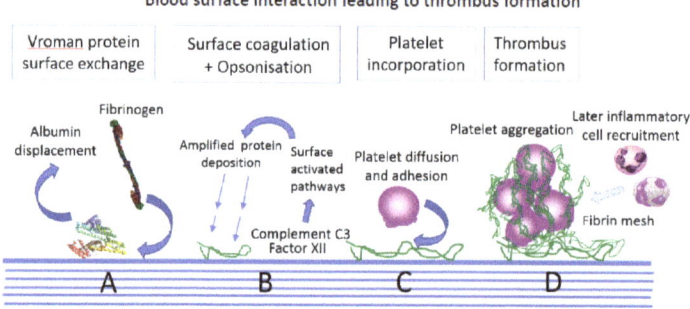

Figure 4. Schematic of surface coagulation sequence. (**A**) Initial rapid protein deposition, in milliseconds, subject complex, competitive displacement/remodeling via the Vroman effect, e.g., fibrinogen displacement of albumin. (**B**) Surface activated C3 and Factor XII trigger complement and coagulation cascades, leading to protein/C3b coating (opsonisation) and fibrin directed at the surface. (**C**) Platelets contact with coated protein sets of adhesion response. (**D**) Platelet adhesion leads to activation and promotion of fibrin clot, later inflammatory cells incorporated.

The platelet is the specialist player of blood that really drives the development of a surface thrombus and it is later one of its most prominent constituents. Its study is difficult because of its environmental reactivity, including to the very surfaces used to handle it, and also its high sensitivity to shear stress. Moreover, its response to the developing surface coagulum is involves specific pathways triggered by specific surface receptor stimulation. The latter lead to dramatic morphological changes in the platelet, including degranulation, shape change to a discoid, and multiple bridging/aggregation. This super-structure of platelets and the entrapped fibrin then add to the growing thrombus [48]. For sensors, although the focus has been on surface chemistry, surface physical profile might also be important. In one proposal, surface roughness of platelet dimensions (~2 µm) was considered to offer a better match for platelet surface contact and, therefore, for thrombus formation than lower dimension roughness giving less platelet purchase [49]. Leukocytes in blood also become surface activated later [50], are then recruited into the thrombus, and promote further coagulation through cytokine release.

4. Tissue Oxygen Electrodes

4.1. Compartmental Difference

The Clark pO_2 sensor has also allowed for continuous monitoring of subcutaneous tissue pO_2 [51]. Such a device has enabled the tracking of peripheral tissue pO_2 during haemorrhagic shock [52], but there are indications that there are compartmental differences between blood and tissue. This is suggested for this study by an inter-sensor agreement that is greater than with venous blood (Figure 5). The measured tissue pO_2 was significantly lower than that of blood at the later part of the shock experiment, a possible outcome of subcutaneous circulation shut down. Blood pO_2 might, alternatively, reflect deeper tissue levels, e.g., of the more protected central organs, but further studies are warranted. Such changes cannot be readily decoupled from sensor drift, but if that was the cause, and then the polyurethane oxygen permeable membranes used would need to be exceptionally lacking in biocompatibility, and two hours had already been allowed for electrode stabilization. Post implantation stabilization periods are considered as artefact and certainly no clear explanation is given over their basis. However, the early drift seen during this run in period may well be a consequencet a tissue functional response, e.g., microvascular changes, to the intrusion. Mechanical tissue damage and microhaemorrhage will certainly also occur, but cannot be the full explanation. Calibration uncertainty in tissue lends uncertainty to true tissue pO_2 values, which, in any case, will show local differences at the micron scale. Venous rather than arterial blood comparison was used for this study, though arterial pO_2 is the benchmark for clinical use. Here, arterial changes were only observed at a very late stage haemorrhage; venous blood, derived from tissue, may more reflect tissue embedded sensor changes.

Gough reported the determinants of tissue pO_2 under non-haemorrhagic conditions using a silicone membrane covered electrode [53,54]. Again, the similar stabilisation delay and uncertainties about tissue O_2 were observed. They attributed variation in output at an array of tissue electrodes to local differences in vascular flow, and also observed slow to rapid fluctuations of tissue pO_2, which they attributed to perfusion variation due to local vasomotor vascular control. The challenge with tissue is that of extracting valid physiological information in the face of an evolving tissue reaction, essentially a wound site. Surface biofouling raises the further uncertainty. A model for oxygen mass transport to the electrode was established [54], which indicated that local mass transport resistance limited the sensor response, whilst more remote oxygen delivery was rapid and associated with blood flows. The high permeability membrane used in this study allowed for the resolution of such extra-sensor effects. Over 13 weeks, these flux sensitive electrodes picked up tissue reactions that led to decay in local tissue O_2 permeability, to ~10% of that in water. Even a collagen fibrous capsule build-up to 5 mm depth apparently did not entirely abolish diffusive transport. Whist such a high permeability experimental membrane can allow investigation of tissue effects, practical monitoring requires diffusion limiting membranes to negate external transport variables. Even here, however, over long time periods, it might prove difficult to achieve this if a substantive fibrous capsule forms. With tendon as a model dense collagen barrier, we found micro-solute diffusion to be just 1% of that in water [55].

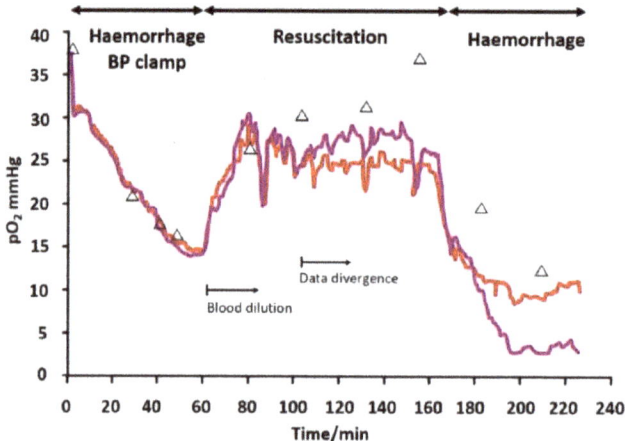

Figure 5. Tissue pO_2 changes monitored in single rat during haemorrhagic shock. Sensors at matched implantation sites in flank. Initial haemorrhage clamped at reduced BP (40 mmHg); saline only resuscitation stabilises BP (60–70 mmHg); haemorrhage to exsanguination with extreme, terminal drop in BP. Resuscitation regimen would lead to cumulative blood dilution, progressively lowering oxygen carrying capacity to peripheral tissue. Adapted from [52].

4.2. Tissue Micro-Heterogeneity

Tissue oxygen delivery distribution at a microscopic level is a field in its own right, and numerous mapping studies of pO_2 have been undertaken using microelectrodes [56,57]. Oxygen micro-heterogeneity is variously a result of cell uptake, vascular delivery, and transport variation across the extracellular compartment. Cerebral tissue has been a particular focus for study because cortical blood vessels are more easily visualised, allowing for combined analysis of vascular organisation and pO_2 distribution. In one study, a <5 µm diameter recess tip electrode with a collodion membrane was used to determine blood pO_2 along a sequence of arteriole, capillary, and venule, together with perivascular tissue oxygen distribution [58]. This showed not only the expected pO_2 reduction along the blood vessel cascade, but steep perivascular oxygen gradients extending ~60 µm into tissue giving pO_2 reductions of up to 80% intravascular values (Figure 6). Muscular arterioles of the CNS are unique in providing through wall tissue oxygenation, so there were also gradients around these vessels. Such work offers insights that may be useful for neurosurgery giving a detailed picture of oxygen profile in CNS tissue. Additionally, the micro-delivery of vasodilator pharmacological agent to a single vessel was examined showing that with resulting relaxation of the arteriolar wall, through wall oxygen delivery was increased.

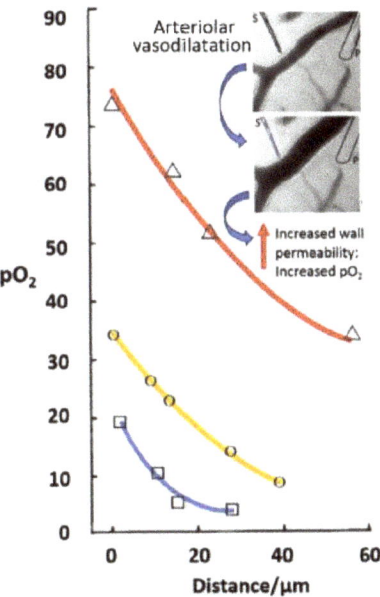

Figure 6. Rat cortical tissue pO$_2$ at varying radial distances from (-) arteriole, (-) capillary, and (-) venule using 4 µm tip oxygen sensor showing exponential reduction with distance from vessel axis. Inset shows microsensor (S) on 30 µm diameter arteriole with subsequent vessel dilatation after delivery from nearby micropipette (P). Adapted from [58].

The question arises as to what macro-electrodes can tell us instead. They offer a 'field of view' extending hundreds of microns and, therefore, a sample aggregate of different tissue oxygen micro-sites, blurring the fractal complexities. The uncertainty is what the exact size of this sampling zone might be. The issue is typically bypassed by setting an empirical in vivo calibration against blood pO$_2$ at the start. The outcome is still meaningful in that trend monitoring of pO$_2$, is obtained for clinical purposes, egduring compromised tissue oxygen delivery [59–61] and in assessing cardiac oxygenation dynamics during ischaemia/reperfusion [62]. A commercial electrode is available for specialist CNS use (Licox, Integra Life Sciences Corporation) [63], but there is, as of yet, no general tissue clinical electrochemical sensor. This commercial system samples oxygen through an extended 18 mm^2 area polyethylene tube so capturing changes across gross tissue regions. Again, uncertainties remain due to interrelationships between microcirculation organisation, flow, vascular distance, and mass transport, all balanced against cell metabolic uptake [64].

Beyond the validation of sensor stability using pre- and-post in vivo use calibration, there is no simple means of establishing the true pO$_2$ experienced by a device [63]. This is where the need for disease correlates and diagnostic benefits diverge from the rigour of measurement science. An example of clinical value is in the case of head injury wheremeasured hypoxia appears to correlates with outcome. Some indication of the degree of uncertainty is shown by reported differences in monitoring output when the principle of measurements is changed, e.g., from electrochemical to optical, but these have been minor.

The CNS is a relatively implant tolerant tissue, a contrast to subcutaneous tissue that demonstrates a florid, cellular inflammation. However, at the opposite extreme is when an inflamed tissue is deliberately monitored. The Licox probe was used to measure the pO$_2$ of inflamed synovial tissue in rheumatoid and psoriatic arthritis patients [65]. Variable degrees of hypoxia were seen, and the degree appeared to correlate with disease severity. There was even the suggestion that hypoxia was a driver of inflammation, as manifested by the degree of oxidative damage coupled with the degree of hypoxia, and through the level of vascular damage and prevalence of T cells and macrophages in the inflammatory field. Consistent with this possibility was the improvement in oxygenation seen following anti-inflammatory therapy; pO_2 doubled from ~20 mmHg in those who responded. A pO_2 correlation was also demonstrated in relation to T cell, but not for B cell, infiltration, suggesting a disease causal link with the former. Any inflammatory milieu remotely like to this at a sensor implant site would radically change measured pO_2, and no longer reflect systemic levels. Despite the oxygenation causal possibilty with the histology, there is also the likelihood that high respiring cells simply caused a low tissue oxygen and the measured levels were a reflection of the respiring cell population.

4.3. Cancer Tissue

Solid cancers grow rapidly and can outgrow their vascular supply, already compromised through disordered, dysfunctional blood vessels. Zones of hypoxia arise, which have a clinical relevance, because hypoxia confers tumour radio-resistance. A range of analytical techniques has been employed to study cancer blood supply and oxygenation and the field has been reviewed [66]. O_2 microelectrodes, in particular, have made it possible to unravel the oxygenation architecture of cancer tissue. A commercial recess tip oxygen electrode (Eppendorf, Hamburg, Germany) housing a 17 μm gold cathode and a PTFE barrier layer enabled sequential tip tracking across tumour tissue at aligned 0.6 mm distances [67].

This generated oxygen distribution data (histography), which when combined for tumours from many patients in the case of cervical tissue, showed a stark oxygen distribution difference for normal vs. cancer. In cervical cancer, for example, median pO_2 was a mere 10mmHg, as compared with normal cervix at 43 mmHg, with also a huge preponderance (>60%) of exceptionally hypoxic microenvironments (Figure 7) [68]. It is notable that, even in normal tissue, sites of near zero oxygen levels do exist. Any hypoxia link to cancer prognosis appeared to be absent, but there is value, nevertheless, to such study of the oxygen state of cancers.

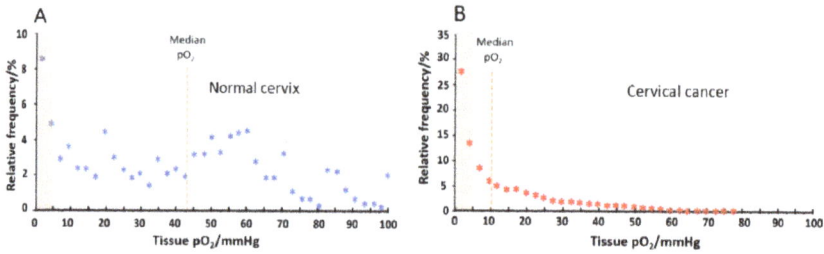

Figure 7. Percentage frequencies of pO_2 measured across (**A**) Normal uterine cervix (seven patients, 432 samples), (**B**) Cervical cancer (150 patients, 13596 samples). Shaded areas highlight percentage prevalence of extreme tissue hypoxia of <5 mmHg. Each set of data represents combined data points from multiple patient samples. Adapted from [68].

Tissue oxygen can also be monitored while using fibreoptic sensors. The principle of operation is fluorescent quenching as with the intravascular oxygen sensor. There is potential for application just as with electrochemical devices, and a widerange of critical care clinical applications have been considered [69]. When brain monitoring was undertaken, the optical affinity system gave slightly lower responses to an electrochemical device (Licox), and no advantage regarding response time was evident [70]. Differences that were seen variously related to device tip geometry, the area of the sensing surface, and how close it was juxtaposed to tissue, but evidently not to sensing principle. A possible discriminator of clinical relevance was the greater accuracy of the electrochemical probe at low pO_2; this could be of value in monitoring brain hypoxia. Stabilisation times is long, and this would affect operational deployment of either device in an emergency situation. Yet again, transduction chemistry is far less important than design.

5. Glucose Electrodes

5.1. Monitoring Needs

Glucose remains the most important target for continuous monitoring. Whilst detailed electrochemical studies have reported on the mechanisms and optimisation of enzyme-electrode electron exchange, for example, these alone cannot alone translate into operational benefit. Commercial development has helped to advance the latter operational aspect, initiated previously through single use strip technology [71]. Diabetes has now escalated in importance since the early days of glucose sensor development, and so glucose sensing has risen up the health priority list, along with this, the interest in continuous monitoring. Diabetes now poses a massive healthcare burden currently affecting 9% of the global adult population. Within this group are the 5-10% Type I diabetics [72] needing insulin who warrant closer monitoring to manage their therapy. They are also liable to marked glucose fluctuations, e.g., during concurrent illness when insulin sensitivity changes. The brittle diabetes sub-population lies within this group, and though small, has highly unpredictable insulin needs and difficult to manage glucose levels; these patients are prone to dangerous hypo-/hyperglycaemia [73]. Accordingly, continuous glucose offers specific benefits to the Type I diabetic, with a reduction of long-term vascular and other complications through improved glucose control.

Sensors when used as single use devices allow for greater design flexibility permitting, for example, genetically modified enzyme use, leachable/soluble mediator, and a host of modifie working electrode surfaces. The major consideration here is mass usage, shelf life stability and calibration-free measurement. Beyond that, survival in blood need only be for a few seconds. An in vivo sensor, by contrast, can be allowed high calibration variabilty pre-use, but, beyond that, stability during use has to be sustained over long periods, and the electrode surface, enzyme component and any incorporated reagent components have to be guaranteed to be safe, or unable to be released. Hence, permitted flexibility over design chemistry is limited. Additionally, intravascular sensors risk microthromus dissemination, so, despite its uncertainty, tissue is the near-universal target for continuous glucose monitoring (CGM).

5.2. Clinical Realities

Measurement based on O_2/H_2O_2 transduction of the glucose oxidase (GOD) reaction forms the basis of all CGMs:

$$\text{Glucose} + O_2 \rightarrow \text{Gluconolactone} + H_2O_2 \tag{3}$$

By avoiding additional leachable chemical components and mediators, thisreagentless approach complies withobligatory requirements for invasive use. Local toxicity, teratogenic, and other adverse effects have also to be avoided based both on the general precautionary principle, and, regulatory compliance. Unfortunately, the glucose K_m for glucose oxidase does not allow for simple application to clinical glucose levels without barrier membranes exerting control over glucose/oxygen access to the enzyme layer. Given the dual substrate kinetics of the enzyme, that such membranes have been developed with any success is an understated achievement. The oxidase not only depends vitally on adequate, freely available, oxygen co-substrate, but its ambient levels then also dictate the apparent glucose K_m. In the discourses on kinetic pO_2 effects at the enzyme, it is often forgotten how precariously limited this oxygen is in terms of actual concentration. At a normal arterial pO_2 range of 80–100 mmHg, to a first approximation, Henry's Law indicates oxygen concentrations of a mere 88–102 µM. This is distinctlyreaction limiting, below V_{max} conditions of oxygen despite the low oxygen

K_m of 180–950 µM (BRENDA) for the enzyme. Apparent glucose K_m is significantly lowered by O_2 limitation under in vivo conditions. Moreover, this can be on a background of a possible down shift in K_m due to a diffusion limitation in the enzyme layer. Barrier membranes are firstly required to diminish glucose access to the enzyme so sensor output is at the lower end of the Michaelis–Menten saturation curve and, therefore, linear. Thereby, the enzyme also becomes less of an oxygen sink with better maintenance of microenvironmental oxygen concentrations. Notwithstanding this, O_2 transport also has to be differentially advantaged, otherwise a reduced response with an unextended glucose K_m results. The membrane design challenge is to achieve this in the face of low transmembrane oxygen gradients. Any benefit here through faster intrinsic oxygen diffusivity is limited; diffusion coefficients in water are glucose 6.7×10^{-6} cm^2/s, oxygen 20×10^{-6} cm^2/s. Membranes with mosaic, composite, or porous strictures are frequently used, accordingly. We are yet to achieve predictive membrane design here despite the bofy of work on membrane modeling and innumerable reports on enzyme kinetics.

As a transduction principle, the second generation glucose biosensor offers a distinct advantage; it removes the key variable of oxygen control. Its signature characteristic is its integral redox mediator, which, when optimised for low potential, mostly also avoids extraneous electrochemical interference. The first generation device, operating at a typical +0.65 V vs. Ag/AgCl, also necessitates a molecular weight discriminating inner membrane as a barrier to any species larger than H_2O_2. This, however, offers added physical protection from electrode poisoning by diffusible biochemical molecules, especially those with thiol moieties. No such protection is possible with the second generation device, and this may matter for long term monitoring. Moreover, leachable or soluble mediator is precluded for in vivo application.

The exception to the in vivo sensor reagentless paradigm has been the osmium electron shuttle that was developed by Heller [74], now used for clinical CGM (FreeStyle Libre, Abbott, Almada, CA, USA). This uses mobile pendent pyridine groups along linear polymer chains for stable retention of osmium (III/II) redox centres, whilst allowing their dynamic interaction to create a relay to the working electrode. The sensor is accepted for 14 days monitoring use, and ultra-low (~2%) drift independent of calibration has been reported [75], considered to be on the basis of a high biocompatibility covering membrane. Even during normal dynamic glucose changes, there was stated to be concordance with blood glucose values. However, this system, enters a new type of unknown into the measurement; devices are already manufacturer calibrated to generate automatic blood equivalence [76]. This brackets a series of known variables, including blood vs. tissue glucose relations within and between patients, glucose dynamic change modulation of this relationship, and also any implant site dependence. Studies have shown that all are variables affecting response, and they should be considered as factors that need to be allowed for on an individual basis. Nevertheless, the clinical value of this approach has been recognised through improved monitoring benefit to the patient.

Generally, in the literature the true measured glucose value in tissue is bypassed, as with oxygen, and the starting point for data recording is after calibrated against blood in vivo. This is really a form of data ratioing across compartments rather than a true calibration, and it provides no actual information regarding the tissue state. Modeling of plasma-tissue exchanges by contrast recognises delayed and variable exchange kinetics and the need to factor in genuine lag times to underpin data correction [77,78].

Added to the physiological uncertainties, there are changes due to the reactivity of tissue at the implant site. As of yet, no material has been able to claim the stealth performance needed to eliminate the disruptive tissue reaction, despite the many sensor design iterations [79–81]. Performance decay is also maximum in the hours following implantation; the so-called run-in period of hours to days, which warrants separate consideration.

5.3. Membranes and Coatings

The coatings and coverings for glucose sensors have mostly used existing materials. The key requirement is low surface fouling and stable glucose and oxygen permeability. Shichiri et al. [82] were the first to demonstrate such packaging in their use of polyurethane in an implanted device. We and others have similarly utilised polyurethane [83,84]. Such repurposing of a medical polymer helps to reduce the unknown risk of a new material and with appropriate porosity and diffusion control enables a response that is sufficiently independent of sample physical properties or oxygen background for the clinical glucose range. However, commercial CGM manufacturers have been able to develop and incorporate new materials, as reviewed by McGarraugh [85]. The Guardian Minimed (Medtronic) employs a block copolymer polyurethane with a glucose permeable hydrophilic diol phase for glucose, balanced against a silicone that would presumably be O_2 only permeable; the DexCom (DexCom Inc., San Diego, CA, USA) uses a hydrophobic/hydrophilic polyurethane mixture for balancing diffusive transport with a presumably similar differential permeability intended. The minimisation of any oxygen diffusion limitation for the enzyme reaction is part of the design goal for materials here. In the absence of a mediator membrane transport selectivity provides an important means of achieving this. The design challenge is the micromolar levels of oxygen concentration in tissue, likely to be below arterial values. The FreeStyle (Abbott) departs from the polyurethane platform and uses a functionalised vinyl pyridine-styrene copolymer, but the mediator based device here is, in any case, O_2 independent. One interesting claim made for the latter was of the unprecedented absence of any tissue encapsulation, even after one year implantation, [80], though the muscle location here might have been a factor.

Membrane innovations have also been reported for CGMs on an experimental basis. Moussey has advanced a range of compositions that have variously included a hydroxypropyl methacrylate hydrogel coating on polyurethane that reduced inflammation and fibrosis [86], humic acid films that provoked less tissue reaction [87], and a structurally robust epoxy-polyurethane, which, though leading to a fibrous capsule, also stimulated vascularisation [88]. In one study, a porous polyvinyl alcohol scaffold was used as a covering matrix over the sensor, and this took up collagen growth from the tissue surround, along with inward growth of new blood vessels. However, the collagen barrier effect countered the blood supply benefit of increased vascularisation [89]. Nafion, a perfluorosulphonic ionomer, has been extensively studied and, as a tissue contacting sensor surface, it has generated a reduced tissue reaction with only a thin fibrous capsule at three months [90]. A comparative study of negatively charged membranes as part of a sol gel layer, espectively, utilised Nafion, dextran sulphate, and polystyrene sulphonate [91]. The results were similar for these, with thin collagen capsules being seen at 12 weeks foreach material. The lack of a difference is a reminder that chemical refinement does not necessarily change the outcome. Neutral polyethylene glycol (PEG) has well recognised antifouling properties and, as a tissue contact surface, provoked less tissue reactivity with a reduced local cellularity and tissue adherence [92].

Phosphoryl choline as an outer cell membrane zwitterion has been used to reduce protein and cell deposition in blood at an intravascular glucose sensor located in the carotid artery of rats [93]. This was an acute study, and long-term outcome would need to be investigated.

Application has been transferred to tissue. Following initial combinatorial screening, a PEG crosslinked phosphoryl choline polymer was used over a commercial CGM sensor in mice and primate studies [94]. Inflammation mitigation by the phosphoryl choline reduced the need for repetitive in vivo calibration. Blood to tissue glucose mismatch was reduced, although there was still late fibrous capsule development. Phosphoryl choline translation from blood to tissue would be a valuable future direction. Here and for other studies, the possibility cannot be excluded of changed surface mechanics, especially with a gel. Tissue is reactive to surface mechanical cues. Whilst not necessarily due to mechanical change, in one study, soft electrospun gelatin coatings on polyurethane fibres over sensors reduced fibrous encapsulation, as compared with non-coated fibre [95].

As an alternative to the registration of H_2O_2 product, Gough has advanced the use of cathodic O_2 measurement. In one study, a surgically fully implanted sensor was operated for a year [96]. A dual sensor arrangement was necessitated here, with a second, non-enzymic, O_2 sensor compensating for background tissue oxygen variation. Glucose oxidase was used in a crosslinked gel and, whilst there was a tissue reaction and a steep response decline with the secondary oxygen sensor (Figure 8), the dual O₂ approach enabled glucose monitoring after two weeks. The wide statistical spread seen in responses for different oxygen sensor implants indicates the variability of the local tissue response. Whilst such a protracted stabilisation delay is an option, it would only seem so if long term implantation is contemplated, and that demands a high level of confidence in a sensor that needs surgical implantation. A subsequent six month human study with this sensor demonstrated stable oxygen compensated glucose tracking [97]. The collagen capsule imposed response delay was of the order of 10 min, so workable for clinical purposes.

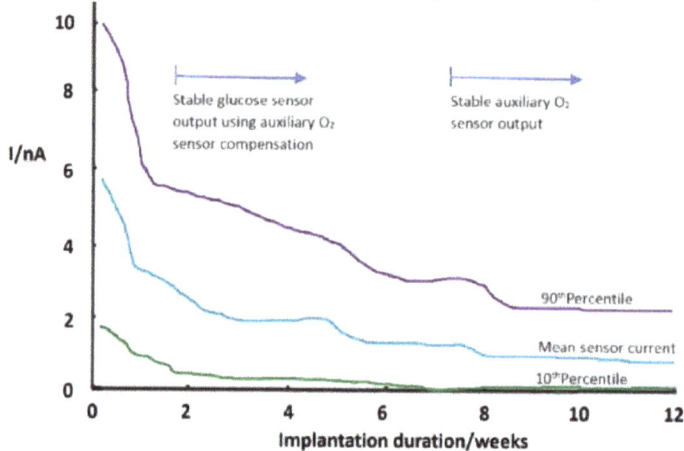

Figure 8. Auxiliary oxygen sensor current decay in subcutaneously implanted glucose sensors in pigs. Data represent one week moving averages of daily mean sampled currents and the spread of data for 60 electrodes. Adapted from [96].

Boronic acid is capable of reversible binding with saccharides and a diboronate system with an attached fluorophore sensitive to glucose binding induced conformational change has been reported for intravascular glucose monitoring [98]. This is an attractive possibility because of the theoretical independence from the need for ongoing glucose mass transport during measurement. However, there was a need for gel containment of the affinity molecule with membrane barriers to prevent access to micro- and macrosolutes into the affinity phase. A special need was for an outer Pt modified membrane to degrade damaging low level peroxides from blood. Response times of over 5 min. would not be particularly slower than most electrochemical sensors. No interference was seen with potential blood constituents, and use of membranes eliminated the effect of blood haematocrit on response. Correlation data, whilst acceptable, appeared no superior to that seen for tissue electrochemical sensors; whether the affinity principle provides for greater reliability and independence from fouling would need detailed comparator studies. However, it is unlikely that an electrochemical glucose sensor would ever be a practical proposition for intravascular use; the intended use of the optical sensor was critical care monitoring. Injectable boronic acid gels with non-invasive optical tracking have been reported for glucose, but these constitute a rather different type of monitoring strategy, subject to the challenges of safety and biocompatibility as well as reliable signal extraction through tissue.

5.4. Bioactive Molecule Release for Biocompatibility

Ahyeon, et al. reviewed drug loaded membranes [99]. Dexamethasone, a high potency steroid, can suppress inflammation and late fibrosis; VEGF (vascular endothelial growth factor) can promote vascularisation to augment glucose delivery. However, biological complexities may becme evident. Vallejo-Heligon et al. [100] found dexamethasone loaded polyurethane to both suppress inflammation and to promote neovascularisation, extending the sensor operating period; however, its combination with VEGF led to the depression of VEGF stimulated neovascularisation [101]. Despite a high early implantation effectiveness of such loaded membranes, eventual reservoir depletion is a potential drawback, especially given the use of thin, low capacity membranes. One study instead delivered VEGF from a cannula while using an osmotic pump [102]. Neovascularisation was demonstrated, evident at least 40 days at a tissue distance of 1.3 cm. Composite material designs might also extend function life time. Dexamethasone when loaded onto poly (lactic-co-glycolic acid) (PLGA) microspheres of two molecular weights provided early and late release with bioactivity retained for six months [103]; this wasconditioned by the differential degradation rates of the two polymers. With microspheres embedded within porous polyvinyl alcohol (PVA), any surface fouling was offset by porosity recovery as the microspheres degraded [104].

Bioactive NO releasing membranes have also been developed [105] with the aim of suppressing pro-inflammatory cytokine release and thereby inflammatory cell recruitment. Stabilised sensor operation depended again on a maintained NO reserve. Interestingly, here, the sensor run-in period was short. A NO pharmacological effect might offer a clue to some altered tissue response, with perhaps a link to cell signalling. NO is electroactive at anodic voltages, but, if its interference is constant, then monitoring would be possible. In regard to this, the authors postulated that the variable generation of endogenous NO by inflammatory tissue might contribute to signal instability.

The widening repertoire of clinical therapeutic agents, including biologics, should provide a new generation of anti-inflammatory agents. Masitinib, a small molecule agent used to treat mast cell tumours, is one example. Mast cells are central to the tissue response; they are immediately deployed in inflammation, undergo ready degranulation, and through this variously release proinflammatory cytokines, serotonin, and histamine, and thus accelerate inflammation. Masitinib works by blocking mast cell receptors, thereby suppressing intracellular kinase signalling and stabilising the cell membrane. However, when used as a released agent in a sensor [106], the protective effect was limited, certainlynot as much as might have been expected from drug potency.

5.5. Tissue Reactivity to Implants

For bulk dependent implant devices, the tissue reaction is of a lower order concern, but for surface response dependent sensors, even a minor reaction can have profound effects. When observed in the opposite sense, tissue is actually the more sophisticated sensor. Thus, it has mobile surveillance through its constituent cells, a strong capacity for recognising, even the smallest of foreign body intrusions as 'non-self' and an ability to resolve shape; whatever way any three-dimensional (3D) object is packaged, or disguised it is readily recognised. From this recognition starting point, a cascade is established, embodied in the Foreign Body Reaction. This is designed to degrade the intrusion, and failing that, to package and isolate it behind a fibrous capsular wall. It is too fundamental a part of the armoury of an organism, linked to its very survival, to be readily countered. While the outcome for the technology is seriously adverse, for the biology it is an unmitigated success. The current state of the art is that, whilst we have dissected the complex response pathways, it is our understanding of these that lags behind [107,108].

On implantation in tissue, as in blood, rapid protein deposition takes place and the device is already packaged by a layer that, though conditioned by the original surface, itself then goes on to condition the subsequent response (Figure 9). Early protein reorganisation, layer accumulation, and denaturation characterise this initial growing protein layer. The tissue cellular tissue response is subsequently affected by specific receptor binding to the adsorbed proteins. The first responders are exploratory polymorphonuclear leucocytes (neutrophils) and, with mast cells, they initiate diffuse chemotactic signalling to attract other phagocytic cells (macrophages). These amplify the directional signalling and recruit even more macrophages. The bioactive factors released include PDGF (platelet derived growth factor), TNF-α (tumor necrosis factor alpha), and IL-6 (interleukin-6). To this mix are added monocytes from blood, together with proinflammatory mediators, replenishing the macrophage pool. If phagocytosis against the device fails, macrophages on the surface fuse to make more effective multinucleated foreign body giant cells, through a trigger that is unknown [109]. There is also an outpouring of degradative diffusible agents with no other purpose than to solubilise the intrusion, which includes acidic cell contents, oxygen free radicals, and hydrolytic enzymes. Ultimately, degradation might be a highly desirable biomaterials outcome if the agent is a surgical suture, but, if it is a sensor membrane, it becomes a clear problem. Preferential degradation of the soft segments of a polyurethane used for glucose sensors illustrates this [110].

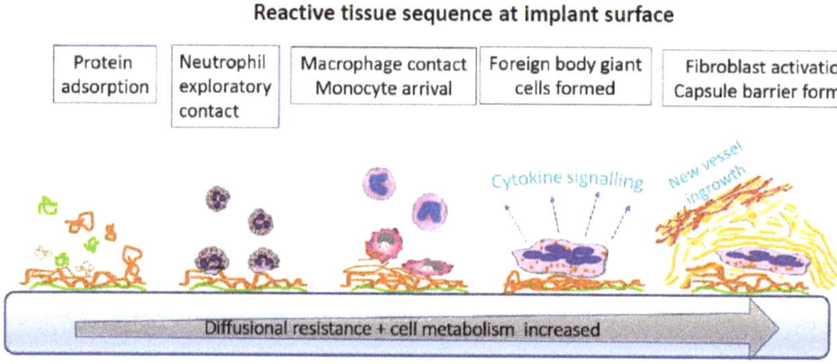

Figure 9. Schematic of tissue foreign body response in sequence: (**1**) Rapid protein deposition masks sensor surface; deposited layer increases. (**2**) Tissue neutrophils sense the surface and send chemotactic signals, mast cells promote inflammatory background. (**3**) Macrophages accumulate with population reinforcement by blood monocytes. (**4**) Failure to degrade surface stimulates more powerful multinucleated giant cell formation from macrophages, with enhanced signalling. (**5**) End stage of more quiescent collagen formation and cumulative barrier formation by fibroblasts with parallel neovasularisation.

If the implant stimulus persists over days or weeks, a more cellularly heterogenous inflammatory cell architecture is built up with added lymphocytes and plasma cells. Healing then ensues if there is no outright toxicity. This progresses behind a cell layer adjacent to the device, and variously hosts a dense network of inwardly directed blood vessels, fibroblasts, and macrophages; this is granulation tissue. This is also the remodeling phase of the response, and a precursor to final collagen capsule deposition by fibroblasts. This sequence of events around a non-toxic implant is pre-programmed constant, and refractory to control with only its quantitative aspects varying across different materials [111] or through suppression regimens.

5.6. Tissue Reaction Implications for Glucose Sensors

Some microhaemorrhage is inevitable during any implantation. The locally released RBCs can then become a sink for glucose, though not for oxygen. After early RBC removal and the entry of more actively metabolising nucleated cells, glucose and oxygen access to the sensor can both become reduced. The direct injection of macrophages to a sensor implant site reproduced this effect [112], but, interestingly, but not lymphocyte injection despite the equivalent metabolic activity of these cells. Within days, a rapid population change occurs with an order of magnitude expansion of neutrophil number followed by decay and a commensurate increase in lymphocyte number. No differences in this tissue response sequence was seen in one study, regardless of whether or not the sensor was operational and releasing toxic H_2O_2 into tissue [113]. What is a constant with all devices is that final fibrous capsule formation is inevitable and it becomes the arbiter of what is then 'seen' by the sensor.

Studies of beyond a week confirm such capsular development and its effect on glucose exchanges, in one example leading to a 24 min. response lag following intravenous glucose [114]. The result was also consistent with the arrival time of injected fluorescent glucose analogue. True physiological glucose exchange between blood and tissue is considered to be quite rapid, requiring < 5 min. for completion, so implanted sensors clearly create an artefactual delay.

The collagen capsule, far from being a simple, static accumulation of collagen fibres, is an evolving structure with its own vascular network and an internal palisade of cells apposed to the sensor. Novak et al. [115] took this structural duality into account in their modeling of glucose transport. They concluded that lag time was determined by capsular thickness, whilst sensitivity was a function of capsular porosity and local vascularity. Additional effects of macrophages and adipocytes metabolism were small here. By contrast, glucose losses due to local cell metabolism were evident in a cell loaded fibrin gel, and further accentuated by exposure to a pro-inflammatory agent [116], again highlighting cellular influence on an inflammatory matrix.

Our own studies on implant sensor stability have led us to a materials independent strategy. Whilst inflammation comprises a hypercellular environment with distinct histological characteristics, it is also a zone of high, histologically silent, fluid influx due to permeabilised capillaries delivering protein rich fluid, passaged then to the lymphatics, and also part returned to the microcirculation. A balance of hydrostatic and osmotic transcapillary pressures drives the fluid flow, as embodied in the Starling mechanism [117]. Our approach here was simply to deliver extra protein-free fluid to the implant site. This utilised an electrode-cannula coaxial arrangement for low volume fluid delivery around the sensor tip (Figure 10A). Subcutaneous tissue has a negative hydrostatic pressure, and the arrangement used pumpless tissue driven flow. The resulting locally reduced protein load both reduced sensor fouling and stabilised response. The response lag time with respect to blood was eliminated and unusually, the measured tissue levels matched blood without in vivo correction.

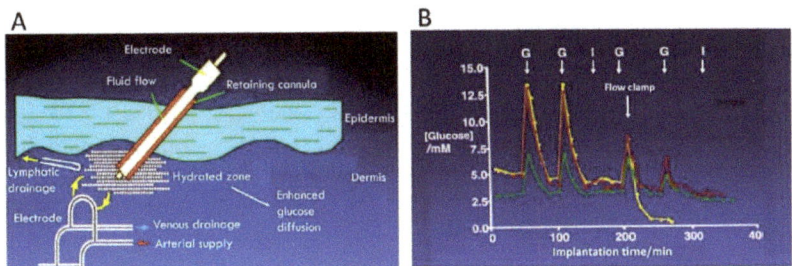

Figure 10. (**A**) Schematic of subcutaneously implanted glucose needle electrode within open ended cannula for delivering fluid around the implanted sensor to create a limited hydrated zone. (**B**) Subcutaneous glucose monitoring in rat (•) venous blood glucose, tissue glucose at 60 µL/h microflow (◦) and at a constrained flow of 10 µL/h (•) showing underestimated glucose and total loss of response with clamped flow. Bolus tail vein administration of glucose (G) and insulin (I). Adapted from [83].

Accordingly, far from local glucose being diluted, glucose access was likely to have been enhanced through the well hydrated, open structure, interstitial tissue space (Figure 10B) [83]. Hence, the conclusion of the approach is that a fluidized zone with rapid glucose exchanges with the blood compartment is obtained. As both a practical and model system, fluid management of the interstitium could provide an alternate means of manipulating the implant inflammatory environment.

6. Lactate

As the end stage metabolite of anaerobic respiration, lactate offers a quantitative measure of hypoxic and shock states where the peripheral tissue O_2 supply is compromised. As such, it has formed a core means of tracking the severity of such states and their response to treatment. It is inevitably subject to rapid change, but, despite this, clinical continuous monitoring is not available. Some experimental work on lactate oxidase sensor based in vivo monitoring in brain has been conducted. Here, oxygen co-substrate limitation at the enzyme could potentially lead to underestimated lactate levels, so an oxygen discriminating membrane has been one option [118]. Alternatively, stoichiometric regeneration of oxygen from H_2O_2 product has been tried using incorporated CeO_2 catalytic particles, and have improved lactate response during brain monitoring in hypoxic rats [119]. As with glucose, without an independent reference measurement in tissue, true tissue lactate level is difficult confirm. Direct validation is possible in the case of an intravascular sensor, and anti-thrombotic NO release from such a sensor for added device stabilisation has been reported with rapid response to both lactate and hypoxia [120]. However, with subcutaneous deployment, a substantially blunted and delayed response was seen. Moreover, in this pig model, upper vs. lower body implantation altered responses. We have also found a blood-tissue discrepancy in experimental shock with subcutaneous electode site dependent output [121].

The scale of the tissue mismatch well exceeds that seen with glucose and suggests that, at least under shock conditions, there might be an added barrier to lactate release from the circulation. This should not occur at the capillaries, which are not selective for micro-solute, but possibly in the interstitial tissue space, which, as a polyelectrolyte, might create an ionomeric barrier to the lactate anion. In diffusion through cartilage, we found the diffusion coefficient for ascorbate anion to be a small percentage of that for similar size neutral molecules [55]. The complexities of blood tissue compartmentalisation were shown in a study of muscle interstitial tissue [122]. Here, microdialysis sampling demonstrated tissue lactate at rest to be double that of plasma water at rest, butlevels converged with plasma during exercise elevations, whilst glucose at rest was about half that of plasma, but again converged with plasma during exercise. The results indicated that muscle can control its extracellular environment. Subcutaneous connective tissue will not have this capability. Our exercise study with subcutaneous tissue microdialysis did not make a similar comparison, but it showed a blunted lactate response, even delayed to the post-exercise period during which tissue glucose appeared to falling [123]. Therefore, it is clear that study of different interstitial locations and comparisons between techniques for lactate are needed to help understand intercompartmental exchange, for if we do not understand these, our understanding of events even for the traditional blood compartment will be limited.

7. Conclusions

Practical in vivo sensors are a unique sub-set of electrochemical sensors, and they constitute a distinct practical offering when compared to the fundamental electrochemistry studies on, say, cell signaling molecules and CNS neurotransmitters. Unlike many sensor types, including industrial, they operate not merely in a hostile environment, but one that is active, reactive, and protean in its nature. This rather counters the idea of biology as a benignmatrix presenting mild solution conditions. The recruitment of high surface activity and destructive cells in high numbers renders the implant site far from representing the normal physiological state locally and provides evidence of a contrived effort at sensor disruption. However, our quest for data immediacy on some variables in the acutely ill patient requires just such sensors.

There also remains a need for repertoire expansion into a broader palette of intermediary metabolites, as these interact dynamically, and they will give added clinical information. Currently, we better understand that, because of compartmental differences, there is an even stronger case for developing the tools for examining these separate entities at different locations.

Electrochemical sensors have been the mainstay of such endeavours, and this review has highlighted the insights that they have given us. This reinforces the need to resolve the generic problem of biocompatibility. It has been an inappropriate quest in many ways to search for the single material or surface that absolves us from this problem—the result has simply been more model systems. The quest needs to be far more deeply rooted in the study of the reactive biology. If nothing else, we have learned that this reactivity is not surface restricted and has a signalling hinterland remote from the surface.

By addressing the right issues, electrochemical sensors will be able to expand their service from physiology to precision medicine. Additionally, future development of closed loop feedback control and autonomous therapeutic management will become feasible. Much of this capability is in place, including sensing chemistry, but it is the biological control of the implant site that remains to be resolved. A far better, multi-parameter, understanding of the individual's dynamic bio-signature might emerge from this, and would be in step with the needs of individualised therapy, currently advancing through genomic profiling.

The implications for future development are that, from our experience of in vivo operation, it is now clear that there are generic material and membrane needs that have to be satisfied to provide the correct contact surfaces in vivo and through this a way of reducing the body's reactivity. This might be only partially achieved, as has been evident with biomaterials more generally, but still tailoring of the device packaging now should be the priority. Success here can help to bring the field forward, as has been evident in the case of implantable electronics. Semi-implantable sensors should be focused on as these can allow ready removal and replacement. Further work with miniaturisation and multiple arrays through microfabrication should also be undertaken by way of creating more reproducible systems and for evolving less intrusive sensors. This will provide for greater measurement confidence through multiple redundancy. What seems less of a need is the invention of ever more chemistries; transduction advances will not deliver practical value without the interfacing and miniaturisation effort. The review has only focused on a few target species, but, from the point of view of critical care, they are entirely sufficient. If, finally, we can achieve a paradigm shift by creating reversible label-free bioaffinity then a whole span of protein, hormone, metabolite and drug species come into scope. The lack of rapidly reversible immunobinding may be a problem for in vitro *assay*, but it is a major drawback to dynamic monitoring. It is, perhaps, time to also look around for rapid reversibility systems that exploit cell membrane receptor principles.

Funding: The authors would like to thank the EPSRC for financial support through Grant EP/H009744/1.

Conflicts of Interest: The author declares no conflict of interest, and neither sponsors nor any other party had a role into the materials collection or interpretation in this review.

References

1. Walker, J.L.; Brown, H.M. Intracellular ionic activity measurements in nerve and muscle. *Physiol. Rev.* **1977**, *57*, 729–778. [CrossRef] [PubMed]
2. Land, S.C.; Porterfield, D.M.; Sanger, R.H.; Smith, P.J.S. The self-referencing oxygen-selective microelectrode: Detection of transmembrane oxygen flux from single cells. *J. Exp. Biol.* **1999**, *202*, 211–218. [PubMed]
3. Wians, E.H. Clinical Laboratory Tests: Which, Why, and What Do The Results Mean? *Labmedicine* **2009**, *40*, 105–113. [CrossRef]
4. Rodenburg, R.J.T.; Schoonderwoerd, G.C.; Tiranti, V.; Taylor, R.W.; Rotig, A.; Valente, L.; Invernizzi, F.; Chretien, D.; He, L.; Backx, G.; et al. A multi-center comparison of diagnostic methods for the biochemical evaluation of suspected mitochondrial disorders. *Mitochondrion* **2013**, *13*, 36–43. [CrossRef]
5. Laje, R.; Agostino, P.V.; Golombek, D.A. The Times of Our Lives: Interaction among Different Biological Periodicities. *Front. Integr. Neurosci.* **2018**, *12*. [CrossRef] [PubMed]
6. Ries, A.L.; Fedullo, P.F.; Clausen, J.L. Rapid changes in arterial blood gas levels after exercise in pulmonary patients. *Chest* **1983**, *83*, 454–456. [CrossRef] [PubMed]
7. Oropello, J.M.; Manasia, A.; Hannon, E.; Leibowitz, A.; Benjamin, E. Continuous fiberoptic arterial and venous blood gas monitoring in hemorrhagic shock. *Chest* **1996**, *109*, 1049–1055. [CrossRef] [PubMed]

8. Heining, M.P.D.; Linton, R.A.F.; Band, D.M. Continuous intravascular monitoring of plasma ionized calcium. In *Ion Measurements in Physiology and Medicine*, 1st ed.; Kessler, M., Harrison, D.K., Hoper, J., Eds.; Springer: Berlin, Germany, 1985; pp. 292–296.
9. Linton, R.A.F.; Lim, M.; Band, D.M. Continuous intravascular monitoring of plasma potassium using potassium-selective electrodes. *Crit. Care Med.* **1982**, *10*, 337–340. [CrossRef]
10. Lim, M.; Linton, R.A.F.; Band, D.M. Continuous intravascular monitoring of ephinephrine-induced changes in plasma potassium. *Anesthesiology* **1982**, *57*, 272–278. [CrossRef] [PubMed]
11. Drake, H.F.; Smith, M.; Corfield, D.R.; Treasure, T. Continuous multi-channel intravascular monitoring of the effects of dopamine and dobutamine on plasma potassium in dogs. *Intensive Care Med.* **1989**, *15*, 446–451. [CrossRef] [PubMed]
12. Band, D.M.; Semple, S.J. Continuous measurement of blood pH with an indwelling arterial glass electrode. *J. Appl. Physiol.* **1967**, *22*, 584–587. [CrossRef]
13. Lang, D.A.; Matthews, D.R.; Peto, J.; Turner, R.C. Cyclic oscillations of basal plasma glucose and insulin concentrations in human beings. *N. Engl. J. Med.* **1979**, *301*, 1023–1027. [CrossRef] [PubMed]
14. Cater, D.B.; Silver, I.A. Quantitative measurements of oxygen tension in normal tissues and in the tumours of patients before and after radiotherapy. *Acta Radiol.* **1960**, *53*, 233–256. [CrossRef]
15. Tracinski, M.; Silver, I.A. Tissue oxygen tension and brain sensitivity to hypoxia. *Respir. Physiol.* **2001**, *128*, 263–276. [CrossRef]
16. Lubbers, D.W.; Baumgartl, H. Heterogeneities and profiles of oxygen pressure in brain and kidney as examples of the pO(2) distribution in the living tissue. *Kidney Int.* **1997**, *51*, 372–380. [CrossRef] [PubMed]
17. Lubbers, D.W.; Baumgartl, H.; Zimelka, W. Heterogeneity and stability of local PO_2 distribution within the brain tissue. *Adv. Exp. Med. Biol. Oxyg. Transp. Tissue XV* **1994**, *345*, 567–574.
18. Hill, J.L.; Gettes, L.S.; Lynch, M.R.; Hebert, N.C. Flexible valinomycin electrodes for online determination of intravascular and myocardial K^+. *Am. J. Physiol.* **1978**, *235*, H455–H459. [PubMed]
19. Hallen, J.; Sejersted, O.M. Intravasal use of pliable K^+-selective electrodes in the femoral vein of humans during exercise. *J. Appl. Physiol.* **1993**, *75*, 2318–2325. [CrossRef]
20. Paterson, D.J.; Estavillo, J.A.; Nye, P.C.G. The effect of hypoxia on plasma potassium concentration and the excitation of the arterial chemoreceptors in the cat. *Q. J. Exp. Physiol. CMS* **1988**, *73*, 623–625. [CrossRef]
21. Webb, S.C.; Canepaanson, R.; Rickards, A.F.; Poole-Wilson, P.A. Myocardial potassium loss after acute coronary occlusion in humans. *J. Am. Coll. Cardiol.* **1987**, *9*, 1230–1234. [CrossRef]
22. Watanabe, I.; Saito, S.; Ozawa, Y.; Hatano, M.; Gettes, L.S. Continuous coronary venous K^+ monitoring in myocardial ischaemia in swine heart. *Jpn. Circ. J.* **1988**, *52*, 1019–1020.
23. Bourdillon, P.D.; Bettmann, M.A.; McCracken, S.; Poole-Wilson, P.A.; Grossman, W. Effects of a new non-ionic and a conventional ionic contrast agent on coronary sinus ionised calcium and left ventricular hemodynamics in dogs. *J. Am. Coll. Cardiol.* **1985**, *6*, 845–853. [CrossRef]
24. Meruva, R.K.; Meyerhoff, M.E. Catheter-Type sensor for potentiometric monitoring of oxygen, pH and carbon dioxide. *Biosens. Bioelectron.* **1998**, *13*, 201–212. [CrossRef]
25. Cobbe, S.M.; Poole-Wilson, P.A. Continuous coronary sinus and arterial pH monitoring during pacing induced ischaemia in coronary artery disease. *Br. Heart J.* **1982**, *47*, 369–374. [CrossRef]
26. Khalid, A.; Peng, L.; Arman, A.; Warren-Smith, S.C.; Schartner, E.P.; Sylvia, G.M.; Hutchinson, M.R.; Ebendorff-Heidepriem, H.; McLaughlin, R.A.; Gibson, B.C.; et al. Silk: A bio-derived coating for optical fibre sensing applications. *Sens. Actuators B Chem.* **2020**, *311*, 127864. [CrossRef]
27. Jin, W.Z.; Jiang, J.J.; Wang, X.; Zhu, X.D.; Wang, G.F.; Song, Y.L.; Bai, C.X. Continuous intra-arterial blood pH monitoring in rabbits with acid-base disorders. *Respir. Physiol. Neurobiol.* **2011**, *177*, 183–188. [CrossRef]
28. Canovas, R.; Sanchez, S.P.; Parrilla, M.; Cuartero, M.; Crespo, G.A. Cytotoxicity Study of Ionophore-Based Membranes: Toward On Body and in Vivo Ion Sensing. *ACS Sens.* **2019**, *4*, 2524–2535. [CrossRef]
29. Jiang, X.J.; Wang, P.; Liang, R.N.; Qin, W. Improving the Biocompatibility of Polymeric Membrane Potentiometric Ion Sensors by Using a Mussel-Inspired Polydopamine Coating. *Anal. Chem.* **2019**, *91*, 6424–6429. [CrossRef]
30. Murphy, S.M.; Hamilton, C.J.; Davies, M.L.; Tighe, B.J. Polymer membranes in clinical sensor applications. 2. The design and fabrication of permselective hydrogels for electrochemical devices. *Biomaterials* **1992**, *13*, 979–990. [CrossRef]

31. Berrocal, M.J.; Johnson, R.D.; Badr, I.H.A.; Liu, M.D.; Gao, D.Y.; Bachas, L.G. Improving the blood compatibility of ion-selective electrodes by employing poly(MPC-co-BMA), a copolymer containing phosphorylcholine, as a membrane coating. *Anal. Chem.* **2002**, *74*, 3644–3648. [CrossRef]
32. Brooks, H.A.; Allen, J.R.; Feldhoff, P.W.; Bachas, L.G. Effect of surface-attached heparin on the response of potassium-selective electrodes. *Anal. Chem.* **1996**, *68*, 1439–1443. [CrossRef] [PubMed]
33. EspadasTorre, C.; Oklejas, V.; Mowery, K.; Meyerhoff, M.E. Thromboresistant chemical sensors using combined nitric oxide release ion sensing polymeric films. *J. Am. Chem. Soc.* **1997**, *119*, 2321–2322. [CrossRef]
34. Vassilev, P.; Koper, M.T.M. Electrochemical reduction of oxygen on gold surfaces: A density functional theory study of intermediates and reaction paths. *J. Phys. Chem. C* **2007**, *111*, 2607–2613. [CrossRef]
35. Bratanow, N.; Polk, K.; Bland, R.; Kram, H.B.; Lee, T.S.; Shoemaker, W.C. Continuous polarographic monitoring of intra-arterial oxygen in the peroperative period. *Crit. Care Med.* **1985**, *13*, 859–860. [CrossRef] [PubMed]
36. Green, G.E.; Hassell, K.T.; Mahutte, C.K. Comparison of arterial blood gas with continuous intra-arterial and trans-cutaneous pO2 sensors in adult critically ill patients. *Crit. Care Med.* **1987**, *15*, 491–494. [CrossRef]
37. Mahutte, C.K. On-line arterial blood gas analysis with optodes: Current status. *Clin. Biochem.* **1998**, *31*, 119–130. [CrossRef]
38. Ren, H.; Coughlin, M.A.; Major, T.C.; Aiello, S.; Pena, A.R.; Bartlett, R.H.; Meyerhoff, M.E. Improved in Vivo Performance of Amperometric Oxygen (PO2) Sensing Catheters via Electrochemical Nitric Oxide Generation/Release. *Anal. Chem.* **2015**, *87*, 8067–8072. [CrossRef]
39. Widness, J.A.; Kulhavy, J.C.; Johnson, K.J.; Cress, G.A.; Kromer, I.J.; Acarregui, M.J.; Feld, R.D. Clinical performance of an in-line point-of-care monitor in neonates. *Pediatrics* **2000**, *106*, 497–504. [CrossRef]
40. Norde, W. My voyage of discovery to proteins in flatland ... and beyond. *Colloid Surf. B* **2008**, *61*, 1–9. [CrossRef]
41. Noh, H.; Vogler, E.A. Volumetric interpretation of protein adsorption: Competition from mixtures and the Vroman effect. *Biomaterials* **2007**, *28*, 405–422. [CrossRef]
42. Brash, J.L. Protein Surface Interactions and Biocompatibility: A Forty Year Perspective. *ACS Symp. Ser. Proteins Interfaces III State Art* **2012**, *1120*, 277–300.
43. Pawlak, M.; Bakker, E. Chemical Modification of Polymer Ion-Selective Membrane Electrode Surfaces. *Electroanalysis* **2014**, *26*, 1121–1131. [CrossRef]
44. Brash, J.L.; Horbett, T.A.; Latour, R.A.; Tengvall, P. The blood compatibility challenge. Part 2: Protein adsorption phenomena governing blood reactivity. *Acta Biomater.* **2019**, *94*, 11–24. [CrossRef] [PubMed]
45. Modinger, Y.; Teixeira, G.Q.; Neidlinger-Wilke, C.; Ignatius, A. Role of the Complement System in the Response to Orthopedic Biomaterials. *Int. J. Mol. Sci.* **2018**, *19*, 3367. [CrossRef] [PubMed]
46. Ekdahl, K.N.; Teramura, Y.; Hamad, O.A.; Asif, S.; Duehrkop, C.; Fromell, K.; Gustafson, E.; Hong, J.; Kozarcanin, H.; Magnusson, P.U.; et al. Dangerous liaisons: Complement, coagulation, and kallikrein/kinin cross-talk act as a linchpin in the events leading to thromboinflammation. *Immunol. Rev.* **2016**, *274*, 245–269. [CrossRef]
47. Sotiri, I.; Robichaud, M.; Lee, D.; Braune, S.; Gorbet, M.; Ratner, B.D.; Brash, J.L.; Latour, R.A.; Reviakine, I. BloodSurf 2017: News from the blood-biomaterial frontier. *Acta Biomater.* **2019**, *87*, 55–60. [CrossRef] [PubMed]
48. Gorbet, M.B.; Sefton, M.V. Biomaterial-associated thrombosis: Roles of coagulation factors, complement, platelets and leukocytes. *Biomaterials* **2004**, *25*, 5681–5703. [CrossRef]
49. Chen, L.; Han, D.; Jiang, L. On improving blood compatibility: From bioinspired to synthetic design and fabrication of biointerfacial topography at micro/nano scales. *Colloid Surf. B* **2011**, *85*, 2–7. [CrossRef]
50. Sundaram, S.; Lim, F.; Cooper, S.L.; Colman, R.W. Role of leucocytes in coagulation induced by artificial surfaces: Investigation of expression of Mac-1, granulocyte elastase release and leucocyte adhesion on modified polyurethanes. *Biomaterials* **1996**, *17*, 1041–1047. [CrossRef]
51. Ward, W.K.; Wood, M.D.; Slobodzian, E.P. Continuous amperometric monitoring of subcutaneous oxygen in rabbit by telemetry. *J. Med. Eng. Technol.* **2002**, *26*, 158–167. [CrossRef]
52. Ward, W.K.; Van Albert, S.; Bodo, M.; Pearce, F.; Gray, R.; Harlson, S.; Rebec, M.V. Design and Assessment of a Miniaturized Amperometric Oxygen Sensor in Rats and Pigs. *IEEE Sens. J.* **2010**, *10*, 1259–1265. [CrossRef]
53. Makale, M.T.; Jablecki, M.C.; Gough, D.A. Mass transfer and gas-phase calibration of implanted oxygen sensors. *Anal. Chem.* **2004**, *76*, 1773–1777. [CrossRef]

54. Kumosa, L.S.; Routh, T.L.; Lin, J.T.; Lucisano, J.Y.; Gough, D.A. Permeability of subcutaneous tissues surrounding long-term implants to oxygen. *Biomaterials* **2014**, *35*, 8287–8296. [CrossRef]
55. Adatia, K.; Raja, M.; Vadgama, P. An electrochemical study of microporous track-etched membrane permeability and the effect of surface protein layers. *Colloids Surf. B* **2017**, *158*, 84–92. [CrossRef]
56. Intaglietta, M.; Johnson, P.C.; Winslow, R.M. Microvascular and tissue oxygen distribution. *Cardiovasc. Res.* **1996**, *32*, 632–643. [CrossRef]
57. Vovenko, E. Distribution of oxygen tension on the surface of arterioles, capillaries and venules of brain cortex and in tissue in normoxia: An experimental study on rats. *Pflug. Arch. Eur. J. Physiol.* **1999**, *437*, 617–623. [CrossRef]
58. Sharan, M.; Vovenko, E.P.; Vadapalli, A.; Popel, A.S.; Pittman, R.N. Experimental and theoretical studies of oxygen gradients in rat pial microvessels. *J. Cerebr. Blood Flow Metab.* **2008**, *28*, 1597–1604. [CrossRef]
59. Finnerty, N.J.; Bolger, F.B. In Vitro development and In Vivo application of a platinum-based electrochemical device for continuous measurements of peripheral tissue oxygen. *Bioelectrochemistry* **2018**, *119*, 124–135. [CrossRef]
60. Russell, D.M.; Garry, E.M.; Taberner, A.J.; Barrett, C.J.; Paton, J.F.R.; Budgett, D.M.; Malpas, S.C. A fully implantable telemetry system for the chronic monitoring of brain tissue oxygen in freely moving rats. *J. Neurosci. Methods* **2012**, *204*, 242–248. [CrossRef]
61. Weltin, A.; Ganatra, D.; Konig, K.; Joseph, K.; Hofmann, U.G.; Urban, G.A.; Kieninger, J. New life for old wires: Electrochemical sensor method for neural implants. *J. Neural Eng.* **2019**, *17*, 016007. [CrossRef]
62. Lee, G.J.; Kim, S.K.; Kang, S.W.; Kim, O.K.; Chae, S.J.; Choi, S.; Shin, J.H.; Park, H.K.; Chung, J.H. Real time measurement of myocardial oxygen dynamics during cardiac ischemia-reperfusion of rats. *Analyst* **2012**, *137*, 5312–5319. [CrossRef]
63. Valadka, A.B.; Gopinath, S.P.; Contant, C.F.; Uzura, M.; Robertson, C.S. Relationship of brain tissue Po-2 to outcome after severe head injury. *Crit. Care Med.* **1998**, *26*, 1576–1581. [CrossRef]
64. Nortje, J.; Gupta, A.K. The role of tissue oxygen monitoring in patients with acute brain injury. *Br. J. Anaesth.* **2006**, *97*, 95–106. [CrossRef]
65. Kennedy, A.; Ng, C.T.; Chang, T.C.; Biniecka, M.; O'Sullivan, J.N.; Heffernan, E.; Fearon, U.; Veale, D.J. Tumor Necrosis Factor Blocking Therapy Alters Joint Inflammation and Hypoxia. *Arthritis Rheum.* **2011**, *63*, 923–932. [CrossRef]
66. Dewhirst, M.W.; Klitzman, B.; Braun, R.D.; Brizel, D.M.; Haroon, Z.A.; Secomb, T.W. Review of methods used to study oxygen transport at the microcirculatory level. *Int. J. Cancer* **2000**, *90*, 237–255. [CrossRef]
67. Collingridge, D.R.; Young, W.K.; Vojnovic, B.; Wardman, P.; Lynch, E.M.; Hill, S.A.; Chaplin, D.J. Measurement of tumor oxygenation: A comparison between polarographic needle electrodes and a time-resolved luminescence-based optical sensor. *Radiat. Res.* **1997**, *147*, 329–334. [CrossRef]
68. Vaupel, P.; Hockel, M.; Mayer, A. Detection and characterization of tumor hypoxia using pO(2) histography. *Antioxid. Redox Signal.* **2007**, *9*, 1221–1235. [CrossRef]
69. De Santis, V.; Singer, M. Tissue oxygen tension monitoring of organ perfusion: Rationale, methodologies, and literature review. *Br. J. Anaesth.* **2015**, *115*, 357–365. [CrossRef]
70. Ngwenya, L.B.; Burke, J.F.; Manley, G.T. Brain Tissue Oxygen Monitoring and the Intersection of Brain and Lung: A Comprehensive Review. *Respir. Care* **2016**, *61*, 1232–1244. [CrossRef]
71. Heller, A.; Feldman, B. Electrochemical glucose sensors and their applications in diabetes management. *Chem. Rev.* **2008**, *108*, 2482–2505. [CrossRef]
72. Menke, A.; Orchard, T.J.; Imperatore, G.; Bullard, K.M.; Mayer-Davis, E.; Cowie, C.C. The Prevalence of Type 1 Diabetes in the United States. *Epidemiology* **2013**, *24*, 773–774. [CrossRef] [PubMed]
73. Vantyghem, M.C.; Press, M. Management strategies for brittle diabetes. *Ann. Enocinol. (Paris)* **2006**, *67*, 287–294. [CrossRef]
74. Mao, F.; Mano, N.; Heller, A. Long tethers binding redox centers to polymer backbones enhance electron transport in enzyme "wiring" hydrogels. *J. Am. Chem. Soc.* **2003**, *125*, 4951–4957. [CrossRef] [PubMed]
75. Hoss, U.; Budiman, E.S.; Liu, H.; Christiansen, M.P. Continuous glucose monitoring in the subcutaneous tissue over a 14-day sensor wear period. *J. Diabetes Sci. Technol.* **2013**, *7*, 1210–1518. [CrossRef]
76. Hoss, U.; Budiman, E.S. Factory-Calibrated Continuous Glucose Sensors: The Science behind the Technology. *Diabetes Technol. Ther.* **2017**, *19*, S44–S50. [CrossRef]

77. Vettoretti, M.; Battocchio, C.; Sparacino, G.; Facchinetti, A. Development of an Error Model for a Factory-Calibrated Continuous Glucose Monitoring Sensor with 10-Day Lifetime. *Sensors* **2019**, *19*, 5320. [CrossRef]
78. Mancini, G.; Berioli, M.G.; Santi, E.; Rogari, F.; Toni, G.; Tascini, G.; Crispoldi, R.; Ceccarini, G.; Esposito, S. Flash Glucose Monitoring: A Review of the Literature with a Special Focus on Type 1 Diabetes. *Nutrients* **2018**, *10*, 992. [CrossRef]
79. Bruen, D.; Delaney, C.; Florea, L.; Diamond, D. Glucose Sensing for Diabetes Monitoring: Recent Developments. *Sensors* **2017**, *17*, 1866. [CrossRef]
80. Chen, C.; Zhao, X.L.; Li, Z.H.; Zhu, Z.G.; Qian, S.H.; Flewitt, A.J. Current and Emerging Technology for Continuous Glucose Monitoring. *Sensors* **2017**, *17*, 182. [CrossRef]
81. Nery, E.W.; Kundys, M.; Jelen, P.S.; Jonsson-Niedziolka, M. Electrochemical Glucose Sensing: Is There Still Room for Improvement? *Anal. Chem.* **2016**, *88*, 11271–11282. [CrossRef]
82. Shichiri, M.; Kawamori, R.; Yamasaki, Y.; Hakui, N.; Abe, H. Wearable artificial endocrine pancreas with needle-type glucose sensor. *Lancet* **1982**, *2*, 1129–1131. [CrossRef]
83. Rigby, G.P.; Crump, P.W.; Vadgama, P. Stabilized needle electrode system for in vivo glucose monitoring based on open flow microperfusion. *Analyst* **1996**, *121*, 871–875. [CrossRef] [PubMed]
84. Rigby, G.P.; Ahmed, S.; Horseman, G.; Vadgama, P. In Vivo glucose monitoring with open microflow—Influences of fluid composition and preliminary evaluation in man. *Anal. Chim. Acta* **1999**, *385*, 23–32. [CrossRef]
85. McGarraugh, G. The Chemistry of Commercial Continuous Glucose Monitors. *Diabetes Technol. Ther.* **2009**, *11*, S17–S24. [CrossRef] [PubMed]
86. Wang, C.; Yu, B.; Knudsen, B.; Harmon, J.; Moussy, F.; Moussy, Y. Synthesis and performance of novel hydrogels coatings for implantable glucose sensors. *Biomacromolecules* **2008**, *9*, 561–567. [CrossRef]
87. Galeska, I.; Hickey, T.; Moussy, F.; Kreutzer, D.; Papadimitrakopoulos, F. Characterization and biocompatibility studies of novel humic acids based films as membrane material for an implantable glucose sensor. *Biomacromolecules* **2001**, *2*, 1249–1255. [CrossRef]
88. Yu, B.Z.; Ju, Y.M.; West, L.; Moussy, Y.; Moussy, F. An investigation of long-term performance of minimally invasive glucose biosensors. *Diabetes Technol. Ther.* **2007**, *9*, 265–275. [CrossRef]
89. Dungel, P.; Long, N.; Yu, B.; Moussy, Y.; Moussy, F. Study of the effects of tissue reactions on the function of implanted glucose sensors. *J. Biomed. Mater. Res. A* **2008**, *85*, 699–706. [CrossRef]
90. Turner, R.F.B.; Harrison, D.J.; Rajotte, R.V. Preliminary in vivo biocompatibility studies on perfluorosulfonic acid polymer membranes for biosensor applications. *Biomaterials* **1991**, *12*, 361–368. [CrossRef]
91. Gerritsen, M.; Kros, A.; Sprakel, V.; Lutterman, J.A.; Nolte, R.J.M.; Jansen, J.A. Biocompatibility evaluation of sol-gel coatings for subcutaneously implantable glucose sensors. *Biomaterials* **2000**, *21*, 71–78. [CrossRef]
92. Quinn, C.A.P.; Connor, R.E.; Heller, A. Biocompatible, glucose-permeable hydrogel for in situ coating of implantable biosensors. *Biomaterials* **1997**, *18*, 1665–1670. [CrossRef]
93. Yang, Y.; Zhang, S.F.; Kingston, M.A.; Jones, G.; Wright, G.; Spencer, S.A. Glucose sensor with improved haemocompatibility. *Biosens. Bioelectron.* **2000**, *15*, 221–227. [CrossRef]
94. Xie, X.; Doloff, J.C.; Yesilyurt, V.; Sadraei, A.; McGarrigle, J.J.; Commis, M.; Veiseh, O.; Farah, S.; Isa, D.; Ghanis, S.; et al. Reduction of measurement noise in a continuous glucose monitor by coating the sensor with a zwitterionic polymer. *Nat. Biomed. Eng.* **2018**, *2*, 894–906. [CrossRef]
95. Burugapalli, K.; Wijesuriya, S.; Wang, N.; Song, W.H. Biomimetic electrospun coatings increase the in vivo sensitivity of implantable glucose biosensors. *J. Biomed. Mater. Res. A* **2018**, *106*, 1072–1081. [CrossRef]
96. Gough, D.A.; Kumosa, L.S.; Routh, T.L.; Lin, J.T.; Lucisano, J.Y. Function of an Implanted Tissue Glucose Sensor for More than 1 Year in Animals. *Sci. Transl. Med.* **2010**, *2*. [CrossRef] [PubMed]
97. Lucisano, J.Y.; Routh, T.L.; Lin, J.T.; Gough, D.A. Glucose Monitoring in Individuals With Diabetes Using a Long-Term Implanted Sensor/Telemetry System and Model. *IEEE Trans. Biomed. Eng.* **2017**, *64*, 1982–1993. [CrossRef] [PubMed]
98. Crane, B.C.; Barwell, N.P.; Gopal, P.; Gopichand, M.; Higgs, T.; James, T.D.; Jones, C.M.; Mackenzie, A.; Mulavisala, K.P.; Paterson, W. The Development of a Continuous Intravascular Glucose Monitoring Sensor. *J. Diabetes Sci. Technol.* **2015**, *9*, 751–761. [CrossRef]
99. Ahyeon, K.; Scott, P.N.; Schoenfisch, M.H. Glucose sensor membranes for mitigating the foreign body response. *J. Diabetes Sci. Technol.* **2011**, *5*, 1052–1059.

100. Vallejo-Heligon, S.G.; Brown, N.L.; Reichert, W.M.; Klitzman, B. Porous, Dexamethasone-loaded polyurethane coatings extend performance window of implantable glucose sensors in vivo. *Acta Biomater.* **2016**, *30*, 106–115. [CrossRef]
101. Norton, L.W.; Koschwanez, H.E.; Wisniewski, N.A.; Klitzman, B.; Reichert, W.M. Vascular endothelial growth factor and dexamethasone release from nonfouling sensor coatings affect the foreign body response. *J. Biomed. Mater. Res. A* **2007**, *81*, 858–869. [CrossRef]
102. Ward, W.K.; Quinn, M.J.; Wood, M.D.; Tiekotter, K.L.; Pidikiti, S.; Gallagher, J.A. Vascularizing the tissue surrounding a model biosensor: How localized is the effect of a subcutaneous infusion of vascular endothelial growth factor (VEGF)? *Biosens. Bioelectron.* **2003**, *19*, 155–163. [CrossRef]
103. Gu, B.; Papadimitrakopoulos, F.; Burgess, D.J. PLGA microsphere/PVA hydrogel coatings suppress the foreign body reaction for 6 months. *J. Control. Release* **2018**, *289*, 35–43. [CrossRef] [PubMed]
104. Vaddiraju, S.; Wang, Y.; Qiang, L.; Burgess, D.J.; Papadimitrakopoulos, F. Microsphere Erosion in Outer Hydrogel Membranes Creating Macroscopic Porosity to Counter Biofouling-Induced Sensor Degradation. *Anal. Chem.* **2012**, *84*, 8837–8845. [CrossRef]
105. Gifford, R.; Batchelor, M.M.; Lee, Y.; Gokulrangan, G.; Meyerhoff, M.E.; Wilson, G.S. Mediation of in vivo glucose sensor inflammatory response via nitric oxide release. *J. Biomed. Mater. Res. A* **2005**, *75*, 755–766. [CrossRef] [PubMed]
106. Avula, M.; Jones, D.; Rao, A.N.; McClain, D.; McGill, L.D.; Grainger, D.W.; Solzbacher, F. Local release of masitinib alters in vivo implantable continuous glucose sensor performance. *Biosens. Bioelectron.* **2016**, *77*, 149–156. [CrossRef] [PubMed]
107. Morais, J.M.; Papadimitrakopoulos, F.; Burgess, D.J. Biomaterials/Tissue Interactions: Possible Solutions to Overcome Foreign Body Response. *AAPS J.* **2010**, *12*, 188–196. [CrossRef]
108. Ratner, B.D. Biomaterials: Been There, Done That, and Evolving into the Future. *Annu. Rev. Biomed. Eng.* **2019**, *21*, 171–191. [CrossRef]
109. Anderson, J.M.; Rodriguez, A.; Chang, D.T. Foreign body reaction to biomaterials. *Semin. Immunol.* **2008**, *20*, 86–100. [CrossRef]
110. Mathur, A.B.; Collier, T.O.; Kao, W.J.; Wiggins, M.; Schubert, M.A.; Hiltner, A.; Anderson, J.M. In Vivo biocompatibility and biostability of modified polyurethanes. *J. Biomed. Mater. Res.* **1997**, *36*, 246–257. [CrossRef]
111. Anderson, J.M. Biological responses to materials. *Annu. Rev. Mater. Res.* **2001**, *31*, 81–110. [CrossRef]
112. Klueh, U.; Frailey, J.T.; Qiao, Y.; Antar, O.; Kreutzera, D.L. Cell based metabolic barriers to glucose diffusion: Macrophages and continuous glucose monitoring. *Biomaterials* **2014**, *35*, 3145–3153. [CrossRef] [PubMed]
113. Henninger, N.; Woderer, S.; Kloetzer, H.M.; Staib, A.; Gillen, R.; Li, L.; Yu, X.L.; Gretz, N.; Kraenzlin, B.; Pill, J. Tissue response to subcutaneous implantation of glucose-oxidase-based glucose sensors in rats. *Biosens. Bioelectron.* **2007**, *23*, 26–34. [CrossRef] [PubMed]
114. McClatchey, P.M.; McClain, E.S.; Williams, I.M.; Malabanan, C.M.; James, F.D.; Lord, P.C.; Gregory, J.M.; Cliffel, D.E.; Wasserman, D.H. Fibrotic Encapsulation Is the Dominant Source of Continuous Glucose Monitor Delays. *Diabetes* **2019**, *68*, 1892–1901. [CrossRef] [PubMed]
115. Novak, M.T.; Yuan, F.; Reichert, W.M. Modeling the relative impact of capsular tissue effects on implanted glucose sensor time lag and signal attenuation. *Anal. Bioanal. Chem.* **2010**, *398*, 1695–1705. [CrossRef]
116. Novak, M.T.; Yuan, F.; Reichert, W.M. Macrophage embedded fibrin gels: An in vitro platform for assessing inflammation effects on implantable glucose sensors. *Biomaterials* **2014**, *35*, 9563–9572. [CrossRef]
117. Wiig, H. Pathophysiology of tissue fluid accumulation in inflammation. *J. Physiol. Lond.* **2011**, *589*, 2945–2953. [CrossRef]
118. Burmeister, J.J.; Palmer, M.; Gerhardt, G.A. L-lactate measures in brain tissue with ceramic-based multisite microelectrodes. *Biosens. Bioelectron.* **2005**, *20*, 1772–1779. [CrossRef]
119. Sardesai, N.P.; Ganesana, M.; Karimi, A.; Leiter, J.C.; Andreescu, S. Platinum-Doped Ceria Based Biosensor for In Vitro and In Vivo Monitoring of Lactate during Hypoxia. *Anal. Chem.* **2015**, *87*, 2996–3003. [CrossRef] [PubMed]
120. Wolf, A.; Renehan, K.; Ho, K.K.Y.; Carr, B.D.; Chen, C.V.; Cornell, M.S.; Ye, M.Y.; Rojas-Pena, A.; Chen, H. Evaluation of Continuous Lactate Monitoring Systems within a Heparinized In Vivo Porcine Model Intravenously and Subcutaneously. *Biosensors* **2018**, *8*, 122. [CrossRef] [PubMed]

121. Rong, Z.M.; Leitao, E.; Popplewell, J.; Alp, B.; Vadgama, P. Needle enzyme electrode for lactate measurement in vivo. *IEEE Sens. J.* **2008**, *8*, 113–120. [CrossRef]
122. MacLean, D.A.; Bangsbo, J.; Saltin, B. Muscle interstitial glucose and lactate levels during dynamic exercise in humans determined by microdialysis. *J. Appl. Physiol.* **1999**, *87*, 1483–1490. [CrossRef] [PubMed]
123. Gowers, S.A.N.; Curto, V.F.; Seneci, C.A.; Wang, C.; Anastasova, S.; Vadgama, P.; Yang, G.Z.; Boutelle, M.G. 3D Printed Microfluidic Device with Integrated Biosensors for Online Analysis of Subcutaneous Human Microdialysate. *Anal. Chem.* **2015**, *87*, 7763–7770. [CrossRef] [PubMed]

© 2020 by the author. Licensee MDPI, Basel, Switzerland. This article is an open access article distributed under the terms and conditions of the Creative Commons Attribution (CC BY) license (http://creativecommons.org/licenses/by/4.0/).

Review

Microfluidics by Additive Manufacturing for Wearable Biosensors: A Review

Mahshid Padash [1,2], Christian Enz [1] and Sandro Carrara [1,*]

1. Laboratory of Integrated Circuits, École Polytechnique Fédérale de Lausanne, CH-2002 Neuchâtel, Switzerland; m.padash@sci.uk.ac.ir or m.padash427@yahoo.com (M.P.); christian.enz@epfl.ch (C.E.)
2. Chemistry Department, Shahid Bahonar University of Kerman, Kerman 76169-13439, Iran
* Correspondence: sandro.carrara@epfl.ch

Received: 29 May 2020; Accepted: 12 July 2020; Published: 29 July 2020

Abstract: Wearable devices are nowadays at the edge-front in both academic research as well as in industry, and several wearable devices have been already introduced in the market. One of the most recent advancements in wearable technologies for biosensing is in the area of the remote monitoring of human health by detection on-the-skin. However, almost all the wearable devices present in the market nowadays are still providing information not related to human 'metabolites and/or disease' biomarkers, excluding the well-known case of the continuous monitoring of glucose in diabetic patients. Moreover, even in this last case, the glycaemic level is acquired under-the-skin and not on-the-skin. On the other hand, it has been proven that human sweat is very rich in molecules and other biomarkers (e.g., ions), which makes sweat a quite interesting human liquid with regards to gathering medical information at the molecular level in a totally non-invasive manner. Of course, a proper collection of sweat as it is emerging on top of the skin is required to correctly convey such liquid to the molecular biosensors on board of the wearable system. Microfluidic systems have efficiently come to the aid of wearable sensors, in this case. These devices were originally built using methods such as photolithographic and chemical etching techniques with rigid materials. Nowadays, fabrication methods of microfluidic systems are moving towards three-dimensional (3D) printing methods. These methods overcome some of the limitations of the previous method, including expensiveness and non-flexibility. The 3D printing methods have a high speed and according to the application, can control the textures and mechanical properties of an object by using multiple materials in a cheaper way. Therefore, the aim of this paper is to review all the most recent advancements in the methods for 3D printing to fabricate wearable fluidics and provide a critical frame for the future developments of a wearable device for the remote monitoring of the human metabolism directly on-the-skin.

Keywords: wearable biosensors; metabolism; remote monitoring; sweat; microfluidic; 3D printing

1. Introduction

Nowadays, the prevention of diseases by monitoring the early stages is considered a very cost-effective approach with respect to treatment costs once the diseases are fully manifested. This new approach also leads to better health outcomes [1,2]. In this endeavor, wearable biosensors have gained considerable attention. The high specificity, portability, fast detection, low-cost, and low-power features of biosensors have made them very suitable as wearable applications. Wearable devices have a considerable role in accomplishing these goals since the collection of crucial information in a continuous and non-invasive manner is easily obtained [3–9]. The USA announced 2015 as "the year of health care for wearables" [10], while The Huffington Post stated wearable technology is "the coming revolution in healthcare" [11]. Wearable biosensors are advancing toward non-invasive monitoring. In this regard, microfluidic systems are very effective and helpful. Due to the important role of microfluidics, manufacturing methods are required to be suitable for these goals. Therefore, an efficient, flexible, fast, and affordable manufacturing method plays a huge role in the future development of wearable biosensors and human health monitoring. Before we get into the manufacturing methods, first we will discuss the wearable technology, its application in health monitoring, and the role of microfluidics in their development.

Wearable technology is often referring to a category of wearable gadgets that can be worn directly by a consumer for fun or just to track their physical activity and fitness. A different category of wearable technology is medical wearable devices that can be worn by patients on their skin, on different parts of the body, and often includes the tracking of the body's physiological information related to health, in some cases, at molecular levels [12–24]. Wearable devices can collect data on a 24-h, seven-day basis, in several environmental settings, as people go through their daily routines at home or at work [25]. Wearable devices are able to relay physiological information as the body evolves over healthy and sick states. They can help persons to monitor themselves without expensive equipment, and neither educated professionals nor teams of expensive medical staff are required [15,26]. Moreover, the characterization of non-invasive and wearable technologies for diagnosis is extremely beneficial for both continuous health monitoring and diagnostics in early and pre-disease states. They also allow a quick access of clinical information by the patients, which encourages people to take more concern in their own health in a more comfortable and cheaper way, which also improves compliance [27,28]. In recent years, several wearable devices to gather the body's physiological data have been proposed by the scientific literature, especially targeting personalized medicine and point-of-care diagnostics [29], as well as home and fitness monitoring. Wearable monitoring is provided by shirts [30], necklaces [31], tattoos [32], lenses [33], headbands [34], smart wristbands [35], watches [36], shoes [37], eyeglasses [38,39], wristbands, and patches [40,41]. Different kinds of wearable sensors perform clinical diagnostics by measuring the major electrolytes, metabolites, ions, acids, heavy metals, alcohols, and toxic gases directly acquired in different body fluids [25,42], as shown in Figure 1.

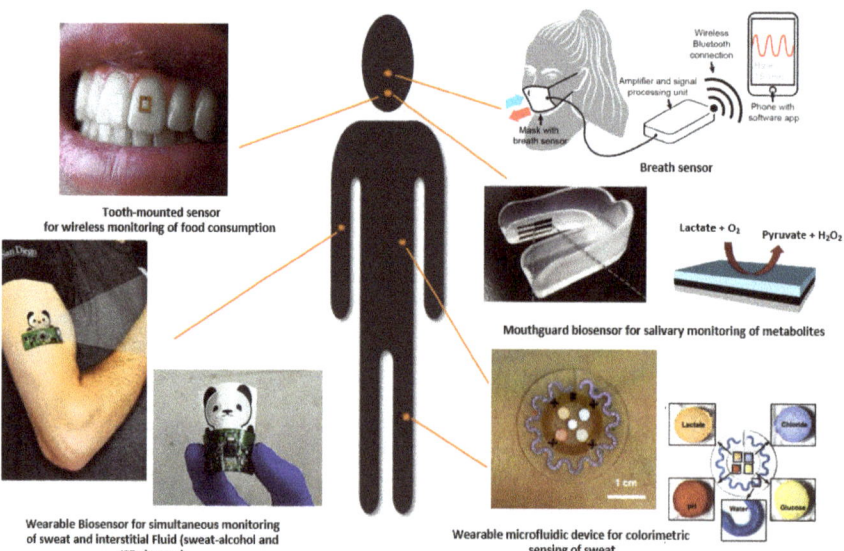

Figure 1. Different shapes of wearable devices for health monitoring (tooth-mounted sensor photo courtesy of Mike Silver, SilkLab, Tufts University, 23 October 2019 (reprinted with permission from [32,43–45])).

There are several candidates as body fluids when looking to sample human molecules in different ways (Figure 2). Blood is the most widely exploited biological fluid for clinical diagnostics. Access to it is usually painful while difficult to reach with non-invasive techniques and more often impossible when trying with wearable platforms [26]. Its collection usually causes bit of pain, and it may also provoke phobias and cause discomfort to patients [28]. Its sampling is usually invasive and it is unsuitable for long-term continuous monitoring [42]. As an alternative to blood sampling, interstitial tissue has been considered, which is another option largely used by commercially available glucometers for diabetic patients. For the sampling of glucose from the interstitial tissues, we still need to typically pass through the skin, therefore, this approach is invasive as well. Blood is not necessarily informative for health monitoring in the cases of some metabolites, so it is possible to transition from blood to other body fluids, such as saliva, sweat, and tears for health monitoring. It provides non-invasive approaches and in-situ monitoring, which is more attractive for the long-term applications of the continuous monitoring of health in daily life [42].

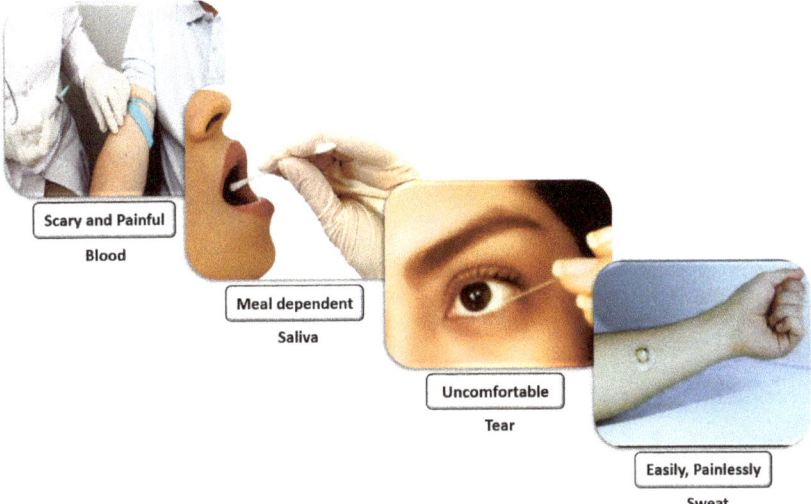

Figure 2. The methods of sampling for several candidate biofluids: Blood collection usually causes a bit of pain and can cause phobias and discomfort to patients. Before sampling saliva for a specific analysis, the time and content of the last meal should be considered. Tears can be uncomfortable or risky to sample. Sweat is easy to collect by painless techniques (reprinted with permission from [43], needle-professional-arm-human-body-blood-skin-647863, Photo by https://pxhere.com/en/photo/647863 is licensed under CC0 1.0, 5 October 2020).

Tears are a promising fluid for protein, lipids, and glucose detection. Tear sampling or its continuous monitoring can be uncomfortable or risky in terms of irritation, which can produce side-effects as well as mislead the sensor readings (e.g., by variations of pH). A capillary micropipette and swab are usually used for tear sampling. When they are being used, the eye usually reacts when coming close to an external object, and also some unwanted contact can cause irritation, which makes sampling uncomfortable. On the other, any irritation with increasing the production of tears can cause a reduction in the biomarker concentration. Saliva includes some markers for several diseases, such as cardiovascular diseases, oral and breast cancer, and human immunodeficiency virus (HIV) [46–48]. Because of the high alteration of the saliva composition from the last meal, it provides limited physiological insight. Instead, sweat is a promising fluid for wearable sensing, providing several analytes such as ions, alcohols, and drugs [15].

Sweat is particularly interesting for non-invasive biosensing because it is an abundant source of information on the inner physiological health, which can be determined from analytes, several of them with potential as biomarkers for diagnosing diseases, e.g., such as ions [49–51], alcohols [18,52], glucose [53,54], lactates [19,55] drugs, and heavy metals [15], as schematically shown in Figure 3.

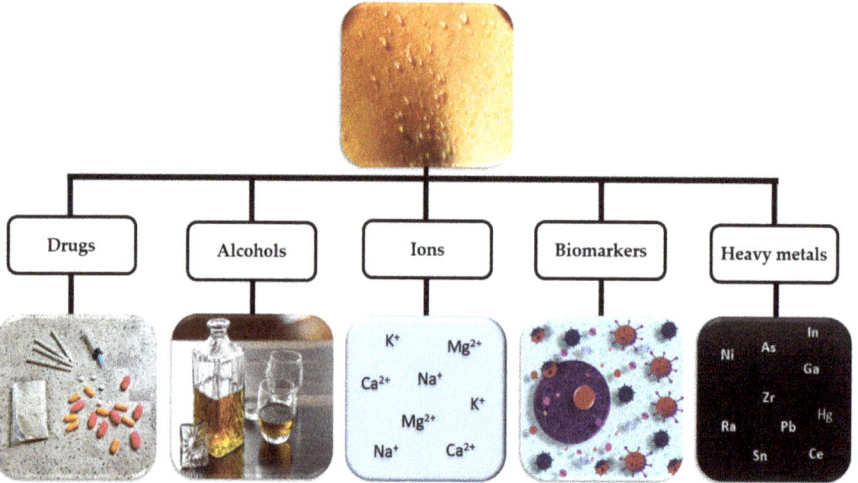

Figure 3. Different kinds of analyte in sweat: biomarkers, ions, alcohols, drugs, and heavy metals (Sweat_body_fitness_sport_fit_training_active_young-552041, Photo by https://pxhere.com/en/photo/552041 is licensed under CC0 1.0, Bottle of whisky and three glasses on wooden table, Photo by Lovely2912 form https://pxhere.com/en/photo/1593175 is licensed under CC0 1.0, 5 October 2020).

Genomics and proteomics play an important role in searching for new biomarkers in sweat. For example, dermcidin (DCD) and prolactin inducible protein (PIP), in addition to blood, have been found in sweat too [56]. DCD and its receptors are present and overexpressed on the cell surfaces of invasive breast carcinomas, and their lymph node metastases as well as in brain neurons. PIP is instead overexpressed in prostate cancer and metastatic breast cancer [56]. Tozser and his group analyzed dermcidin (DCD) and prolactin inducible protein (PIP) in sweat by label-free mass spectrometry [57]. The discovery of new diagnostic molecules in sweat has, of course, pushed further the investigation of new wearable diagnostic devices, which can also be located close to the place of sweat generation, allowing a quick detection before the analytes biodegrade [15,28]. Therefore, and due to this new attention towards sweat analysis, several novel wearable and flexible sweat analyzing platforms have been recently proposed for in situ analyses and continuous health monitoring [12,16,18,19,43,53,54], as shown in Figure 4.

Despite the evident advantage in terms of non-invasive diagnostics, on-body sweat sensing presents several issues including but not limited to liquid evaporation, low sweat-volumes, irregular volumes in the case of non-stimulated sweating, contamination, interfering chemicals from the environment, the need for the continuous sampling of fresh sweat, and biodegradation [26]. An appropriate sweat sampling method is then very crucial to prevent measurement artefacts being evaporated and the contamination of sweat specimens. An effective sweat transport by a fast sampling system can minimize the crossover effect due to the mixing of new and old sweat samples. These undesirable effects are overcome by developing proper microfluidics [16,43]. Typically, the design of microfluidic platforms is carried out with the goal of improving the following functions: (1) collecting body fluids in proper ways, (2) transferring the body fluids to the detection site, and (3) supporting the detection in the right conditions [58]. Figure 5 shows a possible conceptual scheme of such a microfluidic system. Here, the microfluidics typically provides continuous sampling by conveying the sweat along a controlled fluidic channel and then enhancing the sensing in a well-defined encapsulated acquisition chamber. Thin microfluidic layers are typically designed as sweat collectors to bring the sweat on the electrodes, preventing any further re-absorption of the electrodes back to the skin, and also preventing sweat evaporation [59,60]. If built with soft and flexible materials, microfluidics easily addresses some of the requirements for wearable devices, such as being lightweight, comfortable, and conformable [58]. Moreover, the sensing accuracy and reliability is significantly improved by allowing the sampling of precise liquid amounts. This is particularly beneficial since body fluids are often secreted in limited quantities, while the lower sampling volume also reduces the burden on patients. Moreover, microfluidic systems may also integrate little reservoirs for solute storage in case of the need for further injections to the detection part, typically at controlled intervals, for detection purposes, e.g., in case of labelled detections [58]. The primary fabrication methods for microfluidics were photolithographic techniques and chemical etching with rigid materials, such as silicon and glass, which are non-flexible and expensive. Because of the limitations of these methods, three-dimensional (3D) printing methods were considered, which are cheaper and more flexible. In 3D printing processes, materials are solidified with computer control to create a three-dimensional object. In Section 2, we will take a brief look at microfluidics and the original methods of making them and their challenges. Finally, in Sections 3 and 4, we will discuss why the traditional methods have given way to 3D printing methods today, and the details of the 3D printing methods suitable for wearable sensors and their applications.

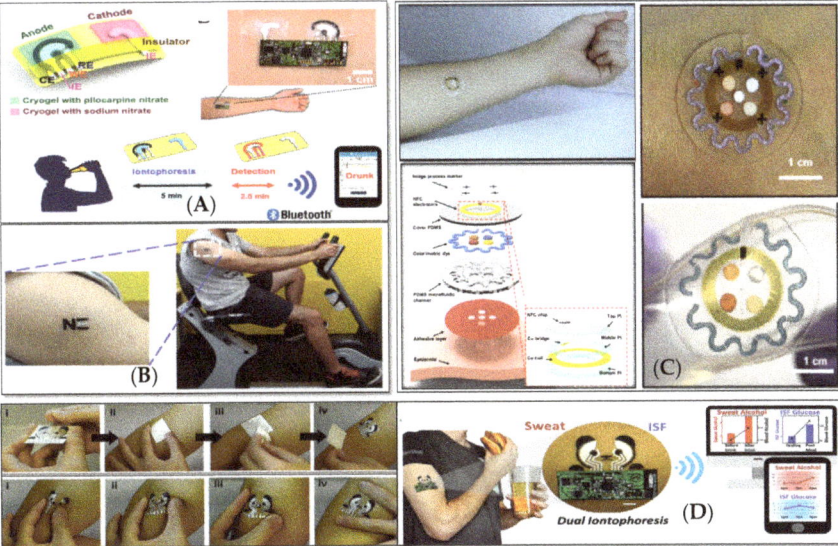

Figure 4. Wearable sweat analyzing platforms: (**A**) a wearable tattoo-based iontophoretic-biosensing system for alcohol monitoring (reprinted with permission from [18]), (**B**) an electrochemical tattoo biosensor for real-time non-invasive lactate monitoring in human perspiration (reprinted with permission from [19]), (**C**) a wearable microfluidic device for the capture, storage, and colorimetric sensing of sweat for markers such as chloride and hydronium ions, glucose, and lactate (reprinted with permission from [43]), and (**D**) a wearable biosensor platform for the simultaneous monitoring of sweat and interstitial fluid for components such as glucose and alcohol, subfigure i–iv shows the tattoo application process and mechanical deformation tests of transferred tattoo respectively (reprinted with permission from [32]).

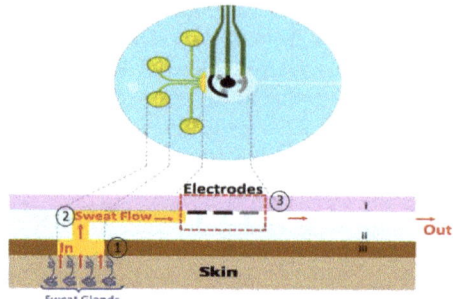

Figure 5. Schematic of wearable microfluidics: (**1**) sampling, (**2**) transferring to the site of detection, (**3**) detection by electrochemical sensors (reprinted with permission from [55]).

2. Microfluidics

Microfluidics is now a well-consolidated field of research, which deals with fabrication on the micron-scale of systems for manipulating fluids. It is most commonly identified by device fabrications with critical sizes of fluidic channels of less than 1 mm [61]. The field of microfluidics has grown rapidly over the last four decades since it emerged in early 1980s [62]. Numerous theoretical studies have been conducted in this field that play an important role in the making of an efficient product [63–65]. In one of these works, a novel continuous flow magnetophoretic microfluidic device for the separation of magnetic microparticles, based on size, is presented [64]. Some of the effective parameters on the motion of microparticles into the microchannel and the associated performance metrics are: the location of the outlets, microchannel height, fluid velocity, and ratio of inlet and of the outlet flow rates. In order to investigate the relationship between the operating and geometric parameters on device performance, a mathematical model is developed. Designers using a mathematical model can choose the parameters for the magnetophoretic microfluidic device with the best performance metrics. In recent years, many of the applications of microfluidic technologies to chemistry, biology, and medicine (e.g., see Figure 6) successfully appeared in the literature mainly due to their great advantages in terms of low volumes, high sensitivity, rapid processing, high spatial resolution, high integration with sensing components, and easy control. Undoubtedly, a short fabrication time, easy prototyping, and simple and cheap methods played an important role for the success of microfluidics [62,66,67].

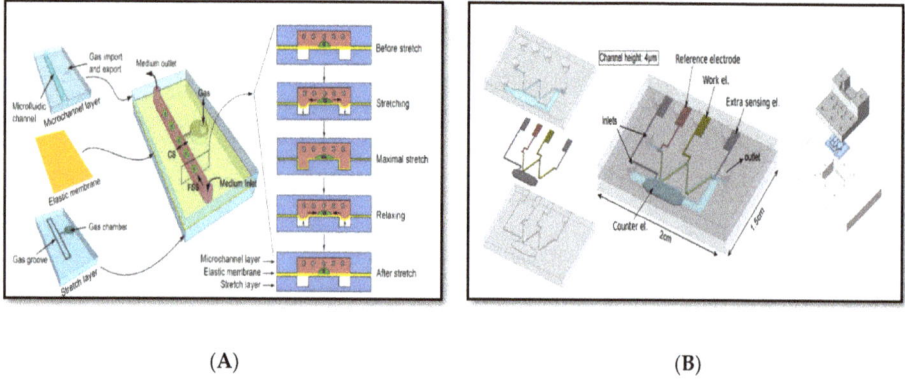

Figure 6. Microfluidic devices: (**A**) artificial organ-on-chip, (**B**) electrochemical detection of drugs (reprinted with permission from [68,69]).

One of the very first microfluidic devices was developed by Andreas Manz and his team in the early 1990s [70]. Such devices were originally fabricated in silicon or glass by a conventional, planar photolithographic technique, and chemical etching, by adapting techniques typically used in microelectronics. These kinds of methods are precise but expensive, non-flexible, and poorly suited to exploratory work by prototyping [71]. As the field has progressed, alternative methods, such as laminate manufacturing methods, various polymer molding technologies (e.g., hot embossing and injection molding) and 3D printing, have emerged for the fabrication of channels with the requisite sizes [61]. Traditional methods have failed to address some of the barriers, while the commercialization of microfluidic devices with 3D printing is able to overcome said barriers. Some of these barriers include a non-standard user interface, complex control system, and the speed and cost of the liquid polymer (i.e., poly(dimethylsiloxane) (PMDS)) modeling. In the 3D printing method, various materials with different properties (e.g., transparent and biocompatible) have been developed that can be used according to the application of the microfluidic device. So, among the several possible methods, this review focuses on additive manufacturing, namely the modern industrial method to create three-dimensional objects, typically in a computer-controlled manner, by progressively adding material, typically in a layer-by-layer approach: the so-called 3D printing.

3. Three-Dimensional Printing Methods

Additive manufacturing, namely 3D printing, has enormous potential for a considerable contribution to the field of microfluidics. In particular, its ability to create truly three-dimensional structures with very complex features in a single step and start from a digital model has obvious attractions for the easy prototyping of very complex microfluidics [72]. This technology was developed originally by Charles Hull in 1986 [73]. In general, 3D printing is an additive manufacturing (AM) technique, proposed for the fabrication of a wide range of structures with computer-controlled processes based on three-dimensional (3D) digital models of the object to print (Figure 7). The process includes printing consecutive layers of materials that are formed on top of each other [73], while digital models provide an extremely flexible way to design and shape the objects. Typical fabrication thicknesses are in the range from 0.001 to 0.1 inches for each printed layer [74]. Additive manufacturing, or 3D printing as it is more often known, has received considerable interest in more recent years in both the academic community and the business society, and it has been mentioned as a third industrial revolution [72,75]. In fact, 3D printing offers many advantages with respect to traditional manufacturing, including an improved versatility, less waste, more freedom in design, a low-cost fabrication, high-automation, and short fabrication cycle time [73,76]. Three-dimensional printing technology allows for the creation of objects with complex internal structures with fewer space requirements [76]. Three-dimensional printing is also suitable for fabricating parts of various sizes from the micro- to macro-scale [73,75].

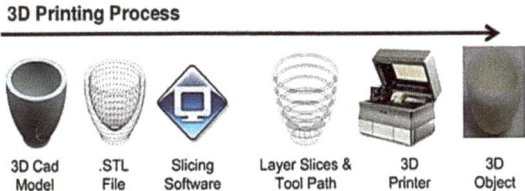

Figure 7. Three-dimensional printing process (3d-printing-a-2014-horizonwatching-trend-summary-report-9-638 by Kholoudabdolqader is licensed under CC-BY-SA-4.0, 1 October 2020).

Product customization has been a challenge for traditional manufacturers, typically due to the high costs in fabricating the mold, especially for small-scale productions of custom-tailored products. On the other hand, 3D printing is able to print small quantities of customized products in plastic (in 3D) with extremely low costs compared to traditional mold-based productions. This is specifically useful in biomedical fields, whereby unique patient-customized products are also required [73].

Actually, the name "3D printing" includes various methods for additive manufacturing (see Figure 8).

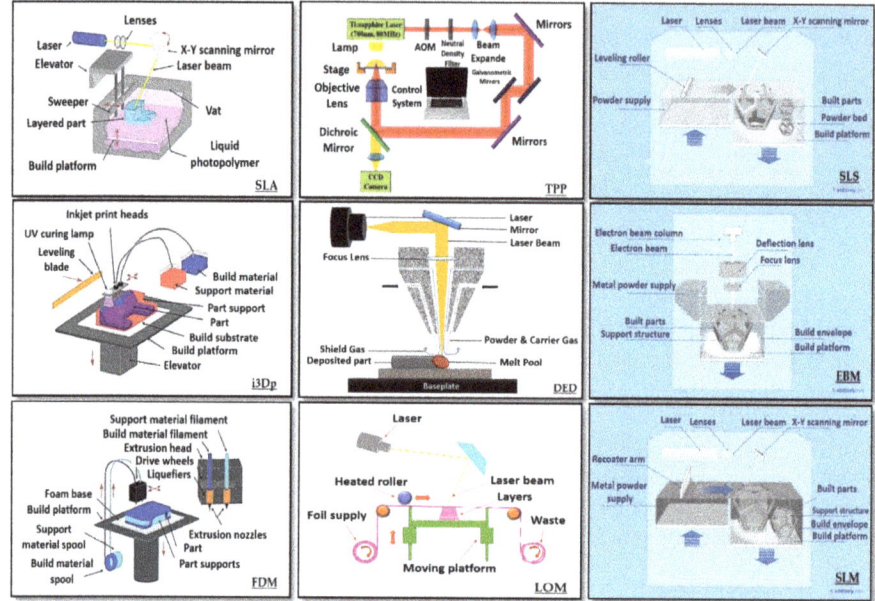

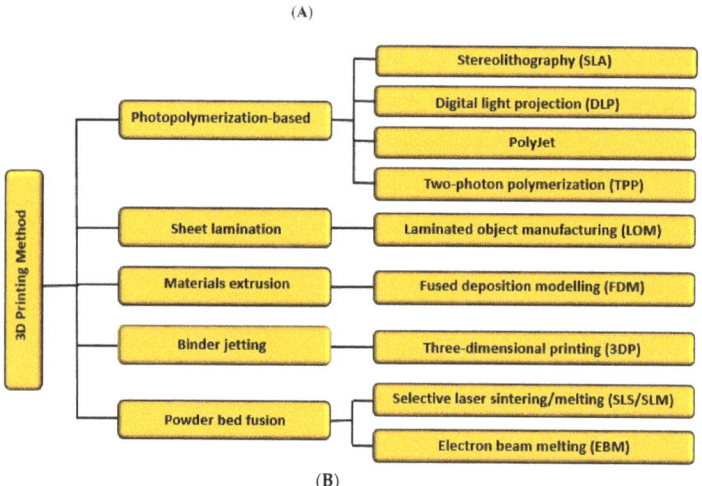

Figure 8. Three-dimensional printing methods: (**A**) various methods for additive manufacturing: stereolithography (SLA), two photon polymerization (TPP), selective laser sintering (SLS), multi-jet modeling (photopolymer inkjet printing (MJM))/ inkjet 3D printing (i3Dp), direct energy deposition (DED), electron beam melting (EBM), fused deposition modeling (FDM), laminated object manufacturing (LOM), selective laser melting (SLM), (**B**) related taxonomy on the base of the product printing way (reprinted with permission from www.additively.com, [77,78]).

By following the new incoming needs of human monitoring (e.g., serious critical concerns about the increasing medical costs related to the aging of the populations in west-countries), an emerging trend in mobile health (mHealth) is about developing new kinds of monitoring systems by the integration of different sensory devices in wearable single-platforms to achieve more complex and efficient monitoring functions for the new concept of digital biomarkers [79]. Of course, several new wearable devices are already on the market with the capability to acquire data, e.g., about temperature, location, physical activity, etc., while very few of these devices already on the market are capable of providing real measurements on molecules that are usually used in diagnostics and possible to find on-the-skin. On the other hand, many examples of possible approaches to realize wearable systems for molecular measurements have already appeared in the scientific literature. These kind of devices typically imply the intimate integration of electrochemical or optical biosensors with proper microfluidics in order to collect sweat from the skin and convey it, as well as its molecular content, to the sensory platform. Among the many different and possible approaches for fabricating such systems, we will focus here on 3D printing techniques and related materials for building wearable microfluidics to be then integrated in biosensors.

All 3D printing techniques are not appropriate for microfluidics. The most widely used 3D printing techniques for microfluidics are selective laser sintering (SLS), fused deposition modeling (FDM), inkjet 3D printing (i3Dp), laminated object manufacturing (LOM), two photon polymerization (TPP) and stereolithography (SLA) [78,80–86]. Since we are pointing out here the work done for wearable biosensors, this review focuses only on progress made towards the use of 3D printing for the fabrication of flexible microfluidics. Polymeric substrates are widely used in flexible devices [58]. The polymeric suitable approaches that have been already successfully to this aim include fused deposition modeling (FDM), inkjet 3D printing (i3Dp) and stereolithography (SLA) [72]. These 3D printing techniques also provide the opportunity for using multi-material to improve the quality of the final printed product [87,88]. Therefore, these three methods for 3D printing of wearable fluidics are discussed more in detail in the following sections of this paper. Table 1 summarizes the materials, benefits, and limitations of the three 3D printing methods suitable for wearable microfluidics.

3.1. Fused Deposition Modeling (FDM)

The fused deposition modeling (FDM) appeared originally in a patent obtained in 1992 and this technology was, and still it is, commercialized by the company Stratasys, funded by Scott Crump, the inventor of this 3D printing approach [76]. This process works by depositing layers through extrusion and fast condensation, in different locations, driven by a computer based on a digital model of the object-to-build [73,89], as schematically shown in Figure 9. The method uses a motor-driven nozzle that moves in three dimensions. A continuous filament of a thermoplastic polymer is extruded on the build platform after heating the nozzle head. After extrusion, the material cools down and solidifies immediately.

Table 1. The materials, benefits, and limitations of the three 3D printing methods suitable for wearable microfluidics.

3D Printing Methods	Materials	Benefits	Drawbacks
Fused deposition modelling (FDM)	Polyethylene terephthalate (PET) Polystyrene (PS) Polycarbonate (PC) Acrylonitrile butadiene styrene (ABS) Polycaprolactone (PCL) Poly-lactic acid (PLA) Polybutylene terephthalate (PBT) Polyglycolic acid (PGA) Polypropylene(pp)	Low cost High speed Simplicity Low-cost Manufacturing of centimeter-sized prototypes Using inexpensive biocompatible polymers	Weak mechanical properties Limited materials (only thermoplastics) Layer-by-layer finish Leakage due to filament bonding Difficulty of removal of support structure for complex internal features Inter-layer distortion
Inkjet printing (i3Dp)	Soft elastomers Liquid metals (i.e., EGaIn) Wax-based inks Liquid suspensions Acrylonitrile butadiene styrene (ABS), Polystyrene (PS), Polypropylene (PP), Polymethylmethacrylate (PMMA), Polycarbonate (PC) Ethylene propylene diene monomer (EPDM) High-impact polystyrene (HIPS)	Layer-by-layer fine structures Fast High resolution smooth surface Low cost Ability to easily print highly complex devices without using lithography Precise control Realizing microfluidics directly on other systems without any bonding steps Absence of sticking agents in between layers	Difficulty in removing the support material Layer-by-layer finish
Stereolithography (SLA)	Epoxy Hybrid resins Acrylate based resin Clear acrylic polymer Elastomers and ceramics Composites of photopolymers Hybrid polymer-ceramics	High quality Smooth surface Use flexible resin Fine resolution (a nometer scale) custom low-cost resins No need for external alignment Ability to directly print the channels Manufacturing complex nanocomposites Making a monolithic structure without the need for bonding	Slow printing Sometimes expensive chemicals Low biochemical adaptability of the resist Limited choice of the materials

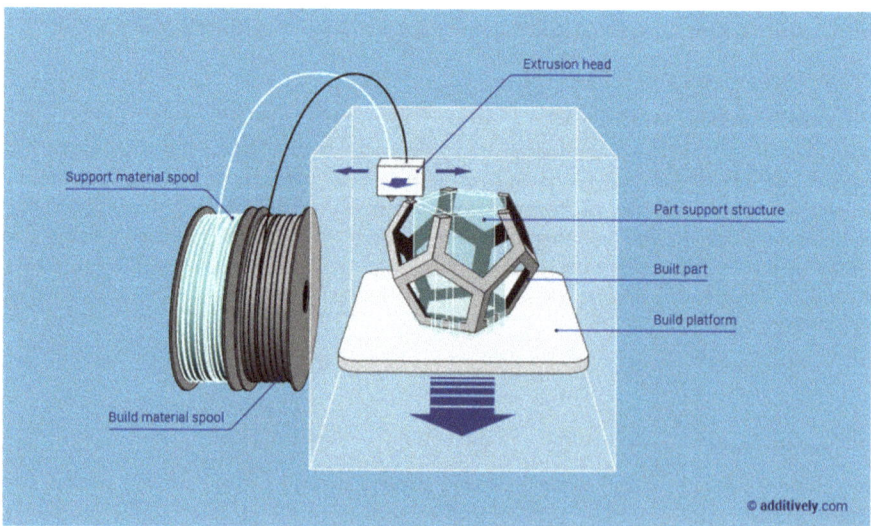

Figure 9. Schematic representation of FDM (reprinted with permission from www.additively.com).

In this method, the thermoplasticity of the polymer is strictly required to obtain the fused filament, which is a key component of the entire fabrication process. Thanks to this property, different polymer filaments may fuse together during printing and then solidify [73]. The mechanical properties of the printed parts are clearly dependent on the processing parameters. The main processing parameters are the width and orientation of the filaments, layer thickness, and air gaps in some of the layers or between the layers [73]. FDM can be successfully used in several applications with inexpensive biocompatible polymers such as polycaprolactone (PCL), polylactic acid (PLA), polybutylene terephthalate (PBT), and polyglycolic acid (PGA) [63,78]. The main benefits of FDM are the low-cost, high speed, and simplicity of the method [73].

Despite the advantages, FDM presents some drawbacks, such as the weak mechanical properties of the final realized solid and evident appearance of the layer-by-layer built structure, a poor surface quality [62], leakage due to the filament bonding when used for fluidics [89], and a limited number of thermoplastic materials are available to use [73]. The main cause of mechanical weakness is related to the inter-layer distortion [73]. Moreover, FDM usually requires support structures, of which removal may be difficult due to the complex internal features of the object [89]. Another drawback of FDM in producing microfluidic systems is the difficulty in appreciating the details of the printed systems. This drawback is because of light diffusion. Optical microscopy is used for the investigation of the systems. The polymers used for layering create light diffusion and make the optical investigation of the realized channels difficult [89]. To solve this problem, the optical transparent windows have been integrated inside the 3D printed microfluidic system [89]. Integrating optical transparent windows is typically possible by stopping the printing process to insert more transparent materials, and then restarting again the printing process. So, this method makes it hard to ensure a leakage-free sealing between the inserted part and the main body of the device [89]. Finally, the resolution of FDM is still on the way for improvement in terms of dealing with fluidics: it is less efficient than other 3D printing systems as compared, for example, to stereolithography (SLA). However, as we will see in the following, SLA presents higher costs for buying both the printer and the materials, while it requires more complex steps to complete a printed device [89]. On the other hand, the fabrication of microchannels with FDM is still a true challenge since the typical required size of a typical channel is usually smaller than the size of the extruded filaments [78]. Fabrications of channels with a well-defined sidewall geometry and straight walls are also difficult to obtain since the process creates rough surfaces [81]. After the extrusion process, the filaments cannot be arbitrarily joined at the channel intersections [62]. Because of the fast hardening of the extruded materials, the adjoining layers are not well fused, causing the low structural strength of FDM-printed products [78]. To increase the intra-layer strength, improvements such as creating covalent bonds for cross-linking between layers by using the thermally reversible Diels–Alder reaction and gamma-irradiation post-printing have been recently proposed for improving the properties of FDM-printed products [78]. or all these several reasons, the FDM remains a challenging technique for the fabrication of microfluidics.

On the other hand, we can successfully use the FDM method to fabricate some other components or parts usually required on integrated wearable systems: e.g., batteries [90], strain sensors on flexible substrates [91], light-emitting diodes (LEDs) [92], antennas on 3D surfaces [93], interconnects [94], electrodes within biological tissue [95], microfluidic pumps for wearable biomedical applications [96], electrochemical detectors [97] and other microfluidic devices [28,88], as shown in Figure 10.

For biomedical applications, inexpensive and biocompatible polymers from spools of filament are possible to be used in FDM [78,80,82]. In particular, the ones mainly used are: acrylonitrile butadiene styrene (ABS, the polymer of Lego), poly (lactic acid (PLA, a biodegradable polymer)), polycarbonate (PC), polyamide, polyethylene terephthalate (PET) polycaprolactone (PCL), polybutylene terephthalate (PBT), polyglycolic acid (PGA), polypropylene (PP), acetoxy silicone polymer and polystyrene (PS). There is an alternative version of FDM that uses, instead, liquid precursors which are extruded through a nozzle without heating. In this way, FDM may extrude a wide range of other materials, such as metallic solutions [80], hydrogels, and cell-based solutions, also including composites to strengthen the mechanical properties of 3D printed objects [73].

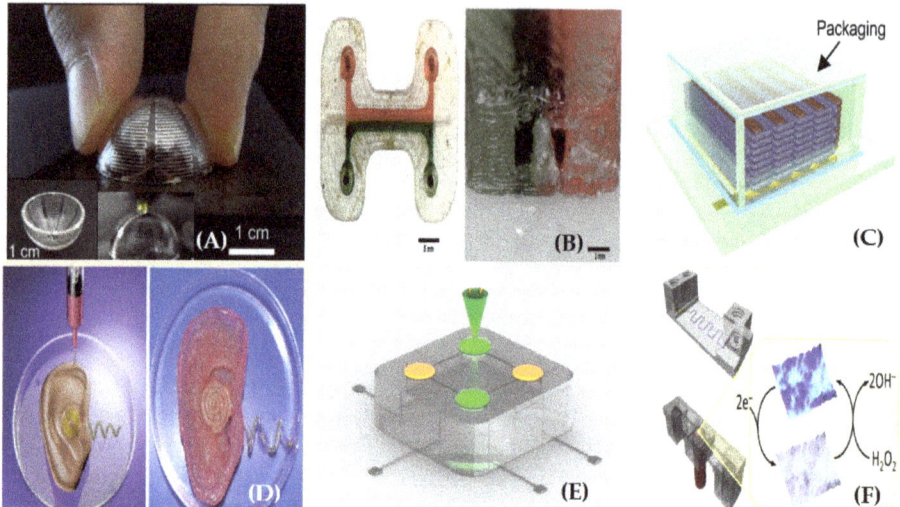

Figure 10. Different wearable devices created with FDM: (**A**) small antenna, (**B**) microfluidic device with an integrated membrane and embedded reagents, (**C**) inter-digitated Li-Ion micro-battery architectures, (**D**) bionic ears, (**E**) quantum dot light-emitting diodes, and (**F**) electrochemical detector (reprinted with permission from [88,90,92,93,95,98]).

3.2. Inkjet 3D Printing (i3Dp)

Inkjet printing is one of the main methods for additive manufacturing, especially used for printing ceramics. In this method, a nozzle and a stable ceramic suspension (such as zirconium oxide powder in water) are used. Via the injection nozzle, the suspension is pumped and deposited in the form of droplets onto the substrate (Figure 11). Then, the droplets form a continuous pattern that solidifies to hold further layers of printed materials. Several factors determine the quality of inkjet-printed objects: the solid content, particle size distribution of ceramics, nozzle size, viscosity of the ink, extrusion rate, and speed of printing [73].

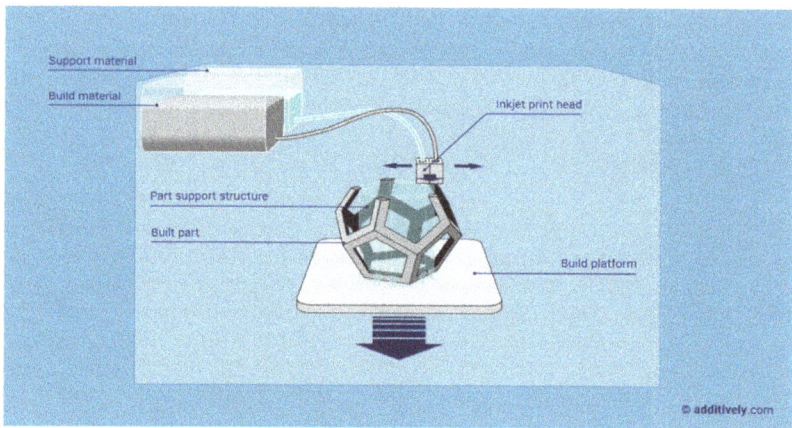

Figure 11. Schematic representation of inkjet 3D printing (i3Dp (reprinted with permission from www.additively.com)).

There are two main types of ceramic inks: wax-based inks, and liquid suspensions. Wax-based inks are melted and deposited onto a cold substrate to solidify, while liquid suspensions are solidified by liquid evaporation [73]. Inkjet technology operates in the i3Dp process either in continuous or in drop-on-demand (DoD) mode [72]. Typically, the inks used in the continuous mode have a lower viscosity that provides a higher drop speed than those typically used for the DoD mode. The DoD mode generates smaller droplets with a higher placement accuracy. Therefore, the DoD is the better choice for 3D microfabrications and leads to finer and more repeatable microfluidic structures [72]. In the DoD technique, the pulse is piezoelectrically or thermoelectrically provided. In the piezoelectric DoD, the deformation of a piezoelectric element generates acoustic pulses. These pulses push droplets of ink from the nozzle. In the thermal DoD, a vapor blob is formed by heating the ink locally and then it is ejected as an ink droplet. In this case, the solvents must be volatile while in piezoelectric DoD, a large variety of organic solvents such as chloroform, dimethyl sulfoxide, and dimethylacetamide can be used.

Inkjet 3D printing is further split into two categories (Figure 12): powder-based and photopolymer-based. In powder-based i3Dp, the powder particles are bonded with a polymeric sticking solution. In this process, a roller initially deposits a layer of ceramic powder, which is spread uniformly on the building stage. Then, the multi-channel printer head sprays droplets of adhesive onto the powder bed at the targeted areas. After completing the first layer, the building platform drops a second powder layer and bounding restarts again with the successive printing of adhesive. This process is repeated until the 3D object is formed. Once the process is completed, the printed object is typically surrounded by a feeble supporting powder that can be easily removed without further post-processing steps. Supporting powders are usually made by a mixture of gypsum, polymer, and silica particles, with adhesives that are composed by glycerol and water-soluble acrylates. These materials are recyclable. So, recycling the unused powder can further lower the costs of printing [72]. The resolution of these kind of printers is determined by some parameters such as the packing density, shape of the printed objects, and particle sizes of the used ceramic. Un-bound particles may increase the surface roughness and reduce the transparency of the printed object. These un-bound particles scatter the light, preventing the microscopy investigations usually required for investigating the printed microfluidics [72]. In the second kind of i3Dp (photopolymer-based), the head deposits small drops for both the build and support materials to create the object in a layer-by-layer process again. The typical materials used are acrylate photopolymers, starting from monomers, oligomers, or photo-initiators, then each printed layer is treated with an ultraviolet (UV) source. At least 100 different composite materials are available in the market, as further developed by a core of 17 primary photopolymers only [72].

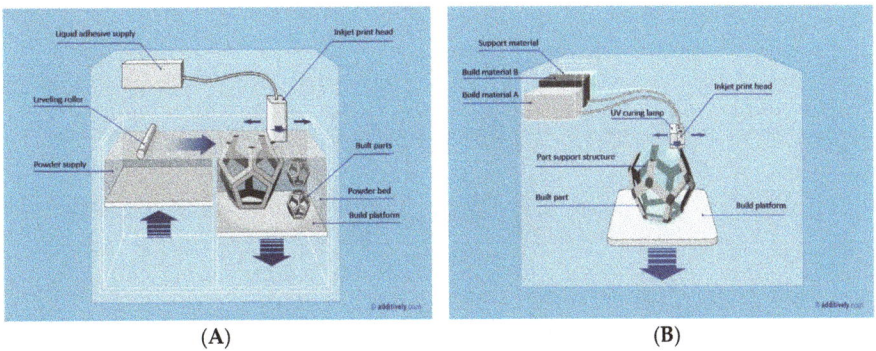

Figure 12. Inkjet 3D printing is split into two further categories: (**A**) powder-based, and (**B**) photopolymer-based (reprinted with permission from www.additively.com).

Theoretically, an XY-resolution of 42 µm and a Z resolution of 16 µm is possible for a high resolution inkjet printing of 600 × 600 dpi [4]. In the case of mass production, a relatively inexpensive alternative way of inkjet printing is also possible for fabricating paper-based microfluidics [99]. Distinct advantages of i3Dp are the high resolution, simplicity, and low-cost [80,99]. This method is fast and efficient for easily printing quite complex structures with a high resolution by avoiding the use of lithography [73]. Inkjet 3D printing (i3Dp) easily modulates the shape and dimensions of the pattern and does not require the use of support structures [15].

With i3Dp, we can also realize the flexible resistive components and sensors by using conductive liquid metals [100]. Although the price is typically higher than that for FDM, i3Dp is probably the most commercially viable 3D printing approach for microfluidics [72]. This method creates microfluidic structures (also flexible) with precise control and manipulation of fluids that are small enough for the intended aim: typically, sub-millimeter or micrometer scales [83]. With this technique, vertically aligned channels usually have size stability and a smooth surface [72]. Interestingly, microfluidics can be realized directly on top of other systems, e.g., transducers or electrodes, without the need of any bonding or assembling steps [100]. In i3Dp, only inks containing weak organic solvents are acceptable [88], since strong organic solvents and hydrophobic materials, such as hexane, heptane, and toluene, damage the cartridges and other components of the printer. So, the main advantages of this method are the absence of sticking agents in between layers, protecting workability, and layer-by-layer fine structures [73]. On the other hand, some disadvantages are the quite high costs from changing materials during printing, removal of the support material (difficult for fully closed structures), and the drying of inks with consequent holes clogging in the print cartridges if the i3Dp system is not regularly used [72].

After the first microfluidic device proposed with i3Dp by McDonald and co-workers in 2002 [72], the development of accurate printers has been extensively improved to fabricate milli- and microfluidics [81]. As well as FDM, i3Dp is also regularly used to print biomedical systems other than wearable biosensors: for example, scaffolds for tissue engineering [73], orthopedic prosthesis [86], cardiac parts [101], and components for intracranial aneurysm surgeries [102]. In microfluidics, devices for flexible, planar, and multilayer microfluidics, membrane, on-chip gel electrophoresis, and wall-jet electrochemical detectors have been published so far [97,100,103–105], as shown in Figure 13.

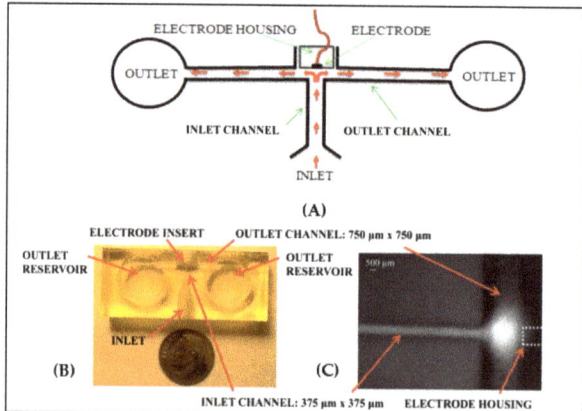

Figure 13. Wearable microfluidics created with i3Dp with a wall-jet electrochemical (WJE) configuration: (**A**) schematic of the WJE device, (**B**) top view of the 3D printed WJE device, (**C**) micrograph of a fluorescein plug hitting the electrode in the WJE design and flowing away from the electrode (reprinted with permission from [105]).

Photopolymer i3Dp is also known by the names of PolyJet or multi-jet modeling (MJM). In MJM, a large range of photopolymers are used, including soft elastomers, liquid metals (i.e., EGaIn), acrylonitrile butadiene styrene (ABS), polystyrene (PS), polypropylene (PP), polymethylmethacrylate (PMMA), polycarbonate (PC), ethylene propylene diene monomer (EPDM), and high-impact polystyrene (HIPS) [72,80].

3.3. Stereolithography (SLA)

Stereolithography (SLA) was proposed as method of additive manufacturing in 1986 by Chuck Hull. He defined SLA as "a method and apparatus for making solid objects by successively 'printing' with thin layers of a curable material, e.g., a UV curable material, one on top of the other". SLA was commercialized in 1988 and became the first commercialized 3D printing system ever proposed [72,82]. SLA uses an energy source (e.g., light or electron beams) for activation (radicalization) of the monomers (mainly acrylic or epoxy-based) in order to obtain polymer chains. After the polymerization, a pattern inside the resin layer is solidified to hold the next layer formation. After the full printing, the unreacted resin is then removed. In some cases, the printed parts need some post-process treatments, such as heating or photo-curing, in order to reach the desired mechanical performance [73]. While Hull described SLA only for material curable by UV, recent advances in resin photochemistry and laser technology achieved polymerization with modern high-intensity lasers or focused LED light sources in the visible wavelength range, using suitable types of photo-initiators [78,80].

In using SLA, the laser spot-size, the pixel resolution, the type and viscosity of the resin must be carefully considered for fabricating microchannels with the minimum cross-sectional area [78]. The thickness of each layer is affected by the energy of the light source and by the exposure time [73]. The laser spot-size and absorption spectra of the photoresins affect the resolution too [78]. SLA is possible in the two most important configurations: the free surface approach (bath configuration), and constrained surface approach (bat configuration). In both the cases, objects are formed by a liquid resin photopolymerized with either a scanning laser or a digital light projector (DLP) by a spatially controlled photopolymerization [72], as schematically shown in Figure 14. In the bath configuration, the vat depth limits the object height, while this limitation does not exist in the bat configuration. The time of curing is faster in the bat configuration because oxygen inhibits the process of photopolymerization, and the reaction happens far from the air–resin interface. For modifications in the constrained surface technique, the bottom plate is made sensitive to oxygen by a controlled oxygen inhibition to the last cured resin layer [78]. The bath configuration is the classical setup for SLA. In this configuration, a substrate is submerged in a tank of photoactive resin and a UV beam affects a two-dimensional (2D) cross-section onto the substrate when the resin polymerizes under illumination. After the completion of the 2D cross-section, the next step is the lowering of the substrate further into the resin by a predefined distance. Then, the next layer is polymerized on top of the previous layer by the UV beam. Working that way ensures that the focus of the UV beam does not change. A blade levels the surface with a further layer of uniform resin before the next step of exposure to UV light. As written, the 'bat' configuration is the name of the constrained surface approach. The reason for choosing this name is the fact that, as shown in Figure 14, the object is created hanging from a movable substrate like a bat from a ceiling. The movable substrate is hanging above the resin tank. The tank has an optically clear bottom and a non-sticking surface, so the printed structure does not stick to the substrate. The light source is typically located under the tank (picture on Figure 14B). The action of gravity on the forming surface, which rests for a certain settling time, refreshes and smooths the surface of the illuminated resin. The object is drawn out of the resin, rather than submersed in it, so only small amounts of resin with low viscosity are needed. Since the illuminated layer is not exposed to the atmosphere, oxygen inhibition is limited. Here, the height of the printed objects is not limited with respect to the bath configuration, and the bat configuration requires minimum cleaning steps. The cured layer is sandwiched between the resin vat and the previous layer. Sometimes, the solidified material strongly sticks to the bottom of vat. In these cases, the object may, unfortunately, break or deform when coming up from the vat [72].

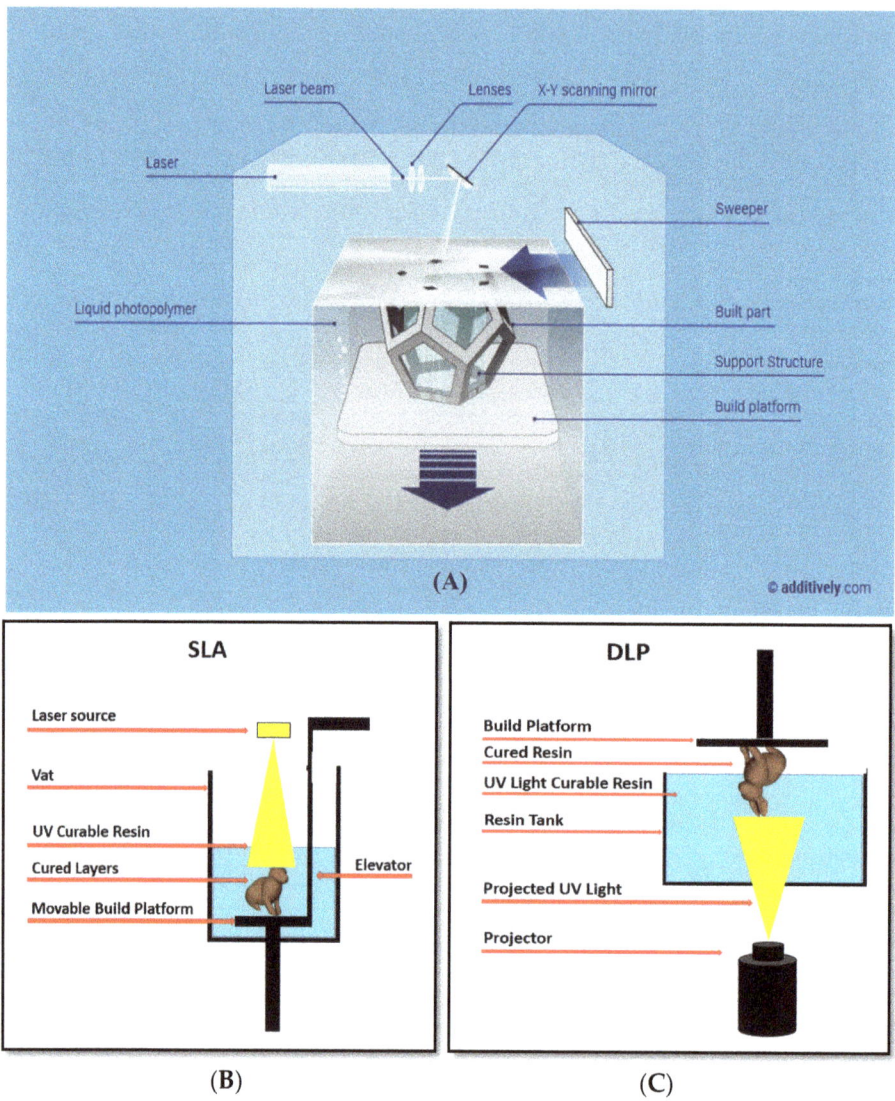

Figure 14. (**A**) stereolithography (SLA): (**B**) the bath, and (**C**) the bat configurations (reprinted with permission from www.additively.com).

SLA produces high-quality objects with a fine resolution down to 10 μm [73], and offers a good balance between the resolution, price, and performance [72]. The fabrication of micro- and nanostructures in a wide variety of shapes is possible as well [83]. There are a considerable number of different choices regarding the material properties, especially regarding the increasing range of the extrusion filaments: it is possible to also use conducting, flexible, and magnetic filaments, as well as a range of different colored polymers [84]. Materials like elastomers and ceramics are also allowed, as well as photoactive resins such as acrylate, clear acrylic, epoxy, hybrid resins and composites of different photopolymers [58,81]. Dispersions of ceramic particles are used to print ceramic–polymer composites or polymer-derived ceramifiable monomers, such as silicon oxycarbide [73]. Recently, Gong et al. investigated the effect of the optical property of SLA's resin on the channel sizes in microfluidic systems. They found that there is fundamental exchange between the critical dose that penetrates into a flow channel during fabrication and the homogeneity of the optical dose within individual layers. In order to obtain a minimum channel size of 60 µm × 108 µm by 10 µm building-layers, they increased the resin absorbance and the XY plane resolution of the DLP illumination [72]. Transparent biocompatible resins are available too for SLA, so the realized microfluidics allow for the characterization of the internal fluid flow as well as the direct observation of the in-situ formation of droplets [106]. Other advantages are the smooth surfaces, custom low-cost resins, lack of need for external alignment, monolithic structures, and direct printing of the fluidic channels [84,107,108].

One of the major limitations of all the current SLA printers is that they are restricted to a single print material at this time. Choi et al. have developed a prototype of a multi-material SLA printer by using four different resin baths. However, the process is too complex and each resin layer requires multiple exposures, making the method quite inefficient [84]. Moreover, removing the uncured resin remains a major challenge in using this multi-material SLA for the printing of microfluidic structures. In SLA, the removal of uncured resin is easier than in i3Dp since the resin is a liquid, but it is still challenging in general [72]. There are several successful examples of using SLA for microfluidics fabrication [82], and SLA is an effective additive method for manufacturing complex nanocomposites, [73]. However, further developments of SLA are still required in order to make it an ideal method of choice for the fabrication of microfluidics for wearable biosensors [72]. Other disadvantages are the slow printing time, sometimes expensive chemicals, low biochemical adaptability of the resin, limited choice of the materials, and a resolution of a few tens of micrometers [73,83]: for example, a number of DLP printers with a resolution in XY and Z of 50 µm have been reported [73]. Even though the stereolithography process was introduced almost 33 years ago, there is still enough room for further improvements. Recently, a novel micro-diamond based composite resin was published to print a thermally conductive prototype for specific applications [72]. Of course, stereolithography is also largely used in other fields other than microfluidics (channels, valves, and pumps) for biosensing, such as organ-on-chip platforms, flexible electronics, micromixers for pKa determinations, and soft robotics [97,109]. Examples of such other systems are shown in Figure 15.

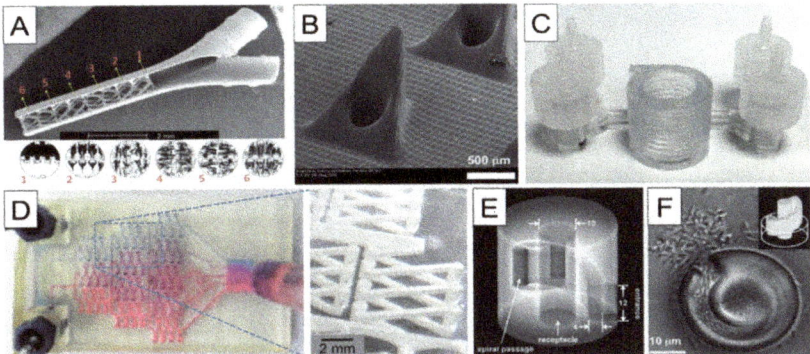

Figure 15. Examples of structures fabricated with SLA: (**A**) a micro-mixer, (**B**) hollow micro-needles, (**C**) Spiral microchannels for size-selective separation of bacterial cells, (**D**) gradient generator, (**E**) a "lobster trap" for bacteria, (**F**) alternative "lobster trap" for colony of *E. coli*.

3.4. Multi-Material Methods

As briefly mentioned at the beginning of Section 3, several multi-material methods have been proposed over the years to improve the quality of printed objects. With this approach, two or more materials can be simultaneously used for the building of a single object [87]. For example, multi-material 3D printing enables the synergic use of soft and rigid polymers with still resolutions in the range of tens of microns [27,103,115]. Several parts and components of microfluidic systems are fabricated with a multi-material method: for example, interconnects [66], membranes, valves, pumps, and multi-flow controllers [103]. All the previously discussed 3D printing methods (i3DP, FDM, and SLA) are possible with multi-materials [87,88]. The multi-material photopolymer inkjet printing method allows up to five different materials with a wide range of properties: from hard to soft plastics, elastomers, and also different colors. This method has a good speed in building an object since the multiple materials can be printed at the same time. However, with the multi-material inkjet printing method, it is difficult to remove the support material from the complex fluidic channels, while FDM printers do not need the support material to create channels. The materials used in the inkjet printing method are limited and their formulations are expensive and proprietary. The selection of the materials is often mostly concentrated on color, while the choice of flexible materials is especially suitable for the fluidics of wearable biosensors. For wearable microfluidics, three substrate interfaces are typically used, including fabric, polymer, and silicone (elastomer/rubber (see Figure 16)) [58]. These kinds of substrates are chosen since they are biocompatible and, therefore, they are particularly suitable for the biosensing systems used in wearable applications. Fabric may be soft, absorbent and breathable. Several different polymers are possible with properties of flexibility, robustness, and a strong resistance to chemicals. Silicone elastomers are stretchable and conformable, with properties of long-term durability, and present an excellent chemical resistance and viscoelasticity [58]. With multi-material methods, valves, pumps, and mixers have been demonstrated with a stronger resistance to deformation [81,87].

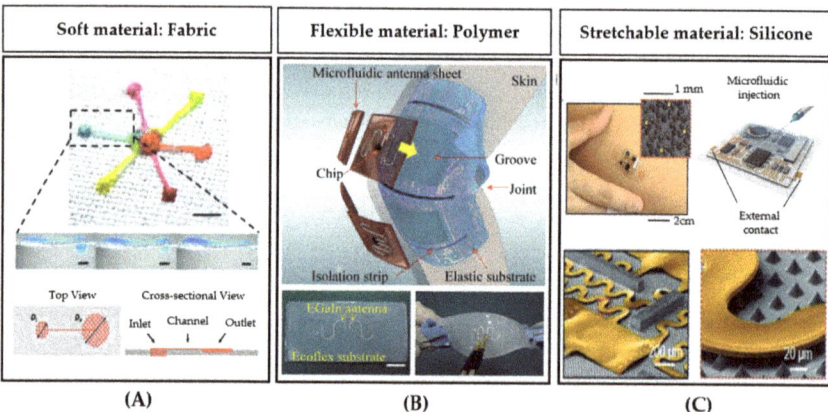

Figure 16. Wearable microfluidic systems realized with multi-material methods: (**A**) interfacial microfluidic transport principle to drive three-dimensional liquid flows on a micropatterned superhydrophobic textile, (**B**) triboelectric pressure sensor integrated with an antenna for data transmission, (**C**) stretchable skin-patch with integrated electronics (reprinted with permission from [116–118]).

Most of the polymers used in multi-material additive manufacturing are split into two main categories: photosensitive polymers and thermoplastic polymers. The first group is widely used in the 3D printing method based on photopolymerization, like SLA, and includes acrylate, epoxy, or hybrid resins. SLA processes also broadly use hybrid resins, such as epoxides with acrylate content. These hybrid resins increase the integrity of the layers during the fabrication and the strength of the finished parts. More importantly, the use of hybrid resins enables the fabrication of transparent and biocompatible microfluidic devices with a high resolution. Thermoplastic polymers include acrylonitrile butadiene styrene (ABS) and polylactic acid (PLA), which are widely employed for extrusion-based methods, such as FDM [119,120]. Extruded materials for microfluidics are essentially polypropylene (PP), ABS, and PLA. PP is used for its high biocompatibility, like polydimethylsiloxane (PDMS), and it is cheaper. It is introduced as an attractive material for the additive fabrications of micro and milli-scale fluidic devices, since it is a robust, flexible, and chemically inert polymer. PP is a semi-crystalline material, which typically does not soften with raising temperatures. It quickly transforms into a low-viscosity liquid and, once solid, it shrinks less in the flow direction than in the transverse direction. PP is extruded in a liquid state, and then it solidifies via crystallization as soon as the temperature goes below the melting point. However, the thermal shrinkage stresses are high during the layer solidification, and lead to very high warping stresses. Therefore, PP is typically only used when a high biocompatibility is required. On the contrary, ABS is an amorphous polymer that can crawl slowly until it cools below the glass point, and starts to warp below the glass point to complete the solidification. Therefore, the thermal stresses above the glass point temperature could be partially compensated. Therefore, it is suitable for the building plates and chambers by minimizing the warping stress at temperatures around the glass point. ABS is a very useful material for many biomedical applications because of its excellent mechanical and processing properties, its versatility and low-cost. Most recently, one of the most common materials considered for 3D printing became PLA, which is an inexpensive, biodegradable, and nontoxic aliphatic polyester [119].

The technical limitations for multi-material printing are still present and typically related to the material's compatibility, the adherence among them, and also the significant differences in the extrusion temperatures. In some cases, it is difficult to remove the support material and the material choice is limited with respect the typical needs in wearable biosensors. Multi-material FDM is cheaper than i3Dp, and easier than SLA. Compared to other printing methods, FDM printing has a good choice of commercially available materials with different properties. FDM typically uses thermoplastics, providing access to a wide variety of cheap and biocompatible materials. Despite these benefits, and as well as those already discussed, FDM still has its own limitations, including a low structural strength of printed objects, lack of structural integrity between the layers, and weak sealing properties when used in microfluidics for wearable biosensors. Furthermore, it provides a lower resolution with respect to SLA and i3Dp [81,87]. With multi-material SLA, the resolution and chemical compatibility is usually much better; the stretchability, gas permeability, and larger heat dissipation are better too. On the other hand, SLA is more expensive than FDM and i3Dp, and the fabrication process is typically slower because there are different resin vats that the printed object must be moved between during the printing process [81].

4. Printing Sensors for Direct Integration

By using the additive manufacturing methods presented above, that are suitable for building the microfluidics for wearable biosensors, it is also possible to print the sensors as well for a direct integration with the fluidics. Stretchable conductors are highly suitable for wearable sensors and electronics. Typical tracks in wearable systems involve electrical conductors, such as carbon nanotubes (CNTs), graphene sheets, metal nanowires (metal NWs), liquid metals, or conductive polymers, which present limitations in these kind of applications because of an increase in resistance upon deformation. Therefore, a new class of conductors is proposed with the aim of a direct integration of sensors by direct printing into the fluidics for a wearable biosensing platform. Ionically conductive materials, mostly hydrogels (Figure 17), ionogels or polyionic elastomers, which use charged ions rather than electrons to transmit electrical signals, are suggested to achieve the aim. Many ionic conductors have the intrinsic properties of a high stretchability, transparency, and biocompatibility. More importantly, they have a good optical transmission and electrical conductance at the same time. The use of such microstructured hydrogel electrodes improves the deformability of the sensor for its application to ionic skins (ionic conductor-based sensors). Printed conductive hydrogels are highly stretchable and elastic, almost fully transparent, highly precise, and stable with regards to their electro-mechanical properties. So, they are very suitable to apply to wearable amperometric biosensors as microstructured current collectors (see Figures 18 and 19). However, it is difficult to obtain a high resolution, especially in the printing of complex structures, typically due to limitations related to the low strength of the used conductive materials.

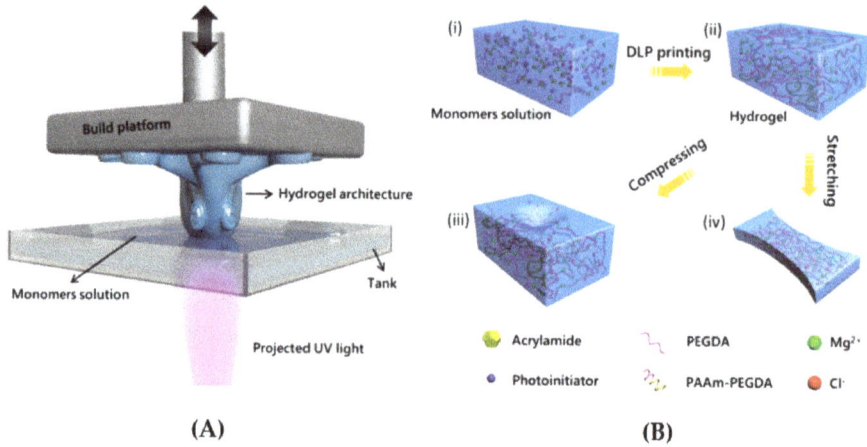

Figure 17. The addictive manufacture of a conductive hydrogel: (**A**) a high resolution fast bottom-up fabrication by a DLP printer, (**B**) the polymerization of a hydrogel network (reprinted with permission from [121]).

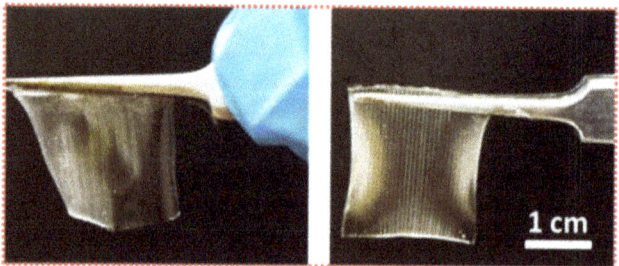

Figure 18. Structured hydrogel films: a total thickness of 400 mm; parallel lines with a deepness of 200 mm, line-to-line spacing of 200 mm (**in the left**) or 400 mm (**in the right** (reprinted with permission from [121])).

To show a good example of this kind of printed sensor, we can mention here the work of Yang et al., who proposed printing a wearable capacitive sensor for detecting both the static and dynamic pressures and strain, with a high sensitivity and low limit of detection [121]. Yang et al. fabricated an elastic and ionically conductive hydrogel by using microstructures in a single printing step with a commercial DLP printer (Figure 17A). Superior capabilities with regards to high-fidelity for the body signal acquisition over multiple skin locations was demonstrated with such a device, including finger bending tracking, pulse waveform monitoring, and larynx vibration tracking.

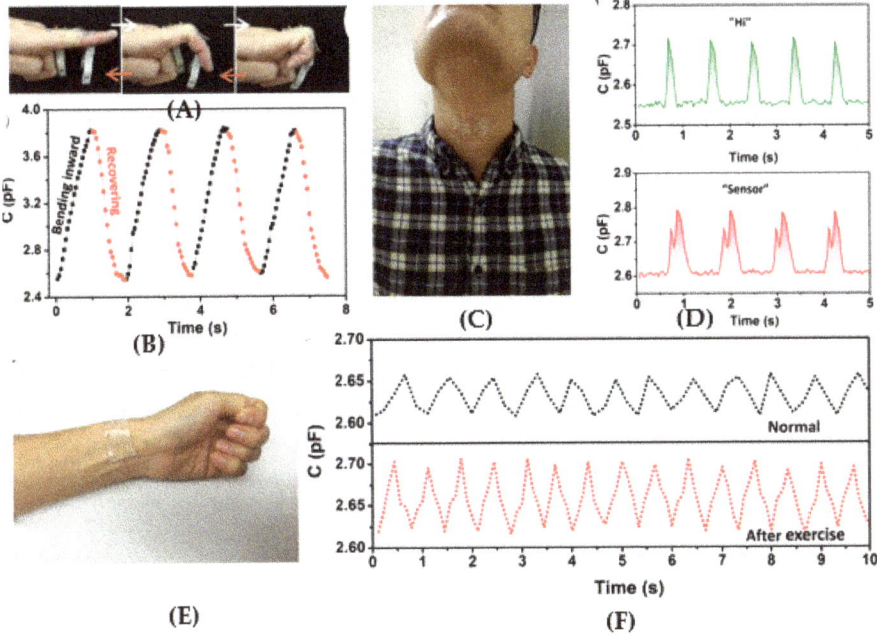

Figure 19. Wearable sensing devices by hydrogel films: (**A**) sensor on-the-skin for the finger bending, (**B**) capacitance as acquired from the finger cyclically bending, (**C**) sensor on-the-skin for throat movements, (**D**) acquired signals when subject says "Hi" and "Sensor", (**E**) sensor on-the-skin for the artery pulse, (**F**) radial artery pulses (reprinted with permission from [121]).

5. Conclusions

This paper provides a review of all the additive manufacturing (3D printing) methods suitable for microfluidics in applications to wearable biosensors. A photolithographic technique and chemical etching were the primary fabrication methods for microfluidics that used silicon and glass. In comparison to traditional methods, 3D printing technology has a short fabrication cycle time. This technology can have a good control of the complex design and quality of the product, with a good resolution and low-cost fabrication Wearable biosensors are required for the analysis of the molecular content of sweat as directly provided on-the-skin. Microfluidics plays an important role in solving some of the challenges for on-the-skin monitoring, for example collecting and conveying a small amount of liquid sample to the detection part. The 3D printing methods suitable for microfluidics are selective laser sintering (SLS), fused deposition modeling (FDM), inkjet 3D printing (i3Dp), laminated object manufacturing (LOM), two photon polymerization (TPP) and stereolithography (SLA). Among them, three suitable methods for flexible devices have been discussed in this paper: fused deposition modeling (FDM), inkjet 3D Printing (i3Dp), and stereolithography (SLA). These methods use polymeric substrates that have wide applications for wearable devices. On the other hand, these methods are able to use multi-material methods which have better control over the quality and flexibility of the product. Each of the discussed methods still present pros and cons when thought for fabricating microfluidics for wearable biosensors. SLA is a suitable method in case of the need for high resolutions, while SLA is not necessarily a good option for the low-cost productions or for fast fabrication processes. FDM is the more appropriate method in the case of inexpensive productions, especially in cases where biocompatibility is strictly required. The i3Dp method is most probably the best method for devices requiring extremely good mechanical properties (e.g., in term of robustness) and to assure fewer rough on-body contact-surfaces. Today, a lot of research is being done to advance these 3D printing methods for their future applications. For example, the inks in the i3Dp method can be modified for certain purposes. The biosensors required for detection on-the-skin are also printable by using additive manufacturing, especially by exploiting conductive properties of modern hydrogels, ionogels, or polyionic elastomers. Therefore, such additive manufacturing methods are definitely suitable for producing microfluidics systems with the right resolution for wearable biosensors in order to develop more advanced and fully non-invasive monitoring of metabolism directly on-the-skin of humans.

This review article will help the reader understand the importance of the role of microfluidics in the development of wearable technology and gain insight into the appropriate ways to fabricate them that will play an important role in future applications of wearable sensors using microfluidics. If a researcher wants to create suitable microfluidics for wearable sensors, by reading this article, they can decide which method is right for them according to their desires and possibilities. This article gives a brief overview of the materials used in these methods and some of their challenges. It is suggested that in future work, the efficiency, limitations, and preference of materials over each other be examined in more detail, separately for each application and method. This critical article can give researchers a good vision of the materials used in a certain method and their challenges. In this case, researchers can choose the most suitable material for the intended purpose from the available materials, and sometimes even decide to change the method by choosing a suitable material for a specific application. This study of materials and their challenges can be considered by researchers who are diligent in overcoming the challenges presented, and progress, and improve the field.

Author Contributions: M.P. wrote the original draft and the manuscript and designed the figures. S.C. worked on the manuscript. All authors contributed for revising the manuscript. All authors have read and agreed to the published version of the manuscript.

Funding: This research received no external funding.

Conflicts of Interest: The authors declare no conflict of interest.

References

1. Adami, H.-O.; Day, N.; Trichopoulos, D.; Willett, W. Primary and secondary prevention in the reduction of cancer morbidity and mortality. *Eur. J. Cancer* **2001**, *37*, 118–127. [CrossRef]
2. Rassi, A.; Dias, J.C.P.; Marin-Neto, J.A. Challenges and opportunities for primary, secondary, and tertiary prevention of Chagas' disease. *Heart* **2009**, *95*, 524–534. [CrossRef]
3. Bagheri-Tadi, S.A.A.T.; Teimouri, F. Features and application of wearable biosensors in medical care. *J. Res. Med. Sci.* **2015**, *20*, 1208–1215. [CrossRef]
4. Appelboom, G.; Camacho, E.; Abraham, M.E.; Bruce, S.S.; Dumont, E.L.; Zacharia, B.E.; D'Amico, R.; Slomian, J.; Reginster, J.-Y.; Bruyere, O.; et al. Smart wearable body sensors for patient self-assessment and monitoring. *Arch. Public Health* **2014**, *72*, 28. [CrossRef] [PubMed]
5. Esteban, M.; Castano, A. Non-invasive matrices in human biomonitoring: A review. *Environ. Int.* **2009**, *35*, 438–449. [CrossRef] [PubMed]
6. Kańtoch, E.; Augustyniak, P. Human activity surveillance based on wearable body sensor network. In Proceedings of the International Conference Computing in Cardiology, Krakow, Poland, 9–12 September 2012; pp. 325–328.
7. Stoppa, M.; Chiolerio, A. Wearable Electronics and Smart Textiles: A Critical Review. *Sensors* **2014**, *14*, 11957–11992. [CrossRef] [PubMed]
8. Tamsin, M. Wearable Biosensor Technologies. *Int. J. Innov. Sci. Res.* **2015**, *13*, 697–703.
9. Zeng, W.; Shu, L.; Li, Q.; Chen, S.; Wang, F.; Tao, X.-M. Fiber-Based Wearable Electronics: A Review of Materials, Fabrication, Devices, and Applications. *Adv. Mater.* **2014**, *26*, 5310–5336. [CrossRef]
10. Tarakci, H.; Kulkarni, S.S.; Ozdemir, Z.D. The impact of wearable devices and performance payments on health outcomes. *Int. J. Prod. Econ.* **2018**, *200*, 291–301. [CrossRef]
11. Afshar, V.; Peterson, D. Wearable technology: The Coming Revolution in Healthcare. Available online: http://www.huffpost.com (accessed on 20 August 2019).

12. Gao, W.; Emaminejad, S.; Nyein, H.Y.Y.; Challa, S.; Chen, K.; Peck, A.; Fahad, H.M.; Ota, H.; Shiraki, H.; Kiriya, D.; et al. Fully integrated wearable sensor arrays for multiplexed in situ perspiration analysis. *Nature* **2016**, *529*, 509–514. [CrossRef]
13. Gao, W.; Nyein, H.Y.Y.; Shahpar, Z.; Fahad, H.M.; Chen, K.; Emaminejad, S.; Gao, Y.; Tai, L.-C.; Ota, H.; Wu, E.; et al. Wearable Microsensor Array for Multiplexed Heavy Metal Monitoring of Body Fluids. *ACS Sens.* **2016**, *1*, 866–874. [CrossRef]
14. Nyein, H.Y.Y.; Gao, W.; Shahpar, Z.; Emaminejad, S.; Challa, S.; Chen, K.; Fahad, H.M.; Tai, L.-C.; Ota, H.; Davis, R.W.; et al. A Wearable Electrochemical Platform for Noninvasive Simultaneous Monitoring of Ca^{2+} and pH. *ACS Nano* **2016**, *10*, 7216–7224. [CrossRef] [PubMed]
15. Bariya, M.; Nyein, H.Y.Y.; Javey, A. Wearable sweat sensors. *Nat. Electron.* **2018**, *1*, 160–171. [CrossRef]
16. Rose, D.P.; Ratterman, M.E.; Griffin, D.K.; Hou, L.; Kelley-Loughnane, N.; Naik, R.R.; Hagen, J.A.; Papautsky, I.; Heikenfeld, J.C. Adhesive RFID Sensor Patch for Monitoring of Sweat Electrolytes. *IEEE Trans. Biomed. Eng.* **2014**, *62*, 1457–1465. [CrossRef]
17. Choi, J.; Kang, D.; Han, S.; Kim, S.B.; Rogers, J.A. Thin, Soft, Skin-Mounted Microfluidic Networks with Capillary Bursting Valves for Chrono-Sampling of Sweat. *Adv. Healthc. Mater.* **2017**, *6*, 1601355. [CrossRef]
18. Kim, J.; Jeerapan, I.; Imani, S.; Cho, T.N.; Bandodkar, A.J.; Cinti, S.; Mercier, P.P.; Wang, J. Noninvasive Alcohol Monitoring Using a Wearable Tattoo-Based Iontophoretic-Biosensing System. *ACS Sens.* **2016**, *1*, 1011–1019. [CrossRef]
19. Jia, W.; Bandodkar, A.J.; Valdés-Ramírez, G.; Windmiller, J.R.; Yang, Z.; Ramírez, J.; Chan, G.; Wang, J. Electrochemical Tattoo Biosensors for Real-Time Noninvasive Lactate Monitoring in Human Perspiration. *Anal. Chem.* **2013**, *85*, 6553–6560. [CrossRef]
20. Bandodkar, A.J.; Jia, W.; Yardımcı, C.; Wang, X.; Ramírez, J.; Wang, J.; Yardimci, C. Tattoo-Based Noninvasive Glucose Monitoring: A Proof-of-Concept Study. *Anal. Chem.* **2014**, *87*, 394–398. [CrossRef]
21. Bandodkar, A.J.; Wang, J. Non-invasive wearable electrochemical sensors: A review. *Trends Biotechnol.* **2014**, *32*, 363–371. [CrossRef]
22. Sonner, Z.; Wilder, E.; Gaillard, T.; Kasting, G.; Heikenfeld, J. Integrated sudomotor axon reflex sweat stimulation for continuous sweat analyte analysis with individuals at rest. *Lab Chip* **2017**, *17*, 2550–2560. [CrossRef]
23. Kim, J.; Imani, S.; De Araujo, W.R.; Warchall, J.; Valdés-Ramírez, G.; Paixão, T.R.L.C.; Mercier, P.P.; Wang, J. Wearable salivary uric acid mouthguard biosensor with integrated wireless electronics. *Biosens. Bioelectron.* **2015**, *74*, 1061–1068. [CrossRef] [PubMed]
24. Emaminejad, S.; Gao, W.; Wu, E.; Davies, Z.A.; Nyein, H.Y.Y.; Challa, S.; Ryan, S.P.; Fahad, H.M.; Chen, K.; Shahpar, Z.; et al. Autonomous sweat extraction and analysis applied to cystic fibrosis and glucose monitoring using a fully integrated wearable platform. *Proc. Natl. Acad. Sci. USA* **2017**, *114*, 4625–4630. [CrossRef] [PubMed]
25. Bariya, M.; Shahpar, Z.; Park, H.; Sun, J.; Jung, Y.; Gao, W.; Nyein, H.Y.Y.; Liaw, T.S.; Tai, L.-C.; Ngo, Q.P.; et al. Roll-to-Roll Gravure Printed Electrochemical Sensors for Wearable and Medical Devices. *ACS Nano* **2018**, *12*, 6978–6987. [CrossRef]
26. Li, G.; Mo, X.; Law, W.-C.; Chan, K.C. Wearable Fluid Capture Devices for Electrochemical Sensing of Sweat. *ACS Appl. Mater. Interfaces* **2018**, *11*, 238–243. [CrossRef] [PubMed]
27. Wu, W.; Haick, H. Materials and Wearable Devices for Autonomous Monitoring of Physiological Markers. *Adv. Mater.* **2018**, *30*, e1705024. [CrossRef]
28. De Castro, L.F.; De Freitas, S.V.; Duarte, L.C.; De Souza, J.A.C.; Paixão, T.R.L.C.; Coltro, W.K. Salivary diagnostics on paper microfluidic devices and their use as wearable sensors for glucose monitoring. *Anal. Bioanal. Chem.* **2019**, *411*, 4919–4928. [CrossRef]
29. Raju, S.P.; Chu, X. Rapid Low-Cost Microfluidic Detection in Point of Care Diagnostics. *J. Med. Syst.* **2018**, *42*, 184. [CrossRef]
30. Chen, M.; Ma, Y.; Song, J.; Lai, C.-F.; Hu, B. Smart Clothing: Connecting Human with Clouds and Big Data for Sustainable Health Monitoring. *Mob. Netw. Appl.* **2016**, *21*, 825–845. [CrossRef]
31. Aldeer, M.; Javanmard, M.; Martin, R.P. A Review of Medication Adherence Monitoring Technologies. *Appl. Syst. Innov.* **2018**, *1*, 14. [CrossRef]

32. Kim, J.; Sempionatto, J.R.; Imani, S.; Hartel, M.C.; Barfidokht, A.; Tang, G.; Campbell, A.S.; Mercier, P.P.; Wang, J. Simultaneous Monitoring of Sweat and Interstitial Fluid Using a Single Wearable Biosensor Platform. *Adv. Sci.* **2018**, *5*, 1800880. [CrossRef]
33. Liao, Y.-T.; Yao, H.; Lingley, A.; Parviz, B.; Otis, B.P. A 3-μW CMOS Glucose Sensor for Wireless Contact-Lens Tear Glucose Monitoring. *IEEE J. Solid State Circuits* **2011**, *47*, 335–344. [CrossRef]
34. Dehghani, M.; Dangelico, R.M. Smart wearable technologies: Current status and market orientation through a patent analysis. In Proceedings of the 2017 IEEE International Conference on Industrial Technology (ICIT), Toronto, ON, Canada, 22–25 March 2017; pp. 1570–1575.
35. Sun, B.; Zhang, Z. Photoplethysmography-Based Heart Rate Monitoring Using Asymmetric Least Squares Spectrum Subtraction and Bayesian Decision Theory. *IEEE Sens. J.* **2015**, *15*, 7161–7168. [CrossRef]
36. Ahanathapillai, V.; Amor, J.D.; Goodwin, Z.; James, C.J. Preliminary study on activity monitoring using an android smart-watch. *Healthc. Technol. Lett.* **2015**, *2*, 34–39. [CrossRef] [PubMed]
37. Jung, P.-G.; Oh, S.; Lim, G.; Kong, K. A Mobile Motion Capture System Based on Inertial Sensors and Smart Shoes. *J. Dyn. Syst. Meas. Control.* **2013**, *136*, 011002. [CrossRef]
38. Amft, O.; Wahl, F.; Ishimaru, S.; Kunze, K. Making Regular Eyeglasses Smart. *IEEE Pervasive Comput.* **2015**, *14*, 32–43. [CrossRef]
39. Holz, C.; Wang, E.J. Glabella: Continuously sensing blood pressure behavior using an unobtrusive wearable device. *Proc. ACM Interact. Mob. Wearable Ubiquitous Technol.* **2017**, *1*, 1–23. [CrossRef]
40. Forrester, J. Securing Patch for Wearable Medical Device. U.S. Patent Application 16/158,954, 18 April 2019.
41. Punjiya, M.; Rezaei, H.; Zeeshan, M.A.; Sonkusale, S. A flexible pH sensing smart bandage with wireless CMOS readout for chronic wound monitoring. In Proceedings of the 2017 19th International Conference on Solid-State Sensors, Actuators and Microsystems (TRANSDUCERS), Kaohsiung, Taiwan, 18–22 June 2017; pp. 1700–1702.
42. Yang, Y.; Gao, W. Wearable and flexible electronics for continuous molecular monitoring. *Chem. Soc. Rev.* **2019**, *48*, 1465–1491. [CrossRef]
43. Koh, A.; Kang, D.; Xue, Y.; Lee, S.; Pielak, R.M.; Kim, J.; Hwang, T.; Min, S.; Banks, A.; Bastien, P.; et al. A soft, wearable microfluidic device for the capture, storage, and colorimetric sensing of sweat. *Sci. Transl. Med.* **2016**, *8*, 366ra165. [CrossRef]
44. Kim, J.; Valdés-Ramírez, G.; Bandodkar, A.J.; Jia, W.; Martinez, A.G.; Ramírez, J.; Mercier, P.P.; Wang, J. Non-invasive mouthguard biosensor for continuous salivary monitoring of metabolites. *Analyst* **2014**, *139*, 1632–1636. [CrossRef]
45. Güder, F.; Ainla, A.; Redston, J.; Mosadegh, B.; Glavan, A.; Martin, T.J.; Whitesides, G.M. Paper-Based Electrical Respiration Sensor. *Angew. Chem. Int. Ed.* **2016**, *55*, 5727–5732. [CrossRef]
46. Boyle, J.O.; Mao, L.; Brennan, J.A.; Koch, W.M.; Eisele, D.W.; Saunders, J.R.; Sidransky, D. Gene mutations in saliva as molecular markers for head and neck squamous cell carcinomas. *Am. J. Surg.* **1994**, *168*, 429–432. [CrossRef]
47. Hu, S.; Arellano, M.; Boontheung, P.; Wang, J.; Zhou, H.; Jiang, J.; Elashoff, D.; Wei, R.; Loo, J.A.; Wong, D.T.W. Salivary proteomics for oral cancer biomarker discovery. *Clin. Cancer Res.* **2008**, *14*, 6246–6252. [CrossRef] [PubMed]
48. Zhang, L.; Xiao, H.; Karlan, S.; Zhou, H.; Gross, J.; Elashoff, D.; Akin, D.; Yan, X.; Chia, D.; Karlan, B.; et al. Discovery and Preclinical Validation of Salivary Transcriptomic and Proteomic Biomarkers for the Non-Invasive Detection of Breast Cancer. *PLoS ONE* **2010**, *5*, e15573. [CrossRef] [PubMed]
49. McCaul, M.; Porter, A.; Barrett, R.; Wallace, G.; Stroiescu, F.; Wallace, G.G.; Diamond, D. Wearable Platform for Real-time Monitoring of Sodium in Sweat. *Chem. Phys. Chem.* **2018**, *19*, 1531–1536. [CrossRef] [PubMed]
50. Bandodkar, A.J.; Molinnus, D.; Mirza, O.; Guinovart, T.; Windmiller, J.R.; Valdés-Ramírez, G.; Andrade, F.J.; Schoening, M.J.; Wang, J. Epidermal tattoo potentiometric sodium sensors with wireless signal transduction for continuous non-invasive sweat monitoring. *Biosens. Bioelectron.* **2014**, *54*, 603–609. [CrossRef]
51. Guinovart, T.; Bandodkar, A.J.; Windmiller, J.R.; Andrade, F.J.; Wang, J. A potentiometric tattoo sensor for monitoring ammonium in sweat. *Analyst* **2013**, *138*, 7031. [CrossRef]
52. Gamella, M.; Campuzano, S.; Manso, J.; De Rivera, G.G.; Lopez-Colino, F.; Reviejo, Á.J.; Pingarrón, J.M. A novel non-invasive electrochemical biosensing device for in situ determination of the alcohol content in blood by monitoring ethanol in sweat. *Anal. Chim. Acta* **2014**, *806*, 1–7. [CrossRef]

53. Lee, H.; Choi, T.K.; Lee, Y.B.; Cho, H.R.; Ghaffari, R.; Wang, L.; Choi, H.J.; Chung, T.D.; Lu, N.; Hyeon, T.; et al. A graphene-based electrochemical device with thermoresponsive microneedles for diabetes monitoring and therapy. *Nat. Nanotechnol.* **2016**, *11*, 566–572. [CrossRef]
54. Lee, H.; Song, C.; Hong, Y.S.; Kim, M.S.; Cho, H.R.; Kang, T.; Shin, K.; Choi, S.H.; Hyeon, T.; Kim, D.-H. Wearable/disposable sweat-based glucose monitoring device with multistage transdermal drug delivery module. *Sci. Adv.* **2017**, *3*, e1601314. [CrossRef]
55. Martin, A.; Kim, J.; Kurniawan, J.; Sempionatto, J.R.; Moreto, J.R.; Tang, G.; Campbell, A.S.; Shin, A.; Lee, M.Y.; Liu, X.; et al. Epidermal Microfluidic Electrochemical Detection System: Enhanced Sweat Sampling and Metabolite Detection. *ACS Sens.* **2017**, *2*, 1860–1868. [CrossRef]
56. Jadoon, S.; Karim, S.; Akram, M.R.; Khan, A.K.; Zia, M.A.; Siddiqi, A.R.; Murtaza, G. Recent Developments in Sweat Analysis and Its Applications. *Int. J. Anal. Chem.* **2015**, *2015*, 164974. [CrossRef] [PubMed]
57. Csősz, É.; Emri, G.; Kalló, G.; Tsaprailis, G.; Tőzsér, J. Highly abundant defense proteins in human sweat as revealed by targeted proteomics and label-free quantification mass spectrometry. *J. Eur. Acad. Dermatol. Venereol.* **2015**, *29*, 2024–2031. [CrossRef]
58. Yeo, J.C.; Kenry; Lim, C.T. Emergence of microfluidic wearable technologies. *Lab Chip* **2016**, *16*, 4082–4090. [CrossRef]
59. Nyein, H.Y.Y.; Tai, L.-C.; Ngo, Q.P.; Chao, M.; Zhang, G.B.; Gao, W.; Bariya, M.; Bullock, J.; Kim, H.; Fahad, H.M.; et al. A Wearable Microfluidic Sensing Patch for Dynamic Sweat Secretion Analysis. *ACS Sens.* **2018**, *3*, 944–952. [CrossRef] [PubMed]
60. Kaya, T.; Liu, G.; Ho, J.; Yelamarthi, K.; Miller, K.; Edwards, J.; Stannard, A.B. Wearable Sweat Sensors: Background and Current Trends. *Electroanalysis* **2018**, *31*, 411–421. [CrossRef]
61. Gale, B.K.; Jafek, A.R.; Lambert, C.J.; Goenner, B.L.; Moghimifam, H.; Nze, U.C.; Kamarapu, S.K. A Review of Current Methods in Microfluidic Device Fabrication and Future Commercialization Prospects. *Inventions* **2018**, *3*, 60. [CrossRef]
62. Yuen, P.K.; Goral, V.N. Low-cost rapid prototyping of flexible microfluidic devices using a desktop digital craft cutter. *Lab Chip* **2010**, *10*, 384–387. [CrossRef]
63. Baghban, A.; Jalali, A.; Shafiee, M.; Ahmadi, M.H.; Ahmadi, M.H. Developing an ANFIS-based swarm concept model for estimating the relative viscosity of nanofluids. *Eng. Appl. Comput. Fluid Mech.* **2018**, *13*, 26–39. [CrossRef]
64. Alnaimat, F.; Mathew, B. Magnetophoretic microdevice for size-based separation: Model-based study. *Eng. Appl. Comput. Fluid Mech.* **2020**, *14*, 738–750. [CrossRef]
65. Naghdloo, A.; Ghazimirsaeed, E.; Shamloo, A. Numerical simulation of mixing and heat transfer in an integrated centrifugal microfluidic system for nested-PCR amplification and gene detection. *Sens. Actuators B Chem.* **2019**, *283*, 831–841. [CrossRef]
66. Cosson, S.; Aeberli, L.G.; Brandenberg, N.; Lutolf, M. Ultra-rapid prototyping of flexible, multi-layered microfluidic devices via razor writing. *Lab Chip* **2015**, *15*, 72–76. [CrossRef] [PubMed]
67. Mi, S.; Du, Z.; Xu, Y.; Sun, W. The crossing and integration between microfluidic technology and 3D printing for organ-on-chips. *J. Mater. Chem. B* **2018**, *6*, 6191–6206. [CrossRef] [PubMed]
68. Zheng, W.; Jiang, B.; Wang, N.; Zhang, W.; Wang, Z.; Jiang, X. A microfluidic flow-stretch chip for investigating blood vessel biomechanics. *Lab Chip* **2012**, *12*, 3441–3450. [CrossRef] [PubMed]
69. Odijk, M.; Baumann, A.; Lohmann, W.; Brink, F.T.G.V.D.; Olthuis, W.; Karst, U.; Berg, A.V.D. A microfluidic chip for electrochemical conversions in drug metabolism studies. *Lab Chip* **2009**, *9*, 1687–1693. [CrossRef]
70. Manz, A.; Verpoorte, E.M.J.; Fettinger, J.C.; Harrison, D.J.; Ludi, H.; Widmer, H.M. Design of Integrated Electroosmotic Pumps and Flow Manifolds for Total Chemical Analysis Systems, Micromechanics Europe, Technical Digest. In Proceedings of the 2nd Workshop on Micromachining, Micromechanics, and Microsystems, Berlin, Germany, 26–27 November 1990; pp. 127–132.
71. Whitesides, G.M.; Stroock, A.D. Flexible Methods for Microfluidics. *Phys. Today* **2001**, *54*, 42–48. [CrossRef]
72. Waheed, S.; Cabot, J.M.; Macdonald, N.P.; Lewis, T.W.; Guijt, R.; Paull, B.; Breadmore, M.C. 3D printed microfluidic devices: Enablers and barriers. *Lab Chip* **2016**, *16*, 1993–2013. [CrossRef]
73. Ngo, T.; Kashani, A.; Imbalzano, G.; Nguyen, Q.T.; Hui, D. Additive manufacturing (3D printing): A review of materials, methods, applications and challenges. *Compos. Part B Eng.* **2018**, *143*, 172–196. [CrossRef]
74. Terry, W.; Tim, C. *Wohlers report 2013: Additive Manufacturing and 3D Printing State of the Industry*; Wohlers Associates: Fort Collins, Colorado, 2013.

75. Lu, B.; Li, D.; Tian, X. Development Trends in Additive Manufacturing and 3D Printing. *Engineering* **2015**, *1*, 85–89. [CrossRef]
76. Marro, A.; Bandukwala, T.; Mak, W.; Information, P.E.K.F.C. Three-Dimensional Printing and Medical Imaging: A Review of the Methods and Applications. *Curr. Probl. Diagn. Radiol.* **2016**, *45*, 2–9. [CrossRef]
77. Lin, Y.; Xu, J. Microstructures Fabricated by Two-Photon Polymerization and Their Remote Manipulation Techniques: Toward 3D Printing of Micromachines. *Adv. Opt. Mater.* **2018**, *6*, 1701359. [CrossRef]
78. Bhattacharjee, N.; Urrios, A.; Kang, S.; Folch, A. The upcoming 3D-printing revolution in microfluidics. *Lab Chip* **2016**, *16*, 1720–1742. [CrossRef] [PubMed]
79. Dagum, P. Digital biomarkers of cognitive function. *NPJ Digit. Med.* **2018**, *1*, 1–3. [CrossRef] [PubMed]
80. Au, A.K.; Huynh, W.; Horowitz, L.F.; Folch, A. 3D-Printed Microfluidics. *Angew. Chem. Int. Ed.* **2016**, *55*, 3862–3881. [CrossRef] [PubMed]
81. Li, F.; Macdonald, N.P.; Guijt, R.; Breadmore, M.C. Increasing the functionalities of 3D printed microchemical devices by single material, multimaterial, and print-pause-print 3D printing. *Lab Chip* **2019**, *19*, 35–49. [CrossRef] [PubMed]
82. Amin, R.; Knowlton, S.; Hart, A.; Yenilmez, B.; Ghaderinezhad, F.; Katebifar, S.; Messina, M.; Khademhosseini, A.; Tasoglu, S. 3D-printed microfluidic devices. *Biofabrication* **2016**, *8*, 022001. [CrossRef]
83. Walczak, R.; Adamski, K. Inkjet 3D printing of microfluidic structures—On the selection of the printer towards printing your own microfluidic chips. *J. Micromech. Microeng.* **2015**, *25*, 85013. [CrossRef]
84. Hür, D.; Say, M.G.; Diltemiz, S.E.; Duman, F.; Ersöz, A.; Say, R. 3D Micropatterned All-Flexible Microfluidic Platform for Microwave-Assisted Flow Organic Synthesis. *ChemPlusChem* **2018**, *83*, 42–46. [CrossRef]
85. Macdonald, N.P.; Cabot, J.M.; Smejkal, P.; Guijt, R.; Paull, B.; Breadmore, M.C. Comparing Microfluidic Performance of Three-Dimensional (3D) Printing Platforms. *Anal. Chem.* **2017**, *89*, 3858–3866. [CrossRef]
86. Ahn, D.-G.; Lee, J.-Y.; Yang, D.-Y. Rapid Prototyping and Reverse Engineering Application for Orthopedic Surgery Planning. *J. Mech. Sci. Technol.* **2006**, *20*, 19–28. [CrossRef]
87. Li, F.; Macdonald, N.P.; Guijt, R.; Breadmore, M.C. Multimaterial 3D Printed Fluidic Device for Measuring Pharmaceuticals in Biological Fluids. *Anal. Chem.* **2018**, *91*, 1758–1763. [CrossRef]
88. Li, F.; Smejkal, P.; Macdonald, N.P.; Guijt, R.; Breadmore, M.C. One-Step Fabrication of a Microfluidic Device with an Integrated Membrane and Embedded Reagents by Multimaterial 3D Printing. *Anal. Chem.* **2017**, *89*, 4701–4707. [CrossRef] [PubMed]
89. Bressan, L.; Adamo, C.B.; Quero, R.F.; De Jesus, D.P.; Da Silva, J.A.F.; Da Silva, J.A.F. A simple procedure to produce FDM-based 3D-printed microfluidic devices with an integrated PMMA optical window. *Anal. Methods* **2019**, *11*, 1014–1020. [CrossRef]
90. Sun, K.; Wei, T.-S.; Ahn, B.Y.; Seo, J.Y.; Dillon, S.J.; Lewis, J.A. 3D Printing of Interdigitated Li-Ion Microbattery Architectures. *Adv. Mater.* **2013**, *25*, 4539–4543. [CrossRef] [PubMed]
91. Muth, J.T.; Vogt, D.M.; Truby, R.L.; Mengüç, Y.; Kolesky, D.B.; Wood, R.J.; Lewis, J.A. Embedded 3D Printing of Strain Sensors within Highly Stretchable Elastomers. *Adv. Mater.* **2014**, *26*, 6307–6312. [CrossRef]
92. Kong, Y.L.; Tamargo, I.A.; Kim, H.; Johnson, B.N.; Gupta, M.K.; Koh, T.-W.; Chin, H.-A.; Steingart, D.A.; Rand, B.P.; McAlpine, M.C. 3D Printed Quantum Dot Light-Emitting Diodes. *Nano Lett.* **2014**, *14*, 7017–7023. [CrossRef]
93. Adams, J.J.; Duoss, E.B.; Malkowski, T.F.; Motala, M.J.; Ahn, B.Y.; Nuzzo, R.G.; Bernhard, J.T.; Lewis, J.A. Conformal printing of electrically small antennas on three-dimensional surfaces. *Adv. Mater.* **2011**, *23*, 1335–1340. [CrossRef]
94. Lopes, A.J.; Macdonald, E.; Wicker, R.B. Integrating stereolithography and direct print technologies for 3D structural electronics fabrication. *Rapid Prototyp. J.* **2012**, *18*, 129–143. [CrossRef]
95. Mannoor, M.S.; Jiang, Z.; James, T.; Kong, Y.L.; Malatesta, K.A.; Soboyejo, W.O.; Verma, N.; Gracias, D.H.; McAlpine, M.C. 3D Printed Bionic Ears. *Nano Lett.* **2013**, *13*, 2634–2639. [CrossRef]
96. Thomas, D.J.; Tehrani, Z.; Redfearn, B. 3-D printed composite microfluidic pump for wearable biomedical applications. *Addit. Manuf.* **2016**, *9*, 30–38. [CrossRef]
97. Chen, C.; Mehl, B.T.; Munshi, A.S.; Townsend, A.D.; Spence, D.M.; Martin, R.S. 3D-printed Microfluidic Devices: Fabrication, Advantages and Limitations—A Mini Review. *Anal. Methods* **2016**, *8*, 6005–6012. [CrossRef]

98. Bishop, G.; Satterwhite, J.E.; Bhakta, S.; Kadimisetty, K.; Gillette, K.M.; Chen, E.; Rusling, J.F. 3D-Printed Fluidic Devices for Nanoparticle Preparation and Flow-Injection Amperometry Using Integrated Prussian Blue Nanoparticle-Modified Electrodes. *Anal. Chem.* **2015**, *87*, 5437–5443. [CrossRef] [PubMed]
99. Xu, Y.; Liu, M.; Kong, N.; Liu, J. Lab-on-paper micro- and nano-analytical devices: Fabrication, modification, detection and emerging applications. *Microchim. Acta* **2016**, *183*, 1521–1542. [CrossRef]
100. Alfadhel, A.; Ouyang, J.; Mahajan, C.G.; Forouzandeh, F.; Cormier, D.; Borkholder, D.A. Inkjet printed polyethylene glycol as a fugitive ink for the fabrication of flexible microfluidic systems. *Mater. Des.* **2018**, *150*, 182–187. [CrossRef] [PubMed]
101. Jywe, W. Progress on Advanced Manufacture for Micro/nano Technology 2005. In Proceedings of the 2005 International Conference on Advanced Manufacture, Taipei, Taiwan, 28 November–2 December 2005.
102. Erbano, B.O.; Opolski, A.C.; Olandoski, M.; Foggiatto, J.A.; Kubrusly, L.F.; Dietz, U.A.; Zini, C.; Marinho, M.M.M.A.; Leal, A.G.; Ramina, R. Rapid prototyping of three-dimensional biomodels as an adjuvant in the surgical planning for intracranial aneurysms. *Acta Cir. Bras.* **2013**, *28*, 756–761. [CrossRef]
103. Ji, Q.; Zhang, J.M.; Liu, Y.; Li, X.; Lv, P.; Jin, D.; Duan, H. A Modular Microfluidic Device via Multimaterial 3D Printing for Emulsion Generation. *Sci. Rep.* **2018**, *8*, 4791. [CrossRef]
104. Walczak, R.; Adamski, K.; Kubicki, W. Inkjet 3D printed modular microfluidic chips for on-chip gel electrophoresis. *J. Micromech. Microeng.* **2019**, *29*, 057001. [CrossRef]
105. Munshi, A.S.; Martin, R.S. Microchip-based electrochemical detection using a 3-D printed wall-jet electrode device. *Analyst* **2016**, *141*, 862–869. [CrossRef]
106. Zhou, Z.; Kong, T.; Mkaouar, H.; Salama, K.N.; Zhang, J.M. A hybrid modular microfluidic device for emulsion generation. *Sens. Actuator A Phys.* **2018**, *280*, 422–428. [CrossRef]
107. Parekh, D.P.; Ladd, C.; Panich, L.; Moussa, K.; Dickey, M.D. 3D printing of liquid metals as fugitive inks for fabrication of 3D microfluidic channels. *Lab Chip* **2016**, *16*, 1812–1820. [CrossRef]
108. Gong, H.; Ramstedt, C.; Woolley, A.T.; Nordin, G.P. 3D Printed Microfluidics. Brigham Young University: Provo, Utah, 2017.
109. Bhattacharjee, N.; Parra-Cabrera, C.; Kim, Y.T.; Kuo, A.; Folch, A. Desktop-Stereolithography 3D-Printing of a Poly(dimethylsiloxane)-Based Material with Sylgard-184 Properties. *Adv. Mater.* **2018**, *30*, e1800001. [CrossRef]
110. Bertsch, A.; Heimgartner, S.; Cousseau, P.; Renaud, P. Static micromixers based on large-scale industrial mixer geometry. *Lab Chip* **2001**, *1*, 56–60. [CrossRef] [PubMed]
111. Miller, P.R.; Gittard, S.D.; Edwards, T.L.; Lopez, D.M.; Xiao, X.; Wheeler, D.R.; Monteiro-Riviere, N.A.; Brozik, S.M.; Polsky, R.; Narayan, R.J. Integrated carbon fiber electrodes within hollow polymer microneedles for transdermal electrochemical sensing. *Biomicrofluidics* **2011**, *5*, 013415. [CrossRef] [PubMed]
112. Shallan, A.I.; Smejkal, P.; Corban, M.; Guijt, R.M.; Breadmore, M.C. Cost-effective three-dimensional printing of visibly transparent microchips within minutes. *Anal. Chem.* **2014**, *86*, 3124–3130. [CrossRef] [PubMed]
113. Nielson, R.; Kaehr, B.; Shear, J.B. Microreplication and design of biological architectures using dynamic-mask multiphoton lithography. *Small* **2009**, *5*, 120–125. [CrossRef] [PubMed]
114. Lee, W.; Kwon, D.; Choi, W.; Jung, G.Y.; Au, A.K.; Folch, A.; Jeon, S. 3D-printed microfluidic device for the detection of pathogenic bacteria using size-based separation in helical channel with trapezoid cross-section. *Sci. Rep.* **2015**, *5*, 7717. [CrossRef]
115. Paydar, O.H.; Paredes, C.; Hwang, Y.; Paz, J.; Shah, N.; Candler, R. Characterization of 3D-printed microfluidic chip interconnects with integrated O-rings. *Sens. Actuators A Phys.* **2014**, *205*, 199–203. [CrossRef]
116. Xing, S.; Jiang, J.; Pan, T. Interfacial microfluidic transport on micropatterned superhydrophobic textile. *Lab Chip* **2013**, *13*, 1937–1947. [CrossRef]
117. Huang, Y.; Wang, Y.; Xiao, L.; Liu, H.; Dong, W.; Yin, Z. Microfluidic serpentine antennas with designed mechanical tunability. *Lab Chip* **2014**, *14*, 4205–4212. [CrossRef]
118. Xu, S.; Zhang, Y.; Jia, L.; Mathewson, K.E.; Jang, K.I.; Kim, J.; Fu, H.; Huang, X.; Chava, P.; Wang, R.; et al. Soft microfluidic assemblies of sensors, circuits, and radios for the skin. *Science* **2014**, *344*, 70–74. [CrossRef]
119. Pranzo, D.; Larizza, P.; Filippini, D.; Percoco, G. Extrusion-Based 3D Printing of Microfluidic Devices for Chemical and Biomedical Applications: A Topical Review. *Micromachines* **2018**, *9*, 374. [CrossRef]

120. Yazdi, A.A.; Popma, A.; Wong, W.; Nguyen, T.; Pan, Y.; Xu, J. 3D printing: An emerging tool for novel microfluidics and lab-on-a-chip applications. *Microfluid. Nanofluid.* **2016**, *20*, 50. [CrossRef]
121. Yin, X.-Y.; Zhang, Y.; Cai, X.; Guo, Q.; Yang, J.; Wang, Z.L. 3D printing of ionic conductors for high-sensitivity wearable sensors. *Mater. Horiz.* **2019**, *6*, 767–780. [CrossRef]

© 2020 by the authors. Licensee MDPI, Basel, Switzerland. This article is an open access article distributed under the terms and conditions of the Creative Commons Attribution (CC BY) license (http://creativecommons.org/licenses/by/4.0/).

Review

Non-Invasive Electrochemical Biosensors Operating in Human Physiological Fluids

Magnus Falk [1], Carolin Psotta [1,2], Stefan Cirovic [1] and Sergey Shleev [1,2,*]

[1] Department of Biomedical Science, Faculty of Health and Society, and Biofilms—Research Center for Biointerfaces, Malmö University, 20506 Malmö, Sweden; magnus.falk@mau.se (M.F.); carolin.psotta@mau.se (C.P.); stefan.cirovic@mau.se (S.C.)
[2] Aptusens AB, 293 94 Kyrkhult, Sweden
* Correspondence: sergey.shleev@mau.se; Tel.: +46-702-351-141

Received: 10 October 2020; Accepted: 4 November 2020; Published: 7 November 2020

Abstract: Non-invasive healthcare technologies are an important part of research and development nowadays due to the low cost and convenience offered to both healthcare receivers and providers. This work overviews the recent advances in the field of non-invasive electrochemical biosensors operating in secreted human physiological fluids, viz. tears, sweat, saliva, and urine. Described electrochemical devices are based on different electrochemical techniques, viz. amperometry, coulometry, cyclic voltammetry, and impedance spectroscopy. Challenges that confront researchers in this exciting area and key requirements for biodevices are discussed. It is concluded that the field of non-invasive sensing of biomarkers in bodily fluid is highly convoluted. Nonetheless, if the drawbacks are appropriately addressed, and the pitfalls are adroitly circumvented, the approach will most certainly disrupt current clinical and self-monitoring practices.

Keywords: non-invasive biosensors; human physiological fluids; tears; sweat; saliva; urine

1. Introduction

Owing to the low cost and convenience shared by healthcare receivers and providers alike, non-invasive healthcare technologies have become increasingly important parts of current research and development [1]. Non-invasive measurements include the use of sweat [2], urine [3], saliva [4], and tears [5], but can also rely on fluid-free technologies. The last option is more attractive because of fast and inexpensive analyses with convenient fluid independent procedures, and the concerns in the scientific community regarding correlations between bioanalyte concentration in blood and in other physiological fluids can be disregarded [6–9]. In addition to traditional fluid-free non-invasive technologies, which are known and have been used for ages, e.g., electrocardiography [10,11], many other non-invasive instruments have been developed, such as cardiovascular diagnostic systems [12], bioimpedance based scales [13,14], and even non-invasive blood analyzers to measure sentinel substances in blood, e.g., hemoglobin [15], oxygen [16], and glucose [17,18]. Hence, fluid-free oximeters are relied on and widely used in acute and critical care [19], but other fluid-free non-invasive blood analyzers, e.g., glucometers and hemoglobinometers, are far from accurate and the readings cannot be trusted [20–22]. Drawing on the contentious performance of current fluid-free analyzers, and since many important bioanalytes cannot be measured using fluid-free technologies, this review is focused on fluid-based biosensors operating in different physiological fluids, viz. tears, saliva, sweat, and urine. Among the variety of fluid-based non-invasive biosensors, this review is focused on electrochemical biodevices.

For all electrochemical techniques, the most common electrode materials are silver, gold, platinum, and carbon, e.g., graphene, graphite, carbon nanotubes, and glassy carbon [23]. Regarding carbon, the working electrode area can be tweaked by selecting from the carbon alternatives [24], and in general, the performance and sensing abilities and can be improved by surface modification using e.g., polymers, nanofibers, or nanoparticles [23,24]. Biomolecules are immobilized on the working area of the electrode surface, and act as recognition elements to generate and transduce an output signal. The selectivity and sensitivity of a particular recognition element can be finely tuned by appropriate helper elements [23,25]. The main components of a complete biosensor for detection of analytes in bodily fluids are the following: bioreceptor, transducer, electronics, and display.

A variety of compounds of clinical relevance are present in secreted physiological fluids, most of which can be converted by different oxidoreductases, and many by oxidases [5–7,9–11]. Hydrogen peroxide (H_2O_2), a common by-product of oxidases, is used as the terminal element in most enzyme-based biosensors and analytical kits, and H_2O_2 can be assessed to estimate the concentration of analytes. In a review, it is impossible to cover all electrochemical techniques used for the detection of analytes in human physiological fluids, and it is equally impossible to even briefly describe all biosensors operating in urine, sweat, sweat, and saliva. Below, we will describe a few examples of a rich variety of biodevices that rely on amperometry, coulometry, cyclic voltammetry, and impedance spectroscopy.

However, prior to the description, the following should be emphasised. On the one hand, exuded human physiological fluids are well suited as information vehicles and sources, provided that the available data can be sampled reliably. Tears, sweat, saliva, and urine are known to carry substances indicative of the health status of an individual. Glucose, lactate, ascorbate, urea, creatinine, as well as their metabolites, and hormones and their metabolites, are all examples of species present in human physiological fluids. By assaying amounts or relative amounts, or, by gauging appearance or disappearance rates, a reasonably accurate appreciation of the health status of an individual can be made. On the other hand, when it comes to non-invasive analysis of bioanalytes in secreted/exuded physiological fluids for diagnostics, it seems pertinent to highlight a major ambiguity, which attenuates the enthusiasm of medical doctors and in general seriously alerts the biomedical community. Unfortunately, data dispersion in the literature, when it comes to amounts, and sometimes even the presence of bioanalytes in physiological fluids, borders on the ridiculous. One can find all possible reports, from no correlation, via certain correlation, up to direct straight dependences between concentrations in different physiological fluids. Neither direct nor indirect reasons for this will be discussed in the current review because of the lack of space and possible loss of focus. It is obvious, however, that the nonsensical dispersion of data definitely calls for proper (i.e., with adequate sampling, taking into account differences in basal, induced, emotional fluids, as well as selective and appropriately sensitive determination, using modern techniques, in full control of possible pitfalls) chemical analysis of main bioanalytes in human physiological fluids naturally released outside the body and simultaneous comparison of blood concentrations.

2. Fluid-Based Biosensors

2.1. Biosensors Operating in Tears

The complex aqueous fluid secreted by lachrymal glands, i.e., human lachrymal liquid or tears, apart from water and electrolytes, carry low molecular weight organic compounds, proteins, enzymes, as well as other biomolecules. Owing to the physiology of the eye, different lachrymal secretions can occur, and three kinds of tear, basal, reflex, and psycho-emotional tears [26], each substantially differing in composition, can be observed. Basal tears are produced in small quantities, 0.5–2.2 μL min^{-1} [27], to maintain a film on the corneal surface, ensuring corneal homeostasis and visual integrity. Reflex tears result from increased lacrimation, 7–23 μL min^{-1} [28], in response to damage to the ocular surface by foreign bodies, including contact lenses [29], chemicals, wounds, and inflammation [26]. Psycho-emotional tears are provoked by cerebral stimuli of psychogenic origin. While basal tears result from spontaneous neuroglandular activity and reflex tears are the result of external sensorial stimulation, and both kind of tears are expletively purposeful, psycho-emotional tears are triggered by cognitive and emotional brain processes, and are of no apparent use for the eye [26].

Several compounds found in tears have diagnostic potential, e.g., glucose, lactate, ascorbate, and neurotransmitters, e.g., dopamine and norepinephrine [30]. Hence, many diseases and ailments can be diagnosed by analyzing the composition of tear fluid, and an adequate summary of this can be found in a recent review [31]. Illustrating the potential of tear-based bioanalyte sensing, glaucoma patients suffer from lower than average tear neurotransmitter concentrations, and determination of catecholamines in tears has been advocated in glaucoma diagnosis [30]. Also, the baseline concentration of the stress marker norepinephrine is high enough for detection, and since the concentration of this is likely to increase during psychological and physical challenges, it should be possible to realize a non-invasive tear-based stress sensor.

Tear sampling is expressively non-trivial and the established methods all suffer from shortcomings. For example, tears can be absorbed using Schirmer strips resting on the lower eyelid, but the procedure tends to collect cellular as well as secreted proteins, and the physical presence of the strip can cause mechanical stimulation of the corneal and conjunctival epithelium, provoking release of reflex tears [32,33]. Consequently, the composition of the samples thus collected most likely differs from that in native basal secretion [34]. If, instead, microcapillary tubes are used to draw tears from the reservoir within the conjunctival sac, while seemingly less invasive than Schirmer strips, collecting basal tears with this method can be tedious and time consuming, and to accumulate volumes adequate for analysis it may be necessary to pool samples [32,35]. In addition, the microcapillary tube method shares the Schirmer strip reflex tear issues. To conclude, the small volumes collected along with low concentrations of tear bioanalytes could explain the large discrepancies in concentration values reported, and it appears that authentic and accurate bioanalyte concentrations in human tears are yet to be ascertained.

In order to bypass tear sampling issues, several research teams have turned their attention to contact lens-based biosensors. As far as can be asserted, the first attempt to design an on-lens electrochemical biosensor coincided with efforts to realize electronically augmented contact lenses for bionic eyesight [36,37]. The lens assembly incorporated a 100 µm thick poly(ethylene terephthalate) (PET) film, fitted with an amperometric sensor, in which the Pt working electrode was connected to an indium-tin oxide (ITO) substrate carrying a self-assembled monolayer of glucose oxidase (GOx). Potential control was achieved using an external Ag|AgCl reference electrode [36,38]. Glucose amounts were indirectly assessed, based on electrochemical oxidation of H_2O_2 (Reaction (1)), formed as a co-product in the GOx catalysed oxidation of glucose (Reaction (2)).

$$H_2O_2 \xrightarrow{Pt} 2H^+ + O_2 + 2e^- \tag{1}$$

$$D\text{-}glucose \xrightarrow{GOx} D\text{-}gluconolactone + H_2O_2 \tag{2}$$

The sensitivity of the first/original sensor was unsatisfactory, and an improved biosensor design was called for [5]. Thus, the second generation sensor was essentially a PET film fitted with a three electrode system with an integrated reference electrode (Figure 1), distinguishing the novel sensor from the original [36]. Bio-modification relied on a GOx/titania sol-gel membrane and the sensor was provided with a Nafion® layer to decrease the influence of other redox active species, like ascorbate, lactate, and urea, present in the lachrymal fluid.

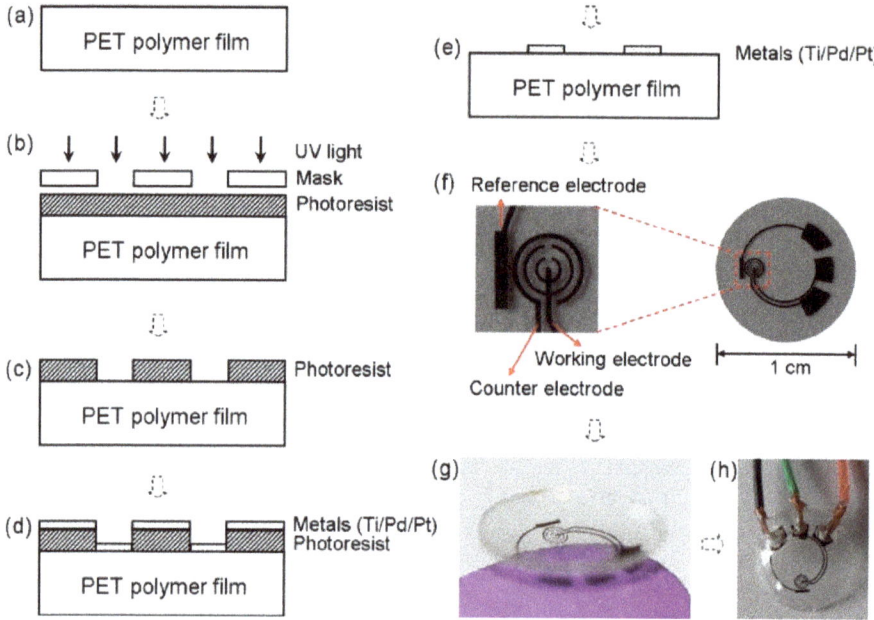

Figure 1. The sensor fabrication process and results: (**a**) a clean PET substrate is prepared; (**b**) the substrate is covered with a photoresist and exposed to UV light through a mask; (**c**) the photoresist is developed; (**d**) thin metal films are evaporated on the sample; (**e**) after lift-off, the metal pattern remains on the surface. After this step, the sensor is cut out of the polymer substrate and heat molded to the contact lens shape and functionalized with enzymes; (**f**) images of a sensor after it has been cut out of the substrate; (**g**) image of a completed sensor after molding held on a finger; (**h**) the sensor hardwired for testing. Reproduced from [5] with permission from Elsevier.

At a constant voltage of 0.4 V, the amperometric sensor showed quite promising performance in glucose containing solutions, and excellent current and glucose concentration linearity in a physiologically relevant glucose concentration range. However, the influence of interfering redox species shifted the detection limit by about one order of magnitude. After storage for two and four days in buffer at 4 °C, the residual current response was 80% and 55%, respectively, of the initial current. In order to minimize the contribution from interfering substances additional control working and counter electrodes were fitted, resulting in a so called dual sensor setup [39]. Control electrodes were assembled using the same approach as for the signal electrodes, but the bio-modification with a GOx/titania sol–gel membrane and electrode coating with Nafion® was omitted. Background currents obtained from the dummy sensor were subtracted from the bioelectrocatalytic sensing electrode currents, and measurements using a physiologically accurate flowing eye model confirmed that the dual sensor did indeed lower the glucose detection limit, but it was still too high for real practical applications. In the on-lens version of the biosensor, wirelessly powered and with an integrated telecommunication circuit, the control electrodes were covered with deactivated GOx to subtract background currents more accurately [40,41]. The sensitivity of the on-lens biosensor, tested on a model human eye, was found to be 18 µA cm^{-2} mM^{-1} with a linear response in the 0–2 mM glucose range in a tear mimicking buffer solution. The artificial tear solution contained redox active species, i.e., ascorbate, lactate, and urea, at concentrations typical for human tears, and representative tear proteins, i.e., lysozyme, mucin, and albumin, all of which may affect the biosensor performance. The residual bioelectrocatalytic current was 97.4% of the initial after 12 h of storage in buffer at 4 °C, and 84.3%, 67.2%, and 54.2% after one, two, and four days, respectively, without any loss of current vs. glucose concentration linearity. The on-lens sensing platform was powered by an on-chip 1.2 V supply, consuming just 3 µW, and could also be wirelessly powered from a distance of 15 cm [40]. Google and Novartis foreshadowed that commercially available glucose monitoring contact lenses could be expected already in 2015 [42]. However, the companies did not deliver on their promise, and to the best of our knowledge, the project is completely terminated. One of the possible reasons for that might be attributed to the partial discharge of reflex tears in response to damage to the ocular surface. Reflex tears have different composition compared to basal lachrymal liquid, and even for one individual, bioanalyte concentrations could vary from one day to another. To mitigate or bypass the problems related to contact lens based biosensors, usage of soft, highly flexible, air breathing materials is suggested.

In addition to glucose detection, the sensor design was adapted to contact lenses with integrated lactate biosensors (Figure 2) [43].

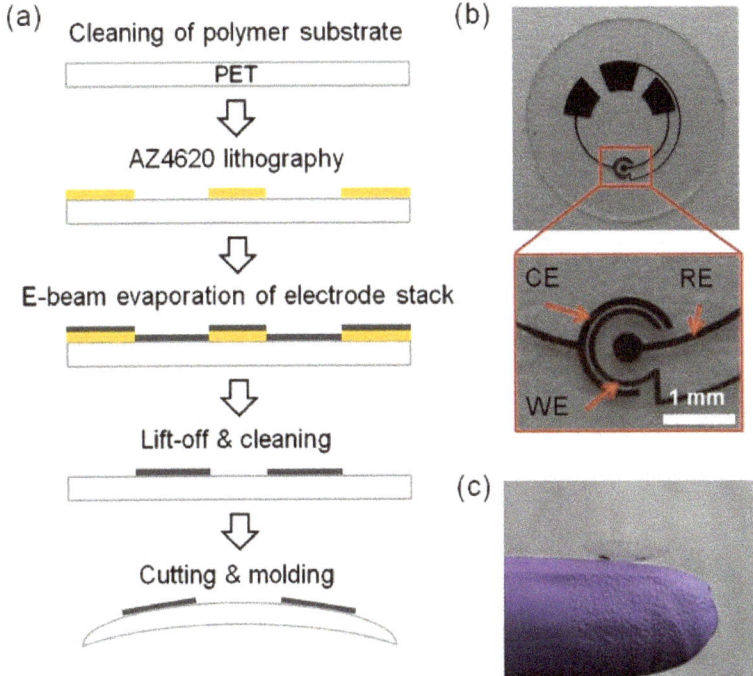

Figure 2. Lactate sensor on a contact lens. (a) Schematic representation of the assembly process for sensors on a transparent PET substrate which is molded to a contact lens. (b) Flat substrate with sensing structure, interconnects and electrode pads for connection to the external potentiostat; WE—working electrode, CE—counter electrode, RE—reference electrode. (c) A completed contact lens sensor held on a finger. Reproduced from [43] with permission from Elsevier.

Lactate was detected based on oxidation of lactate to pyruvate by molecular oxygen, catalyzed by lactate oxidase (LOx) (Reaction (3)), with subsequent electrochemical oxidation of H_2O_2 at a Pt working electrode (see Reaction (1)).

$$\text{L-lactate} + O_2 \xrightarrow{LOx} \text{pyruvate} + H_2O_2 \tag{3}$$

LOx and bovine serum albumin were co-immobilized on the sensing area, and the proteins were simultaneously crosslinked using glutaraldehyde, and to prevent enzyme leakage and to reduce the influence of interfering redox species the electrodes were sequentially covered with polyurethane and Nafion®. In PBS, the sensitivity reported was 53 µA cm^{-2} mM^{-1}, with a response time of 35 s, and a linear detection range of 0–1 mM lactate. An interfering signal owing to direct oxidation of ascorbic acid on the Pt surface was successfully eliminated by using a dual sensor setup (vide supra). The sensor maintained full functionality after 24 h of storage in buffer at room temperature.

Mitsubayashi et al. managed the design of an on-lens glucose biosensor differently [44,45]. Drawing on a two-electrode setup, i.e., a Pt working electrode and a combined Ag|AgCl reference/counter electrode, the amperometric biosensor was affixed to a 70 µm thick polydimethyl siloxane (PDMS) membrane, attached to a soft contact lens made of the same material. GOx was immobilized from a co-polymer of 2-methacryloyloxyethyl phosphorylcholine and 2-ethylhexylmethacrylate (PMEH) mixture, and the GOx-polymer layer was additionally covered with a PMEH membrane to prevent enzyme leakage. Glucose sensing was based on electrochemical oxidation of H_2O_2 on the Pt electrode, vide supra, Reactions (1) and (2), and a linear relation between current output and glucose amount was obtained in buffer solutions with 0.03–5 mM glucose. Testing the sensor in wearable mode using rabbit models gave excellent results and it was also demonstrated that the glucose concentration in tears traces the changes in blood glucose amounts with a delay of approximately 10 min. The sensitivity of biosensors could be substantially improved by building 3D micro-pillar electrodes, which have up to three times higher surface area when compared to the flat analogues [46].

Some non-invasive tear-based biosensors cannot easily be repurposed as wearable devices. A flexible electrochemical microbiosensor has been designed for quantitative analysis of glucose, ascorbate, and dopamine in nanovolumes of human lachrymal fluid (Figure 3) [47]. The biodevice is based on glucose dehydrogenase (GDh), rather than GOx, and ascorbate and dopamine are analyzed electrochemically without pre-enzymatic reactions. GDh was immobilized on a gold microwire modified with carbon nanotubes and an osmium redox polymer. A capillary microcell, with a working volume of 60–100 nL and a sampling deviation of about 7%, was constructed for tear sampling. To check if the microcell was properly filled with buffer or a tear sample, a control electrode was introduced into the construction (Figure 3c). The electrode was used to measure the electrical resistance of a fully filled nanovolume cell. The mechanical flexibility is one of the most important features of the prototype and it allows direct collection of tears with minimal risk of damage to the eye (Figure 3d,e). Based on the experimental results, the authors concluded that the flexible and non-invasive prototype could be converted into a user-friendly microbiosensor, suitable for detection of blood bioanalytes, including glucose, in human lachrymal fluid [47].

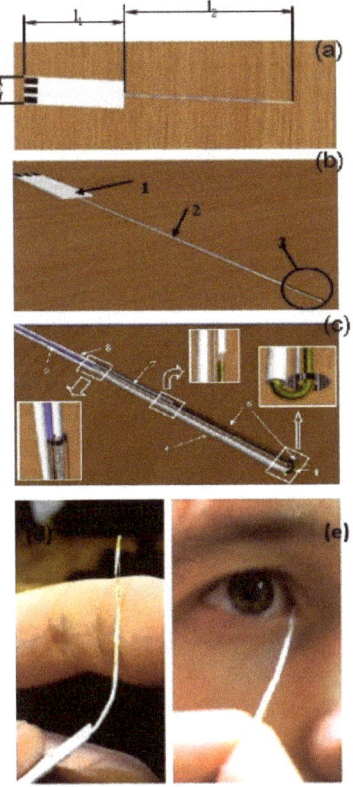

Figure 3. Schematic view of the flexible non-invasive micro-biosensor. (**a**) l1—length of a handle (from 10 mm up to 50 mm, optimum 20–30 mm), l2—length of a flexible sampling part (from 40 mm up to 100 mm, optimum is 50–60 mm), w—width of the handle (from 5 mm up to 10 mm). (**b**) 1—the handle with three electrical contacts, 2—the flexible sampling part (total diameter from 0.05 mm up to 0.5 mm), 3—the flexible part of the device with a microcell. (**c**) Flexible part of the sapling device including a microcell. 4—working electrode (diameter from 0.01 mm up to 0.2), 5—insulated part of the working electrode, 6—polymeric tube (internal diameter from 0.015 mm up to 0.25), external diameter from 0.02 mm up to 0.4, length from 1 up to 40 mm) fused to the insulated part of the working electrode, 7—counter and reference electrode (diameter from 0.01 mm up to 0.2 mm), 8—insulated part of counter and reference electrode, 9—checking/control electrode (diameter from 0.01 up to 0.2 mm). Distance between ends of working and opposite electrodes from 0.01 mm up to 1 mm. (**d**) Photograph demonstrating the flexibility of the biodevice. (**e**) Photograph of authentic sampling. Reproduced from [47] with permission from Springer.

2.2. Biosensors Operating in Saliva

Saliva is an oral fluid that is mainly produced by three pairs of major salivary glands: parotid (inside of the cheeks), sublingual (under the tongue), submandibular (bottom of the mouth), and a large number of minor salivary glands [48–51]. Moreover, saliva is a clear, viscid, complex [52–56], colourless, odourless fluid, with a pH in the 6.6–7.1 range [50,57]. Saliva is a watery substance (99.5% water [58]) that incorporates different elements like bacteria, leukocytes, epithelial cells, crevicular fluid [53,58], hormones, ions [54,56], enzymes, proteins, nucleic acids, antimicrobial constituents, cytokines, and antibodies [56,57]. These different components, originally from the blood, can diffuse through para-cellular or trans-cellular pathways in the oral cavity, adding to the complexity of saliva [49,55]. Additionally, the oral cavity also comprises a large number of bacteria (oral microbiome) [49].

Saliva analysis has a considerable potential regarding general health status monitoring [48,59]. Saliva carries a broad range of biomarkers [49] that can be used for clinical analysis and diagnostic testing of various diseases [49,50,54,56,60]. Moreover, because many biomarkers found in saliva are passed directly from the bloodstream, changes in saliva composition indicate the current health status of the examined person [56,60]. The correlation between the blood and saliva concentration of different biomarkers and metabolites like lactate [48,53,60,61], ethanol [48,53], cholesterol [53,60,61], or glucose [49,54,60–64] has been established. Therefore, saliva analysis gives the opportunity to monitor and surveil the emotional, hormonal, nutritional, and metabolic state of the human body [54,55]. Additionally, utilizing saliva as a diagnostic fluid offers various advantages, e.g., a painless and non-invasive method for diagnostics and monitoring [48,49,52–55,65,66] relying on a simple and fast collection method [48,49,55,67,68]. Moreover, sample collection is conveniently trivial [48,49,55], not privacy invading for the patient [49], and does not require special laboratory equipment or trained medical personnel [49,55,60,65,66]. Additional to that, saliva can be used as an alternative fluid because it is generally safer and has a lower contamination risk compared to blood tests [48,49,54,66]. Therefore, saliva offers the possibility to analyze various biomarkers in an easy accessible, reliable, cost-effective way [48,49,56,65,67] and can realize a multiplex detection of metabolites with high sensitivity and selectivity [61]. On the top, using saliva can realize dynamic measurements of biochemical markers with an read out in real time [54], or a portable biosensing platform for health care monitoring [54,69].

These benefits and the different prospective options led to an increased focus in this research field with the aim of assay developments and technological advancements for the detection of various salivary biomarkers to improve clinical diagnosis, management, and treatment [48]. Normally, detection of salivary biomarkers is performed by using laboratory-based assay methods that involve multiple steps in a time-consuming process. This includes, e.g., collection/transfer of the sample followed by processing in the laboratory [48,65]. Therefore, the demand for fast, simple, inexpensive, reliable, accurate, portable, and on site (point-of-care) quantification of salivary biomarkers through the use of biosensing technology increased and is supported by the progress in nanotechnology [48,60,65].

Various biomarkers or molecules can be detected in saliva, e.g., the cytokines interleukin-6 [70,71], interleukin-8 mRNA [72–74], interleukin-8 protein [72–74], cancer antigens [48] like carcinoembryonic antigen (CEA) [75,76], cancer antigen 125 (CA125) [76] and Her-2/Neu (C-erbB-2) [76,77], VEGF165 [78], TNFα [79], or cytokeratin-19 antigen (Cyfra 21-1) [80]. Moreover, lactate [48,54,58,81–84], glucose [48,60,62,85–96], or hormones [48,66], like cortisol [48,49,51,65,66,97–102], and cortisone [99,103,104] are of interest to detect as well. Furthermore, molecules like uric acid [49,54,105], urea [106], L-tryptophan [107] or different enzymes, like α-amylase [48,66,108–110] or aspartic peptidases [111,112] are found in saliva. Additional to that, the detection of bacteria, viruses, and whole cells is possible, like *Helicobacter pylori*, *Streptococcus sanguinis*, *Escherichia coli*, *Staphylococcus aureus*, *Chikungunya viruses* [113–118] or different antibodies [48,66,114,119], e.g., HIV [48,66,120]. On the top, different drugs [49,51,80,121], like alcohol [53,66] and cocaine [66,122], as well as neurotransmitters [49,66,123] can be determined in saliva. The related diseases to the different biomolecules are cancer [51,70,73–77,79,80,124], oral cancer [56,67,72,108,109], breast cancer [56,67], diabetes [49,60,67,85], and different cardiovascular diseases [68,108,109]. Moreover, diseases of the oral cavity like periodontal infections [56,67,125] or caries risk assessments [56] can be done. Additional to that, infections caused by bacteria or virus-like viral hepatitis A,B,C [51,67], related to infections, or cardiovascular diseases [68,108,109,118] or metabolites, like lactate that can diagnose severe sepsis, septic shock, and many more [83,126].

The correlation between the concentration of glucose in blood and in saliva was established by different groups [49,54,60–64] and, hence, the opportunity was given to realize a biosensing platform for the detection of salivary glucose [48,60,62,85–96]. For the realization of an enzyme-based glucose sensitive sensor, a screen-printed sensor chip was produced by a layer-by-layer assembly process to functionalize the working electrode surface. The multilayer films were composed of single-walled carbon nanotubes functionalized with carboxylic groups and three repeated layers of chitosan/gold nanoparticles/GOx to achieve the best glucose-sensing performance using amperometry [62]. To enable constant monitoring over a specific period of time, an oral biosensor was imbedded in a mouthguard and miniaturized to a detachable 'cavitas sensor' based on an enzyme membrane with GOx (Poly (MPC-co-EHMA)). The device was integrated with a wireless transmitter, based on a platinum and silver/silver chloride electrode, with amperometric read out, and real time measurement of the glucose concentration was achieved [88]. Another group used a bimetallic, bifunctional electrode where a platinum surface was patterned with nanostructured gold fingers with different film thicknesses. The gold fingers were functionalized with GOx by using selective adsorption of a self-assembled monolayer onto gold fingers. GOx on gold acted on glucose and the hydrogen peroxide formed was detected on the platinum sites. The read out was based on cyclic voltammetry and impedance spectroscopy [96]. Another glucose sensor was a disposable saliva nano-biosensor. The platinum working electrode was functionalized with single-walled carbon nanotubes and multilayers of chitosan, gold nanoparticles, and GOx, generated by using a layer-by-layer assembly technique and the final read out relied on amperometry [60]. Last but not least, the detection of salivary glucose can be achieved based on flow injection analysis combined with an amperometric O2 electrode [127]. Glucose sensor presented in Figure 4 consists of transducer constructed of silver anode and golden cathode covered with potassium chloride. The electrode is also covered with Teflon membrane with immobilized GOx, a test solution is injected using a micro syringe [127].

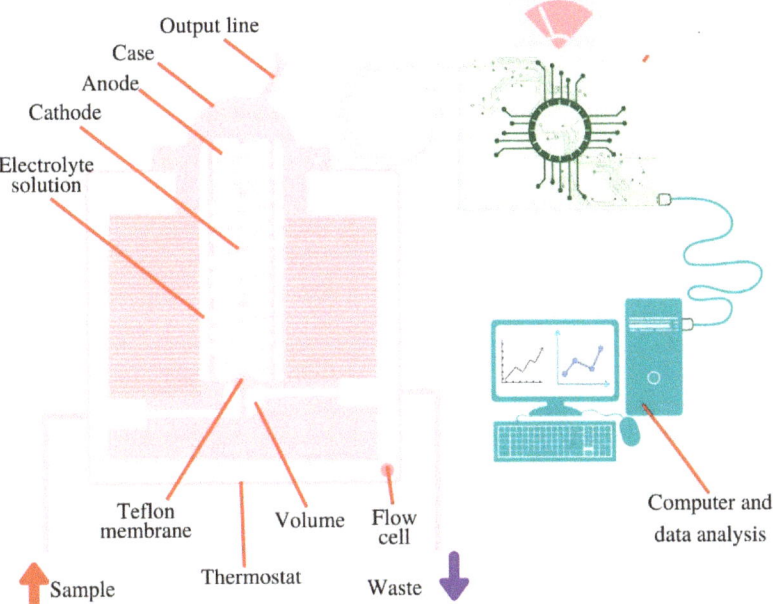

Figure 4. Schematic view of the salivary glucose biosensor inside installed in a 37 °C thermostat.

The correlation between the concentration of cortisol in saliva and blood has been established, providing the opportunity to use the detection of this for the development of various biosensors [98,128,129]. A highly sensitive and non-invasive electrochemical immunosensor for salivary cortisol sensing was developed by using an NiO thin film-based label-free electrochemical immunosensor. For this purpose, cortisol antibodies were immobilized with EDC and NHS on the electrode surface and detected with differential pulse voltammetry [100].

Another electrochemical label-free immunosensor was established with interdigitated microelectrodes and anti-cortisol antibodies, which were covalently immobilized on self-assembled monolayers of dithiobis(succinimidylpropionte) (DTSP). The different concentrations of cortisol were evaluated using cyclic voltammetry [101]. Another electrochemical immunosensor was based on one-dimensional ZnO nanorods and two-dimensional ZnO nanoflakes that were synthesized on gold-coated substrates, followed by immobilizing anti-cortisol antibodies; the detection of cortisol was performed by cyclic voltammetry [102].

For the detection of TNF-α in human saliva and serum samples, a disposable immunosensor based on thin indium tin oxide films, covered by a new semi-conductive conjugated polymer, was developed. The read out was realized by electrochemical impedance spectroscopy [79]. Another electrochemical biosensor platform for TNF-α cytokines detection in both artificial and human saliva was introduced. This fully integrated platform was developed to detect varying cytokine biomarkers by using eight gold working microelectrodes. TNF-α antibodies (anti-TNF-α) were immobilized on gold working electrodes through functionalization with carboxyl diazonium; detection relied on amperometry [59].

Another group developed a graphene-based fully integrated portable nanosensing system for the detection of interleukin-6. This new sensor was based on the permittivity of a HfO_2 dielectric layer in the buried-gate graphene field effect transistor. Moreover, the detection was performed by the immobilization of aptamers and the data transfer was realized by a wireless connection via WiFi [70]. An additional possibility to detect interleukin-6 was based on a magneto immunosensor design. Here, the anti-IL-6 antibodies were immobilized on carboxyl-functionalized magnetic microparticles. This immunoassay created the signal amplification by using poly-HRP-streptavidin conjugates that were immobilized on screen-printed carbon electrodes. The final detection was realized with amperometry [71].

Two biosensors were developed for the simultaneous detection of interleukin-8-protein and interleukin-8-mRNA. One of the sensors was an amperometric magnetosensor which allowed the direct determination of both biomarkers with an antibody and a hairpin DNA. The specific hairpin DNA probe and the antibody were coupled on screen-printed carbon electrodes and used the $HQ/HRP/H_2O_2$ system [72]. The second biosensor was also based on amperometric detection but used biotin and fluorescein dual-labeled hairpin probe for IL-8 mRNA and biotinylated human IL-8 monoclonal antibody on a gold integrated electrode array. Moreover, the group used a conducting polymer as a supporting film to improve sensor performance [73].

The determination of lactate was accomplished with an electrochemical enzyme probe (LOx), by using a dual electrode system (two platinum electrodes), and an enzyme membrane that was imbedded between a cellulose acetate membrane and a polycarbonate membrane. The changes in the lactate concentration were measured with the relating current response, after injecting the salvia sample with PBS and by subtracting the inactive electrode response from the active electrode response [81]. Another approach for detection lactate was to use a cavitas sensor (placed in the oral cavity). This wearable biosensor on a mouthguard is based on an Ag/AgCl reference electrode and contacts (for interfacing the electrochemical analyser) on a flexible PET substrate. Moreover, it was based on the integration of a printable enzymatic electrode, with LOx, on a mouthguard, and detection of the hydrogen peroxide formed. The mouthguard sensor was made of a Prussian-Blue transducer and a poly-orthophenylenediamine (PPD)/LOx layer. The amperometric measurements were realized with the connected PB-PPD-LOx system [82].

The detection of uric acid was realized by a sensor that consisted of an uricase-modified screen-printed electrode system and was also integrated in a mouthguard platform. The whole sensor was based on a flexible PET substrate, including a Prussian-blue transducer and immobilized uricase enzyme. This wearable sensor was connected to a wireless device via Bluetooth for data collection and allowed real-time and continuous measurement of the uric acid concentration with an amperometric read out [105].

Additionally, the detection of orexin A (a neurotransmitter), was accomplished by using a gold field effect transistor-based biosensor that was modified with zinc oxide. For this purpose, a novel peptide recognition element was synthesized and coupled to the electrode surface [123].

For the detection of the protein Pf HRP2 (*Plasmodium falciparum* histidine-rich protein 2) an enzyme-free electrochemical immunosensor was developed. This sensor was based on an immunosandwich format and used a competitive detection principle with methylene blue, hydrazine, and platinum nanoparticles. For this purpose, the specific antibodies were labelled with methylene blue and immobilized on an indium tin oxide electrode. The read out was realized with chronocoulometric measurements [130].

Furthermore, the sensing of tryptophan, a standard amino acid, was reported by a group using two screen printed electrodes modified with multiwall carbon nanotubes on gold electrodes. Additionally, the electrode surface was modified with aptamer molecules to determine tryptophan, followed by impedimetric read out [107]. Another approach was made, by using magnetic multiwalled carbon nanotubes as nanocarrier tags for the detection of human fetuin A with impedimetric read out. Moreover, the electrochemical immunosensor was realized by using the linker molecule diazonium, followed by the immobilization an anti-human fetuin A-antibody and coupled HRP [131].

For the detection of cocaine, a solid-state probe based on an electrochemical aptasensor was developed. The sensor principle was based on a layer-by-layer self-assembled multilayer with ferrocene-appended poly(ethyleneimine) on an indium tin oxide electrode array. Additionally, gold nanoparticles were coupled to the electrode surface. The final read out was realized by using differential pulse voltammetry and measuring the related signal of ferrocene [122].

The development of a biosensor can be also used to evaluate the efficiency of a treatment in the field of pharmacology. For this purpose, another electrochemical aptamer-based sensor was introduced to detect ampicillin. The aptamers were immobilized on a gold electrode, followed by a blocking step with MCH to avoid unspecific binding. When ampicillin is bound by the methylene blue modified aptamer, the aptamer conformation changes; the measurement of the change was realized with the methods of alternating current voltammetry and square wave voltammetry [80].

One group developed a biosensor for the detection of *S. pyogenes*. This sensing platform was based on the immobilization of antibodies on a gold surface. They used screen printed gold electrodes to create a polytyramine (Ptyr)-based immunosensor. Accordingly, NeutrAvidin was coupled to the Ptyr amine group, followed by the immobilization of biotin tagged antibodies against *S. pyogenes* [113].

For the detection of HIV antibodies, an electrochemical peptide-based sensor was introduced. This sensor used the incorporation of extra amino acids that acted as a target recognition element and antifouling agent on gold electrodes. Moreover, the peptide probe was thiolated (coupling to the electrode surface) and methylene blue-modified (detection). With the binding of the HIV antibody, the methylene blue related current decreases and allowed a read out with alternating current voltammetry and cyclic voltammetry [114].

Despite many achievements, a better understanding is needed for the relation between biomarker concentration in blood and in saliva to improve the reliability of various diagnostic platforms, to provide accurate oral monitoring applications, and to develop highly sensitive sensors [54]. In particular, higher sensitivity is an important requirement due to the fact that the concentration of many important biomarkers in saliva are lower than in blood, e.g., proteins with ~30% lower concentration in saliva [54,68]. Additional to that, metabolite measurement in saliva is complicated by the presence of bacteria, epithelial cells and leukocytes [53]. Moreover, when sampling saliva, challenges still remain, e.g., including sampling standardization since the concentration of different compounds in saliva depends on the flow rate and the flow rate varies in response to any pre-sampling stimulation [53]. Additionally, saliva viscosity can vary substantially, which makes it more difficult to provide a reliable testing platform [68]. Furthermore, after the saliva collection, the sample must be put on ice to reduce growing of microorganisms [53]. Moreover, various sensing problems can occur after/before/during food and drink consumption because it can interfere with the analyte sensing [52,54,68]. Furthermore, future development should focus on anti-fouling strategies. These strategies are needed because a high concentration of proteins in saliva, e.g., mucins and proteolytic enzymes, will cause nonspecific adsorption on the electrode surface; the oral microbiome will produce a biofilm, and consequently, lower the life time of sensors in general [52,54,69]. This poses serious challenges, especially for real-time measurements and long-term monitoring of biomarkers since saliva is not only a complex solution but also a dynamically changing one [52,54,68].

Moreover, for development of devices and sensing platforms, different requirements should be met, e.g., on-body/in-vivo compatibility (compatible biomaterials for device and system), including an effective device encapsulation (electronic interface, power supply, wireless communication) and will alleviate the toxicity of the whole sensor [22]. Furthermore, for an in vivo sensing platform the sensor should be mechanically robust and securely fixed in the oral cavity and fit with the mouth anatomy and spatial ranges [55]. This increases the comfort for the wearer and a firmly fixed position will not change with mouth muscle movements [52].

Multiplex sensing platforms have also been introduced. One sensor was developed for the detection of glucose, lactate, and cholesterol, which was based on an organic electrochemical transistor (OECT) microarray integrated with a pumpless "finger-powered" PDMS-based microfluidic system. The group combined a biofunctionalization method and electrically isolating layer between the devices to decrease the background interference and crosstalk for improving the sensing abilities. Additionally, they immobilized GOx, LOx, cholesterol oxidase, and bovine serum albumin (control) on each electrode separately and realized the read out by using chronoamperomerty and the channel current response [61].

Another group developed a flexible organic electrochemical transistor to detect uric acid and glucose. For this purpose, the gate electrodes of the transistor were modified with positively/negatively charged bilayer polymer films and enzymes (uricase and GOx). Additionally, the platinum electrode surface was modified with graphene flakes and Nafion, followed by PANI (polyaniline, a conducting polymer). The whole sensor was based on the selectivity for H_2O_2 and measured the resulting channel current response of the transistor [132].

For future developments, power supply and communication challenges should be addressed. This comprises for instance, to find and implement new solutions/sources for the power that is needed for sensing/detection of biomarkers, data processing/data collection, data transfer/communication. Alternative strategies could include incorporation of batteries, biofuel cells, solar cells, or thermoelectric generation of power [54]. Moreover, communication challenges include the overall integrity of wireless communication and a long-distance data transmission between the sensor and the device for the connected read out. Additionally, data security problems will require a safer collection and storage of the biomedical data because the sensor is monitoring the patients' health status in real time, remotely, and continuously [54]. All in all, considerable validation studies would be necessary to establish sensors in clinical applications and to make is accessible for the general public [54].

2.3. Biosensors Operating in Sweat

Other sensing technologies for measurement of various analytes utilizes the biofluid sweat [23,133]. Sweat contains several compounds that provide helpful medical information about health and metabolic status, physiological state, and disease states [23,24,58,134–137]. The outer skin surface comprises a high number of sweat glands that are densely and widely distributed all over the body [54,138]. These sweat glands produce and excrete an acidic fluid directly to the outer skin surface through microscale pores [23,68]. Moreover, sweat includes various molecules, e.g., lactate [133,138–140], glucose [69,134,137,141,142], cortisol [23,54,135,142–144], testosterone [23], uric acid, as well as larger molecules, like proteins, peptides, and cytokines [24,54,142],

Additionally, the detection of analytes in sweat is feasible due to the given correlation between the blood analyte concentration and the excreted sweat analyte concentration [138,145,146]. For analytes like hormones (cortisol, testosterone), potassium, and different drugs, e.g., alcohol, a strong sweat-blood correlation was proven [23]. Moreover, an understanding has been established about the glucose-level relation between sweat and blood and its potential use in diabetes monitoring [23]. The correlation between sweat and blood concentration for lactate or urea is not so far approved, but lactate/urea measurements in sweat can give an indication of the health status of the examined subject [23].

Therefore, utilizing sweat as a diagnostic biofluid offers the possibility of non-invasive diagnostic platforms [134–138,147] with effortless sample collection [23,53,54], compared to blood [24]. Moreover, sweat biosensors are readily adapted to wearable sensing system [54,133,138] for real-time, dynamic, and continuous measurement of biomarkers [23,54,133]. Over recent years, given the progress in nanotechnology, development of novel sweat-based biosensors has grown [23,24].

When collecting data from sweat, the sensor must be in close contact with the skin, ideally with a planar fit. Moreover, sweat biosensors are mostly produced with a flexible substrate for contact with the skin surface, enabling the option of a wearable sensor platform. Both characteristics, i.e., planar fit and flexibility, ensure proper sweat sample collection and a lower required sample volume [23,24,148]. Additional requirements for a wearable sweat sensor are the following: a fast response time of the detected analyte, high stability, selectivity, and sensitivity under environmental conditions [24]. A fast response time of the sensor can be realized and accomplished by using e.g., electrochemical read out methods, such as amperometry and electrochemical impedance spectroscopy [24]. The integration of electrochemical methods for the sensor development of sweat biosensors has different advantages in terms of low cost, high performance, and device portability [136]. However, a significant drawback when performing quantitative assessments is normalization of the sampled volume. One way to mitigate this is by also incorporating monitoring the sweat flow rate, e.g., by measuring the change in sweat generation rate by skin impedance. However, without a detailed fluidic model between the sweat glands and sensors, the sweat rate does not predict actual biomarker sampling intervals [149].

Enzymes, which are immobilized on the surface of the working electrode via covalent cross linking or bonding [150], are widely used as biorecognition elements in sweat-based biosensors. Some examples include GOx [24,137,141,151,152], LOx [137,139,151,153,154], alcohol oxidase (AOx) [140,155–159]. For an improved enzyme immobilization and at the same time a maximized surface concentration and an increased surface area, nanoparticles with silver [91,160], gold [161], and nickel [162], as well as nanofibers, such as zinc oxide [23], can be used. This modification step results in higher electrode response, faster sensor response time, and a higher selectivity to the investigated analyte [23].

For enzymatic sensing of metabolites, like glucose, lactate, ethanol, and uric acid, an amperometric read out is often used [148,153,157]. The detection of lactate is realized with the immobilization of the enzyme LOx and was demonstrated by different working groups [138–140,154]. Already, in the early nineties, an amperometric H_2O_2 based biosensor was developed to detect lactate and was based on enzyme immobilized between a polycarbonate membrane and a polytetrafluoroethylene (PTFE) blocking membrane [139]. Furthermore, a flexible array patch with LOx and a Prussian blue/gold electrode [137], a temporary tattoo using carbon ink (Figure 5) [140] was introduced to monitor diseases, like pressure ischaemia, peripheral arterial occlusive disease, or hypoxia [137,140,151]. Another accomplishment drew on two electrocardiogram electrodes combined with the detection of lactate (three electrode system). The so called "Chem-phys"-patch is based on a biocatalytic layer with LOx and modified Prussian blue on a polyester sheet [154].

The possible sensing of the metabolite glucose in sweat was demonstrated by different working groups for the detection of diabetes mellitus [24,137,141,151,152]. This included, e.g., a flexible array patch with immobilized GOx on a Prussian blue/Au electrode [137], multi-analyte glasses based on GOx attached to a gold electrode [151], or a graphene based stretchable patch using GOx on a Prussian blue/graphene-Au electrode [152]. For another sensing platform gold and platinum alloy nanoparticles were electrochemically deposited on a reduced graphene oxide surface and chitosan-GOx composites were integrated onto the modified surface [141].

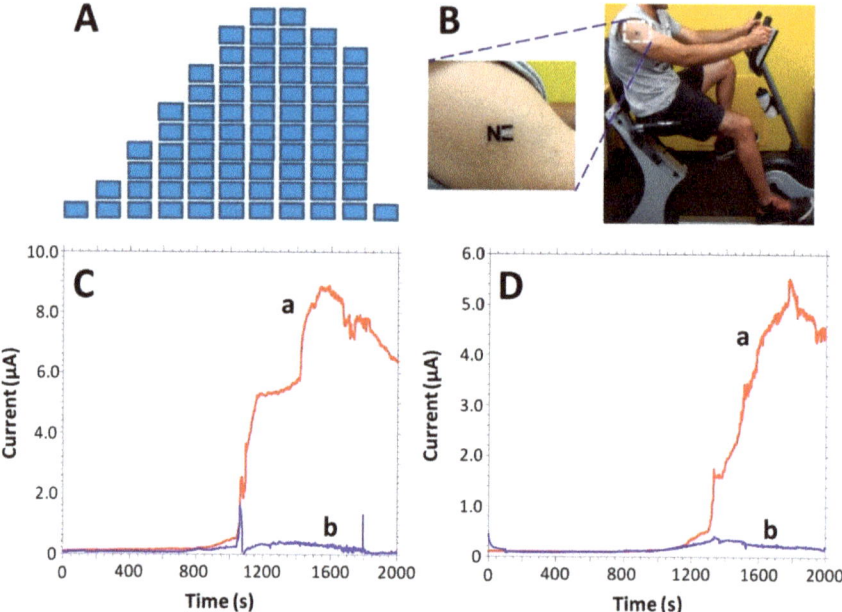

Figure 5. Monitoring of sweat lactate during 33 min of cycling exercise while adjusting the work intensity. (**A**) Exercise resistance profile on a stationary cycle. Subjects were asked to maintain a constant cycling rate while the resistance was increased every 3 min for a total evaluation of 30 min. A 3-min cool down period followed the exercise. (**B**) An "NE" lactate biosensor applied to a male volunteer's deltoid; (**C,D**) Response of the LOx- (a) and enzyme-free (b) tattoo biosensors during the exercise regimen (shown in part A) using two representative subjects. Constant potential, 0.05 V (vs. Ag|AgCl); measurement intervals, 1 s. Reproduced from [140] with permission from the American Chemical Society.

Moreover, the detection of another metabolite, ethanol, was successfully accomplished, using a skin surface-based sensing device for determining the blood's ethanol content by monitoring transdermal alcohol concentration. For this purpose, two enzymes were used, GOx/horseradish peroxidase (GOx/HRP), immobilized on a graphite-Teflon electrode [158]. Other developments were a temporary tattoo with AOx immobilized either on Prussian blue [157] or a wearable patch using platinum electrodes [159].

Impedance-based sensors were developed to detect different metabolites in sweat, e.g., glucose [135,143], lactate [163], or biomarkers, like Interleukin-6 [142] or cortisol [135,142,144]. For the detection of cortisol, MoS2 nanosheets were functionalized with cortisol antibodies to create a non-faradaic label-free cortisol biosensor (Figure 6) [144]. One working group developed flexible, wearable, nanoporous tunable electrical double layer biosensors with a bio-functionalized area of Zinc oxide (active region) to detect cortisol in sweat due the changes of impedance caused by the modulation of the double layer capacitance [135]. The same working group introduced a lancet-free and label-free diagnostic platform to detect glucose and cortisol in sweat. They used again zinc oxide based flexible bioelectronics of stacked metal/metal-oxide (gold/zinc oxide) thin films within porous polyamide substrates. Additionally, antibodies specific to GOx and for cortisol were attached to the zinc oxide region [143]. Moreover, this group also developed a sensor which enhanced the stability of biomolecules by using room temperature ionic liquids. In this paper they used sensors on nanoporous, flexible polymer membranes functionalized with antibodies to detect interleukin-6 (IL-6) and cortisol in human sweat [142].

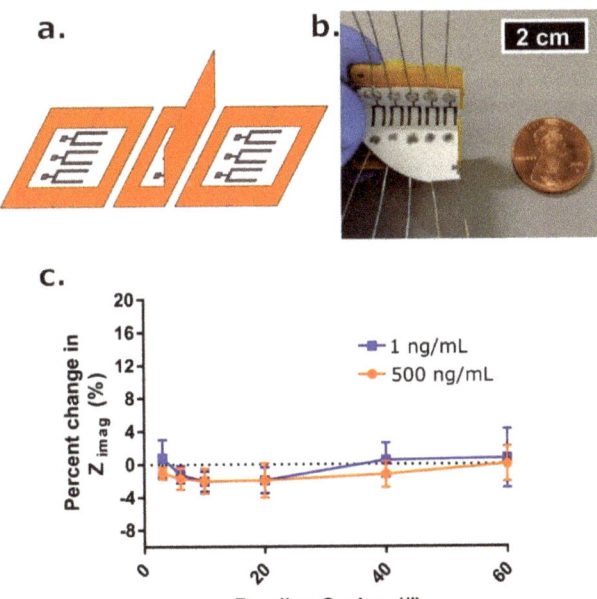

Figure 6. (**a**) Schematic drawing for one complete bending cycle of the sensor. The cycle is comprised of (left) unbent state, (middle) 90° flexion motion bend, and (right) return to unbent state at which point a measurement occurs. (**b**) Picture of bending apparatus with an affixed sensor array affixed. Penny for reference. (**c**) Percent change in Zimag impedance with respect to the initial measurement post-cortisol dosing and 7-min incubation time (blue box—1 ng/mL, red circle—500 ng/mL) after # of bending cycles ($n = 3$). Error bars are standard error of the mean. Reproduced from [144] with permission from Springer Nature Limited.

For a more practical use in health status monitoring, different working groups developed multiplex analysis platforms. One interesting approach was to measure simultaneously glucose, lactate, sodium, potassium, and skin temperature in one fully integrated sensor array [137]. The patch-type sensor was flexible, wearable, and made of a PET substrate (Figure 7). The metabolites were detected by using GOx and LOx which were imbedded in a chitosan film (amperometric read out). Moreover, the analysis of sodium and potassium were realized by integrating ion selective electrodes (potentiometric read out). The skin temperature measurement was based on a chromium/gold metal microwire. Additionally, the electronic parts were sealed and covered with insulating parylene.

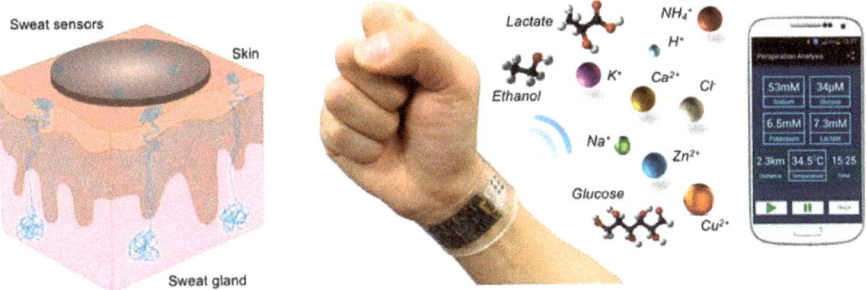

Figure 7. Wearable sweat biosensors which continuously measure a variety of sweat components for health monitoring. Reproduced from [164] with permission from IEEE.

Another approach was made for the simultaneous detection of ethanol, glucose, and lactate with a low sample volume of sweat based on non-faradaic chronoamperometric read out [165]. For this purpose, nanotextured zinc oxide films were integrated on a flexible porous membrane. The specific enzymes were immobilized in the active zinc oxide region by using a linker molecule (DSP). A streptavidin biotinylated AOx was used for the detection of alcohol, whereas a glucose antibody and coupled GOx, and LOx was used for glucose and lactate, respectively. The measured changes in current were associated with interactions of the target biomarkers with their specific enzyme and the relating analyte.

2.4. Biosensors Operating in Urine

Urine, a typically sterile liquid by-product, is often used as a diagnostic tool for many disease conditions. An adult human produces around two liters of urine per day within about seven urinations. The number of urinations depends on state of hydration, activity level, environmental factors, weight, and the individual's health. About 95% of the urine consists of water, but the fluid also contains different inorganic and organic, low and high molecular weight compounds [166]. A variety of compounds with clinical relevance are present in urine, such as glucose, lactate, urate, ascorbate, cholesterol and oxalate, all of which can be converted by different oxidoreductases. The occurrence of glucose in urine is associated with diabetes [167,168], whereas lactate is a prognostic marker for various disorders, and urinary lactate have been shown to correlate with blood lactate [169]. Significant amounts of cholesterol were detected in the urine of nephritic patients, whereas renal excretion of the bioanalyte is unsubstantial in healthy individuals [170]. Additionally, there are reports showing correlations between other biomarker levels in blood/plasma and urine, e.g., ascorbate and oxalate. For instance, oxalate concentrations in regular urine and blood range from about 160 µM to about 550 µM [171], and from about 17 µM to about 39 µM, respectively [172]. Ascorbate concentrations in plasma and urine of apparently healthy volunteers were found to be 76.50 ± 8.88 µM and 5.94 ± 1.43 µM, respectively [173]. Moreover, it was shown that increased renal excretion of ascorbate because of certain illnesses, e.g., sickle cell anaemia, might result in decreased plasma levels.

Biosensors based on oxidases are one of the well-known biosensors for the detection of bioanalytes in urine. These sensors detect the H_2O_2 generated, which can be used to estimate the concentration of the particular analyte [174]. The most common enzyme used for the detection of glucose in urine is, as in other cases, GOx [175]. One of the methods used for the detection of glucose in urine using amperometric biosensors is exploitation of conductometric biotransducer, which gives a binary response, when the analyte is present in urine. The working principle is based on a Prussian blue-cellulose acetate layer modified with GOx. When the substrate is present, H_2O_2 is formed, reacting with the layer (Figure 8). The reaction leads to the change of conductivity of Prussian blue-cellulose acetate layer making it possible to estimate the presence of glucose using a wireless biosensor [176]. Another reported method used for determination of glucose in urine is based on amperometry by using redox mediators and a bi-enzyme system. The measurement can be achieved by compressing electrically conductive carbon with the strip of a biosensor simultaneously having two redox mediators, i.e., an enzyme system for the oxidation of glucose, and silver/silver chloride reference electrode. The analysis readout can be achieved by applying a drop of urine on a sensor, where the result is compared with a standard calibration curve or by converting the current flow to some units of urine glucose levels [177].

Another important bioanalyte present in urine is urea. The analysis of urea in urine is mostly based on the measurement of NH_4^+ and HCO_3^-, which are the hydrolysis products. For detection, potentiometric, amperometric, optical, thermal, piezo-electric, and conductometric sensors are used [178]. The most reliable sensor for measuring urea in urine is the amperometric urease biosensor, which is relatively simple and offers a low cost analysis. Results obtained with this sensor are directly associated with hydrolysis of urea on the electrode surface [179]. One of the first potentiometric urea sensitive biosensors designed in 1969, was also based on urease [180].

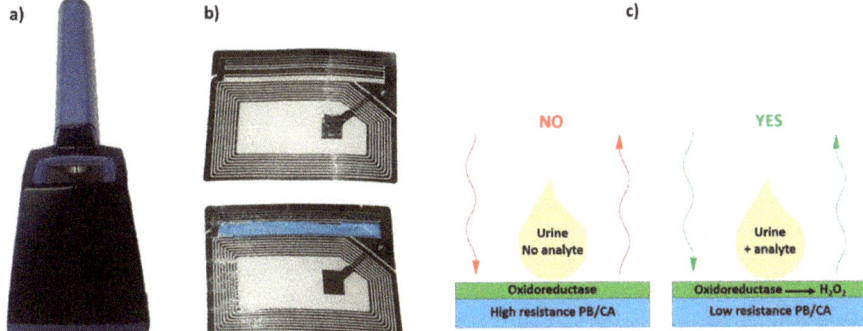

Figure 8. (a) Photographs of the portable monitoring unit, (b) sensor circuit developed for moisture detection (top) and modified sensor circuit modified with PB/CA layer (bottom), (c) Schematics of wireless biosensing. Reproduced from [176] with permission from IEEE.

One of the most common bacterial infections, which poses a significant healthcare problem, is urinary tract infection (UTI). The standard culture-based diagnosis of UTI has a typical lead time of several days, and in the absence of microbiological diagnosis at point of care facilities, physicians frequently initiate broad-spectrum antibiotic treatment, thus contributing to the emergence of resistant pathogens. The powerful diagnostic platforms for infectious diseases are based on biosensors. For instance, an interesting example of a biosensor for uropathogen identification is the UTI Sensor Array (Figure 9) [181,182]. An electrochemical sensor array customized with bacteria specific DNA probes as recognition elements represents the sensor-platform. Each of the 16 sensors is modified with a self-assembled monolayer, which allows versatility in surface modification and, simultaneously, reduces background noise [183]. On the surface of the sensor, a library of DNA probes targeting the most common uropathogens is immobilized [184,185]. The detection protocol is based on conversion of hybridization events into quantifiable electrochemical signals.

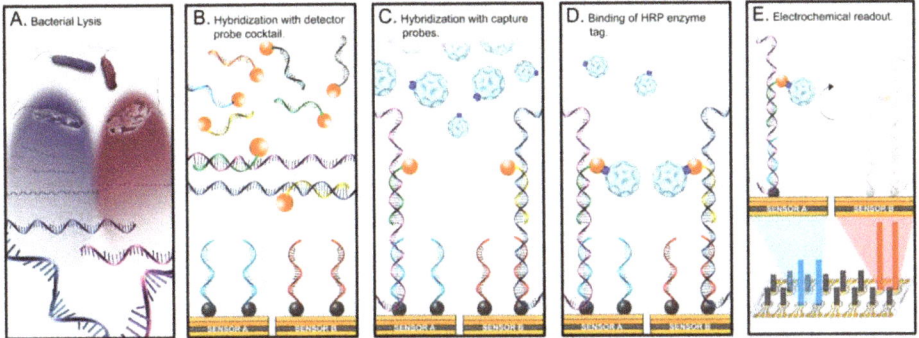

Figure 9. Multiplex pathogen detection scheme using an array of 16 electrochemical biosensors (UTI sensor array). Reproduced from Reproduced from [181] with permission from the US National Library of Medicine.

Currently, biosensor diagnosis for UTI has moved beyond the proof-of-concept stage into the validation phase, with authentic clinical samples, and development of assays for rapid molecular pathogen identification and antimicrobial susceptibility testing. UTI sensor arrays offer a promising technology platform without the need for nucleic acid amplification. Removal of the technology bottlenecks still remaining, i.e., sample preparation and system integration, is crucial for the technology to be used in decentralized settings such as clinicians' offices and emergency departments.

A novel innovative approach draws on a single platform strategy, i.e., a so-called universal electrode, that incorporates the central microfluidics of molecular analyses, i.e., pumping, mixing, washing, and sensing [186]. In an important illustration of the validity of the universal electrode platform, a bacterial phylogenetic marker was detected, promoting the rapid diagnosis of urinary tract infections. Since the platform is operated with electronic interfaces only, not only will it be possible to streamline systems integration and thus unlock the potential of microfluidics in molecular diagnostics at point of care sites, but also offer advanced biosensing in uncustomary health care settings [186].

Last but not least, recent efforts have been directed towards the development of wireless bacteria sensitive biosensors based on near field communication and radio frequency identification tags [187,188]. The approaches were realized by coupling a biosensor electrode as a part of the tag antenna. The transduction mechanism of these wireless biosensors exploits biological redox reactions. Specifically, the reactions change the impedance of the tag antennas, which is then wirelessly monitored by vector network analyzer or mobile phone. Future development of these wireless biosensors tags will target monitoring of UTI, as well as monitoring of bacterial growth in hygiene and medical products.

3. Conclusions and Outlook

Utilizing biological fluids for health monitoring offers the opportunity for non-invasive measurements and straightforward sample collection [55,134,135,137,138,147]. However, for successful sensor development, full and detailed knowledge is needed of the biological and chemical characteristics of sweat, saliva, tears, and urine, as well as the required technology for sensor realization [54]. Therefore, further efforts and research are needed to recognize the full diagnostic potential, in order to bypass the remaining challenges regarding sample collection, measurement, and sensing [147]. As an example, controlled and reproducible sampling is essential to improve the reliability of the results [54]. As regards sweat and tears, sample collection methods are wanting, and separate collection and analysis stages are in use [23,47]. Moreover, perspiration needs to be generated by exercising, heating, stress, or iontophoretic stimulation [54]. Additionally, variations in environmental and personal conditions, like temperature or pH, the individual skin composition, state of the oral cavity, or type of collected tears, fluid contamination, or mixing old and new fluid during sensing/sample collection impede reliable results [69]. Other challenges thwart the improvement of sensor characteristics, e.g., stability, biofouling, sensitivity, selectivity, robustness, accuracy, and power supply [24]. For instance, physiological fluids are complex solutions, provoking the integrity of the working electrode [142], and stability enhancing measures need to be taken in order achieve long term, continuous monitoring and measurement [54]. Moreover, electrodes should be usable without pre-treatment [24] or calibration, or storage in conditioning solutions (ion selective electrodes) [23,24]. On a final note, the power supply and packaging is one of the challenges for using sensors with an electro(-chemical) read out. Additionally, the packaging and integrated electronics should be in one comfortable, reliable, and safe platform [24].

After this presentation of a somewhat motley collection of vehicles/fluids, analytes, sampling techniques, sensors and sensor technology, power supplies, communication and data processing, it should be apparent to the reader that the field of non-invasive sensing of biomarkers in bodily fluid is highly convoluted. Nonetheless, if the drawbacks are appropriately addressed, and the pitfalls are adroitly circumvented, the approach will most certainly disrupt current clinical and self-monitoring practices.

As an example, the formation rate of all four fluids, affecting the availability of analytes, is varying in ways that are difficult to appreciate and control. However, by adhering to strict sampling protocols, relevant to clinical or point-of-care settings, or by relying on continuous measurements over an extended time, relevant to self-monitoring settings, much of the uncertainty emanating from varying fluid formation rates can be removed. Analogously, the blood concentration of some important biomarker targets is not exactly mirrored in the fluids under investigation. Depending on the particular target, the blood/fluid concentration discrepancy can be temporal or permanent. If the real (i.e., blood) value is critically needed, individual benchmarking using blood analysis combined with long term data collection usually featured by non-invasive sensors can be employed. If, on the other hand, the real value is non-critical, the aforementioned data collection features can identify positive or negative trends or abrupt baseline changes.

Funding: This research was funded by the Swedish Knowledge Foundation, viz. "ComBine" Industrial PhD school and "Biobarriers" profile projects, by Malmö University, viz. MultiSens platform at the Faculty of Health and Society, as well as by the European Commission, viz. Marie Curie "ImplantSens" ITN project.

Acknowledgments: The authors thank Zoltan Blum (Malmö University) for critical reading and helpful suggestions.

Conflicts of Interest: The authors declare no conflict of interest.

References

1. Yamani:, A.Z.; Alqahtani, F.M.; Alshahrani, N.S.; Alzamanan, R.M.; Aslam, N.; Algherairy, A.S. A proposed noninvasive point-of-care technique for measuring hemoglobin concentration. In Proceedings of the 2019 International Conference on Computer and Information Sciences (ICCIS), Aljouf, Saudi Arabia, 10–11 April 2019; pp. 1–4.
2. Olarte, O.; Chilo, J.; Pelegri-Sebastia, J.; Barbe, K.; Van Moer, W. Glucose detection in human sweat using an electronic nose. In Proceedings of the 35th Annual International Conference of the IEEE Engineering in Medicine and Biology Society, Osaka, Japan, 3–7 July 2013; pp. 1462–1465.
3. Okutucu, B.; Onal, S. Molecularly imprinted polymers for separation of various sugars from human urine. *Talanta* **2011**, *87*, 74–79. [CrossRef] [PubMed]
4. Aydin, S. A comparison of ghrelin, glucose, alpha-amylase and protein levels in saliva from diabetics. *J. Biochem. Mol. Biol.* **2007**, *40*, 29–35. [CrossRef] [PubMed]
5. Yao, H.F.; Shum, A.J.; Cowan, M.; Lahdesmaki, I.; Parviz, B.A. A contact lens with embedded sensor for monitoring tear glucose level. *Biosens. Bioelectron.* **2011**, *26*, 3290–3296. [CrossRef] [PubMed]
6. Moyer, J.; Wilson, D.; Finkelshtein, I.; Wong, B.; Potts, R. Correlation between sweat glucose and blood glucose in subjects with diabetes. *Diabetes Technol.* **2012**, *14*, 398–402. [CrossRef] [PubMed]
7. Morris, L.R.; McGee, J.A.; Kitabchi, A.E. Correlation between plasma and urine glucose in diabetes. *Ann. Intern. Med.* **1981**, *94*, 469–471. [CrossRef] [PubMed]
8. Puttaswamy, K.A.; Puttabudhi, J.H.; Raju, S. Correlation between salivary glucose and blood glucose and the implications of salivary factors on the oral health status in type 2 diabetes mellitus patients. *J. Int. Soc. Prev. Community Dent.* **2017**, *7*, 28–33. [CrossRef]
9. Lane, J.D.; Krumholz, D.M.; Sack, R.A.; Morris, C. Tear glucose dynamics in diabetes mellitus. *Curr. Eye Res.* **2006**, *31*, 895–901. [CrossRef]
10. Howell, J.D. Diagnostic technologies - X-rays, electrocardiograms, and cat-scans. *South. Calif. Law Rev.* **1991**, *65*, 529–564.
11. Fye, W.B. A history of the origin, evolution and impact of electrocardiography. *Am. J. Cardiol.* **1994**, *73*, 937–949. [CrossRef]
12. Bonetti, P.O.; Pumper, G.M.; Higano, S.T.; Holmes, D.R.; Kuvin, J.T.; Lerman, A. Noninvasive identification of patients with early coronary atherosclerosis by assessment of digital reactive hyperemia. *J. Am. Coll. Cardiol.* **2004**, *44*, 2137–2141. [CrossRef] [PubMed]
13. Lukaski, H.C.; Johnson, P.E.; Bolonchuk, W.W.; Lykken, G.I. Assessment of fat-free mass using bioelectrical impedance measurements of the human body. *Am. J. Clin. Nutr.* **1985**, *41*, 810–817. [CrossRef] [PubMed]
14. Khalil, S.F.; Mohktar, M.S.; Ibrahim, F. The theory and fundamentals of bioimpedance analysis in clinical status monitoring and diagnosis of diseases. *Sensors* **2014**, *14*, 10895–10928. [CrossRef]
15. Barker, S.J.; Shander, A.; Ramsay, M.A. Continuous noninvasive hemoglobin monitoring: A measured response to a critical review. *Anesth. Analg.* **2016**, *122*, 565–572. [CrossRef] [PubMed]
16. Taylor, M.B.; Whitwam, J.G. The current status of pulse oximetry—Clinical value of continuous noninvasive oxygen-saturation monitoring. *Anaesthesia* **1986**, *41*, 943–949. [CrossRef]

17. Tierney, M.J.; Tamada, J.A.; Potts, R.O.; Jovanovic, L.; Garg, S.; Cygnus Res, T. Clinical evaluation of the GlucoWatch (R) biographer: A continual, non-invasive glucose monitor for patients with diabetes. *Biosens. Bioelectron.* **2001**, *16*, 621–629. [CrossRef]
18. Caduff, A.; Dewarrat, F.; Talary, M.; Stalder, G.; Heinemann, L.; Feldman, Y. Non-invasive glucose monitoring in patients with diabetes: A novel system based on impedance spectroscopy. *Biosens. Bioelectron.* **2006**, *22*, 598–604. [CrossRef]
19. Ahrens, T.; Tucker, K. *Pulse Oximetry*; United States FIELD Citation; Department of Critical Care, Barnes-Jewish Hospital: St. Louis, MO, USA, 1999; pp. 87–98.
20. Smith, J.L. *The Pursuit of Noninvasive Glucose: "Hunting the deceitful Turkey" Seventh Edition*; NIVG Consulting LLC: Portland, OR, USA, 2020.
21. Yuan, J.; Ding, H.; Gao, H.; Lu, Q. Research on improving the accuracy of near infrared non-invasive hemoglobin detection. *Infrared Phys. Technol.* **2015**, *72*, 117–121. [CrossRef]
22. Adel, A.; Awada, W.; Abdelhamid, B.; Omar, H.; Hasanin, A.; Rady, A.; Awada, W.; Abd El Dayem, O.; Hasanin, A. Accuracy and trending of non-invasive hemoglobin measurement during different volume and perfusion statuses. *J. Clin. Monit. Comput.* **2018**, *32*, 1025–1031. [CrossRef] [PubMed]
23. Chung, M.; Fortunato, G.; Radacsi, N. Wearable flexible sweat sensors for healthcare monitoring: A review. *J. R. Soc. Interface* **2019**, *16*, 20190217. [CrossRef]
24. Bandodkar, A.J.; Jeang, W.J.; Ghaffari, R.; Rogers, J.A. Wearable Sensors for Biochemical Sweat Analysis. *Annu. Rev. Anal. Chem.* **2019**, *12*, 1–22. [CrossRef]
25. Chambers, J.; Arulanandam, B.; Matta, L.; Weis, A.; Valdes, J. Biosensor recognition elements. *Curr. Issues Mol. Biol.* **2008**, *10*, 1–12. [PubMed]
26. Murube, J. Basal, reflex, and psycho-emotional tears. *Ocul. Surf.* **2009**, *7*, 60–66. [CrossRef]
27. Berman, E.R. *Biochemistry of the Eye*; Plenum Press: New York, NY, USA, 1991.
28. van Haeringen, N.J.; Glasius, E. Collection method dependent concentrations of some metabolites in human tear fluid, with special reference to glucose in hyperglycaemic conditions. *Graefe's Arch. Clin. Exp. Ophthalmol.* **1977**, *202*, 1–7. [CrossRef] [PubMed]
29. Phillips, A. *Contact Lenses*; Elsevier: New York, NY, USA, 2007; p. 218.
30. Agarwal, S.; Agarwal, A.; Buratto, L.; Apple, D.J.; Ali, J.L. *Textbook of Ophthalmology*; Jaypee Brothers Publishers: New Delhi, India, 2002; p. 3000.
31. Farandos, N.M.; Yetisen, A.K.; Monteiro, M.J.; Lowe, C.R.; Yun, S.H. Contact lens sensors in ocular diagnostics. *Adv. Healthc. Mater.* **2015**, *4*, 792–810. [CrossRef]
32. Green-Church, K.B.; Nichols, K.K.; Kleinholz, N.M.; Zhang, L.; Nichols, J.J. Investigation of the human tear film proteome using multiple proteomic approaches. *Mol. Vis.* **2008**, *14*, 456–470.
33. Li, S.; Sack, R.; Vijmasi, T.; Sathe, S.; Beaton, A.; Quigley, D.; Gallup, M.; McNamara Nancy, A. Antibody protein array analysis of the tear film cytokines. *Optom. Vis. Sci. Off. Publ. Am. Acad. Optom.* **2008**, *85*, 653–660. [CrossRef]
34. Markoulli, M.; Papas, E.; Petznick, A.; Holden, B. Validation of the flush method as an alternative to basal or reflex tear collection. *Curr. Eye Res.* **2011**, *36*, 198–207. [CrossRef]
35. Sack, R.; Conradi, L.; Beaton, A.; Sathe, S.; McNamara, N.; Leonardi, A. Antibody array characterization of inflammatory mediators in allergic and normal tears in the open and closed eye environments. *Exp. Eye Res.* **2007**, *85*, 528–538. [CrossRef]
36. Shum, A.J.; Cowan, M.; Lahdesmaki, I.; Lingley, A.; Otis, B.; Parviz, B.A. Functional modular contact lens. *Proc. SPIE* **2009**, *7397*, 73970K.
37. Parviz, B.A. For your eye only. *IEEE Spectr.* **2009**, *46*, 36–41. [CrossRef]
38. Fang, A.P.; Ng, H.T.; Li, S.F.Y. A high-performance glucose biosensor based on monomolecular layer of glucose oxidase covalently immobilised on indium-tin oxide surface. *Biosens. Bioelectron.* **2003**, *19*, 43–49. [CrossRef]
39. Yao, H.; Afanasiev, A.; Lahdesmaki, I.; Parviz, B.A. A dual microscale glucose sensor on a contact lens, tested in conditions mimicking the eye. In Proceedings of the 2011 IEEE 24th International Conference on Micro Electro Mechanical Systems, Cancun, Mexico, 23–27 January 2011; pp. 25–28.
40. Liao, Y.T.; Yao, H.F.; Lingley, A.; Parviz, B.; Otis, B.P. A 3-mu W CMOS glucose sensor for wireless contact-lens tear glucose monitoring. *IEEE J. Solid-St. Circ.* **2012**, *47*, 335–344. [CrossRef]

41. Yao, H.; Liao, Y.; Lingley, A.R.; Afanasiev, A.; Lahdesmaki, I.; Otis, B.P.; Parviz, B.A. A contact lens with integrated telecommunication circuit and sensors for wireless and continuous tear glucose monitoring. *J. Micromech. Microeng.* **2012**, *22*. [CrossRef]
42. Morse, A. *Novartis and Google to Work on Smart Contact Lenses*; The Wall Steeet Journal: New York, NY, USA, 2014.
43. Thomas, N.; Lahdesmaki, I.; Parviz, B.A. A contact lens with an integrated lactate sensor. *Sens. Actuators B* **2012**, *162*, 128–134. [CrossRef]
44. Chu, M.X.; Miyajima, K.; Takahashi, D.; Arakawa, T.; Sano, K.; Sawada, S.; Kudo, H.; Iwasaki, Y.; Akiyoshi, K.; Mochizuki, M.; et al. Soft contact lens biosensor for in situ monitoring of tear glucose as non-invasive blood sugar assessment. *Talanta* **2011**, *83*, 960–965. [CrossRef]
45. Chu, M.X.; Shirai, T.; Takahashi, D.; Arakawa, T.; Kudo, H.; Sano, K.; Sawada, S.; Yano, K.; Iwasaki, Y.; Akiyoshi, K.; et al. Biomedical soft contact-lens sensor for in situ ocular biomonitoring of tear contents. *Biomed Microdevices* **2011**, *13*, 603–611. [CrossRef]
46. Patel, J.N.; Gray, B.L.; Kaminska, B.; Gates, B.D. Flexible three-dimensional electrochemical glucose sensor with improved sensitivity realized in hybrid polymer microelectromechanical systems technique. *J. Diabetes Sci. Technol.* **2011**, *5*, 1036–1043. [CrossRef]
47. Andoralov, V.; Shleev, S.; Arnebrant, T.; Ruzgas, T. Flexible micro(bio)sensors for quantitative analysis of bioanalytes in a nanovolume of human lachrymal liquid. *Anal. Bioanal. Chem.* **2013**, *405*, 3871–3879. [CrossRef]
48. Malon, R.S.P.; Sadir, S.; Balakrishnan, M.; Córcoles, E.P. Saliva-Based Biosensors: Noninvasive Monitoring Tool for Clinical Diagnostics. *Biomed Res. Int.* **2014**, *2014*, 962903. [CrossRef]
49. Ilea, A.; Andrei, V.; Feurdean, C.N.; Babtan, A.M.; Petrescu, N.B.; Campian, R.S.; Bosca, A.B.; Ciui, B.; Tertis, M.; Sandulescu, R.; et al. Saliva, a Magic Biofluid Available for Multilevel Assessment and a Mirror of General Health-A Systematic Review. *Biosensors* **2019**, *9*, 27. [CrossRef]
50. Lee, Y.-H.; Wong, D.T. Saliva: An emerging biofluid for early detection of diseases. *Am. J. Dent.* **2009**, *22*, 241–248.
51. Forde, M.D.; Koka, S.; E Eckert, S.; Carr, A.B.; Wong, D.T. Systemic Assessments Utilizing Saliva: Part 1 General Considerations and Current Assessments. *Int. J. Prosthodont.* **2006**, *19*, 43–52.
52. Yang, Y.; Gao, W. Wearable and flexible electronics for continuous molecular monitoring. *Chem. Soc. Rev.* **2019**, *48*, 1465–1491. [CrossRef] [PubMed]
53. Guilbault, G.G.; Palleschi, G.; Lubrano, G. Non-invasive biosensors in clinical analysis. *Biosens. Bioelectron.* **1995**, *10*, 379–392. [CrossRef]
54. Kim, J.; Campbell, A.S.; de Avila, B.E.; Wang, J. Wearable biosensors for healthcare monitoring. *Nat. Biotechnol.* **2019**, *37*, 389–406. [CrossRef]
55. Bandodkar, A.J.; Wang, J. Non-invasive wearable electrochemical sensors: A review. *Trends Biotechnol.* **2014**, *32*, 363–371. [CrossRef]
56. Khan, R.S.; Khurshid, Z.; Yahya Ibrahim Asiri, F. Advancing Point-of-Care (PoC) Testing Using Human Saliva as Liquid Biopsy. *Diagnostics* **2017**, *7*, 39. [CrossRef]
57. Liu, J.; Geng, Z.; Fan, Z.; Liu, J.; Chen, H. Point-of-care testing based on smartphone: The current state-of-the-art (2017–2018). *Biosens. Bioelectron.* **2019**, *132*, 17–37. [CrossRef] [PubMed]
58. Yang, X.; Cheng, H. Recent Developments of Flexible and Stretchable Electrochemical Biosensors. *Micromachines* **2020**, *11*, 243. [CrossRef]
59. Bellagambi, F.G.; Baraket, A.; Longo, A.; Vatteroni, M.; Zine, N.; Bausells, J.; Fuoco, R.; Di Francesco, F.; Salvo, P.; Karanasiou, G.S.; et al. Electrochemical biosensor platform for TNF-α cytokines detection in both artificial and human saliva: Heart failure. *Sens. Actuators B Chem.* **2017**, *251*, 1026–1033. [CrossRef]
60. Zhang, W.; Du, Y.; Wang, M.L. Noninvasive glucose monitoring using saliva nano-biosensor. *Sens. Bio-Sens. Res.* **2015**, *4*, 23–29. [CrossRef]
61. Pappa, A.-M.; Curto, V.F.; Braendlein, M.; Strakosas, X.; Donahue, M.J.; Fiocchi, M.; Malliaras, G.G.; Owens, R.M. Organic Transistor Arrays Integrated with Finger-Powered Microfluidics for Multianalyte Saliva Testing. *Adv. Healthc. Mater.* **2016**, *5*, 2295–2302. [CrossRef] [PubMed]
62. Du, Y.; Zhang, W.; Wang, M.L. An On-Chip Disposable Salivary Glucose Sensor for Diabetes Control. *J. Diabetes Sci. Technol.* **2016**, *10*, 1344–1352. [CrossRef] [PubMed]

63. Gupta, S.; Nayak, M.T.; Sunitha, J.D.; Dawar, G.; Sinha, N.; Rallan, N.S. Correlation of salivary glucose level with blood glucose level in diabetes mellitus. *J. Oral Maxillofac. Pathol.* **2017**, *21*, 334–339.
64. Kumar, S.; Padmashree, S.; Jayalekshmi, R. Correlation of salivary glucose, blood glucose and oral candidal carriage in the saliva of type 2 diabetics: A case-control study. *Contemp. Clin. Dent.* **2014**, *5*, 312–317. [CrossRef]
65. Justino, C.I.L.; Duarte, A.C.; Rocha-Santos, T.A.P. Critical overview on the application of sensors and biosensors for clinical analysis. *Trac. Trends Anal. Chem.* **2016**, *85*, 36–60. [CrossRef]
66. Yamaguchi, M. Salivary Sensors in Point-of-Care Testing. *Sens. Mater.* **2010**, *22*, 143–153.
67. Bhalinge, P.; Kumar, S.; Jadhav, A.; Suman, S.; Gujjar, P.; Perla, N. Biosensors: Nanotools of Detection—A Review. *Int. J. Healthc. Biomed. Res.* **2016**, *4*, 26–39.
68. Corrie, S.R.; Coffey, J.W.; Islam, J.; Markey, K.A.; Kendall, M.A. Blood, sweat, and tears: Developing clinically relevant protein biosensors for integrated body fluid analysis. *Analyst* **2015**, *140*, 4350–4364. [CrossRef]
69. Kim, J.; Jeerapan, I.; Sempionatto, J.R.; Barfidokht, A.; Mishra, R.K.; Campbell, A.S.; Hubble, L.J.; Wang, J. Wearable Bioelectronics: Enzyme-Based Body-Worn Electronic Devices. *Acc. Chem. Res.* **2018**, *51*, 2820–2828. [CrossRef] [PubMed]
70. Hao, Z.; Pan, Y.; Shao, W.; Lin, Q.; Zhao, X. Graphene-based fully integrated portable nanosensing system for on-line detection of cytokine biomarkers in saliva. *Biosens. Bioelectron.* **2019**, *134*, 16–23. [CrossRef]
71. Ojeda, I.; Moreno-Guzmán, M.; González-Cortés, A.; Yáñez-Sedeño, P.; Pingarrón, J.M. Electrochemical magnetoimmunosensor for the ultrasensitive determination of interleukin-6 in saliva and urine using poly-HRP streptavidin conjugates as labels for signal amplification. *Anal. Bioanal. Chem.* **2014**, *406*, 6363–6371. [CrossRef]
72. Torrente-Rodríguez, R.M.; Campuzano, S.; Ruiz-Valdepeñas Montiel, V.; Gamella, M.; Pingarrón, J.M. Electrochemical bioplatforms for the simultaneous determination of interleukin (IL)-8 mRNA and IL-8 protein oral cancer biomarkers in raw saliva. *Biosens. Bioelectron.* **2016**, *77*, 543–548. [CrossRef]
73. Wei, F.; Patel, P.; Liao, W.; Chaudhry, K.; Zhang, L.; Arellano-Garcia, M.; Hu, S.; Elashoff, D.; Zhou, H.; Shukla, S.; et al. Electrochemical sensor for multiplex biomarkers detection. *Clin. Cancer Res.* **2009**, *15*, 4446–4452. [CrossRef]
74. Topkaya, S.N.; Azimzadeh, M.; Ozsoz, M. Electrochemical Biosensors for Cancer Biomarkers Detection: Recent Advances and Challenges. *Electroanalysis* **2016**, *28*, 1402–1419. [CrossRef]
75. Yu, Q.; Wang, X.; Duan, Y. Capillary-Based Three-Dimensional Immunosensor Assembly for High-Performance Detection of Carcinoembryonic Antigen Using Laser-Induced Fluorescence Spectrometry. *Anal. Chem.* **2014**, *86*, 1518–1524. [CrossRef]
76. Jokerst, J.V.; Raamanathan, A.; Christodoulides, N.; Floriano, P.N.; Pollard, A.A.; Simmons, G.W.; Wong, J.; Gage, C.; Furmaga, W.B.; Redding, S.W.; et al. Nano-bio-chips for high performance multiplexed protein detection: Determinations of cancer biomarkers in serum and saliva using quantum dot bioconjugate labels. *Biosens. Bioelectron.* **2009**, *24*, 3622–3629. [CrossRef] [PubMed]
77. Chen, J.; Lin, J.; Zhang, X.; Cai, S.; Wu, D.; Li, C.; Yang, S.; Zhang, J. Label-free fluorescent biosensor based on the target recycling and Thioflavin T-induced quadruplex formation for short DNA species of c-erbB-2 detection. *Anal. Chim. Acta* **2014**, *817*, 42–47. [CrossRef] [PubMed]
78. Cho, H.; Yeh, E.-C.; Sinha, R.; Laurence, T.A.; Bearinger, J.P.; Lee, L.P. Single-Step Nanoplasmonic VEGF165 Aptasensor for Early Cancer Diagnosis. *ACS Nano* **2012**, *6*, 7607–7614. [CrossRef] [PubMed]
79. Aydın, E.B.; Aydın, M.; Sezgintürk, M.K. A highly sensitive immunosensor based on ITO thin films covered by a new semi-conductive conjugated polymer for the determination of TNFα in human saliva and serum samples. *Biosens. Bioelectron.* **2017**, *97*, 169–176. [CrossRef]
80. Song, C.K.; Oh, E.; Kang, M.S.; Shin, B.S.; Han, S.Y.; Jung, M.; Lee, E.S.; Yoon, S.-Y.; Sung, M.M.; Ng, W.B.; et al. Fluorescence-based immunosensor using three-dimensional CNT network structure for sensitive and reproducible detection of oral squamous cell carcinoma biomarker. *Anal. Chim. Acta* **2018**, *1027*, 101–108. [CrossRef] [PubMed]
81. Palleschi, G.; Faridnia, M.H.; Lubrano, G.J.; Guilbault, G.G. Determination of lactate in human saliva with an electrochemical enzyme probe. *Anal. Chim. Acta* **1991**, *245*, 151–157. [CrossRef]
82. Kim, J.; Valdés-Ramírez, G.; Bandodkar, A.J.; Jia, W.; Martinez, A.G.; Ramírez, J.; Mercier, P.; Wang, J. Non-invasive mouthguard biosensor for continuous salivary monitoring of metabolites. *Analyst* **2014**, *139*, 1632–1636. [CrossRef]

83. Ballesta Claver, J.; Valencia Mirón, M.C.; Capitán-Vallvey, L.F. Disposable electrochemiluminescent biosensor for lactate determination in saliva. *Analyst* **2009**, *134*, 1423–1432. [CrossRef]
84. Yao, Y.; Li, H.; Wang, D.; Liu, C.; Zhang, C. An electrochemiluminescence cloth-based biosensor with smartphone-based imaging for detection of lactate in saliva. *Analyst* **2017**, *142*, 3715–3724. [CrossRef]
85. Liu, C.; Sheng, Y.; Sun, Y.; Feng, J.; Wang, S.; Zhang, J.; Xu, J.; Jiang, D. A glucose oxidase-coupled DNAzyme sensor for glucose detection in tears and saliva. *Biosens. Bioelectron.* **2015**, *70*, 455–461. [CrossRef]
86. Ye, D.; Liang, G.; Li, H.; Luo, J.; Zhang, S.; Chen, H.; Kong, J. A novel nonenzymatic sensor based on CuO nanoneedle/graphene/carbon nanofiber modified electrode for probing glucose in saliva. *Talanta* **2013**, *116*, 223–230. [CrossRef]
87. Li, Z.; Chen, Y.; Xin, Y.; Zhang, Z. Sensitive electrochemical nonenzymatic glucose sensing based on anodized CuO nanowires on three-dimensional porous copper foam. *Sci. Rep.* **2015**, *5*, 16115. [CrossRef]
88. Arakawa, T.; Kuroki, Y.; Nitta, H.; Chouhan, P.; Toma, K.; Sawada, S.-i.; Takeuchi, S.; Sekita, T.; Akiyoshi, K.; Minakuchi, S.; et al. Mouthguard biosensor with telemetry system for monitoring of saliva glucose: A novel cavitas sensor. *Biosens. Bioelectron.* **2016**, *84*, 106–111. [CrossRef]
89. Soni, A.; Jha, S.K. Smartphone based non-invasive salivary glucose biosensor. *Anal. Chim. Acta* **2017**, *996*, 54–63. [CrossRef]
90. Dominguez, R.B.; Orozco, M.A.; Chávez, G.; Márquez-Lucero, A. The Evaluation of a Low-Cost Colorimeter for Glucose Detection in Salivary Samples. *Sensors* **2017**, *17*, 2495. [CrossRef]
91. Anderson, K.P.B.; Dudgeon, J.; Li, S.-E.; Ma, X. A Highly Sensitive Nonenzymatic Glucose Biosensor Based on the Regulatory Effect of Glucose on Electrochemical Behaviors of Colloidal Silver Nanoparticles on MoS2. *Sensors* **2017**, *17*, 1807. [CrossRef]
92. Bell, C.; Nammari, A.; Uttamchandani, P.; Rai, A.; Shah, P.; Moore, A.L. Flexible electronics-compatible non-enzymatic glucose sensing via transparent CuO nanowire networks on PET films. *Nanotechnology* **2017**, *28*, 245502. [CrossRef] [PubMed]
93. Velmurugan, M.; Karikalan, N.; Chen, S.-M. Synthesis and characterizations of biscuit-like copper oxide for the non-enzymatic glucose sensor applications. *J. Colloid Interface Sci.* **2017**, *493*, 349–355. [CrossRef]
94. Kim, D.-M.; Moon, J.-M.; Lee, W.-C.; Yoon, J.-H.; Choi, C.S.; Shim, Y.-B. A potentiometric non-enzymatic glucose sensor using a molecularly imprinted layer bonded on a conducting polymer. *Biosens. Bioelectron.* **2017**, *91*, 276–283. [CrossRef]
95. Santana-Jiménez, L.A.; Márquez-Lucero, A.; Osuna, V.; Estrada-Moreno, I.; Dominguez, R.B. Naked-Eye Detection of Glucose in Saliva with Bienzymatic Paper-Based Sensor. *Sensors* **2018**, *18*, 1071. [CrossRef]
96. Raymundo-Pereira, P.A.; Shimizu, F.M.; Coelho, D.; Piazzeta, M.H.O.; Gobbi, A.L.; Machado, S.A.S.; Oliveira, O.N. A Nanostructured Bifunctional platform for Sensing of Glucose Biomarker in Artificial Saliva: Synergy in hybrid Pt/Au surfaces. *Biosens. Bioelectron.* **2016**, *86*, 369–376. [CrossRef] [PubMed]
97. Tlili, C.; Myung, N.V.; Shetty, V.; Mulchandani, A. Label-free, chemiresistor immunosensor for stress biomarker cortisol in saliva. *Biosens. Bioelectron.* **2011**, *26*, 4382–4386. [CrossRef] [PubMed]
98. Mitchell, J.S.; Lowe, T.E.; Ingram, J.R. Rapid ultrasensitive measurement of salivary cortisol using nano-linker chemistry coupled with surface plasmon resonance detection. *Analyst* **2009**, *134*, 380–386. [CrossRef]
99. Frasconi, M.; Mazzarino, M.; Botrè, F.; Mazzei, F. Surface plasmon resonance immunosensor for cortisol and cortisone determination. *Anal. Bioanal. Chem.* **2009**, *394*, 2151–2159. [CrossRef]
100. Dhull, N.; Kaur, G.; Gupta, V.; Tomar, M. Highly sensitive and non-invasive electrochemical immunosensor for salivary cortisol detection. *Sens. Actuators B Chem.* **2019**, *293*, 281–288. [CrossRef]
101. Pasha, S.K.; Kaushik, A.; Vasudev, A.; Snipes, S.A.; Bhansali, S. Electrochemical Immunosensing of Saliva Cortisol. *J. Electrochem. Soc.* **2013**, *161*, B3077–B3082. [CrossRef]
102. Vabbina, P.K.; Kaushik, A.; Pokhrel, N.; Bhansali, S.; Pala, N. Electrochemical cortisol immunosensors based on sonochemically synthesized zinc oxide 1D nanorods and 2D nanoflakes. *Biosens. Bioelectron.* **2015**, *63*, 124–130. [CrossRef]
103. Pires, N.M.; Dong, T. Measurement of salivary cortisol by a chemiluminescent organic-based immunosensor. *Biomed. Mater. Eng.* **2014**, *24*, 15–20. [CrossRef]
104. Usha, S.P.; Shrivastav, A.M.; Gupta, B.D. A contemporary approach for design and characterization of fiber-optic-cortisol sensor tailoring LMR and ZnO/PPY molecularly imprinted film. *Biosens. Bioelectron.* **2017**, *87*, 178–186. [CrossRef] [PubMed]

105. Kim, J.; Imani, S.; de Araujo, W.R.; Warchall, J.; Valdés-Ramírez, G.; Paixão, T.R.L.C.; Mercier, P.P.; Wang, J. Wearable salivary uric acid mouthguard biosensor with integrated wireless electronics. *Biosens. Bioelectron.* **2015**, *74*, 1061–1068. [CrossRef]
106. Soni, A.; Surana, R.K.; Jha, S.K. Smartphone based optical biosensor for the detection of urea in saliva. *Sens. Actuators B Chem.* **2018**, *269*, 346–353. [CrossRef]
107. Majidi, M.R.; Omidi, Y.; Karami, P.; Johari-Ahar, M. Reusable potentiometric screen-printed sensor and label-free aptasensor with pseudo-reference electrode for determination of tryptophan in the presence of tyrosine. *Talanta* **2016**, *150*, 425–433. [CrossRef]
108. Lee, M.-H.; Thomas, J.L.; Tseng, H.-Y.; Lin, W.-C.; Liu, B.-D.; Lin, H.-Y. Sensing of Digestive Proteins in Saliva with a Molecularly Imprinted Poly(ethylene-co-vinyl alcohol) Thin Film Coated Quartz Crystal Microbalance Sensor. *ACS Appl. Mater. Interfaces* **2011**, *3*, 3064–3071. [CrossRef]
109. Attia, M.S.; Zoulghena, H.; Abdel-Mottaleb, M.S.A. A new nano-optical sensor thin film cadmium sulfide doped in sol–gel matrix for assessment of α-amylase activity in human saliva. *Analyst* **2014**, *139*, 793–800. [CrossRef]
110. Shetty, V.; Zigler, C.; Robles, T.F.; Elashoff, D.; Yamaguchi, M. Developmental validation of a point-of-care, salivary α-amylase biosensor. *Psychoneuroendocrinology* **2011**, *36*, 193–199. [CrossRef] [PubMed]
111. Gorodkiewicz, E.; Breczko, J.; Sankiewicz, A. Surface Plasmon Resonance Imaging biosensor for cystatin determination based on the application of bromelain, ficin and chymopapain. *Surf. Plasmon Reson. Imaging Biosens. Cystatin Determ. Based Appl. BromelainFicin Chymopapain* **2012**, *50*, 130–136. [CrossRef]
112. Gorodkiewicz, E.; Regulska, E.; Wojtulewski, K. Development of an SPR imaging biosensor for determination of cathepsin G in saliva and white blood cells. *Microchim. Acta* **2011**, *173*, 407–413. [CrossRef]
113. Ahmed, A.; Rushworth, J.V.; Wright, J.D.; Millner, P.A. Novel Impedimetric Immunosensor for Detection of Pathogenic Bacteria Streptococcus pyogenes in Human Saliva. *Anal. Chem.* **2013**, *85*, 12118–12125. [CrossRef]
114. Zaitouna, A.J.; Maben, A.J.; Lai, R.Y. Incorporation of extra amino acids in peptide recognition probe to improve specificity and selectivity of an electrochemical peptide-based sensor. *Anal. Chim. Acta* **2015**, *886*, 157–164. [CrossRef]
115. Zilberman, Y.; Sonkusale, S.R. Microfluidic optoelectronic sensor for salivary diagnostics of stomach cancer. *Biosens. Bioelectron.* **2015**, *67*, 465–471. [CrossRef]
116. Mannoor, M.S.; Tao, H.; Clayton, J.D.; Sengupta, A.; Kaplan, D.L.; Naik, R.R.; Verma, N.; Omenetto, F.G.; McAlpine, M.C. Graphene-based wireless bacteria detection on tooth enamel. *Nat. Commun.* **2012**, *3*, 763. [CrossRef]
117. Priye, A.; Bird, S.W.; Light, Y.K.; Ball, C.S.; Negrete, O.A.; Meagher, R.J. A smartphone-based diagnostic platform for rapid detection of Zika, chikungunya, and dengue viruses. *Sci. Rep.* **2017**, *7*, 44778. [CrossRef]
118. Blicharz, T.M.; Siqueira, W.L.; Helmerhorst, E.J.; Oppenheim, F.G.; Wexler, P.J.; Little, F.F.; Walt, D.R. Fiber-Optic Microsphere-Based Antibody Array for the Analysis of Inflammatory Cytokines in Saliva. *Anal. Chem.* **2009**, *81*, 2106–2114. [CrossRef]
119. Campuzano, S.; Yáñez-Sedeño, P.; Pingarrón, J.M. Electrochemical bioaffinity sensors for salivary biomarkers detection. *Trac Trends Anal. Chem.* **2017**, *86*, 14–24. [CrossRef]
120. Zachary, D.; Mwenge, L.; Muyoyeta, M.; Shanaube, K.; Schaap, A.; Bond, V.; Kosloff, B.; de Haas, P.; Ayles, H. Field comparison of OraQuick® ADVANCE Rapid HIV-1/2 antibody test and two blood-based rapid HIV antibody tests in Zambia. *BMC Infect. Dis.* **2012**, *12*, 183. [CrossRef]
121. Machini, W.B.S.; Teixeira, M.F.S. Analytical development of a binuclear oxo-manganese complex bio-inspired on oxidase enzyme for doping control analysis of acetazolamide. *Biosens. Bioelectron.* **2016**, *79*, 442–448. [CrossRef]
122. Du, Y.; Chen, C.; Yin, J.; Li, B.; Zhou, M.; Dong, S.; Wang, E. Solid-State Probe Based Electrochemical Aptasensor for Cocaine: A Potentially Convenient, Sensitive, Repeatable, and Integrated Sensing Platform for Drugs. *Anal. Chem.* **2010**, *82*, 1556–1563. [CrossRef] [PubMed]
123. Hagen, J.; Lyon, W.; Chushak, Y.; Tomczak, M.; Naik, R.; Stone, M.; Kelley-Loughnane, N. Detection of Orexin A Neuropeptide in Biological Fluids Using a Zinc Oxide Field Effect Transistor. *ACS Chem. Neurosci.* **2013**, *4*, 444–453. [CrossRef]
124. Eftekhari, A.; Hasanzadeh, M.; Sharifi, S.; Dizaj, S.M.; Khalilov, R.; Ahmadian, E. Bioassay of saliva proteins: The best alternative for conventional methods in non-invasive diagnosis of cancer. *Int. J. Biol. Macromol.* **2019**, *124*, 1246–1255. [CrossRef]

125. Wignarajah, S.; Suaifan, G.A.R.Y.; Bizzarro, S.; Bikker, F.J.; Kaman, W.E.; Zourob, M. Colorimetric Assay for the Detection of Typical Biomarkers for Periodontitis Using a Magnetic Nanoparticle Biosensor. *Anal. Chem.* **2015**, *87*, 12161–12168. [CrossRef] [PubMed]
126. Calabria, D.; Caliceti, C.; Zangheri, M.; Mirasoli, M.; Simoni, P.; Roda, A. Smartphone–based enzymatic biosensor for oral fluid L-lactate detection in one minute using confined multilayer paper reflectometry. *Biosens. Bioelectron.* **2017**, *94*, 124–130. [CrossRef]
127. Yamaguchi, M.; Mitsumori, M.; Kano, Y. Noninvasively Measuring Blood Glucose Using Saliva. *IEEE Eng. Med. Biol. Mag.* **1998**, *17*, 59–63. [CrossRef]
128. Dorn, L.D.; Lucke, J.F.; Loucks, T.L.; Berga, S.L. Salivary cortisol reflects serum cortisol: Analysis of circadian profiles. *Ann. Clin. Biochem.* **2007**, *44*, 281–284. [CrossRef]
129. Gozansky, W.S.; Lynn, J.S.; Laudenslager, M.L.; Kohrt, W.M. Salivary cortisol determined by enzyme immunoassay is preferable to serum total cortisol for assessment of dynamic hypothalamic–pituitary–adrenal axis activity. *Clin. Endocrinol.* **2005**, *63*, 336–341. [CrossRef]
130. Dutta, G.; Nagarajan, S.; Lapidus, L.J.; Lillehoj, P.B. Enzyme-free electrochemical immunosensor based on methylene blue and the electro-oxidation of hydrazine on Pt nanoparticles. *Biosens. Bioelectron.* **2017**, *92*, 372–377. [CrossRef]
131. Sánchez-Tirado, E.; González-Cortés, A.; Yáñez-Sedeño, P.; Pingarrón, J.M. Magnetic multiwalled carbon nanotubes as nanocarrier tags for sensitive determination of fetuin in saliva. *Biosens. Bioelectron.* **2018**, *113*, 88–94. [CrossRef] [PubMed]
132. Liao, C.; Mak, C.; Zhang, M.; Chan, H.L.W.; Yan, F. Flexible Organic Electrochemical Transistors for Highly Selective Enzyme Biosensors and Used for Saliva Testing. *Adv. Mater.* **2015**, *27*, 676–681. [CrossRef]
133. Nyein, H.Y.; Gao, W.; Shahpar, Z.; Emaminejad, S.; Challa, S.; Chen, K.; Fahad, H.M.; Tai, L.C.; Ota, H.; Davis, R.W.; et al. A Wearable Electrochemical Platform for Noninvasive Simultaneous Monitoring of Ca(2+) and pH. *ACS Nano* **2016**, *10*, 7216–7224. [CrossRef]
134. Chandran, K.; Kalpana, B.; Robson, B. *Biosensors and Bioelectronics*; Elsevier: Amsterdam, The Netherlands, 2015; p. 344.
135. Munje, R.D.; Muthukumar, S.; Panneer Selvam, A.; Prasad, S. Flexible nanoporous tunable electrical double layer biosensors for sweat diagnostics. *Sci. Rep.* **2015**, *5*, 14586. [CrossRef]
136. Bandodkar, A.J.; Molinnus, D.; Mirza, O.; Guinovart, T.; Windmiller, J.R.; Valdés-Ramírez, G.; Andrade, F.J.; Schöning, M.J.; Wang, J. Epidermal tattoo potentiometric sodium sensors with wireless signal transduction for continuous non-invasive sweat monitoring. *Biosens. Bioelectron.* **2014**, *54*, 603–609. [CrossRef]
137. Gao, W.; Emaminejad, S.; Nyein, H.Y.Y.; Challa, S.; Chen, K.; Peck, A.; Fahad, H.M.; Ota, H.; Shiraki, H.; Kiriya, D.; et al. Fully integrated wearable sensor arrays for multiplexed in situ perspiration analysis. *Nature* **2016**, *529*, 509–514. [CrossRef]
138. Campbell, A.S.; Kim, J.; Wang, J. Wearable Electrochemical Alcohol Biosensors. *Curr. Opin. Electrochem.* **2018**, *10*, 126–135. [CrossRef]
139. Faridnia, M.H.; Palleschi, G.; Lubrano, G.J.; Guilbault, G.G. Amperometric biosensor for determination of lactate in sweat. *Anal. Chim. Acta* **1993**, *278*, 35–40. [CrossRef]
140. Jia, W.; Bandodkar, A.J.; Valdes-Ramirez, G.; Windmiller, J.R.; Yang, Z.; Ramirez, J.; Chan, G.; Wang, J. Electrochemical tattoo biosensors for real-time noninvasive lactate monitoring in human perspiration. *Anal. Chem.* **2013**, *85*, 6553–6560. [CrossRef]
141. Xuan, X.; Yoon, H.S.; Park, J.Y. A wearable electrochemical glucose sensor based on simple and low-cost fabrication supported micro-patterned reduced graphene oxide nanocomposite electrode on flexible substrate. *Biosens. Bioelectron.* **2018**, *109*, 75–82. [CrossRef] [PubMed]
142. Munje, R.D.; Muthukumar, S.; Jagannath, B.; Prasad, S. A new paradigm in sweat based wearable diagnostics biosensors using Room Temperature Ionic Liquids (RTILs). *Sci. Rep.* **2017**, *7*, 1950. [CrossRef]
143. Munje, R.D.; Muthukumar, S.; Prasad, S. Lancet-free and label-free diagnostics of glucose in sweat using Zinc Oxide based flexible bioelectronics. *Sens. Actuators B Chem.* **2017**, *238*, 482–490. [CrossRef]
144. Kinnamon, D.; Ghanta, R.; Lin, K.C.; Muthukumar, S.; Prasad, S. Portable biosensor for monitoring cortisol in low-volume perspired human sweat. *Sci. Rep.* **2017**, *7*, 13312. [CrossRef]
145. Karpova, E.V.; Karyakina, E.E.; Karyakin, A.A. Wearable non-invasive monitors of diabetes and hypoxia through continuous analysis of sweat. *Talanta* **2020**, *215*, 120922. [CrossRef] [PubMed]

146. Karpova, E.V.; Laptev, A.I.; Andreev, E.A.; Karyakina, E.E.; Karyakin, A.A. Relationship between sweat and blood lactate levels during exhaustive physical exercise. *ChemElectroChem* **2020**, *7*, 191–194. [CrossRef]
147. Zhao, J.; Guo, H.; Li, J.; Bandodkar, A.J.; Rogers, J.A. Body-Interfaced Chemical Sensors for Noninvasive Monitoring and Analysis of Biofluids. *Trends Chem.* **2019**, *1*, 559–571. [CrossRef]
148. Gao, W.; Nyein, H.Y.Y.; Shahpar, Z.; Fahad, H.M.; Chen, K.; Emaminejad, S.; Gao, Y.; Tai, L.-C.; Ota, H.; Wu, E.; et al. Wearable Microsensor Array for Multiplexed Heavy Metal Monitoring of Body Fluids. *ACS Sens.* **2016**, *1*, 866–874. [CrossRef]
149. Sonner, Z.; Wilder, E.; Heikenfeld, J.; Kasting, G.; Beyette, F.; Swaile, D.; Sherman, F.; Joyce, J.; Hagen, J.; Kelley-Loughnane, N.; et al. The microfluidics of the eccrine sweat gland, including biomarker partitioning, transport, and biosensing implications. *Biomicrofluidics* **2015**, *9*, 031301. [CrossRef]
150. Datta, S.; Christena, L.R.; Rajaram, Y.R.S. Enzyme immobilization: An overview on techniques and support materials. *3 Biotech* **2013**, *3*, 1–9. [CrossRef]
151. Sempionatto, J.R.; Nakagawa, T.; Pavinatto, A.; Mensah, S.T.; Imani, S.; Mercier, P.; Wang, J. Eyeglasses based wireless electrolyte and metabolite sensor platform. *Lab Chip* **2017**, *17*, 1834–1842. [CrossRef] [PubMed]
152. Lee, H.; Choi, T.K.; Lee, Y.B.; Cho, H.R.; Ghaffari, R.; Wang, L.; Choi, H.J.; Chung, T.D.; Lu, N.; Hyeon, T.; et al. A graphene-based electrochemical device with thermoresponsive microneedles for diabetes monitoring and therapy. *Nat. Nanotechnol.* **2016**, *11*, 566–572. [CrossRef]
153. Pribil, M.M.; Laptev, G.U.; Karyakina, E.E.; Karyakin, A.A. Noninvasive hypoxia monitor based on gene-free engineering of lactate oxidase for analysis of undiluted sweat. *Anal. Chem.* **2014**, *86*, 5215–5219. [CrossRef]
154. Imani, S.; Bandodkar, A.J.; Mohan, A.M.; Kumar, R.; Yu, S.; Wang, J.; Mercier, P.P. A wearable chemical-electrophysiological hybrid biosensing system for real-time health and fitness monitoring. *Nat. Commun.* **2016**, *7*, 11650. [CrossRef]
155. Kim, J.; Campbell, A.S.; Wang, J. Wearable non-invasive epidermal glucose sensors: A review. *Talanta* **2018**, *177*, 163–170. [CrossRef]
156. Anastasova, S.; Crewther, B.; Bembnowicz, P.; Curto, V.; Ip, H.M.D.; Rosa, B.; Yang, G.-Z. A wearable multisensing patch for continuous sweat monitoring. *Biosens. Bioelectron.* **2017**, *93*, 139–145. [CrossRef] [PubMed]
157. Kim, J.; Jeerapan, I.; Imani, S.; Cho, T.N.; Bandodkar, A.; Cinti, S.; Mercier, P.P.; Wang, J. Noninvasive Alcohol Monitoring Using a Wearable Tattoo-Based Iontophoretic-Biosensing System. *ACS Sens.* **2016**, *1*, 1011–1019. [CrossRef]
158. Gamella, M.; Campuzano, S.; Manso, J.; Gonzalez de Rivera, G.; Lopez-Colino, F.; Reviejo, A.J.; Pingarron, J.M. A novel non-invasive electrochemical biosensing device for in situ determination of the alcohol content in blood by monitoring ethanol in sweat. *Anal. Chim. Acta* **2014**, *806*, 1–7. [CrossRef]
159. Hauke, A.; Simmers, P.; Ojha, Y.R.; Cameron, B.D.; Ballweg, R.; Zhang, T.; Twine, N.; Brothers, M.; Gomez, E.; Heikenfeld, J. Complete validation of a continuous and blood-correlated sweat biosensing device with integrated sweat stimulation. *Lab Chip* **2018**, *18*, 3750–3759. [CrossRef]
160. Amjadi, M.; Pichitpajongkit, A.; Lee, S.; Ryu, S.; Park, I. Highly Stretchable and Sensitive Strain Sensor Based on Silver Nanowire–Elastomer Nanocomposite. *ACS Nano* **2014**, *8*, 5154–5163. [CrossRef] [PubMed]
161. Hasegawa, Y.; Shikida, M.; Ogura, D.; Suzuki, Y.; Sato, K. Fabrication of a wearable fabric tactile sensor produced by artificial hollow fiber. *J. Micromech. Microeng.* **2008**, *18*, 085014. [CrossRef]
162. Lam Po Tang, S. Recent developments in flexible wearable electronics for monitoring applications. *Trans. Inst. Meas. Control* **2007**, *29*, 283–300. [CrossRef]
163. Zaryanov, N.V.; Nikitina, V.N.; Karpova, E.V.; Karyakina, E.E.; Karyakin, A.A. Nonenzymatic Sensor for Lactate Detection in Human Sweat. *Anal. Chem.* **2017**, *89*, 11198–11202. [CrossRef]
164. Gao, W.; Nyein, H.; Shahpar, Z.; Tai, L.-C.; Wu, E.; Bariya, M.; Ota, H.; Fahad, H.; Chen, K.; Javey, A. Wearable Sweat Biosensors. In Proceedings of the 2016 IEEE International Electron Devices Meeting (IEDM), San Francisco, CA, USA, 3–7 December 2016.
165. Bhide, A.; Cheeran, S.; Muthukumar, S.; Prasad, S. Enzymatic low volume passive sweat based assays for mlti-bomarker detection. *Biosensors* **2019**, *9*, 13. [CrossRef]
166. Guyton, A.C.; Hall, J.E. *Textbook of Medical Physiology*; Elsevier Saunders: Philadelphia, PA, USA, 2006; p. 1116.

167. Gerich, J.E. Role of the kidney in normal glucose homeostasis and in the hyperglycaemia of diabetes mellitus: Therapeutic implications. *Diabet. Med.* **2010**, *27*, 136–142. [CrossRef]
168. Walker, H.K.; Hall, W.D.; Hurst, J.W. *Clinical Methods: The History, Physical, and Laboratory Examinations*, 3rd ed.; Butterworths: Boston, MA, USA, 1990.
169. Hagen, T.; Korson, M.S.; Wolfsdorf, J.I. Urinary Lactate Excretion to Monitor the Efficacy of Treatment of Type I Glycogen Storage Disease. *Mol. Genet. Metab.* **2000**, *70*, 189–195. [CrossRef]
170. Bloch, E.; Sobotka, H. Urinary cholesterol in cancer. *J. Biol. Chem.* **1938**, *124*, 567–572.
171. Kobos, R.K.; Ramsey, T.A. Enzyme electrode system for oxalate determination utilizing oxalate decarboxylase immobilized on a carbon dioxide sensor. *Anal. Chim. Acta* **1980**, *121*, 111–118. [CrossRef]
172. Rubinstein, I.; Martin, C.R.; Bard, A.J. Electrogenerated chemiluminescent determination of oxalate. *Anal. Chem.* **1983**, *55*, 1580–1582. [CrossRef]
173. Westerman, M.P.; Zhang, Y.; McConnell, J.P.; Chezick, P.A.; Neelam, R.; Freels, S.; Feldman, L.S.; Allen, S.; Baridi, R.; Feldman, L.E.; et al. Ascorbate levels in red blood cells and urine in patients with sickle cell anemia. *Am. J. Hematol.* **2000**, *65*, 174–175. [CrossRef]
174. Rocchitta, G.; Spanu, A.; Babudieri, S.; Latte, G.; Madeddu, G.; Galleri, G.; Nuvoli, S.; Bagella, P.; Demartis, M.I.; Fiore, V.; et al. Enzyme Biosensors for Biomedical Applications: Strategies for Safeguarding Analytical Performances in Biological Fluids. *Sensors* **2016**, *16*, 780. [CrossRef]
175. Wilson, R.T.; Turner, A.P.F. Glucose oxidase: An ideal enzyme. *Biosens. Bioelectron.* **1992**, *7*, 165–185. [CrossRef]
176. Falk, M.; Cirovic, S.; Falkman, P.; Sjoholm, J.; Hydbom, O.; Shleev, S. Wireless biosensing of analytes in human urine: Towards smart electronic diapers. *J. Biosens. Biomark. Diagn.* **2019**, *4*, 1–6.
177. Shieh, P. Non-Invasive Glucose Biosensor: Determination of Glucose in Urine. U.S. Patent No 5,876,952, 8 December 1999.
178. Hao, W.; Das, G.; Yoon, H.H. Fabrication of an amperometric urea biosensor using urease and metal catalysts immobilized by a polyion complex. *J. Electroanal. Chem.* **2015**, *747*, 143–148. [CrossRef]
179. Velichkova, Y.; Ivanov, Y.; Marinov, I.; Ramesh, R.; Kamini, N.R.; Dimcheva, N.; Horozova, E.; Godjevargova, T. Amperometric electrode for determination of urea using electrodeposited rhodium and immobilized urease. *J. Mol. Catal. B-Enzym.* **2011**, *69*, 168–175. [CrossRef]
180. Guilbault, G.G.; Montalvo, J.G., Jr. A urea-specific enzyme electrode. *J. Am. Chem. Soc.* **1969**, *91*, 2164–2165. [CrossRef]
181. Mach, K.E.; Wong, P.K.; Liao, J.C. Biosensor diagnosis of urinary tract infections: A path to better treatment? *Trends Pharmacol. Sci.* **2011**, *32*, 330–336. [CrossRef]
182. Pan, Y.; Sonn, G.A.; Sin, M.L.Y.; Mach, K.E.; Shih, M.-C.; Gau, V.; Wong, P.K.; Liao, J.C. Electrochemical immunosensor detection of urinary lactoferrin in clinical samples for urinary tract infection diagnosis. *Biosens. Bioelectron.* **2010**, *26*, 649–654. [CrossRef]
183. Chaki, N.K.; Vijayamohanan, K. Self-assembled monolayers as a tunable platform for biosensor applications. *Biosens. Bioelectron.* **2002**, *17*, 1–12. [CrossRef]
184. Liao, J.C.; Mastali, M.; Gau, V.; Suchard, M.A.; Moller, A.K.; Bruckner, D.A.; Babbitt, J.T.; Li, Y.; Gornbein, J.; Landaw, E.M.; et al. Use of electrochemical DNA biosensors for rapid molecular identification of uropathogens in clinical urine specimens. *J. Clin. Microbiol.* **2006**, *44*, 561–570. [CrossRef]
185. Mach, K.E.; Du, C.B.; Phull, H.; Haake, D.A.; Shih, M.-C.; Baron, E.J.; Liao, J.C. Multiplex pathogen identification for polymicrobial urinary tract infections using biosensor technology: A prospective clinical study. *J. Urol.* **2009**, *182 Pt 1*, 2735–2741. [CrossRef]
186. Sin, M.L.Y.; Gau, V.; Liao, J.C.; Wong, P.K. A universal electrode approach for automated electrochemical molecular analyses. *J. Microelectromech. Syst.* **2013**, *22*, 1126–1132. [CrossRef]
187. Larpant, N.; Pham, A.D.; Shafaat, A.; Gonzalez-Martinez, J.F.; Sotres, J.; Sjoeholm, J.; Laiwattanapaisal, W.; Faridbod, F.; Ganjali, M.R.; Arnebrant, T.; et al. Sensing by wireless reading Ag/AgCl redox conversion on RFID tag: Universal, battery-less biosensor design. *Sci. Rep.* **2019**, *9*, 1–9. [CrossRef]

188. Ruzgas, T.; Larpant, N.; Shafaat, A.; Sotres, J. Wireless, battery-less biosensors based on direct electron transfer reactions. *ChemElectroChem* **2019**, *6*, 5167–5171. [CrossRef]

Publisher's Note: MDPI stays neutral with regard to jurisdictional claims in published maps and institutional affiliations.

© 2020 by the authors. Licensee MDPI, Basel, Switzerland. This article is an open access article distributed under the terms and conditions of the Creative Commons Attribution (CC BY) license (http://creativecommons.org/licenses/by/4.0/).

Perspective

Recent Advances in Noninvasive Biosensors for Forensics, Biometrics, and Cybersecurity

Leif K. McGoldrick [1] and Jan Halámek [2,*]

1. Department of Chemistry, University at Albany, State University of New York, 1400 Washington Ave, Albany, NY 12222, USA; lmcgoldrick@albany.edu
2. Department of Environmental Toxicology (ENTX), Texas Tech University, 1207 Gilbert Drive, Lubbock, TX 79416, USA
* Correspondence: jan.halamek@ttu.edu

Received: 27 September 2020; Accepted: 20 October 2020; Published: 22 October 2020

Abstract: Recently, biosensors have been used in an increasing number of different fields and disciplines due to their wide applicability, reproducibility, and selectivity. Three large disciplines in which this has become relevant has been the forensic, biometric, and cybersecurity fields. The call for novel noninvasive biosensors for these three applications has been a focus of research in these fields. Recent advances in these three areas has relied on the use of biosensors based on primarily colorimetric assays based on bioaffinity interactions utilizing enzymatic assays. In forensics, the use of different bodily fluids for metabolite analysis provides an alternative to the use of DNA to avoid the backlog that is currently the main issue with DNA analysis by providing worthwhile information about the originator. In biometrics, the use of sweat-based systems for user authentication has been developed as a proof-of-concept design utilizing the levels of different metabolites found in sweat. Lastly, biosensor assays have been developed as a proof-of-concept for combination with cybersecurity, primarily cryptography, for the encryption and protection of data and messages.

Keywords: biosensors; forensics; biometrics; cybersecurity; fingerprints; sweat; blood; cipher

---

1. Background

Biosensors are widely used in multiple processes today. These include, but are not limited to, clinical diagnostics [1–10], environmental processes [11–13], the food industry [13–18], and devices for military use [15,19,20]. More recently, the use of biosensors has been noted in other disciplines, namely forensics, biometrics, and cybersecurity. As biosensors are devices that employ sensing techniques relying on biorecognition elements, they are able to provide specific, rapid results pertaining to bioaffinity-based reactions. The use of biosensors in forensics enables investigators to have another source of information in addition to DNA analysis that also provides worthwhile information quickly for them to narrow down their investigation in a timely manner. Biometrics and biosensors are becoming more closely related as the technology improves in that field. The differentiation of people with more noninvasive biosensors, biosensors that do not involve intrusive procedures, is exceedingly useful. The main procedures used here involve electrochemical and enzymatic assays for analysis. Lastly, with the advent of computers, the use of biosensors in unconventional computing [21] and the combination of computing with chemistry, biology, and physics have become another facet for biosensors.

Forensics

In the realm of forensic science, there is an increasing need for new technologies to aid investigators and lab scientists in the pursuit of gathering worthwhile information from evidence. There are many subsections of evidence that are pertinent in the forensic field, but recent research has focused on three: Fingerprints, blood samples, and sweat-based field testing for ethanol and other drugs.

In the history of forensics, fingerprints have been essential in addition to being a widely emblematic feature of the forensic field in pop culture. However, fingerprints are mainly used in the field as a comparative means of identification [22], and if a print cannot be utilized for this pictorial comparison based on ridge structure, size, and shape, it is treated as exclusionary evidence [23]. This can be viewed as a large limitation on the amount of data that one can gain from this relevant piece of evidence. By analyzing the content of a print, namely the amino acids, one can gain some understanding of who the donor of that print was and be able to narrow down the search for the investigators. This is due to the metabolic [24] differences [25–28] in people due to their gender, age, medications, and lifestyle. By analyzing these types of biomarkers in a fingerprint sample, it would not only allow for one to gain much needed information that would provide additional context for investigators, but would also lead to the reduction in the need to wait for the lengthy analysis of DNA that causes a backlog [29], if any was recovered. According to the NIJ, a backlog is defined as any evidence that was not analyzed for at least 30 days after submission to a laboratory. By analyzing the content of a fingerprint instead of the pictorial fingerprint commonly relied on, it allows for smudged or partial prints, the prints that would not provide ample evidence for comparison, to have value for investigators. The chemical content of fingerprints has been examined as well, mainly focused on laboratory-based equipment such as mass spectrometric (MS) techniques that focus on total fingerprint content [24,30,31], drugs of abuse [32,33], and fatty acids to differentiate individuals based on age [34]. In addition to MS, there were optical techniques, as well using spectrophotometric instrumentation with age differences based on lipids [35], visual representation [36], and explosive content found on prints [37], as well as combined techniques such as desorption electrospray ionization (DESI) and direct analysis in real-time mass spectrometry (DART-MS) for total content [38–40] and for pictorial [41–43]. Further analysis of these techniques can be found in a review in Trends in Analytical Chemistry [44].

In addition to fingerprints, blood is another matrix that is commonly obtained as evidence for forensic investigators. In addition to the commonly known DNA matching with a database [45,46], bloodstains are also used for splatter analysis [47–49], and there are even techniques to differentiate if there are multiple overlapping stains [50]. An important quality relating to the bloodstain that was missing was the time since it was deposited onto the surface, utilizing a technique that is practical and did not require great sample prep or laboratory instrumentation. This is a vital piece of information as it would allow for corroboration of stories told by possible witnesses and would enable the reduction in irrelevant and unnecessary lab work to be done with blood that is too fresh or too old at a particular scene. Another lapse in analysis of bloodstains is the determination of the age of the person the blood is from, as this can be done with DNA but is a lengthy process [51]. Similar studies using biosensors were to find biological sex and ethnicity [52,53]. Other lab techniques were attempted to be able to perform age deduction; however, they were not applicable, due to serious flaws in the techniques [54].

The third forensic matrix is sweat. Sweat is a viable forensic sample for multiple reasons, as it contains a small amount of DNA [55–58], and other metabolites and compounds [59–61], and people leave traces of it upon contact of surfaces with their skin [62]. As fingerprints contain sweat in addition to other components, it is a comparable matrix to what has been discussed previously. Sweat can be detected in the field utilizing one of the main components in sweat—lactate [63–67]. Lactate is prominent in sweat samples and, even though there are techniques that can be used in order to detect it [68–71], they require complex laboratory instrumentation and are not viable methodologies for on-site deployment. Some of these techniques rely on a tattoo-like sensing device that consists of a potentiometric sensor with a wireless receiver [70,71], involving the use of a screen-printed electrode on tattoo paper with a microfluidic channel for sweat collection and detection of analytes. By being able to find sweat on surfaces directly at a crime scene, investigators would have additional samples of viable evidence. Building upon this, emerging research has also looked at analysis involving ethanol. Multiple studies have shown that ethanol is excreted into sweat [72,73] and that sweat cannot be tampered with, similarly to how people "trick" breathalyzers. There are wearable technologies that allow for biosensing of metabolites in sweat [74,75]; however, they have a long delay, some up to two hours, that makes them useless for on-site Driving While Intoxicated (DWI) analysis. Similar research has looked into saliva for the identification of tetrahydrocannabinol (THC) [76] and an overall approach that analyzes multiple types of legal and illegal compounds in sweat [77].

Biometrics

The second main discipline has been the use of biosensors for strictly biometric purposes. Namely, current research has aimed to use bioassays for the identification and differentiation of individuals. As mentioned previously, the content of different metabolites in sweat can be quantified, which can be used to find differences between people. This has implications in both the prior forensics section, biometrics in general, and even cybersecurity, which will be explained below. This method can be used to provide an alternate to DNA testing similar to the other techniques being developed for forensics. This kind of technology is similar to emerging research in biosensors and bioelectronics [78,79]. As mentioned previously, the content within a person's sweat is a result of metabolic processes [24–27,59,80] related to what can be a person's identifying factors such as age, biological sex, diet, and activity level. By taking measurements of and comparing results for multiple biomarkers, one can differentiate a person from others with these fluctuating factors. This can be applied for the unlocking of smart devices as the technology is moving in this direction as well [81–83].

As mentioned previously, sweat has been an emerging source of information in both forensics and biometrics. Many studies have been done for the advancement of various methodologies for different compounds found in sweat [84–88]. Further descriptions of some of the recent methodologies that have been developed for sweat analysis can be seen in Table 1 below.

Table 1. Recent biosensor analysis techniques for sweat.

Protocol	Description	Technique	Analyte	LOD/Range	Ref.
Tattoo	Wearable skin tattoo for wireless signal transduction	Potentiometry	Sodium	0.1–100 mM	[71]
Tattoo	Wearable skin tattoo for pH monitoring	Potentiometry	pH	pH 3–7	[89]
Tattoo	Wearable skin tattoo for lactate monitoring	Potentiometry	Lactate	1–20 mM	[70]
Tattoo	Wearable skin tattoo for monitoring	Potentiometry	Ammonium	10^{-4}–0.1 M	[90]
Superwettable bands	Multiplex method for on-body sampling	Colorimetric	pH chloride glucose calcium	pH 4.5–7 0–100 mM 0–15 mM 0–15 mM	[91]
Screen-printed electrode	Monitoring of cystic fibrosis patients	Potentiometry	Chloride	2.7×10^{-5} mol/L	[92]
Janus textile bands	Multiplex method for on body sampling	Potentiometry	Glucose Lactate Potassium Sodium	18–40 µM 10 mM 0.3–6.3 mM 60 mM	[93]
Wearable sensor	Stretchable, skin-attachable sweat sensor	Potentiometry, Carbon nanotubes, gold nanosheets	Glucose pH	10.89 µA mM^{-1} cm^{-2} 71.44 mV pH^{-1}	[94]
Graphene electrochemical	Diabetes monitoring	Gold-doped graphene	Glucose	10 µM–0.7 mM	[95]
Microfluidic wearable sensor	Multiplex analysis for sensing in sweat	Colorimetric	Lactate Chloride Creatine pH Glucose	0–100 mM 0–mM 0–1000 µM pH 5–8.5 0–25 mM	[96]
liquigel	Organic electrochemical transistor	Transistor	Lactate	0.3–1.3 mM	[97]
Direct iontophoresis	Sweat extraction and electrochemical analysis using smartphone	Potentiometry	Glucose Chloride	0–100 µM 20–80 mM	[98]

Table 1. Cont.

Protocol	Description	Technique	Analyte	LOD/Range	Ref.
Free amino acid analysis	Eccrine sweat amino acid composition	Cation chromatography and amino acid analyzer, GC-MS	Amino acids	-	[99]
Wearable Sensor	Chemical electrocardiogram and simultaneous metabolite monitoring	Amperometry	Lactate	0–28 mM	[100]
microfluidic	Sweat collection and analysis for kidney disorders	Colorimetric	Creatine Urea pH	0–0.5 mM 0–250 mM pH 5–7	[101]
Wearable sensor	Integrated multiplex array for sweat analysis	Amperometry	Sodium Potassium Glucose lactate	20–120 mM 2–16 mM 0–200 µM 2–30 mM	[102]
Wearable sensor	Monitoring for cystic fibrosis patients for sodium concentration	Potentiometry Atomic Absorption	Sodium	20–100 mM	[103]
Watch sensor	Monitoring of sodium levels	Potentiometry	Sodium	10^{-4}–10^{-1} M	[104]
Tattoo	Wearable skin tattoo for alcohol monitoring in sweat	Amperometry	Alcohol	0–36 mM	[74]
Wearable sensor	Drug monitoring via differential pulse	Voltammetry	Caffeine	0–40 × µM	[105]
Wearable sensor	Detection of THC and Alcohol	Voltammetry Amperometry	THC Alcohol	0.5 µM 0.1–1 mM	[76]

Cybersecurity

The third and final discipline has been the use of biosensors for cybersecurity purposes. This is a small transition from biometrics into cybersecurity as they are closely related. This research can be applied in two different areas of cybersecurity—authentication and cryptography. Both are important in our world with the advent of the digital age, so there is a call for innovative and worthwhile technologies across all disciplines to innovate and advance the novel research into cyber technologies. For authentication, biometrics that were mentioned previously can be applied. Ideally, if one can differentiate people, the same technology can be used to identify a person, or at the very least, dismiss an imposter. Cryptography is the use of codes and cyphers in order to encrypt data to keep them safe, either in transmission between people as a message or safekeeping in storage [106–108]. Many multidisciplinary researchers have been applying their research to encryption, including, but not limited to, fluorescence [109–117], nuclear magnetic resonance (NMR) [118], bacteria [119], antibodies [120], and molecular computing systems [121–124], with the heaviest research in DNA applications [125–133].

2. Research

Forensics

First, in the forensic field, emerging research has focused on the three areas of study above with fingerprints [44,134–137], blood [138,139], sweat [140,141], and one general review on the use of biocomputing in forensics [142]. These five fingerprint papers provide novel applications of biorecognition elements that can be used for future biosensing devices. These papers are centralized on the idea that people have different levels of certain L-amino acids, which are related to their metabolism and different traits, allowing them to be differentiated into groups. The blood papers use the degradation of certain enzymes in order to determine the time that a blood spot has been outside the body and to identify the age of the originator from the level of a separate enzyme. The sweat papers use the levels of different compounds in sweat in order to identify sweat and to be able to provide an alternative technique for alcohol intoxication. The methodologies within these papers use different biosensors, some via enzyme assays, some with chemical reactions. A generic enzymatic assay diagram can be seen in Figure 1, where specific substrates for the enzyme are used in the assay to produce byproducts, one of which is a recognition element. The last paper is a review on other trending types of fingerprint analysis such as the use of mass spectrometry, spectroscopy, nanotechnology, and combinatorial methods [44].

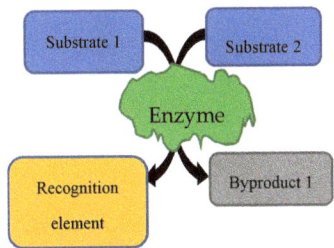

Figure 1. Generic enzymatic assay example.

A fingerprint paper from 2015 uses an enzymatic cascade utilizing L-amino acid oxidase (L-AAO) and horseradish peroxidase (HRP) in the presence of L-amino acids to produce a visible color by oxidizing the redox dye o-dianisidine, which results in a color that may be analyzed at 436 nm [134]. Due to the fact that women produce a higher amount of these amino acids than men [26–28], this assay would allow for the determination of a person's biological sex. First, the assay was performed utilizing 50 mimicked samples, 25 male and 25 female, containing 23 target amino acids in concentrations based on reported values calculated in R-project software. This resulted in the area under the curve (AUC) of a receiver operating characteristic (ROC) curve of 99%, showing a high probability for this methodology to differentiate between the two sexes. This was then repeated with three male and three female volunteers testing both their right and left thumbs in an acid extraction methodology using polyurethane film as a medium that was developed in the same article. These real samples also show definitive differentiation between the male and female prints. Different surfaces around the lab, such as a computer screen and doorknob, were also tested in this manuscript to show viability on different surfaces. An additional paper from 2016 follows the same principles as the L-AAO/HRP assay in order to find an alternative methodology to determine biological sex, this time utilizing a chemical assay [135]. This assay uses ninhydrin, which produces a colorimetric reaction with alpha-amino (α-NH$_2$) acids [143,144]. Ninhydrin is commonly used in forensics already as it produces Ruhemann's purple as a product in the presence of all α-NH$_2$ acids. The process in this paper is similar to the last: 50 total mimicked samples with designated concentrations of the 23 amino acids from R-project software are analyzed photometrically at 570 nm. The area under the ROC curve in this case was at 94%, showing a high probability to correctly distinguish the two sexes. After this, five male and five female volunteers were analyzed using this technique, resulting in a 91% area under the curve for authentic samples.

Following this trend, a paper from 2017 also uses a chemical assay for the determination of biological sex with the focus being on the Bradford reagent, Coomassie Brilliant Blue G-250 dye [136]. Bradford is commonly used for quantifying proteins and is less affected by reagents and nonprotein components of samples than other commonly used reagents [145]. The goal of utilizing this assay is that it only targets six specific amino acids with which to form a colorimetric complex, thus enabling a more focused approach to the determination of biological sex with the ideal being the determination of a single amino acid assay to differentiate the sexes. Contrary to the previous papers, only authentic fingerprints were sampled from 50 authentic fingerprint samples from volunteers—25 males and 25 females. This resulted in an area under the curve of the ROC graph of 99%, showing the highest probability of the three methods to correctly identify the biological sex of the fingerprint donor.

The final fingerprint paper to be examined goes one step further, by using two separate tests, each targeting a specific amino acid, in order to differentiate on the basis of biological sex [137]. The two methods used are an enzymatic cascade targeting alanine and a chemical assay that targets arginine. The alanine-targeting assay consists of a three-enzyme cascade with alanine transaminase, pyruvate oxidase, and horseradish peroxidase [146]. In the presence of alanine in addition to the other substrates necessary for the assay, a redox dye is oxidized by HRP and can be spectrophotometrically measured. Following the previously established standard, even though the mimicked samples produced the lowest AUC of the ROC curve with a value of 82%, the authentic samples of 50 total individuals was vastly improved at 99.8%. The second technique in this research consisted of the application of the Sakaguchi Test [147], which involves α-naphthol, NaOH, and sodium hypobromite in order to form a red-colored complex. As in the previous experiments, both mimicked and authentic samples were tested, resulting in both AUCs being 100%.

In order to apply these types of biosensor techniques to other bodily fluids that are forensically relevant, blood is another focus of research. The main topic that research on blood is centered around is the estimation of the time since deposition, TSD, of a blood spot. The first paper focuses on this by measuring the levels of two biomarkers in blood, citrate kinase (CK) and alanine transaminase (ALT), which denature with the passage of time of up to 5 days [138]. The CK assay involves creatine and adenosine triphosphate as substrates for the first enzyme, CK, in a three-enzyme cascade utilizing pyruvate kinase (PK) and lactate dehydrogenase (LDH) as the other two enzymes. LDH, the third enzyme in the cascade, is the enzyme involved with production of the recognition element β-nicotinamide adenine dinucleotide from β-nicotinamide adenine dinucleotide reduced, which produces a reduction in signal at 340 nm. ALT is a two-enzyme assay that also utilizes LDH to allow for simultaneous determination. ALT recognizes the substrates alanine and α-ketoglutaric acid. By using a two-analyte system, this provides a more reliable system of determination as it has parallel markers being analyzed compared to a single marker. Building upon this, the technique outlined in the second paper uses one assay to determine not only the TSD, but also an estimation of the age of the source individual [139]. By looking at alkaline phosphatase (ALP), this research achieves both goals. This is due to the fact that ALP is a commonly used biomarker in clinical diagnostics for bone growth that relates directly to the age of the individual. To measure this, ALP converts the substrate p-nitrophenol phosphate into p-nitrophenol, which also acts as the recognition element as p-nitrophenol is observable at 405 nm. As ALP is a biomarker for bone growth, and that it degenerates over time when out of the body, one can obtain data about the relative age of the blood donor and the TSD up to 2 days. For the analysis of this methodology, 100 samples were prepared via the R-project software mentioned previously, split evenly between young and old, males and females. The samples provided an AUC for the ROC curve of 99% for males and 100% for the female group in differentiating between old and young.

The third forensic medium that is being researched currently is sweat. As previously mentioned, sweat has many forensic applications but is difficult to identify at a crime scene. To this end, a novel methodology in order to identify sweat by use of a biosensor strip based on the detection of lactate was developed [140]. This method utilized an enzymatic assay in order to detect lactate, a major component in sweat. The assay used involved a two-enzyme cascade of lactate oxidase (LOx) and horseradish peroxidase (HRP). LOx involves the substrates lactate and oxygen, which are used to produce hydrogen peroxide. This hydrogen peroxide is then used by HRP as mentioned previously with a redox dye to produce a signal. This methodology was able to detect sweat with minimal decay for up to two weeks and at low amounts of sweat: Around 50 nL. This technique was even applied onto a paper strip modified with polystyrene for use as a field-deployable device. This optical strip provides a binary YES/NO for the presence of sweat via a color change, which is ideal for preliminary detection of sweat that can itself be analyzed further. Additionally, sweat was further examined in a noninvasive testing methodology for ethanol sensing on the surface of one's skin [141]. This conceptualizes an alternative method to breathalyzers by relying on an enzymatic assay involving alcohol oxidase (AOx) and HRP that is quantified not only by UV-Vis spectrophotometry but also an optical camera. AOx uses ethanol and oxygen as substrates to produce hydrogen peroxide, which is used by HRP as mentioned previously. This research shows that there is a correlation between both techniques and the currently used breathalyzer. The data were achieved from a 26-volunteer drinking study with people of different ages, biological sexes, and food habits. The sweat samples were obtained through pilocarpine electrophoresis similar to the Gibson and Cooke method [148], which allowed the sweat to be collected in gauze pads and analyzed. A minimum of 3 µL of sweat was required for this method.

These advances with biosensors in the field of forensics have produced a viable way for investigators to receive some information to pursue leads if DNA evidence is backlogged or not applicable using sweat and blood evidence found at the crime scene. Multiple enzymatic assays were developed for the differentiation of biological sex of an individual from fingerprint content, enabling an alternative or additional analysis for fingerprints depending on the clarity of the print for pictorial analysis. Blood was examined, and provides a viable methodology in order to show the time since deposition, TSD, and also an estimated age of the originator of the blood spot. Additionally, a field-deployable testing strip was developed for the determination of sweat, supplying clarity for a difficult-to-detect bodily fluid at crime scenes. Sweat was also tested in a laboratory setting for an alternative to breathalyzers for the detection and quantification of alcohol in sweat utilizing enzymatic assays and a colorimetric response. In the future, different drugs and illicit substances, such as THC, in addition to a more broad analysis of metabolites characteristic to certain habits or biological features can be examined for their use in forensics for providing a more deterministic and rapid analysis for forensic and law enforcement personnel in a sweeping suite of biosensor devices.

Biometrics

Currently, biometrics is another avenue of interest for biosensors [149]. This paper from Hair et al. uses the levels of three metabolites found in sweat in order to differentiate people. This analysis is performed using three enzymatic assays that each target the metabolites: Lactate, urea, and glutamate. The assay for lactate is the same one that was outlined in the sweat paper involving the paper strips. The assay for urea involves the enzymes urease and glutamate dehydrogenase. The final assay that is used for glutamate involves only glutamate dehydrogenase. These assays could be measured spectrophotometrically using either a redox dye in LOx/HRP or with conversions involving NADH and NAD^+ in the other two. First, 50 mimicked sweat samples were run and compared, which showed that this method was viable as there was no overlap between the samples. Additionally, 25 authentic samples were analyzed, where the sweat was collected according to the same procedure as the Gibson and Cooke method mentioned above [148]. A multivariate analysis of variance statistic test (MANOVA), was performed for both the authentic and mimicked sample sets to determine if the combination of the three analytes were truly unique. Both sets produced p-values of <0.001 each. In addition, six ANOVA tests were performed between each analyte for the mimicked and authentic samples. All six of these tests also produced p-values of <0.001. These statistical values show that there is not only a significant statistical difference between individuals in the combination of the three analytes, but also a significant difference between each individual for a single analyte. The analysis shows that these three metabolites can be used in order to produce an individual's "sweat profile," enabling differentiation between individuals. The implications for this research are multidisciplinary as there are many forensic, cybersecurity, and point-of-care diagnostic applications that would benefit from this method.

This use of sweat content for biometric purposes is progressing but further research needs to be done for biometrics to be a reliable form of authentication using biosensor methodologies. The main future aspect that would need to be studied is a long-term study relating to the monitoring of the levels of the chosen metabolites in people and how the levels fluctuate over time relating to different factors such as stress, diet, and other day-to-day habits. For higher security when used for authentication, especially for higher-security systems and cybersecurity, additional metabolites would need to also be concurrently measured as well. This monitoring process would not only assist in the future of biometrics but also in existing disciplines such as clinical diagnostics.

Cybersecurity

Lastly, the use of biosensors for cybersecurity is a growing trend. The use of sensors for authentication of an individual and a novel methodology related to cryptography were recently developed [150,151]. The first paper represents a review with the aim of introducing a multi-assay wearable biosensor that would provide continuous tracking of a person's sweat metabolites for authentication purposes. This review looks at many of the assays previously mentioned: ALT/LDH, ALT/POx/HRP, and GlDH, and some that were not mentioned: Alanine and glutamate assay with ALT, glutamate oxidase, and HRP; aspartate using the aspartate transaminase enzyme; and a combined version with all three of these new analytes in a single assay. By monitoring these assays, one can produce output data that would be beneficial in the authentication of a person with many cybersecurity applications. The second paper illustrates the use of three enzyme assays in order to encrypt a short message using a basic cipher [151]. The three enzymatic assays used involved HRP, lysozyme, and ALP. HRP and ALP were used as previously mentioned. Lysozyme breaks down cell walls from cells added during the assay to produce a reduction in signal at 450 nm, acting as the recognition element. The data resulting from the colorimetric assays are used in the encryption of a message. Provided that the receiver of the message performs the same experiment under the same conditions, the message will be properly decrypted. The data from these enzymatic assays act as "keys" that one use in order to lock and unlock data in relation to encryption.

This brief combination of cryptology and biosensors can have a large impact on the future of user authentication, cryptography, and unconventional computing as a whole. The processes outlined here can be combined with biometrics for user authentication, which is considered just as, or more, important compared to data security through encryption. In cryptography, the further research of other biosensor systems combined with stronger and more robust encryption methods can lead to the advancement of these systems to be used instead of random number generators, which have been controversial for use in cryptography since their inception [152]. Going even further, biosensors can be further researched for direct encryption of data to provide an alternative to the widely researched encryption via DNA [94–102,122–126,153–157].

3. Conclusions

Current research in biosensors has led to advancement in the use of biosensors on three fronts: Forensics, biometrics, and cybersecurity. In the field of forensics, the use of fingerprint material has been demonstrated to be capable of being used to determine the biological sex of a person through multiple methods. These methods relied on the detection of certain amino acids, some methods consisted of a broad detection of 23 amino acids, and some methods were much more selective and targeted down to a single amino acid. Additionally, the use of another medium, blood, allows one to deterministically estimate the amount of time that the blood has been outside the body for up to 2 days by analyzing the degradation of enzymes present in blood. Additionally, it was found that a single assay was able to not only estimate this time since deposition, but to also estimate the age of the originator. Sweat has been the third major medium in forensic endeavors, building upon the fingerprint analysis as fingerprints contain sweat. A novel methodology for a lactate-detecting on-site testing strip was developed in order to identify the presence of sweat due to the high concentration of lactate in sweat. This method was highly sensitive and required extremely small volumes of sweat to produce a tangible response. Sweat has also been examined for other purposes, showing that one can detect the presence of ethanol in sweat, providing another method of determining intoxication levels besides a breathalyzer or invasive blood techniques, in addition to detecting other drugs of abuse. Sweat was further analyzed for biometric purposes by comparing the levels of three metabolites found in sweat to differentiate individuals. Lastly, methods involving biosensors for both the authentication of individuals and for cryptography were developed, benefitting two major establishments of cybersecurity.

In the future, biosensors can further fulfill the expansion of these three fields with additional research. In the field of forensics, a wider array of metabolites may be examined for use in a device that would analyze a certain body fluid and provide more information relating to the habits and identifying information of the originator in addition to the biological sex and age mentioned previously. Additionally, more research can be performed in order to provide for a broad testing kit with higher accuracy and precision for various compounds, illicit and legal, for use in roadside testing to aid law enforcement officers. Biometrics and biosensors are closely related, as shown by the research seen here. Further analysis utilizing monitoring and other metabolite tracking will reinforce not only the strengths in the use of this methodology but also possibly reduce or remove the current unknowns and limitations for this method. Lastly, the use of different bioaffinity-based assay systems for cryptography for the use of different cipher systems will provide a reliable alternative to the random number generator systems used in cryptography today. This work in cybersecurity can also be combined with biometrics for user authentication for digital and evolving systems. Biosensors have been an important facet in the fields of clinical diagnostics, environmental processes, and military devices, and is a strong emerging technique in the fields of forensic science, biometrics, and cybersecurity. In these three fields, biosensors have produced considerable results thus far and have an auspicious future for further research.

Funding: This research was funded by NIJ grant number 2016-DN-BX-0188.

Conflicts of Interest: The authors declare no conflict of interest.

References

1. D'Orazio, P. Biosensors in clinical chemistry—2011 update. *Clin. Chim. Acta* **2011**, *412*, 1749–1761. [CrossRef] [PubMed]
2. Burcu Bahadir, E.; Kemal Sezgintürk, M. Applications of electrochemical immunosensors for early clinical diagnostics. *Talanta* **2015**, *132*, 162–174. [CrossRef] [PubMed]
3. Bianchi, V.; Guerra, C.; De munari, I.; Ciampolini, P. Wearable sensors for behavioral assessment. *Gerontechnology* **2016**. [CrossRef]
4. Guilbault, G.G.; Palleschi, G.; Lubrano, G. Non-invasive biosensors in clinical analysis. *Biosens. Bioelectron.* **1995**, *10*, 379–392. [CrossRef]
5. Guilbault, G.G.; Pravda, M.; Kreuzer, M.; O'Sullivan, C.K. Biosensors—42 Years and counting. *Anal. Lett.* **2004**, *37*, 1481–1496. [CrossRef]
6. Ferapontova, E.E. DNA Electrochemistry and Electrochemical Sensors for Nucleic Acids. *Annu. Rev. Anal. Chem.* **2018**, *11*, 197–218. [CrossRef]
7. Sanghavi, B.J.; Wolfbeis, O.S.; Hirsch, T.; Swami, N.S. Nanomaterial-based electrochemical sensing of neurological drugs and neurotransmitters. *Microchim. Acta* **2015**, *182*, 1–41. [CrossRef]
8. Shen, Y.; Tran, T.T.; Modha, S.; Tsutsui, H.; Mulchandani, A. A paper-based chemiresistive biosensor employing single-walled carbon nanotubes for low-cost, point-of-care detection. *Biosens. Bioelectron.* **2019**, *130*, 367–373. [CrossRef]
9. García-Arellano, H.; Fink, D.; Muñoz Hernández, G.; Vacík, J.; Hnatowicz, V.; Alfonta, L. Nuclear track-based biosensors with the enzyme laccase. *Appl. Surf. Sci.* **2014**, *310*, 66–76. [CrossRef]
10. Tzouvadaki, I.; De Micheli, G.; Carrara, S. Memristive Biosensors for Ultrasensitive Diagnostics and Therapeutics. In *Springer Series in Advanced Microelectronics*; Springer: Singapore, 2020.
11. Volkov, A.G.; Volkova-Gugeshashvili, M.I.; Osei, A.J. Plants as environmental biosensors: Non-invasive monitoring techniques. In Proceedings of the Appropriate Technologies for Environmental Protection in the Developing World—Selected Papers from ERTEP 2007, Ghana, Africa, 17–19 July 2007.
12. Somerset, V. *Environmental Biosensors*; InTech: Rijeka, Croatia, 2011; ISBN 978-953-307-486-3.
13. Van Dorst, B.; Mehta, J.; Bekaert, K.; Rouah-Martin, E.; De Coen, W.; Dubruel, P.; Blust, R.; Robbens, J. Recent advances in recognition elements of food and environmental biosensors: A review. *Biosens. Bioelectron.* **2010**, *26*, 1178–1194. [CrossRef]
14. Bollella, P.; Hibino, Y.; Kano, K.; Gorton, L.; Antiochia, R. Highly Sensitive Membraneless Fructose Biosensor Based on Fructose Dehydrogenase Immobilized onto Aryl Thiol Modified Highly Porous Gold Electrode: Characterization and Application in Food Samples. *Anal. Chem.* **2018**, *90*, 12131–12136. [CrossRef] [PubMed]

15. Bahadir, E.B.; Sezgintürk, M.K. Applications of commercial biosensors in clinical, food, environmental, and biothreat/biowarfare analyses. *Anal. Biochem.* **2015**, *478*, 107–120. [CrossRef]
16. Alocilja, E.C.; Radke, S.M. Market analysis of biosensors for food safety. *Biosens. Bioelectron.* **2003**, *18*, 841–846. [CrossRef]
17. Thakur, M.S.; Ragavan, K.V. Biosensors in food processing. *J. Food Sci. Technol.* **2013**, *50*, 625–641. [CrossRef] [PubMed]
18. Siepenkoetter, T.; Salaj-Kosla, U.; Magner, E. The Immobilization of Fructose Dehydrogenase on Nanoporous Gold Electrodes for the Detection of Fructose. *ChemElectroChem* **2017**, *4*, 905–912. [CrossRef]
19. Shi, H.; Zhao, H.; Liu, Y.; Gao, W.; Dou, S.C. Systematic analysis of a military wearable device based on a multi-level fusion framework: Research directions. *Sensors (Switzerland)* **2019**, *19*, 2651. [CrossRef] [PubMed]
20. Pohanka, M.; Jun, D.; Kuca, K. Amperometric biosensors for real time assays of organophosphates. *Sensors* **2008**, *8*, 5303–5312. [CrossRef]
21. Andrew, A.M. Unconventional computing. *Kybernetes* **2012**, *41*, 518–522. [CrossRef]
22. Yamashita, B.; French, M.; Bleay, S.; Cantu, A.; Inlow, V.; Ramotowski, R.; Sears, V.; Wakefield, M. Latent Print Development-Chapter 7 Fingerprint Sourcebook. In *The Fingerprint Sourcebook*; U.S. Department of Justice: Washington, DC, USA, 2010.
23. Francese, S.; Bradshaw, R.; Ferguson, L.S.; Wolstenholme, R.; Clench, M.R.; Bleay, S. Beyond the ridge pattern: Multi-informative analysis of latent fingermarks by MALDI mass spectrometry. *Analyst* **2013**, *138*, 4215–4228. [CrossRef]
24. Thody, A.J.; Shuster, S. Control and function of sebaceous glands. *Physiol. Rev.* **1989**, *59*, 383–416. [CrossRef]
25. Beskaravainy, P.M.; Molchanov, M.V.; Suslikov, A.V.; Paskevich, S.I.; Kutyshenko, V.P.; Vorob'ev, S.I. NMR study of human biological fluids for detection of pathologies. *Biomeditsinskaya Khim.* **2015**, *61*, 141–149. [CrossRef] [PubMed]
26. Hier, S.W.; Cornbleet, T.; Bergeim, O. The amino acids of human sweat. *J. Biol. Chem.* **1946**, *166*, 327–333. [PubMed]
27. Coltman, C.A.; Rowe, N.J.; Atwell, R.J. The amino acid content of sweat in normal adults. *Am. J. Clin. Nutr.* **1966**, *18*, 373–378. [CrossRef] [PubMed]
28. Croxton, R.S.; Baron, M.G.; Butler, D.; Kent, T.; Sears, V.G. Variation in amino acid and lipid composition of latent fingerprints. *Forensic Sci. Int.* **2010**, *199*, 93–102. [CrossRef]
29. NIJ DNA Backlog. Available online: https://www.ncjrs.gov/App/Publications/abstract.aspx?ID=259066 (accessed on 15 September 2020).
30. Hartzell-Baguley, B.; Hipp, R.E.; Morgan, N.R.; Morgan, S.L. Chemical composition of latent fingerprints by gas chromatography-mass spectrometry. An experiment for an instrumental analysis course. *J. Chem. Educ.* **2007**, *84*, 689. [CrossRef]
31. Archer, N.E.; Charles, Y.; Elliott, J.A.; Jickells, S. Changes in the lipid composition of latent fingerprint residue with time after deposition on a surface. *Forensic Sci. Int.* **2005**, *154*, 224–239. [CrossRef]
32. Jacob, S.; Jickells, S.; Wolff, K.; Smith, N. Drug Testing by Chemical Analysis of Fingerprint Deposits from Methadone- Maintained Opioid Dependent Patients Using UPLC-MS/MS. *Drug Metab. Lett.* **2008**. [CrossRef]
33. Goucher, E.; Kicman, A.; Smith, N.; Jickells, S. The detection and quantification of lorazepam and its 3-O-glucuronide in fingerprint deposits by LC-MS/MS. *J. Sep. Sci.* **2009**. [CrossRef]
34. Michalski, S.; Shaler, R.; Dorman, F.L. The Evaluation of Fatty Acid Ratios in Latent Fingermarks by Gas Chromatography/Mass Spectrometry (GC/MS) Analysis. *J. Forensic Sci.* **2013**. [CrossRef]
35. Antoine, K.M.; Mortazavi, S.; Miller, A.D.; Miller, L.M. Chemical differences are observed in children's versus adults' latent fingerprints as a function of time. *J. Forensic Sci.* **2010**. [CrossRef]
36. Ricci, C.; Phiriyavityopas, P.; Curum, N.; Chan, K.L.A.; Jickells, S.; Kazarian, S.G. Chemical imaging of latent fingerprint residues. *Appl. Spectrosc.* **2007**. [CrossRef] [PubMed]
37. Mou, Y.; Rabalais, J.W. Detection and Identification of Explosive Particles in Fingerprints Using Attenuated Total Reflection-Fourier Transform Infrared Spectromicroscopy. *J. Forensic Sci.* **2009**. [CrossRef] [PubMed]
38. Morelato, M.; Beavis, A.; Kirkbride, P.; Roux, C. Forensic applications of desorption electrospray ionisation mass spectrometry (DESI-MS). *Forensic Sci. Int.* **2013**, *226*, 10–21. [CrossRef] [PubMed]
39. Ifa, D.R.; Jackson, A.U.; Paglia, G.; Cooks, R.G. Forensic applications of ambient ionization mass spectrometry. *Anal. Bioanal. Chem.* **2009**, *394*, 1995–2008. [CrossRef] [PubMed]

40. Hazarika, P.; Jickells, S.M.; Wolff, K.; Russell, D.A. Multiplexed detection of metabolites of narcotic drugs from a single latent fingermark. *Anal. Chem.* **2010**. [CrossRef] [PubMed]
41. van Dam, A.; Aalders, M.C.G.; de Puit, M.; Gorré, S.M.; Irmak, D.; van Leeuwen, T.G.; Lambrechts, S.A.G. Immunolabeling and the compatibility with a variety of fingermark development techniques. *Sci. Justice* **2014**. [CrossRef] [PubMed]
42. Bradshaw, R.; Rao, W.; Wolstenholme, R.; Clench, M.R.; Bleay, S.; Francese, S. Separation of overlapping fingermarks by Matrix Assisted Laser Desorption Ionisation Mass Spectrometry Imaging. *Forensic Sci. Int.* **2012**. [CrossRef]
43. Wolstenholme, R.; Bradshaw, R.; Clench, M.R.; Francese, S. Study of latent fingermarks by matrix-assisted laser desorption/ionisation mass spectrometry imaging of endogenous lipids. *Rapid Commun. Mass Spectrom.* **2009**. [CrossRef]
44. Huynh, C.; Halámek, J. Trends in fingerprint analysis. *TrAC—Trends Anal. Chem.* **2016**, *82*, 328–336. [CrossRef]
45. An, J.H.; Shin, K.J.; Yang, W.I.; Lee, H.Y. Body fluid identification in forensics. *BMB Rep.* **2012**, *45*, 545–553. [CrossRef]
46. Gill, P.; Jeffreys, A.J.; Werrett, D.J. Forensic application of DNA "fingerprints". *Nature* **1985**. [CrossRef]
47. Flight, C.; Jones, M.; Ballantyne, K.N. Determination of the maximum distance blood spatter travels from a vertical impact. *Forensic Sci. Int.* **2018**. [CrossRef]
48. Brettell, T.A.; Butler, J.M.; Saferstein, R. Forensic science. *Anal. Chem.* **2005**, *77*, 3839–3860. [CrossRef] [PubMed]
49. Erbisti, P.C.F.; Gardner, R.M. *Bloodstain Pattern Analysis with an Introduction to Crime Scene Reconstruction*; CRC Press: Boca Raton, FL, USA, 2008.
50. Elkins, K. *Forensic DNA Biology*, 3rd ed.; CRC Press: Boca Raton, FL, USA, 2013; ISBN 9780123945853.
51. Bocklandt, S.; Lin, W.; Sehl, M.E.; Sánchez, F.J.; Sinsheimer, J.S.; Horvath, S.; Vilain, E. Epigenetic predictor of age. *PLoS ONE* **2011**, *6*, e41821. [CrossRef] [PubMed]
52. Kramer, F.; Halámková, L.; Poghossian, A.; Schöning, M.J.; Katz, E.; Halámek, J. Biocatalytic analysis of biomarkers for forensic identification of ethnicity between Caucasian and African American groups. *Analyst* **2013**, *138*, 6251–6257. [CrossRef] [PubMed]
53. Bakshi, S.; Halámková, L.; Halámek, J.; Katz, E. Biocatalytic analysis of biomarkers for forensic identification of gender. *Analyst* **2014**, *139*, 559–563. [CrossRef]
54. Meissner, C.; Ritz-Timme, S. Molecular pathology and age estimation. *Forensic Sci. Int.* **2010**, *203*, 34–43. [CrossRef]
55. Pizzamiglio, M.; Mameli, A.; Maugeri, G.; Garofano, L. Identifying the culprit from LCN DNA obtained from saliva and sweat traces linked to two different robberies and use of a database. *Int. Congr. Ser.* **2004**. [CrossRef]
56. Pizzamiglio, M.; Marino, A.; Portera, G.; My, D.; Bellino, C.; Garofano, L. Robotic DNA extraction system as a new way to process sweat traces rapidly and efficiently. *Int. Congr. Ser.* **2006**. [CrossRef]
57. Sikirzhytski, V.; Sikirzhytskaya, A.; Lednev, I.K. Multidimensional Raman spectroscopic signature of sweat and its potential application to forensic body fluid identification. *Anal. Chim. Acta* **2012**. [CrossRef]
58. Stouder, S.; Reubush, K.; Hobson, D.; Smith, J. Trace Evidence Scrapings: A Valuable Source of DNA? *Forensic Sci. Commun.* **2001**, *4*, 4.
59. Verde, T.; Shephard, R.J.; Corey, P.; Moore, R. Sweat composition in exercise and in heat. *J. Appl. Physiol. Respir. Environ. Exerc. Physiol.* **1982**. [CrossRef] [PubMed]
60. Patterson, M.J.; Galloway, S.D.R.; Nimmo, M.A. Variations in regional sweat composition in normal human males. *Exp. Physiol.* **2000**. [CrossRef] [PubMed]
61. Costa, F.; Calloway, D.H.; Margen, S. Regional and total body sweat composition of men fed controlled diets. *Am. J. Clin. Nutr.* **1969**. [CrossRef]
62. Pinson, E.A. Evaporation from human skin with sweat glands inactivated. *Am. J. Physiol. Content* **1942**. [CrossRef]
63. Åstrand, I. Lactate Content in Sweat. *Acta Physiol. Scand.* **1963**. [CrossRef]
64. Buono, M.J.; Lee, N.V.L.; Miller, P.W. The relationship between exercise intensity and the sweat lactate excretion rate. *J. Physiol. Sci.* **2010**. [CrossRef]
65. Kondoh, Y.; Kawase, M.; Ohmori, S. D-Lactate concentrations in blood, urine and sweat before and after exercise. *Eur. J. Appl. Physiol. Occup. Physiol.* **1992**. [CrossRef]
66. Derbyshire, P.J.; Barr, H.; Davis, F.; Higson, S.P.J. Lactate in human sweat: A critical review of research to the present day. *J. Physiol. Sci.* **2012**, *62*, 429–440. [CrossRef]
67. Meyer, F.; Laitano, O.; Bar-Or, O.; McDougall, D.; Heingenhauser, G.J.F. Effect of age and gender on sweat lactate and ammonia concentrations during exercise in the heat. *Braz. J. Med. Biol. Res.* **2007**. [CrossRef] [PubMed]

68. Ghamouss, F.; Ledru, S.; Ruillé, N.; Lantier, F.; Boujtita, M. Bulk-modified modified screen-printing carbon electrodes with both lactate oxidase (LOD) and horseradish peroxidase (HRP) for the determination of l-lactate in flow injection analysis mode. *Anal. Chim. Acta* **2006**. [CrossRef] [PubMed]
69. Ballesta Claver, J.; Valencia Mirón, M.C.; Capitán-Vallvey, L.F. Disposable electrochemiluminescent biosensor for lactate determination in saliva. *Analyst* **2009**. [CrossRef] [PubMed]
70. Jia, W.; Bandodkar, A.J.; Valdés-Ramírez, G.; Windmiller, J.R.; Yang, Z.; Ramírez, J.; Chan, G.; Wang, J. Electrochemical tattoo biosensors for real-time noninvasive lactate monitoring in human perspiration. *Anal. Chem.* **2013**. [CrossRef] [PubMed]
71. Bandodkar, A.J.; Molinnus, D.; Mirza, O.; Guinovart, T.; Windmiller, J.R.; Valdés-Ramírez, G.; Andrade, F.J.; Schöning, M.J.; Wang, J. Epidermal tattoo potentiometric sodium sensors with wireless signal transduction for continuous non-invasive sweat monitoring. *Biosens. Bioelectron.* **2014**. [CrossRef] [PubMed]
72. Buono, M.J. Sweat ethanol concentrations are highly correlated with co-existing blood values in humans. *Exp. Physiol.* **1999**. [CrossRef]
73. Nyman, E.; Palmlöv, A. The Elimination of Ethyl Alcohol in Sweat. *Skand. Arch. Physiol.* **1936**. [CrossRef]
74. Kim, J.; Jeerapan, I.; Imani, S.; Cho, T.N.; Bandodkar, A.; Cinti, S.; Mercier, P.P.; Wang, J. Noninvasive Alcohol Monitoring Using a Wearable Tattoo-Based Iontophoretic-Biosensing System. *ACS Sens.* **2016**. [CrossRef]
75. Swift, R. Transdermal alcohol measurement for estimation of blood alcohol concentration. *Alcohol. Clin. Exp. Res.* **2000**, *24*, 422–423. [CrossRef]
76. Mishra, R.K.; Sempionatto, J.R.; Li, Z.; Brown, C.; Galdino, N.M.; Shah, R.; Liu, S.; Hubble, L.J.; Bagot, K.; Tapert, S.; et al. Simultaneous detection of salivary Δ9-tetrahydrocannabinol and alcohol using a Wearable Electrochemical Ring Sensor. *Talanta* **2020**. [CrossRef]
77. Teymourian, H.; Parrilla, M.; Sempionatto, J.R.; Montiel, N.F.; Barfidokht, A.; Van Echelpoel, R.; De Wael, K.; Wang, J. Wearable Electrochemical Sensors for the Monitoring and Screening of Drugs. *ACS Sens.* **2020**. [CrossRef]
78. Jeerapan, I.; Sempionatto, J.R.; Wang, J. On-Body Bioelectronics: Wearable Biofuel Cells for Bioenergy Harvesting and Self-Powered Biosensing. *Adv. Funct. Mater.* **2020**. [CrossRef]
79. Sempionatto, J.R.; Jeerapan, I.; Krishnan, S.; Wang, J. Wearable Chemical Sensors: Emerging Systems for On-Body Analytical Chemistry. *Anal. Chem.* **2019**, *92*, 378–396. [CrossRef] [PubMed]
80. Mickelsen, O.; Keys, A. The composition of sweat, with special reference to vitamins. *J. Biol. Chem.* **1943**, *149*, 479–490.
81. Lemonick, S. You Could Unlock Your Phone with Sweat. *Forbes* **2017**. Available online: https://www.forbes.com/sites/samlemonick/2017/11/15/you-could-unlock-your-phone-with-sweat/#7225db7849ae (accessed on 15 September 2020).
82. French, L. Virtual Case Notes: Sweat, Skin Secretions Could Contain Chemical 'Password' for Future Mobile Authentication. *Forensic Magazine*, 16 November 2017.
83. Murphy, M. Sweat Could Soon Unlock Your Phone. Available online: https://nypost.com/2017/11/14/sweat-could-soon-unlock-your-phone/ (accessed on 15 September 2020).
84. Nagamine, K.; Nomura, A.; Ichimura, Y.; Izawa, R.; Sasaki, S.; Furusawa, H.; Matsui, H.; Tokito, S. Printed organic transistor-based biosensors for non-invasive sweat analysis. *Anal. Sci.* **2020**. [CrossRef] [PubMed]
85. Bandodkar, A.J.; Jeang, W.J.; Ghaffari, R.; Rogers, J.A. Wearable Sensors for Biochemical Sweat Analysis. *Annu. Rev. Anal. Chem.* **2019**, *12*, 1–22. [CrossRef]
86. Gao, W.; Nyein, H.Y.Y.; Shahpar, Z.; Tai, L.C.; Wu, E.; Bariya, M.; Ota, H.; Fahad, H.M.; Chen, K.; Javey, A. Wearable sweat biosensors. In Proceedings of the Technical Digest—International Electron Devices Meeting, IEDM, San Francisco, CA, USA, 3–7 December 2016.
87. Bandodkar, A.J.; Wang, J. Non-invasive wearable electrochemical sensors: A review. *Trends Biotechnol.* **2014**, *32*, 363–371. [CrossRef]
88. Heikenfeld, J. Non-invasive Analyte Access and Sensing through Eccrine Sweat: Challenges and Outlook circa 2016. *Electroanalysis* **2016**, *28*, 1242–1249. [CrossRef]
89. Bandodkar, A.J.; Hung, V.W.S.; Jia, W.; Valdés-Ramírez, G.; Windmiller, J.R.; Martinez, A.G.; Ramírez, J.; Chan, G.; Kerman, K.; Wang, J. Tattoo-based potentiometric ion-selective sensors for epidermal pH monitoring. *Analyst* **2013**. [CrossRef]
90. Guinovart, T.; Bandodkar, A.J.; Windmiller, J.R.; Andrade, F.J.; Wang, J. A potentiometric tattoo sensor for monitoring ammonium in sweat. *Analyst* **2013**. [CrossRef]

91. He, X.; Xu, T.; Gu, Z.; Gao, W.; Xu, L.P.; Pan, T.; Zhang, X. Flexible and Superwettable Bands as a Platform toward Sweat Sampling and Sensing. *Anal. Chem.* **2019**. [CrossRef]
92. Hauke, A.; Oertel, S.; Knoke, L.; Fein, V.; Maier, C.; Brinkmann, F.; Jank, M.P.M. Screen-Printed Sensor for Low-Cost Chloride Analysis in Sweat for Rapid Diagnosis and Monitoring of Cystic Fibrosis. *Biosensors* **2020**, *10*, 123. [CrossRef]
93. He, X.; Yang, S.; Pei, Q.; Song, Y.; Liu, C.; Xu, T.; Zhang, X. Integrated Smart Janus Textile Bands for Self-Pumping Sweat Sampling and Analysis. *ACS Sens.* **2020**. [CrossRef] [PubMed]
94. Oh, S.Y.; Hong, S.Y.; Jeong, Y.R.; Yun, J.; Park, H.; Jin, S.W.; Lee, G.; Oh, J.H.; Lee, H.; Lee, S.S.; et al. Skin-Attachable, Stretchable Electrochemical Sweat Sensor for Glucose and pH Detection. *ACS Appl. Mater. Interfaces* **2018**. [CrossRef] [PubMed]
95. Lee, H.; Choi, T.K.; Lee, Y.B.; Cho, H.R.; Ghaffari, R.; Wang, L.; Choi, H.J.; Chung, T.D.; Lu, N.; Hyeon, T.; et al. A graphene-based electrochemical device with thermoresponsive microneedles for diabetes monitoring and therapy. *Nat. Nanotechnol.* **2016**. [CrossRef] [PubMed]
96. Koh, A.; Kang, D.; Xue, Y.; Lee, S.; Pielak, R.M.; Kim, J.; Hwang, T.; Min, S.; Banks, A.; Bastien, P.; et al. A soft, wearable microfluidic device for the capture, storage, and colorimetric sensing of sweat. *Sci. Transl. Med.* **2016**. [CrossRef]
97. Khodagholy, D.; Curto, V.F.; Fraser, K.J.; Gurfinkel, M.; Byrne, R.; Diamond, D.; Malliaras, G.G.; Benito-Lopez, F.; Owens, R.M. Organic electrochemical transistor incorporating an ionogel as a solid state electrolyte for lactate sensing. *J. Mater. Chem.* **2012**. [CrossRef]
98. Emaminejad, S.; Gao, W.; Wu, E.; Davies, Z.A.; Nyein, H.Y.Y.; Challa, S.; Ryan, S.P.; Fahad, H.M.; Chen, K.; Shahpar, Z.; et al. Autonomous sweat extraction and analysis applied to cystic fibrosis and glucose monitoring using a fully integrated wearable platform. *Proc. Natl. Acad. Sci. USA* **2017**. [CrossRef]
99. Mark, H.; Harding, C.R. Amino acid composition, including key derivatives of eccrine sweat: Potential biomarkers of certain atopic skin conditions. *Int. J. Cosmet. Sci.* **2013**. [CrossRef]
100. Imani, S.; Bandodkar, A.J.; Mohan, A.M.V.; Kumar, R.; Yu, S.; Wang, J.; Mercier, P.P. A wearable chemical-electrophysiological hybrid biosensing system for real-time health and fitness monitoring. *Nat. Commun.* **2016**. [CrossRef]
101. Zhang, Y.; Guo, H.; Kim, S.B.; Wu, Y.; Ostojich, D.; Park, S.H.; Wang, X.; Weng, Z.; Li, R.; Bandodkar, A.J.; et al. Passive sweat collection and colorimetric analysis of biomarkers relevant to kidney disorders using a soft microfluidic system. *Lab Chip* **2019**. [CrossRef]
102. Gao, W.; Emaminejad, S.; Nyein, H.Y.Y.; Challa, S.; Chen, K.; Peck, A.; Fahad, H.M.; Ota, H.; Shiraki, H.; Kiriya, D.; et al. Fully integrated wearable sensor arrays for multiplexed in situ perspiration analysis. *Nature* **2016**. [CrossRef] [PubMed]
103. Schazmann, B.; Morris, D.; Slater, C.; Beirne, S.; Fay, C.; Reuveny, R.; Moyna, N.; Diamond, D. A wearable electrochemical sensor for the real-time measurement of sweat sodium concentration. *Anal. Methods* **2010**. [CrossRef]
104. Glennon, T.; O'Quigley, C.; McCaul, M.; Matzeu, G.; Beirne, S.; Wallace, G.G.; Stroiescu, F.; O'Mahoney, N.; White, P.; Diamond, D. 'SWEATCH': A Wearable Platform for Harvesting and Analysing Sweat Sodium Content. *Electroanalysis* **2016**. [CrossRef]
105. Tai, L.C.; Gao, W.; Chao, M.; Bariya, M.; Ngo, Q.P.; Shahpar, Z.; Nyein, H.Y.Y.; Park, H.; Sun, J.; Jung, Y.; et al. Methylxanthine Drug Monitoring with Wearable Sweat Sensors. *Adv. Mater.* **2018**. [CrossRef] [PubMed]
106. Tanenbaum, A.S.; Wetherall, D.J. Computer Networks. *World Wide Web Internet Web Inf. Syst.* **2011**. [CrossRef]
107. Schneier, B. *Secrets and Lies: Digital Security in a Networked World*; John Wiley & Sons, Inc.: New York, NY, USA, 2000.
108. Kaufman, C.; Perlman, R.; Spencer, M. *Network Security Private Communication in a Public World*; Prentice Hall: Westford, MA, USA, 2002; ISBN 0-13-046019-2.
109. Kishimura, A.; Yamashita, T.; Yamaguchi, K.; Aida, T. Rewritable phosphorescent paper by the control of competing kinetic and thermodynamic self-assembling events. *Nat. Mater.* **2005**. [CrossRef]
110. Mutai, T.; Satou, H.; Araki, K. Reproducible on-off switching of solid-state luminescence by controlling molecular packing through heat-mode interconversion. *Nat. Mater.* **2005**. [CrossRef]
111. Perruchas, S.; Goff, X.F.L.; Maron, S.; Maurin, I.; Guillen, F.; Garcia, A.; Gacoin, T.; Boilot, J.P. Mechanochromic and thermochromic luminescence of a copper iodide cluster. *J. Am. Chem. Soc.* **2010**. [CrossRef]

112. Yoon, S.J.; Chung, J.W.; Gierschner, J.; Kim, K.S.; Choi, M.G.; Kim, D.; Park, S.Y. Multistimuli two-color luminescence switching via different slip-stacking of highly fluorescent molecular sheets. *J. Am. Chem. Soc.* **2010**. [CrossRef]
113. Yan, D.; Lu, J.; Ma, J.; Wei, M.; Evans, D.G.; Duan, X. Reversibly thermochromic, fluorescent ultrathin films with a supramolecular architecture. *Angew. Chem. Int. Ed.* **2011**. [CrossRef]
114. Li, K.; Xiang, Y.; Wang, X.; Li, J.; Hu, R.; Tong, A.; Tang, B.Z. Reversible photochromic system based on rhodamine B salicylaldehyde hydrazone metal complex. *J. Am. Chem. Soc.* **2014**. [CrossRef]
115. Sun, H.; Liu, S.; Lin, W.; Zhang, K.Y.; Lv, W.; Huang, X.; Huo, F.; Yang, H.; Jenkins, G.; Zhao, Q.; et al. Smart responsive phosphorescent materials for data recording and security protection. *Nat. Commun.* **2014**. [CrossRef] [PubMed]
116. Wu, Y.; Xie, Y.; Zhang, Q.; Tian, H.; Zhu, W.; Li, A.D.Q. Quantitative photoswitching in bis(dithiazole)ethene enables modulation of light for encoding optical signals. *Angew. Chem. Int. Ed.* **2014**, *53*, 2090–2094. [CrossRef] [PubMed]
117. Sarkar, T.; Selvakumar, K.; Motiei, L.; Margulies, D. Message in a molecule. *Nat. Commun.* **2016**, *7*, 1–9. [CrossRef] [PubMed]
118. Ratner, T.; Reany, O.; Keinan, E. Encoding and processing of alphanumeric information by chemical mixtures. *ChemPhysChem* **2009**, *10*, 3303–3309. [CrossRef]
119. Palacios, M.A.; Benito-Pena, E.; Manesse, M.; Mazzeo, A.D.; LaFratta, C.N.; Whitesides, G.M.; Walt, D.R. InfoBiology by printed arrays of microorganism colonies for timed and on-demand release of messages. *Proc. Natl. Acad. Sci. USA* **2011**. [CrossRef]
120. Kim, K.W.; Bocharova, V.; Halámek, J.; Oh, M.K.; Katz, E. Steganography and encrypting based on immunochemical systems. *Biotechnol. Bioeng.* **2011**, *108*, 1100–1107. [CrossRef]
121. Shoshani, S.; Piran, R.; Arava, Y.; Keinan, E. A molecular cryptosystem for images by DNA computing. *Angew. Chem. Int. Ed.* **2012**, *51*, 2883–2887. [CrossRef]
122. Poje, J.E.; Kastratovic, T.; Macdonald, A.R.; Guillermo, A.C.; Troetti, S.E.; Jabado, O.J.; Fanning, M.L.; Stefanovic, D.; Macdonald, J. Visual displays that directly interface and provide read-outs of molecular states via molecular graphics processing units. *Angew. Chem. Int. Ed.* **2014**. [CrossRef]
123. Ling, J.; Naren, G.; Kelly, J.; Moody, T.S.; De Silva, A.P. Building pH Sensors into Paper-Based Small-Molecular Logic Systems for Very Simple Detection of Edges of Objects. *J. Am. Chem. Soc.* **2015**, *137*, 3763–3766. [CrossRef]
124. Ling, J.; Naren, G.; Kelly, J.; Fox, D.B.; Prasanna De Silva, A. Small molecular logic systems can draw the outlines of objects via edge visualization. *Chem. Sci.* **2015**. [CrossRef] [PubMed]
125. Zhang, Y.; Liu, X.; Sun, M. DNA based random key generation and management for OTP encryption. *BioSystems* **2017**. [CrossRef] [PubMed]
126. Clelland, C.T.; Risca, V.; Bancroft, C. Hiding Data in DNA Microdots. *Nature* **1999**, *399*, 533–534. [CrossRef] [PubMed]
127. Lustgarten, O.; Carmieli, R.; Motiei, L.; Margulies, D. A Molecular Secret Sharing Scheme. *Angew. Chem. Int. Ed.* **2019**, *58*, 184–188. [CrossRef]
128. Srilatha, N.; Murali, G. Fast three level DNA Cryptographic technique to provide better security. In Proceedings of the 2016 2nd International Conference on Applied and Theoretical Computing and Communication Technology, iCATccT 2016, Bangalore, India, 21–23 July 2016.
129. Malathi, P.; Manoaj, M.; Manoj, R.; Raghavan, V.; Vinodhini, R.E. Highly Improved DNA Based Steganography. *Procedia Comput. Sci.* **2017**, *115*, 651–659.
130. Cherian, A.; Raj, S.R.; Abraham, A. A Survey on different DNA Cryptographic Methods. *Int. J. Sci. Res. (IJSR)* **2013**, *2*, 167–169.
131. Roy, P.; Dey, D.; De, D.; Sinha, S. DNA cryptography. In *Cyber Security and Threats: Concepts, Methodologies, Tools, and Applications*; IGI Global: Hershey, PA, USA, 2018; ISBN 9781522556350.
132. Halvorsen, K.; Wong, W.P. Binary DNA Nanostructures for Data Encryption. *PLoS ONE* **2012**, *7*, e44212. [CrossRef]
133. Marwan, S.; Shawish, A.; Nagaty, K. DNA-based cryptographic methods for data hiding in DNA media. *BioSystems* **2016**, *150*, 110–118. [CrossRef]
134. Huynh, C.; Brunelle, E.; Halámková, L.; Agudelo, J.; Halámek, J. Forensic Identification of Gender from Fingerprints. *Anal. Chem.* **2015**, *87*, 11531–11536. [CrossRef]

135. Brunelle, E.; Huynh, C.; Le, A.M.; Halámková, L.; Agudelo, J.; Halámek, J. New Horizons for Ninhydrin: Colorimetric Determination of Gender from Fingerprints. *Anal. Chem.* **2016**. [CrossRef]
136. Brunelle, E.; Le, A.M.; Huynh, C.; Wingfield, K.; Halámková, L.; Agudelo, J.; Halámek, J. Coomassie Brilliant Blue G-250 Dye: An Application for Forensic Fingerprint Analysis. *Anal. Chem.* **2017**. [CrossRef] [PubMed]
137. Brunelle, E.; Huynh, C.; Alin, E.; Eldridge, M.; Le, A.M.; Halámková, L.; Halámek, J. Fingerprint Analysis: Moving Toward Multiattribute Determination via Individual Markers. *Anal. Chem.* **2018**. [CrossRef]
138. Agudelo, J.; Huynh, C.; Halámek, J. Forensic determination of blood sample age using a bioaffinity-based assay. *Analyst* **2015**, *140*, 1411–1415. [CrossRef] [PubMed]
139. Agudelo, J.; Halámková, L.; Brunelle, E.; Rodrigues, R.; Huynh, C.; Halámek, J. Ages at a Crime Scene: Simultaneous Estimation of the Time since Deposition and Age of Its Originator. *Anal. Chem.* **2016**, *88*, 6479–6484. [CrossRef] [PubMed]
140. Huynh, C.; Brunelle, E.; Agudelo, J.; Halámek, J. Bioaffinity-based assay for the sensitive detection and discrimination of sweat aimed at forensic applications. *Talanta* **2017**. [CrossRef]
141. Hair, M.E.; Gerkman, R.; Mathis, A.I.; Halámková, L.; Halámek, J. Noninvasive Concept for Optical Ethanol Sensing on the Skin Surface with Camera-Based Quantification. *Anal. Chem.* **2019**. [CrossRef] [PubMed]
142. Brunelle, E.; Halámek, J. Biocomputing approach in forensic analysis. *Int. J. Parallel Emergent Distrib. Syst.* **2017**. [CrossRef]
143. Moore, S.; Stein, W.H. Photometric ninhydrin method for use in the chromatography of amino acids. *J. Biol. Chem.* **1948**, *178*, 367–388.
144. Meyer, H. The ninhydrin reaction and its analytical applications. *Biochem. J.* **1957**. [CrossRef]
145. Walker, J.M. *Protein Protocols Handbook*; Humana Press: Totowa, NJ, USA, 2002.
146. Halámek, J.; Bocharova, V.; Chinnapareddy, S.; Windmiller, J.R.; Strack, G.; Chuang, M.C.; Zhou, J.; Santhosh, P.; Ramirez, G.V.; Arugula, M.A.; et al. Multi-enzyme logic network architectures for assessing injuries: Digital processing of biomarkers. *Mol. Biosyst.* **2010**, *6*, 2554–2560. [CrossRef]
147. Sakaguchi, S. A new method for the colorimetric determination of arginine. *J. Biochem.* **1950**. [CrossRef]
148. Gibson, L.E.; Cooke, R.E. A test for concentration of electrolytes in sweat in cystic fibrosis of the pancreas utilizing pilocarpine by iontophoresis. *Pediatrics* **1959**, *23*, 545–549. [PubMed]
149. Hair, M.E.; Mathis, A.I.; Brunelle, E.K.; Halámková, L.; Halámek, J. Metabolite Biometrics for the Differentiation of Individuals. *Anal. Chem.* **2018**, *90*, 5322–5328. [CrossRef] [PubMed]
150. Agudelo, J.; Privman, V.; Halámek, J. Promises and Challenges in Continuous Tracking Utilizing Amino Acids in Skin Secretions for Active Multi-Factor Biometric Authentication for Cybersecurity. *ChemPhysChem* **2017**, *18*, 1714–1720. [CrossRef]
151. McGoldrick, L.K.; Weiss, E.A.; Halámek, J. Symmetric-Key Encryption Based on Bioaffinity Interactions. *ACS Synth. Biol.* **2019**. [CrossRef]
152. Park, S.K.; Miller, K.W. Random number generators: Good ones are hard to find. *Commun. ACM* **1988**. [CrossRef]
153. Leier, A.; Richter, C.; Banzhaf, W.; Rauhe, H. Cryptography with DNA binary strands. *BioSystems* **2000**. [CrossRef]
154. Cox, J.P.L. Long-term data storage in DNA. *Trends Biotechnol.* **2001**, *19*, 247–250. [CrossRef]
155. Gahlaut, A.; Bharti, A.; Dogra, Y.; Singh, P. DNA based cryptography. In *Communications in Computer and Information Science*; Springer: Singapore, 2017; Volume 750.
156. Cui, G.; Wang, Y.; Han, D.; Wang, Y.; Wang, Z.; Wu, Y. An encryption scheme based on DNA microdots technology. *Commun. Comput. Inf. Sci.* **2014**. [CrossRef]
157. Zhang, Q.; Guo, L.; Wei, X. Image encryption using DNA addition combining with chaotic maps. *Math. Comput. Model.* **2010**. [CrossRef]

Publisher's Note: MDPI stays neutral with regard to jurisdictional claims in published maps and institutional affiliations.

© 2020 by the authors. Licensee MDPI, Basel, Switzerland. This article is an open access article distributed under the terms and conditions of the Creative Commons Attribution (CC BY) license (http://creativecommons.org/licenses/by/4.0/).

Review

Development Perspective of Bioelectrocatalysis-Based Biosensors

Taiki Adachi, Yuki Kitazumi, Osamu Shirai and Kenji Kano *,†

Division of Applied Life Sciences, Graduate School of Agriculture, Kyoto University, Sakyo, Kyoto 606-8502, Japan; adachi.taiki.62s@st.kyoto-u.ac.jp (T.A.); kitazumi.yuki.7u@kyoto-u.ac.jp (Y.K.); shirai.osamu.3x@kyoto-u.ac.jp (O.S.)
* Correspondence: kano.kenji.5z@kyoto-u.ac.jp
† Present address: Center for Advanced Science and Innovation, Kyoto University, Gokasho, Uji, Kyoto 611-0011, Japan.

Received: 19 July 2020; Accepted: 25 August 2020; Published: 26 August 2020

Abstract: Bioelectrocatalysis provides the intrinsic catalytic functions of redox enzymes to nonspecific electrode reactions and is the most important and basic concept for electrochemical biosensors. This review starts by describing fundamental characteristics of bioelectrocatalytic reactions in mediated and direct electron transfer types from a theoretical viewpoint and summarizes amperometric biosensors based on multi-enzymatic cascades and for multianalyte detection. The review also introduces prospective aspects of two new concepts of biosensors: mass-transfer-controlled (pseudo)steady-state amperometry at microelectrodes with enhanced enzymatic activity without calibration curves and potentiometric coulometry at enzyme/mediator-immobilized biosensors for absolute determination.

Keywords: current–potential curve; multi-enzymatic cascades; multianalyte detection; mass-transfer-controlled amperometric response; potentiometric coulometry

---

1. Introduction

Electron transfer reactions such as photosynthesis, respiration and metabolisms play an important role in all living things. A huge variety of redox enzymes catalyze the oxidation and reduction of couples of two inherent substrates. Usually, the electrons are transferred between the two substrates through a cofactor(s) that is covalently or non-covalently bound to the redox enzymes. The most important ones are pyridine nucleotide coenzymes (NAD(P)(H)). The coenzymes shuttle back and forth between NAD(P)-dependent enzymes and solution and mediate hydride ion transfers (single step two-electron one-proton redox reactions) between two organic substrates in the biologic system without generating any intermediate radicals. On the other hand, there are a variety of metallic ion cofactors such as hemes, iron–sulfur clusters, copper ion, nickel ion and manganese ion, in redox enzymes. These metallic ion cofactors undergo single or multistep single-electron transfers (SETs). Flavin cofactors ($FAD(H_2)$ $FMN(H_2)$) and quinone cofactors—including pyrroloquinoline quinone (PQQ)—can undergo both hydride ion transfers and SETs. Molybdopterin cofactors (Mocos) have similar properties. Therefore, flavin and quinone cofactors and Mocos have very important roles to mediate between the hydride ion transfer and the SET systems.

The coupling of redox enzymatic reactions with electrochemical reactions has received worldwide medical and scientific interests [1–14]. The coupled reaction is called bioelectrocatalysis. Since enzymatic reactions show very high performance in the selectivity, catalytic activity, uniformity and enormous chemical versatility, the coupling provides those redox–enzymatic characteristics to nonspecific electrode reactions. Therefore, several bioelectrochemical devices have been developed based on bioelectrocatalysis, such as biosensors [9,15], biologic fuel cells [16–20], bioelectrochemical reactors [21,22] and biosupercapacitors [23].

Most of redox–enzymatic reactions can be coupled with electrode reactions via redox mediators [3,24–26]. This reaction is called mediated electron transfer (MET)-type bioelectrocatalysis. Since electrode reactions are not hydride ion transfers, but single or multistep SETs, NAD(P)-dependent enzymatic reactions must be coupled to electrode reactions by using redox mediators that can transfer both hydride ion and single electrons, such as flavins, quinones (especially o-quinones) and phenothiazines (such as Meldola's blue). For other redox enzymes (i.e., flavoenzymes, quinoenzymes, metal-containing enzymes), a large variety of redox compounds (that undergo SETs)—including metallic ion complexes (such ferrocenes, ferrocyanide, osmium complexes)—can be used as mediators.

On the other hand, it is known that some metal-containing enzymes and flavoenzymes can directly exchange electrons with electrodes in the absence of any mediators in the catalytic reaction. Such reactions are referred to as direct electron transfer (DET)-type bioelectrocatalysis [14,27–31]. In theory, DET-type bioelectrocatalysis can simplify the fabrication process of bioelectrodes and can minimize the thermodynamic overpotential in a coupled reaction. However, redox enzymes that can provide clear DET-type bioelectrocatalytic waves are limited in number, and DET-type reactions are very susceptible to the chemical properties and structure of the electrode surface.

On the basis of bioelectrocatalysis, amperometric biosensors are analytical devices used to detect specific target analytes (substrates) and have been widely used in various fields such as medical care, environmental monitoring, food safety and industrial bioprocess monitoring [12,13,32,33]. Generally, there are many compounds (analytes) in the target fluid. Therefore, multianalyte biosensors that may combine cost-effective and rapid analysis with reducing the volume of samples have been improved [34]. When one enzyme reaction cannot be effectively coupled with electrode reactions, multiple enzymes that work in a cascade mode are required. In this case, immobilization and colocalization of multiple enzymes are necessary for sensor fabrication, and various methods for immobilization and colocalization of multi-enzyme have been reported [35–37].

However, the response of an amperometric sensor depends on detection time, which limits practical application [4,9,11–13,26,32,33]. It is necessary for amperometric sensors to obtain steady-state currents. If one can use biosensors without calibration, it is easy to monitor the concentration of target analytes. In this sense, microelectrode detection was introduced to develop amperometric biosensors [38–44]. Nonlinear diffusion of the substrate at microelectrodes causes diffusion-controlled steady-state currents. In addition, microelectrode-type biosensors are suitable for miniaturization and reducing the volume of samples. On the other hand, coulometry that can determine the absolute quantity is an absolute analytical method. However, non-Faradaic background electricity (background integrated current during absolute electrolysis) becomes too large to be ignored, compared with Faradaic electricity to be measured at decreased concentrations of analytes. In order to overcome this issue, potentiometric coulometry based on MET-type bioelectrocatalysis has been proposed, in which the total electricity of an analyte in a small value of test solution is transferred to the mediator immobilized on an electrode, and the change in the redox state of the mediator is potentiometrically detected [45,46].

In this review, we describe fundamental concepts of MET- and DET-type bioelectrocatalytic reactions, practical use of biosensors such as multi-enzyme biosensor and multianalyte biosensor and prospective biosensors that can be utilized without calibration.

2. Fundamentals of Bioelectrocatalytic Sensors

Bioelectrocatalysis is the core reaction in electrochemical biosensors constructed with redox enzymes and electrodes. The effective coupling between the enzymatic and electrode reactions improves high performance for biosensors. In this section, the basic principle of the bioelectrocatalysis is introduced.

2.1. Theory of Steady-State Catalytic Currents in Met-Type Bioelectrocatalysis

2.1.1. MET-Type Bioelectrocatalysis in Homogeneous System

In the MET-type bioelectrocatalysis (for the oxidation of a substrate (S) to a product (P)), the enzymatic reaction catalyzed by an enzyme (E) and the electrode reaction of a mediator (M) are coupled as follows:

$$S + \frac{n_S}{n_M} M_O \xrightarrow{\text{Enzyme}} P + \frac{n_S}{n_M} M_R, \tag{1}$$

and

$$M_R \xrightarrow{\text{Electrode}} M_O + n_M e^-, \tag{2}$$

where n_X is the number of electrons of X, M_O and M_R are the oxidized and reduced forms of the mediator, respectively. In the presence of the enzyme, the mediator and the substrate in a quiescent solution, a 1D reaction–diffusion equation of the MET-type bioelectrocatalytic reaction under the steady-state conditions is expressed as follows:

$$\frac{\partial c_{M_O}}{\partial t} = D_M \frac{\partial^2 c_{M_O}}{\partial x^2} - \frac{n_S}{n_M} \frac{k_c c_E}{1 + \frac{K_M}{c_{M_O}} + \frac{K_S}{c_S}} = 0, \tag{3}$$

where c_X is the concentration of X, t is the time, x is the distance from the electrode surface, D_M is the diffusion coefficient of the mediator, k_c is the catalytic constant, and K_X is the Michaelis constant for X, respectively. When the overpotential is high enough and the excess amount of S is present in solution ($c_S \gg K_S$), the limiting steady-state current ($i_{s,\text{lim}}$) can be obtained and is expressed by solving Equation (3) [47]:

$$i_{s,\text{lim}} = FA \sqrt{2n_S n_M D_M k_c K_M c_E \left\{ \frac{c_M}{K_M} - \ln\left(1 + \frac{c_M}{K_M}\right) \right\}}, \tag{4}$$

where F is the Faraday constant, A is the electrode surface area, and $c_M \equiv c_{M_O} c_{M_R}$, respectively. Furthermore, Equation (4) is simplified in two limited cases of c_M; when $c_M \ll K_M$:

$$i_{s,\text{lim}} = FAc_M \sqrt{n_S n_M D_M \frac{k_c}{K_M} c_E}, \tag{5}$$

and when $c_M \gg K_M$:

$$i_{s,\text{lim}} = FA \sqrt{2n_S n_M D_M k_c c_E c_M}. \tag{6}$$

2.1.2. Reaction-Layer Model at Enzyme/Mediator-Immobilized Electrodes

Enzymes and mediators are often immobilized on the electrode surface for the application to biosensors. In such situations, an enzyme/mediator-co-immobilized layer and a reaction layer must be considered.

When the enzyme/mediator-co-immobilized layer is sufficiently smaller in thickness than the reaction layer ($L \ll \mu$, L and μ being the thickness of immobilized and reaction layers, respectively), reaction (1) becomes a rate-determining step of the total reaction and the concentration polarization of the mediator and the substrate becomes negligible. Introducing a reaction layer theory [48] under the conditions of large overpotentials and $c_S \gg K_S$, $i_{s,\text{lim}}$ is expressed as follows:

$$i_{s,\text{lim}} = FAn_S D_M \frac{k_c c_E}{1 + \frac{K_M}{c_{M_O}}} L. \tag{7}$$

when $L \gg \mu$ or $L \approx \mu$, $i_{s,\text{lim}}$ is expressed as follows:

$$i_{s,\text{lim}} = FA\sqrt{2n_Sn_MD_Mk_cc_EB}\tanh\left\{\frac{L}{1+\frac{K_M}{c_M}}\sqrt{\frac{n_S}{n_M}\frac{k_cc_E}{2D_MK_MB}}\right\}, \quad (8)$$

where $B = \frac{c_M}{K_M} - \ln\left(1 + \frac{c_M}{K_M}\right)$. Equation (8) is simplified in two limited cases of c_M, and μ is expressed as follows: When $c_M \ll K_M$,

$$i_{s,\text{lim}} = FAc_M\sqrt{n_Sn_MD_M\frac{k_c}{K_M}c_E}\tanh\left(\frac{L}{\mu}\right), \quad (9)$$

$$\mu = \sqrt{\frac{n_S}{n_M}\frac{k_cc_E}{D_MK_M}}; \quad (10)$$

and when $c_M \gg K_M$,

$$i_{s,\text{lim}} = FA\sqrt{2n_Sn_MD_Mk_cc_Ec_M}\tanh\left(\frac{L}{\mu}\right), \quad (11)$$

$$\mu = \sqrt{\frac{n_S}{n_M}\frac{k_cc_E}{2D_Mc_M}}. \quad (12)$$

It is essential to optimize the value of L, in order to construct biosensors with large values of $i_{s,\text{lim}}/A$ using minimum amounts of enzymes and mediators.

2.1.3. Serial Resistance Model

The steady-state MET-type bioelectrocatalysis can be explained based on a serial resistance model. In this model, we assume a set of series reactions (Figure 1): (1) mass transfer, (2) membrane permeation, (3) enzymatic reaction and (4) electrode reaction. The steady-state current (i_s) can be expressed as follows:

$$\frac{1}{i_s} \cong \frac{1}{i_{mt}} + \frac{1}{i_{perm}} + \frac{1}{i_{enz}} + \frac{1}{i_{elec}}, \quad (13)$$

where i_{mt}, i_{perm}, i_{enz} and i_{elec} are the limiting steady-state currents controlled by mass-transfer, permeation, enzymatic reaction and electrode reaction, respectively. The value of i_{mt} is given by Levich equation at a rotating disk electrode (RDE) or by Equation (15) at a microdisk electrode:

$$i_{mt,\text{RDE}} = \pm 0.620 n_S FAD_S^{\frac{2}{3}} \nu^{-\frac{1}{6}} \omega^{\frac{1}{2}} c_S, \quad (14)$$

$$i_{mt,\text{microdisk}} = \pm 4 n_S FD_S r c_S, \quad (15)$$

where D_S is the diffusion coefficient of the substrate, ν is a kinetic viscosity of the buffer solution, ω is an angular rotation rate of the RDE, and r is a radius of the microdisk electrode. The plus and minus signs in the current indicate the oxidation and reduction currents, respectively. The term i_{perm} can be expressed as follows:

$$i_{perm} = \pm n_S FAP_m c_S, \quad (16)$$

where P_m is a permeation coefficient of a membrane. The value of i_{enz} is given by a Michaelis–Menten-type equation in the presence of an excess amount of a mediator as follows:

$$i_{enz} = \pm n_S FA\frac{k_c\Gamma_E}{1+\frac{K_S}{c_S}}, \quad (17)$$

where Γ_E is the surface concentration of an enzyme. The value of i_{elec} is given by a Butler–Volmer-type equation as follows:

$$i_{elec} = n_S F A k_M^o c_S \exp\left\{\frac{n_{M,rds}F}{RT}(E - E^{o\prime}_{M,rds})\right\}^{1-\alpha_{M,rds}}, \quad (18)$$

where k_M^o is the standard rate constant in the rate-determining step (rds) of the interfacial electron transfer of the mediator at the electrode, R is the gas constant, T is the absolute temperature, $n_{M,rds}$ is the number of electrons in the rds of the mediator (generally $n_{M,rds} = 1$), E is the electrode potential, $E^{o\prime}_{M,rds}$ is the formal potential of the rds of the mediator, and $\alpha_{M,rds}$ is the transfer coefficient in the rds of the mediator.

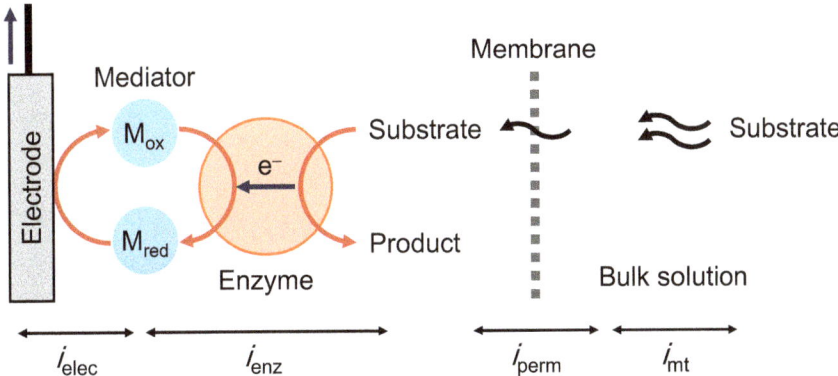

Figure 1. Schematic view of mediated electron transfer (MET)-type bioelectrocatalysis based on a serial resistance model.

Under the condition of $E \gg E^{o\prime}_{M,rds}$, i_{elec} becomes sufficiently larger than i_{mt}, i_{perm}, and i_{enz} to give a steady-state limiting current. On the other hand, i_{perm} depends on an outer membrane and i_{enz} is directly affected by $k_c\Gamma_E$ that often varies under measurement conditions such as temperature and pH. Thus, a diffusion-controlled steady-state condition ($i_{mt} \ll i_{perm}$, i_{enz}, and i_{elec}) is ideal for amperometric biosensors, in which no calibration curve is required for determination.

2.2. Theory of Steady-State DET-Type Bioelectrocatalysis

In DET-type bioelectrocatalysis, an enzyme reacts with both a substrate and an electrode as follows:

$$S + E_O \xrightarrow{\text{Enzyme}} P + E_R, \quad (19)$$

$$E_R \xrightleftharpoons{\text{Electrode}} E_O + n_E e^-, \quad (20)$$

In this situation, i_s can be expressed by the following equation based on the serial resistance model with a series of the mass-transfer process and the DET-type bioelectrocatalysis on the electrode surface (electro–enzyme reaction):

$$\frac{1}{i_s} = \frac{1}{i_{mt}} + \frac{1}{i_{elec-enz}}, \quad (21)$$

where i_{mt} is given by Equations (14) and (15) for RDE and microdisk electrode experiments, respectively. $i_{elec-enz}$ (the electro–enzyme reaction-controlled current) can be expressed as follows [49]:

$$i_{elec-enz} = \pm \frac{n_S F A k_{c,DET} \Gamma_E}{1 + \frac{k_{c,DET}}{k_f} + \frac{k_b}{k_f}}, \tag{22}$$

where $k_{c,DET}$ is the catalytic constant in DET-type bioelectrocatalysis, k_f and k_b are the interfacial electron transfer rate constants of the forward and backward reactions, respectively, which are given by the Butler–Volmer equations as follows: For oxidation,

$$k_f = k_E^o \eta_E^{1-\alpha_{E,rds}} \tag{23}$$

$$k_b = k_E^o \eta_E^{-\alpha_{E,rds}}, \tag{24}$$

and for reduction,

$$k_f = k_E^o \eta_E^{-\alpha_{E,rds}} \tag{25}$$

$$k_b = k_E^o \eta_E^{1-\alpha_{E,rds}} \tag{26}$$

where K_E^o is the standard rate constant in the rds of the interfacial electron transfer of the enzyme at the electrode, $\eta_E = \exp\left\{\frac{n_E' F}{RT}(E - E_E^{o'})\right\}$, n_E' is the number of electrons in the rds of the heterogeneous electron transfer of the enzyme (generally $n_E' = 1$), $E_E^{o'}$ is the formal potential in the rds of the enzyme, and $\alpha_{E,rds}$ is the transfer coefficient in the rds of the enzyme. When $E \gg E_E^{o'}$, $i_{elec-enz}$ is limited to $\pm n_S F A k_{c,DET} \Gamma_E$.

In addition, K_E^o decreases exponentially with an increase in the distance between the electrode surface and an electrode-active site of the enzyme (d) [50–52]. Thus, DET-type bioelectrocatalysis is often improved by using mesoporous electrode materials on which enzymes adsorb in increased probability of orientations favorable for DET-type reactions. Suitable modification of the electrode surface with chemical substances leads to electrostatic or specific attractive interaction between the electrode surface and the enzyme surface close to the electrode-active site [14,30].

2.3. Examples of MET/DET-Type Biosensors

The most popular enzymatic biosensors are those for self-monitoring of blood glucose (SMBG); blood glucose concentration being an important index in the treatment of diabetes. Various types of amperometric glucose biosensors have been reported [53–59]. SMBG sensors involve FAD-dependent glucose oxidase (GOD) [60,61], bacterial and fungal FAD-dependent glucose dehydrogenase (FAD-GDH) [62,63], PQQ-dependent soluble GDH (PQQ–sGDH) [64–66] and NAD-dependent GDH (NAD-GDH) [67]. Characteristics of the enzymes are summarized in a review [68].

Kakehi et al. reported a biofuel cell-type glucose biosensor using bacterial flavohemo-GDH complex containing a cytochrome subunit as well as an FAD subunit. The authors proposed that the enzyme proceeded DET-type bioelectrocatalysis at the cytochrome subunit and the open-circuit potential was used as a measure of the glucose concentration in the range of 0.5 mM to 6 mM [69].

Fructose biosensors are also useful in food analyses and diagnoses of kidney function. The sensors can be utilized for determining the inulin clearance, a difference of the intake and discharge of inulin filtered in the kidney, inulin being hydrolyzed into fructose and glucose by inulinase [70]. A membrane-bound flavohemo enzyme, D-fructose dehydrogenase (FDH), catalyzes two-electron oxidation of D-fructose, shows high activity of DET-type bioelectrocatalysis and is often mounted on fructose biosensors [71]. Both DET- and MET-type fructose biosensors are summarized in a review [72].

Cellobiose dehydrogenase (CDH) is useful for lactose biosensing [73]. CDH comprises an FAD-containing larger catalytic dehydrogenase domain and a heme b-containing smaller cytochrome domain that directly communicates with electrodes [73]. As a mimic of CDH, in addition, the cytochrome domain was introduced into some other flavoenzymes by protein engineering methods and the engineered enzymes realized DET-type bioelectrocatalysis. Ito et al. designed a cytochrome domain-linked fungal FAD-GDH and constructed a DET-type amperometric glucose biosensor [74].

3. Multi-Enzymatic Cascades

3.1. Diaphorase/NAD(P)$^+$-Dependent Enzymes

NAD(P)$^+$/NAD(P)H is a natural coenzyme involved in a large variety of redox enzymes. Usually, NAD(P)(H) shuttle back and forth between NAD(P)-dependent enzymes and solution to transfer the hydride ion. However, NAD(P) is an unfavorable mediator in MET-type reactions for NAD(P)-dependent enzymes, because the direct electrochemical reaction of NAD(P)(H) at electrodes requires high overpotentials, because NAD(P) is not a two-SET carrier, but a hydride ion carrier; SET characteristics are essential for electrode reactions. Therefore, other additional mediators, such as Meldola's blue (MB) that can undergo both hydride ion transfer and SETs, were used as mediators. MB shows large values of the second-order reaction rate constant with NAD(P)H (hydride ion transfer) and also the heterogeneous electron transfer rate constant with electrodes (SET) [75]. Avramescu et al. constructed a D-lactate biosensor using NAD-dependent D-lactate dehydrogenase (D-LDH), NAD(H) and MB, with a detection limit of 0.05 mM, a linear range of 0.1–1 mM and a sensitivity of 1.2 µA cm^{-2} Mm^{-1} [76].

Furthermore, FMN-containing diaphorase (DI) that catalyzes a redox reaction between NAD(P)(H) and an artificial mediator can be utilized to build up more efficient mediated systems and was introduced in MET-type biosensors using NAD-dependent enzymes [77–80]. Takagi et al. analyzed bienzymatic MET-type bioelectrocatalysis of NAD-dependent L-lactate dehydrogenase (L-LDH) and DI using several mediators (Figure 2A) [77]. The bienzyme system realized an interconversion (two-way conversion) between L-lactate and pyruvate. Nikitina et al. reported an amperometric formaldehyde biosensor with NAD-dependent formaldehyde dehydrogenase, DI and an osmium redox polymer, with a detection limit of 32 µM, a linear range of 50–500 µM and a sensitivity of 2.2 µA cm^{-2} mM^{-1} [79].

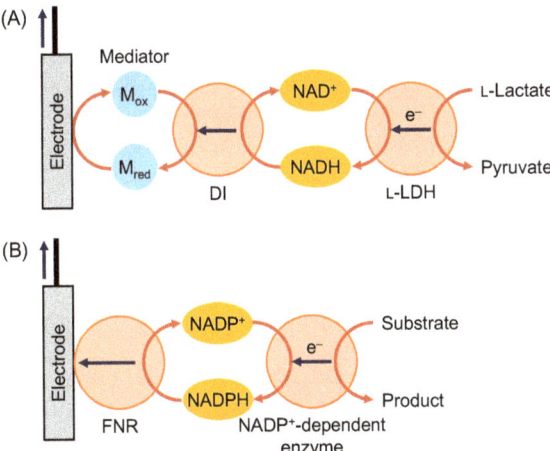

Figure 2. (**A**) Schematic view of MET-type bioelectrocatalysis of L-lactate dehydrogenase (L-LDH) and FMN-containing diaphorase (DI); (**B**) schematic view of bienzymatic bioelectrocatalysis using an NADP$^+$-dependent enzyme and ferredoxin-NADP$^+$ reductase (FNR).

On the other hand, Siritanaratkul et al. reported a DET-type interconversion of NADP$^+$/NADPH catalyzed by ferredoxin-NADP$^+$ reductase (FNR) [81]. They also demonstrated L-glutamate synthesis from 2-oxoglutarate and NH$_4$$^+$ by the coupled reactions of FNR and NADP-dependent glutamate dehydrogenase (GLDH) [81]. Multi-enzymatic biosensors using FNR have not yet been reported, but a DET-type NADP$^+$/NADPH interconversion by FNR is potentially important in constructing NADP-dependent enzymatic biosensors without any additional mediators (Figure 2B).

3.2. Peroxidase/Oxidases

Several oxidases (ODs) were used as bioelectrocatalysts of 1st generation biosensors, which can detect and quantify target compounds by direct electrochemical oxidation of the enzymatically generated hydrogen peroxide (H_2O_2) to dioxygen (O_2) [82–84]. However, direct oxidation of H_2O_2 at the electrode surface requires high overpotentials, and thus these biosensors are sensitive to interference by coexisting reductants [83,84].

In order to overcome this issue, H_2O_2 generated in an OD reaction was reductively detected with a help of horseradish peroxidase (HRP) that catalyzes two-electron reduction of H_2O_2 to H_2O. The HRP reaction was frequently coupled with electrode reactions via MET-type bioelectrocatalysis. The two-enzyme reaction coupled electrode may be called OP-type biosensors, here. In OP-type biosensors, the following reactions proceed:

$$S + O_2 \quad \xrightarrow{OD} \quad P + H_2O_2, \tag{27}$$

$$H_2O_2 + 2H^+ + 2e^- \quad \xrightarrow{HRP/electrode} \quad 2H_2O, \tag{28}$$

By introducing HRP, H_2O_2 detection at lower potentials was realized and the interference from other substances was reduced. The ODs mounted on OP-type biosensors reported so far are as follows: GOD [85,86], galactose oxidase (GalOD) [87], D- and L-amino acid oxidase [85,88,89], L-glutamate oxidase [90], amine oxidase [91], alcohol oxidase [85,91–95], urate oxidase [96], zinc superoxide dismutase [97], etc. Castillo et al. also investigated the bioelectrocatalytic characteristics of sweet potato peroxidase (SPP), and SPP-based OP-type biosensors showed higher performance than HRP-based ones [91].

In addition, multi-enzymatic biosensors were constructed by incorporating upstream enzymatic reactions in OP-type biosensors. Tkáč et al. constructed a glucose-non-interfering lactose biosensor by introducing β-galactosidase that hydrolyzes lactose to galactose and glucose, into the GalOD-HRP system [98]. Several groups reported cholesterol oxidase/cholesterol esterase/HRP co-immobilized biosensors, which enabled to monitor total cholesterol [99–101]. On the other hand, Nieh et al. reported a multi-enzymatic creatinine biosensor, as shown in Figure 3 [102].

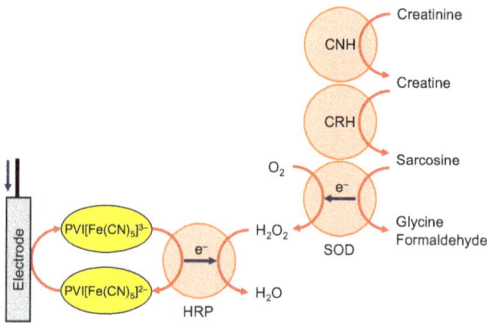

Figure 3. Schematic view of multi-enzymatic creatinine biosensor. CNH, CRH, SOD and PVI[Fe(CN)$_5$]$^{2-/3-}$ indicate creatinine amidohydrolase, creatine amidohydrolase, sarcosine oxidase and pentacyanoferrate-bound poly(1-vinylimidazole), respectively.

Recently, it was found that HRP showed a DET-type bioelectrocatalytic activity at mesoporous carbon and nanostructured gold electrodes [103,104] and third generation OP-type biosensors using the DET-type reaction of HRP were reported. Xia et al. reported glucose and putrescine biosensors using GOD and putrescine oxidase, respectively [41,103]. In addition, Kawai et al. constructed an analytical model of an OP-type pyruvate biosensor based on the serial resistance concept [44].

4. Multianalyte Detection

Real samples are mixtures containing multiple components. In particular, biologic samples have complex compositions and interrelation among the compositions remains unclear. In addition, there are many diseases of which the specific biomarker has not been found out. Status of the human health is often evaluated on the basis of the balance of the compositions of the body fluid. Therefore, the multianalyte sensing and monitoring of biologic samples, that is, big data in human health attracted attention [105–107]. Similarly, multianalyte detection in environmental samples is also an important subject [108]. Electrochemical sensors are easily miniaturized and compatible with such multianalyte detection [109–111]. The development of printing technologies reduces the cost of the fabrication of bioelectrochemical sensors and realize a disposable sensor array system with a complicated structure. In this section, some notes in simultaneous multicomponent measurements are outlined.

4.1. Crosstalk among Amperometric Biosensors

Electrical circuits of the amperometric sensors are on the basis of potentiostat. In the simplest construction of potentiostat, the working electrode set to common potential [112]. The circuit functions to maintain the potential difference between the working and reference electrodes at the setting potential by current flowing. Therefore, it is basically impossible to employ multiple working electrodes and a single pair of reference and counter electrodes by multiple potentiostats. Moreover, when the multiple electrochemical systems locate in a sample solution, the unnoticed electric connection of the potentiostats (for example, sharing ground) possibly causes the current flow between the working electrodes. In order to avoid these problems in simultaneous measurements, it is necessary to employ a special apparatus (it is called multipotentiostat).

The desorption of enzymes and mediators from the electrode surface also becomes the origin of crosstalk among the sensing devices in biosensors for multianalyte detection. The immobilization of enzymes and mediators and employment of the permeable membrane are effective to reduce this type of crosstalk. In the disposable biosensor array, the design and layout of the sensors on the sensing device are important to prevent the cross-contamination within the measurement period.

4.2. Absolute and Relative Concentration

In order to reduce the damage of the subjects in sampling, discharged body fluids such as saliva, sweat, tear and urine are better samples than blood. The homeostasis of the blood is important in the living things, while the water contents in such discharged body fluids change readily with time, movement and uptake, etc. Therefore, determination of the relative concentration of the components to the internal standard becomes more important than the determination of the absolute concentration of the components in the body fluids. Here, the homeostasis of the internal standard will become problems because its concentration affects all over the measurements. Furthermore, the calibration of the individual sensors is a fatal problem because the calibration process of the sensor for the multiple-analyte is incompatible with the disposable sensors. Calibration of the sensors is required if the sensitivities of sensors are unstable. Therefore, the stability of the sensor is important in disposable sensors.

4.3. Examples of the Internal Standard in Body Fluids

The most widely used internal standard in the urine sample is creatinine. Creatinine is a metabolite of creatine in muscle. The produced creatinine transfers from blood to urine through the kidney.

The steady-state production of creatinine in the human body is widely accepted. Therefore, in the blood of healthy people, the concentration of creatinine is practically constant, and the amount of creatinine transferred into urine is also constant. In urine analysis, creatinine is usually employed for calibration of other components. However, since renal dysfunction causes the variation of creatinine concentration in urine, creatinine is not a perfect standard [113].

The water content in sweat is easily varied. The suggested internal standards in sweat are chloride ion [114] and sodium ion [115]. Because it is difficult to incorporate those ions into the redox reaction, potentiometry is suitable to detect these ions.

In the metabolism of human bodies, L-lactate is a normal product. However, bacteria can produce both D- and L-lactates. Therefore, D-lactate in body fluids is a marker for bacterial infection [116]. In order to remove the dilution effect of body fluid, the most suitable reference material is L-lactate. Therefore, the simultaneous detection of lactate enantiomers is the most effective construction. In order to determine the D/L ratio of lactate in body fluid, the combination of the diffusion-limited amperometric D- and L-lactate sensors was reported based on MET-type bioelectrocatalysis of corresponding NAD-dependent enzymes and MB [43].

5. Prospective Biosensors without Calibration

Biosensors are fundamentally fragile, due to enzyme properties. The functions of biomaterials are easily suppressed by acid, base, oxidation, heat, dehydration and other external factors. Because the sensitivities of biosensors are generally labile to change, frequent calibrations of biosensors may be required to guarantee accuracy. In this section, prospective biosensors without calibration are introduced.

5.1. Significance of Mass-Transfer Controlling

As mentioned above, the current response of an amperometric biosensor is expressed as a serial combination of each reaction step. A simple case of the substrate reductive MET-type biosensing employing rotating disk electrode is considered here.

As mentioned in Section 2, the response of biosensors is affected by many physical quantities. However, the stability or reproducibility of these quantities is different from each other in the constructed biosensors. The features of the selectable or controllable values are summarized in Table 1. Generally, the electrode area (A) is easy to control except for nanoelectrodes. The formal potential of the mediator ($E_M^{o\prime}$) is possible to control by the selection of the mediator. Since the stability of redox mediators are limited due to oxygen and light damage, it is difficult to define the rigid redox state of mediators. Since the stability of the reference electrode is frequently poor, the value of E is unstable. The lack of uniformity of permeable membranes causes the poor reproducibility in P_m. The rotating speed is easy to control. The most unstable component in the biosensor is enzymes. The value of $k_c \Gamma_E$ decreases by the denaturation of the enzyme with time. Therefore, the mass transfer process is the most stable and controllable in amperometric biosensors. The stabilization of biosensors must be achieved by setting the mass transfer process as the rate-determining step.

Table 1. Physical quantities affected to the amperometric biosensing.

Physical Quantities	Stability	Reproducibility	Controllability
Surface area of electrode (A)	Good	good	good
Standard redox potential ($E_M^{o\prime}$)	good	good	poor
Concentration of mediator (c_M)	poor	good	good
Electrode potential (E)	poor	poor	good
Permeability of membrane (P_m)	good	poor	poor
Rotating speed (ω)	good	good	good
Enzyme activity ($k_c \Gamma_E$)	poor	poor	poor

5.2. Bioelectrocatalysis at Microelectrodes

As described in the above (Section 2), the steady-state mass transfer from the solution to an electrode is realized by rotating electrodes or microelectrodes. However, since the rotating electrode requires a special apparatus, the application in the biosensors for real sample measurements seems to be difficult. Therefore, microelectrodes are more suitable for construction for amperometric biosensors. Another advantage of microelectrodes is a high signal–noise ratio because of the relatively high Faradaic current density against the charging current density.

Microelectrodes are classified into some types based on the symmetry of the shape, such as microsphere, microdisk, microcylinder and microband [112]. The microdisk electrode is the most available one. The time-dependence of the limited current of a microdisk electrode (i_{md}) with a radius of r is given as follows [117],

$$i_{md} = 4n_S F D_S c_S r \left\{ 0.7854 + 0.8862\sqrt{\tau_d} + 0.2146 \exp\left(-\frac{0.7823}{\sqrt{\tau_d}}\right) \right\}, \tag{29}$$

where τ_d is the dimensionless time for the microdisk and is defined as follows:

$$\tau_d = \frac{4 D_S t}{r^2}. \tag{30}$$

At the long-time limitation ($t \to \infty$), the current reaches a steady-state value given by Equation (15); the mass-transfer-limited current density at the microdisk electrode increases with a decrease in the radius of the electrode. On the other hand, the enzymatic reaction-limited current density would be constant at a constant surface density of the enzyme. Therefore, it would become difficult to satisfy the requirement for diffusion-controlled biosensors ($i_{enz} \gg i_{mt,microdisk}$) at microdisk electrodes with extremely small values of r. A similar situation is expected for other types of microelectrodes. In other words, most of the microelectrode-type biosensors are often controlled by the enzymatic kinetics (or the permeability of the outer membrane, if any), and therefore the response would be labile to change. However, if one can realize extremely fast enzymatic reactions on microelectrodes, one may satisfy the condition: $i_{enz} \gg i_{mt,microdisk}$.

The simplest case is extremely fast DET-type bioelectrocatalysis at a microdisk electrode. In this situation, the substrate concentration at the microelectrode surface is regarded as zero, that is, diffusion-controlled limiting current conditions are realized. Some of multicopper-oxidases (MCOs) reduce O_2 to water with sufficiently high activity at the surface of mesoporous electrodes by DET-type bioelectrocatalysis. Mass-transfer-controlled DET-type bioelectrocatalysis was realized at MCO-modified porous gold microdisk electrodes [118]. The porous electrode was employed to increase the effective surface area and the probability of orientations suitable for DET-type bioelectrocatalysis thanks to curvature effects [19,119,120]. Figure 4A shows voltammograms recorded at a Cu-efflux oxidase-modified porous gold microelectrode (solid line) with and (dotted line) without oxygen. The sigmoidal curve indicates the DET-type bioelectrocatalysis of O_2 reduction. A clear potential-independent limiting current was observed. The limiting current linearly increased with the bulk concentration of O_2, as shown in Figure 4B. The solid line in Figure 4B shows the theoretical sensitivity calculated with Equation (15) using the literature value of D_S and the microscopically measured r. The agreement between the experimental and theoretical sensitivities verifies that the response of the constructed O_2 sensor is truly controlled by the mass transfer of O_2. Since the sensitivity of the sensor is independent of the activity of CueO, the sensor shows excellent reproducibility and stability at a given temperature.

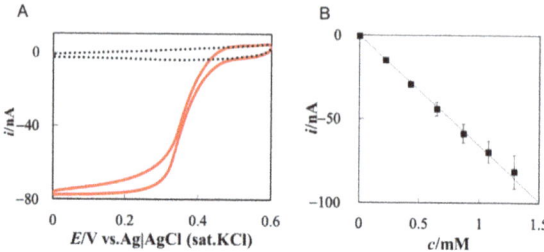

Figure 4. (**A**) Cyclic voltammogram of CueO-modified porous gold microdisk electrode recorded with $r = 20$ µm (solid line) with and (dotted line) without oxygen; (**B**) (squares) experimental and (solid line) theoretical calibration curve for the oxygen biosensor. Reprinted from ref. [118], Copyright (2020), with permission from Elsevier.

The situation of MET-type bioelectrocatalysis at a microelectrode is more complicated than that of DET-type bioelectrocatalysis due to the diffusion of the mediator. However, when a large amount of active enzymes exist in the system, it is possible to realize the substrate–diffusion-controlled MET-type bioelectrocatalysis at a microdisk electrode. According to this concept, a MET-type glucose sensor was constructed using a microdisk electrode, 1,2-benzoquinone and FAD-GDH [39]. The current response quickly reached a steady-state (Figure 5). The linear range of the sensor was from 0 mM to 3 mM of glucose while the concentration of the mediator was 1 mM in the system (inset in Figure 5). Obviously, the linear range was beyond the endpoint situation. Moreover, the current response of the sensor agreed with the diffusion-controlled value of glucose.

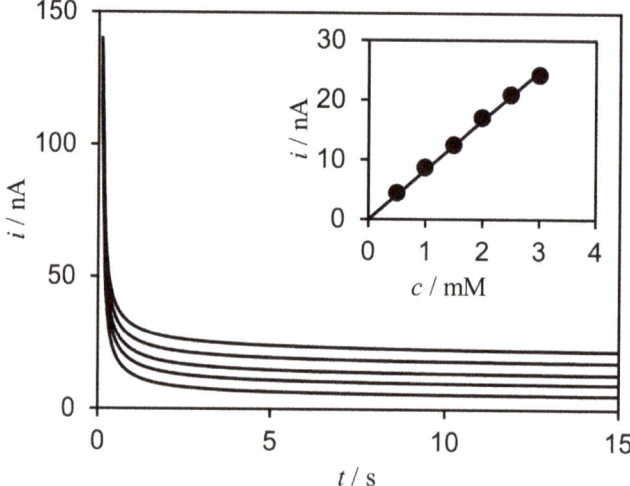

Figure 5. Chronoamperometric response for glucose oxidation in a buffer solution containing 0.21-mM FAD-dependent glucose dehydrogenase (FAD-GDH) and 1-mM 1,2-benzoquinone at a microelectrode with diameter of 50 µm. Inset shows the calibration curve based on the current at 10 s. Reprinted with permission from ref. [39], Copyright (2013) Japan Society for Analytical Chemistry.

The situation where the MET-type bioelectrocatalysis occurs around a microelectrode was simulated by the finite element method [40]. The concentration profiles of the substrate, reduced form of the mediator, and reduced enzyme are given in Figure 6 under the conditions that the enzymatic reaction is extremely fast. While the concentration gradient of the substrate spreads hemispherically, the concentration gradient of the mediator is located only in the vicinity of the electrode surface. The concentration profile of the enzyme clearly shows that the enzymatic reaction occurs only in the thin region (as a reaction plane) where the enzyme concentration gradient exists. The flux of the substrate is quickly converted to that of the mediator in the thin region. Therefore, the surface of the thin region plays the role of a virtual electrode that selectively reacts with the substrate. The increase in the substrate concentration leads to a decrease in the distance of the thin region and then increases the current density.

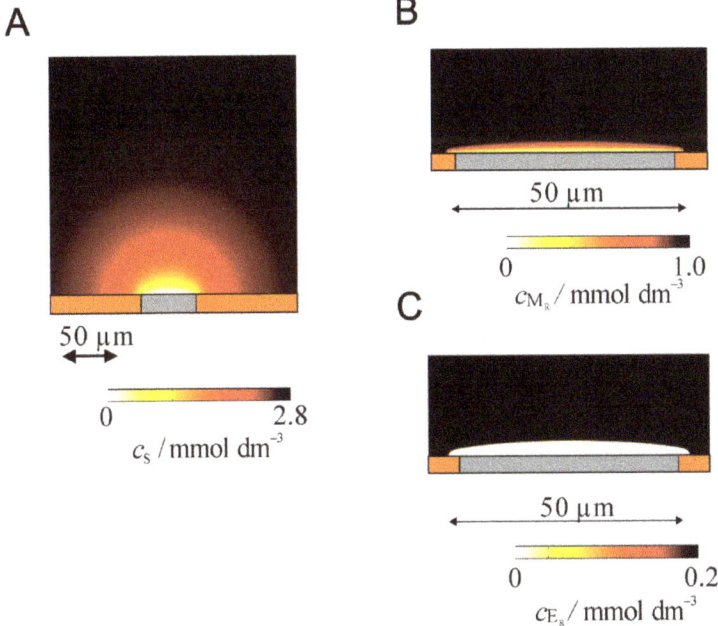

Figure 6. Concentration profiles of (**A**) substrate, (**B**) reduced mediator and (**C**) reduced form of enzyme around an microdisk electrode with r = 25 µm at a substrate concentration of 4 mmol dm^{-3}, a mediator concentration of 1 mmol dm^{-3}, an enzyme concentration of 0.2 mmol dm^{-3}. Profiles calculated for the situation of 20 s after the potential step at the limiting current conditions. Reprinted from ref. [40] with permission from the PCCP Owner Societies.

5.3. Pseudo-Steady-State Response

Although microdisk electrodes provide clear steady-state current, the current is quite small. In order to increase the current, microdisk electrode array without overlapping of the diffusional concentration gradients (scattered microdisk electrode array) is one of the solutions [38,121–123]. However, the fabrication of such a scattered microdisk electrode array is a technically challenging issue. Another solution to increase the current may be the employment of microband electrodes. Ultrathin ring electrodes can be considered as rod electrodes [124,125]. The time-dependence of the limiting current at a microband electrode (i_{mb}) with a length of l and a thickness of w is given by [126]:

$$i_{mb} = n_S F D_S c_S l \left[\frac{\pi \exp\left(-\frac{2}{5}\sqrt{\pi \tau_b}\right)}{4\sqrt{\pi \tau_b}} + \frac{\pi}{\ln\left\{64\tau_b \exp(-0.5772156) + \exp\left(\frac{5}{3}\right)\right\}} \right], \quad (31)$$

where τ_b is the nondimensional time for the band electrode and is defined as follows,

$$\tau_b = \frac{D_S t}{w^2}. \tag{32}$$

According to Equation (31), i_{mb} easily magnifies with an increase in l. On the other hand, i_{mb} becomes 0 at $t \to \infty$. Therefore, the exact analysis of i_{mb} based on the steady-state current is not possible. However, the decay of i_{mb} is so slow that the current is practically indistinguishable from the so-called steady-state current. In actual electrochemical measurements, a quasi-steady-state response is acceptable for practical use. For example, MET-type bioelectrocatalysis of FAD-GDH at microband electrodes provided the values very close to the diffusion-limited response [42].

Since the current density at a microelectrode is quite large, the mass-transfer-limited bioelectrocatalysis is fundamentally difficult. Therefore, a relatively large microelectrode ($r = 1.5$ mm) is promising to realize the mass-transfer-limited bioelectrocatalysis. In Equation (29), time is a dimensionless quantity and the scale is determined by r. Therefore, Equation (29) is more suitable to define the limiting current decay at a disk electrode than the Cottrell equation. The current decay given by Equation (29) is slower than that of the Cottrell equation. Even if the electrode radius is few mm, the limiting current reaches quasi-steady-state values. The quasi-steady-state characteristics of a disk electrode ($r = 1.5$ mm) realized diffusion-controlled lactate biosensing [43].

5.4. Potentiometric Coulometry

Coulometry is one of the absolute quantification methods and the most accurate electroanalytical methods in theory. In coulometry, the charge (electricity) due to the objective redox reaction is measured. Coulometry coupled with bioelectrocatalysis is a familiar technique [127–131]. However, small Faradaic current compared with the non-Faradaic current at porous electrodes causes a decrease in the accuracy of bioelectrocatalytic coulometry.

On the other hand, potentiometry is expected to avoid the effect of non-Faradaic current. When redox mediators exist in the solution, the redox enzymatic reaction with the mediator changes the ratio of the oxidized and reduced forms of the mediator. The change of the ratio changes the equilibrated solution potential based on the Nernst equation. When the redox mediator is immobilized at the electrode surface in a thin layer, the change of the ratio of the oxidized and reduced forms of the mediator (Γ_O/Γ_R) changes the equilibrated electrode potential (E). If the total amount of immobilized mediator ($A\Gamma_T$) is constant, the accumulated charge (Q_S) is evaluated directly from the initial value and equilibrated value of E (E_i and E_f, respectively) as follows:

$$Q_S = \frac{-n_M F A \Gamma_T}{1 + \exp\left[\frac{n_M F}{RT}(E_f - E^{\circ\prime})\right]} + \frac{n_M F A \Gamma_T}{1 + \exp\left[\frac{n_M F}{RT}(E_i - E^{\circ\prime})\right]}. \tag{33}$$

However, in the case of osmium complex polymer as a mediator immobilized at the electrode surface, the relationship between E and Γ_O/Γ_R was deviated from the Nernst equation [45]. The most possible cause is strong electrostatic interactions between the redox sites in the polymer. An increase in the ionic strength in the medium will decrease the electrostatic interaction between the redox species at the electrode surface. In order to increase the ionic strength, a redox active thin liquid film containing a highly concentrated electrolyte solution was fabricated on the electrode surface [46]. Figure 7 shows the construction of the liquid film-modified electrode. The components of the thin liquid film were a hydrophobic ionic liquid (1-ethyl-3-methylimidazolium bis(nonafluorobutanesulfonyl)imide), organic medium (dibutyl phthalate) and a redox mediator (ferrocene). Ferrocene is enzymatically oxidized at the surface of the liquid film by H_2O_2 with peroxidase (POD). Since the electrostatic interaction was effectively suppressed, the liquid film-modified electrode played as a reversible and ideal surface-confined system. Since the value of $A\Gamma_T$ is controlled in the fabrication of the modified electrode, the value of Q_S by the bioelectrocatalysis can be estimated from Equation (33) without any calibration.

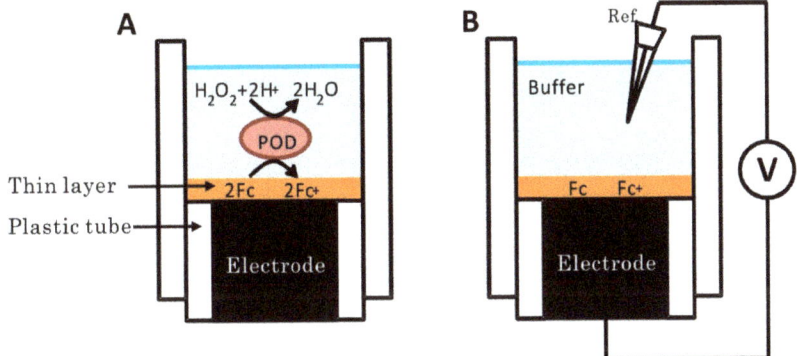

Figure 7. Schematic illustrations of (**A**) accumulation process and (**B**) potentiometric measurement of a liquid-film-modified electrode. Reprinted from ref. [46], Copyright (2015), with permission from Elsevier.

The sensitivity of the potentiometric coulometry could be regulated on the basis of the value of $A\Gamma_T$. Since no current flows across the system, the interference due to the non-Faradaic processes could be eliminated by potentiometric coulometry.

6. Conclusions

Over the past few decades in the field of bioelectrochemistry, several attempts have been made to utilize redox enzymes as electrocatalysts and to develop novel bioelectrochemical systems. By comparison with inorganic catalysts, redox enzymes have distinctive characteristics such as high activity, extremely large size, identity by regeneration, uniformity, versatility and fragility. In this review, we provide an overview and additional insights and discuss the recent progress on the practical use of amperometric biosensors such as multi-enzyme biosensors, multianalyte biosensors. The characteristics of each type of bioelectrocatalysis are summarized in Table 2. Prospective diffusion-controlled biosensors based on DET- and MET-type reactions and potentiometric coulometry based on MET-type reaction may be utilized without calibration. All these efforts may be useful for constructing bioelectrochemical sensors for practical use.

Table 2. Advantages and disadvantages of the bioelectrocatalytic systems for biosensing.

Type of Bioelectrocatalysis	Advantage	Disadvantage
MET-type	Easy coupling of enzyme reaction High loading of enzyme and mediator per projected area	Leakage of mediator and/or enzyme Stability of mediator Low thermodynamic efficiency
DET-type	Crosstalk-free High thermodynamic efficiency	Limited amounts of effective enzyme per projected area Limited number of enzymes Interference from strongly adsorbing substances
Multi-enzymatic cascade	Flexibility in sensor design	Instability due to series reactions

Funding: This research received no external funding.

Conflicts of Interest: The authors declare no conflict of interest.

References

1. Bartlett, P.N. *Bioelectrochemistry: Fundamentals, Experimental Techniques and Applications*; John Wiley & Sons: Chichester, UK, 2008.
2. Wilson, G.S.; Johnson, M.A. In-vivo Electrochemistry: What Can We Learn about Living Systems? *Chem. Rev.* **2008**, *108*, 2462–2481. [CrossRef] [PubMed]
3. Heller, A. Electrical Connection of Enzyme Redox Centers to Electrodes. *J. Phys. Chem.* **1992**, *96*, 3579–3587. [CrossRef]
4. Thévenot, D.R.; Tóth, K.; Durst, R.A.; Wilson, G. Electrochemical Biosensors: Recommended Definitions and Classification. *Pure Appl. Chem.* **1999**, *71*, 2333–2348. [CrossRef]
5. Armstrong, F.A.; Hinst, J. Reversibility and Efficiency in Electrocatalytic Energy Conversion and Lessons from Enzymes. *Proc. Natl. Acad. Sci. USA* **2011**, *108*, 14049–14054. [CrossRef] [PubMed]
6. de Poulpiquet, A.; Ranava, D.; Monsalve, K.; Giudici-Orticoni, M.-T.; Lojou, E. Biohydrogen for a New Generation of H_2/O_2 Biofuel Cells: A Sustainable Energy Perspective. *ChemElectroChem* **2014**, *1*, 1724–1750. [CrossRef]
7. Fourmond, V.; Léger, C. Modelling the Voltammetry of Adsorbed Enzymes and Molecular Catalysts. *Curr. Opin. Electrochem.* **2017**, *1*, 110–120. [CrossRef]
8. Martinkova, P.; Kostelnik, A.; Valek, T.; Pohanska, M. Main Streams in the Construction of Biosensors and Their Applications. *Int. J. Electrochem. Sci.* **2017**, *12*, 7386–7403. [CrossRef]
9. Bollella, P.; Gorton, L. Enzyme Based Amperometric Biosensors. *Curr. Opin. Electrochim.* **2018**, *10*, 157–173. [CrossRef]
10. Kano, K. Fundamentals and Applications of Redox Enzyme-Functionalized Electrode Reactions. *Electrochemistry* **2019**, *87*, 301–311. [CrossRef]
11. Kucherenko, I.S.; Soldatkin, O.O.; Kucherenko, D.Y.; Soldatkina, O.V.; Dzyadevych, S.V. Advances in Nanomaterial Application in Enzyme-Based Electrochemical Biosensors. *Nanoscale Adv.* **2019**, *1*, 4560–4577. [CrossRef]
12. Ngyen, H.H.; Lee, S.H.; Lee, U.J.; Fermin, C.D.; Kim, M. Immobilized Enzymes in Biosensor Applications. *Materials* **2019**, *12*, 121. [CrossRef] [PubMed]
13. Pinyou, P.; Blay, V.; Muresan, L.M.; Noguer, T. Enzyme-Modified Electrodes for Biosensors and Biofuel Cells. *Mater. Horiz.* **2019**, *6*, 1336–1358. [CrossRef]
14. Bollella, P.; Katz, E. Enzyme-Based Biosensors: Tackling Electron Transfer Issues. *Sensors* **2020**, *20*, 3517. [CrossRef] [PubMed]
15. Willner, I.; Katz, E.; Willner, B. Electrical Contact of Redox Enzyme Layers Associated with Electrodes; Routes to Amperometric Biosensors. *Electroanalysis* **1997**, *9*, 965–977. [CrossRef]
16. Barton, S.C.; Gallaway, J.; Atanassov, P. Enzymatic Biofuel Cells for Implantable and Microscale Devices. *Chem. Rev.* **2004**, *104*, 4867–4886. [CrossRef]
17. Cracknell, J.A.; Vincent, K.A.; Armstrong, F.A. Enzymes as Working or Inspirational Electrocatalysts for Fuel Cells and Electrolysis. *Chem. Rev.* **2008**, *108*, 2439–2461. [CrossRef]
18. Meredith, M.T.; Minteer, S.D. Biofuel Cells; Enhanced Enzymatic Bioelectrocatalysis. *Annu. Rev. Anal. Chem.* **2012**, *5*, 157–179. [CrossRef]
19. Mazurenko, I.; de Poulpiquet, A.; Lojou, E. Recent Developments in High Surface Area Bioelectrodes for Enzymatic Fuel Cells. *Curr. Opin. Electrochem.* **2017**, *5*, 74–84. [CrossRef]
20. Mano, N.; de Poulpiquet, A. O_2 Reduction in Enzymatic Biofuel Cells. *Chem. Rev.* **2018**, *118*, 2392–2468. [CrossRef]
21. Krieg, T.; Sydow, A.; Schröder, U.; Schrader, J.; Holtmann, D. Reactor Concepts for Bioelectrochemical Syntheses and Energy Conversion. *Trends Biotechnol.* **2014**, *32*, 645–655. [CrossRef]
22. Milton, R.D.; Minteer, S.D. Enzymatic Bioelectrosynthetic Ammonia Production: Recent Electrochemistry of Nitrogenase, Nitrate Reductase, and Nitrite Reductase. *ChemPlusChem* **2017**, *82*, 513–521. [CrossRef] [PubMed]
23. Shleev, S.; González-Arribas, E.; Falk, M. Biosupercapacitors. *Curr. Opin. Electrochem.* **2017**, *5*, 226–233. [CrossRef]
24. Dutton, P.L. Redox Potentiometry: Determination of Midpoint Potentials of Oxidation-Reduction Components of Biological Electron-Transfer Systems. *Methods Enzymol.* **1978**, *54*, 411–435. [PubMed]

25. Scheller, F.; Kirstein, D.; Kirstein, L.; Schubert, F.; Wollenberger, U.; Olsson, B.; Gorton, L.; Johansson, G. Enzyme Electrodes and Their Application. *Phil. Trans. R. Lond. B* **1987**, *316*, 85–94.
26. Scheller, F.; Schubert, F.; Pfeiffer, D.; Hintsche, R.; Dransfeld, I.; Renneberg, R.; Wollenberger, U.; Riedel, K.; Pavlova, M.; Kühn, M.; et al. Research and Development of Biosensors. *Analyst* **1989**, *114*, 653–662. [CrossRef]
27. Freire, R.S.; Pessoa, C.A.; Mello, L.D.; Kubota, L.T. Direct Electron Transfer: An Approach for Electrochemical Biosensors with Higher Selectivity and Sensitivity. *J. Braz. Chem. Soc.* **2003**, *14*, 230–243. [CrossRef]
28. Léger, C.; Bertrand, P. Direct Electrochemistry of Redox Enzymes as Tool for Mechanistic Studies. *Chem. Rev.* **2008**, *108*, 2379–2438. [CrossRef]
29. Milton, R.D.; Minteer, S.D. Direct Enzymatic Bioelectrocatalysis: Differentiating between Myth and Reality. *J. R. Soc. Interface* **2017**, *14*, 20170253. [CrossRef]
30. Adachi, T.; Kitazumi, Y.; Shirai, O.; Kano, K. Direct Electron Transfer-Type Bioelectrocatalysis of Redox Enzymes at Nanostructured Electrodes. *Catalysts* **2020**, *10*, 236. [CrossRef]
31. Kitazumi, Y.; Shirai, O.; Kano, K. *Significance of Nanostructures of an Electrode Surface in Direct Electron Transfer-Type Bioelectrocatalysis of Redox Enzymes, in Novel Catalyst Materials for Bioelectrochemical Systems: Fundamentals and Applications*; ACS Publications: Washington, DC, USA, 2020.
32. Hooda, V.; Gahlaut, A.; Gothwal, A.; Hooda, V. Recent Trends and Perspectives in Enzyme Based Biosensor Development for the Screening of Triglycerides: A Comprehensive Review. *Artif. Cells Nanomed. Biotechnol.* **2018**, *46*, 626–635. [CrossRef]
33. Semenova, D.; Gernaey, K.V.; Morgan, B.; Silina, Y.E. Towards One-Step Design of Tailored Enzymatic Nanobiosensors. *Analyst* **2020**, *145*, 1014–1024. [CrossRef] [PubMed]
34. Pilas, J.; Yazici, Y.; Selmer, T.; Keusgen, M.; Schöning, M.J. Application of a Portable Multi-Analyte Biosensor for Organic Acid Determination in Silage. *Sensors* **2018**, *18*, 1470. [CrossRef] [PubMed]
35. Zhu, L.; Yang, R.; Zhai, J.; Tian, C. Bienzymatic Glucose Biosensor Based on Co-Immobilization of Peroxidase and Glucose Oxidase on a Carbon Nanotubes Electrode. *Biosens. Bioelectron.* **2007**, *23*, 528–535. [CrossRef] [PubMed]
36. Jia, F.; Narasimhan, B.; Mallapragada, S. Materials-Based Strategies for Multi-Enzyme Immobilization and Co-Localization. *Biotechnol. Bioeng.* **2013**, *111*, 209–222. [CrossRef]
37. Serafín, V.; Hernández, P.; Agüí, L.; Yáñez-Sedeño, P.; Pingarrón, J.M. Electrochemical Biosensor for Creatinine Based on the Immobilization of Creatininase, Creatinase and Sarcosine Oxidase onto a Ferrocene/Horseradish Peroxidase/Gold Nanoparticles/Multi-Walled Carbon Nanotubes/Teflon Composite Electrode. *Electrochim. Acta* **2013**, *97*, 175–183. [CrossRef]
38. Noda, T.; Hamamoto, K.; Tsutumi, M.; Tsujimura, S.; Shirai, O.; Kano, K. Bioelectrocatalytic Endpoint Assays Based on Steady-State Diffusion Current at Microelectrode Array. *Electrochem. Commun.* **2010**, *12*, 839–842. [CrossRef]
39. Noda, T.; Wanibuchi, M.; Kitazumi, Y.; Tsujimura, S.; Shirai, O.; Yamamoto, M.; Kano, K. Diffusion-Controlled Detection of Glucose with Microelectrodes in Mediated Bioelectrocatalytic Oxidation. *Anal. Sci.* **2013**, *29*, 279–281. [CrossRef]
40. Kitazumi, Y.; Noda, T.; Shirai, O.; Yamamoto, M.; Kano, K. Characteristics of Fast Mediated Bioelectrocatalytic Reaction near Microelectrodes. *Phys. Chem. Chem. Phys.* **2014**, *16*, 8905–8910. [CrossRef]
41. Xia, H.-Q.; Kitazumi, Y.; Shirai, O.; Ohta, H.; Kurihara, S.; Kano, K. Putrescine Oxidase/Peroxidase-Co-Immobilized and Mediator-Less Mesoporous Microelectrode for Diffusion-Controlled Steady-State Amperometric Detection of Putrescine. *J. Electroanal. Chem.* **2017**, *804*, 128–132. [CrossRef]
42. Matsui, Y.; Hamamoto, K.; Kitazumi, Y.; Shirai, O.; Kano, K. Diffusion-Controlled Mediated Electron Transfer-Type Bioelectrocatalysis Using Ultrathin-Ring and Microband Electrodes as Ultimate Amperometric Glucose Sensors. *Anal. Sci.* **2017**, *33*, 845–851. [CrossRef]
43. Matsui, Y.; Kitazumi, Y.; Shirai, O.; Kano, K. Simultaneous Detection of Lactate Enantiomers Based on the Diffusion-Controlled Bioelectrocatalysis. *Anal. Sci.* **2018**, *34*, 1137–1142. [CrossRef] [PubMed]
44. Kawai, H.; Kitazumi, Y.; Shirai, O.; Kano, K. Performance Analysis of an Oxidase/Peroxidase-Based Mediatorless Amperometric Biosensor. *J. Electroanal. Chem.* **2019**, *841*, 73–78. [CrossRef]
45. Nieh, C.-H.; Kitazumi, Y.; Shirai, O.; Yamamoto, M.; Kano, K. Potentiometric Coulometry Based on Charge Accumulation with a Peroxidase/Osmium Polymer-Immobilized Electrode for Sensitive Determination of Hydrogen Peroxide. *Electrochem. Commun.* **2013**, *33*, 135–137. [CrossRef]

46. Katsube, R.; Kitazumi, Y.; Shirai, O.; Yamamoto, M.; Kano, K. Potentiometric Coulometry Using a Liquid-Film-Modified Electrode as a Reversible Surface-Confined System. *J. Electroanal. Chem.* **2016**, *780*, 114–118. [CrossRef]
47. Matsumoto, R.; Kano, K.; Ikeda, T. Theory of Steady-State Catalytic Current of Mediated Bioelectrocatalysis. *J. Electroanal. Chem.* **2002**, *535*, 37–40. [CrossRef]
48. Albery, W.J.; Cass, A.E.G.; Shu, Z.X. Inhibited Enzyme Electrodes. Part 1: Theoretical Model. *Biosens. Bioelectron.* **1990**, *5*, 367–378. [CrossRef]
49. Tsujimura, S.; Nakagawa, T.; Kano, K.; Ikeda, T. Kinetic Study of Direct Bioelectrocatalysis of Dioxygen Reduction with Bilirubin Oxidase at Carbon Electrodes. *Electrochemistry* **2004**, *72*, 437–439. [CrossRef]
50. Moser, C.C.; Keske, J.M.; Warncke, K.; Farid, R.S.; Dutton, P.L. Nature of Biological Electron Transfer. *Nature* **1992**, *355*, 796–802. [CrossRef]
51. Marcus, R.A.; Sutin, N. Electron Transfers in Chemistry and Biology. *BBA Rev. Bioenerg.* **1985**, *811*, 265–322. [CrossRef]
52. Marcus, R.A. Electron Transfer Reactions in Chemistry: Theory and Experiment. *Angew. Chem. Int. Ed.* **1993**, *32*, 1111–1121. [CrossRef]
53. Wang, J. Electrochemical Glucose Biosensors. *Chem. Rev.* **2008**, *108*, 814–825. [CrossRef] [PubMed]
54. Heller, A.; Feldman, B. Electrochemical Glucose Sensors and Their Applications in Diabetes Management. *Chem. Rev.* **2008**, *108*, 2482–2505. [CrossRef] [PubMed]
55. Chen, C.; Xie, Q.; Yang, D.; Xiao, H.; Fu, Y.; Tan, Y.; Yao, S. Recent Advances in Electrochemical Glucose Biosensors. *RSC Adv.* **2013**, *3*, 4473–4491. [CrossRef]
56. Zhu, Z.; Garcia-Gancedo, L.; Flewitt, A.J.; Xie, H.; Moussy, F.; Milne, W.I. A Critical Review of Glucose Biosensors Based on Carbon Nanomaterials: Carbon Nanotubes and Graphene. *Sensors* **2012**, *12*, 5996–6022. [CrossRef] [PubMed]
57. Lai, J.; Yi, Y.; Zhu, P.; Shen, J.; Wu, K.; Zhang, L.; Liu, J. Polyaniline-Based Glucose Biosensor. *J. Electroanal. Chem.* **2016**, *782*, 138–153. [CrossRef]
58. Rahman, M.M.; Ahammad, A.J.S.; Jin, J.-H.; Ahn, S.J.; Lee, J.-J. A Comprehensive Review of Glucose Biosensors Based on Nanostructured Metal-Oxides. *Sensors* **2010**, *10*, 4855–4886. [CrossRef]
59. Rahman, G.; Mian, S.A. Recent Trends in the Development of Electrochemical Glucose Biosensors. *Int. J. Biosen. Bioelectron.* **2017**, *3*, 210–213. [CrossRef]
60. Gregg, B.A.; Heller, A. Cross-Linked Redox Gels Containing Glucose Oxidase for Amperometric Biosensor Applications. *Anal. Chem.* **1990**, *62*, 258–263. [CrossRef]
61. Trifonov, A.; Stemmer, A.; Tel-Vered, R. Enzymatic Self-Wiring in Nanopores and Its Application in Direct Electron Transfer Biofuel Cells. *Nanoscale Adv.* **2019**, *1*, 347–356. [CrossRef]
62. Yamashita, Y.; Ferri, S.; Huynh, M.L.; Shimizu, H.; Yamaoka, H.; Sode, K. Direct Electron Transfer Type Disposable Sensor Strip for Glucose Sensing Employing an Engineered FAD Glucose Dehydrogenase. *Enzym. Microb. Technol.* **2013**, *52*, 123–128. [CrossRef]
63. Tsujimura, S.; Kojima, S.; Kano, K.; Ikeda, T.; Sato, M.; Sanada, H.; Omura, H. Novel FAD-Dependent Glucose Dehydrogenase for a Dioxygen-Insensitive Glucose Biosensor. *Biosci. Biotechnol. Biochem.* **2006**, *70*, 654–659. [CrossRef] [PubMed]
64. Malinauskas, A.; Kuzmarskyte, J.; Meskys, R.; Ramanavicius, A. Bioelectrochemical Sensor Based on PQQ-Dependent Glucose Dehydrogenase. *Sens. Actuators B* **2004**, *100*, 387–394. [CrossRef]
65. Flexer, V.; Durand, F.; Tsujimura, S.; Mano, N. Efficient Direct Electron Transfer of PQQ-Glucose Dehydrogenase on Carbon Cryogel Electrodes at Neutral pH. *Anal. Chem.* **2011**, *83*, 5721–5727. [CrossRef] [PubMed]
66. Bollella, P.; Lee, I.; Blaauw, D.; Katz, E. A Microelectronic Sensor Device Powered by a Small Implantable Biofuel Cell. *ChemPhysChem* **2020**, *21*, 120–128. [CrossRef]
67. Mie, Y.; Yasutake, Y.; Ikegami, M.; Tamura, T. Anodized Gold Surface Enables Mediator-Free and Low-Overpotential Electrochemical Oxidation of NADH: A Facile Method for the Development of an NAD^+-Dependent Enzyme Biosensor. *Sens. Actuators B* **2019**, *288*, 512–518. [CrossRef]
68. Ferri, S.; Kojima, K.; Sode, K. Review of Glucose Oxidases and Glucose Dehydrogenases: A Bird's Eye View of Glucose Sensing Enzymes. *J. Diabetes Sci. Technol.* **2011**, *5*, 1068–1076. [CrossRef]
69. Kakehi, N.; Yamazaki, T.; Tsugawa, W.; Sode, K. A Novel Wireless Glucose Sensor Employing Direct Electron Transfer Principle Based Enzyme Fuel Cell. *Biosens. Bioelectron.* **2007**, *22*, 2250–2255. [CrossRef]

70. Roe, J.H.; Epstein, J.H.; Goldstein, N.P. A Photometric Method for the Determination of Inulin in Plasma and Urine. *J. Biol. Chem.* **1949**, *178*, 839–845.
71. Bollella, P.; Hibino, Y.; Kano, K.; Gorton, L.; Antiochia, R. Highly Sensitive Membraneless Fructose Biosensor Based on Fructose Dehydrogenase Immobilized onto Aryl Thiol Modified Highly Porous Gold Electrode: Characterization and Application in Food Samples. *Anal. Chem.* **2018**, *90*, 12131–12136. [CrossRef]
72. Adachi, T.; Kitazumi, Y.; Shirai, O.; Kano, K. Bioelectrocatalytic Performance of D-Fructose Dehydrogenase. *Bioelectrochemistry* **2019**, *129*, 1–9. [CrossRef]
73. Bollella, P.; Ludwig, R.; Gorton, L. Cellobiose Dehydrogenase: Insights on the Nanostructuration of Electrodes for Improved Development of Biosensors and Biofuel Cells. *Appl. Mater. Today* **2017**, *9*, 319–332. [CrossRef]
74. Ito, K.; Okuda-Shimazaki, J.; Mori, K.; Kojima, K.; Tsugawa, W.; Ikebukuro, K.; Lin, C.-E.; La Belle, J.; Yoshida, H.; Sode, K. Designer Fungus FAD Glucose Dehydrogenase Capable of Direct Electron Transfer. *Biosens. Bioelectron.* **2019**, *123*, 114–123. [CrossRef] [PubMed]
75. Gorton, L.; Torstensson, A.; Jaegfeldt, H.; Johansson, G. Electrocatalytic Oxidation of Reduced Nicotinamide Coenzymes by Graphite Electrodes Modified with an Adsorbed Phenoxazinium Salt, Meldola Blue. *J. Electroanal. Chem.* **1984**, *161*, 103–120. [CrossRef]
76. Avramescu, A.; Noguer, T.; Magearu, V.; Marty, J.-L. Chronoamperometric Determination of D-Lactate Using Screen-Printed Enzyme Electrodes. *Anal. Chim. Acta* **2001**, *433*, 81–88. [CrossRef]
77. Takagi, K.; Kano, K.; Ikeda, T. Mediated Bioelectrocatalysis Based on NAD-Related Enzymes with Reversible Characteristics. *J. Electroanal. Chem.* **1998**, *445*, 211–219. [CrossRef]
78. Antiochia, R.; Gallina, A.; Lavagnini, I.; Magno, F. Kinetic and Thermodynamic Aspects of NAD-Related Enzyme-Linked Mediated Bioelectrocatalysis. *Electroanalysis* **2002**, *14*, 1256–1261. [CrossRef]
79. Nikitina, O.; Shleev, S.; Gayda, G.; Demkiv, O.; Gonchar, M.; Gorton, L.; Csöregi, E.; Nistor, M. Bi-Enzyme Biosensor Based on NAD$^+$- and Glutathione-Dependent Recombinant Formaldehyde Dehydrogenase and Diaphorase for Formaldehyde Assay. *Sens. Actuators B* **2007**, *125*, 1–9. [CrossRef]
80. Lobo, M.J.; Miranda, A.J.; Tuñón, P. Amperometric Biosensors Based on NAD(P)-Dependent Dehydrogenase Enzymes. *Electroanalysis* **1997**, *9*, 191–201. [CrossRef]
81. Siritanaratkul, B.; Megarity, C.F.; Roberts, T.G.; Samuels, T.O.M.; Winkler, M.; Warner, J.H.; Happe, T.; Armstrong, F.A. Transfer of Photosynthetic NADP$^+$/NADPH Recycling Activity to a Porous Metal Oxide for Highly Specific, Electrochemically-Driven Organic Synthesis. *Chem. Sci.* **2017**, *8*, 4579–4586. [CrossRef]
82. Xu, C.X.; Marzouk, S.A.M.; Cosofret, V.V.; Buck, R.P.; Neuman, M.R.; Sprinkle, R.H. Development of a Diamine Biosensor. *Talanta* **1997**, *44*, 1625–1632. [CrossRef]
83. Carsol, M.-A.; Mascini, M. Diamine Oxidase and Putrescine Oxidase Immobilized Reactors in Flow Injection Analysis: A Comparison in Substrate Specificity. *Talanta* **1999**, *50*, 141–148. [CrossRef]
84. Nagy, L.; Nagy, G.; Gyurcsanyi, R.E.; Neuman, M.R.; Lindner, E. Development and Study of an Amperometric Biosensor for the in Vitro Measurement of Low Concentration of Putrescine in Blood. *J. Biochem. Biophys. Methods* **2002**, *53*, 165–175. [CrossRef]
85. Gorton, L.; Jönsson-Petterson, G.; Csöregi, E.; Johansson, K.; Domínguez, E.; Marko-Varga, G. Amperometric Biosensors Based on an Apparent Direct Electron Transfer between Electrodes and Immobilized Peroxidases. Plenary Lecture. *Analyst* **1992**, *117*, 1235–1241. [CrossRef]
86. Matsumoto, R.; Mochizuki, M.; Kano, K.; Ikeda, T. Unusual Response in Mediated Biosensors with an Oxidase/Peroxidase Bienzyme System. *Anal. Chem.* **2002**, *74*, 3297–3303. [CrossRef]
87. Tkáč, J.; Gemeiner, P.; Šturdík, E. Rapid and Sensitive Galactose Oxidase-Peroxidase Biosensor for Galactose Detection with Prolonged Stability. *Biotechnol. Tech.* **1999**, *13*, 931–936. [CrossRef]
88. Nieh, C.-H.; Kitazumi, Y.; Shirai, O.; Kano, K. Sensitive D-Amino Acid Biosensor Based on Oxidase/Peroxidase System Mediated by Pentacyanoferrate-Bound Polymer. *Biosens. Bioelectron.* **2013**, *47*, 350–355. [CrossRef]
89. Kacaniklic, V.; Johansson, K.; Marko-Varga, G.; Gorton, L.; Jönsson-Petterson, G.; Csöregi, E. Amperometric Biosensors for Detection of L- and D-Amino Acids Based on Coimmobilized Peroxidase and L- and D-Amino Acid Oxidases in Carbon Paste Electrodes. *Electroanalysis* **1994**, *6*, 381–390. [CrossRef]
90. Yoshida, S.; Kanno, H.; Watanabe, T. Glutamate Sensors Carrying Glutamate Oxidase/Peroxidase Bienzyme System on Tin Oxide Electrode. *Anal. Sci.* **1995**, *11*, 251–256. [CrossRef]
91. Castillo, J.; Gáspár, S.; Sakharov, I.; Csöregi, E. Bienzyme Biosensors for Glucose, Ethanol and Putrescine Built on Oxidase and Sweet Potato Peroxidase. *Biosens. Bioelectron.* **2003**, *18*, 705–714. [CrossRef]

92. Vijayakumar, A.R.; Csöregi, E.; Heller, A.; Gorton, L. Alcohol Biosensors Based on Coupled Oxidase-Peroxidase Systems. *Anal. Chim. Acta* **1996**, *327*, 223–234. [CrossRef]
93. Hasunuma, T.; Kuwabata, S.; Fukusaki, E.; Kobayashi, A. Real-Time Quantification of Methanol in Plants Using a Hybrid Alcohol Oxidase–Peroxidase Biosensor. *Anal. Chem.* **2004**, *76*, 1500–1506. [CrossRef] [PubMed]
94. Johansson, K.; Jönsson-Pettersson, G.; Gorton, L.; Marko-Varga, G.; Csöregi, E. A Reagentless Amperometric Biosensor for Alcohol Detection in Column Liquid Chromatography Based on Co-Immobilized Peroxidase and Alcohol Oxidase in Carbon Paste. *J. Biotechnol.* **1993**, *31*, 301–316. [CrossRef]
95. Smutok, O.; Ngounou, B.; Pavlishko, H.; Gayda, G.; Gonchar, M.; Schuhmann, W. A Reagentless Bienzyme Amperometric Biosensor Based on Alcohol Oxidase/Peroxidase and an Os-Complex Modified Electrodeposition Paint. *Sens. Actuators B* **2006**, *113*, 590–598. [CrossRef]
96. Akyilmaz, E.; Sezgintürk, M.K.; Dinçkaya, E. A Biosensor Based on Urate Oxidase–Peroxidase Coupled Enzyme System for Uric Acid Determination in Urine. *Talanta* **2003**, *61*, 73–79. [CrossRef]
97. Dharmapandian, P.; Rajesh, S.; Rajasingh, S.; Rajendran, A.; Karunakaran, C. Electrochemical Cysteine Biosensor Based on the Selective Oxidase–Peroxidase Activities of Copper, Zinc Superoxide Dismutase. *Sens. Actuators B* **2010**, *148*, 17–22. [CrossRef]
98. Tkáč, J.; Šturdík, E.; Gemeiner, P. Novel Glucose Non-Interference Biosensor for Lactose Detection Based on Galactose Oxidase–Peroxidase with and without Co-Immobilised β-Galactosidase. *Analyst* **2000**, *125*, 1285–1289. [CrossRef]
99. Crumbliss, A.L.; Stonehuerner, J.G.; Henkens, R.W.; Zhao, J.; O'Daly, J.P. A Carrageenan Hydrogel Stabilized Colloidal Gold Multi-Enzyme Biosensor Electrode Utilizing Immobilized Horseradish Peroxidase and Cholesterol Oxidase/Cholesterol Esterase to Detect Cholesterol in Serum and Whole Blood. *Biosens. Bioelectron.* **1993**, *8*, 331–337. [CrossRef]
100. Li, G.; Liao, J.M.; Hu, G.Q.; Ma, N.Z.; Wu, P.J. Study of Carbon Nanotube Modified Biosensor for Monitoring Total Cholesterol in Blood. *Biosens. Bioelectron.* **2005**, *20*, 2140–2144. [CrossRef]
101. Singh, S.; Solanki, P.R.; Pandey, M.K.; Malhotra, B.D. Cholesterol Biosensor Based on Cholesterol Esterase, Cholesterol Oxidase and Peroxidase Immobilized onto Conducting Polyaniline Films. *Sens. Actuators B* **2006**, *115*, 534–541. [CrossRef]
102. Nieh, C.-H.; Kitazumi, Y.; Shirai, O.; Kano, K. Amperometric Biosensor Based on Reductive H_2O_2 Detection Using Pentacyanoferrate-Bound Polymer for Creatinine Determination. *Anal. Chim. Acta* **2013**, *767*, 128–133. [CrossRef]
103. Xia, H.-Q.; Kitazumi, Y.; Shirai, O.; Kano, K. Direct Electron Transfer-Type Bioelectrocatalysis of Peroxidase at Mesoporous Carbon Electrodes and Its Application for Glucose Determination Based on Bienzyme System. *Anal. Sci.* **2017**, *33*, 839–844. [CrossRef]
104. Sakai, K.; Kitazumi, Y.; Shirai, O.; Kano, K. Nanostructured Porous Electrodes by the Anodization of Gold for an Application as Scaffolds in Direct-Electron-Transfer-Type Bioelectrocatalysis. *Anal. Sci.* **2018**, *34*, 1317–1322. [CrossRef]
105. Sadik, O.A.; Aluoch, A.O.; Zhou, A. Status of Biomolecular Recognition Using Electrochemical Techniques. *Biosens. Bioelectron.* **2009**, *24*, 2749–2765. [CrossRef]
106. Andreu-Perez, J.; Poon, C.C.Y.; Merrifield, R.D.; Wong, S.T.C.; Yang, G.-Z. Big Data for Health. *IEEE J. Biomed. Health* **2015**, *19*, 1193–1208. [CrossRef]
107. Tu, J.; Torrente-Rodríguez, R.M.; Wang, M.; Gao, W. The Era of Digital Health: A Review of Portable and Wearable Affinity Biosensors. *Adv. Funct. Mater.* **2020**, *30*, 1906713. [CrossRef]
108. March, G.; Nguyen, T.D.; Piro, B. Modified Electrodes Used for Electrochemical Detection of Metal Ions in Environmental Analysis. *Biosensors* **2015**, *5*, 241–275. [CrossRef]
109. Wang, J.; Katz, E. Digital Biosensors with Built-In Logic for Biomedical Applications–Biosensors Based on a Biocomputing Concept. *Anal. Bioanal. Chem.* **2010**, *398*, 1591–1603. [CrossRef]
110. Guz, N.; Halámek, J.; Rusling, J.F.; Katz, E. A Biocatalytic Cascade with Several Output Signals–Towards Biosensors with Different Levels of Confidence. *Anal. Bioanal. Chem.* **2014**, *406*, 3365–3370. [CrossRef]
111. Katz, E.; Poghossian, A.; Schöning, M.J. Enzyme-Based Logic Gates and Circuits–Analytical Applications and Interfacing with Electronics. *Anal. Bioanal. Chem.* **2017**, *409*, 81–94. [CrossRef]
112. Bard, A.J.; Faulkner, L.R. *Electrochemical Methods: Fundamentals and Applications*, 2nd ed.; Wiley: New York, NY, USA, 2001.

113. Waikar, S.S.; Sabbisetti, V.S.; Bonventre, J.V. Normalization of Urinary Biomarkers to Creatinine during Changes in Glomerular Filtration Rate. *Kidney Int.* **2010**, *78*, 486–494. [CrossRef]
114. Ganguly, A.; Rice, P.; Lin, K.-C.; Muthukumar, S.; Prasad, S. A Combinatorial Electrochemical Biosensor for Sweat Biomarker Benchmarking. *SLAS Technol.* **2020**, *25*, 25–32. [CrossRef]
115. Appenzeller, B.M.R.; Schummer, C.; Rodrigues, S.B.; Wennig, R. Determination of the Volume of Sweat Accumulated in a Sweat-Patch Using Sodium and Potassium as Internal Reference. *J. Chromatogr. B* **2007**, *852*, 333–337. [CrossRef]
116. Smith, S.M.; Eng, R.H.K.; Buccini, F. Use of D-Lactic Acid Measurements in the Diagnosis of Bacterial Infections. *J. Infect. Dis.* **1986**, *154*, 658–664. [CrossRef]
117. Shoup, D.; Szabo, A. Chronoamperometric Current at Finite Disk Electrodes. *J. Electroanal. Chem.* **1982**, *140*, 237–245. [CrossRef]
118. Miyata, M.; Kitazumi, Y.; Shirai, O.; Kataoka, K.; Kano, K. Diffusion-Limited Biosensing of Dissolved Oxygen by Direct Electron Transfer-Type Bioelectrocatalysis of Multi-Copper Oxidases Immobilized on Porous Gold Microelectrodes. *J. Electroanal. Chem.* **2020**, *860*, 113895. [CrossRef]
119. Sugimoto, Y.; Takeuchi, R.; Kitazumi, Y.; Shirai, O.; Kano, K. Significance of Mesoporous Electrodes for Noncatalytic Faradaic Process of Randomly Oriented Redox Proteins. *J. Phys. Chem. C* **2016**, *120*, 26270–26277. [CrossRef]
120. Sugimoto, Y.; Kitazumi, Y.; Shirai, O.; Kano, K. Effects of Mesoporous Structures on Direct Electron Transfer-Type Bioelectrocatalysis: Facts and Simulation on a Three-Dimensional Model of Random Orientation of Enzymes. *Electrochemistry* **2017**, *85*, 82–87. [CrossRef]
121. Lee, H.J.; Beriet, C.; Ferrigno, R.; Girault, H.H. Cyclic Voltammetry at a Regular Microdisc Electrode Array. *J. Electroanal. Chem.* **2001**, *502*, 138–145. [CrossRef]
122. Ordeig, O.; del Campo, J.; Muñoz, F.X.; Banks, C.E.; Compton, R.G. Electroanalysis Utilizing Amperometric Microdisk Electrode Arrays. *Electroanalysis* **2007**, *19*, 1973–1986. [CrossRef]
123. Guo, J.; Lindner, E. Cyclic Voltammograms at Coplanar and Shallow Recessed Microdisk ElectrodeAarrays: Guidelines for Design and Experiment. *Anal. Chem.* **2009**, *81*, 130–138. [CrossRef]
124. Bixler, J.W.; Fifield, M.; Poler, J.C.; Bond, A.M.; Thormann, W. An Evaluation of Ultrathin Ring and Band Microelectrodes as Amperometric Sensors in Electrochemical Flow Cells. *Electroanalysis* **1989**, *1*, 23–33. [CrossRef]
125. Kitazumi, Y.; Hamamoto, K.; Noda, T.; Shirai, O.; Kano, K. Fabrication and Characterization of Ultrathin-Ring Electrodes for Pseudo-Steady-State Amperometric Detection. *Anal. Sci.* **2015**, *31*, 603–607. [CrossRef]
126. Szabo, A.; Cope, D.K.; Tallman, D.E.; Kovach, P.M.; Wightman, R.M. Chronoamperometric Current at Hemicylinder and Band Microelectrodes: Theory and Experiment. *J. Electroanal. Chem.* **1987**, *217*, 417–423. [CrossRef]
127. Uchiyama, S.; Kato, S.; Suzuki, S.; Hamamoto, O. Biocoulometry of Cholesterol Using Porous Carbon Felt Electrodes. *Electroanalysis* **1991**, *3*, 59–62. [CrossRef]
128. Morris, N.A.; Cardosi, M.F.; Birch, B.J.; Turner, A.P.F. An Electrochemical Capillary Fill Device for the Analysis of Glucose Incorporating Glucose Oxidase and Ruthenium (III) Hexamine as Mediator. *Electroanalysis* **1992**, *4*, 1–9. [CrossRef]
129. Fukaya, M.; Ebisuya, H.; Furukawa, K.; Akita, S.; Kawamura, Y.; Uchiyama, S. Self-Driven Coulometry of Ethanol, D-Glucose and D-Fructose Using Ubiquinone-Dependent Dehydrogenase Reactions. *Anal. Chim. Acta* **1995**, *306*, 231–236. [CrossRef]
130. Tsujimura, S.; Nishina, A.; Kamitaka, Y.; Kano, K. Coulometric D-Fructose Biosensor Based on Direct Electron Transfer Using D-Fructose Dehydrogenase. *Anal. Chem.* **2009**, *81*, 9383–9387. [CrossRef] [PubMed]
131. Mizutani, F.; Kato, D.; Kurita, R.; Mie, Y.; Sato, Y.; Niwa, O. Highly-Sensitive Biosensors with Chemically-Amplified Responses. *Electrochemistry* **2008**, *76*, 515–521. [CrossRef]

© 2020 by the authors. Licensee MDPI, Basel, Switzerland. This article is an open access article distributed under the terms and conditions of the Creative Commons Attribution (CC BY) license (http://creativecommons.org/licenses/by/4.0/).

Review

Enzyme-Based Biosensors: Tackling Electron Transfer Issues

Paolo Bollella * and Evgeny Katz

Department of Chemistry and Biomolecular Science, Clarkson University, Potsdam, New York, NY 13699-5810, USA; ekatz@clarkson.edu
* Correspondence: pbollell@clarkson.edu

Received: 30 May 2020; Accepted: 19 June 2020; Published: 21 June 2020

Abstract: This review summarizes the fundamentals of the phenomenon of electron transfer (ET) reactions occurring in redox enzymes that were widely employed for the development of electroanalytical devices, like biosensors, and enzymatic fuel cells (EFCs). A brief introduction on the ET observed in proteins/enzymes and its paradigms (e.g., classification of ET mechanisms, maximal distance at which is observed direct electron transfer, etc.) are given. Moreover, the theoretical aspects related to direct electron transfer (DET) are resumed as a guideline for newcomers to the field. Snapshots on the ET theory formulated by Rudolph A. Marcus and on the mathematical model used to calculate the ET rate constant formulated by Laviron are provided. Particular attention is devoted to the case of glucose oxidase (GOx) that has been erroneously classified as an enzyme able to transfer electrons directly. Thereafter, all tools available to investigate ET issues are reported addressing the discussions toward the development of new methodology to tackle ET issues. In conclusion, the trends toward upcoming practical applications are suggested as well as some directions in fundamental studies of bioelectrochemistry.

Keywords: enzyme-based biosensors; direct electron transfer (DET); redox enzymes; nanostructured electrodes; protein film voltammetry (PFV)

---

1. Introduction

Redox enzymes are defined as proteins that facilitate biological electron transfer (ET) processes, acquitting for multiple essential biological functions like photosynthesis, respiration, nucleic acid biosynthesis, etc. [1–3]. Redox cofactors within the enzymes exhibit different ET thermodynamics and kinetics [4–6]. Moreover, redox cofactors exhibit different formal potentials (E°) spread over a potential window of approximately 1.5 V [3,7–15], which is wider, especially compared to the water thermodynamic stability window, Figure 1, considering hydrogen ions reduction to molecular hydrogen ($E^{0'}_{2H^+/H_2}$ = −0.41 V vs. standard hydrogen electrode (SHE) at pH 7) and oxygen reduction to water ($E^{0'}_{O_2/H_2O}$ = +0.82 V vs. SHE at pH 7) normally occurring in biological systems [16,17]. Recently, it was demonstrated how the potential of redox cofactors is affected by the redox center architecture and the surrounding protein structure [18–20].

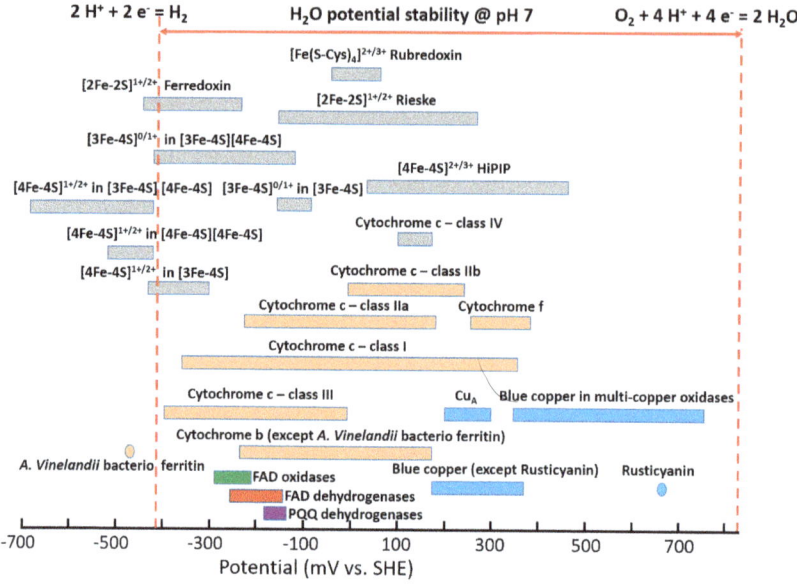

Figure 1. Standard potentials (E°) of various redox proteins and enzymes measured at pH 7.0 and recalculated vs. standard hydrogen electrode (SHE) reference. The potentials spread over range of values for the species originating from different biological sources. The data (except FAD oxidases/dehydrogenases and PQQ dehydrogenases) is adopted from [3] with permission. The potentials of FAD oxidases originate from refs. [7,8]; the potentials of FAD dehydrogenases originate from refs. [9–11]; the potentials of PQQ dehydrogenases originate from [12–15].

The driving force in the investigation of biological redox molecules is mainly related to understanding the biochemical reactions being molecular bases of life [21]. Redox enzymes are extensively employed in the production of biofuels (e.g., hydrogen, methane, cellulose breakdown, etc.) [22–24]. However, they have also been used to develop new biocatalysts to solve challenging synthetic problems, to capture atmospheric CO_2 [25]. Despite the great achievements in synthetic biology and green energy production, redox enzymes, being able to convert biological stimuli into electronic signals, are widely exploited in the development of electrochemical biosensors [26–28]. Among different sensing applications, the most famous example is certainly about blood glucose sensing, which greatly improved the life of billions of people worldwide [29–31].

In this research frame, most of bioelectrochemists have focused their attention on possible solutions to tackle direct electron transfer (DET) issues mainly for the development of sensitive, selective and stable biosensors [32–34]. The electronic coupling between redox enzymes and electrodes for the development of biosensors and biofuel cells can be accomplished according to three mechanisms, denoted as first-, second-, and third-generation biosensors. [35,36] Notably, first-generation biosensors are based on the electroactivity of a substrate or product of the enzymatic reaction [37] (Figure 2A). Second-generation biosensors based on mediated electron transfer (MET) use redox mediators (relays), which are small electroactive molecules shuttling electrons between the enzyme active sites and an electrode [38], Figure 2B. These can be freely diffusing mediators or bound to side chains of flexible redox polymers. In this class, we certainly include all enzymes that are using freely diffusing nicotinamide dinucleotide (NAD^+) as primary electron acceptor, which later needs an immobilized catalyst (e.g., phenothiazines or quinones, particularly including pyrroloquinoline quinone (PQQ) [39], etc.) to reoxidize (recycle) NADH [40,41], Figure 2C.

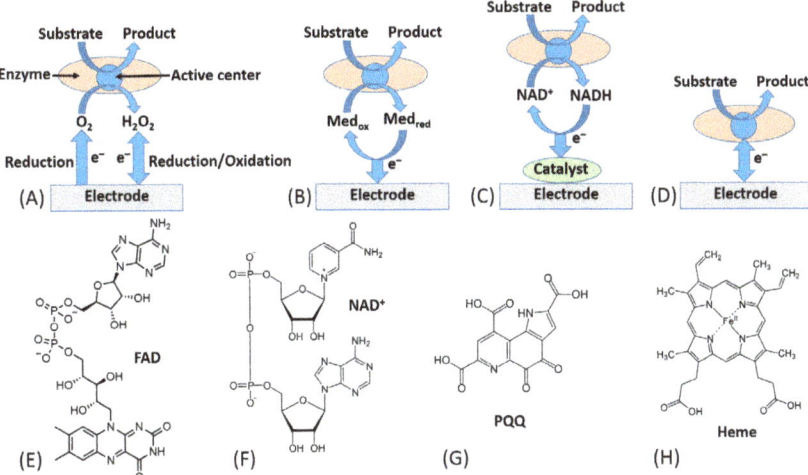

Figure 2. Different ways of electronic communication between redox enzymes and conductive electrodes: (**A**) Electrical communication through electrochemical transformations of enzyme substrate or product (exemplified with reduction of O_2 and reduction/oxidation of H_2O_2 typical for oxidases). (**B**) Electrical communication using electron transfer mediators (relays) cyclic between oxidized (Med_{ox}) and reduced (Med_{red}) states (exemplified with an enzyme oxidizing a substrate and reducing a mediator, which is electrochemically re-oxidized and recycled back to Med_{ox}). (**C**) Electrical communication using NAD^+/NADH cofactor re-oxidized and recycled electrocatalytically (exemplified with an enzyme oxidizing a substrate and reducing NAD^+ yielding NADH). (**D**) Electrical communication via direct electron transfer (DET) from an enzyme active center to an electrode (exemplified with an enzyme oxidizing a substrate and generating anodic current at an electrode). (**E–H**) Structures of the most typical enzyme redox cofactors: flavin adenine dinucleotide (FAD), nicotinamide adenine dinucleotide (NAD^+), pyrroloquinoline quinone (PQQ) and heme.

The mediated ET has been achieved in systems of different complexity [1], ranging from very simple diffusion-operating soluble electron transfer mediators to very sophisticated molecular "machines" shuttling electrons between redox active centers of immobilized enzymes and a conductive electrode support [42]. A very efficient and at the same time simple construct was based on a redox enzyme (e.g., glucose oxidase, GOx) immobilized in a polymer matrix with pendant redox mediator units [43,44]. The enzyme was physically entrapped into the polymer matrix (Figure 3A) or covalently bound to the polymer chain (Figure 3B). The electrocatalytic (ET) current [44] (Figure 3C) has been achieved with random electron hopping from a mediator site to another site, finally reaching an electrode surface. This approach, pioneered by Adam Heller (Figure 3D), was one of the first effective electronic coupling of redox enzymes with electrodes. Another approach has been developed using redox groups tethered to an enzyme backbone, then operating as electron-transporting stations through quasi-diffusional conformational changes in the linker, if the linker was long enough to provide flexibility and mobility of the bound redox mediator [45]. Importantly, the location of the linker should be close to the catalytically active enzyme center. When amino groups of lysine residues are used for the covalent binding of the mediator, their position in different enzymes is important [46] (Figure 4A–C). The mediator-functionalized enzymes first operated in a solution [45] (Figure 4D,E) and then were immobilized at an electrode surface [47]. The tethered mediator facilitated ET from oxidizing enzymes to electrodes (providing anodic current) [20,45] (Figure 4D,F) or to reducing enzymes (providing cathodic flow of electrons) [48,49] (Figure 4E,G), depending on the type of the enzyme and appropriate redox potential of the bound mediator.

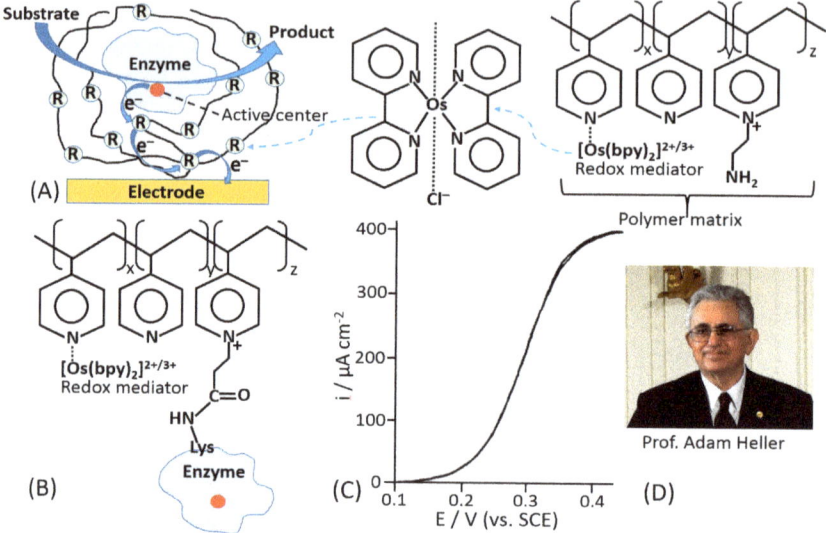

Figure 3. (**A**) An enzyme physically entrapped in a polymer operating as a redox matrix providing the enzyme electrical wiring by electron hopping through redox sites finally reaching an electrode conducting support. The redox-mediator sites are represented with [Os-(2,2′-bipyridine)$_2$]$^{2+/3+}$-complex pendant at poly(vinylpyridine) matrix. (**B**) A similar system where the enzyme is covalently bound to the redox polymer. The systems have been pioneered by Prof. Adam Heller. They are exemplified with an enzyme oxidizing a substrate and generating anodic current mediated by the redox polymer irrespective of the enzyme orientation. (**C**) A cyclic voltammogram, showing a 400 µA cm^{-2} glucose diffusion limited current density reached at 40 mM glucose concentration with the wired-enzyme shown schematically in (**B**). The scan rate is 5 mV/s. (**D**) Prof. Adam Heller – the pioneer in the enzyme wiring according to many various approaches, particularly including systems exemplified in (**A**,**B**). (Part C was adopted from [44]; (**D**) the photo was adopted from Wikipedia, public domain).

Third generation biosensors or DET-based biosensors are realized with the direct electronic connection between the redox center of the enzyme and the electrode surface, which is working as a signal transducer [50], Figure 2D.

From the perspective of biosensing application, the third-generation electrode platform based on DET mechanism shows important advantages compared to MET, considering both soluble and immobilized mediator, and first generation. First of all, the absence of mediators and electroactive substrates/products allows a higher selectivity because the biosensor can operate at a potential closer to the E° of the redox enzyme, thus reducing possible interfering reactions. Second, both soluble/immobilized mediators and electroactive substrates/products may also facilitate unspecific reactions. Next, the absence of a reagent in the reaction sequence makes the device easier to realize. However, as mentioned above adsorbed/immobilized mediators allow the realization of reagentless biosensors (no freely diffusing mediator in solution), which is an obvious advantage compared to other second-generation biosensors that rely on the addition of mediators to sensing solution [2,34].

Today, an efficient ET connection between a variety of electrodes and a wide range of redox enzymes has been accomplished for many redox enzymes (e.g., flavin adenine dinucleotide (FAD), nicotinamide dinucleotide (NAD$^+$), pyrroloquinoline quinone (PQQ), or heme-based redox enzymes) [51]. The chemical structures of their cofactors are shown in Figure 2E–H.

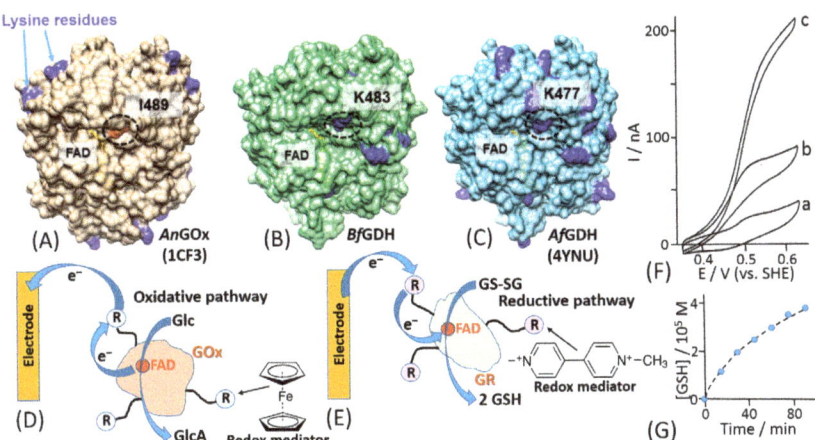

Figure 4. (**A**–**C**) Comparison of the positions of lysine residues in *Aspergillus niger* derived glucose oxidase (*An*GOx) (PDB ID: 1CF3), *Botryotinia fuckeliana* derived glucose dehydrogenase (*Bf*GDH) (model), and *A.flavus* derived GDH (*Af*GDH) (PDB ID: 4YNU). Lysine residues are shown in dark blue. In *Bf*GDH and *Af*GDH, lysine residues (K483, K477, circled) are located at the entrance of what appears to be a pathway to the active center. In *An*GOx, an isoleucine residue (I489, circled) is located at this position. (**D**) The electron transfer from soluble GOx to a Au electrode mediated by ferrocene redox relay (R) species covalently tethered to the enzyme with long flexible chains. Note that ferrocene has a positive redox potential needed to mediate the oxidative biocatalytic process. The biocatalytic reaction results in glucose (Glc) oxidation and gluconic acid (GlcA) formation. (**E**) The electron transfer from a Au electrode to soluble glutathione reductase (GR) mediated by viologen redox relay species covalently tethered to the enzyme with long flexible chains. Note that viologen has a very negative redox potential needed to mediate the reductive biocatalytic process. The biocatalytic reaction results in transformation of the oxidized glutathione (GS-SG) to the reduced glutathione (G-SH). (**F**) Cyclic voltammograms obtained with a bare (unmodified) Au electrode (a disk of 1.5 mm diameter) measured in the presence of a ferrocene-functionalized GOx (12 ferrocene electron relays per a GOx molecule, shown schematically in (**D**); 10 mg/mL): (a) in the absence of glucose; (b) and (c) in the presence of 0.8 mM and 5 mM glucose, respectively. A phosphate buffer solution (0.085 M, pH 7.0) was used as a background electrolyte applied under N_2 atmosphere. Scan rate was 2 mV/s. (**G**) Formation of reduced glutathione bioelectrocatalyzed by GR functionalized with viologen mediator units tethered to the enzyme with long flexible chains (see experimental details in [48]. (Part A is adopted from ref. [46] with permission; part F is adopted from [20] with permission.)

Based on the previous literature, electrostatic compatibility between an electrode and protein surface (part of the surface responsible for ET) seems to play a key role in order to establish an efficient DET, thus showing fully reversible or quasi-reversible cyclic voltammograms in non-turnover conditions (in the absence of an enzyme substrate) [52–54]. Thereafter, the reversibility of non-turnover cyclic voltammograms (depending on the ET rate) will affect the catalytic current produced in the presence of an enzyme substrate. Moreover, the polarity of redox enzyme/electrode interfaces is dramatically affecting the enzyme molecules adsorption and orientation onto the electrode, thus sometimes not facilitating DET processes or even hindering their adsorption, which impedes any biological ET without relying on redox mediators [55,56] (Figure 5A). In this regard, another important aspect, that has been deeply investigated about the enzyme–enzyme and enzyme–interface interactions, is the ability of small multivalent cations to promote the ET between negatively charged proteins and electrodes (e.g., Mg^{2+} or Ca^{2+}, which are ubiquitous in nature) [57]. In this regard, Schulz and his co-workers have been able to increase the catalytic activity of cellobiose dehydrogenase (CDH) by the addition of $CaCl_2$ to the buffer [58,59] (Figure 6). Cellobiose dehydrogenase (CDH, EC 1.1.99.18) is an extracellular monomeric redox enzyme that consists of a catalytically active dehydrogenase domain (CDH$_{DH}$), connected through a flexible linker region to a cytochrome domain (CDH$_{CYT}$). During the catalytic process, carbohydrates (e.g., cellobiose, lactose or glucose) undergo two electrons oxidation at the FAD cofactor (CDH$_{DH}$) subsequently transferring electrons from CDH$_{DH}$ to CDH$_{CYT}$ by an internal electron transfer (IET) process. The reduced CDH$_{CYT}$ further transfers electrons to one-electron molecule acceptors, like cytochrome c (Cyt c), in a biological process or to a macroscopic electrode in a bioelectrochemical process [60–62]. The observed effects of especially divalent Ca^{2+} on the catalytic currents (increased up to five times) can be ascribed to a modified interaction between CDH$_{CYT}$ and the electrode and/or between CDH$_{CYT}$ and CDH$_{DH}$. Regarding the IET, most probably Ca^{2+} ions are complexed by the carboxyl groups of aspartic and glutamic acid at the interface of the CDH$_{DH}$ and CDH$_{CYT}$ domains, thus resulting in a closer domain interaction and a higher IET rate. This concept has been recently demonstrated also for fructose dehydrogenase (FDH), which exhibits a similar structure compared to CDH [63].

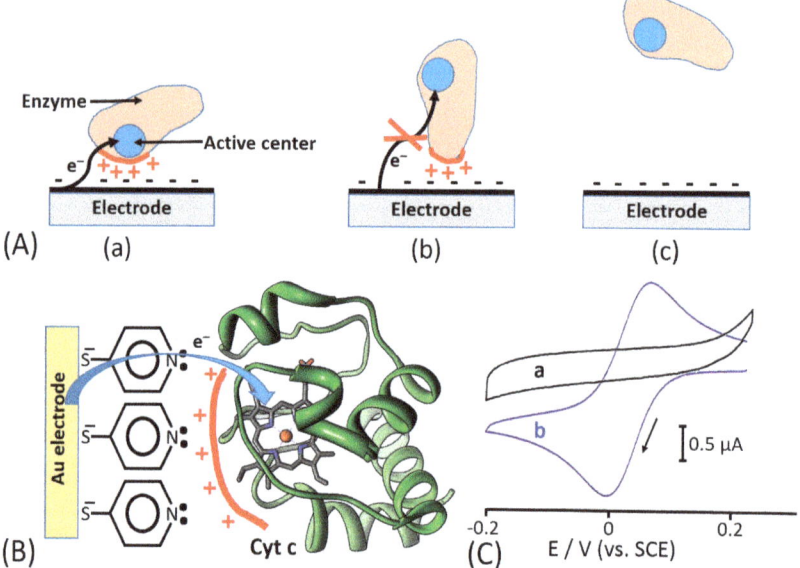

Figure 5. (A) Non-specific protein adsorption outcomes: (a) Electrostatic attraction of oppositely charged protein residues and electrode surface results in immobilization of the protein in an electroactive orientation, facilitating direct electron transfer between a redox center and the electrode. (b) Protein becomes adsorbed in an orientation that does not facilitate direct electron transfer. (c) Protein does not adsorb to the electrode surface and the direct electron transfer is not possible. (B) Alignment of cytochrome c (Cyt c) at a Au electrode surface functionalized with a promoter self-assembled monolayer. The alignment results in a short distance between the heme active center facilitating the direct electron transfer. (C) The cyclic voltammograms obtained in the presence of Cyt c (0.1 mM): (a) at a bare Au electrode without the protein alignment and with no direct electron transfer; (b) at a modified electrode (as shown in (B)) with the alignment facilitating the direct electron transfer. Potential scan rate is 50 mV s^{-1}. Phosphate buffer (1 mM, pH 7.0) under Ar was used as a background electrolyte.

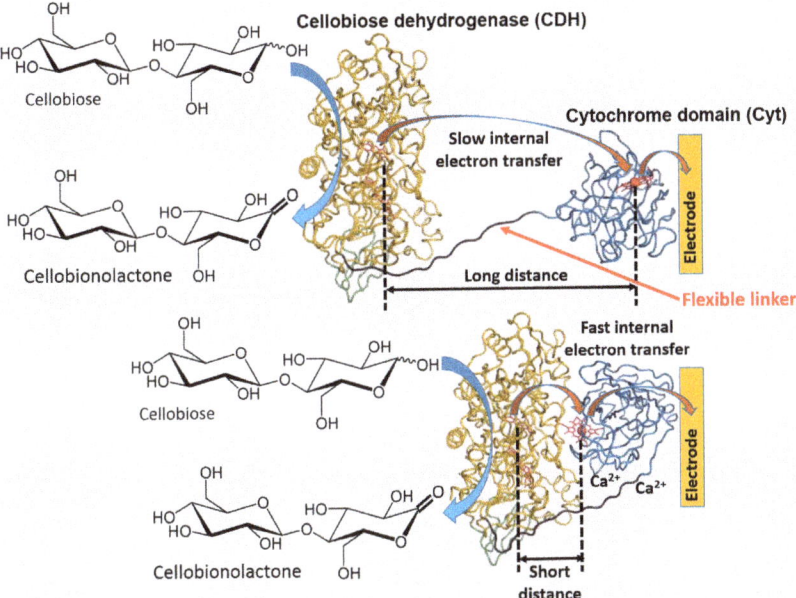

Figure 6. Direct electron transfer from the immobilized cellobiose dehydrogenase (CDH) catalytic domain to an electrode via a covalently linked cytochrome (Cyt) domain. The bioelectrocatalytic current depends on the conformation of the flexible linker. The short electron transfer path resulting in facilitation of the current was realized in the presence of Ca^{2+} cations.

Electrostatic interactions or hydrogen bond formation between redox proteins (e.g., cytochrome c) and monolayer-functionalized electrode surfaces have been used to align proteins at an electrode surface providing a short distance between the redox cofactor and conducting interface (Figure 5B), thus allowing reversible electron transfer [64,65], which was impossible without the orientation effect for the protein molecules (Figure 5C).

In addition to the electrostatic forces that are affecting the enzyme orientation, DET efficiency is also affected by the internal electron tunneling distances. On this specific aspect, Harry Gray and his co-workers demonstrated that electron tunneling distances play a key role in the ET between electron-donor and electron-acceptor partner redox-active centers, thus affecting the ET rate [66–68]. According to Guo and Hill theory [69], the enzymes can be classified in intrinsic enzymes, in which there are no pathways for the electron tunneling because of the absence of appropriate redox sites, and extrinsic enzymes, in which there is a redox acceptor allowing the electron tunneling toward the electrode. Moreover, Dutton et al. established a simple and practical rule that within metalloenzyme structures, high ET rates are supported by an electron tunneling distance of less than 14 Å between redox active sites and electrodes avoiding limiting steps in the redox catalysis. Ideally, in order to allow efficient ET, all enzyme molecules would adsorb and orient on the electrode with the same sub-14 Å distance between the redox cofactor and the electrode [70].

This review aims at summarizing all findings about DET of redox enzymes with a special focus on theoretical (e.g., Marcus theory) and practical aspects (e.g., electrochemical techniques used to study DET and (bio)engineering approaches used to tackle DET issues). A particular attention will be devoted to the case of glucose oxidase (GOx, E.C. 1.1.3.4) from *Aspergillus niger* that has been widely and wrongly used to develop DET based biosensors. Despite the huge number of publications on this subject, that unfortunately accounts for thousands of citations, there is no solid evidence to support DET in GOx based on a stunning statement made by George Wilson: "based on recent experimental results, the observed electrochemical signal corresponds to the FAD cofactor non-covalently bound to the enzyme scaffold that comes out from the redox enzyme upon application of potential, getting adsorbed onto the electrode surface" [71].

2. Theoretical Aspects of Electron Transfer (ET) Processes

2.1. Marcus Theory

In 1992, Rudolph A. Marcus was awarded with the Nobel Prize in Chemistry for his contribution to the development of the ET theory in chemical systems [72,73] (Figure 7A). The theory takes into consideration changes in the structure of the reacting molecules and the solvent's molecules. Based on changes in the energy of the molecular systems, the rate of chemical reactions can be calculated [74].

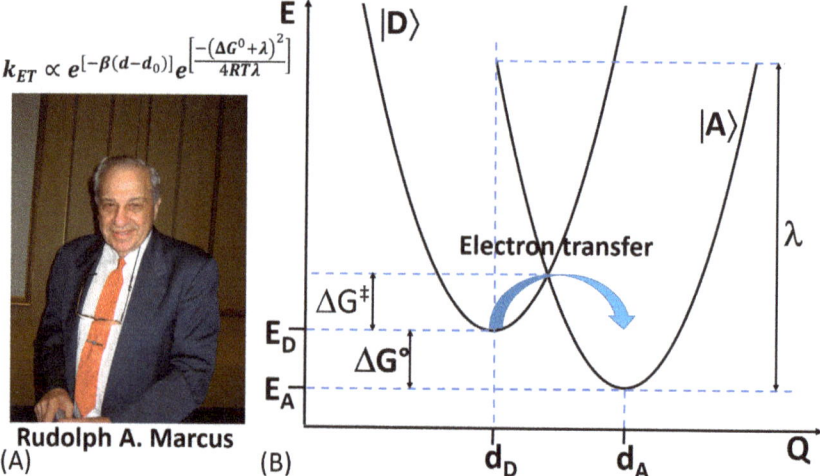

Figure 7. (A) Rudolph A. Marcus, the Nobel Prize in Chemistry (1992) recipient "for his contributions to the theory of electron transfer reactions in chemical systems". (B) Profile of potential energy (E) surfaces vs. nuclear coordinates (Q) for a reactant, an electron donor ($|D\rangle$), and a product, electron acceptor ($|A\rangle$), corresponding to an electron transfer reaction: d_D and d_A—coordinates corresponding to the energy minimum of the reactant and product, respectively; E_D and E_A minimum energies (redox potentials) of the reactant and product, respectively; $\Delta G°$ and $\Delta G^\ddagger$—free Gibbs energy change and activation energy, respectively, in the course of the electron transfer reaction; λ—reorganization energy upon transition from the reactant to the product. The equation is the theoretical expression derived by R. A. Marcus for the electron transfer reaction rate dependence on the energy parameters and electron transfer distance (see explanations for all parameters in [75]). (The photo is adopted from Wikipedia, public domain.)

In an ET reaction, we must first define the electron donor species (D) and the electron accepting ones (A). To enhance the coupling probability of their electronic orbitals, D and A (reactants) should be as close as possible. On the other hand, both vibrational and orientational (affected by surrounding environment) coordinates are varying around the equilibrium values mainly due to charge transfer occurring during the ET process [75]. The potential energy of D and A as reactants and products is expressed as a function of their nuclear coordinates, which can be represented in a multidimensional potential-energy surface (Figure 7B). It should be emphasized that biological ET shares certain features in common with the ET in chemical systems [76]. However, despite the similarities in the ET in chemical and biological systems, we should also consider some substantial differences typical for biochemical reactions—(i) lack of self-exchange reactions, (ii) slightly less available structural information compared to chemical systems, (iii) less homogeneous environment for the ET in biological systems, (iv) lack of free energy (ΔG^0) data for biological systems, (v) ΔG^0 data are dependent on electric potential across the membrane of biological systems, (vi) protein conformational changes may precede or follow the electron transfer in biological systems, thus the binding free energy might differ between the oxidized and reduced form of redox cofactor, and (vii) hindering the possibility of any contacts between donator and acceptor redox center due to physical constrains (e.g., redox centers should have locked position in biological systems) [77].

In biological systems, the dependence of the ET rate on the distance between D and A has been widely elucidated both theoretically and experimentally. In this regard, the ET theory for biological systems, named afterwards "Marcus Theory", is able to predict the ET rate constant value as given by Equation (1):

$$k_{ET} \propto e^{[-\beta(d-d_0)]} e^{[\frac{-(\Delta G^0 + \lambda)^2}{4RT\lambda}]} \tag{1}$$

where β corresponds to the decay or attenuation factor (about 10 nm^{-1} for proteins), ΔG^0 and λ correspond to the free Gibbs energy and reorganization energy accompanying the ET process; d_0 and d are the Van der Waals distance and actual distance between redox active sites; while R and T have their usual meanings [78].

Considering biological D and A reactants similar to those involved in a long range non-biological intramolecular ET, the ET rate constant can be approximated as $e^{-\beta r}$, so exponentially dependent on the distance (β_r) between D and A (as reactants). In the aforementioned formula, the ET rate constant is also dependent on intrinsic (λ) and thermodynamic (ΔG^0) factors as well as dependent from the mutual orientation of the reactants. Although this is the universal rule to study and improve DET connection between redox enzymes and electrodes, other formulas and models have been proposed to calculate the ET rate at an electrode surface.

2.2. Other Theoretical Aspects

Today, many bioelectrochemists are using the model proposed by Laviron to compute the ET rate constant valid for diffusionless (surface-confined) electrochemical systems. The model was derived at first considering linear sweep voltammetry measurements, and it can be applied for any degree of reversibility of the electrochemical reactions [79,80]. However, the main constrain of the model is that both the oxidized and reduced forms of redox species should be strongly adsorbed (immobilized) onto the electrode surface. The first theoretical approximation was derived with the assumption that the adsorbed species do not interact with each other (Langmuir isotherm) [79]. The ET rate constant can be calculated considering the trend between the variation of peak potentials (both anodic and cathodic) toward the logarithm of potential scan rates. The heterogeneous electron transfer rate constant (k_s) for adsorbed (and eventually monolayer immobilized) species can be calculated as follows (Equation (2)):

$$\log k_s = \alpha \log(1-\alpha) + (1-\alpha)\log\alpha - \log\left(\frac{RT}{nFv}\right) - \alpha(1-\alpha)nF\frac{\Delta E_p}{2.3RT} \tag{2}$$

where α denotes the electron transfer coefficient, k_s is the standard rate constant of the surface reaction, v represents the potential scan rate, n is the number of electrons transferred, RT is the gas constant and absolute temperature (K), and ΔE_p is the peak-to-peak separation. The experimentally determined ΔE_p (the difference between the anodic and cathodic peaks; $\Delta E_p = E_{pa} - E_{pc}$) can be found from linear sweep or cyclic voltammetry experiments. The calculated electron transfer rate constant, k_s, can be only a rough estimated value because of many assumptions (mostly assuming Langmuir isotherm for the adsorbed redox species) used in the first theoretical approximation [79]. The second approximation developed by Laviron and Roullier [80] partially solved this problem taking into account possible interactions (attractive or repulsing) of the redox species in the monolayer. However, this theoretical treatment included many parameters which are usually unknown and difficult to find experimentally. Thus, the second Laviron's approximation was rather useless for practical calculations.

Overall, the experimental procedure required for the Laviron's estimation of the rate constant corresponding to the interfacial electron transfer usually includes cyclic voltammetry performed with different potential scan rates, then finding ΔE_p as a function of the logarithm of the potential scan rate. All other parameters can be found from independent experiments, thus allowing use of the Laviron's equation (Equation (2)).

3. Why Glucose Oxidase (GOx) Cannot Undergo DET?

In 1962, Clark and Lyons reported for the first time the employment of glucose oxidase (GOx) from *Aspergillus niger* (E.C. 1.1.3.4) for the development of an enzyme-based electrode [81]. This report has always been recognized by the scientific community as the year of birth of biosensors. Afterward, GOx was widely studied as a redox enzyme for many bioelectrochemical applications (e.g., biosensors, enzymatic fuel cells (EFCs), etc.) [31,82,83]. There are several reasons for its popularity [84]. It is commercially available at relatively cheap costs. Moreover, it is highly active, very stable, and robust as an enzyme [85]. However, the most important reason is its ability to oxidize β-D(+)-glucose, thus allowing monitoring of β-D(+)-glucose for clinical applications like the management of diabetes [29,86,87].

Since Clark's initial paper, an enormous number of papers on electrochemical glucose biosensors have been published [88–93]. As earlier reported in this review, also amperometric glucose biosensors can be divided into three classes based on the type of ET mechanism. While the pioneering work of Clark was highly important for development of electrochemical biosensors, much more practically important results have been obtained with the second-generation biosensors based on mediated electron transfer (MET), greatly contributed by Anthony Cass and Adam Heller [43,44,94] who used ferrocene mediating electron transfer from GOx and included enzymes in polymeric redox matrices, respectively [43,44,94,95]. These works are certainly the most significant advancement in the topic of biosensors [96]. About the "third generation" glucose biosensors, there has always been an open debate on the possibility for GOx to transfer electrons directly from its FAD-cofactor to an electrode surface [71,97]. In this regard, Bartlett et al. recently managed to prove experimentally that there is no evidence to support DET of GOx, thus the vast majority of publications in the literature about DET of GOx are claiming misleading results [98]. This paper raised considerable attention of the scientific community, especially after a "strong" statement made by George Wilson (published as editorial of *Biosensors and Bioelectronics* in 2016 [71]), and today, it has been acknowledged and endorsed by all bioelectrochemists.

The main explanation is based on a deep and comprehensive analysis of the GOx enzymatic structure. The enzyme is a homodimer (composed of two identical units containing flavin adenine dinucleotide (FAD) active sites) highly glycosylated with a molecular weight (MW) of about 160 kDa (the MW is dependent on the level of glycosylation) [99]. The GOx enzyme exhibits high specificity for oxidation of β-anomer of D(+)-glucose and the reaction occurs through a "ping-pong" mechanism, where one of the oxidized FAD in the homodimer reacts with the substrate to give the reduced flavin (FADH$_2$) and the product gluconolactone (which undergoes a subsequent hydrolysis in neutral solution to gluconic acid) [100], as shown in Equations (3a) and (3b).

$$\beta\text{-D(+)-glucose} + \text{GOx(FAD)} \rightarrow \text{D(+)-glucono-1,5-lactone} + \text{GOx(FADH}_2\text{)} \quad (3a)$$

$$\text{D(+)-glucono-1,5-lactone} + \text{H}_2\text{O} \rightarrow \text{gluconic acid} \quad (3b)$$

when the GOx-catalyzed reaction proceeds in an artificial system (not in a native biological environment) the reduced active centers, GOx(FADH$_2$), in the homodimers are then oxidized by reaction with oxygen, regenerating the initial oxidized form GOx(FAD) and producing H$_2$O$_2$ as a byproduct (Equation (4)).

$$\text{GOx(FADH}_2\text{)} + \text{O}_2 \rightarrow \text{GOx(FAD)} + \text{H}_2\text{O}_2 \quad (4)$$

Unfortunately, the majority of the early studies reported on glucose biosensors were performed without any knowledge of the enzyme crystal structure, thus without the knowledge of the distance separating the FAD cofactor and an electrode surface. Notably, the 3D structure of GOx was published by Hecht in 1993 [101–103] (Figure 8A). Based on the analysis of the GOx crystal structure, the two flavin active sites are deeply buried within the enzyme body, thus hindering any redox communication between the two dimeric units of the enzyme. From the crystal structure, it is possible to observe a distance of about 17–22 Å between the active sites and the enzyme surface. Actually, this distance was estimated for a deglycosylated enzyme; thus, in reality, distances might be larger considering that the molecular weight and size of the "native" (glycosylated) GOx are higher by 16-25% compared to the deglycosylated species [104–106]. Therefore, the DET for GOx is rather unlikely. Moreover, it is possible to observe the presence of a channel at the interface between the GOx homodimers. This structure actually hinders any possibility of non-specific electron transfer to electron acceptors available in biological systems and controls the local environment around the FAD cofactor ensuring high selectivity, not oxidizing any closely related carbohydrates. The turnover rate for the β-anomer of D(+)-glucose is 150 times higher than that for the α-anomer [107].

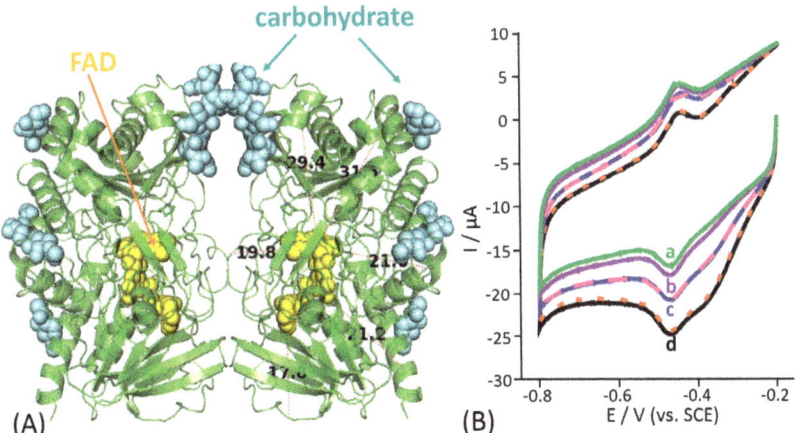

Figure 8. (**A**) Representation of the secondary structure of GOx. The image was obtained with PyMol software, PyMol visualizations are based on the crystal structure of GOx from *Aspergillus niger*, PDB code 1gal http://www.rcsb.org/pdb/explore/explore.do?structureId=1GAL. (**B**) Cyclic voltammograms showing oxygen reduction on a glassy carbon electrode modified with multi-walled carbon nanotubes and loaded with GOx: (a) in the absence of glucose, and in the presence of (b) 2 mM, (c) 4 mM and (d) 8 mM glucose. The experiment was performed in 0.1 M phosphate buffer, pH 6.8, in the presence of oxygen (in equilibrium with air) with the potential scan rate of 60 mV s^{-1}. Note that this and similar cyclic voltammograms were erroneously reported as the proof of the DET with GOx. (The figure is adopted from ref. [98] with permission.)

In most previous papers, GOx has been immobilized on carbon-based nanomaterials (e.g., carbon nanotubes, graphene, graphite, etc.) or metal-based nanomaterials (e.g., gold nanoparticles, porous gold, etc.) [108–114]. Concerning carbon nanomaterials, the claimed DET of GOx is often attributed to some "special" but not clearly specified, properties of the carbon nanomaterials or possibly some particular interactions of the enzyme and the carbon nanotubes that allow to access the active site, thus enabling the charge transfer between the FAD cofactor and carbon nanotubes [115–118].

In a typical cyclic voltammogram, the data reported to support these claims of DET for glucose oxidase are (*i*) a pair of peaks with the shape characteristic of surface-confined redox species at around −0.26 V vs. SHE at pH 7, assumed to be the FAD in the active enzyme center, and (*ii*) the catalytic current upon addition of glucose that correlates with the glucose concentration. Figure 8B shows the typical results usually considered as evidence for DET of GOx, without taking into account the real scientific meaning of the results. First, the current response shows a reductive "catalytic" current that cannot be considered as oxidation response expected from the enzymatically catalyzed oxidation of glucose. Second, the surface redox peaks around −0.26 V vs. SHE are unchanged by addition of D(+)-glucose with both oxidation and reduction peaks clearly present throughout and simply moving with the changing background. Third, the change in the current upon glucose addition starts at approximately 0 V vs. SHE, which does not match with the thermodynamic potential of FAD. To analyze and understand the results reported in Figure 8B, we should consider the fact that oxygen is dissolved in the solution. Most likely, the reductive "catalytic" wave correlates with the reduction of oxygen on carbon nanotubes electrode that starts at 0 V vs. SHE. The reductive "catalytic" current decreases because oxygen is serving as electron acceptor for the reaction catalyzed by the enzyme, thus being less available for its reduction at the electrode surface upon addition of different concentration of glucose [119]. In other words, the observed effect of glucose originates not from the glucose oxidation but rather from O_2 depletion. Next, we should also consider that the reduction of oxygen produces mainly H_2O_2 in that potential window [120]. Moreover, H_2O_2 exhibits a large overpotential window on carbon nanotubes being thus unavailable for its further reduction to water. However, the reaction can be easily catalyzed in the presence of metal nanoparticles incorporated into carbon nanotubes on the electrode surface, therefore consuming the H_2O_2 produced from the enzymatic reaction [121–123]. It should be noted that the redox peaks observed ($E^{0\prime}$ = −0.26 V vs. SHE in pH 6.8) in non-turnover conditions cannot be taken as the evidence for the DET from the flavin in the active site of GOx to the electrode surface. Conversely, they most likely arise from free FAD adsorbed directly on the electrode surface. The free FAD can either be present as impurity of the enzyme sample or dissociate from the enzyme during the incubation of the electrode with the enzyme [98]. These experiments unequivocally prove that GOx immobilized on an electrode surface cannot undergo DET reaction mechanism. Unfortunately, many of these claims were usually supported by the sentence "the data presented are similar to previously reported literature", without considering the reliability and the "correct" scientific meaning or interpretation of the results.

4. Methods to Investigate DET Issues

For bioelectrochemists, one of the most intriguing and recent challenges has been the construction of an electrode platform based on the DET mechanisms. Notably, to realize the DET mechanism and then to prove it many methods have been developed. Herein, we sort them in two classes: biochemical methods facilitating the DET (e.g., deglycosylation, enzyme mutation, etc.) and electrochemical methods investigating the process and providing the prove of the DET (e.g., cyclic voltammetry, amperometry, protein film voltammetry, etc.) [16,97,124,125].

4.1. Biochemical Methods

This section is resuming the main biochemical methods that have been widely employed to distinguish DET bioelectrocatalysis from other catalytic signals, originating possibly from a dissociated cofactor, like in the case with GOx. At first, we should take into account that small conformational changes in the enzyme may occur upon its immobilization at an electrode, thus, lowering also the enzyme activity. From the biochemical point of view, one control experiment cannot confirm the DET mechanism of the enzyme, thus multiple experiments are indeed required to prove the ET pathway of the enzyme immobilized on the electrode and to support the conclusion on the DET.

For example, as the first experiment, we might consider determining the formation of the expected product, which does not give any confirmation for the ET mechanism, but it only tells that the reaction is actually taking place (it might occur at a dissociated cofactor or other reasons not related to the enzyme activity). Recently, Duca et al. co-immobilized nitrogenase with a noble metal catalyst in order to perform the stepwise reduction of nitrate to ammonia through its intermediate, namely nitrite [126]. Initially, nitrate reduction to nitrite was assayed by using the well-established Griess method [127] to detect nitrite, also confirmed by computing the theoretical nitrite amount produce during the bioelectrocatalytic process (the charge passed through the electrode during the catalytic process was correlated with the concentration of nitrite by using Coulomb's law). After proving the intermediate, the authors used a fluorescent compound, namely (o-phthalaldehyde), to quantify the ammonia produced during the further reaction step catalyzed by the noble metal catalyst. This method is particularly useful when the whole biocatalytic process proceeds through several intermediate steps [128,129].

Besides the product analysis, we should consider the importance of enzyme tertiary and quaternary structure for its activity. Recently, it has been proved that estimating the activity loss occurring upon denaturation processes (e.g., heating to elevated temperature from 80 °C to 100 °C for a short time or by treating the enzyme with proteases, like trypsin, etc.), it is possible to differentiate between a DET bioelectroctalytic mechanism and a cofactor that dissociates from the enzyme and undergo only electrocatalysis (without any contribution from the enzymatic structure to selectivity and turnover) [130,131]. Additionally, if the enzyme exhibits a complex ET mechanism it is possible to add different inhibitors during bioelectrocatalytic measurements, thus determining all the steps and ET mechanisms contributing to the enzyme turnover [132–134].

Although previously reported methods are easy to perform, one of the most appealing method to confirm possible DET still remains the generation of mutated or modified oxidoreductases that exhibit altered catalytic properties, sometimes depending on the orientation and immobilization at the electrode surface [135]. These mutations certainly include single-point mutations, the cleavage of component subunits or even the deglycosylation of enzymes [136]. Consequently, these might alter the apparent kinetics parameters like maximum reaction rate, V_{MAX}, or the Michaelis constant, K_M. Mutations can also induce modification of the ET mechanism or internal electron-tunneling pathway. For example, Léger et al., who altered the ligation of the distal [4Fe–4S] cluster in a hydrogenase, by replacing a histidine residue with a glycine residue. The DET bioelectrocatalysis was significantly facilitated as the result of this change in the protein backbone [137]. However, the generation of mutant enzymes is neither straightforward nor trivial in the majority of the cases.

Finally, the potential applied to the electrode may indicate whether direct bioelectrocatalysis of active enzyme is occurring. This assumption is based on the fact that the reduction potential of the enzyme's cofactor has been predetermined and that its reduction potential is not mildly altered upon any small conformational changes that may occur by the enzyme immobilization.

4.2. Electrochemical Methods

Classical electrochemical methods like cyclic voltammetry, linear sweep voltammetry, differential pulse voltammetry and chronoamperometry have been widely employed to investigate the catalytic properties of many enzyme-modified electrodes giving important insight on their apparent kinetics properties and also on the ET mechanism [138].

However, the main achievements were reported after introduction of a new investigation methodology reported by Fraser Armstrong (Figure 9A) as a protein-film voltammetry (PFV) [139–141]. In this technique, redox enzymes are "wired" directly to the electrochemical analyzer, which is able to activate and measure the redox behavior of the enzyme. A redox enzyme can be likened to an intriguing electronic device of which we would like to know all electronic features. To investigate its properties, we would plug the device to an electronic probe to measure all the parameters about its relays, switches, gates, etc., that have their representatives within the complex machinery of multi-centered redox enzymes as redox active sites [142]. The PFV provides a powerful way to investigate how ET processes occurs between active sites and electrodes and how the catalytic ET through an enzyme is controlled. The enzyme is adsorbed on a suitable electrode as a stable mono-/sub-monolayer film of molecules that are oriented to ease ET process [143], Figure 9B. This approach allows to overcome the problems of sluggish protein diffusion and kinetics limitations of the protein at the electrode. Therefore, PFV allows the detection and quantification of the complex and redox coupled chemical reactions that occur at the active sites [144].

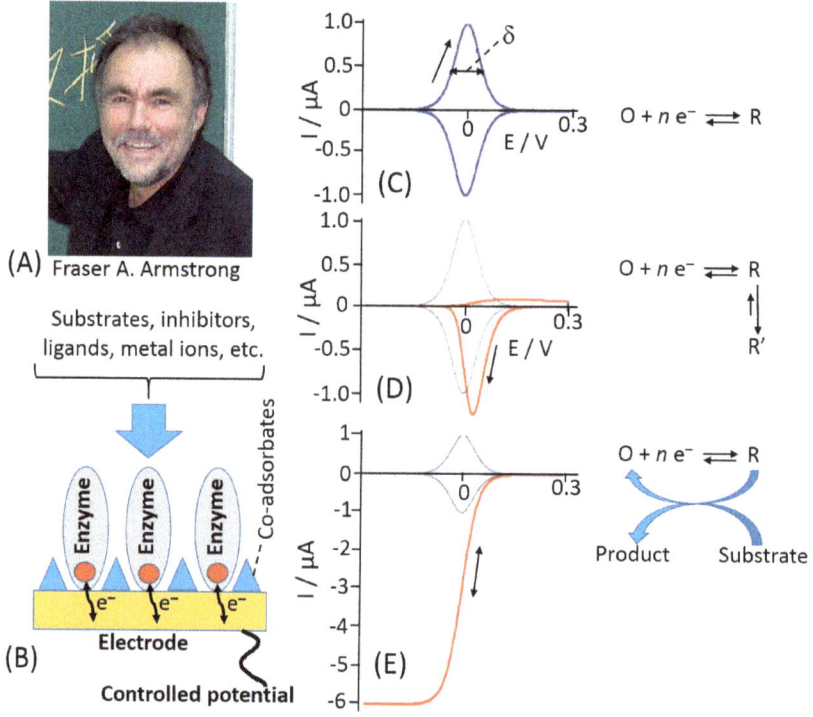

Figure 9. (A) Fraser A. Armstrong. (B) Cartoon showing an adsorbed monolayer of protein molecules on an electrode. Electron transfer accompanying a biocatalytic process is shown schematically. (C–E) Voltammograms expected for adsorbed redox couples displaying different types of ET coupling: (C) Reversible ET. (D) Orange trace shows ET coupled to a spontaneous chemical reaction of the reduced form; on this timescale, the reverse chemical process gates electron transfer. Blue trace shows uncoupled ET for comparison. (E) Orange trace shows ET coupled to catalytic regeneration of an oxidized form. Blue trace shows uncoupled ET. (The Armstrong photo is adopted from Wikipedia, public domain; Parts (C–E) were adopted from ref. [140] with permission.)

Notably, the PFV exhibits several advantages in the investigation of the protein ET mechanism compared to other electrochemical techniques—(*i*) redox active centers are fully controlled through the electrode potential, thus allowing fine-tuning of the enzyme redox properties; (*ii*) well-defined curves especially considering the ideal case of a reversible ET mechanism that would give a couple of redox peaks with a Nerstian half-height width $90.6/n$ mV (where n is a number of electrons in the ET process) (Figure 9C)—by integrating the peak area, it is possible to determine the number of active sites within the protein layer; (*iii*) little amount of samples is needed to form a monolayer based on the assumption that the electrode surface can accommodate about 10^{-12}–10^{-11} moles cm^{-2} of redox protein (other techniques require much bigger samples to get similar information); (*iv*) high sensitivity, considering the little amount of enzyme deposited onto the electrode; (*v*) high rate for ET reactions because they are not limited by diffusion. If the ET rate constant determined is at least 500 s^{-1} or higher, a chemical step with a very short half-life (milliseconds) can be coupled to the main ET reaction [145].

A completely different cyclic voltammogram from the one reported in Figure 9C can be obtained if an ET is followed by a spontaneous chemical process resulting in a product, which is reoxidized in a relatively slow reverse chemical process, showing an irreversible cyclic voltammogram, Figure 9D. Normally, the effect of catalytic turnover on the voltammetry depends primarily on how much mass transport of a substrate is limiting the reaction rate (current). This will be true in case of a macro-electrode coated with a high coverage of a very active enzyme. As the coverage or activity decreases, or if the electrode is a micro-electrode or one that is rotating at high speed, the current will more likely be determined by ET or properties of the enzyme. The catalytic turnover causes the peak-like signal to convert to a sigmoidal wave, where current is directly correlated to turnover rate, Figure 9E. Some of the well-studied examples include cytochrome P450, other heme-containing proteins, like hemoglobin, myoglobin, cytochrome c, and furthermore PS-I and PS-II photosystems, the proteins from the electron transfer chain, several hydrogenases, some Mo-containing proteins and various Fe–S, and other metal-containing proteins (mainly with Ni–Fe, Mn, or Cu as redox centers) [125,139,143].

In the last two decades, the PFV has been largely exploited to investigate the ET of many redox enzymes, that otherwise would remain unknown considering other techniques. Recently, the PFV has been coupled with special spectrophotometric techniques in order to monitor the variation in absorbance of a redox site while applying a specific potential at the electrode [146].

5. Different Approaches to Tackle DET Issues

Considering the advantages of the DET mechanism for the construction of many electrochemical biosensors, multiple different approaches to tackle the DET issues have been proposed over the past thirty years, like apo-enzyme reconstitution at an electrode surface, enzyme bioengineering, deglycosylation, site-oriented immobilization, and electrode nano-structuration [147–154].

Since the distance separating enzyme redox active catalytic centers and electrodes is the main problem for the DET, several approaches have been reported for decreasing this distance and facilitating the ET. One of the methods is based on plugging-in electronically conducting nanospecies, such as small Au nanoparticles or carbon nanotubes. This nano-size conducting bridges electronically connecting enzyme active centers and electrodes represent a nanotechnological approach to the electrical "wiring" of redox enzymes. It should be noted that the ultra-small size of the nano-bridges is critically important to allow their insertion into the protein globule for the efficient electrical contacting with the redox active centers located inside the protein.

Willner et al. proposed reconstitution of an apo-flavoenzyme, namely apo-GOx, on a small gold nanocluster formed by 55 Au atoms (1.4 nm diameter) functionalized with the FAD cofactor [155] (Figure 10A). The gold nanoparticles were immobilized onto a gold electrode using a bifunctional thiol linker, benzene-1,4-dithiol, readily chemisorbed on the gold electrode producing a self-assembled monolayer, then attached to the Au nanoparticles with the second thiol group. The FAD derivative bound to the Au nanoparticles included an additional amino group linked to the cofactor unit through a short spacer (Figure 10C), which allowed the FAD cofactor to be kept separate from the nanoparticle to allow its reconstitution but still at a distance that provided the efficient electron transfer. In a control experiment, when the reconstitution of the apo-GOx with the FAD-functionalized Au nanoparticles was performed in a solution, scanning transmission electron microscopy (STEM) demonstrated a single Au nanoparticle bound to a GOx molecule (Figure 10B). Importantly, the reconstitution method resulted in the specific positioning of the Au nanoparticle near the active center of the reconstituted enzyme being partially embedded into the protein body, thus allowing efficient electron transfer from the FAD active center to the Au nanoparticle and then to the electrode support. Cyclic voltammetry measurements have demonstrated an electrocatalytic current corresponding to the glucose oxidation with the electron transfer through the Au nanoparticle operating as a conducting bridge (Figure 10D). This approach allowed obtaining an ET rate constant of approximately 5000 s^{-1}. Despite the fact that the turnover number for the reconstituted enzyme was impressive and the ET was organized through the conducting bridge, the anodic electrocatalytic current was observed only with a very large overpotential starting at approximately +0.3 V (vs. SCE), while the FAD potential is ca. −0.45V (pH 7.0; SCE), thus requiring at least ca. 750 mV overpotential for the oxidation process (much bigger overpotential for the increased anodic current). A similar approach using carbon nanotubes functionalized with the same amino-FAD derivative for reconstituting apo-GOx has demonstrated an ET over a very long distance [156] (Figure 11). Both systems, where Au nanoparticles or carbon nanotubes have been used as conducting bridges, demonstrated quasi-direct ET transfer from the enzyme active centers to the electrode supports. The ET in these systems was not mediated by chemical redox species but rather provided by electronically conducting nano-wires. The major disadvantages of the reported systems [155,156] are the following: (i) a large overpotential originating from distances separating the nano-bridges and FAD cofactor and between the nano-bridges and the electrode surface and (ii) the use of the artificial (synthetic) FAD derivative (note that its synthesis is extremely complicated [148]). Both disadvantages resulted in low practical importance of the systems despite their scientific novelty. Particularly, the large overpotential for the anodic process did not allow use of these biocatalytic electrodes in biofuel cells.

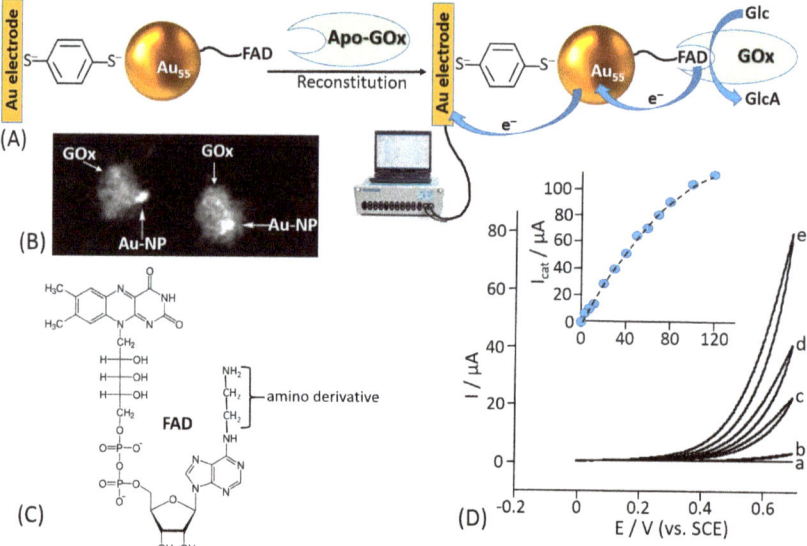

Figure 10. (A) Reconstitution of apo-GOx on FAD-functionalized Au nanoparticles operating as an electronically conducting nano-bridges facilitating electron transfer from the reduced FAD active center to the electrode support. (B) The scanning transmission electron microscopy (STEM) image showing binding of a single Au nanoparticle per the GOx enzyme molecule. (C) The synthetic FAD amino-derivative structure (note an additional amino group connected to the adenine with a spacer composed of two methylene groups). (D) Cyclic voltammograms corresponding to the bioelectrocatalyzed oxidation of glucose by the reconstituted GOx in the presence of different glucose concentrations: (a) 0 mM, (b) 1 mM, (c) 10 mM, (d) 20 mM, and (e) 50 mM. Results were recorded in 0.1 M phosphate buffer (pH 7.0), under Ar, potential scan rate 5 mV s^{-1}. Inset: Calibration plot derived from the cyclic voltammograms at $E = 0.6$ V vs. SCE. (Part D was adopted from [155] with permission.)

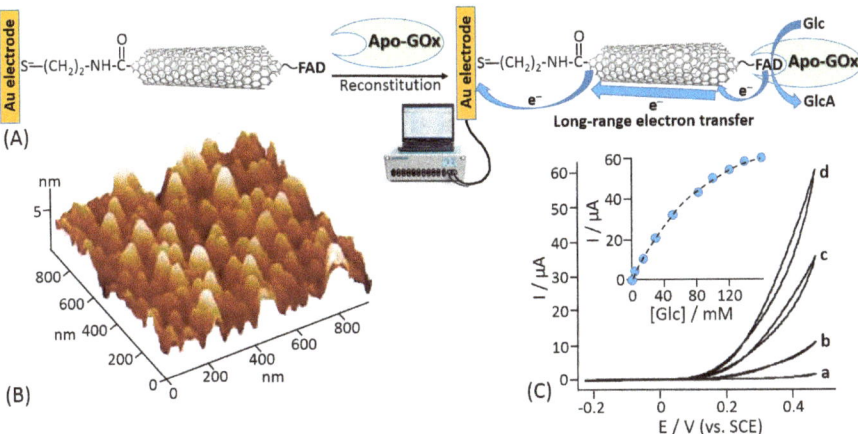

Figure 11. (**A**) Reconstitution of apo-GOx on FAD-functionalized carbon nanotubes operating as an electronically conducting nano-bridges facilitating long-range electron transfer from the reduced FAD active center to the electrode support. (**B**) Atomic force microscope (AFM) image of the GOx reconstituted on the FAD-functionalized carbon nanotubes monolayer associated with the Au electrode surface. (**C**) Cyclic voltammograms corresponding to the electrocatalyzed oxidation of different concentrations of glucose by the GOx reconstituted on the 25 nm-long FAD-functionalized carbon nanotube assembly: (a) 0, (b) 20, (c) 60 and (d) 160 mM glucose. Data recorded in phosphate buffer, 0.1 M, pH 7.4, scan rate 5 mV s^{-1}. Inset: Calibration curve corresponding to the amperometric responses of the reconstituted GOx-electrode at E = 0.45 V in the presence of different concentrations of glucose. (Parts B and C were adopted from ref. [156] with permission.)

Banta and Atanassov et al. proposed a different approach to optimize the electrical communication between GOx and an electrode surface [157]. They introduced cysteine residues offering thiol groups into the GOx protein backbone by genetic engineering substituting natural amino acids, Figure 12A. Depending on the position of the newly introduced cysteine, the distance from its thiol group to the FAD cofactor was different ranging from 13.8 Å to 28.5 Å. The orientation of the GOx molecules at the electrode surface was controlled by the position of the artificially introduced cysteine used for the immobilization. Among all prepared GOx mutants, only H447C-mutant showed the DET activity as reported by the cyclic voltammetry that is displaying a small catalysis in the presence of glucose, in turnover conditions, Figure 12C. This might be ascribed to the low surface coverage of the enzyme. For this reason, the authors linked the mutated enzyme to a gold nanoparticle in order to facilitate the DET and increase the enzyme surface coverage. The artificially introduced thiol group was reacted with a maleimide-functionalized Au nanoparticles resulting in their covalent binding to the GOx protein backbone at the specific site, Figure 12B. Then, the Au-GOx mutant conjugate was bound to the electrode surface. The electrode modified with the H447C-Au nanoparticle showed a great catalytic current in the presence of the glucose substrate (Figure 12C). Unfortunately, the published cyclic voltammogram does not show clear redox peaks for FAD (in the absence of glucose, in non-turnover conditions), but the catalytic wave is starting at the potential close to the thermodynamic potential of FAD embedded in GOx, thus proving the DET features of the modified electrode.

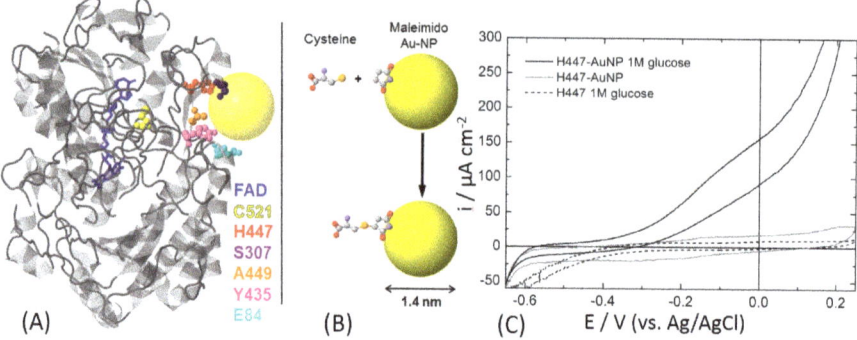

Figure 12. (**A**) Ribbon diagram of a GOx monomer (from *A. niger*) with the FAD molecule shown in blue. The amino acid residues targeted for mutagenesis are highlighted as space-filling models: cysteine (yellow), histidine (red), serine (purple), alanine (orange), tyrosine (pink), and glutamate (light blue). The yellow sphere represents an idealized Au nanoparticle (Au-NP) on the same scale as GOx. (**B**) Schematic drawing of the covalent-binding chemistry of cysteine to a maleimide-modified Au nanoparticle. The molecules are displayed as ball-and-stick: carbon (gray), oxygen (red), nitrogen (blue), and sulfur (yellow). (**C**) Cyclic voltammograms of H447C-Au-NP conjugates on a gold electrode in the presence (black line) and absence (gray line) of 1 M glucose (N_2-saturated buffer, pH 7, 10 mV s^{-1}). The cyclic voltammogram for unconjugated H44C is shown as a dotted line. The H447C-Au-NP conjugates in the presence of glucose exhibit enzymatic glucose oxidation starting at ca. −400 mV. The definition of all used mutants and their abbreviate names can be found in ref. [157]. (The figure was adopted from ref. [157] with permission.)

Recently, Gorton et al. proposed deglycosylation as a very innovative approach to tackle DET issues, especially based on their promising results previously obtained for horseradish and tobacco peroxidase, where the distance between the prosthetic group (*heme b*) and the electrode surface was effectively reduced, thus enhancing the DET rate. They proposed the same approach to enhance the ET rate of cellobiose dehydrogenase (CDH) [158]. In particular, they studied the effect of deglycosylation on two of most representative variants of CDH, namely CDH from *Phanerochaete chrysosporium* (PcCDH) and *Ceriporiopsis subvermispora* (CsCDH). Note that these enzymes are composed of two covalently linked domains, the catalytic dehydrogenase domain (CDH$_{DH}$) and electron transfer cytochrome domain (CDH$_{Cyt}$). The electron transfer proceeds as an internal interdomain process. Indeed, the study demonstrated that deglycosylation improves the catalytic current density, I_{max}, and the sensitivity for lactose, as a substrate, which could be ascribed to a higher number of the electroactive CDH molecules at the electrode surface due to the downsizing of the enzyme's dimensions and a facilitated DET due to the deglycosylation, which reduces the ET distance. Although the DET rate between CDH$_{CYT}$ and the electrodes was increased, no DET between CDH$_{DH}$ and the electrodes has been observed. The increased current density observed with the deglycosylated CDH-modified electrodes originates certainly from the decreased size of the deglycosylated CDHs. However, deglycosylation was also affecting the intrinsic kinetic parameters of the enzyme. The main drawbacks of this approach are the high cost and the impossibility to scale-up the process to industrial level for the production of very sensitive DET-based biosensors. However, the same approach has been used for pyranose dehydrogenase and other highly glycosylated enzymes [159–161].

Nowadays, one of the most used approaches to optimize DET rate or tackle DET issues is the site-oriented immobilization of redox enzymes through site-directed mutagenesis. Different specific protocols have been proposed by different research groups working in bioelectrochemistry. In order to obtain a productive orientation of the enzyme onto the electrode surface, it is needed a deep knowledge on the enzyme structure [162–165].

For example, Bartlett et al. reported on the covalent coupling between a surface-exposed cysteine residue and maleimide groups to immobilize different variants of *Myriococcum thermophilum* cellobiose dehydrogenase (*Mt*CDH) at multiwall carbon nanotube electrodes [166,167] (Figure 13E). By placing individual cysteine residues around the surface of the CDH_{DH} domain of the enzyme, they were able to immobilize the different variants with different orientations (Figure 13A–D). Notably, it was shown that DET occurs exclusively through the *heme b* cofactor and that the redox potential of the cofactor is unaffected by the orientation of the enzyme. This immobilization approach also resulted in an increased amount by 4–5 times of the electrically contacted (active) enzyme immobilized onto the electrode compared to not site-oriented immobilization. The current generated by the enzyme-modified electrodes in the presence of the cellobiose was dependent on the site-specific orientation of the enzyme-mutants (Figure 13F). In a similar approach [168], the same enzyme was immobilized onto a gold electrode by placing cysteine residues only around the CDH_{DH} domain in order to study the influence of CDH_{CYT} domain mobility on the ET rate (Figure 14A,B). For DET, the CDH_{CYT} domain needs to move from the closed-state conformation, where it obtains an electron from the catalytic CDH_{DH} to the open state where it can donate an electron to the electrode. Except for the optimal enzyme orientation (both domains on the side with the CDH_{CYT} in proximity of the electrode), CDH is not able to swing back the closed conformation, thus not allowing an efficient DET (Figure 14C,D). However, this approach does not necessarily require a site-directed mutagenesis as proposed in many papers. In some unique systems, a cysteine residue might be present at the optimum location near a redox active site already in natural structures, as it was the case for oriented immobilization of bacterial photosynthetic reaction centers at a modified electrode surface [169].

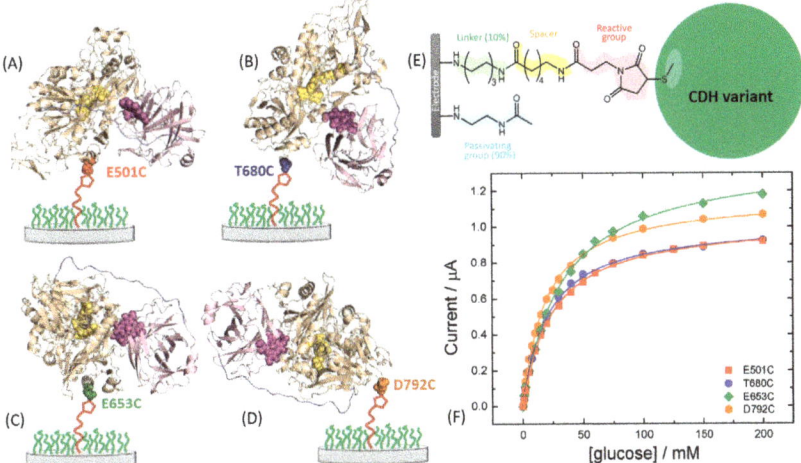

Figure 13. Cartoon representations of the structures of the four different *Mt*CDH variants attached to the electrode surface in different orientations through the cysteine-maleimide bond. The cytochrome domain is shown in purple and the dehydrogenase domain in pale brown. (**A**) E501C, substrate channel close to electrode (front on); (**B**) T680C, top of enzyme facing electrode (top on); (**C**) E653C, right side of substrate channel facing electrode (right side on); and (**D**) D792C, C-terminus close to electrode (bottom on). The images were obtained with PyMol software based on the crystal structure of *Mt*CDH, PDB code 4QI6. (**E**) Chemical structure of the whole electrode modification used in this work to immobilize *Mt*CDH variants with a single surface exposed cysteine, with the different components in different colors. (**F**) Background-corrected catalytic currents at 0.0 V vs. SCE for the four *Mt*CDH variants from parts (**A**–**D**) plotted as a function of the glucose concentration. The definition of all used mutants and their abbreviate names can be found in ref. [167]. (The figure was adopted from ref. [167] with permission.)

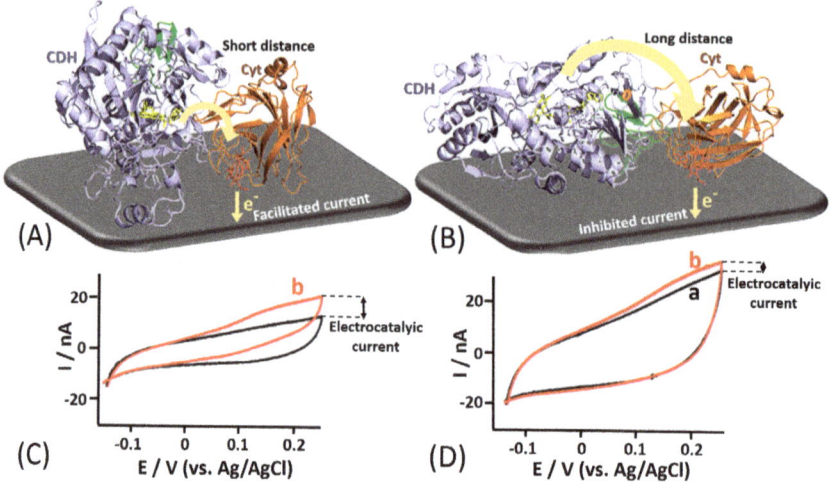

Figure 14. Direct electron transfer anisotropy of a site-specifically immobilized cellobiose dehydrogenase. (**A**) The enzyme immobilization with a short internal electron transfer path from a catalytic cellobiose dehydrogenase (CDH) domain to a cytochrome (Cyt) mediator domain connected with a flexible spacer. (**B**) The enzyme immobilization with a long electron transfer path. (**C,D**) Cyclic voltammograms obtained in the absence (a) and presence (b) of lactose (10 mM) for the enzyme immobilization with the short and long internal electron transfer distances, respectively. (The figure adopted from ref. [168] with permission.)

Different methods based on non-covalent enzyme binding have been applied for the oriented immobilization of enzymes in order to facilitate the DET. For example, Armstrong and his co-workers proposed the immobilization of laccase through a hydrophobic pocket nearby one of the metal centers included in the enzyme, namely T1 copper (T1Cu), in order to enhance the DET rate [170]. Notably, laccases belong to the family of multicopper oxidases (MCOs), where the ET proceeds through the following three steps: (*i*) the reduction of the T1Cu site through the electrons transferred from a substrate (or electrode considering the immobilized enzyme), (*ii*) the internal electron transfer (IET) or tunneling between the T1Cu and trinuclear copper cluster (TNC) proceeding through the Cys-(His)$_2$ bridge over a distance of 13 Å, and (*iii*) O_2 reduction taking place at TNC [171,172]. The authors proposed the electrodeposition of diazonium salts of a wide group of aryl amines, thus producing a highly aromatic electrode surface that would be able to access the hydrophobic pocket of the enzyme (Figure 15A). The cyclic voltammogram for an unmodified carbon electrode with randomly adsorbed laccase showed a very small catalytic wave (Figure 15B, curve a) that was doubled upon electrode incubation with additional amount of the enzyme (Figure 15B, curve b), thus showing a partial coverage of the electrode. On the other hand, using the site-oriented immobilization the DET rate was greatly improved by at least six-fold (Figure 15B, curve c.) Upon further incubation with an additional amount of enzyme, an overlapping cyclic voltammogram was recorded (Figure 15B, curve d), meaning that a full enzyme coverage was achieved. This approach to the enzyme immobilization has been further used mainly for the biofuel cell development because of the minimization of overpotential needed to activate the reduction of O_2 at the cathode surface [173–177].

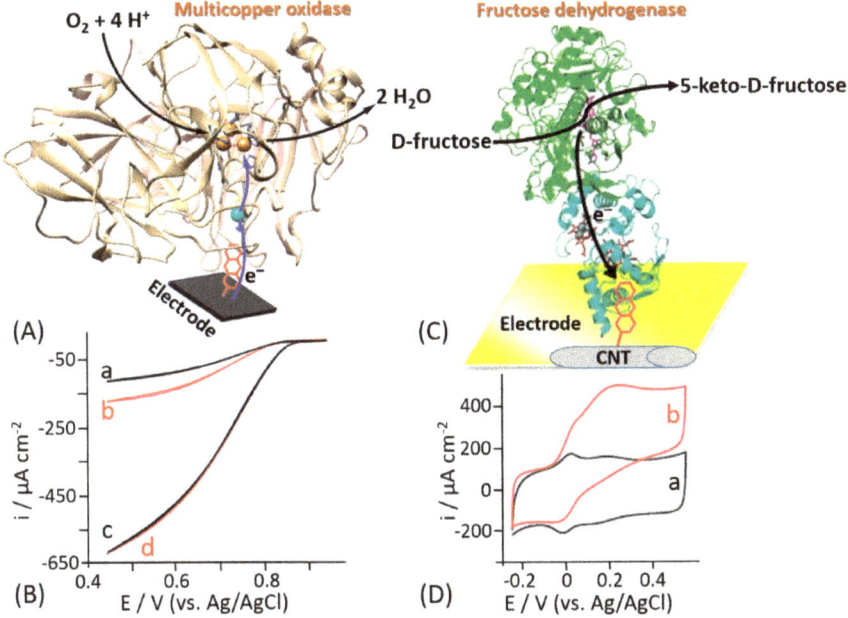

Figure 15. (**A**) Laccase (multicopper oxidase) site-specific immobilization due to the enzyme binding to the surface-located aromatic species. (**B**) The electrocatalytic activity of a film of *Pycnoporus cinnabarinus* laccase lcc3-1 (PcL) on an electrode without aromatic species and random orientation of the adsorbed enzyme (a–b) and on an electrode functionalized with the aromatic species providing orientation of the enzyme favorable for the DET (c–d). Cyclic voltammograms (a) and (c) correspond to the catalytic waves immediately after spotting on laccase solution. Cyclic voltammograms (b) and (d) show the catalytic waves after additional treatment of the modified electrode with a new portion of the enzyme. Potential scan rate was 5 mV s^{-1}. (**C**) Fructose dehydrogenase site-specific immobilization due to the enzyme binding to the surface-located aromatic species. (**D**) Cyclic voltammograms measured with fructose dehydrogenase-electrode modified according the scheme shown in (**C**); in the absence (a) and presence of 10 mM D-fructose (b). The background electrolyte was 50 mM acetic buffer, pH 4.5; potential scan rate was 10 mV s^{-1}. (Part A was adopted from [3] with permission; part B was adopted from [170] with permission; part D was adopted from [178] with permission.)

The same approach has been reported by Bollella et al. for the immobilization of a carbohydrate oxidizing enzyme, namely fructose dehydrogenase (FDH) [178]. An efficient DET reaction pathway between FDH and a carbon nanotube-modified electrode further grafted with an aromatic compound has been reported (Figure 15C). Anthracene molecules have been deposited onto single walled carbon nanotubes (SWCNTs) by electrochemical reduction of 2-aminoanthracene diazonium. Cyclic voltammograms measured in the absence of D-fructose with the FDH-modified electrode revealed two couples of redox waves attributed to *heme* c_1 and *heme* c_3 of the cytochrome domain (Figure 15D, curve a). The addition of 10 mM D-fructose, which is a substrate of FDH, resulted in two catalytic waves correlated with *heme* c_1 and *heme* c_3 with a maximum current density of 485 ± 21 µA cm^{-2} (Figure 15D, curve b). Conversely, only one couple of redox peaks and one catalytic wave in the absence and presence of D-fructose, respectively, were observed for the plain carbon nanotube-modified electrode. The difference has been explained by different orientation of the FDH enzyme molecules at the electrode surface. Indeed, the FDH molecules are randomly adsorbed at the electrode surface in the absence of the anthracene grafted at SWCNTS. On the other hand, the hydrophobic pocket close to the *heme* groups of the cytochrome domain interact with the grafted anthracene due to the π-π interactions with the aromatic side chains of the amino acids present in the hydrophobic pocket of FDH [179,180]. This interaction results in the oriented deposition of the FDH molecules facilitating redox transformations for both *heme* groups according to the DET mechanism.

Although the discussed approaches allowed to enhance the DET rate for many enzymes, they exhibit as main drawbacks the high cost and complexity of the modified electrode preparation, which do not allow them to scale-up to an industrial level. Therefore, the efficient ET between enzymes and electrodes is still recognized by the scientific community as the major challenge. Nanotechnological approaches facilitating the ET, also increasing the enzyme load, remain as active research directions [27,110,181,182]. Among all kinds of electrode materials, metal nanoparticles and carbon-based nanomaterials play an important role in the electrode modification, because of their high surface area-to-volume ratios and high surface energy, which facilitate immobilization of enzymes, allowing them to act as electron conducting pathways between the prosthetic groups of the enzymes and electrode surfaces. Due to the high research activity in this area, the advances in this research direction have been extensively reviewed [183–185].

6. Conclusions and Future Perspectives

The investigation of the electrochemical properties of biological materials has gained a solid foundation over the past 80 years. From the very first observation of the protein electrochemical activity in polarographic measurements reported by Brdička in 1930s [186–188], great achievements [189] have been progressed in electrochemical studies of redox proteins, enzymes, and whole biological cells (e.g., bacterial cells, yeast cells, etc.). The success in bioelectrochemistry has been achieved by using carefully designed chemically modified electrode surfaces and particular experimental conditions [51]. Recently, many bioelectrochemists successfully attempted to tackle ET issues for various redox enzymes with detailed understanding of the ET mechanism. Despite the general progress on the ET mechanism elucidation for many redox enzymes, the ET mechanism of GOx still remains an open debate based on experimental evidences reported in the literature from both sides (pro and quo). On the one hand, many researchers are claiming that GOx immobilized on an electrode surface cannot undergo DET reaction mechanism, without considering the reliability and the "correct" scientific meaning or interpretation of the results. On the other hand, other scientists are claiming the absence of FAD dissociation from the enzyme structure, thus suggesting glucose can be determined directly either by the redox process of the co-enzyme FAD or by oxygen consumption (competitive mechanism) [190–193].

In the future, the development of all these technologies will be certainly scaled-up to the industrial level allowing to take great advantages of fundamental studies in bioelectrochemistry [194,195]. The future developments in bioelectrochemistry are related to understanding the ET mechanism of more complex biological species (e.g., whole cells, etc.), possibly extending the numbers of enzymes that could be connected in DET to an electrode surface. However, bioelectrochemistry is also connected to the developments in different scientific and technological areas, like bioengineering and materials science. The former is mainly important for the biological mutation of redox enzymes, while the latter is important for the synthesis or electrosynthesis of new nano-catalysts used to support the DET of many redox enzymes [196]. Although the analytical aspects of the ET mechanisms have not been herein reviewed, as the development of biosensors, we summarized all the fundamentals needed for newcomers in order to enrich their knowledge about bioelectrochemistry [197,198]. Besides biosensing applications [199,200], bioelectrochemisty can also find applications for the development of special power-generating systems (e.g., enzyme-based [201–206] and microbial [207–209] biofuel cells) or employed as a biocomponent in unconventional information processing systems [210,211].

Author Contributions: P.B. and E.K. wrote and revised the manuscript. All authors have read and agreed to the published version of the manuscript.

Funding: This work was supported in part by Human Frontier Science Program, project grant RGP0002/2018 to E.K.

Conflicts of Interest: The authors declare no conflict of interest.

References

1. Willner, I.; Katz, E. Integration of layered redox proteins and conductive supports for bioelectronic applications. *Angew. Chem. Int. Ed.* **2000**, *39*, 1180–1218. [CrossRef]
2. Bollella, P.; Gorton, L. Enzyme based amperometric biosensors. *Curr. Opin. Electrochem.* **2018**, *10*, 157–173. [CrossRef]
3. Yates, N.D.; Fascione, M.A.; Parkin, A. Methodologies for "wiring" redox proteins/enzymes to electrode surfaces. *Chem. Eur. J.* **2018**, *24*, 12164–12182. [CrossRef]
4. Willner, I.; Katz, E.; Willner, B. Electrical contact of redox enzyme layers associated with electrodes: Routes to amperometric biosensors. *Electroanalysis* **1997**, *9*, 965–977. [CrossRef]
5. Davidson, V.L. Electron transfer in quinoproteins. *Arch. Biochem. Biophys.* **2004**, *428*, 32–40. [CrossRef]
6. Kennedy, M.L.; Gibney, B.R. Metalloprotein and redox protein design. *Curr. Opin. Struct. Biol.* **2001**, *11*, 485–490. [CrossRef]
7. Mewies, M.; McIntire, W.S.; Scrutton, N.S. Covalent attachment of flavin adenine dinucleotide (FAD) and flavin mononucleotide (FMN) to enzymes: The current state of affairs. *Protein Sci.* **1998**, *7*, 7–20. [CrossRef] [PubMed]
8. Gorton, L. Carbon paste electrodes modified with enzymes, tissues, and cells. *Electroanalysis* **1995**, *7*, 23–45. [CrossRef]
9. Schulz, C.; Kittl, R.; Ludwig, R.; Gorton, L. Direct electron transfer from the FAD cofactor of cellobiose dehydrogenase to electrodes. *ACS Catal.* **2016**, *6*, 555–563. [CrossRef]
10. Hunt, J.; Massey, V.; Dunham, W.R.; Sands, R.H. Redox potentials of milk xanthine dehydrogenase. Room temperature measurement of the FAD and 2Fe/2S center potentials. *J. Biol. Chem.* **1993**, *268*, 18685–18691.
11. Kawai, S.; Yakushi, T.; Matsushita, K.; Kitazumi, Y.; Shirai, O.; Kano, K. The electron transfer pathway in direct electrochemical communication of fructose dehydrogenase with electrodes. *Electrochem. Commun.* **2014**, *38*, 28–31. [CrossRef]
12. Duine, J.A. Quinoproteins: Enzymes containing the quinonoid cofactor pyrroloquinoline quinone, topaquinone or tryptophan-tryptophan quinone. *Eur. J. Biochem.* **1991**, *200*, 271–284. [CrossRef] [PubMed]
13. Tkac, J.; Svitel, J.; Vostiar, I.; Navratil, M.; Gemeiner, P. Membrane-bound dehydrogenases from *Gluconobacter sp.*: Interfacial electrochemistry and direct bioelectrocatalysis. *Bioelectrochemistry* **2009**, *76*, 53–62. [CrossRef]
14. Anthony, C. Quinoprotein-catalysed reactions. *Biochem. J.* **1996**, *320*, 697–711. [CrossRef]
15. Oubrie, A. Structure and mechanism of soluble glucose dehydrogenase and other PQQ-dependent enzymes. *Biochim. Biophys. Acta* **2003**, *1647*, 143–151. [CrossRef]
16. Léger, C.; Bertrand, P. Direct electrochemistry of redox enzymes as a tool for mechanistic studies. *Chem. Rev.* **2008**, *108*, 2379–2438. [CrossRef] [PubMed]
17. Rusling, J.F. (Ed.) *Biomolecular Films: Design, Function, and Applications*; Marcel Dekker: New York, NY, USA, 2003.
18. Ghindilis, A.L.; Atanasov, P.; Wilkins, E. Enzyme-catalyzed direct electron transfer: Fundamentals and analytical applications. *Electroanalysis* **1997**, *9*, 661–674. [CrossRef]
19. Gorton, L.; Lindgren, A.; Larsson, T.; Munteanu, F.D.; Ruzgas, T.; Gazaryan, I. Direct electron transfer between heme-containing enzymes and electrodes as basis for third generation biosensors. *Anal. Chim. Acta* **1999**, *400*, 91–108. [CrossRef]
20. Degani, Y.; Heller, A. Direct electrical communication between chemically modified enzymes and metal electrodes. I. Electron transfer from glucose oxidase to metal electrodes via electron relays, bound covalently to the enzyme. *J. Phys. Chem.* **1987**, *91*, 1285–1289. [CrossRef]
21. Rutherford, A.W.; Osyczka, A.; Rappaport, F. Back-reactions, short-circuits, leaks and other energy wasteful reactions in biological electron transfer: Redox tuning to survive life in O_2. *FEBS Lett.* **2012**, *586*, 603–616. [CrossRef]

22. Quinlan, R.J.; Sweeney, M.D.; Leggio, L.L.; Otten, H.; Poulsen, J.-C.N.; Johansen, K.S.; Krogh, K.B.; Jørgensen, C.I.; Tovborg, M.; Anthonsen, A. Insights into the oxidative degradation of cellulose by a copper metalloenzyme that exploits biomass components. *Proc. Natl. Acad. Sci. USA* **2011**, *108*, 15079–15084. [CrossRef] [PubMed]

23. Phillips, C.M.; Beeson, W.T.; Cate, J.H.; Marletta, M.A. Cellobiose dehydrogenase and a copper-dependent polysaccharide monooxygenase potentiate cellulose degradation by *Neurospora crassa*. *ACS Chem. Biol.* **2011**, *6*, 1399–1406. [CrossRef] [PubMed]

24. Nishio, N.; Nakashimada, Y. Recent development of anaerobic digestion processes for energy recovery from wastes. *J. Biosci. Bioeng.* **2007**, *103*, 105–112. [CrossRef] [PubMed]

25. Srikanth, S.; Maesen, M.; Dominguez-Benetton, X.; Vanbroekhoven, K.; Pant, D. Enzymatic electrosynthesis of formate through CO_2 sequestration/reduction in a bioelectrochemical system (BES). *Bioresour. Technol.* **2014**, *165*, 350–354. [CrossRef]

26. Schuhmann, W. Amperometric enzyme biosensors based on optimised electron-transfer pathways and non-manual immobilisation procedures. *Rev. Mol. Biotechnol.* **2002**, *82*, 425–441. [CrossRef]

27. Wang, J. Carbon-nanotube based electrochemical biosensors: A review. *Electroanalysis* **2005**, *17*, 7–14. [CrossRef]

28. Calvo, E.J.; Danilowicz, C. Amperometric enzyme electrodes. *J. Braz. Chem. Soc.* **1997**, *8*, 563–574. [CrossRef]

29. Wang, J. Electrochemical glucose biosensors. *Chem. Rev.* **2008**, *108*, 814–825. [CrossRef]

30. Heller, A.; Feldman, B. Electrochemical glucose sensors and their applications in diabetes management. *Chem. Rev.* **2008**, *108*, 2482–2505. [CrossRef]

31. Wang, J. Glucose biosensors: 40 years of advances and challenges. *Electroanalysis* **2001**, *13*, 983–988. [CrossRef]

32. Wu, Y.; Hu, S. Biosensors based on direct electron transfer in redox proteins. *Microchim. Acta* **2007**, *159*, 1–17. [CrossRef]

33. Murphy, L. Biosensors and bioelectrochemistry. *Curr. Opin. Chem. Biol.* **2006**, *10*, 177–184. [CrossRef] [PubMed]

34. Bollella, P.; Gorton, L.; Antiochia, R. Direct electron transfer of dehydrogenases for development of 3rd generation biosensors and enzymatic fuel cells. *Sensors* **2018**, *18*, 1319. [CrossRef] [PubMed]

35. Habermüller, K.; Mosbach, M.; Schuhmann, W. Electron-transfer mechanisms in amperometric biosensors. *Fresenius J. Anal. Chem.* **2000**, *366*, 560–568. [CrossRef]

36. Lötzbeyer, T.; Schuhmann, W.; Schmidt, H.-L. Electron transfer principles in amperometric biosensors: Direct electron transfer between enzymes and electrode surface. *Sens. Actuators B* **1996**, *33*, 50–54. [CrossRef]

37. Karyakin, A.A.; Gitelmacher, O.V.; Karyakina, E.E. Prussian blue-based first-generation biosensor. A sensitive amperometric electrode for glucose. *Anal. Chem.* **1995**, *67*, 2419–2423. [CrossRef]

38. Scheller, F.W.; Schubert, F.; Neumann, B.; Pfeiffer, D.; Hintsche, R.; Dransfeld, I.; Wollenberger, U.; Renneberg, R.; Warsinke, A.; Johansson, G. Second generation biosensors. *Biosens. Bioelectron.* **1991**, *6*, 245–253. [CrossRef]

39. Katz, E.; Lötzbeyer, T.; Schlereth, D.D.; Schuhmann, W.; Schmidt, H.-L. Electrocatalytic oxidation of reduced nicotinamide coenzymes at gold and platinum electrode surfaces modified with a monolayer of pyrroloquinoline quinone. Effect of Ca^{2+} cations. *J. Electroanal. Chem.* **1994**, *373*, 189–200. [CrossRef]

40. Gorton, L.; Domínguez, E. Electrocatalytic oxidation of NAD(P)H at mediator-modified electrodes. *Rev. Mol. Biotechnol.* **2002**, *82*, 371–392. [CrossRef]

41. Gorton, L. Chemically modified electrodes for the electrocatalytic oxidation of nicotinamide coenzymes. *J. Chem. Soc. Faraday Trans. 1* **1986**, *82*, 1245–1258. [CrossRef]

42. Katz, E.; Lioubashevsky, O.; Willner, I. Electromechanics of a redox-active rotaxane in a monolayer assembly on an electrode. *J. Am. Chem. Soc.* **2004**, *126*, 15520–15532. [CrossRef]

43. Heller, A. Electrical connection of enzyme redox centers to electrodes. *J. Phys. Chem.* **1992**, *96*, 3579–3587. [CrossRef]

44. Heller, A. Electrical wiring of redox enzymes. *Acc. Chem. Res.* **1990**, *23*, 128–134. [CrossRef]

45. Schuhmann, W.; Ohara, T.J.; Schmidt, H.-L.; Heller, A. Electron transfer between glucose oxidase and electrodes via redox mediators bound with flexible chains to the enzyme surface. *J. Am. Chem. Soc.* **1991**, *113*, 1394–1397. [CrossRef]

46. Suzuki, N.; Lee, J.; Loew, N.; Takahashi-Inose, Y.; Okuda-Shimazaki, J.; Kojima, K.; Mori, K.; Tsugawa, W.; Sode, K. Engineered glucose oxidase capable of quasi-direct electron transfer after a quick-and-easy modification with a mediator. *Int. J. Mol. Sci.* **2020**, *21*, 1137. [CrossRef] [PubMed]
47. Willner, I.; Riklin, A.; Shoham, B.; Rivenzon, D.; Katz, E. Development of novel biosensor enzyme electrodes: Glucose oxidase multilayer arrays immobilized onto self-assembled monolayers on electrodes. *Adv. Mater.* **1993**, *5*, 912–915. [CrossRef]
48. Willner, I.; Katz, E.; Riklin, A.; Kasher, R. Mediated electron transfer in glutathione reductase organized in self-assembled monolayers on Au electrodes. *J. Am. Chem. Soc.* **1992**, *114*, 10965–10966. [CrossRef]
49. Willner, I.; Lapidot, N.; Riklin, A.; Kasher, R.; Zahavy, E.; Katz, E. Electron transfer communication in glutathione reductase assemblies: Electrocatalytic, photocatalytic and catalytic systems for the reduction of oxidized glutathione. *J. Am. Chem. Soc.* **1994**, *116*, 1428–1441. [CrossRef]
50. Freire, R.S.; Pessoa, C.A.; Mello, L.D.; Kubota, L.T. Direct electron transfer: An approach for electrochemical biosensors with higher selectivity and sensitivity. *J. Braz. Chem. Soc.* **2003**, *14*, 230–243. [CrossRef]
51. Bartlett, P.N. (Ed.) *Bioelectrochemistry: Fundamentals, Experimental Techniques and Applications*; Wiley: Chichester, UK, 2008.
52. Frew, J.E.; Hill, H.A.O. Electron-transfer biosensors. *Philos. Trans. R. Soc. B* **1987**, *316*, 95–106.
53. Frew, J.E.; Hill, H.A.O. Direct and indirect electron transfer between electrodes and redox proteins. *Eur. J. Biochem.* **1988**, *172*, 261–269. [CrossRef] [PubMed]
54. Armstrong, F.A.; Hill, H.A.O.; Walton, N.J. Reactions of electron-transfer proteins at electrodes. *Q. Rev. Biophys.* **1985**, *18*, 261–322. [CrossRef] [PubMed]
55. Léger, C.; Jones, A.K.; Albracht, S.P.; Armstrong, F.A. Effect of a dispersion of interfacial electron transfer rates on steady state catalytic electron transport in [NiFe]-hydrogenase and other enzymes. *J. Phys. Chem. B* **2002**, *106*, 13058–13063. [CrossRef]
56. Lopez, R.J.; Babanova, S.; Ulyanova, Y.; Singhal, S.; Atanassov, P. Improved interfacial electron transfer in modified bilirubin oxidase biocathodes. *ChemElectroChem* **2014**, *1*, 241–248. [CrossRef]
57. Hill, H.A.O.; Walton, N.J. Investigation of some intermolecular electron transfer reactions of cytochrome c by electrochemical methods. *J. Am. Chem. Soc.* **1982**, *104*, 6515–6519. [CrossRef]
58. Schulz, C.; Ludwig, R.; Micheelsen, P.O.; Silow, M.; Toscano, M.D.; Gorton, L. Enhancement of enzymatic activity and catalytic current of cellobiose dehydrogenase by calcium ions. *Electrochem. Commun.* **2012**, *17*, 71–74. [CrossRef]
59. Kracher, D.; Zahma, K.; Schulz, C.; Sygmund, C.; Gorton, L.; Ludwig, R. Inter-domain electron transfer in cellobiose dehydrogenase: Modulation by pH and divalent cations. *FEBS J.* **2015**, *282*, 3136–3148. [CrossRef]
60. Ludwig, R.; Ortiz, R.; Schulz, C.; Harreither, W.; Sygmund, C.; Gorton, L. Cellobiose dehydrogenase modified electrodes: Advances by materials science and biochemical engineering. *Anal. Bioanal. Chem.* **2013**, *405*, 3637–3658. [CrossRef]
61. Ludwig, R.; Harreither, W.; Tasca, F.; Gorton, L. Cellobiose dehydrogenase: A versatile catalyst for electrochemical applications. *ChemPhysChem* **2010**, *11*, 2674–2697. [CrossRef]
62. Bollella, P.; Ludwig, R.; Gorton, L. Cellobiose dehydrogenase: Insights on the nanostructuration of electrodes for improved development of biosensors and biofuel cells. *Appl. Mater. Today* **2017**, *9*, 319–332. [CrossRef]
63. Bollella, P.; Hibino, Y.; Kano, K.; Gorton, L.; Antiochia, R. The influence of pH and divalent/monovalent cations on the internal electron transfer (IET), enzymatic activity, and structure of fructose dehydrogenase. *Anal. Bioanal. Chem.* **2018**, *410*, 3253–3264. [CrossRef] [PubMed]
64. Taniguchi, I.; Toyosawa, K.; Yamaguchi, H.; Yasukouchi, K. Reversible electrochemical reduction and oxidation of cytochrome c at a bis(4-pyridyl) disulphide-modified gold electrode. *J. Chem. Soc. Chem. Commun.* **1982**, *18*, 1032–1033. [CrossRef]
65. Armstrong, F.A.; Hill, H.A.O.; Walton, N.J. Direct electrochemistry of redox proteins. *Acc. Chem. Res.* **1988**, *21*, 407–413. [CrossRef]
66. Gray, H.B.; Winkler, J.R. Electron transfer in proteins. *Annu. Rev. Biochem.* **1996**, *65*, 537–561. [CrossRef] [PubMed]
67. Winkler, J.R.; Malmström, B.G.; Gray, H.B. Rapid electron injection into multisite metalloproteins: Intramolecular electron transfer in cytochrome oxidase. *Biophys. Chem.* **1995**, *54*, 199–209. [CrossRef]
68. Winkler, J.R.; Gray, H.B. Long-range electron tunneling. *J. Am. Chem. Soc.* **2014**, *136*, 2930–2939. [CrossRef]

69. Guo, L.-H.; Hill, H.A.O. Direct electrochemistry of proteins and enzymes. *Adv. Inorg. Chem.* **1991**, *36*, 341–375.
70. Page, C.C.; Moser, C.C.; Chen, X.; Dutton, P.L. Natural engineering principles of electron tunnelling in biological oxidation–reduction. *Nature* **1999**, *402*, 47–52. [CrossRef]
71. Wilson, G.S. Native glucose oxidase does not undergo direct electron transfer. *Biosens. Bioelectron.* **2016**, *82*, Vii–Viii. [CrossRef]
72. Marcus, R.A.; Sutin, N. Electron transfers in chemistry and biology. *Biochim. Biophys. Acta* **1985**, *811*, 265–322. [CrossRef]
73. Marcus, R.A. Electron transfer reactions in chemistry: Theory and experiment (Nobel lecture). *Angew. Chem. Int. Ed. Engl.* **1993**, *32*, 1111–1121. [CrossRef]
74. Marcus, R.A. Electron transfer reactions in chemistry. Theory and experiment. *Rev. Mod. Phys.* **1993**, *65*, 599–610. [CrossRef]
75. Marcus, R.A. Chemical and electrochemical electron-transfer theory. *Annu. Rev. Phys. Chem.* **1964**, *15*, 155–196. [CrossRef]
76. Adams, D.M.; Brus, L.; Chidsey, C.E.; Creager, S.; Creutz, C.; Kagan, C.R.; Kamat, P.V.; Lieberman, M.; Lindsay, S.; Marcus, R.A. Charge transfer on the nanoscale: Current status. *J. Phys. Chem. B* **2003**, *107*, 6668–6697. [CrossRef]
77. Bendall, D.S. (Ed.) *Protein Electron Transfer*; BIOS Scientific Pub.: Oxford, UK, 1996.
78. Katz, E.; Shipway, A.N.; Willner, I. The electrochemical and photochemical activation of redox-enzymes. In *Electron Transfer in Chemistry. Volume 4: Heterogeneous Systems, Solid State Systems, Gas Phase Systems. Section 1: Catalysis of Electron Transfer*; Balzani, V., Piotrowiak, P., Rodgers, M.A.J., Eds.; Wiley-VCH: Weinheim, Germany, 2001; pp. 127–201.
79. Laviron, E. General expression of the linear potential sweep voltammogram in the case of diffusionless electrochemical systems. *J. Electroanal. Chem. Interfacial Electrochem.* **1979**, *101*, 19–28. [CrossRef]
80. Laviron, E.; Roullier, L. General expression of the linear potential sweep voltammogram for a surface redox reaction with interactions between the adsorbed molecules: Applications to modified electrodes. *J. Electroanal. Chem. Interfacial Electrochem.* **1980**, *115*, 65–74. [CrossRef]
81. Clark, L.C., Jr.; Lyons, C. Electrode systems for continuous monitoring in cardiovascular surgery. *Ann. N. Y. Acad. Sci.* **1962**, *102*, 29–45.
82. Falk, M.; Blum, Z.; Shleev, S. Direct electron transfer based enzymatic fuel cells. *Electrochim. Acta* **2012**, *82*, 191–202. [CrossRef]
83. Barton, C.S.; Gallaway, J.; Atanassov, P. Enzymatic biofuel cells for implantable and microscale devices. *Chem. Rev.* **2004**, *104*, 4867–4886. [CrossRef]
84. Wilson, R.; Turner, A.P.F. Glucose oxidase: An ideal enzyme. *Biosens. Bioelectron.* **1992**, *7*, 165–185. [CrossRef]
85. Newman, J.D.; Turner, A.P.F. Home blood glucose biosensors: A commercial perspective. *Biosens. Bioelectron.* **2005**, *20*, 2435–2453. [CrossRef] [PubMed]
86. Yoo, E.-H.; Lee, S.-Y. Glucose biosensors: An overview of use in clinical practice. *Sensors* **2010**, *10*, 4558–4576. [CrossRef] [PubMed]
87. Zhu, Z.; Garcia-Gancedo, L.; Flewitt, A.J.; Xie, H.; Moussy, F.; Milne, W.I. A critical review of glucose biosensors based on carbon nanomaterials: Carbon nanotubes and graphene. *Sensors* **2012**, *12*, 5996–6022. [CrossRef] [PubMed]
88. Gregg, B.A.; Heller, A. Cross-linked redox gels containing glucose oxidase for amperometric biosensor applications. *Anal. Chem.* **1990**, *62*, 258–263. [CrossRef]
89. Quinn, C.A.; Connor, R.E.; Heller, A. Biocompatible, glucose-permeable hydrogel for in situ coating of implantable biosensors. *Biomaterials* **1997**, *18*, 1665–1670. [CrossRef]
90. Chen, C.; Xie, Q.; Yang, D.; Xiao, H.; Fu, Y.; Tan, Y.; Yao, S. Recent advances in electrochemical glucose biosensors: A review. *RSC Adv.* **2013**, *3*, 4473–4491. [CrossRef]
91. Bollella, P.; Gorton, L.; Ludwig, R.; Antiochia, R. A third generation glucose biosensor based on cellobiose dehydrogenase immobilized on a glassy carbon electrode decorated with electrodeposited gold nanoparticles: Characterization and application in human saliva. *Sensors* **2017**, *17*, 1912. [CrossRef]
92. Bollella, P.; Sharma, S.; Cass, A.E.G.; Tasca, F.; Antiochia, R. Minimally invasive glucose monitoring using a highly porous gold microneedles-based biosensor: Characterization and application in artificial interstitial fluid. *Catalysts* **2019**, *9*, 580. [CrossRef]

93. Bollella, P.; Sharma, S.; Cass, A.E.G.; Antiochia, R. Minimally-invasive microneedle-based biosensor array for simultaneous lactate and glucose monitoring in artificial interstitial fluid. *Electroanalysis* **2019**, *31*, 374–382. [CrossRef]
94. Cass, A.E.G.; Davis, G.; Francis, G.D.; Hill, H.A.O.; Aston, W.J.; Higgins, I.J.; Plotkin, E.V.; Scott, L.D.; Turner, A.P.F. Ferrocene-mediated enzyme electrode for amperometric determination of glucose. *Anal. Chem.* **1984**, *56*, 667–671. [CrossRef]
95. Mao, F.; Mano, N.; Heller, A. Long tethers binding redox centers to polymer backbones enhance electron transport in enzyme "wiring" hydrogels. *J. Am. Chem. Soc.* **2003**, *125*, 4951–4957. [CrossRef] [PubMed]
96. Turner, A.P.F. Biosensors–sense and sensitivity. *Science* **2000**, *290*, 1315–1317. [CrossRef] [PubMed]
97. Milton, R.D.; Minteer, S.D. Direct enzymatic bioelectrocatalysis: Differentiating between myth and reality. *J. R. Soc. Interface* **2017**, *14*, 20170253. [CrossRef] [PubMed]
98. Bartlett, P.N.; Al-Lolage, F.A. There is no evidence to support literature claims of direct electron transfer (DET) for native glucose oxidase (GOx) at carbon nanotubes or graphene. *J. Electroanal. Chem.* **2018**, *819*, 26–37. [CrossRef]
99. Bankar, S.B.; Bule, M.V.; Singhal, R.S.; Ananthanarayan, L. Glucose oxidase—an overview. *Biotechnol. Adv.* **2009**, *27*, 489–501. [CrossRef] [PubMed]
100. Leskovac, V.; Trivić, S.; Wohlfahrt, G.; Kandrač, J.; Peričin, D. Glucose oxidase from *Aspergillus niger*: The mechanism of action with molecular oxygen, quinones, and one-electron acceptors. *Int. J. Biochem. Cell Biol.* **2005**, *37*, 731–750. [CrossRef] [PubMed]
101. Hecht, H.J.; Schomburg, D.; Kalisz, H.; Schmid, R.D. The 3D structure of glucose oxidase from *Aspergillus niger*. Implications for the use of GOD as a biosensor enzyme. *Biosens. Bioelectron.* **1993**, *8*, 197–203. [CrossRef]
102. Hecht, H.J.; Kalisz, H.M.; Hendle, J.; Schmid, R.D.; Schomburg, D. Crystal structure of glucose oxidase from *Aspergillus niger* refined at 2.3 Å resolution. *J. Mol. Biol.* **1993**, *229*, 153–172. [CrossRef]
103. Alvarez-Icaza, M.; Kalisz, H.M.; Hecht, H.J.; Aumann, K.-D.; Schomburg, D.; Schmid, R.D. The design of enzyme sensors based on the enzyme structure. *Biosens. Bioelectron.* **1995**, *10*, 735–742. [CrossRef]
104. Pazur, J.H.; Kleppe, K.; Ball, E.M. The glycoprotein nature of some functional carbohydrases. *Arch. Biochem. Biophys.* **1963**, *103*, 515–516. [CrossRef]
105. Swoboda, B.E.; Massey, V. Purification and properties of the glucose oxidase from *Aspergillus niger*. *J. Biol. Chem.* **1965**, *240*, 2209–2215. [PubMed]
106. Kalisz, H.M.; Hecht, H.-J.; Schomburg, D.; Schmid, R.D. Effects of carbohydrate depletion on the structure, stability and activity of glucose oxidase from *Aspergillus niger*. *Biochim. Biophys. Acta* **1991**, *1080*, 138–142. [CrossRef]
107. Raba, J.; Mottola, H.A. Glucose oxidase as an analytical reagent. *Crit. Rev. Anal. Chem.* **1995**, *25*, 1–42. [CrossRef]
108. Bollella, P.; Fusco, G.; Tortolini, C.; Sanzò, G.; Favero, G.; Gorton, L.; Antiochia, R. Beyond graphene: Electrochemical sensors and biosensors for biomarkers detection. *Biosens. Bioelectron.* **2017**, *89*, 152–166. [CrossRef] [PubMed]
109. Mazzei, F.; Favero, G.; Bollella, P.; Tortolini, C.; Mannina, L.; Conti, M.E.; Antiochia, R. Recent trends in electrochemical nanobiosensors for environmental analysis. *Int. J. Environ. Health* **2015**, *7*, 267–291. [CrossRef]
110. Yanez-Sedeno, P.; Pingarron, J.M. Gold nanoparticle-based electrochemical biosensors. *Anal. Bioanal. Chem.* **2005**, *382*, 884–886. [CrossRef] [PubMed]
111. Pingarrón, J.M.; Yanez-Sedeno, P.; González-Cortés, A. Gold nanoparticle-based electrochemical biosensors. *Electrochim. Acta* **2008**, *53*, 5848–5866. [CrossRef]
112. Sánchez-Obrero, G.; Cano, M.; Ávila, J.L.; Mayén, M.; Mena, M.L.; Pingarrón, J.M.; Rodríguez-Amaro, R. A gold nanoparticle-modified PVC/TTF-TCNQ composite amperometric biosensor for glucose determination. *J. Electroanal. Chem.* **2009**, *634*, 59–63. [CrossRef]
113. Yáñez-Sedeño, P.; Pingarrón, J.M.; Riu, J.; Rius, F.X. Electrochemical sensing based on carbon nanotubes. *TrAC Trends Anal. Chem.* **2010**, *29*, 939–953. [CrossRef]
114. Bollella, P. Porous gold: A new frontier for enzyme-based electrodes. *Nanomaterials* **2020**, *10*, 722. [CrossRef]
115. Agüí, L.; Yáñez-Sedeño, P.; Pingarrón, J.M. Role of carbon nanotubes in electroanalytical chemistry: A review. *Anal. Chim. Acta* **2008**, *622*, 11–47. [CrossRef] [PubMed]

116. Das, P.; Das, M.; Chinnadayyala, S.R.; Singha, I.M.; Goswami, P. Recent advances on developing 3rd generation enzyme electrode for biosensor applications. *Biosens. Bioelectron.* **2016**, *79*, 386–397. [CrossRef] [PubMed]
117. Harper, A.; Anderson, M.R. Electrochemical glucose sensors—developments using electrostatic assembly and carbon nanotubes for biosensor construction. *Sensors* **2010**, *10*, 8248–8274. [CrossRef] [PubMed]
118. Luong, J.H.; Glennon, J.D.; Gedanken, A.; Vashist, S.K. Achievement and assessment of direct electron transfer of glucose oxidase in electrochemical biosensing using carbon nanotubes, graphene, and their nanocomposites. *Microchim. Acta* **2017**, *184*, 369–388. [CrossRef]
119. Bollella, P.; Katz, E. Bioelectrocatalysis at carbon nanotubes. *Methods Enzymol.* **2020**, *630*, 215–247.
120. Byers, J.C.; Güell, A.G.; Unwin, P.R. Nanoscale electrocatalysis: Visualizing oxygen reduction at pristine, kinked, and oxidized sites on individual carbon nanotubes. *J. Am. Chem. Soc.* **2014**, *136*, 11252–11255. [CrossRef]
121. Wang, L.; Pumera, M. Residual metallic impurities within carbon nanotubes play a dominant role in supposedly "metal-free" oxygen reduction reactions. *Chem. Commun.* **2014**, *50*, 12662–12664. [CrossRef]
122. Lehman, J.H.; Terrones, M.; Mansfield, E.; Hurst, K.E.; Meunier, V. Evaluating the characteristics of multiwall carbon nanotubes. *Carbon* **2011**, *49*, 2581–2602. [CrossRef]
123. Mello, P.A.; Rodrigues, L.F.; Nunes, M.A.; Mattos, J.C.P.; Müller, E.I.; Dressler, V.L.; Flores, E.M. Determination of metal impurities in carbon nanotubes by direct solid sampling electrothermal atomic absorption spectrometry. *J. Braz. Chem. Soc.* **2011**, *22*, 1040–1049. [CrossRef]
124. Eddowes, M.J.; Hill, H.A.O. Investigation of electron-transfer reactions of proteins by electrochemical methods. *Biosci. Rep.* **1981**, *1*, 521–532. [CrossRef]
125. Léger, C.; Elliott, S.J.; Hoke, K.R.; Jeuken, L.J.; Jones, A.K.; Armstrong, F.A. Enzyme electrokinetics: Using protein film voltammetry to investigate redox enzymes and their mechanisms. *Biochemistry* **2003**, *42*, 8653–8662. [CrossRef] [PubMed]
126. Duca, M.; Weeks, J.R.; Fedor, J.G.; Weiner, J.H.; Vincent, K.A. Combining noble metals and enzymes for relay cascade electrocatalysis of nitrate reduction to ammonia at neutral pH. *ChemElectroChem* **2015**, *2*, 1086–1089. [CrossRef]
127. Moorcroft, M.J.; Davis, J.; Compton, R.G. Detection and determination of nitrate and nitrite: A review. *Talanta* **2001**, *54*, 785–803. [CrossRef]
128. Burgess, B.K.; Lowe, D.J. Mechanism of molybdenum nitrogenase. *Chem. Rev.* **1996**, *96*, 2983–3012. [CrossRef] [PubMed]
129. Milton, R.D.; Abdellaoui, S.; Khadka, N.; Dean, D.R.; Leech, D.; Seefeldt, L.C.; Minteer, S.D. Nitrogenase bioelectrocatalysis: Heterogeneous ammonia and hydrogen production by MoFe protein. *Energy Environ. Sci.* **2016**, *9*, 2550–2554. [CrossRef]
130. Holade, Y.; Yuan, M.; Milton, R.D.; Hickey, D.P.; Sugawara, A.; Peterbauer, C.K.; Haltrich, D.; Minteer, S.D. Rational combination of promiscuous enzymes yields a versatile enzymatic fuel cell with improved coulombic efficiency. *J. Electrochem. Soc.* **2016**, *164*, H3073. [CrossRef]
131. Havaux, M. Characterization of thermal damage to the photosynthetic electron transport system in potato leaves. *Plant Sci.* **1993**, *94*, 19–33. [CrossRef]
132. Tortolini, C.; Bollella, P.; Antiochia, R.; Favero, G.; Mazzei, F. Inhibition-based biosensor for atrazine detection. *Sens. Actuators B* **2016**, *224*, 552–558. [CrossRef]
133. De Poulpiquet, A.; Kjaergaard, C.H.; Rouhana, J.; Mazurenko, I.; Infossi, P.; Gounel, S.; Gadiou, R.; Giudici-Orticoni, M.T.; Solomon, E.I.; Mano, N. Mechanism of chloride inhibition of bilirubin oxidases and its dependence on potential and pH. *ACS Catal.* **2017**, *7*, 3916–3923. [CrossRef]
134. Bollella, P.; Fusco, G.; Tortolini, C.; Sanzò, G.; Antiochia, R.; Favero, G.; Mazzei, F. Inhibition-based first-generation electrochemical biosensors: Theoretical aspects and application to 2, 4-dichlorophenoxy acetic acid detection. *Anal. Bioanal. Chem.* **2016**, *408*, 3203–3211. [CrossRef]
135. Lo, K.K.-W.; Wong, L.-L.; Hill, H.A.O. Surface-modified mutants of cytochrome P450cam: Enzymatic properties and electrochemistry. *FEBS Lett.* **1999**, *451*, 342–346. [CrossRef]
136. Campàs, M.; Prieto-Simón, B.; Marty, J.-L. A review of the use of genetically engineered enzymes in electrochemical biosensors. In Proceedings of the Seminars in Cell & Developmental Biology; Elsevier: Amsterdam, The Netherlands, 2009; pp. 3–9.

137. Dementin, S.; Belle, V.; Bertrand, P.; Guigliarelli, B.; Adryanczyk-Perrier, G.; De Lacey, A.L.; Fernandez, V.M.; Rousset, M.; Léger, C. Changing the ligation of the distal [4Fe4S] cluster in NiFe hydrogenase impairs inter- and intramolecular electron transfers. *J. Am. Chem. Soc.* **2006**, *128*, 5209–5218. [CrossRef]
138. Armstrong, F.A. Recent developments in dynamic electrochemical studies of adsorbed enzymes and their active sites. *Curr. Opin. Chem. Biol.* **2005**, *9*, 110–117. [CrossRef] [PubMed]
139. Armstrong, F.A. Protein film voltammetry: Revealing the mechanisms of biological oxidation and reduction. *Russ. J. Electrochem.* **2002**, *38*, 49–62. [CrossRef]
140. Armstrong, F.A. Insights from protein film voltammetry into mechanisms of complex biological electron-transfer reactions. *J. Chem. Soc. Dalton Trans.* **2002**, 661–671. [CrossRef]
141. Gulaboski, R.; Mirčeski, V.; Bogeski, I.; Hoth, M. Protein film voltammetry: Electrochemical enzymatic spectroscopy. A review on recent progress. *J. Solid State Electrochem.* **2012**, *16*, 2315–2328. [CrossRef]
142. Hirst, J. Elucidating the mechanisms of coupled electron transfer and catalytic reactions by protein film voltammetry. *Biochim. Biophys. Acta* **2006**, *1757*, 225–239. [CrossRef]
143. Angove, H.C.; Cole, J.A.; Richardson, D.J.; Butt, J.N. Protein film voltammetry reveals distinctive fingerprints of nitrite and hydroxylamine reduction by a cytochrome c nitrite reductase. *J. Biol. Chem.* **2002**, *277*, 23374–23381. [CrossRef]
144. Armstrong, F.A.; Heering, H.A.; Hirst, J. Reaction of complex metalloproteins studied by protein-film voltammetry. *Chem. Soc. Rev.* **1997**, *26*, 169–179. [CrossRef]
145. Wijma, H.J.; Jeuken, L.J.; Verbeet, M.P.; Armstrong, F.A.; Canters, G.W. Protein film voltammetry of copper-containing nitrite reductase reveals reversible inactivation. *J. Am. Chem. Soc.* **2007**, *129*, 8557–8565. [CrossRef]
146. Best, S.P. Spectroelectrochemistry of hydrogenase enzymes and related compounds. *Coord. Chem. Rev.* **2005**, *249*, 1536–1554. [CrossRef]
147. Willner, I.; Blonder, R.; Katz, E.; Stocker, A.; Bückmann, A.F. Reconstitution of apo-glucose oxidase with a nitrospiropyran-modified FAD cofactor yields a photoswitchable biocatalyst for amperometric transduction of recorded optical signals. *J. Am. Chem. Soc.* **1996**, *118*, 5310–5311. [CrossRef]
148. Katz, E.; Riklin, A.; Heleg-Shabtai, V.; Willner, I.; Bückmann, A.F. Glucose oxidase electrodes via reconstitution of the apo-enzyme: Tailoring of novel glucose biosensors. *Anal. Chim. Acta* **1999**, *385*, 45–58. [CrossRef]
149. Zayats, M.; Katz, E.; Willner, I. Electrical contacting of glucose oxidase by surface-reconstitution of the apo-protein on a relay-boronic acid-FAD cofactor monolayer. *J. Am. Chem. Soc.* **2002**, *124*, 2120–2121. [CrossRef] [PubMed]
150. Gorton, L.; Jönsson-Pettersson, G.; Csöregi, E.; Johansson, K.; Domínguez, E.; Marko-Varga, G. Amperometric biosensors based on an apparent direct electron transfer between electrodes and immobilized peroxidases. Plenary lecture. *Analyst* **1992**, *117*, 1235–1241. [CrossRef]
151. Zhang, W.; Li, G. Third-generation biosensors based on the direct electron transfer of proteins. *Anal. Sci.* **2004**, *20*, 603–609. [CrossRef] [PubMed]
152. Bistolas, N.; Wollenberger, U.; Jung, C.; Scheller, F.W. Cytochrome P450 biosensors—A review. *Biosens. Bioelectron.* **2005**, *20*, 2408–2423. [CrossRef]
153. Bollella, P.; Mazzei, F.; Favero, G.; Fusco, G.; Ludwig, R.; Gorton, L.; Antiochia, R. Improved DET communication between cellobiose dehydrogenase and a gold electrode modified with a rigid self-assembled monolayer and green metal nanoparticles: The role of an ordered nanostructuration. *Biosens. Bioelectron.* **2017**, *88*, 196–203. [CrossRef] [PubMed]
154. Zappi, D.; Masci, G.; Sadun, C.; Tortolini, C.; Antonelli, M.L.; Bollella, P. Evaluation of new cholinium-amino acids based room temperature ionic liquids (RTILs) as immobilization matrix for electrochemical biosensor development: Proof-of-concept with *Trametes Versicolor* laccase. *Microchem. J.* **2018**, *141*, 346–352. [CrossRef]
155. Xiao, Y.; Patolsky, F.; Katz, E.; Hainfeld, J.F.; Willner, I. "Plugging into enzymes": Nanowiring of redox enzymes by a gold nanoparticle. *Science* **2003**, *299*, 1877–1881. [CrossRef] [PubMed]
156. Patolsky, F.; Weizmann, Y.; Willner, I. Long-range electrical contacting of redox enzymes by SWCNT connectors. *Angew. Chem. Int. Ed.* **2004**, *43*, 2113–2117. [CrossRef] [PubMed]
157. Holland, J.T.; Lau, C.; Brozik, S.; Atanassov, P.; Banta, S. Engineering of glucose oxidase for direct electron transfer via site-specific gold nanoparticle conjugation. *J. Am. Chem. Soc.* **2011**, *133*, 19262–19265. [CrossRef] [PubMed]

158. Ortiz, R.; Matsumura, H.; Tasca, F.; Zahma, K.; Samejima, M.; Igarashi, K.; Ludwig, R.; Gorton, L. Effect of deglycosylation of cellobiose dehydrogenases on the enhancement of direct electron transfer with electrodes. *Anal. Chem.* **2012**, *84*, 10315–10323. [CrossRef] [PubMed]

159. Courjean, O.; Gao, F.; Mano, N. Deglycosylation of glucose oxidase for direct and efficient glucose electrooxidation on a glassy carbon electrode. *Angew. Chem. Int. Ed.* **2009**, *48*, 5897–5899. [CrossRef] [PubMed]

160. Killyéni, A.; Yakovleva, M.E.; MacAodha, D.; Conghaile, P.Ó.; Gonaus, C.; Ortiz, R.; Leech, D.; Popescu, I.C.; Peterbauer, C.K.; Gorton, L. Effect of deglycosylation on the mediated electrocatalytic activity of recombinantly expressed *Agaricus meleagris* pyranose dehydrogenase wired by osmium redox polymer. *Electrochim. Acta* **2014**, *126*, 61–67. [CrossRef]

161. Yakovleva, M.E.; Killyéni, A.; Ortiz, R.; Schulz, C.; MacAodha, D.; Conghaile, P.Ó.; Leech, D.; Popescu, I.C.; Gonaus, C.; Peterbauer, C.K. Recombinant pyranose dehydrogenase—A versatile enzyme possessing both mediated and direct electron transfer. *Electrochem. Commun.* **2012**, *24*, 120–122. [CrossRef]

162. Liu, Y.; Yu, J. Oriented immobilization of proteins on solid supports for use in biosensors and biochips: A review. *Microchim. Acta* **2016**, *183*, 1–19. [CrossRef]

163. Wang, C.; Feng, B. Research progress on site-oriented and three-dimensional immobilization of protein. *Mol. Biol.* **2015**, *49*, 1–20. [CrossRef]

164. Al-Lolage, F.A.; Bartlett, P.N.; Gounel, S.; Staigre, P.; Mano, N. Site-Directed Immobilization of Bilirubin Oxidase for Electrocatalytic Oxygen Reduction. *ACS Catal.* **2019**, *9*, 2068–2078. [CrossRef]

165. Hitaishi, V.P.; Clement, R.; Bourassin, N.; Baaden, M.; De Poulpiquet, A.; Sacquin-Mora, S.; Ciaccafava, A.; Lojou, E. Controlling redox enzyme orientation at planar electrodes. *Catalysts* **2018**, *8*, 192. [CrossRef]

166. Al-Lolage, F.A.; Meneghello, M.; Ma, S.; Ludwig, R.; Bartlett, P.N. A flexible method for the stable, covalent immobilization of enzymes at electrode surfaces. *ChemElectroChem* **2017**, *4*, 1528–1534. [CrossRef]

167. Meneghello, M.; Al-Lolage, F.A.; Ma, S.; Ludwig, R.; Bartlett, P.N. Studying direct electron transfer by site-directed immobilization of cellobiose dehydrogenase. *ChemElectroChem* **2019**, *6*, 700–713. [CrossRef] [PubMed]

168. Ma, S.; Laurent, C.V.; Meneghello, M.; Tuoriniemi, J.; Oostenbrink, C.; Gorton, L.; Bartlett, P.N.; Ludwig, R. Direct electron-transfer anisotropy of a site-specifically immobilized cellobiose dehydrogenase. *ACS Catal.* **2019**, *9*, 7607–7615. [CrossRef]

169. Katz, E. Application of bifunctional reagents for immobilization of proteins on a carbon electrode surface: Oriented immobilization of photosynthetic reaction centers. *J. Electroanal. Chem.* **1994**, *365*, 157–164. [CrossRef]

170. Blanford, C.F.; Heath, R.S.; Armstrong, F.A. A stable electrode for high-potential, electrocatalytic O_2 reduction based on rational attachment of a blue copper oxidase to a graphite surface. *Chem. Commun.* **2007**, *17*, 1710–1712. [CrossRef]

171. Solomon, E.I.; Sundaram, U.M.; Machonkin, T.E. Multicopper oxidases and oxygenases. *Chem. Rev.* **1996**, *96*, 2563–2606. [CrossRef]

172. Solomon, E.I.; Augustine, A.J.; Yoon, J. O_2 Reduction to H_2O by the multicopper oxidases. *Dalton Trans.* **2008**, *30*, 3921–3932. [CrossRef]

173. Gentil, S.; Rousselot-Pailley, P.; Sancho, F.; Robert, V.; Mekmouche, Y.; Guallar, V.; Tron, T.; Le Goff, A. Efficiency of site-specific clicked laccase-carbon nanotubes biocathodes towards O_2 reduction. *Chem. Eur. J.* **2020**, *26*, 4798–4804. [CrossRef]

174. Neto, S.A.; Da Silva, R.G.; Milton, R.D.; Minteer, S.D.; De Andrade, A.R. Hybrid bioelectrocatalytic reduction of oxygen at anthracene-modified multi-walled carbon nanotubes decorated with $Ni_{90}Pd_{10}$ nanoparticles. *Electrochim. Acta* **2017**, *251*, 195–202. [CrossRef]

175. Rasmussen, M.; Abdellaoui, S.; Minteer, S.D. Enzymatic biofuel cells: 30 years of critical advancements. *Biosens. Bioelectron.* **2016**, *76*, 91–102. [CrossRef]

176. Wang, T.; Milton, R.D.; Abdellaoui, S.; Hickey, D.P.; Minteer, S.D. Laccase inhibition by arsenite/arsenate: Determination of inhibition mechanism and preliminary application to a self-powered biosensor. *Anal. Chem.* **2016**, *88*, 3243–3248. [CrossRef] [PubMed]

177. Bollella, P.; Fusco, G.; Stevar, D.; Gorton, L.; Ludwig, R.; Ma, S.; Boer, H.; Koivula, A.; Tortolini, C.; Favero, G. A glucose/oxygen enzymatic fuel cell based on gold nanoparticles modified graphene screen-printed electrode. Proof-of-concept in human saliva. *Sens. Actuators B* **2018**, *256*, 921–930. [CrossRef]

178. Bollella, P.; Hibino, Y.; Kano, K.; Gorton, L.; Antiochia, R. Enhanced direct electron transfer of fructose dehydrogenase rationally immobilized on a 2-aminoanthracene diazonium cation grafted single-walled carbon nanotube based electrode. *ACS Catal.* **2018**, *8*, 10279–10289. [CrossRef]
179. Bollella, P.; Hibino, Y.; Kano, K.; Gorton, L.; Antiochia, R. Highly sensitive membraneless fructose biosensor based on fructose dehydrogenase immobilized onto aryl thiol modified highly porous gold electrode: Characterization and application in food samples. *Anal. Chem.* **2018**, *90*, 12131–12136. [CrossRef]
180. Bollella, P.; Hibino, Y.; Conejo-Valverde, P.; Soto-Cruz, J.; Bergueiro, J.; Calderón, M.; Rojas-Carrillo, O.; Kano, K.; Gorton, L. The influence of the shape of Au nanoparticles on the catalytic current of fructose dehydrogenase. *Anal. Bioanal. Chem.* **2019**, *411*, 7645–7657. [CrossRef] [PubMed]
181. Calitri, G.; Bollella, P.; Ciogli, L.; Tortolini, C.; Mazzei, F.; Antiochia, R.; Favero, G. Evaluation of different storage processes of passion fruit (*Passiflora edulis* Sims) using a new dual biosensor platform based on a conducting polymer. *Microchem. J.* **2020**, *154*, 104573. [CrossRef]
182. Bollella, P.; Schulz, C.; Favero, G.; Mazzei, F.; Ludwig, R.; Gorton, L.; Antiochia, R. Green synthesis and characterization of gold and silver nanoparticles and their application for development of a third generation lactose biosensor. *Electroanalysis* **2017**, *29*, 77–86. [CrossRef]
183. Silva, T.A.; Moraes, F.C.; Janegitz, B.C.; Fatibello-Filho, O. Electrochemical biosensors based on nanostructured carbon black: A review. *J. Nanomater.* **2017**, *2017*, 4571614. [CrossRef]
184. Yáñez-Sedeño, P.; Campuzano, S.; Pingarrón, J.M. Carbon nanostructures for tagging in electrochemical biosensing: A review. *J. Carbon Res.* **2017**, *3*, 3. [CrossRef]
185. Taurino, I.; Sanzò, G.; Antiochia, R.; Tortolini, C.; Mazzei, F.; Favero, G.; De Micheli, G.; Carrara, S. Recent advances in third generation biosensors based on Au and Pt nanostructured electrodes. *Trends Anal. Chem.* **2016**, *79*, 151–159. [CrossRef]
186. Brdička, R. Etudes polarographiques des proteines du serum et leur signification pour le diagnostic du cancer. *C. R. Soc. Biol.* **1938**, *128*, 54–56.
187. Müller, O.H. Polarographic analysis of proteins, amino acids, and other compounds by means of the Brdička reaction. In *Methods of Biochemical Analysis*; Glick, D., Ed.; John Wiley & Sons: Easton, PA, USA, 1963; pp. 329–403.
188. Brabec, V. Polarography of cytochrome c in ammoniacal buffers containing cobalt ions. The effect of the protein conformation. *Gen. Physiol. Biophys.* **1985**, *4*, 609–623. [PubMed]
189. Blanford, C.F. The birth of protein electrochemistry. *Chem. Commun.* **2013**, *49*, 11130–11132. [CrossRef] [PubMed]
190. Haghighi, B.; Tabrizi, M.A. Direct electron transfer from glucose oxidase immobilized on a nano-porous glassy carbon electrode. *Electrochim. Acta* **2011**, *56*, 10101–10106. [CrossRef]
191. Bai, Y.F.; Xu, T.B.; Luong, J.H.; Cui, H.F. Direct electron transfer of glucose oxidase-boron doped diamond interface: A new solution for a classical problem. *Anal. Chem.* **2014**, *86*, 4910–4918. [CrossRef] [PubMed]
192. Wen, Z.; Ye, B.; Zhou, X. Direct electron transfer reaction of glucose oxidase at bare silver electrodes and its application in analysis. *Electroanalysis* **1997**, *9*, 641–644. [CrossRef]
193. Ghica, M.E.; Pauliukaite, R.; Fatibello-Filho, O.; Brett, C.M.A. Application of functionalised carbon nanotubes immobilised into chitosan films in amperometric enzyme biosensors. *Sens. Actuators B* **2009**, *142*, 308–315. [CrossRef]
194. Wu, F.; Yu, P.; Mao, L. Bioelectrochemistry for in vivo analysis: Interface engineering toward implantable electrochemical biosensors. *Curr. Opin. Electrochem.* **2017**, *5*, 152–157. [CrossRef]
195. Desmet, C.; Marquette, C.A.; Blum, L.J.; Doumèche, B. Paper electrodes for bioelectrochemistry: Biosensors and biofuel cells. *Biosens. Bioelectron.* **2016**, *76*, 145–163. [CrossRef]
196. Ruff, A. Redox polymers in bioelectrochemistry: Common playgrounds and novel concepts. *Curr. Opin. Electrochem.* **2017**, *5*, 66–73. [CrossRef]
197. Putzbach, W.; Ronkainen, N.J. Immobilization techniques in the fabrication of nanomaterial-based electrochemical biosensors: A review. *Sensors* **2013**, *13*, 4811–4840. [CrossRef]
198. Gooding, J.J.; Gonçales, V.R. Recent advances in the molecular level modification of electrodes for bioelectrochemistry. *Curr. Opin. Electrochem.* **2017**, *5*, 203–210. [CrossRef]
199. Pandey, C.M.; Malhotra, B.D. *Biosensors: Fundamentals and Applications*; Walter de Gruyter: Shawbury, UK, 2019.
200. Tiwari, A.; Turner, A.P.F. (Eds.) *Biosensors Nanotechnology*; Wiley-VCH: Hoboken, NJ, USA, 2014.

201. Luckarift, H.R.; Atanassov, P.; Johnson, G.R. *Enzymatic Fuel Cells–From Fundamentals to Applications*; Wiley: Hoboken, NJ, USA, 2014.
202. Davis, F.; Higson, S.P.J. Biofuel cells-Recent advances and applications. *Biosens. Bioelectron.* **2007**, *22*, 1224–1235. [CrossRef] [PubMed]
203. Bullen, R.A.; Arnot, T.C.; Lakeman, J.B.; Walsh, F.C. Biofuel cells and their development. *Biosens. Bioelectron.* **2006**, *21*, 2015–2045. [CrossRef] [PubMed]
204. Luz, R.A.S.; Pereira, A.R.; de Souza, J.C.P.; Sales, F.C.P.F.; Crespilho, F.N. Enzyme biofuel cells: Thermodynamics, kinetics and challenges in applicability. *ChemElectroChem* **2014**, *1*, 1751–1777. [CrossRef]
205. Yu, E.H.; Scott, K. Enzymatic biofuel cells–Fabrication of enzyme electrodes. *Energies* **2010**, *3*, 23–42. [CrossRef]
206. Shleev, S. Quo vadis, implanted fuel cell? *ChemPlusChem* **2017**, *82*, 522–539. [CrossRef]
207. Slate, A.J.; Whitehead, K.A.; Brownson, D.A.C.; Banks, C.E. Microbial fuel cells: An overview of current technology. *Renew. Sustain. Energy Rev.* **2019**, *101*, 60–81. [CrossRef]
208. Santoro, C.; Arbizzani, C.; Erable, B.; Ieropoulos, I. Microbial fuel cells: From fundamentals to applications. A review. *J. Power Sources* **2017**, *356*, 225–244. [CrossRef]
209. Du, Z.; Li, H.; Gu, T. A state of the art review on microbial fuel cells: A promising technology for wastewater treatment and bioenergy. *Biotechnol. Adv.* **2007**, *25*, 464–482. [CrossRef]
210. Katz, E. (Ed.) *Biomolecular Information Processing: From Logic Systems to Smart Sensors and Actuators*; Wiley-VCH: Weinheim, Germany, 2012.
211. Katz, E. *Enzyme-Based Computing Systems*; Wiley-VCH: Weinheim, Germany, 2019.

© 2020 by the authors. Licensee MDPI, Basel, Switzerland. This article is an open access article distributed under the terms and conditions of the Creative Commons Attribution (CC BY) license (http://creativecommons.org/licenses/by/4.0/).

Review

Electrochemical Biosensors Based on Membrane-Bound Enzymes in Biomimetic Configurations

Julia Alvarez-Malmagro, Gabriel García-Molina and Antonio López De Lacey *

Instituto de Catálisis y Petroleoquímica, CSIC, c/Marie Curie 2, 28049 Madrid, Spain; j.malmagro@csic.es (J.A.-M.); gabriel.garcia.m@csic.es (G.G.-M.)
* Correspondence: alopez@icp.csic.es

Received: 27 May 2020; Accepted: 14 June 2020; Published: 16 June 2020

Abstract: In nature, many enzymes are attached or inserted into the cell membrane, having hydrophobic subunits or lipid chains for this purpose. Their reconstitution on electrodes maintaining their natural structural characteristics allows for optimizing their electrocatalytic properties and stability. Different biomimetic strategies have been developed for modifying electrodes surfaces to accommodate membrane-bound enzymes, including the formation of self-assembled monolayers of hydrophobic compounds, lipid bilayers, or liposomes deposition. An overview of the different strategies used for the formation of biomimetic membranes, the reconstitution of membrane enzymes on electrodes, and their applications as biosensors is presented.

Keywords: biosensor; biomimetic membranes; membrane-bound enzymes; electrodes

1. Introduction

The attachment of enzymes to electrodes has been in the last decades a powerful strategy for the development of efficient biosensors. It couples the high specificity and turnover of enzymatic catalysis, thus assuring selective target detection and signal amplification, with the versatility, fast-response, sensitivity, and simplicity of electrochemical transduction [1,2].

Although a high proportion of redox enzymes in nature are membrane-bound ones, either associated to the external part or trans-membrane, most of the developed enzymatic biosensors are based on soluble ones. The main reasons are the higher structural complexity and lower stability of the purified membrane-bound enzymes. In fact, in most electrochemical studies that involve membrane-bound enzymes, their hydrophobic subunits are dissociated, which generally results in diminished activity and stability [3]. In order to take full advantage of the use of membrane-bound enzymes in electrochemical devices, it is necessary to design immobilization procedures that favor maintaining their natural configuration by stabilizing their hydrophobic regions. The formation of biomimetic membranes over electrode surfaces that are based on phospholipid bilayers is a versatile and powerful option for the reconstitution of membrane-bound enzymes [4,5]. Other methods that have been successfully used for developing electrochemical biosensors based on this type of enzymes have been the deposition of liposomes or formation of monolayers of hydrophobic compounds on the electrode surface [6,7].

In this review, we provide an overview of the strategies developed for the reconstitution of membrane-enzymes on biomimetic membranes over electrodes and, in particular, focusing on works that aimed at their application as electroenzymatic biosensors.

2. Biomimetic Membranes on Electrodes

Lipid biomimetic membranes that formed on solid materials are the key to model the properties of cell membrane processes, opening new research opportunities for surface electrochemistry [8]. These architectures provide simple systems, where it is possible to study in a systematically way fundamental membrane-related processes while preserving the essential characteristics of the membrane, such as fluidity or electrical sealing properties [9]. Indeed, gold substrates show a particular interest, because it is a common material that is employed for biomedical devices, such as electrochemical biosensors.

Lipid bilayers over gold surfaces are normally built by combination of Langmuir–Blodgett (LB) and Langmuir–Schaefer (LS) transfer methods or by vesicle fusion (VF) [10]. The LB technique (Figure 1a) allows for forming a lipid monolayer on gold substrates [11]. In a typical LB deposition, the gold electrode is immersed in a water subphase under an air/water interphase, in which a lipid solution in chloroform is spread. After the evaporation of the chloroform a lipid monolayer is formed in the air/water interphase. Subsequently, the monolayer is slowly compressed to a controlled surface pressure. Finally, the deposition is accomplished by raising the gold substrate from the subphase through the compressed monolayer. The subphase pressure of the monolayer should be in the range from 10 to 40 mN m^{-1}, and the monolayer temperature needs to be controlled, ensuring that the organic film is in a condensed and stable state, to obtain a good transfer ratio (τ) between 0.9 and 1.1. The microstructure and the packing density of this monolayer (the inner layer) affects the structure of the successive monolayer (the outer layer) deposited on the already modified gold electrode [12]. Drying completely the first transferred monolayer has been shown to improve the deposition process of the second layer [10]. The outlet layer of the bilayer is deposited by the LS technique (Figure 1a) [11]. The substrate horizontally touches a compressed lipid monolayer in a subphase and it is immediately withdrawn slowly. As a result, a Y-type bilayer, in which the membrane is in tail-to-tail arrangement, on the gold substrate is obtained. Because the deposition of each layer is an independent process, the LB–LS deposition method allows for constructing asymmetric bilayers. Moreover, this strategy produces more stable and better ordered bilayers.

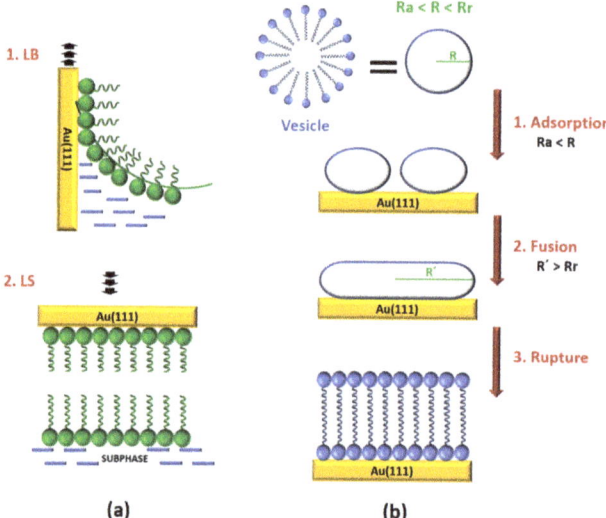

Figure 1. Schematic representation of the preparation of lipid bilayers on gold substrates by (**a**) a combination of Langmuir–Blodgett and Langmuir–Schaefer (LB-LS) methods and (**b**) vesicle fusion (VF). Adapted with permission from [11,13]. Copyright (2000 and 2007) American Chemical Society.

The VF method is an easier strategy for preparing bilayers that include integral membrane proteins from proteoliposomes [13,14]. A vesicle is a biological structure that consists of liquid enclosed by a lipid bilayer. Briefly, the VF process on gold substrates involves several steps (Figure 1b). Firstly, small unilamellar vesicles (20–50 nm in diameter), which are in an aqueous vesicle dispersion, are adhered to the substrate in highly ordered stripe-like domains. Subsequently, the fusion takes place giving a hemimicelle film. As a result, a lipid bilayer is formed by rupturing of the vesicles, unrolling, and spreading onto gold substrate. It is assumed that all of these steps depend on the vesicle size [12,13,15,16]. The VF process requires that the initial vesicle radius (R) is higher than the critical adsorption radius (Ra) and lower that the minimum rupture radius (Rr). If the radius of the vesicle is lower than Ra, the adsorption process does not take place, whereas, if the radius is higher than the Rr, then the vesicle can rupture and directly form a single bilayer disk [13]. However, for a successful rupturing of the adhered vesicles in the second step the new vesicles radius (R´) need to be higher than Rr. Moreover, temperature, presence of cations, surface charge, surface roughness, ionic strength, and solution pH should be also taken into account to obtain a good bilayer [17]. Alternatively, if gold electrodes are previously modified with relatively low hydrophobic self-assembled monolayers (SAMs) a lipid bilayer can be built by the rapid solvent exchange technique reported by Cornell et al. [18]. In this method, the SAM-coated gold electrode is incubated in a lipid solution in ethanol, followed by a fast transfer to an aqueous buffer solution. As a result, more reproducible bilayers with less defects are obtained.

The use of gold electrodes as substrate allows for obtaining molecular level of information from the bilayer formed on top. In the recent years, a combination of traditional electrochemical techniques and surface sensitive methods, such as spectroscopy, neutron scattering, and microscopic methods, have been employed to understand the behavior and structure of these biomimetic systems. On gold surfaces, it is possible to build different types of biomimetic bilayers (Figure 2), such as metal supported bilayer lipid membranes (sBLMs), tethered bilayer lipid membranes (tBLMs), or floating bilayer lipid membranes (fBLMs) [19,20].

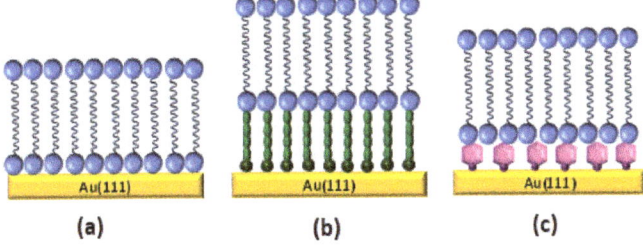

Figure 2. Cartoon of some type of biomimetic membranes at gold surface: (**a**) supported bilayer lipid membranes (sBLM), (**b**) tethered bilayer lipid membranes (tBLM), and (**c**) floating bilayer lipid membranes (fBLM). Adapted from [19] with permission from Elsevier.

Metal supported lipid membranes (sBLMs) (Figure 2a) were the first generation of lipid bilayer system used to mimic and understand the cell membrane processes. In the sBLMs, the inner layer is directly adsorbed onto the gold substrate [12]. Lipkowski's group was pioneer in successfully building sBLMs on gold electrodes either by VF [21–23] or a combination of LB-LS techniques [11,24]. The quality of the bilayer can be improved by the addition of 30% of cholesterol. The lipid membrane becomes more fluid and stress within the membrane is released, which is closer to a real biological cell membrane [25]. Atomic force microscopy (AFM) [16], scanning tunneling microscope (STM) [16,26,27], and electrochemical scanning tunneling microscope EC-STM [28] techniques have been employed to obtain molecular imaging information and molecular level resolution about the electrode surface. sBLMs on gold electrodes behave as an ideal capacitor over a large potential range. The effect of the electric field has effectively been studied with fundamental electrochemical techniques, such as differential capacitance and charge density. These results allow for the characterization of sBLMs providing information regarding the quality, compactness, and defect level of supported bilayer lipid membranes [19]. Electrochemical impedance spectroscopy (EIS) analysis provides interesting quantitative information about the interphase. A model that considers the sBLMs systems, like a series of constant phase elements (CPE) and an ohmic resistance, was employed to simulate the electrical properties of the interphase for a better understanding of interfacial processes, such as the incorporation of a carrier into real cell membranes and drug release [29,30]. In sBLMs, each side of the bilayer is exposed to different environments. The outer layer is exposed to the electrolyte while the inner leaflet is physically adsorbed on the gold subphase. From this point of view, electrochemical measurements provide valuable information regarding average properties, such as the effect of the electric field in the orientation and conformation in each layer of the membrane. However, it is not possible to obtain insight into the structure of the membrane at molecular level. The polarization modulation infrared reflection absorption spectroscopy (PM-IRRAS) technique, developed in Lipkowski´s group [31], has been successfully employed to evaluate the structure and organization of sBLMs. It demonstrated that the inner leaflet is more ordered than the outer leaflet [24], which agree with the physical and kinetic properties of the membrane constituents [32,33]. sBLMs have been used as a model system to study biological processes, like peptide incorporation [34], ligand-receptor interactions [35], or drug delivery process [36]. However, studies involving the inclusion of large transmembrane proteins are not possible with sBLM, because a water reservoir (1–2 nm thick water-rich layer) is needed between the bilayer and the gold electrode to avoid denaturation or the alteration of the functionality of the proteins [32,37].

To address this issue, tethered bilayer lipid membranes (tBLMs) (Figure 2b), where the bilayer is covalently bound to the gold electrode via hydrophilic tethering molecules [38], have been employed. These platforms show high mechanical and chemical stability. Traditionally, a tBLM is built in two separated steps [39]. First, the inner layer is self-assembled through a covalent bond with the gold electrode. This layer usually consists on flexible disulphide or thiolipid derivatives with a hydrophilic part [9] attached to the electrode by a covalent Au-S bond. The hydrophilic part provides an ion and water reservoir underneath the bilayer. In the second step, the outer layer is anchored by VF [9,40] or rapid solvent exchange [41]. Alternatively, the composition of the lipid tethered bilayer can be modified by vesicle exchange [42]. PM-IRRAS [43], neutron reflectivity experiments [44], and surface enhanced infrared absorption spectroscopy (SEIRAS) [45] have been used to evaluate the densely packed state of the SAM component, as well as the relation with the amount of water present between the bilayer and the electrode. These studies predicted that the insertion of protein and peptides is more favorable into sparsely than densely packed tBLMs [45]. In terms of electrical properties, tBLMs are stable model membrane systems [18,31,46]. EIS provides unique information regarding the electrical parameters of these biomimetic membranes. A model that was developed by Valincius et al. considering membrane defects has been employed to evaluate changes in capacitance and membrane resistance induced by the incorporation of protein or peptide molecules in the bilayer [47,48]. AFM imaging has allowed for determining the surface morphology of the tBLMs [9] as well as to provide visual images concerning the insertion of the protein or peptide [48]. Despite all of the advantages when compared to sBLMs, tBLMs still do not display long term stability and present a restrictive mobility. These are two important disadvantages to employ these platforms for biosensors development.

Floating bilayer lipid membranes (fBLMs) (Figure 2c), which interact with the gold substrate by physical interactions, are able to mimic the quasi natural environment of real membranes. They show better lateral mobility of the bilayer [48] and reduce the risk of protein denaturalization, thus preserving it activity. fBLMs are composed by a bilayer that floats ~2.4 nm over a supporting layer on the surface [21,49] or a monolayer of S-layer protein [50]. The inner leaflet should be a water rich lubricant layer that usually is a water rich polymer [16,51–53] or a hydrogel film [54]. The outer bilayer is deposited by VF [45] or a combination of LB-LS [37,55]. AFM images [37] confirmed that the lipid molecules in fBLM are tightly packed. Moreover, PM-IRRAS studies demonstrated that the lipid bilayer in fBLMS is separated from the gold surface by a water region, which allows for packing the lipid in a zigzag configuration [55]. Recently, SEIRAS [56] has been employed to probe that water molecules are a more ordered structure in the sub-membrane region of a fBLM than in a bulk solution. This strategy is a potential tool for obtaining the molecular level of information related with the hydration of fBLM and with the changes induced when a protein is incorporated.

3. Reconstitution of Membrane-Bound Enzyme on Electrodes

Reconstituted membrane enzymes play an important role in several fields, such as medicine, analytical chemistry, alternatives energies, and materials development [57]. One of the main problems for the application of these enzymes is their denaturation and loss of catalytic activity when they are not in the native-like environments [3]. Coupling the catalytic function of membrane enzymes to an electrode requires the optimization of their immobilization process, so that their in vivo structure is preserved. The best strategies for this purpose involve creating a biomimetic environment of the membrane enzyme that is attached to the electrode surface. Electronic communication between the enzymatic active site and the electrode can then be established by direct electron transfer (DET) or through the incorporation of a redox compound for achieving mediated electron transfer (MET).

Membrane enzymes have high structural complexity, thus both electrostatic and hydrophobic forces contribute greatly during the processes of adsorption onto electrodes and the subsequent electronic transfer. When the enzyme has certain surface regions in which charged amino acid zones predominate, then the adequate modification of the electrode surface can modulate the orientation of the immobilized enzyme molecules [58–60]. As this kind of enzymes have hydrophilic and hydrophobic domains, surfactants are necessary for the solubility of the membrane enzymes after their purification. It has to be taken into account that the surfactant might adsorb to the electrode, therefore affecting DET or even suppressing it completely [61]. Kawai et al. studied the effects of Triton® X-100 (a non-ionic surfactant) on the DET process of membrane-bound formate dehydrogenase (FDh) from *Gluconobacter japonicus* adsorbed onto gold electrodes modified with different thiol SAMs [61]. Changes in the frequency observed in Quartz Crystal Microbalance (QCM) measurements, performed in order to monitor the adsorption of Triton® X-100 over the SAMs, showed the formation of a surfactant monolayer (−40 Hz) in the electrodes modified with mercaptoethane (MEtn) and a bilayer (−100 Hz) in the case of 2-mercaptoethanol (MetOH) (Figure 3a). The surfactant monolayer interacted strongly over the hydrophobic MEtn SAM, thus preventing DET of the adsorbed FDh, which would have its redox centers to far away from the electrode surface. Under the same conditions, but with a hydrophilic MetOH SAM on the electrode surface, the surfactant formed a bilayer over the SAM interacting very weakly with it and allowed the insertion of FDH into the surfactant bilayer for DET with the electrode (Figure 3b).

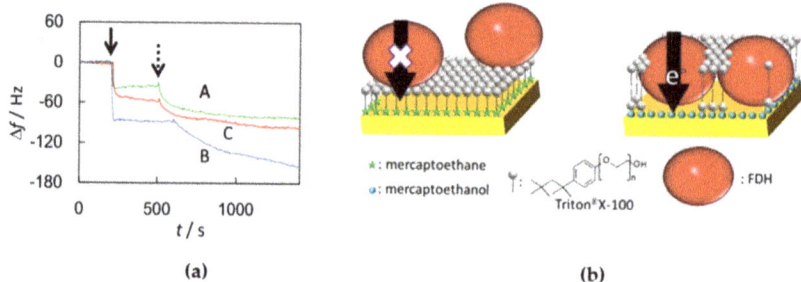

Figure 3. (a) Frequency changes measured by Quartz Crystal Microbalance (QCM) on the addition of 1% Triton® X-100 (at the solid arrow) and fructose dehydrogenase (FDh) (at the dashed arrow) at mercaptoethane (MEtn)-modified (green line), 2-mercaptoethanol (MEtOH)-modified (blue line), and bare Au electrodes (red line). (b) Proposed scheme of the adsorption of fructose dehydrogenase (FDh) and Triton® X-100 to hydrophobic (left) and hydrophilic electrodes (right). Reprinted from [61] with permission from Elsevier.

On the other hand, Lojou and co-workers found that remains of the detergent n-Dodecyl β-D-maltoside (DDM) strongly attached around the hydrophobic zones surrounding the distal 4Fe4S cluster (the redox site for electron exchange) of the membrane-bound NiFe hydrogenase (Hase) from *Aquifex aeolicus* when adsorbed on modified gold electrodes. This effect modifies the hydrophobicity of this areas, which makes them more hydrophilic. Therefore, in the case of the enzyme adsorption on an electrode modified with hydrophobic SAMs, the enzyme molecules always oriented with the distal cluster region on the opposite side to the SAM, too far for establishing DET with the electrode. In the case of hydrophilic SAMs on the electrode, there was no preferential enzyme orientation during adsorption, so the MET and DET processes had the same incidence with a catalytic current ratio of $I_{DET}/I_{DET+MET}$ around 0.5 for H_2 oxidation [59].

Cytochrome p450 (CyP) and human flaving containing monooxygenase 3 (hFMO3) are membrane-bound redox enzymes that have also been studied for optimizing their DET with electrodes modified with hydrophobic SAMs. In the case of the CyPs, its active center is an iron protoheme. The natural compounds that supply electrons to CyPs for their catalytic activity, NADPH, is very expensive and time consuming. Immobilizing CyPs on electrodes can replace these products [62]. Microsomes (lipid membranes) containing CyP and CyP reductase (CPR) and deposited on electrodes modified with hydrophobic SAMs of aromatic compounds benzenethiolate (BT) and naphtalene thiolate (NT) gave good current intensities by DET with reduction peaks around −0.4 V vs Ag/AgCl. The electroenzymatic system was tested for testosterone metabolization, measuring by high performance liquid chromatography (HPLC) a production of 270 pmol of 6 β-hydroxytestosterones [63]. hFMO3 is a liver protein that belongs to the second most important class of phase-1 drug-metabolizing enzymes [64–66]. Castrignano et al. reported the immobilization of hFMO3 on glassy carbon/graphite oxide (GO) modified with di-dodecyl di-methylammonium bromide (DDAB), which mimicked the enzyme's native environment. By HPLC, they measured the products obtained from the electroenzymatic N-oxidation of benzydamine (a nonsteroidal anti-inflamatory) and tamoxifen (an antiestrogenic used in therapies against breast cancer and chemoprotection) [66].

Quinone oxidoreductases are a type of membrane-bound enzymes that catalyze redox processes of the quinone pool in cell membranes. They can be reconstituted on sBLMs formed over electrodes, in which lipophilic quinones that are embedded in the sBLM act as redox mediators with the electrode [67]. Jeuken and co-workers studied this strategy for an ubiquinol oxidase (cytochrome bo3 from *Escherichia coli*), which couples the oxidation of ubiquinol to ubiquinone with the reduction of O_2 to H_2O [68]. Furthermore, they used the layer by layer (LBL) technique with bacterial membrane extracts on gold electrodes to create multilayers of ubiquinol oxidase. In this strategy, poly-l-lysine was used as an electrostatic polymer for connecting the different lipid bilayers. The same strategy was studied with an oxygen tolerant membrane bound Hase (MBH, Figure 4a). This type of NiFe-Hases have has three subunits, one of them is a hydrophobic one that is inserted in the cell membrane and ensures the electron transfer between the other Hase subunits and the pool of quinones of the respiratory chain [69]. Measurements by fluorescence recovery after bleaching were performed to determine the lateral diffusion for the base bilayer and for the interconnections between bilayers, being 0.6 ± 0.1 µm^2 s^{-1} and 0.7 ± 0.2 µm^2 s^{-1}, respectively [68]. Thus, these values are indistinguishable from each other, indicating that the membrane stacks are interconnected via lipid phases. These interconnections can be maintained throughout the layers, creating diffusion routes throughout the multilayer, not only of lipids, but also of lipophilic quinones within the bilayer, such as ubiquinone-10 (UQ10) and menoquinone-7 (MQ7). An important property of quinones is their hydrophobicity, which restricts them within the bilayer, but also allows them to diffuse freely within it, allowing for MET-based electoenzymatic catalysis. The cyclic voltammograms (CVs) performed showed a linear increase in the MQ7 current and in its peak area with the number of lipid bilayers deposited on the electrode, indicating an increase in the number of quinones that interact with the electrode. The same behavior was observed for UQ10 [68]. In the CVs, electrocatalytic H$_2$ oxidation coincided with the ubiquinol oxidation peak, confirming that the electronic transfer between MBH and the electrode was mediated by the quinone pool (Figure 4b), with a considerable stability of the immobilized membrane enzyme (Figure 4c).

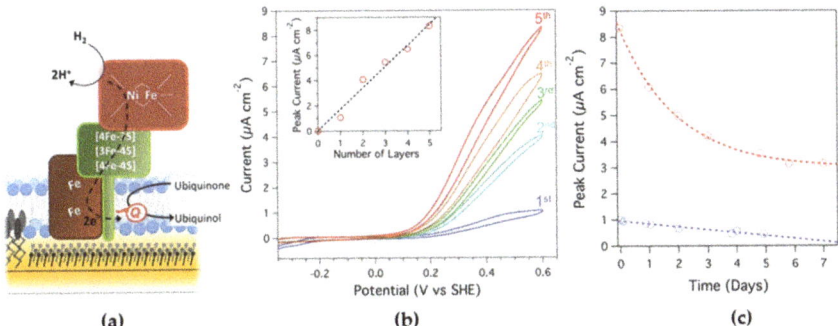

Figure 4. (a) Scheme of immobilized MBH activity on a biomimetic membrane-modified electrode. Ubiquinone is reduced by the enzyme to ubiquinol using electrons generated from hydrogen oxidation. The ubiquinone is reoxidized at the electrode. (b) Cyclic voltammograms (CVs) for 1–5 lipid bilayers containing MBH. Inset: Catalytic current (measured at 600 mV) as a function of the number of lipid bilayers. (c) Peak current (obtained at 600 mV) as a function of time (up to a week) for a five-layered MBH multilayer (red) and a single-MBH bilayer (blue) as measured by CV. Reproduced from [68] under the terms and conditions of the Creative Commons Attribution (CC BY) license.

Another Hase, the NiFeSe one from *Desulfovibrio vulgaris*, is attached to the periplasmic cell membrane in a different way to that of the O$_2$-tolerant Hases. Instead of having a hydrophobic third subunit inserted into the membrane, it is peripherally attached via a lipid tail at its N-terminus, which is in the opposite region of the distal FeS cluster [70]. Gutierrez-Sánchez et al. studied the reconstitution of the membrane-bound enzyme over gold electrodes with two different configurations characterized by AFM and electrochemical measurements [60]. In the first, the enzyme was attached through its lipid tail after the formation of a fBML over a gold electrode that was modified with a 4-aminothiophenol (4-APh) SAM. In this case, catalytic current for H$_2$ oxidation was only measured by MET when adding 0.16 mm of methyl viologen (MV), because the Hase's distal FeS cluster was facing the solution, not the electrode. This result also showed the existence of defects in the membrane allowing permeability of MV. The second configuration was obtained immobilizing the Hase together with liposomes and BioBeads (to remove detergent excess) in a single step over the SAM-modified gold electrode. The positive charges at the electrode surface oriented the immobilized Hase molecules by electrostatic interactions with the negatively charged region surrounding the distal FeS cluster, leaving the lipid tail towards the solution that allowed for the formation of a fBLM on the top of the Hase layer. With this last system, a clear catalytic current of H_2 oxidation was observed by DET [60], as well as the generation of a proton gradient across the fBLM [71]. The AFM characterization indicated that the fBLM thickness was the expected one of approximately 5 nm [60], whereas the SEIRA studies confirmed the two different structural configurations that were obtained with these two reconstitution strategies [72].

Another step further was the coupling of the ability of this system to generate a proton gradient across a biomimetic membrane over an electrode with the activity of another membrane-bound enzyme, F_1-F_0 ATP-synthase from *Escherichia coli* (ATPase). This enzyme uses the proton gradient across the membrane as a driving force for the synthesis of adenosine triphosphate (ATP) [73–75]. ATPase was inserted into liposomes and a fBLM was formed by the fusion of the proteoliposomes over the NiFeSe Hase monolayer covalently attached to the electrode surface. In the presence of 500 µm of adenosine diphosphate (ADP) and phosphate in the solution, 40 µg of ATP was synthesized in 2 h [76]. The AFM images indicated that 30–40% of the surface was covered by enzyme, thus the amount of the ATPase on the gold surface was estimated to be around 350 ng cm^{-2}. It was further reported that ATPase proteoliposomes could be directly fused over the gold electrode modified with the 4-APh SAM to form a fBLM. Two types of proteoliposomes were studied for ATP hydrolysis: with and without poly (ethylene glycol) 5000 MW (PEG) (Figure 5). AFM and electrochemical measurements indicated that more reproducible and stable results were obtained when PEG was included as spacer between the fBLM and the electrode surface. This improvement was attributed to an increase of the hydrophilic boundary, allowing for more translocation of protons across the fBLM [77].

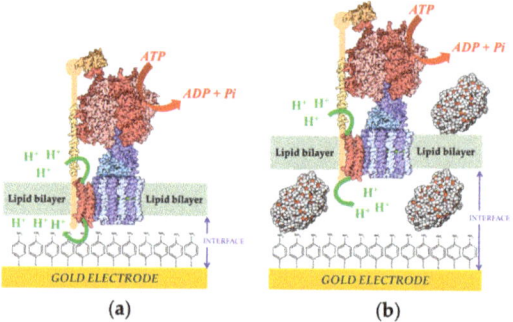

Figure 5. Adenosin triphosphate-synthase from *Escherichia coli* (ATPase) reconstitution in (**a**) fBLM (**b**) fBLM-polyethylene glycol (fBLM-PEG) over gold electrodes modified with a 4-APh self-assembled monolayer (SAM). Reprinted from [77] with permission from Elsevier.

A fBLM on gold electrodes modified with 4-APh has also been used to reconstitute the multienzimatic complex I (CpI) [78], which plays a fundamental role in the production of cellular energy. It contributes to the stabilization and maintenance of the transmembrane electrochemical potential difference necessary for ATP synthesis, transport, and mobility. Deficiencies in this enzyme can lead to some neurodegenerative diseases, such as Leber's hereditary optic neuropathy, Parkinson's disorders, and dystonia [79]. The hydrophilic peripheral subunits contain the prosthetic groups (FeS clusters and FMN) for NADH oxidation and electron transport [80], while the hydrophobic subunits inserted in the cell membrane part are involved in quinone reduction and charge translocation [81]. The system allowed for the electrochemical study of electron transfer and proton translocation by CpI. 2,3-dimethyl-1,4-naphthoquinone (DMN) was incorporated into the bilayer as electron acceptor from the enzyme after NADH oxidation and as redox mediator at the electrode. Without DMN in the system the catalytic current was negligible. The AFM study indicated the presence of protuberances of 6–8 nm coming out from the fBLM, which were attributed to the hydrophilic components of reconstituted CpI molecules [78]. In a later work, the two functions of CpI were monitored by SEIRA spectroscopy. It was found that changing the way of constructing the system affected the amide I/amide II infrared bands intensity ratio, which might indicate different orientations or arrangements of the enzyme on the electrode. The SEIRA experiments showed that CpI was preferably incorporated into the fBLM with the catalytic hydrophilic arm towards the solution, which made it catalytically active on NADH oxidation and translocation of protons, thus acidifying the electrode/fBLM interface [82].

4. Applications as Biosensors

The immobilization of membrane-bound enzymes on tailored electrodes that mimic their natural environment allows for enhancing the electrocatalytic properties by preserving their optimal structural integrity, as indicated in the previous section. In consequence, more sensitive and stable biosensors that are based on this type of enzymes can be developed using these strategies. Furthermore, the presence of a biomimetic membrane over the electrode can prevent the fouling of its surface by the medium or the non-desired signals due to redox-active interferents.

An early example of a biosensor based on a membrane-enzyme co-immobilized with a biomimetic membrane was reported by Kinnear and Monbouquette [83]. Membrane-bound fructose dehydrogenase (FDh) was reconstituted with a mix of two phospholipids (dioleoyl-L-phosphatidyl ethanolamine and dioleoyl-L-phosphatidyl choline) and its natural cofactor ubiquinone-6. Subsequently, it was deposited on a gold electrode modified with a mixed SAM of thiols. The SAM contained a hydrophobic long-chain thiol together with two polar short-chain ones, in order to facilitate both electrostatic and hydrophobic interactions with the charged groups of the phospholipids and the largely lipophilic FDh, respectively. The enzymatic electrode was studied for amperometric detection of fructose, in which the ubiquinone-6 co-immobilized in the mixed thiolate/phospholipid layer acted as redox mediator. The calibration curve was linear up to 5 mm fructose with a detection limit of 10 µm. The biosensor was highly selective by showing no appreciable response to other sugars, which is due to the specificity of FDh activity towards fructose. The fructose biosensor was tested in apple and orange juice with low interference by ascorbate, which was attributed to the blocking effect of the hydrophobic layer on the electrode. The storage stability of the biosensor improved in comparison to the FDh-modified electrode without the biomimetic membrane, which was due to the decreased leaching of the ubiquinone redox mediator.

Darder et al. also immobilized FDh to gold electrodes that were modified with a mixed SAM of hydrophobic and hydrophilic thiols in order to simultaneously provide electrostatic interactions with charged residues of the enzyme and hydrophobic interactions with its largely lipophilic regions [84]. Sodium hydroxymethyl ferrocene in solution was used as redox mediator. The detection limit for fructose was 20 µM and the lineal range reached up to 0.7 mm, having similarly low ascorbate interference, as reported by Kinnear and Monbouquette [83]. The mixed thiol SAM was also tested as an immobilization platform for another membrane-bound enzyme, D-gluconate dehydrogenase, giving good results for gluconate detection. On the other hand, no electrocatalytic responses were obtained when soluble redox enzymes, such us glucose oxidase (GOx) or horse radish peroxidase (HRP), were deposited on gold electrodes that were modified in the same way. The QCM measurements indicated that such enzymes did not immobilize on the surface, which indicated that the mixed SAM was only adequate for attaching lipophilic enzymes [84].

More recently, a FDh-based amperometric biosensor has been reported in which the membrane-bound enzyme was entrapped in liquid-crystalline lipidic cubic phase, an adequate matrix for lipophilic enzymes. The encapsulated FDh was then deposited on a glassy carbon electrode modified with single wall carbon nanotubes (SWCNTs) to allow for DET with the heme group of the enzyme. In this way, very large electroacatalytic currents were measured in the presence of fructose without requiring the addition of a redox mediator, thus obtaining a third-generation biosensor [85]. The linear range of the fructose biosensor was 1–10 mM. Furthermore, the operational stability of the enzymatic electrode was high, as no appreciable loss of the amperometric signal was observed during 10 h of continuous cycling. Indeed, this high stability of the immobilized FDh was attributed to the favorable environment that was provided by the lipidic matrix, preventing enzyme leaching or degradation.

Cholesterol oxidase (COx) is a flavoenzyme that catalyzes the oxidation of cholesterol while reducing O_2 to H_2O_2 and presents a hydrophobic side chain that inserts into the lipid membrane in vivo [86]. The measurement of cholesterol levels in blood is a very important biomedical parameter for coronary heart diseases, arteriosclerosis, and cerebral thrombosis [87]. Wicklein et al. studied the immobilization of this enzyme on different biomimetic interfaces that were based on phosphatidyl choline (PC) assembled on to silicate sepiolite [88]. The best results were obtained when a PC bilayer was used, to which COx bound by inserting a side loop as it does in its natural environment. The activity of the enzyme in the presence of cholesterol was measured amperometrically by the oxidation of the H_2O_2 produced at the working electrode. The lineal range and sensitivity obtained were 0–5 µm and 154 mA M^{-1}, respectively. The same PC bilayer on sepiolite was used for immobilizing urease, an enzyme that peripherally binds to the cellular membrane, to build an urea biosensor on top of a gold electrode modified with a SAM of 5,5'-dithiobis (2-nitrobenzoic acid) (Figure 6a). The electrochemical transduction was based on the pH change at the electrode surface due to the urea hydrolysis activity of the immobilized enzyme. The SAM on the electrode acted as potentiometric sensor, because its redox potential shifted 59 mV per pH unit (Figure 6b). Fast detection of urea was measured with high sensitivity (30.8 ± 0.7 V M^{-1}) and low interference from ascorbic acid. The biosensor could be stored during at least six months without sensitivity loss, which indicated the excellent compatibility of the sepiolite/PC bilayer assembly with the enzyme [88].

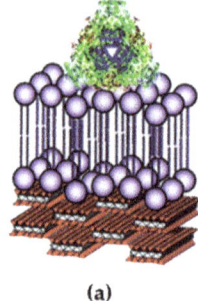
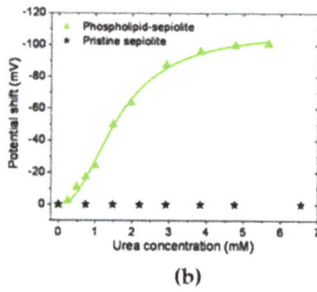

(a) (b)

Figure 6. (a) Schematic representation of urease attached to a sBLM on sepiolite fiber. (b) Dependence on urea concentration of the potential shift of the redox probe as a result of pH increase due to the catalytic activity of urease immobilized on pristine and sBLM-modified sepiolite. Reproduced with permission from [88]. Copyright (2011) American Chemical Society.

Psychoios et al. reported another cholesterol biosensor based on COx immobilized on an electrode in a biomimetic configuration [89]. The enzyme was encapsulated in a lipid film that was polymerized over a matrix of ZnO nanowalls, which acted as transducer by providing a potentiometric signal related to the change of the double layer charge on their surface caused by the enzymatic reaction. The biosensor was tested in blood serum and urine, having a sensitivity of 57 mV per decade of cholesterol concentration, a broad logarithmic detection range of several orders of magnitude, a limit of detection of 0.7 µm, low interferences from albumin, and excellent storage stability.

Very recently, Moura and co-workers have reported an interesting third-generation nitric oxide biosensor based on the membrane-bound enzyme nitric oxide reductase [90]. NO is involved in many biological processes, but it is very reactive, thus its quantification requires detection times between 5 and 15 s [91]. The biosensing device was formed by a mix of the enzyme, BLMs, and SWCNTs on a pyrolytic graphite electrode (Figure 7). The SWCNTs served as wires for DET from the electrode to the enzyme and were functionalized with carboxylic groups to favor electrostatic interactions with the predominantly positively charged phospholipids in the bilayer composition. A polyethylene glycol (PEG)-modified phospholipid was included to avoid liposome formation, while favoring the stabilization of the membrane-bound enzyme within the SWCNTs-BLM network. Square wave voltammetry showed DET of nitric oxide reductase and the peak current increased linearly with NO concentration in the 0.4–1.0 µm range. A limit of detection of 0.13 µm was measured, which is adequate for measuring NO evolution in biological processes. The storage stability was good with 83.5% of the initial response was retained after five weeks, which indicated sufficient binding between the three components that were deposited on the electrode that minimized leaching [90].

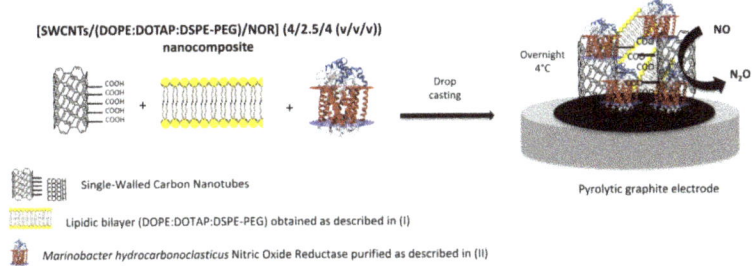

Figure 7. Scheme of nitric oxide biosensor construction. Reprinted from [90] with permission from Elsevier.

There are many studies of electrochemical biosensors that are based on transmembrane proteins reconstituted on phospholipid bilayers supported on electrodes, although the vast majority comprises ion-transport proteins or olfactory receptors, not enzymes [57]. However, a very interesting transmembrane enzyme for biosensing is ATP-synthase or ATPase. The measurement of adenosin triphosphate (ATP) concentration is of great interest for studying cell metabolism [92] and for the detection of microbial contamination on surfaces [93]. In order to develop a sensitive and versatile biosensor for ATP detection, the ATPase from *Escherichia coli* was reconstituted on a floating phospholipid bilayer over a gold electrode modified with a 4-APh SAM (Figure 5). The role of the thiol SAM was not only to favor the formation of the biomimetic membrane over the electrode, but also to serve as redox probe of pH changes at the electrode/fBLM interface after oxidation and dimerization [71]. In the presence of ATP in solution, the ATPase hydrolyzes the compound while translocating protons across the membrane. The potential of the redox probe on the gold surface is pH-dependent, thus by differential pulse voltammetry a positive shift of the peak potential was measured that was proportional to the logarithm of ATP concentration in the bulk solution in the 0.001–1 mM range. The potentiometric ATP biosensor was tested for monitoring the presence of microbial contamination. ATP extracted from *Escherichia coli* cultures with different cells concentration were analyzed with the biosensor obtaining an electrochemical signal proportional to the cells concentration with a time response of 5 min [77].

A very original electrochemical biosensor design has been reported for the detection of aldolase activity. In this case the biosensor does not integrate a membrane-bound enzyme; instead the catalytic activity of the soluble enzyme induces the formation of a tBLM on the electrode surface that decreases the amperometric response of the redox probe ferricyanide. Aldolase is a disease marker for cancer diagnosis and it catalyzes the reversible conversion of fructose-1,6-bisphosphate to glyceraldehyde 3-phosphate and dihydroxyacetone. Both products of the enzyme reaction have carbonyl groups that acted as linkers between the SAM thiol on the Au electrode and a PEG-phospholipid, thus driving the formation of a tBLM that blocked ferricyanide reduction at the electrode. The aldolase activity was analyzed over a broad linear detection range from 5 mU L^{-1} to 100 U L^{-1}, measuring a 1 mU L^{-1} detection limit [94].

Liposomes can be also used to stabilize a membrane-bound enzyme on an electrode for improvement of biosensor performance. Besides providing a biomimetic environment to the enzyme, the liposomes give protection from proteases action and improve confinement on the electrode surface. Guan et al. developed an electrochemical biosensor for organophosphorus pesticides formed by layer-by-layer deposition of chitosan (a biocompatible polymer) and spherical shell liposomes containing acetylcholinesterase (AChE) [95]. AChE is a transmembrane enzyme that is part of the nervous system, in which it maintains the level of the neurotransmitter acetylcholine by catalyzing its hydrolysis into thiocholine [57]. Porins were added to the liposomes that were deposited on the electrode to allow the entrance of AChE substrate and pesticide inhibitor. Dichlorvos was used as a model pesticide for studying the biosensing properties of the enzymatic electrode. Amperometric transduction was performed by measuring the decrease of the oxidation current of the thiocholine product due to the inhibition of AChE by the action of the pesticide. The amperometric biosensor was quite sensitive to dichlorvos with a linear range of 0.25–10 µm and a detection limit of 0.9 ± 0.1 µg L^{-1}. The storage stability of the biosensor was studied, retaining full activity after storing for 15 days, whereas, in comparison, the enzyme in solution lost 80% of its activity [95]. In a subsequent work, the same authors increased the sensitivity of the biosensor by including multiwalled carbon nanotubes in the layer-by-layer system to improve the electrochemical detection of thiocholine. The optimal construction involved six complete layers, in which the detection limit for dichlorvos was 0.68 ± 0.076 µg L^{-1}. In addition, the storage stability improved by retaining full activity after 30 days [96].

5. Perspectives

The number of reported electrochemical biosensors based on membrane-bound enzymes is still quite low when compared to those based on soluble enzymes, even if, in nature, the amount of enzymes of the first kind that catalyze reactions of interest in sensing is large [3,5,57]. Therefore, we think that there is a great potential for increasing research in this particular field. A very important issue for this purpose is to develop reliable methods for immobilizing membrane-enzymes onto electrodes while maintaining their native structure.

Biomimetic chemistry comes to aid for affronting this challenge. In the last decades, there has been large progress on the formation and characterization of biomimetic lipid bilayers on electrodes for fundamental studies [8,19]. Besides stabilizing the membrane-bound enzyme, the lipid bilayer over the electrode might provide protection from surface fouling or from redox-active interferents. More studies should be done in this direction in order to obtain more stable and specific biosensors for analysis in complex samples. Although sBLMs and tBLMs have shown to be very valuable as model systems for fundamental studies of cell membrane processes [9,25,29,45,48], we think that fBLMs are more adequate for biosensor development. The presence of a water cushion in the interface between the fBLM and the electrode facilitates the insertion of transmembrane proteins in the lipid bilayer and gives higher flexibility and mobility within the biomimetic construction [77,78,82]. Moreover, the presence of an aqueous interphase facilitates charge transfer process across the membrane, which can help in coupling the biological recognition event to the electrochemical transductor [77,78]. Thus, we think that there is great potential for the development of biosensors with potentiometric or impedance detection. Not only BLMs are valid for providing a suitable biomimetic immobilization of membrane enzymes on electrodes. Other alternatives, such as formation of mixed SAMs with controlled hydrophobic/hydrophilic composition [83,84], entrapment in liquid-crystalline lipidic cubic phase [85], or deposition of liposomes on the electrode surface have been shown to be successful for different biosensor applications [95,96].

Several surface characterization techniques are now available for electrochemical research, such as AFM/STM, SERS, SEIRAS, and PM-IRRAS, which allow for the complete characterization of the biomimetic constructions and immobilized membrane enzymes on the electrode [8]. The use of these surface characterization techniques combined with the diverse panoply of existing electrochemical methods permits correlating the configuration of the immobilized membrane enzyme in the biomimetic construction with its electrocatalytical properties [7,69,72,82]. In our opinion, this kind of studies should, in the future, considerably aid the optimization of biosensors design based on biomimetic systems.

Author Contributions: Conceptualization, A.L.D.L.; writing—original draft preparation, J.A.-M., G.G.-M., and A.L.D.L.; project administration, A.L.D.L.; funding acquisition, A.L.D.L. All authors have read and agreed to the published version of the manuscript.

Funding: This research was funded by the Spanish MCIU/AEI and European Union FEDER, grant number RTI2018-095090-B-I00.

Acknowledgments: G.G.-M thanks the Spanish MCIU/AEI and European Union FEDER for a predoctoral contract BES-2016-078815.

Conflicts of Interest: The authors declare no conflict of interest.

References

1. Wang, J. Nanomaterial-based electrochemical biosensors. *Analyst* **2005**, *130*, 421–426. [CrossRef]
2. Heller, A.; Feldman, B. Electrochemical glucose sensors and their applications in diabetes management. *Chem. Rev.* **2008**, *108*, 2482–2505. [CrossRef]
3. Jeuken, L.J.C. Electrodes for integral membrane enzymes. *Nat. Prod. Rep.* **2009**, *26*, 1234–1240. [CrossRef]
4. Laftsoglou, T.; Jeuken, L.J.C. Supramolecular electrode assemblies for bioelectrochemistry. *Chem. Commun.* **2017**, *53*, 3801–3809. [CrossRef] [PubMed]
5. Saboe, P.O.; Conte, E.; Farell, M.; Bazan, G.C.; Kumar, M. Biomimetic and bioinspired approaches for wiring enzymes to electrode interfaces. *Energy Environ. Sci.* **2017**, *10*, 14–42. [CrossRef]
6. Marchal, D.; Pantigny, M.; Laval, J.M.; Moiroux, J.; Bourdillon, C. Rate constants in two dimensions of electron transfer between pyruvate oxidase, a membrane enzyme, and ubiquinone (coezyme Q_8), its water-insoluble electron carrier. *Biochemistry* **2001**, *40*, 1248–1256. [CrossRef] [PubMed]
7. Infossi, P.; Lojou, E.; Chauvin, J.P.; Herbette, G.; Brugna, M.; Giudici-Orticoni, M.T. Aquifex aeolicus membrane hydrogenase of hydrogen biooxidation: Role of lipids and physiological partners in enzyme stability and activity. *Int. J. Hydrogen Energy* **2010**, *35*, 10778–10789. [CrossRef]
8. Lipkowski, J. Biomimetics: A new research opportunity for surface electrochemistry. *J. Solid State Electrochem.* **2020**. [CrossRef]
9. Junghans, A.; Köper, I. Structural analysis of tethered bilayer lipid membranes. *Langmuir* **2010**, *26*, 11035–11040. [CrossRef]
10. Kycia, A.H.; Su, Z.; Brosseau, C.L.; Lipkowski, J. In situ PM–IRRAS studies of biomimetic membranes supported at gold electrode surfaces. In *Vibrational Spectroscopy at Electrified Interfaces*; Wieckowski, A., Korzeniewski, C., Braunschweig, B., Eds.; John Wiley & Sons: Hoboken, NJ, USA, 2013; pp. 345–417.
11. Zawisza, I.; Bin, X.; Lipkowski, J. Potential-driven structural changes in Langmuir–Blodgett DMPC bilayers determined by in situ spectroelectrochemical PM IRRAS. *Langmuir* **2007**, *23*, 5180–5194. [CrossRef]
12. Sackmann, E. Supported membranes: Scientific and practical applications. *Science* **1996**, *271*, 43–48. [CrossRef] [PubMed]
13. Reviakine, I.; Brisson, A. Formation of supported phospholipid bilayers from unilamellar vesicles investigated by atomic force microscopy. *Langmuir* **2000**, *16*, 1806–1815. [CrossRef]
14. Leonenko, Z.V.; Carnini, A.; Cramb, D.T. Supported planar bilayer formation by vesicle fusion: The interaction of phospholipid vesicles with surfaces and the effect of gramicidin on bilayer properties using atomic force microscopy. *BBA Biomembr.* **2000**, *1509*, 131–147. [CrossRef]
15. Lipowsky, R.; Seifert, U. Adhesion of vesicles and membranes. *Mol. Cryst. Liq. Cryst.* **1991**, *202*, 17–25. [CrossRef]
16. Pawłowski, J.; Juhaniewicz, J.; Güzeloğlu, A.; Sek, S. Mechanism of lipid vesicles spreading and bilayer formation on a Au(111) Surface. *Langmuir* **2015**, *31*, 11012–11019. [CrossRef] [PubMed]
17. Hardy, G.J.; Nayak, R.; Zauscher, S. Model cell membranes: Techniques to form complex biomimetic supported lipid bilayers via vesicle fusion. *Curr. Opin. Colloid Interface Sci.* **2013**, *18*, 448–458. [CrossRef]
18. Cornell, B.A.; Braach-Maksvytis, V.L.B.; King, L.G.; Osman, P.D.J.; Raguse, B.; Wieczorek, L.; Pace, R.J. A biosensor that uses ion-channel switches. *Nature* **1997**, *387*, 580–583. [CrossRef]

19. Su, Z.; Leitch, J.J.; Lipkowski, J. Electrode-supported biomimetic membranes: An electrochemical and surface science approach for characterizing biological cell membranes. *Curr. Opin. Electrochem.* **2018**, *12*, 60–72. [CrossRef]
20. Juhaniewicz, J.; Sek, S. Atomic force microscopy and electrochemical studies of melittin action on lipid bilayers supported on gold electrodes. *Electrochim. Acta* **2015**, *162*, 53–61. [CrossRef]
21. Bin, X.; Zawisza, I.; Goddard, J.D.; Lipkowski, J. Electrochemical and PM-IRRAS studies of the effect of the static electric field on the structure of the DMPC bilayer supported at a Au(111) electrode surface. *Langmuir* **2005**, *21*, 330–347. [CrossRef]
22. Zawisza, I.; Lachenwitzer, A.; Zamlynny, V.; Horswell, S.L.; Goddard, J.D.; Lipkowski, J. Electrochemical and photon polarization modulation infrared reflection absorption spectroscopy study of the electric field driven transformations of a phospholipid bilayer supported at a gold electrode surface. *Biophys. J.* **2003**, *85*, 4055–4075. [CrossRef]
23. Horswell, S.L.; Zamlynny, V.; Li, H.Q.; Merrill, A.R.; Lipkowski, J. Electrochemical and PM-IRRAS studies of potential controlled transformations of phospholipid layers on Au (111) electrodes. *Faraday Discuss.* **2002**, *121*, 405–422. [CrossRef] [PubMed]
24. Garcia-Araez, N.; Brosseau, C.L.; Rodriguez, P.; Lipkowski, J. Layer-by-layer PMIRRAS characterization of DMPC bilayers deposited on a Au(111) electrode surface. *Langmuir* **2006**, *22*, 10365–10371. [CrossRef] [PubMed]
25. Brosseau, C.L.; Leitch, J.; Bin, X.; Chen, M.; Roscoe, S.G.; Lipkowski, J. Electrochemical and PM-IRRAS a glycolipid-containing biomimetic membrane prepared using Langmuir–Blodgett/Langmuir–Schaefer deposition. *Langmuir* **2008**, *24*, 13058–13067. [CrossRef] [PubMed]
26. Shimizu, H.; Matsunaga, S.; Yamada, T.; Kobayashi, T.; Kawai, M. Formation of ordered phospholipid monolayer on a hydrophilically modified Au (111) substrate. *ACS Nano* **2016**, *10*, 7811–7820. [CrossRef] [PubMed]
27. Sek, S.; Xu, S.; Chen, M.; Szymanski, G.; Lipkowski, J. STM studies of fusion of cholesterol suspensions and mixed 1,2-dimyristoyl-sn-glycero-3-phosphocholine (DMPC)/cholesterol vesicles onto a Au (111) electrode surface. *J. Am. Chem. Soc.* **2008**, *130*, 5736–5743. [CrossRef]
28. Xu, S.; Szymanski, G.; Lipkowski, J. Self-assembly of phospholipid molecules at a Au (111) electrode surface. *J. Am. Chem. Soc.* **2004**, *126*, 12276–12277. [CrossRef]
29. Prieto, F.; Rueda, M.; Naitlho, N.; Vázquez-González, M.; González-Rodríguez, M.L.; Rabasco, A.M. Electrochemical characterization of a mixed lipid monolayer supported on Au (111) electrodes with implications for doxorubicin delivery. *J. Electroanal. Chem.* **2018**, *815*, 246–254. [CrossRef]
30. Alvarez-Malmagro, J.; Jablonowska, E.; Nazaruk, E.; Szwedziak, P.; Bilewicz, R. How do lipid nanocarriers – cubosomes affect electrochemical properties of DMPC bilayers deposited on gold (111) electrodes? *Bioelectrochemistry* **2020**, *134*, 107516. [CrossRef]
31. Zamlynny, V.; Zawisza, I.; Lipkowski, J. PM FTIRRAS studies of potential-controlled transformations of a monolayer and a bilayer of 4-pentadecylpyridine, a model surfactant, adsorbed on a Au(111) electrode surface. *Langmuir* **2003**, *19*, 132–145. [CrossRef]
32. Groves, J.T.; Ulman, N.; Boxer, S.G. Micropatterning fluid lipid bilayers on solid supports. *Science* **1997**, *275*, 651–653. [CrossRef]
33. Groves, J.T.; Boxer, S.G. Micropattern formation in supported lipid membranes. *Acc. Chem. Res.* **2002**, *35*, 149–157. [CrossRef] [PubMed]
34. Su, Z.F.; Shodiev, M.; Jay Leitch, J.; Abbasi, F.; Lipkowski, J. In situ electrochemical and PM-IRRAS studies of alamethicin ion channel formation in model phospholipid bilayers. *J. Electroanal. Chem.* **2018**, *819*, 251–259. [CrossRef]
35. Yang, T.; Baryshnikova, O.K.; Mao, H.; Holden, M.A.; Cremer, P.S. Investigations of bivalent antibody binding on fluid-supported phospholipid membranes: The effect of hapten density. *J. Am. Chem. Soc.* **2003**, *125*, 4779–4784. [CrossRef] [PubMed]
36. Alvarez-Malmagro, J.; Matyszewska, D.; Nazaruk, E.; Szwedziak, P.; Bilewicz, R. PM-IRRAS study on the effect of phytantriol-based cubosomes on DMPC bilayers as model lipid membranes. *Langmuir* **2019**, *35*, 16650–16660. [CrossRef]

37. Kycia, A.H.; Wang, J.; Merrill, A.R.; Lipkowski, J. Atomic force microscopy studies of a floating-bilayer lipid membrane on a Au(111) surface modified with a hydrophilic monolayer. *Langmuir* **2011**, *27*, 10867–10877. [CrossRef]
38. Knoll, W.; Köper, I.; Naumann, R.; Sinner, E.K. Tethered bimolecular lipid membranes. A novel model membrane platform. *Electrochim. Acta* **2008**, *53*, 6680–6689. [CrossRef]
39. Vockenroth, I.K.; Rossi, C.; Shah, M.R.; Köper, I. Formation of tethered bilayer lipid membranes probed by various surface sensitive techniques. *Biointerphases* **2009**, *4*, 19–26. [CrossRef]
40. Ragaliauskas, T.; Mickevicius, M.; Rakovska, B.; Penkauskas, T.; Vanderah, D.J.; Heinrich, F.; Valincius, G. Fast formation of low-defect-density tethered bilayers by fusion of multilamellar vesicles. *Biochim. Biophys. Acta Biomembr.* **2017**, *1859*, 669–678. [CrossRef]
41. McGillivray, D.J.; Valincius, G.; Vanderah, D.J.; Febo-Ayala, W.; Woodward, J.T.; Heinrich, F.; Kasianowicz, J.J.; Lösche, M. Molecular-scale structural and functional characterization of sparsely tethered bilayer lipid membranes. *Biointerphases* **2007**, *2*, 21–33. [CrossRef]
42. Budvytyte, R.; Mickevicius, M.; Vanderah, D.J.; Heinrich, F.; Valincius, G. Modification of tethered bilayers by phospholipid exchange with vesicles. *Langmuir* **2013**, *29*, 4320–4327. [CrossRef]
43. Leitch, J.; Kunze, J.; Goddard, J.D.; Schwan, A.L.; Faragher, R.J.; Naumann, R.; Knol, W.; Dutcher, J.R.; Lipkowski, J. In situ PM-IRRAS studies of an archaea analogue thiolipid assembled on a Au(111) electrode surface. *Langmuir* **2009**, *25*, 10354–10363. [CrossRef] [PubMed]
44. Vockenroth, I.K.; Ohm, C.; Robertson, J.W.; McGillivray, D.J.; Lösche, M.; Köper, I. Stable insulating tethered bilayer lipid membranes. *Biointerphases* **2008**, *3*, FA68–FA73. [CrossRef] [PubMed]
45. Forbrig, E.; Staffa, J.K.; Salewski, J.; Mroginski, M.A.; Hildebrandt, P.; Kozuch, J. Monitoring the orientational changes of alamethicin during incorporation into bilayer lipid membranes. *Langmuir* **2018**, *34*, 2373–2385. [CrossRef] [PubMed]
46. Schiller, S.M.; Naumann, R.; Lovejoy, K.; Kunz, H.; Knoll, W. Archaea analogue thiolipids for tethered bilayer lipid membranes on ultrasmooth gold surfaces. *Ang. Chem. Int. Ed.* **2003**, *42*, 208–211. [CrossRef]
47. Valincius, G.; Meškauskas, T.; Ivanauskas, F. Electrochemical impedance spectroscopy of tethered bilayer membranes. *Langmuir* **2012**, *28*, 977–990. [CrossRef]
48. Valincius, G.; Mickevicius, M.; Penkauskas, T.; Jankunec, M. Electrochemical impedance spectroscopy of tethered bilayer membranes: An effect of heterogeneous distribution of defects in membranes. *Electrochim. Acta* **2016**, *222*, 904–913. [CrossRef]
49. Girard-Egrot, A.P.; Blum, L. Langmuir-Blodgett technique for synthesis of biomimetic lipid membranes. In *Nanobiotechnology of Biomimetic Membranes*; Ferrari, M., Martin, D.K., Eds.; Springer: Boston, MA, USA, 2007; pp. 23–74.
50. Györvary, E.; Wetzer, B.; Sleytr, U.B.; Sinner, A.; Offenhäusser, A.; Knoll, W. Lateral diffusion of lipids in silane-, dextran-, and s-layer-supported mono-and bilayers. *Langmuir* **1999**, *15*, 1337–1347. [CrossRef]
51. Purrucker, O.; Förtig, A.; Jordan, R.; Tanaka, M. Supported membranes with well-defined polymer tethers-Incorporation of cell receptors. *Chem. Phys. Chem.* **2004**, *5*, 327–335. [CrossRef]
52. Tanaka, M.; Sackmann, E. Polymer-supported membranes as models of the cell surface. *Nature* **2005**, *437*, 656–663. [CrossRef]
53. Hertrich, S.; Stetter, F.; Rühm, A.; Hugel, T.; Nickel, B. Highly hydrated deformable polyethylene glycol-tethered lipid bilayers. *Langmuir* **2014**, *30*, 9442–9447. [CrossRef] [PubMed]
54. Kibrom, A.; Roskamp, R.F.; Jonas, U.; Menges, B.; Knoll, W.; Paulsen, H.; Naumann, R.L. Hydrogel-supported protein-tethered bilayer lipid membranes: A new approach toward polymer-supported lipid membranes. *Soft Matter* **2011**, *7*, 237–246. [CrossRef]
55. Su, Z.; Jiang, Y.; Velázquez-Manzanares, M.; Leitch, J.J.; Kycia, A.; Lipkowski, J. Electrochemical and PM-IRRAS studies of floating lipid bilayers assembled at the Au(111) electrode pre-modified with a hydrophilic monolayer. *J. Electroanal. Chem.* **2013**, *688*, 76–85. [CrossRef]
56. Su, Z.; Juhaniewicz-Debinska, J.; Sek, S.; Lipkowski, J. Water structure in the submembrane region of a floating lipid bilayer: The effect of an ion channel formation and the channel blocker. *Langmuir* **2020**, *36*, 409–418. [CrossRef]
57. Ryu, H.; Fuwad, A.; Yoon, S.; Jang, H.; Lee, J.C.; Kim, S.M.; Jeon, T.J. Biomimetic membranes with transmembrane proteins: State-of-the-art in transmembrane protein applications. *Int. J. Mol. Sci.* **2019**, *20*, 1437. [CrossRef]

58. Casero, E.; Darder, M.; Pariente, F.; Lorenzo, E.; Martín-Benito, J.; Vázquez, L. Thiol-functionalized gold surfaces as a strategy to induce order in membrane-bound enzyme immobilization. *Nano Lett.* **2002**, *2*, 577–582. [CrossRef]
59. Ciaccafava, A.; Infossi, P.; Ilbert, M.; Guiral, M.; Lecomte, S.; Giudici-Orticoni, M.T.; Lojou, E. Electrochemistry, AFM, and PM-IRRA spectroscopy of immobilized hydrogenase: Role of a hydrophobic helix in enzyme orientation for efficient H_2 oxidation. *Angew. Chem. Int. Ed.* **2012**, *51*, 953–956. [CrossRef]
60. Gutiérrez-Sánchez, C.; Olea, D.; Marques, M.; Fernández, V.M.; Pereira, I.A.C.; Vélez, M.; De Lacey, A.L. Oriented immobilization of a membrane-bound hydrogenase onto an electrode for direct electron transfer. *Langmuir* **2011**, *27*, 6449–6457. [CrossRef]
61. Kawai, S.; Yakushi, T.; Matsushita, K.; Kitazumi, Y.; Shirai, O.; Kano, K. Role of a non-ionic surfactant in direct electron transfer-type bioelectrocatalysis by fructose dehydrogenase. *Electrochim. Acta* **2015**, *152*, 19–24. [CrossRef]
62. Sultana, N.; Schenkman, J.B.; Rusling, J.F. Protein film electrochemistry of microsomes genetically enriched in human cytochrome P450 monooxygenases. *J. Am. Chem. Soc.* **2005**, *127*, 13460–13461. [CrossRef]
63. Mie, Y.; Suzuki, M.; Komatsu, Y. Electrochemically driven drug metabolism by membranes containing human cytochrome P450. *J. Am. Chem. Soc.* **2009**, *131*, 6646–6647. [CrossRef] [PubMed]
64. Cashman, J.R. The implications of polymorphisms in mammalian flavin-containing monooxygenases in drug discovery and development. *Drug Discov. Today* **2004**, *9*, 574–581. [CrossRef]
65. Cashman, J.R. Role of flavin-containing monooxgenase in drug development. *Expert Opin. Drug Metab. Toxicol.* **2008**, *4*, 1507–1521. [CrossRef] [PubMed]
66. Castrignano, S.; Gilardi, G.; Sadeghi, S.J. Human flavin-containing monooxygenase 3 on graphene oxide for drug metabolism screening. *Anal. Chem.* **2015**, *87*, 2974–2980. [CrossRef]
67. Weiss, S.A.; Jeuken, L.J.C. Lipid-membrane modified electrodes to study quinone oxidoreductases. *Biochem. Soc. Trans.* **2009**, *37*, 707–712. [CrossRef]
68. Heath, G.R.; Li, M.; Rong, H.; Radu, V.; Frielingsdorf, S.; Lenz, O.; Butt, J.N.; Jeuken, L.J.C. Multilayered lipid membrane stacks for biocatalysis using membrane enzymes. *Adv. Funct. Mater.* **2017**, *27*, 1606265. [CrossRef]
69. Sezer, M.; Frielingsdorf, S.; Millo, D.; Heidary, N.; Utesch, T.; Mroginski, M.-A.; Friedrich, B.; Hildebrandt, P.; Zebger, I.; Weidinger, I.M. Role of the HoxZ subunit in the electron transfer pathway of the membrane-bound [NiFe]-hydrogenase from *Ralstonia eutropha* immobilized on electrodes. *J. Phys. Chem. B* **2011**, *115*, 10368–10374. [CrossRef]
70. Valente, F.M.A.; Pereira, P.M.; Venceslau, S.S.; Regalla, M.; Coelho, A.V.; Pereira, I.A.C. The [NiFeSe] hydrogenase from *Desulfovibrio vulgaris* Hildenborough is a bacterial lipoprotein lacking a typical lipoprotein signal peptide. *FEBS Lett.* **2007**, *581*, 3341–3344. [CrossRef]
71. Gutiérrez-Sanz, O.; Tapia, C.; Marques, M.C.; Zacarias, S.; Vélez, M.; Pereira, I.A.C.; De Lacey, A.L. Induction of a proton gradient across a gold-supported biomimetic membrane by electroenzymatic H_2 oxidation. *Angew. Chem. Int. Ed.* **2015**, *54*, 2684–2687. [CrossRef]
72. Gutiérrez-Sanz, O.; Marques, M.; Pereira, I.A.C.; De Lacey, A.L.; Lubitz, W.; Rüdiger, O. Orientation and function of a membrane-bound enzyme monitored by electrochemical surface-enhanced infrared absorption spectroscopy. *J. Phys. Chem. Lett.* **2013**, *4*, 2794–2798. [CrossRef]
73. Shi, Y. Common folds and transport mechanisms of secondary active transporters. *Annu. Rev. Biophys.* **2013**, *42*, 51–72. [CrossRef] [PubMed]
74. Junge, W.; Nelson, N. ATP synthase. *Annu. Rev. Biochem.* **2015**, *84*, 631–657. [CrossRef] [PubMed]
75. Kojima, S. Dynamism and regulation of the stator, the energy conversion complex of the bacterial flagellar motor. *Curr. Opin. Microbiol* **2015**, *28*, 66–71. [CrossRef] [PubMed]
76. Gutiérrez-Sanz, O.; Natale, P.; Márquez, I.; Marques, M.C.; Zacarias, S.; Pita, M.; Pereira, I.A.C.; López-Montero, I.; De Lacey, A.L.; Vélez, M. H_2-fueled ATP synthesis on an electrode: Mimicking cellular respiration. *Angew. Chem. Int. Ed.* **2016**, *55*, 6216–6220. [CrossRef] [PubMed]
77. García-Molina, G.; Natale, P.; Valenzuela, L.; Alvarez-Malmagro, J.; Gutiérrez-Sánchez, C.; Iglesias-Juez, A.; López-Montero, I.; Vélez, M.; Pita, M.; De Lacey, A.L. Potentiometric detection of ATP based on the transmembrane proton gradient generated by ATPase reconstituted on a gold electrode. *Bioelectrochemistry* **2020**, *133*, 107490. [CrossRef]

78. Gutiérrez-Sanz, O.; Olea, D.; Pita, M.; Batista, A.P.; Alonso, A.; Pereira, M.M.; Vélez, M.; De Lacey, A.L. Reconstitution of respiratory complex I on a biomimetic membrane supported on gold electrodes. *Langmuir* **2014**, *30*, 9007–9015. [CrossRef]
79. Schapira, A. Human complex I defects in neurodegenerative diseases. *Biochim. Biophys. Acta-Bioenerg.* **1998**, *1364*, 261–270. [CrossRef]
80. Hunte, C.; Zickermann, V.; Brandt, U. Functional modules and structural basis of conformational coupling in mitochondrial complex I. *Science* **2010**, *329*, 448–451. [CrossRef]
81. Baradaran, R.; Berrisford, J.M.; Minhas, G.S.; Sazanov, L.A. Crystal structure of the entire respiratory complex I. *Nature* **2013**, *494*, 443–448. [CrossRef]
82. Gutiérrez-Sanz, O.; Forbrig, E.; Batista, A.P.; Pereira, M.M.; Salewski, J.; Mroginski, M.A.; Götz, R.; De Lacey, A.L.; Kozuch, J.; Zebger, I. Catalytic activity and proton translocation of reconstituted respiratory complex I monitored by surface-enhanced infrared absorption spectroscopy. *Langmuir* **2018**, *34*, 5703–5711. [CrossRef]
83. Kinnear, K.T.; Monbouquette, H.G. An amperometric fructose biosensor based on fructose dehydrogenase immobilized in a membrane layer on gold. *Anal. Chem.* **1997**, *69*, 1771–1775. [CrossRef]
84. Darder, M.; Casero, E.; Pariente, F.; Lorenzo, E. Biosensors based on membrane-bound enzymes immobilized in 5-(octyldithio)-2-nitrobenzoic acid layer on gold electrodes. *Anal. Chem.* **2000**, *72*, 3784–3792. [CrossRef] [PubMed]
85. Nazaruk, E.; Landau, E.M.; Bilewicz, R. Membranne bound enzyme hosted in liquid crystalline cubic phase for sensing and fuel cells. *Electrochim. Acta* **2014**, *140*, 96–100. [CrossRef]
86. Kreit, J.; Sampson, N.S. Cholesterol oxidase: Physiological functions. *FEBS J.* **2009**, *276*, 6844–6856. [CrossRef] [PubMed]
87. Nauck, M.; Marz, W.; Wieland, H. Is lipoprotein(a) cholesterol a significant indicator of cardiovascular risk? *Clin. Chem.* **2000**, *46*, 436–437. [CrossRef] [PubMed]
88. Wicklein, B.; Darder, M.; Aranda, P.; Ruiz-Hitzky, E. Phospholipid-sepiolite biomimetic interfaces for the immobilization of enzymes. *ACS Appl. Mater. Interfaces* **2011**, *3*, 4339–4348. [CrossRef]
89. Psychoyios, V.N.; Nikoleli, G.P.; Tzamtzis, N.; Nikolelis, D.P.; Psaroudakis, N.; Danielsson, B.; Qadir Israr, M.; Willander, M. Potentiometric cholesterol biosensor based on ZnO nanowalls and stabilized polymerized lipid film. *Electroanalysis* **2013**, *25*, 367–372. [CrossRef]
90. Gomes, F.O.; Maia, L.B.; Loureiro, J.A.; Pereira, M.C.; Delerue-Matos, C.; Moura, I.; Moura, J.J.G.; Morais, S. Biosensor for direct bioelectrocatalysis detection of nitric oxide using nitric oxide reductase incorporated in carboxylated single-walled carbon nanotubes/lipidic 3 bilayer nanocomposite. *Bioelectrochemistry* **2019**, *127*, 76–86. [CrossRef]
91. Calabrese, V.; Cornelius, C.; Rizarelli, E.; Owen, J.B.; Dinkova-Kostova, A.T.; Butterfield, D.A. Nitric oxide in in cell survival: A janus molecule. *Antioxid. Redox Signal* **2009**, *11*, 2717–2739. [CrossRef]
92. Ziller, C.; Lin, J.; Knittel, P.; Friedrich, L.; Andronescu, C.; Pöller, S.; Schuhmann, W.; Kranz, C. Poly(benzooxacine) as an immobilization matrix for miniaturized ATP and glucose biosensors. *ChemElectroChem* **2017**, *4*, 864–871. [CrossRef]
93. Moretro, T.; Normann, M.A.; Saebo, H.R.; Langsrud, S. Evaluation of ATP bioluminiscence-based methods for hygienic assessment in fish industry. *J. Appl. Microbiol.* **2019**, *127*, 186–195. [CrossRef] [PubMed]
94. Zhang, J.; Wang, X.; Chen, T.; Feng, C.; Li, G. Electrochemical analysis of enzyme based on the self-assembly of lipid bilayer on an electrode surface mediated by hydrazone chemistry. *Anal. Chem.* **2017**, *89*, 13245–13251. [CrossRef] [PubMed]
95. Guan, H.; Zhang, F.; Yu, J.; Chi, D. The novel acetylcholinesterase biosensor based on liposome bioreactors-chitosan nanocomposite film for detection of organophosphates pesticides. *Food Res. Int.* **2012**, *49*, 15–21. [CrossRef]
96. Yan, J.; Guan, H.; Yu, J.; Chi, D. Acetylcholinesterase biosensor based on assembly of multiwall carbon natubes onto liposome bioreactors for detection of organophosphates pesticides. *Pesticide Biochem. Physiol.* **2013**, *105*, 197–202. [CrossRef]

© 2020 by the authors. Licensee MDPI, Basel, Switzerland. This article is an open access article distributed under the terms and conditions of the Creative Commons Attribution (CC BY) license (http://creativecommons.org/licenses/by/4.0/).

Review

Enhancement of Biosensors by Implementing Photoelectrochemical Processes

Melisa del Barrio [1,2], Gabriel Luna-López [1] and Marcos Pita [1,*]

1. Departamento de Biocatálisis, Instituto de Catálisis y Petroleoquímica, CSIC. C/Marie Curie, 2. L10, 28049 Madrid, Spain; melisa.delbarrio@uam.es (M.d.B.); gabriel.luna@csic.es (G.L.-L.)
2. Departamento de Química Analítica y Análisis Instrumental, Universidad Autónoma de Madrid, Campus de Cantoblanco, 28049 Madrid, Spain
* Correspondence: marcospita@icp.csic.es

Received: 7 May 2020; Accepted: 6 June 2020; Published: 9 June 2020

Abstract: Research on biosensors is growing in relevance, taking benefit from groundbreaking knowledge that allows for new biosensing strategies. Electrochemical biosensors can benefit from research on semiconducting materials for energy applications. This research seeks the optimization of the semiconductor-electrode interfaces including light-harvesting materials, among other improvements. Once that knowledge is acquired, it can be implemented with biological recognition elements, which are able to transfer a chemical signal to the photoelectrochemical system, yielding photo-biosensors. This has been a matter of research as it allows both a superior suppression of background electrochemical signals and the switching ON and OFF upon illumination. Effective electrode-semiconductor interfaces and their coupling with biorecognition units are reviewed in this work.

Keywords: biosensors; bioelectrochemistry; photo-biosensors; enzyme; biocatalysis

---

1. Introduction

Humankind is on a permanent quest for better ways to extract relevant information from the environment. Many devices, known as sensors, have been created, designed and perfected since the ancient days aiming at increasing the knowledge and therefore allowing better decisions. Therefore, sensors have been implemented in most facets of life and have involved an incommensurable panoply of processes and systems to provide meaningful information.

Regardless of their implementation area, some common features are desired to define efficient sensors: easiness to use, reliability and fast response are the most desired characteristics. The technological revolution accomplished since the late 20th Century has had a huge impact in the sensing sector. Sensors have improved thanks to the evolution of optical and electronic transducers. Sensors have implemented new catalytic processes and used more reliable materials to achieve a faster response, among other improved properties [1]. Biosensors are a particular case worth research, in which a chemical reaction catalyzed by a biological entity, mainly an enzyme, triggers the process to inform about the presence and concentration of a specific molecule. Enzymes provide great features to biosensors, such as selectivity and specificity, which help to ease the biosensing process by avoiding purification steps or matrix effects. Among all kinds of enzymes, redox ones are particularly well suited to being linked with electrochemical methods, as electroactive surfaces may transfer electrons from and to the enzymes; this transfer substitutes that of one of the substrates and directly correlates the enzymatic activity to electrochemical signals. This combination allows for easy, affordable and reliable processes for sensing specific analytes. Some examples of amperometric biosensors have been developed to sense key biomolecules such as adenosine triphosphate (ATP) [2] or general substrates like oxygen [3]. Enzymatic reactions can also be combined with each other to perform simultaneous analysis of biochemicals, emulating logic operations [4,5].

Classically there have been three generations of biosensors [6]. In the first generation, the enzymatic reaction takes place and the product is directly measured with the electrode. The second generation substitutes the enzyme's substrate that is not sensed by a mediator, which accomplishes the electron transfer to the electrode and adds a catalytic effect able to increase the signal. The third generation can be achieved with enzymes, the active site of which is available for direct electron transfer. The immobilization of these enzymes on the electrode surface allows a direct electrochemical measurement of the substrate. Regardless of the biosensor architecture, there are common challenges to overcome. Selectivity is a major issue for efficient sensors to avoid false positives, which is provided by the enzymes in the case of biosensors. Another one is the sensitivity, which is related to the noise level measured in the absence of substrate. There have been many attempts to study the noise level in bioelectrochemical systems. The inclusion of a single enzyme system [7,8] or cascade-concatenated biochemical reactions [9,10] and use of strategies like the incorporation of chemical filters to suppress or delay the background signal [11] are some examples of these efforts. A way to reduce the noise level is to include a semiconductor between the electrode and the biosensing structures [12]. Moreover, semiconductors often can harvest light energy and become an electric conductor upon illumination with visible light, which has even been used for water splitting [13–15]. Such building allows a huge noise reduction while adding a switch system to the biosensor, improving the device performance, and yielding photo-biosensors as a new tool for better sensors [16]. Early reports of photobioelectrochemical (PEC) sensors were published more than 20 years ago [17], but it has been in more recent years when the field has blossomed into many systems for different applications. In this review we will focus on photo-biosensors for relevant analytes such as glucose, lactate, protein kinase or Acetyl Choline Esterase (AChE), among other examples.

2. Semiconductors Used in Photo-Biosensors

The main characteristic of photo-biosensors with respect to amperometric biosensors is the addition of a semiconducting material acting as a switch upon illumination, which can be triggered or modulated when the analyte is present. Key aspects to develop such devices are how to combine the electroactive surface, the semiconductor and the sensing biocatalysts; another key aspect is which materials are more suitable, depending on the application desired. Conventional bioelectrochemistry requires that the redox biocatalysts are connected to the electrode in some fashion. Most enzymes have the catalytic site buried inside the protein structure, so the use of mediators for transferring the electrons is a common strategy. However, there are enzymes that can be oriented on the electrode surface and achieve direct electron transfer. Photo-assisted electrochemistry adds higher complexity to the electron transfer, because semiconductors are generally not as reversible as conductors. Their p- or n-type semiconducting behavior marks the main flux of electrons, so depending on the reaction biocatalyzed only one of the semiconductor types will offer successful photo-biosensor constructions. This characteristic must be considered when selecting the semiconductors.

A very early approach utilized fused-silica optical fibers covered with gold [17]. This sensor was designed to detect catalase activity while providing its substrate, H_2O_2, generated from existing O_2 upon illumination. Other successful interfaces have included quantum dots (QDs) on the surface of electrodes. Some of them use gold as electrode and load it with CdS to detect formaldehyde [18], although gold has not been the most common electrode material used for these purposes. Another alternative is to use conductive liquid contacts as electroactive material, such as a eutectic mixture of gallium and indium, which is liquid at room temperature. This has been used to contact silicon wafers etched with HF [19].

One of the most successful electrodes that has been used in many works is indium-doped tin oxide (ITO). Because tin oxide is itself a semiconductor, when doped with other ions like indium (III) or fluoride it yields a conductive surface with high transparency. These have become the most common electroactive materials for photoelectrodes, which are the bases of photo-biosensors. On the top of ITO many semiconductors have been tested, such as BiOI nanoflakes [20]; layers of graphene, chitosan-Cd^{2+} [21]; Bi_2S_3 [22]; CdS QDs [23]; ZnS nanoparticles [24]; TiO_2 covered with

QDs made of CdSeTe@CdS@ZnS [25]; layers of NiO and CuInS$_2$ [26]; WS$_2$ and gold nanoparticles [27]; laser-induced TiO$_2$-decorated graphene (LITG) [28]; carbon nitride with gold nanoparticles [29]; or NiWO$_4$ nanostructures [30]. It would be worth comparing these materials' performance with the same biosensing system, but to date they have been tested for different applications, so such study is beyond today's reported knowledge.

Many electroactive surfaces have been developed besides ITO-based electrodes, such as silicon (111) working as substrate for InGaN/GaN nanowire growth [31], which served to detect reduced nicotinamide adenine dinucleotide (NADH). Graphene combined with TiO$_2$ nanowires has also been used as substrate on the top of polymer nanosponges, becoming conductive and photoactive. Another strategy is an oxidizing etching-annealing of a metal foil's surface, which has been applied to copper, yielding a CuO nanotubes' coverage [32], and titania, forming a TiO$_2$ cover [33]; the latter has been also used in combination with chitosan to favor the cross-linking of enzymes like horseradish peroxidase (HRP) [34]. Photoactive polymers like polythiophene derivatives have also been tested [35] combined with CdS QDs. Another example of photoelectrode for biosensing has been demonstrated using a Ni:FeOOH/BiVO$_4$ photoanode [36].

Overall, the combination of conductive materials with photoactive materials is a key aspect to look at when developing photo-biosensors. There are many options to suit specific needs, and it is still a field that remains open to new composites to be developed and/or discovered.

3. Enzymes Immobilization and Performance

Enzymes play a major role in any kind of biosensors, as they are the recognition unit responsible to provide the information to the transducing system. The enzymatic reaction should take place close enough to the photoelectrode surface to transfer the chemical information either via direct electron transfer or by means of a mediator, so its immobilization in the surface or close to it is a very common strategy. Many enzymes have been used for different sensors. Very common analytical targets are glucose, lactate, kinase-like proteins or acetylcholine esterases, although they are not the only ones.

3.1. Glucose Detection

Glucose biosensors have become important devices in the medical field due to their contribution against diabetes mellitus. Many different strategies have been developed to address this problem. In the following section, some of the newer approaches and strategies will be reviewed. Furthermore, the performance of the resulting photo-biosensors is summarized in Table 1.

Table 1. Performance of various glucose photo-biosensors.

Electrode	Limit of Detection	Linear Range	Ref.
Pt/ZnO/GOx	5.6 µM	N.A.	[37]
Pt/CdS/GOx	1 µM	1 µM–2.5 mM	[38]
ITO/CdSe@CdS/GOx	0.05 mM	1–8 nM	[39]
FTO/CdTe QD/GOx	0.04 mM	0.1–11 mM	[40]
ITO/ZnS/GOx	0.02 mM	0.1–5.5 mM	[41]
FTO/TiO$_2$ NW/GOx	0.9 nM	N.A.	[42]
FTO/ZnO IOPC/GOx	N.A.	0.4–4.5 mM	[43]
GCE/p-HT/GDH	1.5 µM	5 µM–1 mM	[44]
GCE/ZnS-CdS QD/GDH	4.0 µM	0.010–2.0 mM	[45]
PGE/ZnS-CdS QD/GDH	0.05 mM	0.2–8.0 mM	[46]
ITO/PbS QD/GOx	0.3 µM	1 µM–10 mM	[47]
ITO/g-C$_3$N$_4$/GOx	0.01 mM	0.05–15 mM	[48]
ITO/Co$_3$O$_4$/CNT/GOx	0.20 µM	0–4 mM	[49]
FTO/α-Fe$_2$O$_3$/NB/GDH	25.2 µM	0.25–2.0 mM	[50]
FTO/R/A TiO$_2$/GOx	0.019 mM	1–20 mM	[51]
ITO/NDC-TiO$_2$ NPs/GOx	13 nM	0.05–10 µM	[52]
GCE/CoP-PCN/GOx	1.1 µM	0.05–0.7 mM	[53]
GCE/g-C$_3$N$_4$/ZnIn$_2$S$_4$/GOx/HRP	0.28 µM	1 µM–10 mM	[54]
FTO/BiVO$_4$/GOx	0.73 µM	1–400 µM	[55]
ITO/Au NP/MoS$_2$/GOx	1.2 µM	4 µM–1.75 mM	[56]
3D hollow-out TiO$_2$ NWc/GOx	8.7 µM	0–2 mM	[57]

N.A: No data available.

Ren et al. used ZnO nanoparticles bound to glucose oxidase (GOx) due to its biocompatibility, photoconductivity, photocatalytic activity and high electron transfer capacity [37]. These ZnO nanoparticles were multigrain and the hexagonal phase of ZnO suited best for GOx immobilization, probably due to its larger surface to volume ratio. The adsorption process did not significantly affect the secondary structure of the macrobiomolecule. The effects of ZnO nanoparticles were monitored by amperometric measurements. A control experiment lacking ZnO nanoparticles yielded a current response of 0.82 µAcm^{-2}, while ZnO presence provided up to 21 µAcm^{-2}. The larger surface of the ZnO nanoparticles intensely enhances the current activity of the electrode through a better adsorption of GOx. The optimal conditions for these biosensors were pH 6.8 and 45 °C, although the system was also successful at 35 °C, which allows compatibility with the human body. They also discovered that the current density increased upon irradiating UV light to the sample, reaching 27 µAcm^{-2}. Further experiments showed that the current increased up to 30% more when irradiated, but the photo-bioelectrode inactivated after a long exposure to light, probably due to denaturation of the enzyme.

Sun et al. manufactured a new photoelectrochemical biosensor based on CdS nanoparticles [38]. Polyamidoamine (PAMAM) dendrimer was used as inner template to synthesize CdS nanoparticles. GOx was immobilized on Pt electrodes together with the CdS nanoparticles through layer-by-layer (LbL) technique. This immobilization method consisted of adsorbing sequentially charged macromolecules, where PAMAM acted as scaffold for the CdS nanoparticles to grow. Along with GOx it formed the glucose-detecting electrode in aqueous solution. Platinum nanoparticles were used as charge separator. Nafion was used both as ion exchange matrix and as interference barrier. The electrodes' performance improved under UV light irradiation at 350 nm. They have achieved twice the current response under UV light compared to dark tests. The stability of the sensor, checked every 2 days for a month, showed that the GOx immobilization on the electrode was highly effective. Another photoelectrochemical biosensor was based on a TiO$_2$CdSe@CdS QDs nanocomposite electrode (Figure 1) [39]. Despite using a different kind of quantum dots, glucose was detected with GOx and the biosensor was assembled through a LbL process. The junction of TiO$_2$ with CdS QDs improved the charge separation, and therefore increased the photocurrent. Furthermore, the addition of electron mediators such as [Co(Phen)$_3$]$^{2+}$ enhanced the photocurrent by suppressing the electron-hole recombination process. The sensors built this way showed a linear range from 1 to 8 nM of glucose and the lowest concentration of glucose detected was 0.05 mM, although that range and limit of detection could be improved through the optimization of the number of [Co(Phen)$_3$]$^{2+}$ and GOx bilayers. Regarding stability, this sensor was able to retain 95% of its activity after 3 weeks of storage at −20 °C.

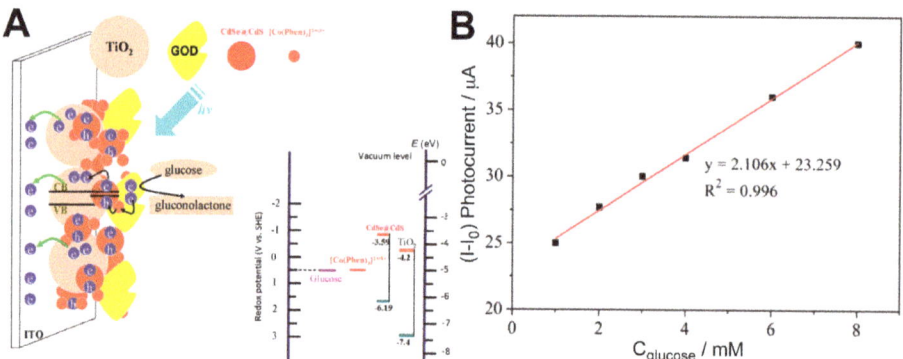

Figure 1. (**A**) Schematic of the photo-bioelectrode building based on QDs and GOx, and the energy levels of its components. (**B**) Linear response of the sensor. Reprinted with permission from ref. [39].

Wang et al. developed a photoelectrochemical biosensor built around CdTe QDs capable of detecting glucose [40]. The CdTe QDs are settled on a fluorine-doped tin oxide (FTO) electrode. Then GOx is covalently attached to CdTe via amide reaction, making it able to work under visible light. This sensor showed a broad linear range and a high upper detection limit. The QDs exhibited an emission peak at 650 nm and an important absorbance shoulder at 627 nm. The experimental conditions set for this sensor were −0.2V vs. SCE, pH 7, and room temperature. The wavelength of the incident light was set to 505 nm. The accuracy and detection limits of the sensor make it promising for measuring glucose in blood in the future.

Another glucose biosensor was developed combining GOx, ITO and ZnS nanoparticles [41]. The main goal was to be able to replace CdS as the main semiconductor in the sensor due to its toxicity and potential harm to the environment. On the surface of the ITO electrode, ZnS nanoparticles were electrodeposited. Then the enzyme was immobilized on the electrode via sol-gel method. The authors proved that the immobilized GOx maintained its tertiary structure and catalytic activity. Photoelectrochemical activity from ZnS nanoparticles resulted in an improved sensibility and lower detection limit when irradiated as compared to experiments without any source of light. In addition, the sensor proved to be stable and retained its activity throughout time.

A photoelectrochemical biosensor based on TiO_2 nanowires and GOx was developed by Tang et al. [42]. Their single-crystalline rutile-phased TiO_2 was hydrothermally grown on an FTO electrode. Then GOx was attached to its surface through silane/glutaraldehyde linkage. Experiments carried out with commercial TiO_2 nanoparticles revealed that the sensitivity obtained with TiO_2 nanowires-GOx sensors was clearly higher. Therefore, the sensitivity enhancement was directly related to the TiO_2 nanostructure. They also evaluated the effect of interference molecules (metal ions, amino acids, glucose analogues, etc.) and their influence proved to be minimal. Finally, they tested the biosensor performance with mice serum, with remarkable results.

A novel photobiosensor built with ZnO inverse opal photonic crystals (IOPCs) was developed by Xia et al. [43]. ZnO IOPCs have a uniform porous distribution and a massive surface area due to their structure. These crystals were obtained via sol-gel method using polymethylmethacrylate (PMMA) as a building scaffold and an FTO electrode. Then Nafion and GOx were attached to the surface and the lining of ZnO. This type of biosensor harnesses the "slow light effect" and multiple scattering from ZnO IOPCs to increase light absorption. This sensor layout proved to be highly selective, sensitive and reproducible.

Dilgin et al. developed a glucose biosensor composed of electropolymerized hematoxylin (p-HT) film on PAMAM dendrimers that were adsorbed on a glassy carbon electrode (GCE) [44]. Then glucose dehydrogenase (GDH) was immobilized onto the whole ensemble. The electrode depends on the electrocatalytic oxidation reaction of NADH. They used a halogen lamp and flow injection analysis (FIA) to carry out the experiments. The main advantages of this technique are: low sample consumption, fast analysis and suitability for the analysis of species that would involve arduous operations of separation and chemical conversions. This sensor electrode is remarkably sensitive, selective and durable, and has proven limit of detection enhancement under irradiation.

Ertek et al. proposed photo-biosensors based on GDH and electrodeposited ZnS-CdS QDs on both multiwalled carbon nanotube modified GCE [45] and pencil graphite electrodes (PGE) [46]. Cyclic voltammetry and FIA were employed to assess the performance of the biosensors under visible radiation generated by a 250 W halogen lamp. The GCE-based photo-biosensor showed a narrower linear range (from 0.010 to 2.0 mM of glucose) compared to that of PGE-based system (from 0.2 to 8.0 mM of glucose), but a lower detection limit (4.0 µM); the latter electrode offered a higher detection limit (0.05 mM). These results suggest that these two biosensors could work complementarily.

One of the main advantages of self-powered biosensor is that there is no need to apply any voltage to the cathode and the anode. Dai et al. [47] designed a self-powered cathodic photo-biosensor focused on a hybrid PbS QDs/nanoporous NiO film nanostructure. They used ITO electrodes upon which they hydrothermally built 3D NiO nanostructures and then attached thioglycolic acid (TGA) capped PbS QDs to form a p-type heterostructure. P-type semiconductors are less prone to react with reductive interference substances. Finally, GOx was immobilized on the electrode via succinimide coupling reaction between NH_2 groups in the enzyme and COOH groups on the surface of TGA-capped PbS

QDs. The resulting biosensor proved to be highly selective, stable and sensitive and provided a fast response.

Liu et al. synthesized a composite comprising g-C_3N_4 and TiO_2 bidimensional nanosheets [48]. Each of these components compensates the flaws of the other. g-C_3N_4 enhances the mediocre visible light excitation of TiO_2 and the latter delays the otherwise rapid charge recombination from g-C_3N_4. They constructed a biosensor to evaluate the performance of this new composite in combination with ITO electrodes and GOx. Nafion was used as a binding agent to secure the enzyme to the electrode. The photoelectrochemical efficiency rose to 350% when compared to the g-C_3N_4 or TiO_2 alone.

Çakıroğlu et al. developed a self-powered biosensor [49] (Figure 2). Co_3O_4 and carbon nanotubes (CNT) were deposited on an ITO electrode coated with TiO_2 anatase, creating a p-n junction. Then CNTs were functionalized with 1-pyrenic boronic acid so that a covalent bound between this moiety and the carbohydrate groups of GOx could take place through an esterification reaction. Since it is a self-powered biosensor, no external potential is needed. While normally TiO_2 would need UV light for electrons to overcome its wide band gap, Co_3O_4 enhanced its photoelectrical capabilities under visible light. A linear range from 0.2 µM to 4 mM glucose concentration and a limit of detection of 0.20 µM were reported at 0 V.

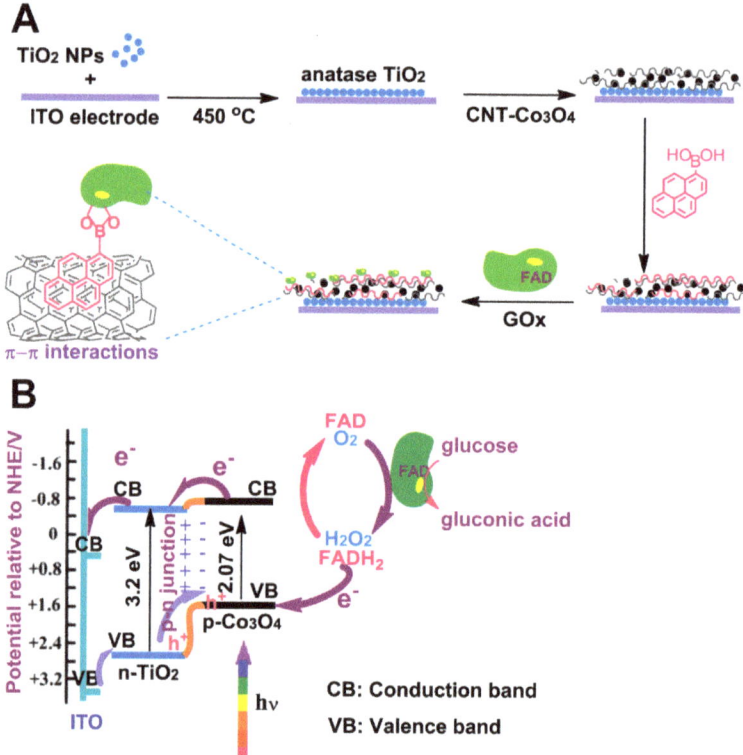

Figure 2. (**A**) Schematics of the photo-biosensor synthesized step by step. (**B**) Energy transfer diagram for the work showing the valence bands and conduction bands of the photo-biosensor components. Taken with permission from ref. [49].

Ryu et al. proposed a photoelectrochemical platform based on hematite (α-Fe_2O_3), a visible light active catalyst which was immobilized upon a FTO electrode through a two-step annealing process [50]. Polydopamine (PDA) was used to immobilize the redox mediator Nile Blue (NB) for the electrocatalytic NADH oxidation. The resulting platform was suitable for biosensing of glucose, ethanol or lactate by selecting the appropriate enzyme. GDH was used as biocatalyst for the sensor. The photo- biosensor showed a reasonable limit of detection and sensitivity, great selectivity and promising future applications.

In another example, GOx was successfully entrapped in Nafion and deposited a on a rutile nanorod/anatase nanowire TiO_2/FTO photoelectrode [51]. The use of these TiO_2 phases (rutile and anatase) improved the PEC glucose biosensor performance as this structure facilitated the loading of the enzyme and favored the transport of electrons from the conduction band on anatase TiO_2 to that of rutile TiO_2. Glucose could be detected in a relatively broad concentration range (1–20 mM) and with low limit of detection (0.019 mM).

Atchudan et al. developed an ultrasensitive PEC biosensor that comprised a novel nanocomposite of nitrogen-doped carbon sheets (NDC) wrapped titanium dioxide nanoparticles (NDC-TiO_2 NPs) and GOx covalently immobilized on it [52]. Interestingly, the NDC-TiO NPs were synthetized from peach extract by a new green method. The energy levels of both the valence and conduction bands of NDC are at a higher level than those of TiO_2 NPs, which favored the migration of generated electrons and holes and minimized their recombination. Regarding the PEC sensing mechanism, the H_2O_2 molecules—formed on the photo-bioelectrode surface from O_2 during the GOx-catalyzed oxidation of the analyte—acted as electron donors and those electrons were transferred to the ITO electrode, while the photogenerated holes migrated from the valence band of TiO_2 NPs to that of NDC. The H_2O_2 oxidation photocurrent increased linearly upon addition of glucose in the range from 50 nM to 10 µM. The PEC biosensor showed excellent selectivity, reproducibility, stability, and durability. The detection limit was as low as 13 nM. Moreover, the biosensor was capable of analyzing glucose levels in real human serum.

A nanocomposite of porous carbon nitride modified with cobalt phosphide nanoparticles (CoP/PCN) was as well proposed as photo-electroactive material and support for GOx [53]. The CoP were employed to increase the PEC response upon visible light—given that it is a good photosensitizer—and also served as electron acceptors to accelerate charge separation. In their approach, the decrease in the concentration of dissolved O_2, which is consumed during the enzymatic reaction, and the subsequent loss of O_2 reduction photocurrent were used for the determination of glucose. Their PEC biosensor showed a linear response in the range from 0.05 to 0.7 mM and a detection limit of 1.1 µM.

Zhang et al. proposed a bi-enzymatic glucose sensor based on graphitic carbon nitride and $ZnIn_2S_4$ composites (g-C_3N_4/$ZnIn_2S_4$) and a biocatalyzed precipitation reaction [54]. GOx and horseradish peroxidase (HRP) were immobilized, with the aid of gold nanoparticles, on a glassy carbon/g-C_3N_4/$ZnIn_2S_4$ photoelectrode, in which GOx catalyzed the oxidation of glucose to generate H_2O_2. In the presence of 4-chloro-1-naphthol (4-CN), HRP used H_2O_2 to catalyze the oxidation of 4-CN to form an insoluble compound (benzo-4-chlorohexadienone). The formed precipitate acted as a barrier towards electron transfer between g-C_3N_4/$ZnIn_2S_4$ and an electron donor (L-cysteine, which trapped the photogenerated holes of the semiconductors), accelerated the carrier recombination and, as a consequence, the oxidation photocurrent of the electron donor decreased. They found a linear relationship between the photocurrent and the logarithm of glucose concentration in the range 1–10,000 µM and a low detection limit of 0.28 µM. This novel methodology was also applied for the determination of glucose in diluted human serum.

Furthermore, a study by Chen et al. concerning the photocurrent switching effect of $BiVO_4$ semiconductors (i.e., the p-type semiconductor behavior of this typical n-type semiconductor at a bias potential) led to the design of a PEC glucose sensor [55]. The sensing strategy was based on the measurement of the reduction photocurrent of H_2O_2 produced during the GOx-catalyzed glucose oxidation. Their FTO/$BiVO_4$/GOx photoelectrode exhibited excellent selectivity, high sensitivity (the detection limit was 0.73 µM) and was barely affected by oxygen level fluctuations.

In a very recent study, Çakıroğlu et al. constructed a mesoporous TiO_2 ($MTiO_2$) structure with enhanced surface area which improved both GOx and gold NPs immobilization [56]. MoS_2 was added for visible light harvesting, while gold NPs aimed to improve the photonic efficiency of the PEC system. The multiple heterojunctions of the $MTiO_2$-gold NPs-MoS_2 system enhanced the PEC response towards glucose of the biosensor, which exhibited a detection limit of 1.2 µM and a broad linear range (0.004–1.75 mM). Because $MTiO_2$ was synthesized by using tannic acid—a green and cheap material—as a template, their results encourage sustainable strategies for porous material preparation. Another recent work also proposed the use of an advanced TiO_2-based material for GOx immobilization [57]. More specifically, the authors aimed to provide the biosensor with a higher number of exposed enzyme active sites by means of 3-dimensional (3D) hollow-out titanium dioxide (TiO_2) nanowire clusters (NWc) on a Ti wire mesh, as illustrated in Figure 3A. The enzymatic reaction could occur with high efficiency on the resulting mesh electrode, which allowed excellent diffusion of glucose and products around the immobilized enzyme. As a result, glucose could be detected with ultrahigh sensitivity in the range between 0 and 2 mM (Figure 3C,D). Moreover, the sensor showed remarkable short- and long-term stability.

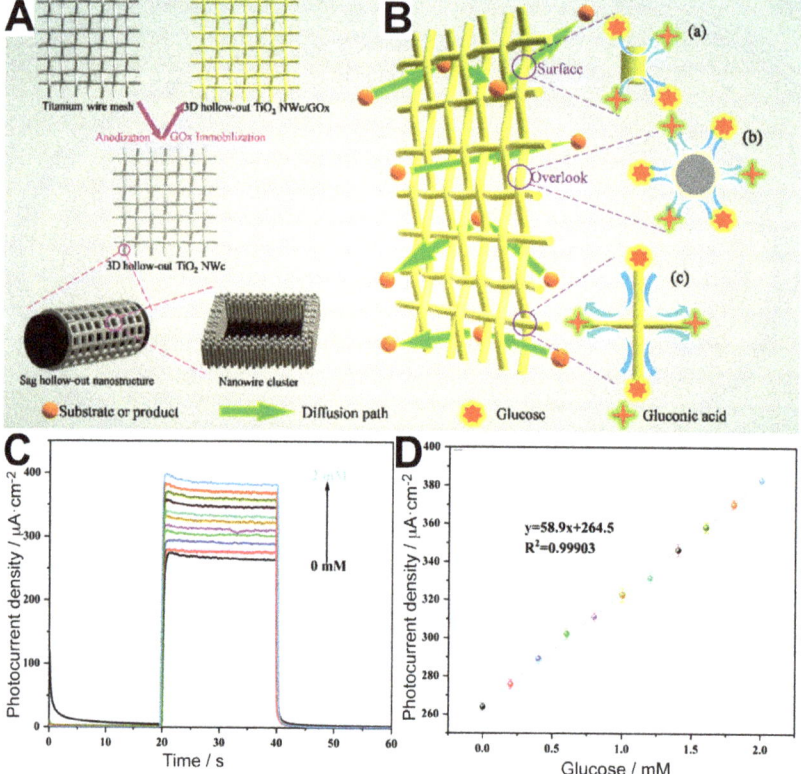

Figure 3. Schematic of (**A**) the preparation of 3D hollow-out TiO_2 NWc/GOx electrode and (**B**) glucose detection on the mesh electrode. Insets (a–c) show that the 3D network structure allows GOx to perform catalysis on a high surface. (**C**) Photocurrent density response over time of 3D hollow-out TiO_2 NWc/GOx in the presence of increasing concentrations of glucose. (**D**) Calibration plot. Adapted from [57] with permission from The Royal Society of Chemistry.

3.2. Acetylcholine Esterase

Acetylcholine esterase (AChE) is a hydrolase involved in the termination of nerve impulses that catalyzes the hydrolysis of the neurotransmitter acetylcholine to acetate and choline. Its activity is affected by various inhibitors, such as organophosphorus and carbamate compounds used as pesticides and nerve agents (because they lead to the accumulation of acetylcholine, disrupting the neurotransmission). As enzymatic sensors can detect not only the substrate but also the enzyme inhibitors, biosensors based on AChE inhibition can be used for the detection of different analytes by measuring the relative difference between the response in the absence and in the presence of the inhibitor [58,59]. Regarding photoelectrochemical sensors, since Pardo-Yissar et al. demonstrated that AChE could be combined with semiconductor QDs for photoelectrochemical biosensing of an enzyme inhibitor [60], many biosensors using new photoactive hybrid materials have been reported for the detection of organophosphate pesticides (OPs) [20,35,61], aflatoxin B_1 [62], and for AChE activity studies linked to the investigation of neurodegenerative diseases [63–65].

Gong et al. integrated AChE within a nanostructured porous network of crossed bismuth oxyiodide BiOI nanoflake arrays (BiOINFs) in the design of a highly sensitive biosensor for the detection of an organophosphate pesticide [20]. A 3D network of BiOINFs turned out as an excellent matrix for the enzyme immobilization, which enhanced mass transport and AChE loading on the photoelectrode. Moreover, BiOI exhibits good visible light harvesting properties. As depicted in Figure 4A, the photocurrent of AChE-BiOINFs/ITO electrodes increased in presence of the enzyme substrate acetylthiocholine (ATCl) as a consequence of the hole scavenging properties of the product of the enzymatic reaction (thiocholine) upon irradiation; when an organophosphate pesticide (methyl parathion) was added, an irreversible inhibition effect impaired the enzymatic production of the hole scavenger and that was reflected as a decrease in the photocurrent. The relative difference between the photocurrent values in the absence and presence of the inhibitor methyl parathion (MP) was proportional to its concentration in the ranges 0.001–0.08 µg mL^{-1} and 0.3–1.0 µg mL^{-1} (Figure 4B,C). A detection limit of 0.04 ng mL^{-1} was reported. On the same enzyme inhibition strategy, the use of CdSe@ZnS QDs and graphene nanocomposites [61] and laser-induced TiO$_2$-decorated graphene (LITG) [28] was also proposed for the determination of OPs. The simple and scalable preparation method of the latter photoelectrode by direct-laser-writing of LITG on ITO, in which graphene greatly improved the photoresponse of the semiconductor (detection limit of chlorpyrifos: 5.4 pg mL^{-1}), could be very promising for PEC assays, although the immobilization of the enzyme was not considered.

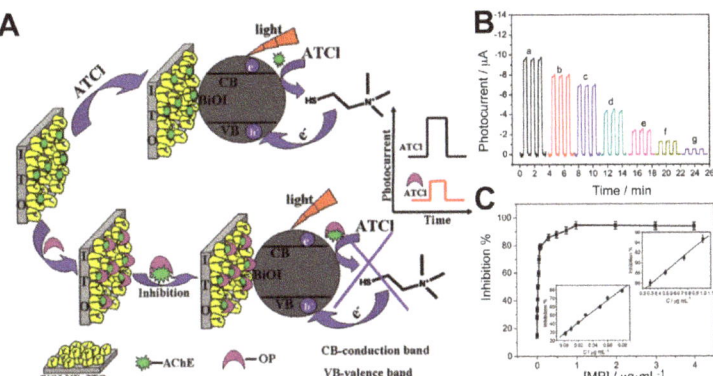

Figure 4. (**A**) Schematic of the photobioelectrochemical (PEC) biosensing principle using Acetyl Choline Esterase (AChE)-bismuth oxyiodide BiOI nanoflake arrays (BiOINFs)/indium-doped tin oxide (ITO) photoelectrodes. (**B**) Photocurrent response over time for increasing concentrations of methyl parathion. (**C**) Inhibition as a function of methyl parathion concentration and linear calibration plots. Adapted with permission from ref. [20].

A highly sensitive and selective self-powered PEC biosensor for OPs based on an enzymatic fuel cell was also reported [35]. They used (PEDOT)-sensitized CdS QDs forming a bilayer heterojunction—which promoted electron-hole separation and prevented charge recombination—as the photoanode and AChE immobilization platform. For the biocathode, they employed multiwalled carbon nanotubes, gold nanoparticles and bilirubin oxidase (BOx). The electrons generated in the photoanode by the enzyme product thiocholine could be transferred to the biocathode, where O_2 was reduced by BOx, and as a consequence a high open circuit voltage (E^{OCV}) was produced. The variation in the E^{OCV} in the presence of different concentrations of the inhibitor chlorpyrifos was used for the determination of the OP. Their PEC biosensor showed a wide linear range (0.00005 to 0.1 µg mL^{-1}) and a detection limit of 0.012 ng mL^{-1}.

In a different approach, Zhao et al. proposed a PEC enzymatic sensor aiming at preserving the optimal activity of the enzyme in the absence of the inhibitor by the use of antibodies [66]. AChE antibodies (anti-AChE) were immobilized, instead of the enzyme, on the surface of a photoelectrode consisting of a BiOI nanoflakes/TiO_2 nanoparticles p-n heterojunction. The presence of the inhibitor methyl parathion in a sample solution containing AChE and acetylthiocholine, in which the photoelectrode was immersed for the immunoreaction with anti-AChE, led to a decrease in the photocurrent that allowed the inhibitor determination (limit of detection: 0.015 ng mL^{-1}). This strategy could be extended to the study of the enzymatic activity or inhibition of other enzymes on condition that antibodies are appropriately immobilized.

The inhibitory effect of aflatoxin B_1 on AChE activity was also exploited for its detection, although in a great lesser extent than for OPs sensing. One PEC biosensor based on TiO_2 nanotubes, gold nanoparticles and AChE immobilized by crosslinking with glutaraldehyde was reported for the determination of this toxin (Yuan coatings 2018). AFB_1 competitively inhibited the enzyme and could be determined in the range 1–6 nM with a detection limit of 0.33 nM. The performance of their biosensor competes well with more costly methods for AFB_1 detection.

Furthermore, the study of AChE inhibition may be crucial in neurodegenerative disease research, because the dysfunction of this enzyme disturbs the cholinergic neurotransmission (i.e., involving neurotransmitter acetylcholine), which is related to the pathogenesis of neurodegenerative disorders such as Parkinson's disease (PD). A few PEC biosensors have been proposed as simple and sensitive platforms for the study of AChE activity in the presence of neurotoxins or Cd^{2+} ions. As an example, a hybrid photoelectrode for the evaluation of AChE inhibition by two endogenous neurotoxins ((R)-Sal and (R)-NMSal)—which have been believed to play a role in PD—was constructed by using nitrogen and fluorine co-doped TiO_2 nanotubes (TNs), Ag nanoparticles and AChE [63]. By measuring the photocurrent variations observed as a result of AChE inhibition and the subsequent decrease of thiocholine concentration (the product of acetylthiocholine hydrolysis which acts as an electron donor to scavenge the holes in the valence band of TNs), (R)-Sal and (R)-NMSal could be determined with a detection limit of 0.1 nM and 0.2 nM, respectively. Their results showed that the inhibition by these endogenous neurotoxins was reversible and mixed (competitive-uncompetitive). The inhibition constants were also calculated (K_i = 0.35 µM for (R)-Sal and 0.12 µM for (R)-NMSal). The same group reported the preparation of a nanomaterial composed of TNs modified with ZnO nanorods for the immobilization of AChE and the investigation of the effect of Cd^{2+} ions on its activity [64]. The results obtained with their PEC system revealed that Cd^{2+} had an activation effect on AChE activity at low concentration, whereas it had an inhibitory effect at high concentration.

3.3. Protein Kinases

Kinases are enzymes that catalyze the transfer of phosphate groups from ATP to other biomolecules such as amino acids in substrate peptides or proteins, sugars, nucleotides or lipids; the case of protein kinases (PKs) have been matter of photoelectrochemical biosensors. Abnormalities in protein kinase activity and the phosphorylation process are related to many diseases, including cancer, diabetes and

Alzheimer's disease [67]. With the aim of determining kinase activities in a simple, rapid and sensitive way and of screening its inhibitors, various photochemical biosensors have been proposed.

Most PEC biosensors for PK activity employed graphite-like carbon nitride (g-C$_3$N$_4$) as photoactive material [29,68,69], occasionally combined with TiO$_2$ to facilitate the effective charge separation and for recognition of the phosphorylated peptide after the PK-catalyzed reaction [70,71]. Yin et al. developed a visible-light activated PEC biosensor based on g-C$_3$N$_4$, the specific recognition molecule Phos-tag-biotin and avidin modified alkaline phosphatase (streptavidin-ALP) for signal amplification [29]. They constructed a g-C$_3$N$_4$-AuNPs-ITO electrode whereby a substrate peptide could bind (via the AuNPs and -SH groups of the peptide residues) and then, protein kinase A (PKA) transferred one phosphate group from ATP to the peptide. Subsequently, Phos-tag-biotin identified the phosphate group and the streptavidin-ALP was further captured on the electrode surface through the highly specific interaction between avidin and biotin. Finally, the immobilized ALP catalyzed the conversion of L-ascorbic acid-2-phosphate trisodium salt (AAP) into ascorbic acid (AA), which acted as electron donor and provided one electron to capture the photo-generated hole of g-C$_3$N$_4$, resulting in an increase of the photocurrent. PKA was thus selectively and sensitively detected (the detection limit was 0.015 U mL^{-1}) through the relationship between the photocurrent and PKA concentration. In another work from the same group, the interaction between phosphorylated g-C$_3$N$_4$ (P-g-C$_3$N$_4$) nanoparticles and a PKA-induced phosphorylated peptide (P-peptide) triggered by Zr^{4+} ion coordination [68] was proposed as a simple method for PKA activity biosensing. However, the detection limit reported could be improved (0.077 U mL^{-1}) and the preparation of P-g-C$_3$N$_4$ was time-consuming. Moreover, they developed a PEC biosensor that used a g-C$_3$N$_4$-TiO$_2$ composite, as both photoactive material and P-peptide conjugation platform, and a signal amplification strategy triggered by a polyamidoamine (PAMAM) dendrimer and ALP (which catalyzes the production of AA, an electron donor for the generation of the photoelectrochemical response) [70]. The PKA-catalyzed phosphorylation was performed in solution, instead of on the electrode surface, to simplify the experimental procedure and improve the contact between the reactants. Nevertheless, the separation process carried out before the capture of the P-peptide on the electrode (by the use of magnetic beads and carboxypeptidase Y for the hydrolysis and release of the P-peptide) was tedious and actually made the detection process more complicated.

The specific biotin-streptavidin interaction proposed in ref. [29] had been also used by Zhou et al. [72] in a simple label-free PEC biosensing method. Kinase-induced phosphopeptides, previously immobilized on a Bi$_2$S$_3$-AuNPs-ITO electrode, could bind to a biotinylated Phos-tag in the presence of Zn^{2+} and then streptavidin could be captured on the electrode surface, resulting in a decrease in the photocurrent due to the blocking of streptavidin towards the electron donor AA diffused to the Bi$_2$S$_3$ surface. The response was related to the phosphorylation extent and therefore to the PKA activity (Figure 5A,B), which could be detected with a detection limit of 0.017 U mL^{-1}. Furthermore, the sensor showed good selectivity when tested with other protein kinases and acceptable stability (Figure 5C,D).

Yan et al. further improved the sensitivity achieved by PK sensors and developed a highly sensitive PEC biosensor for PKA activity detection based on Au NPs localized surface plasmon resonance (LSPR) enhancement and dye sensitization [67]. They constructed a TiO$_2$-ITO electrode for the immobilization of the peptide and subsequent phosphorylation catalyzed by PKA. Then DNA was conjugated onto AuNPs and specifically coordinated to the P-peptides on the electrode via Zr^{4+} ions. [Ru(bipy)$_3$]$^{2+}$ was intercalated into the DNA grooves and harvested visible light to produce excited electrons that injected into TiO$_2$ conduction band, resulting in a strong photocurrent. The LSPR and fast electron transfer kinetics provided by AuNPs further improved the photocurrent efficiency and amplified the response. Their biosensor showed extremely low background signals and a detection limit of 0.005 U mL^{-1}.

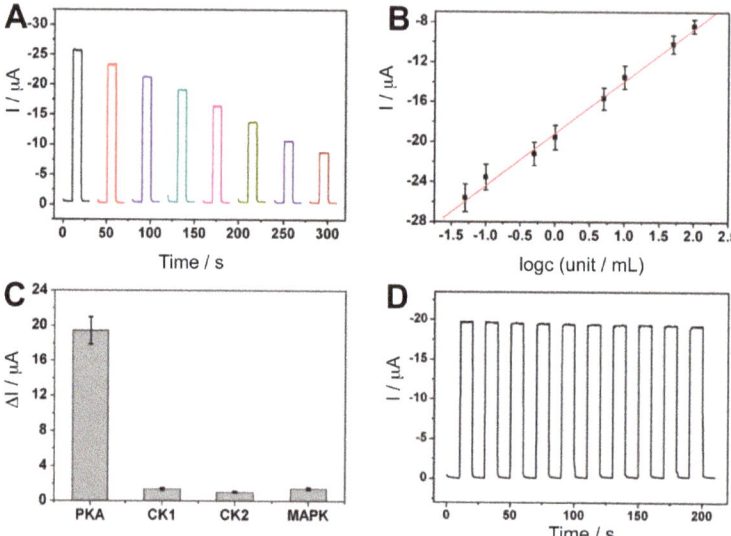

Figure 5. (**A**) Photocurrent response of the PEC biosensor based on biotinylated Phos-tag specific recognition for protein kinase A (PKA)-phosphorylated peptides in the presence of increasing concentrations of PKA (from left to right). (**B**) Photocurrent vs. logarithm of PKA concentration calibration plot. (**C**) Biosensor selectivity after incubation of the peptide-Bi_2S_3-AuNPs-ITO electrodes with different protein kinases. (**D**) Stability of the photocurrent response upon chopped-irradiation. Reprinted with permission from ref. [72].

Metal-organic frameworks (MOFs), a class of organic-inorganic hybrid crystalline porous materials, were also used to improve the sensitivity of PK PEC biosensors. For instance, Zr-based metal-organic frameworks (UiO-66) containing $[Ru(bipy)_3]^{2+}$ in the pores were selected in the design of a biosensor based on surface defect recognition and multiple signal amplification [73]. The surface defects on the ZrO clusters in UiO-66 enabled the binding of the phosphate groups of the peptide previously immobilized on a TiO_2-ITO electrode. Moreover, the high surface and porosity of the UiO-66 enhanced the amount of $[Ru(bipy)_3]^{2+}$, which injected excited electrons into the TiO_2 semiconductor; therefore, that increased the photocurrent and the sensitivity of the biosensor, which presented a detection limit as low as 0.0049 U mL^{-1} and a linear range from 0.005 to 0.0625 U mL^{-1}. As a final example, Wang et al. employed gold nanoparticle-decorated zeolitic imidazolate frameworks (Au-ZIF-8) for the immobilization of the substrate peptide on a ITO electrode modified with carbon microspheres [71]. Then, a g-C_3N_4-TiO_2 nanocomposite specifically interacted with the PKA phosphorylated peptide and provided a strong PEC response under visible light. The sensitivity of their biosensor was poorer than that of the UiO-66-based system previously described (0.02·U mL^{-1}) but the detection range was significantly wider (0.05 and 50 U mL^{-1}).

The PEC biosensors introduced here also performed the detection of PKA activities in cell lysates, which is promising for drug discovery applications, disease diagnosis and evaluation of therapeutic efficiency. Furthermore, these detection and inhibition screening methodologies can be extended to other kinases by changing the substrate peptide.

3.4. Lactate Detection

Lactate monitoring is of great importance in medical diagnosis and sports medicine. For instance, lactate is an indicator of traumatic brain injury [74,75] and its levels inform on the training status of athletes. State-of-the art lactate biosensors are moving towards non-invasive point-of-care (POC) detection and wearable systems. To improve the selectivity, both lactate oxidase (LOx) or lactate dehydrogenase (LDH) enzymes are used [76]. The scarce number of PEC enzymatic sensors for lactate detection that have been developed over the last decade uses the latter enzyme. However, wearable POC systems based on LDH and photoelectrochemical principles have not been reported yet for non-invasive lactate monitoring.

The first PEC LDH biosensor that demonstrated its practical applicability in real samples used a TiO$_2$ nanoparticle-multiwall carbon nanotube composite as immobilization matrix for LDH [77]. The system showed that the LDH co-substrate nicotinamide adenine dinucleotide (NAD$^+$) can be regenerated from the NADH produced during the biocatalytic reaction at a moderate potential (0.2 V vs. Ag/AgCl) by the photoexcited holes of the composite. The biosensor exhibited good long-term stability, high selectivity, a dynamic range of 0.5–120 µM, a sensitivity of 0.0242 µA µM^{-1}, and a detection limit of 0.1 µM.

Zhu et al. immobilized LDH, NAD$^+$ and a ternary composite onto ITO electrodes to develop a PEC biosensor that showed enhanced performance compared to other electrochemical biosensors for lactate [78]. The composite consisted of TiO$_2$ nanotubes (TiONTs), gold nanoparticles (GNPs)—which provided a surface plasmon resonance effect (SPR)—and polyaniline (PANI) with excellent electrochromic properties (Figure 6A). This system allowed the efficient regeneration of NAD$^+$ and the amplification of the photocurrent response, as depicted in Figure 6B, and it responded to a broad range of lactate concentrations (Figure 6C). The linear range, sensitivity, and detection limit of their method were 0.5–210 µM, 0.0401 µA µM^{-1}, and 0.15 µM, respectively.

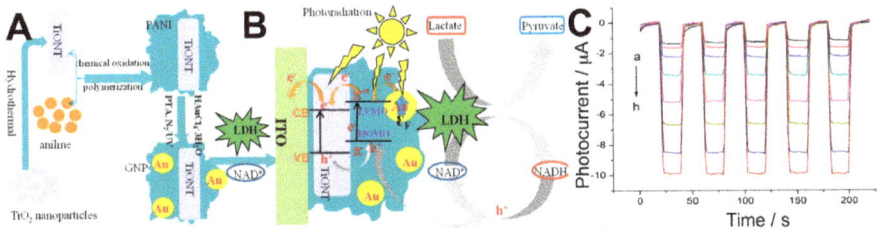

Figure 6. (**A**) Schematic of the preparation of TiO$_2$ nanotubes (TiONT)-polyaniline (PANI)-gold nanoparticles (GNPs) composites. (**B**) SPR-enhanced lactate biosensing mechanism of the TiONT-PANI-GNPs/lactate dehydrogenase (LDH)/nicotinamide adenine dinucleotide (NAD$^+$)/ITO system comprising PEC cosubstrate regeneration. (**C**) Photocurrent response of the PEC biosensor in the presence of increasing concentrations of lactate. Adapted with permission from [78]. Copyright (2016) American Chemical Society.

Furthermore, lactate dehydrogenase can use redox proteins, such as cytochrome c (cyt c), as electron acceptors. A platform that coupled cyt c and pyridine-functionalized CdS nanoparticles was combined with cytochrome-dependent lactate dehydrogenase and allowed the detection of millimolar concentrations of lactate [79]. The system generated oxidation photocurrents (at 0 V vs. SCE and λ = 420 nm) that were enhanced in the presence of increasing concentrations of lactate, as a result of the enzymatic regeneration of reduced cyt c.

3.5. Photo-Biosensors for Specific Applications

Besides the examples shown for the most common enzymatic sensors, there have been some other enzymatic systems developed for photoelectrochemical biosensors. A branch of sensors has focused on monitoring enzymes and their activity, in addition to protein kinases and AChE, already mentioned earlier. The very early optical fiber-based example was designed to detect the presence of catalase [17]. The optical fiber was modified by a partial chemical etching, allowing an interstitial space for the solution containing a sensitizer. The external layer of gold was exposed to a solution with tris(2,2'-bipyridine)ruthenium, afterwards the optical fiber was irradiated with an Ar laser and the internal silica layer provided the photons to excite the electrons of Ru(bpy)$_3^{2+}$, which reduced O_2 present in the solution to H_2O_2. The gold ring was then used to detect the in-situ generated H_2O_2. The sensor was able to quantify catalase activity by measuring the decrease of H_2O_2 concentration. Other photoelectrochemical systems have been designed for detecting protease activity, specifically tyrosinase and thrombin, which selectively cleaves arginine-glycine amide bonds [23]. The photoelectrode comprised ITO glasses modified with multiple layers of the sequence CdS and mercaptopropionic acid. After the final CdS layer was deposited a peptide ended in 4-phenyl was immobilized on the surface to serve as protease sensor. Tyrosinase oxidizes the radical to an ortho-quinone derivative, and thrombin cleaves the peptide chain. Both modifications impact the photoelectrochemical response. Detection of tyrosinase limited at 1.5 µg·mL^{-1} and yielded a linear range from 2.6 to 32 µg·mL^{-1}; whereas thrombin was detected at 1.9 µg·mL^{-1} and gave a linear range from 4.5 to 100 µg·mL^{-1}. Another example where photo-biosensors are used to detect enzymes relied on ITO electrodes as foundation for NiWO$_4$ nanostructures, which comprised the photoelectrodes [30] and showed a suitable ability to oxidize uric acid, which served as sacrificial electron donor and allowed the photocurrent. An immunosensor specific for neuron-specific enolase (NSE) was placed on the top of the nanostructures. Upon the presence of the NSE the surface of the electrode gets blocked by the immunoreceptors activity and the photocurrent is hindered. This structure yielded a sensitivity of 0.12 ng mL^{-1}. Another enzyme worth monitoring is a cancer marker like type IV collagenase, which is related to liver, breast, colon, lung carcinomas, and leukemia [80]. To do so, an ITO electrode was layer by layer modified with CdTe QDs and a synthetic peptide containing the specific sequence Gly-Pro-Ala, which is the cleavage target of collagenase. To one end of the sequence arginine amino acids were added to be modified with silver nanoparticles. The detection took place by a photoelectrochemical current increase caused by the cleavage of the peptide, which released the silver nanoparticles and allowed a larger radiation to reach the CdTe-ITO electrodes. The photocurrent increase was coherent with the variation in the impedance spectra of the surface. The photosensor yielded a limit of detection of 96 ng mL^{-1} and a linear detection range from 0.5 to 50 µg mL^{-1}.

Enzymes are not the only biomolecules that have been a target for photo-biosensors; proteins, peptides and key oligomers have also been matter of research due to its huge relevance in diagnosis. A first example is carcinogenic biomarkers, which have also been addressed with this technology. An ITO electrode modified with a layer of graphene oxide and a second layer of chitosan-Cd^{2+} has been used against a carcinoembryonic antigen (CEA), which served as recognition molecule [21]. The target molecule was the antibody for CEA. When the target molecule is not present, an artificial antibody loaded with horseradish peroxidase (HRP) linked to the surface. The authors promoted this way the reduction of sodium sulfite with H_2O_2 catalyzed by HRP, yielding sodium sulfide, which formed QDs with the existing cadmium in the chitosan layer. The appearance of CdS allowed the photooxidation of ascorbic acid, added as revealing agent. A GOx-based photo-biosensor has been used to detect the cancer biomarker α-fetoprotein (AFP) [81]. AFP is a glycoprotein which excess flags a high probability of hepatic carcinoma or endodermic sinus tumor. The sensor consisted in TiO$_2$ coupled with an AFP-CdTe-GOx conjugate that includes labels antibodies for AFT and GOx attached to CdTe QDs so its signal can be amplified. The electrode was coated with a layer of chitosan to provide a biocompatible matrix suitable for AFP antibody binding. CdTe QDs improve visible light absorption, thus avoiding the irradiation of the electrode with UV light, which is harmful against enzymes. In addition, quick electron transfer grants enhanced charge separation due to matching energy levels between CdTe and TiO$_2$, improving photocurrent response. Furthermore, in the presence of glucose, GOx catalyzes the production of H_2O_2 that acts as an electron donor and scavenges the photogenerated electron holes in CdTe QDs valence band, which enhances the photocurrent response even more. The electrodes tested turned out to be long-lasting, highly reproducible and with good sensitivity.

Nucleic acids' related activity has also been a matter for photo-biosensors. ITO electrodes modified with Bi$_2$S$_3$ and antibodies specific for methylated DNA have been used to detect the enzyme DNA methyltransferase [22], as the malfunction of this enzyme is related to several diseases and cancer development. The activity of DNA methyltransferase was detected by treating a DNA palindrome single stranded probe, which was methylated by the enzyme and later on linked to a biotinylated complementary sequence. The double-stranded modified sequence was trapped on the photoactive surface loaded with antibodies, and the exposed biotin was used to link alkaline phosphatase. The addition of the revealing probe, phosphorylated ascorbic acid, allowed for the photodetection of the resulting ascorbic acid, which enhanced the photoelectrochemical current if present. Malfunction of the DNA methyltransferase enzyme yielded no immunorecognition, no alkaline phosphatase, and no photooxidation of ascorbic acid. The revealing system was used for peptide detection also on different photoactive materials [25]. In this case an ITO electrode modified with macroporous TiO_2 loaded with complex quantum dots CdSeTe@CdS@ZnS was used. The photoelectrode was loaded with polyethyleneimine (PEI) and later with a biotinylated peptide for leukemia recognition. The signal was transduced with an equivalent alkaline phosphatase and ascorbic acid oxidation system. In another example, a photo-biosensor designed for hydroxymethylated DNA was presented by using ITO electrodes, which were modified with WS_2 and gold nanoparticles [27]. On the top of the surface, a DNA probe was immobilized to match the methylated DNA target. This system yielded a linear response in the photocurrent from 0.01 to 100 nM concentrations. Further on, the DNA-modified electrode could also be used to detect glycosyl transferase activity, since this enzyme can use the hydroxymethyl derivative of the DNA and substitute it by a sugar derivative, which can be detected with boronic acid-terminated quantum dots.

Finally, some other biosensors are devoted to detection of small biomolecules that usually act as biomarkers for several conditions and their monitoring can maintain or improve our health. A formaldehyde biosensor based on formaldehyde dehydrogenase was proven by replacing the NAD cofactor by a CdS-covered gold electrode [18], although this work focused on optimizing the enzyme-CdS interface and its light-dependence rather than developing the analytical conditions of the biosensor. Nitrite is an important analyte to monitor in environmental and food chemistry. An example of nitrite detection used a cytochrome C as recognition molecule deposited on a nanosponge modified with graphene-TiO_2 nanowires. Upon illumination, the detection limit of nitrite was 0.225 mM and a linear range from 0.5 µM to 9 mM was achieved. Uric acid is a biomolecule that works as biomarker of purine metabolism, and when it is out of range can anticipate gout or other cardiovascular condition [24]. The sensor was built on an ITO electrode coated with several kinds of ZnS nanostructures and the enzyme uricase, which oxidizes the uric acid to allantoin, CO_2 and H_2O_2. The best performing electrode was modified with ZnS urchin-like nanostructures, which upon illumination showed a limit of determination of 45 nM and a linear range from 0.01 to 0.54 mM. Another sensor [32] was prepared using semiconducting CuO nanotubes by oxidation of copper foil in two steps: a wet etching and further annealing. The work was focused on photoelectrode development using the enzyme xanthine oxidase as model reaction for guanine detection.

Lactose determination in dairy products and in particular in those called "lactose-free" is drawing more and more attention because of lactose intolerance problems. Very recently Çakıroğlu et al. have investigated an effective PEC strategy for lactose detection for the first time [82]. They developed a PEC biosensor for glucose and lactose consisting of TiO_2 modified with gold nanoparticles and a layer of MnO_2/g-C_3N_4 for the co-immobilization of glucose oxidase and β-galactosidase. Lactose measurements could be performed at low potential (−0.4 V vs. Ag/AgCl) with good sensitivity (detection limit of 0.23 µM) and linear range (0.008–2.50 mM).

N-methylglycine, also known as sarcosine, is a natural amino acid present in many organisms and plays a role in some metabolic paths like glycine synthesis or degradation. In addition, it may serve as biomarker of prostate cancer [26]. A photoactive biosensor built on ITO electrodes covered subsequently with layers of NiO, $CuInS_2$ and sarcosine oxidase was tested. It should be noted that the interface NiO-$CuInS_2$ is a p-p type heterojunction. The system worked by reducing the photocurrent due to the enzymatic activity of sarcosine oxidase, which competed with the photoelectrode for O_2 and depleted the substrate for the photoelectrode. The photoelectrode offered a limit of detection of

0.008 mM and a linear range from 0.01 to 1 mM. Typical interferences were tested successfully, which can be attributed to the selectivity of sarcosine oxidase. Hydrogen peroxide is also a very interesting molecule for detection, as it is a byproduct of oxidases, substrate of peroxidases and intermediate of many enzymatic cascades. Li el al. [33] prepared arrays of TiO_2 nanotubes by anodic oxidation of Ti foil and temperature crystallization for 2 h at 450 °C in aerobic conditions. The nanotubes were coated with polydopamine-HRP mix by incubation in solutions of dopamine and later on HRP. The HRP activity oxidized the dopamine, yielding an insoluble product that decreased the photocurrent. The sensor offered a 0.7 nM detection limit and a linear range from 1 nM to 50 µM. This is not the only example of photo-biosensors using TiO_2 nanotubes synthesized from titanium foil; another example has covered it with chitosan and cross-linked HRP to it with glutaraldehyde, by a double Schiff-base imine formation [34]. In this case, the sensor works by feeding HRP with H_2O_2, sodium sulfite and cadmium ytride, which produces CdS that precipitates and increases the photocurrent. The presence of the herbicide asulam avoids the formation of sulfide from sulfite, decreasing the photocurrent. Finally, a very recent work devoted to carcinoembryonic antigen demonstrates a dual electrochromic biosensor where a photoelectrochemical cell plays a role for both powering and sensing the device [36]. The photoanode comprised Ni:FeOOH/BiVO$_4$ nanocomposites and was set to power a cathode made of Prussian Blue (Figure 7). The detection of the carcinoembryonic antigen was performed with specific antibodies loaded with glucose oxidase set to form a sandwich structure when the antigen appears. Then glucose was fed into the anodic chamber. If the secondary antibody is linked, which happens only when the antigen has been trapped, H_2O_2 is produced by the glucose oxidase enzymatic reaction. H_2O_2 then acts as electron donor and is oxidized in the photoelectrode, increasing the current upon illumination. The sensor needed for dual optimization in time incubation, setting that it needs 45 min for immunologic process and 35 min for glucose oxidation by GOx (Figure 7C). The photocurrent generated fed the cathode, where the Prussian Blue was reduced to Prussian White, making the blue color disappear (Figure 7D).

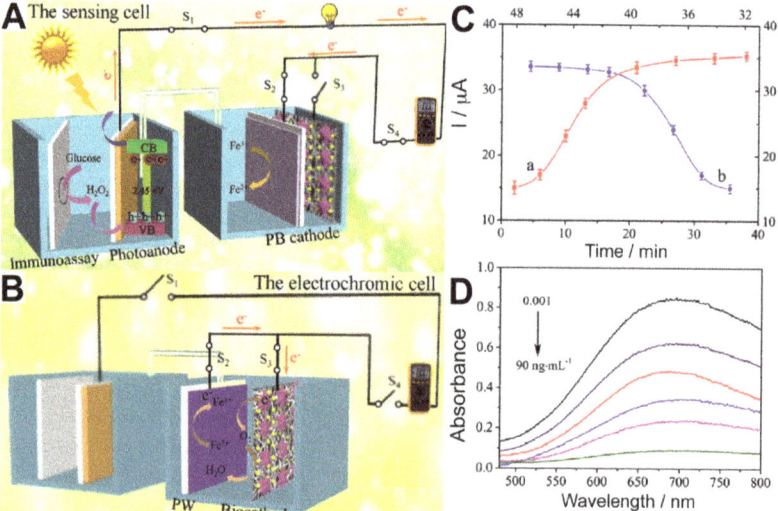

Figure 7. (**A**) Analytical principle of the photoelectrochemical sensing cell triggered by an immunoreaction and connected to a digital multimeter readout. (**B**) Analytical principle of the readout electrochromic cell. (**C**) Time optimization of (a) glucose oxidation and (b) immunosensing processes. (**D**) Electrochromic sensor decreasing absorbance of Prussian Blue, reduced to Prussian White. Adapted from ref. [36].

4. Conclusions

Photoelectrochemical biosensors have shown a very recent development, which came later than other biosensors due to their higher complexity and difficulty to achieve. Nevertheless, they provide better performance with respect to background signal suppression and the ability to switch the sensor ON and OFF. Moreover, the larger number of components opens the door to tailor on-demand sensors that are not only better suited to specific applications than regular biosensors, but have also shown new ways to create and analyze a sensing signal. Future work will focus on new interfaces electrode-semiconductor and semiconductor-macrobiomolecule—where the optimization can be carried out—while increasing the sensitivity and selectivity of the sensor. Another research trend that allows the inclusion of light-dependent sensors is the selection of the excitation wavelength, which opens plenty of possibilities of multi-sensing platforms by a mere decomposition of white light.

Author Contributions: This review has been conceptualized by M.P. and M.d.B. The selection of articles to review was performed by M.P. and M.d.B. Writing and original draft preparation was performed by M.P., M.d.B. and G.L.-L. Editing was performed by M.P., M.d.B. and G.L.-L. Project administration was performed by M.P. and M.d.B. Funding acquisition was performed by M.P. and M.d.B. All authors have read and agreed to the published version of the manuscript.

Funding: M. del Barrio acknowledges funding from the European Union's Horizon 2020 research and innovation programme under the Marie Skłodowska-Curie grant agreement No. 713366. G. Luna-López wishes to thank to "Comunidad de Madrid" and European Structural Funds for their financial support to FotoArt-CM project (S2018/NMT-4367). M. Pita thanks the Retos MCIU/AEI/FEDER, EU for funding project RTI2018-095090-B-I00.

Conflicts of Interest: The authors declare no conflict of interest.

References

1. Katz, E.; Willner, I.; Wang, J. Electroanalytical and Bioelectroanalytical Systems Based on Metal and Semiconductor Nanoparticles. *Electroanalysis* **2004**, *16*, 19–44. [CrossRef]
2. García-Molina, G.; Natale, P.; Valenzuela, L.; Alvarez-Malmagro, J.; Gutiérrez-Sánchez, C.; Iglesias-Juez, A.; López-Montero, I.; Vélez, M.; Pita, M.; De Lacey, A.L. Potentiometric Detection of ATP Based on the Transmembrane Proton Gradient Generated by ATPase Reconstituted on a Gold Electrode. *Bioelectrochemistry* **2020**, *133*, 107490. [CrossRef] [PubMed]
3. Pita, M.; Gutierrez-Sanchez, C.; Toscano, M.D.; Shleev, S.; De Lacey, A.L. Oxygen Biosensor Based on Bilirubin Oxidase Immobilized on a Nanostructured Gold Electrode. *Bioelectrochemistry* **2013**, *94*, 69–74. [CrossRef] [PubMed]
4. Pita, M. Switchable Biofuel Cells Controlled by Biomolecular Computing Systems. *Int. J. Unconv. Comput.* **2012**, *8*, 391–417.
5. Katz, E.; Pita, M. Biofuel Cells Controlled by Logically Processed Biochemical Signals: Towards Physiologically Regulated Bioelectronic Devices. *Chem. Eur. J.* **2009**, *15*, 12554–12564. [CrossRef] [PubMed]
6. Zhao, W.-W.; Xu, J.-J.; Chen, H.-Y. Photoelectrochemical enzymatic biosensors. *Biosens. Bioelectron.* **2017**, *92*, 294–304. [CrossRef] [PubMed]
7. Privman, V.; Pedrosa, V.; Melnikov, D.; Pita, M.; Simonian, A.; Katz, E. Enzymatic AND-Gate Based on Electrode-Immobilized Glucose-6-Phosphate Dehydrogenase: Towards Digital Biosensors and Biochemical Logic Systems with Low Noise. *Biosens. Bioelectron.* **2009**, *25*, 695–701. [CrossRef]
8. Pedrosa, V.; Melnikov, D.; Pita, M.; Halámek, J.; Privman, V.; Simonian, A.; Katz, E. Enzymatic Logic Gates with Noise-Reducing Sigmoid Response. *Int. J. Unconv. Comput.* **2009**, *6*, 451–460.
9. Privman, V.; Strack, G.; Solenov, D.; Pita, M.; Katz, E. Optimization of Enzymatic Biochemical Logic for Noise Reduction and Scalability: How Many Biocomputing Gates Can Be Interconnected in a Circuit? *J. Phys. Chem. B* **2008**, *112*, 11777–11784. [CrossRef]
10. Privman, V.; Arugula, M.A.; Halámek, J.; Pita, M.; Katz, E. Network Analysis of Biochemical Logic for Noise Reduction and Stability: A System of Three Coupled Enzymatic AND Gates. *J. Phys. Chem. B* **2009**, *113*, 5301–5310. [CrossRef]

11. Pita, M.; Privman, V.; Arugula, M.A.; Melnikov, D.; Bocharova, V.; Katz, E. Towards Biochemical Filters with a Sigmoidal Response to pH Changes: Buffered Biocatalytic Signal Transduction. *Phys. Chem. Chem. Phys.* **2011**, *13*, 4507–4513. [CrossRef]
12. Krämer, M.; Pita, M.; Zhou, J.; Ornatska, M.; Poghossian, A.; Schöning, M.J.; Katz, E. Coupling of Biocomputing Systems with Electronic Chips: Electronic Interface for Transduction of Biochemical Information. *J. Phys. Chem. C* **2009**, *113*, 2573–2579. [CrossRef]
13. Jarne, C.; Paul, L.; Conesa, J.C.; Shleev, S.; De Lacey, A.L.; Pita, M. Underpotential Photoelectrooxidation of Water by SnS_2 −Laccase Co-catalysts on Nanostructured Electrodes with Only Visible-Light Irradiation. *ChemElectroChem* **2019**, *6*, 2755–2761. [CrossRef]
14. Tapia, C.; Shleev, S.; Conesa, J.C.; De Lacey, A.L.; Pita, M. Laccase-Catalyzed Bioelectrochemical Oxidation of Water Assisted with Visible Light. *ACS Catal.* **2017**, *7*, 4881–4889. [CrossRef]
15. Tapia, C.; Berglund, S.P.; Friedrich, D.; Dittrich, T.; Bogdanoff, P.; Liu, Y.; Levcenko, S.; Unold, T.; Conesa, J.C.; De Lacey, A.L.; et al. Synthesis and Characterization of V-Doped β-In_2S_3 Thin Films on FTO Substrates. *J. Phys. Chem. C* **2016**, *120*, 28753–28761. [CrossRef]
16. Devadoss, A.; Sudhagar, P.; Terashima, C.; Nakata, K.; Fujishima, A. Photoelectrochemical Biosensors: New Insights into Promising Photoelectrodes and Signal Amplification Strategies. *J. Photochem. Photobiol. C Photochem. Rev.* **2015**, *24*, 43–63. [CrossRef]
17. Cohen, C.B.; Weber, S.G. Photoelectrochemical Sensor for Catalase Activity Based on the in Situ Generation and Detection of Substrate. *Anal. Chem.* **1993**, *65*, 169–175. [CrossRef]
18. Curri, M.L.; Agostiano, A.; Leo, G.; Mallardi, A.; Cosma, P.; Della Monica, M. Development of a Novel Enzyme/Semiconductor Nanoparticles System for Biosensor Application. *Mater. Sci. Eng. C* **2002**, *22*, 449–452. [CrossRef]
19. Tantra, R.; Hutton, R.S.; Williams, D.E. A Biosensor Based on Transient Photoeffects at a Silicon Electrode. *J. Electroanal. Chem.* **2002**, *538–539*, 205–208. [CrossRef]
20. Gong, J.; Wang, X.; Li, X.; Wang, K. Highly Sensitive Visible Light Activated Photoelectrochemical Biosensing of Organophosphate Pesticide Using Biofunctional Crossed Bismuth Oxyiodide Flake Arrays. *Biosens. Bioelectron.* **2012**, *38*, 43–49. [CrossRef]
21. Zeng, X.; Tu, W.; Li, J.; Bao, J.; Dai, Z. Photoelectrochemical Biosensor Using Enzyme-Catalyzed in Situ Propagation of CdS Quantum Dots on Graphene Oxide. *ACS Appl. Mater. Interfaces* **2014**, *6*, 16197–16203. [CrossRef] [PubMed]
22. Yang, Z.; Wang, F.; Wang, M.; Yin, H.; Ai, S. A Novel Signal-on Strategy for *M.SssI* Methyltransfease Activity Analysis and Inhibitor Screening Based on Photoelectrochemical Immunosensor. *Biosens. Bioelectron.* **2015**, *66*, 109–114. [CrossRef] [PubMed]
23. Chen, J.; Liu, Y.; Zhao, G.-C. A Novel Photoelectrochemical Biosensor for Tyrosinase and Thrombin Detection. *Sensors* **2016**, *16*, 135. [CrossRef]
24. Zhao, Y.; Wei, X.; Peng, N.; Wang, J.; Jiang, Z. Study of ZnS Nanostructures Based Electrochemical and Photoelectrochemical Biosensors for Uric Acid Detection. *Sensors* **2017**, *17*, 1235. [CrossRef]
25. Zhou, S.; Wang, Y.; Zhao, M.; Jiang, L.-P.; Zhu, J.-J. CdSeTe@CdS@ZnS Quantum-Dot-Sensitized Macroporous TiO_2 Film: A Multisignal-Amplified Photoelectrochemical Platform. *ChemPhysChem* **2015**, *16*, 2826–2835. [CrossRef]
26. Jiang, X.-Y.; Zhang, L.; Liu, Y.-L.; Yu, X.-D.; Liang, Y.-Y.; Qu, P.; Zhao, W.-W.; Xu, J.-J.; Chen, H.-Y. Hierarchical $CuInS_2$-based heterostructure: Application for photocathodic bioanalysis of sarcosine. *Biosens. Bioelectron.* **2018**, *107*, 230–236. [CrossRef]
27. Sui, C.; Wang, T.; Zhou, Y.; Yin, H.; Meng, X.; Zhang, S.; Waterhouse, G.I.N.; Xu, Q.; Zhuge, Y.; Ai, S. Photoelectrochemical biosensor for hydroxymethylated DNA detection and T4-β-glucosyltransferase activity assay based on WS_2 nanosheets and carbon dots. *Biosens. Bioelectron.* **2019**, *127*, 38–44. [CrossRef]
28. Ge, L.; Hong, Q.; Li, H.; Li, F. A Laser-Induced TiO_2-Decorated Graphene Photoelectrode for Sensitive Photoelectrochemical Biosensing. *Chem. Commun.* **2019**, *55*, 4945–4948. [CrossRef]
29. Yin, H.; Sun, B.; Dong, L.; Li, B.; Zhou, Y.; Ai, S. A signal "on" photoelectrochemical biosensor for assay of protein kinase activity and its inhibitor based on graphite-like carbon nitride, Phos-tag and alkaline phosphatase. *Biosens. Bioelectron.* **2015**, *64*, 462–468. [CrossRef]

30. Soomro, R.A.; Kalwar, N.H.; Avci, A.; Pehlivan, E.; Hallam, K.R.; Willander, M. In-Situ Growth of NiWO$_4$ Saw-Blade-Like Nanostructures and Their Application in Photo-Electrochemical (PEC) Immunosensor System Designed for the Detection of Neuron-Specific Enolase. *Biosens. Bioelectron.* **2019**, *141*, 111331. [CrossRef]
31. Riedel, M.; Hölzel, S.; Hille, P.; Schörmann, J.; Eickhoff, M.; Lisdat, F. InGaN/GaN nanowires as a new platform for photoelectrochemical sensors—Detection of NADH. *Biosens. Bioelectron.* **2017**, *94*, 298–304. [CrossRef]
32. Zhang, L.; Zhu, Y.-C.; Liang, Y.-Y.; Zhao, W.-W.; Xu, J.-J.; Chen, H.-Y. Semiconducting CuO Nanotubes: Synthesis, Characterization, and Bifunctional Photocathodic Enzymatic Bioanalysis. *Anal. Chem.* **2018**, *90*, 5439–5444. [CrossRef]
33. Li, J.; Li, X.; Zhao, Q.; Jiang, Z.; Tadé, M.; Wang, S.; Liu, S. Polydopamine-assisted decoration of TiO$_2$ nanotube arrays with enzyme to construct a novel photoelectrochemical sensing platform. *Sens. Actuators B Chem.* **2018**, *255*, 133–139. [CrossRef]
34. Tian, J.; Li, Y.; Dong, J.; Huang, M.; Lu, J. Photoelectrochemical TiO$_2$ nanotube arrays biosensor for asulam determination based on in-situ generation of quantum dots. *Biosens. Bioelectron.* **2018**, *110*, 1–7. [CrossRef] [PubMed]
35. Gai, P.; Zhang, S.; Yu, W.; Li, H.; Li, F. Light-Driven Self-Powered Biosensor for Ultrasensitive Organophosphate Pesticide Detection *via* Integration of the Conjugated Polymer-Sensitized Cds and Enzyme Inhibition Strategy. *J. Mater. Chem. B* **2018**, *6*, 6842–6847. [CrossRef]
36. Sun, J.; Cui, K.; Li, L.; Zhang, L.; Yu, J. Visible-light-driven renewable photoelectrochemical/synchronous visualized sensing platform based on Ni:FeOOH/BiVO$_4$ photoanode and enzymatic cascade amplification for carcinoembryonic antigen detection. *Sens. Actuators B Chem.* **2020**, *304*, 127301. [CrossRef]
37. Ren, X.; Chen, D.; Meng, X.; Tang, F.; Hou, X.; Han, D.; Zhang, L. Zinc oxide nanoparticles/glucose oxidase photoelectrochemical system for the fabrication of biosensor. *J. Colloid Interface Sci.* **2009**, *334*, 183–187. [CrossRef]
38. Sun, J.; Zhu, Y.; Yang, X.; Li, C. Photoelectrochemical glucose biosensor incorporating CdS nanoparticles. *Particuology* **2009**, *7*, 347–352. [CrossRef]
39. Zheng, M.; Cui, Y.; Li, X.; Liu, S.; Tang, Z. Photoelectrochemical sensing of glucose based on quantum dot and enzyme nanocomposites. *J. Electroanal. Chem.* **2011**, *656*, 167–173. [CrossRef]
40. Wang, W.; Bao, L.; Lei, J.; Tu, W.; Ju, H. Visible light induced photoelectrochemical biosensing based on oxygen-sensitive quantum dots. *Anal. Chim. Acta* **2012**, *744*, 33–38. [CrossRef]
41. Du, J.; Yu, X.; Wu, Y.; Di, J. ZnS nanoparticles electrodeposited onto ITO electrode as a platform for fabrication of enzyme-based biosensors of glucose. *Mater. Sci. Eng. C* **2013**, *33*, 2031–2036. [CrossRef] [PubMed]
42. Tang, J.; Wang, Y.; Li, J.; Da, P.; Geng, J.; Zheng, G. Sensitive enzymatic glucose detection by TiO$_2$ nanowire photoelectrochemical biosensors. *J. Mater. Chem. A* **2014**, *2*, 6153–6157. [CrossRef]
43. Xia, L.; Song, J.; Xu, R.; Liu, D.; Dong, B.; Xu, L.; Song, H. Zinc oxide inverse opal electrodes modified by glucose oxidase for electrochemical and photoelectrochemical biosensor. *Biosens. Bioelectron.* **2014**, *59*, 350–357. [CrossRef]
44. Dilgin, D.G.; Gökçel, H.İ. Photoelectrochemical glucose biosensor in flow injection analysis system based on glucose dehydrogenase immobilized on poly-hematoxylin modified glassy carbon electrode. *Anal. Methods* **2015**, *7*, 990–999. [CrossRef]
45. Ertek, B.; Dilgin, Y. Photoamperometric flow injection analysis of glucose based on dehydrogenase modified quantum dots-carbon nanotube nanocomposite electrode. *Bioelectrochemistry* **2016**, *112*, 138–144. [CrossRef]
46. Ertek, B.; Akgül, C.; Dilgin, Y. Photoelectrochemical glucose biosensor based on a dehydrogenase enzyme and NAD$^+$/NADH redox couple using a quantum dot modified pencil graphite electrode. *RSC Adv.* **2016**, *6*, 20058–20066. [CrossRef]
47. Dai, W.-X.; Zhang, L.; Zhao, W.-W.; Yu, X.-D.; Xu, J.-J.; Chen, H.-Y. Hybrid PbS Quantum Dot/Nanoporous NiO Film Nanostructure: Preparation, Characterization, and Application for a Self-Powered Cathodic Photoelectrochemical Biosensor. *Anal. Chem.* **2017**, *89*, 8070–8078. [CrossRef]
48. Liu, P.; Huo, X.; Tang, Y.; Xu, J.; Liu, X.; Wong, D.K.Y. A TiO$_2$ nanosheet-g-C$_3$N$_4$ composite photoelectrochemical enzyme biosensor excitable by visible irradiation. *Anal. Chim. Acta* **2017**, *984*, 86–95. [CrossRef]

49. Çakıroğlu, B.; Özacar, M. A self-powered photoelectrochemical glucose biosensor based on supercapacitor Co_3O_4-CNT hybrid on TiO_2. *Biosens. Bioelectron.* **2018**, *119*, 34–41. [CrossRef]
50. Ryu, G.M.; Lee, M.; Choi, D.S.; Park, C.B. A hematite-based photoelectrochemical platform for visible light-induced biosensing. *J. Mater. Chem. B* **2015**, *3*, 4483–4486. [CrossRef]
51. Yan, B.; Zhuang, Y.; Jiang, Y.; Xu, W.; Chen, Y.; Tu, J.; Wang, X.; Wu, Q. Enhanced photoeletrochemical biosensing performance from rutile nanorod/anatase nanowire junction array. *Appl. Surface Sci.* **2018**, *458*, 382–388. [CrossRef]
52. Atchudan, R.; Muthuchamy, N.; Edison, T.N.J.I.; Perumal, S.; Vinodh, R.; Park, K.H.; Lee, Y.R. An ultrasensitive photoelectrochemical biosensor for glucose based on bio-derived nitrogen-doped carbon sheets wrapped titanium dioxide nanoparticles. *Biosens. Bioelectron.* **2019**, *126*, 160–169. [CrossRef]
53. Liang, D.; Luo, J.; Huang, Y.; Liang, X.; Qiu, X.; Wang, J.; Yang, M. A porous carbon nitride modified with cobalt phosphide as an efficient visible-light harvesting nanocomposite for photoelectrochemical enzymatic sensing of glucose. *Microchim. Acta* **2019**, *186*, 856. [CrossRef]
54. Zhang, X.Y.; Liu, S.G.; Zhang, W.J.; Wang, X.H.; Han, L.; Ling, Y.; Li, N.B.; Luo, H.Q. Photoelectrochemical platform for glucose sensing based on $g-C_3N_4$/$ZnIn_2S_4$ composites coupled with bi-enzyme cascade catalytic in-situ precipitation. *Sens. Actuators B Chem.* **2019**, *297*, 126818. [CrossRef]
55. Chen, L.; Chen, Y.; Miao, L.; Gao, Y.; Di, J. Photocurrent switching effect on $BiVO_4$ electrodes and its application in development of photoelectrochemical glucose sensor. *J. Solid State Electrochem.* **2020**, *24*, 411–420. [CrossRef]
56. Çakıroğlu, B.; Özacar, M. A Photoelectrochemical Biosensor Fabricated using Hierarchically Structured Gold Nanoparticle and MoS_2 on Tannic Acid Templated Mesoporous TiO_2. *Electroanalysis* **2020**, *32*, 166–177. [CrossRef]
57. Yang, W.; Wang, X.; Hao, W.; Wu, Q.; Peng, J.; Tu, J.; Cao, Y. 3D hollow-out TiO_2 nanowire cluster/GOx as an ultrasensitive photoelectrochemical glucose biosensor. *J. Mater. Chem. B* **2020**, *8*, 2363–2370. [CrossRef]
58. Arduini, F.; Amine, A.; Moscone, D.; Palleschi, G. Biosensors Based on Cholinesterase Inhibition for Insecticides, Nerve Agents and Aflatoxin B1 Detection (review). *Microchim. Acta* **2010**, *170*, 193–214. [CrossRef]
59. Pohanka, M. Electrochemical Biosensors based on Acetylcholinesterase and Butyrylcholinesterase. A Review. *Int. J. Electrochem. Sci.* **2016**, 7440–7452. [CrossRef]
60. Pardo-Yissar, V.; Katz, E.; Wasserman, J.; Willner, I. Acetylcholine Esterase-Labeled CdS Nanoparticles on Electrodes: Photoelectrochemical Sensing of the Enzyme Inhibitors. *J. Am. Chem. Soc.* **2003**, *125*, 622–623. [CrossRef]
61. Li, X.; Zheng, Z.; Liu, X.; Zhao, S.; Liu, S. Nanostructured Photoelectrochemical Biosensor for Highly Sensitive Detection of Organophosphorous Pesticides. *Biosens. Bioelectron.* **2015**, *64*, 1–5. [CrossRef]
62. Yuan, Q.; He, C.; Mo, R.; He, L.; Zhou, C.; Hong, P.; Sun, S.; Li, C. Detection of AFB1 via TiO2 Nanotubes/Au Nanoparticles/Enzyme Photoelectrochemical Biosensor. *Coatings* **2018**, *8*, 90. [CrossRef]
63. Huang, Q.; Chen, H.; Xu, L.; Lu, D.; Tang, L.; Jin, L.; Xu, Z.; Zhang, W. Visible-Light-Activated Photoelectrochemical Biosensor for the Study of Acetylcholinesterase Inhibition Induced by Endogenous Neurotoxins. *Biosens. Bioelectron.* **2013**, *45*, 292–299. [CrossRef]
64. Huang, Q.; Wang, Y.; Lei, L.; Xu, Z.; Zhang, W. Photoelectrochemical Biosensor for Acetylcholinesterase Activity Study Based on Metal Oxide Semiconductor Nanocomposites. *J. Electroanal. Chem.* **2016**, *781*, 377–382. [CrossRef]
65. Zhu, W.; An, Y.-R.; Luo, X.-M.; Wang, F.; Zheng, J.-H.; Tang, L.-L.; Wang, Q.-J.; Zhang, Z.-H.; Zhang, W.; Jin, L.-T. Study on Acetylcholinesterase Inhibition Induced by Endogenous Neurotoxin with an Enzyme–Semiconductor Photoelectrochemical System. *Chem. Commun.* **2009**, 2682. [CrossRef]
66. Zhao, W.-W.; Shan, S.; Ma, Z.-Y.; Wan, L.-N.; Xu, J.-J.; Chen, H.-Y. Acetylcholine Esterase Antibodies on BiOI Nanoflakes/TiO2 Nanoparticles Electrode: A Case of Application for General Photoelectrochemical Enzymatic Analysis. *Anal. Chem.* **2013**, *85*, 11686–11690. [CrossRef]
67. Yan, Z.; Wang, Z.; Miao, Z.; Liu, Y. Dye-Sensitized and Localized Surface Plasmon Resonance Enhanced Visible-Light Photoelectrochemical Biosensors for Highly Sensitive Analysis of Protein Kinase Activity. *Anal. Chem.* **2016**, *88*, 922–929. [CrossRef]

68. Li, X.; Zhou, Y.; Xu, Y.; Xu, H.; Wang, M.; Yin, H.; Ai, S. A novel photoelectrochemical biosensor for protein kinase activity assay based on phosphorylated graphite-like carbon nitride. *Anal. Chim. Acta* **2016**, *934*, 36–43. [CrossRef]
69. Sui, C.; Liu, F.; Tang, L.; Li, X.; Zhou, Y.; Yin, H.; Ai, S. Photoelectrochemical determination of the activity of protein kinase A by using g-C$_3$N$_4$ and CdS quantum dots. *Microchim. Acta* **2018**, *185*, 541. [CrossRef]
70. Li, X.; Zhu, L.; Zhou, Y.; Yin, H.; Ai, S. Enhanced Photoelectrochemical Method for Sensitive Detection of Protein Kinase A Activity Using TiO$_2$/g-C$_3$N$_4$, PAMAM Dendrimer, and Alkaline Phosphatase. *Anal. Chem.* **2017**, *89*, 2369–2376. [CrossRef]
71. Wang, Y.; Li, X.; Waterhouse, G.I.N.; Zhou, Y.; Yin, H.; Ai, S. Photoelectrochemical biosensor for protein kinase A detection based on carbon microspheres, peptide functionalized Au-ZIF-8 and TiO$_2$/g-C$_3$N$_4$. *Talanta* **2019**, *196*, 197–203. [CrossRef] [PubMed]
72. Zhou, Y.; Wang, M.; Yang, Z.; Yin, H.; Ai, S. A Phos-tag-based photoelectrochemical biosensor for assay of protein kinase activity and inhibitors. *Sens. Actuators B Chem.* **2015**, *206*, 728–734. [CrossRef]
73. Wang, Z.; Yan, Z.; Wang, F.; Cai, J.; Guo, L.; Su, J.; Liu, Y. Highly sensitive photoelectrochemical biosensor for kinase activity detection and inhibition based on the surface defect recognition and multiple signal amplification of metal-organic frameworks. *Biosens. Bioelectron.* **2017**, *97*, 107–114. [CrossRef]
74. Pita, M.; Zhou, J.; Manesh, K.M.; Halámek, J.; Katz, E.; Wang, J. Enzyme logic gates for assessing physiological conditions during an injury: Towards digital sensors and actuators. *Sens. Actuators B Chem.* **2009**, *139*, 631–636. [CrossRef]
75. Manesh, K.M.; Halámek, J.; Pita, M.; Zhou, J.; Tam, T.K.; Santhosh, P.; Chuang, M.-C.; Windmiller, J.R.; Abidin, D.; Katz, E.; et al. Enzyme Logic Gates for the Digital Analysis of Physiological Level Upon Injury. *Biosens. Bioelectron.* **2009**, *24*, 3569–3574. [CrossRef]
76. Alam, F.; RoyChoudhury, S.; Jalal, A.H.; Umasankar, Y.; Forouzanfar, S.; Akter, N.; Bhansali, S.; Pala, N. Lactate biosensing: The emerging point-of-care and personal health monitoring. *Biosens. Bioelectron.* **2018**, *117*, 818–829. [CrossRef]
77. Liu, X.; Yan, R.; Zhu, J.; Huo, X.; Wang, X. Development of a Photoelectrochemical Lactic Dehydrogenase Biosensor Using Multi-Wall Carbon Nanotube -TiO$_2$ Nanoparticle Composite as Coenzyme Regeneration Tool. *Electrochim. Acta* **2015**, *173*, 260–267. [CrossRef]
78. Zhu, J.; Huo, X.; Liu, X.; Ju, H. Gold Nanoparticles Deposited Polyaniline–TiO$_2$ Nanotube for Surface Plasmon Resonance Enhanced Photoelectrochemical Biosensing. *ACS Appl. Mater. Interfaces* **2016**, *8*, 341–349. [CrossRef]
79. Katz, E.; Zayats, M.; Willner, I.; Lisdat, F. Controlling the direction of photocurrents by means of CdS nanoparticles and cytochrome c-mediated biocatalytic cascades. *Chem. Commun.* **2006**, 1395. [CrossRef]
80. Chen, J.; Zhao, G.-C.; Wei, Y.; Feng, D. A signal-on photoelectrochemical biosensor for detecting cancer marker type IV collagenase by coupling enzyme cleavage with exciton energy transfer biosensing. *Anal. Methods* **2019**, *11*, 5880–5885. [CrossRef]
81. Li, Y.-J.; Ma, M.-J.; Zhu, J.-J. Dual-Signal Amplification Strategy for Ultrasensitive Photoelectrochemical Immunosensing of α-Fetoprotein. *Anal. Chem.* **2012**, *84*, 10492–10499. [CrossRef] [PubMed]
82. Çakıroğlu, B.; Demirci, Y.C.; Gökgöz, E.; Özacar, M. A photoelectrochemical glucose and lactose biosensor consisting of gold nanoparticles, MnO$_2$ and g-C$_3$N$_4$ decorated TiO$_2$. *Sens. Actuators B Chem.* **2019**, *282*, 282–289. [CrossRef]

© 2020 by the authors. Licensee MDPI, Basel, Switzerland. This article is an open access article distributed under the terms and conditions of the Creative Commons Attribution (CC BY) license (http://creativecommons.org/licenses/by/4.0/).

Review

Electrochemical Biosensors Employing Natural and Artificial Heme Peroxidases on Semiconductors

Bettina Neumann and Ulla Wollenberger *

Institute of Biochemistry and Biology, University of Potsdam, 14476 Potsdam, Germany; bettina.neumann@uni-potsdam.de
* Correspondence: uwollen@uni-potsdam.de

Received: 4 June 2020; Accepted: 27 June 2020; Published: 1 July 2020

Abstract: Heme peroxidases are widely used as biological recognition elements in electrochemical biosensors for hydrogen peroxide and phenolic compounds. Various nature-derived and fully synthetic heme peroxidase mimics have been designed and their potential for replacing the natural enzymes in biosensors has been investigated. The use of semiconducting materials as transducers can thereby offer new opportunities with respect to catalyst immobilization, reaction stimulation, or read-out. This review focuses on approaches for the construction of electrochemical biosensors employing natural heme peroxidases as well as various mimics immobilized on semiconducting electrode surfaces. It will outline important advances made so far as well as the novel applications resulting thereof.

Keywords: electrochemical biosensors; heme; peroxidases; semiconductors; peroxidase mimics

1. Introduction

Heme peroxidases are popular tools for a variety of bioanalytical techniques where they serve e.g., as reporter enzymes in affinity-based assays or as recognition elements in biosensors [1]. The redox enzymes are especially attractive biocatalysts for the construction of electrochemical biosensors for the detection of peroxides, phenolic compounds, or aromatic amines [2]. Phenols and derivatives are highly abundant toxic wastewater contaminants of e.g., the plastic, paper, and pharmaceutical industries [3,4], while the determination of peroxide concentrations is of high relevance for e.g., the food, pharmaceutical, paper, and textile industries [5–7]. As by-product of many oxidases like glucose oxidase (GOx) or cholesterol oxidase, its detection is further important for diagnostics [2]. In addition, the in vivo relevance of hydrogen peroxide as a signaling molecule [8], cell toxin, and disease indicator [9] has gained more attention, invoking the need for highly sensitive biosensors.

Electrochemical biosensors employing heme peroxidases immobilized on classical electrode materials like carbon or noble metals have been extensively reviewed in the past [2,10–12]. Semiconducting materials, on the other hand, have not gained that much attention in this field, although they have been employed already since decades as substrates for the immobilization of enzymes [13–18]. The ability of semiconductors to control charge accumulation and release by potential, light or heat cannot only be exploited for photovoltaic systems, solar energy harvesting, and conversion [19,20], but can also open up new possibilities for the application of enzymes like heme peroxidases in photoelectrocatalytic devices [21].

Despite their wide use and large potential, the structural properties of heme peroxidases interfere with their production on high scales and, more importantly, can impede their performance in electrochemical biosensors. By the use of chemical and biological engineering techniques and computational methods, researchers have therefore developed various alternative chemical and biological molecules that could serve as mimics for heme peroxidases, catalyzing the same or even more reactions, but outperforming the natural enzymes in terms of electrocatalytic activity or stability under harsh conditions. In combination with the development of new electrode surfaces and immobilization techniques, new efficient biosensors based on peroxidase reactions have been designed.

This review compiles approaches for the development of electrochemical biosensors employing natural heme peroxidases as well as various mimics—ranging from nature-derived heme-peptide complexes to fully synthetic heme derivatives—immobilized on semiconducting electrode surfaces. It will outline important advances made so far as well as the novel applications resulting thereof.

2. Heme Peroxidases and Their Mimics

2.1. Biochemistry of Heme Peroxidases

Heme peroxidases, EC 1.11.1.7, catalyze the oxidation of a broad variety of reductants (AH_2) by peroxides, usually hydrogen peroxide, following Equation (1) [22,23]:

$$H_2O_2 + 2AH_2 \longrightarrow 2H_2O + 2AH\bullet. \tag{1}$$

Heme peroxidases are found in all domains of life. Yet, for bioanalytical applications secretory plant peroxidases, constituting family III of the superfamily of peroxidases-catalases [24], are mostly used. The heme peroxidase from horseradish (HRP) is probably the best-studied member of this family and therefore, the most popular peroxidase for the construction of electrochemical biosensors. However, systems employing other heme peroxidases e.g., from soybean [25–27], tobacco [28–30], as well as peanut and sweet potato [28,31], have been reported.

Figure 1 shows the overall structure of HRP isoenzyme C1A as well as the active site arrangement which is common in all plant peroxidases. HRP C is a monomeric and mainly α-helical glycoprotein with a molecular weight of 44 kDa. Furthermore, HRP contains four disulfide bonds and two calcium ions, both of which are important for the enzyme's structural stability [32,33]. Figure 1b shows a close-up of the active site highlighting the cofactor and three amino acids, which have been identified as highly conserved and essential for the high activity of heme peroxidases. His170 coordinates the heme iron on the proximal side and anchors it in the center of the enzyme. The sixth coordination site is the substrate binding site and vacant in the enzyme's resting state. During catalysis, the distal amino acids Arg38 and His42 facilitate the efficient heterolytic cleavage of the bound peroxide by abstracting and donating protons and by stabilizing leaving groups, as depicted in Figure 1c [34,35]. Upon reduction of hydrogen peroxide, water is released and Compound I, a high-valent reaction intermediate with an Fe-oxoferryl center in oxidation state +IV and a porphyrin-based cationic radical, is formed. The three main reaction steps of heme peroxidase catalysis are described with Equations (2)–(4) [32,35]:

$$\text{Peroxidase} + H_2O_2 \longrightarrow \text{Compound I} + H_2O, \tag{2}$$

$$\text{Compound I} + AH_2 \longrightarrow \text{Compound II} + AH\bullet, \tag{3}$$

$$\text{Compound II} + AH_2 \longrightarrow \text{Peroxidase} + AH\bullet + H_2O. \tag{4}$$

In two consecutive one electron transfer steps, the resting state is restored from Compound I via formation of the second intermediate Compound II. The latter is structurally similar to Compound I, but lacks the porphyrin-based radical. Single-electron or hydrogen atom donors serve as electron sources and are oxidized to radicals, which in turn can form dimers or higher oligomers.

2.2. Peroxidase Reactions in Electrochemical Biosensors

Figure 2 illustrates the two different measuring modes of electrochemical biosensors for the detection of hydrogen peroxide by heme peroxidases. In the direct approach, the intermediates Compound I and II formed upon reaction of the enzyme with hydrogen peroxide are both reduced at the electrode via direct electron transfer (Figure 2a). This way, the resting state is restored and a new turnover can be initiated. The reduction potentials of the redox couples Compound I/Compound II and Compound II/resting state were shown to be as high as +700 mV vs. Ag/AgCl at pH 7 [2,36]. Therefore, the electrocatalytic reduction of hydrogen peroxide by heme peroxidases can be detected at highly positive potentials.

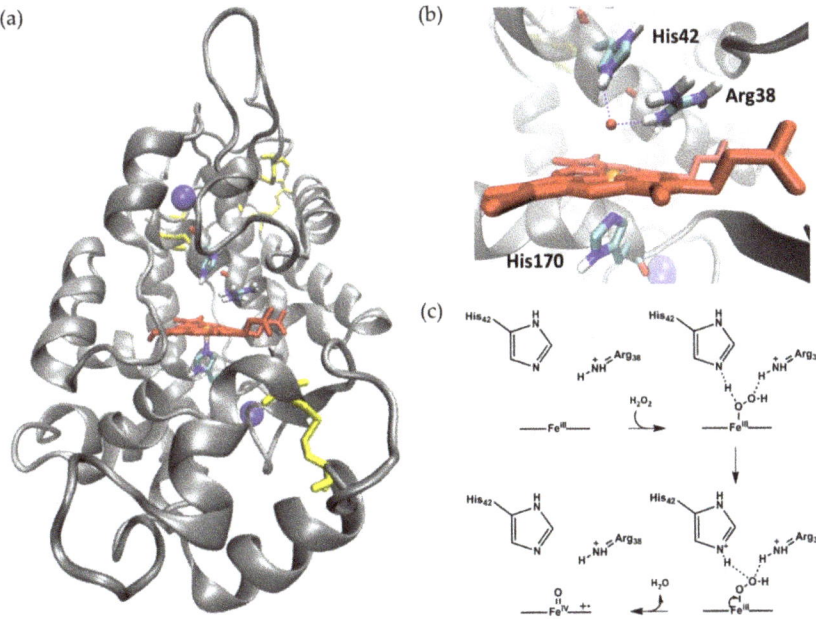

Figure 1. Crystal structure of HRP and mechanism for hydrogen peroxide reduction. (**a**) Overall globular structure of HRP C1A. The polypeptide chain is shown as the grey cartoon, the heme cofactor, disulfide bonds, and selected amino acids as red, yellow, and multi-colored sticks, respectively. The two calcium ions are shown as blue spheres. (**b**) Active site of HRP C1A with the heme cofactor depicted as red sticks with the iron center as orange sphere. The essential amino acids Arg38, His42, and His170 are shown as sticks colored by element. Structures were visualized using PDB 1ATJ [33] and VMD 1.9.3 [37]. From Neumann, 2019 [38]. (**c**) Proposed mechanism for hydrogen peroxide reduction by heme peroxidases and concomitant Compound I formation. Adapted with permission from Rodríguez-López et al., 2001 [34]. Copyright (2001) American Chemical Society.

However, often electrocatalysis is reported to occur at much more negative potentials close to the formal potential of the $Fe^{2+/3+}$ transition. In these cases, the reaction most likely proceeds via a Fenton-like mechanism involving first the reduction of the ferric resting state by the electrode followed by a reaction of the ferrous enzyme with hydrogen peroxide to form a hydroxyl anion and a hydroxy radical (Figure 2a) [39]. Alternatively, the ferrous peroxidase could react with hydrogen peroxide to form water and Compound II which is subsequently reduced by the electrode [22]. In both cases, the reaction is initiated by generation of ferrous heme which usually requires very negative working potentials [23]. This has the disadvantage that many compounds and background reactions, especially oxygen reduction reactions, can interfere with the detection. Therefore, biosensors based on the direct reduction of Compounds I and II are preferred as they can operate at moderate potentials. The mediated approach, on the other hand, employs additional compounds that shuttle electrons between the electrode and the reaction centers of Compounds I and II (Figure 2b). Thus, they can increase the sensitivity for peroxide while the working potential of the biosensor can be tuned according to the reduction potential of the mediator. The reactivity of peroxidases with mediators thereby varies for enzymes from different sources and can be even tuned by enzyme engineering [40]. As many substrates of peroxidases, like phenolic compounds or aromatic amines, are oxidized to redox active products, also their concentration can be determined via their reduction at the working electrode following the scheme in Figure 2b [2].

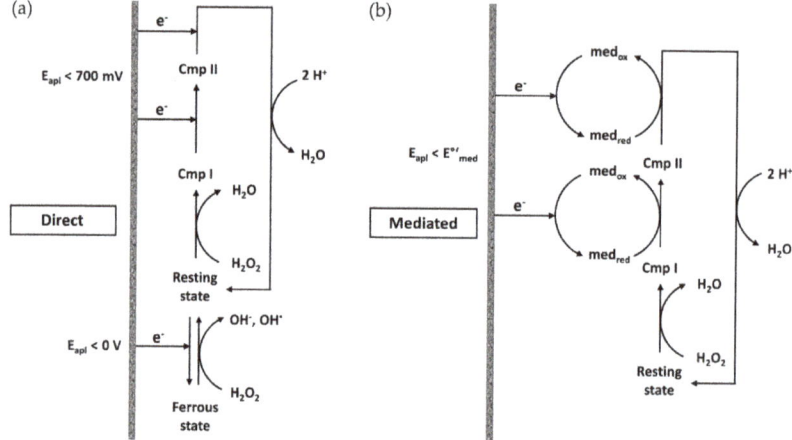

Figure 2. Electrochemical transduction modes of peroxidase-based biosensors. (**a**) In the direct approach, the reaction intermediates Compound I and II are directly reduced at the electrode at high potentials. The Fenton-type reaction route is initiated by conversion of the heme cofactor from the ferric (Fe^{3+}) to the ferrous (Fe^{2+}) state at lower potentials. (**b**) In the mediated approach, an additional compound is used to mediate electron transfer between working electrode and Compounds I and II. Peroxidase substrates can act similar to mediators when their products can be recycled at the electrode.

2.3. Engineering of Heme Peroxidase Mimics

In most common heme peroxidases from plants, the heme cofactor is centered in the middle of the protein matrix and the glycosylation shell can hinder an efficient direct electron transfer to the heme active site, which may lead to a significant decrease in sensitivity or even completely prevent a direct detection of peroxides. In a number of groundbreaking studies, it was reported that the rate of direct heterogeneous electron transfer of HRP on polycrystalline gold could be increased from 1 to up to 400 s^{-1} when the deglycosylated form of the enzyme (dgHRP) was used and engineered surface cysteines and tags were employed for its immobilization [41,42]. Consequently, the sensitivity of dgHRP towards hydrogen peroxide at high potentials was reported to be more than 100 times higher compared to the glycosylated enzyme [43]. These findings demonstrated that a size reduction as well as an oriented surface immobilization of the enzyme significantly influence its electrocatalytic activity as both factors lead to a decrease of the electron transfer distance between the redox site and the electrode.

The heme cofactor itself, the smallest catalytic unit of heme peroxidases, exhibits efficient direct electron transfer on various surfaces and its inherent peroxidatic activity has been exploited for the construction of manifold biosensors. However, compared to HRP, its enzymatic activity is around 1000 times lower due to the lack of amino acids essential for efficient catalysis [44]. The possibility to chemically modify the heme cofactor e.g., by introduction of new functional groups, led to engineering approaches for a variety of heme derivatives. Introduction of pyrrole-, thienyl-, and phenoxy-groups to the porphyrin scaffold, for instance, enabled its polymerization and formation of films with electrocatalytic activity [45–48]. Furthermore, by coupling a rigid linker to the macrocycle, functional groups were positioned in the second coordination sphere of the iron center, as shown in Figure 3a. These so-called Hangman-porphyrins constitute a group of simplified mimics of the active site of heme peroxidases. It was reported that an Fe-Hangman porphyrin bearing an acidic group in the second coordination sphere showed a three orders of magnitude higher activity towards hydrogen peroxide reduction compared to the Fe-porphyrin lacking the hanging group [49]. It was proposed that the proton abstracting/donating properties of the hanging group facilitate the heterolytic cleavage of bound hydrogen peroxide and formation of a reaction intermediate similar to Compound I of heme peroxidases.

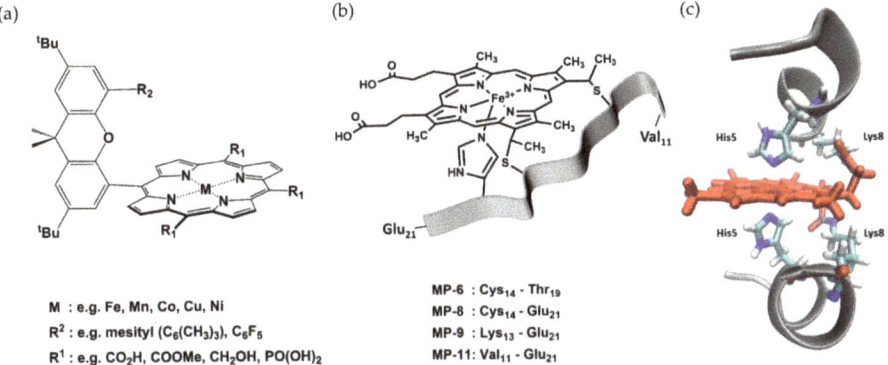

M : e.g. Fe, Mn, Co, Cu, Ni
R^2 : e.g. mesityl ($C_6(CH_3)_3$), C_6F_5
R^1 : e.g. CO_2H, COOMe, CH_2OH, $PO(OH)_2$

MP-6 : Cys_{14} - Thr_{19}
MP-8 : Cys_{14} - Glu_{21}
MP-9 : Lys_{13} - Glu_{21}
MP-11: Val_{11} - Glu_{21}

Figure 3. Structures of selected peroxidase mimics. (**a**) Chemical structure of a Hangman porphyrin with a xanthene linker. Various combinations of meso-substituents (R1), metal centers (M), and hanging groups (R2) have been reported. Examples for each functionality are given below. Taken from [38]. (**b**) Illustration of microperoxidases with the heme in black lines and the polypeptide chain as a grey ribbon. The respective polypeptide segments of the different microperoxidases are noted below. (**c**) Crystal structure of Co-mimochrome IV with polypeptide chains as grey cartoon, the heme cofactor, and selected amino acids as red and multi-colored sticks, respectively. PDB 1PYZ [50] visualized with VMD 1.9.3. [38].

Additionally, various biomolecules were shown to increase the activity of the heme cofactor upon complex formation by preventing its aggregation and providing acid-base functionalities for catalysis. The interaction partners range from G-quadruplexes (DNAzymes) [51] and monoclonal antibodies (Hemoabzymes) [52] to short peptides, including amyloid β peptides involved in Alzheimer's disease [53,54]. Presumably, the most prominent examples of peroxidase mimics though are microperoxidases. These mini-enzymes with peroxidatic activity are prepared via proteolytic digestion of cytochrome c, leading to formation of heme-peptide complexes with a polypeptide chain of typically six to eleven amino acids (MP-6 to MP-11), as shown in Figure 3b [55]. The recently reported fully synthetic approach for the synthesis of microperoxidases further opened up new possibilities for the design of customized heme-peptides incorporating even non-natural groups [56]. Microperoxidases were shown not only to possess a peroxidase-like activity, but also to catalyze dehalogenation reactions [57] and oxygen-transfer reactions similar to cytochrome P450 enzymes [58], making them attractive tools for the construction of biosensors. Nastri et al. also pursued an approach for the rational design of heme-peptide conjugates. By using the β-chain of human deoxyhemoglobin as the template, they designed mimochrome I, a complex composed of two helical peptide chains and a deuteroporphyrin arranged in a helix-heme-helix sandwich structure [59]. Figure 3c shows exemplarily the crystal structure of Co-mimochrome IV, one of the subsequently created mimochrome variants incorporating different metal centers and polypeptide chains [50]. The prototype Fe-mimochrome I was characterized by two symmetrical peptide chains with histidine coordinating the central metal ion on both sides, and thus did not show significant peroxidase activity. Fe-mimochrome VI on the other hand, obtained after iterative optimization, has an asymmetric structure and a vacant coordination site at the iron center [60,61]. The most recent mimochrome was reported to be highly active towards oxidation of classic peroxidase substrates as well as nitration of phenols, with the reaction proceeding via formation of Compound I [62].

3. Peroxidase Reactions on Semiconductors for Electrochemical Biosensing

3.1. Semiconductors as Electrode Materials for Biosensors

The use of inorganic or organic semiconductors in potentiometric sensors like field-effect transistors is well-established. These can be easily miniaturized employing microelectronics or screen-printing technologies, enabling, for example, the fabrication of integrated sensor arrays and even their deposition on flexible surfaces [63,64]. The coupling of semiconductors and illumination is widely used for photovoltaic systems with either the semiconductor itself as light-active component or in combination with an immobilized photosensitizer as a light-harvesting unit [20,65]. Photoswitches constitute another group of possible applications for semiconductors in combination with biomolecules. In enzyme photoswitches, for example, the activation of charge carriers of the semiconductor by internal or external irradiation induces redox changes in an immobilized enzyme and thereby initiates a substrate conversion [21]. The photocurrents resulting from these photo(electro)catalytic processes can then be detected in dependency of the substrate concentration.

The comparatively low conductivity of semiconductors can result in slow direct electron transfers and high background currents. However, their conductivity can be tuned by varying the material composition e.g., by metal ion doping. In addition, a general prerequisite for the construction of an electrochemical biosensor is an appropriate surface that facilitates a productive immobilization and stabilization of the target biomolecules. Analogous to other materials, the surface of semiconductors can be designed according to the requirements given by the properties of the catalyst. Transparent conducting oxides (TCOs) form a group of optically transparent semiconductors including e.g., the widely used titanium dioxide or indium tin oxide (ITO) [66]. They provide a biocompatible matrix for catalyst immobilization that enables the combination of electrochemical and spectroscopic transmission measurements in the visible range of the spectrum. This is explicitly attractive for the analysis of surface-confined heme peroxidases or their mimics as the heme cofactor exhibits a strong absorbance in this range. The absorbance wavelength is sensitive to its redox state as well as its immediate environment, thus allowing for example the identification of reaction intermediates [22]. Additionally, the oxidation products of the peroxidase reaction are often characterized by a high fluorescence or pronounced absorbance in the visible range, thus enabling a facile spectroscopic monitoring of the heme peroxidase activity. Enzymes were adsorbed to planar or porous TCO-substrates as well as to TCO-nanomaterials [18,67,68]. For covalent coupling surfaces can be further modified with different functionalities e.g., by formation of silane- [1,69], phosphonic acid- [69,70], or aryl diazonium salt-based self-assembled monolayers (SAMs) [71,72]. The sol-gel process is a well-established procedure for the preparation of TCOs based on the hydrolysis of metal alkoxide precursors like tetramethoxysilane [73]. Sol-gel materials offer a tunable porosity and biocompatibility and are characterized by a high optical transparency as well as mechanical, chemical, and thermal stability and negligible swelling in various solvents. Moreover, refinements in the experimental conditions with respect to the used solvents, pH, as well as the temperature required for the final drying step paved the way for the addition of biomolecules during the sol or gel formation, thus enabling their encapsulation during fabrication [74].

Organic semiconductors, polymers like polythiophene and polypyrrole (PPy), have also been widely used for the construction of peroxidase-based biosensors. Electropolymerization often serves as a method for a highly controllable deposition of these polymers onto conductive surfaces. Given suitable conditions, the electropolymerization performed in presence of the enzyme leads to its entrapment and thus enables a one-step synthesis of electrochemical biosensors [75]. Besides their physical entrapment into electropolymerized films or the subsequent adsorption, approaches for covalent electropolymerization of biomolecules bearing polymerizable groups have been reported.

These range from DNA fragments [76] to whole enzymes as GOx, where its copolymerization via pyrrole groups led to a higher activity in the film than achieved by entrapment [77].

3.2. Biosensors with Natural Heme Peroxidases

In 1989, Tatsuma et al. reported the first immobilization of HRP on tin oxide coated glass [14]. Amino groups were introduced on the semiconductor's surface via silanization with (3-aminopropyl)triethoxysilane (APTES) followed by covalent coupling of the enzyme. The thus obtained biosensor detected hydrogen peroxide at +150 mV vs. Ag/AgCl with ferrocenecarboxylic acid as mediator. By additional coupling of GOx, the sensor was further successfully employed for the detection of glucose. Afterwards, plenty of publications on electrochemical biosensors employing peroxidases on TCO materials followed. Wu et al. published the first encapsulation of HRP in a sol-gel matrix and reported the preservation of its enzymatic activity inside the glass demonstrated by dibenzothiophene oxidation [78]. Studies for hydrogen peroxide detection via chemiluminescence [79] as well as cholesterol detection via co-entrapment of HRP and cholesterol oxidase [80] followed. Lloyd et al. transferred the sol-gel system to 96-well microplates and demonstrated the protective effect of the enzyme encapsulation at high hydrogen peroxide/HRP ratios during 3,3′,5,5′-tetramethylbenzidine oxidation [81]. However, the spectrum of application of sol-gels in biosensors was quickly expanded from only optical to electrochemical read-out. At first, most approaches focused on the deposition of sol-gels on non-transparent materials like carbon with HRP encapsulated in, adsorbed on or covered with the sol-gel matrix for fabrication of biosensors for hydrogen peroxide either in absence [82–84] or presence [85–87] of a mediator. Chen et al. additionally doped a silica based sol-gel with multi-walled carbon nanotubes, leading to a four times higher sensitivity towards hydrogen peroxide compared to the matrix lacking the nanotubes [88]. Furthermore, TCO nanomaterials like various zinc oxide nanostructures [89,90], antimony oxide bromide nanorods [91], as well as iron and cobalt oxide nanoparticles [92,93] have been employed for HRP immobilization on non-TCO materials. Additionally, titanium dioxide has been used in the form of soluble nanoparticles and nanotubes [94,95], but also as nanotube arrays directly grown on titanium foil via anodic oxidation [96–99]. Kumar et al. reported that the introduction of such a nanoporous oxide layer enabled a direct electrochemical communication between adsorbed HRP and the electrode which was not observed on titanium alone [99]. Also for a mediated approach a significant increase in the hydrogen peroxide reduction by HRP was obtained when the enzyme was immobilized in a graphite composite with mesoporous TiO_2 rather than non-porous TiO_2 [100]. Alternative attempts to improve this communication included the incorporation of gold nanoparticles in between the enzyme and the TCO substrate [101]. However, in the afore-mentioned systems, a direct reduction of hydrogen peroxide was only observed at negative potentials down to -0.6 V vs. Ag/AgCl, thus indicating a Fenton-type reaction at the electrode and not the electrocatalytic behavior expected for HRP.

In 2009, Astuti et al. reported on the direct spectroelectrochemistry of HRP and cytochrome c peroxidase immobilized on mesoporous TiO_2 as well as polylysine modified mesoporous SnO_2 employed for electrocatalytic measurements [102]. Here, the authors were able to confirm the formation of reaction intermediates Compound I and II on the electrode surface by following spectroscopic changes as well as by the high onset potentials of the cathodic reduction of hydrogen peroxide. However, they also reported that HRP showed a much less favorable heterologous electron transfer than cytochrome c peroxidase due to its glycosylation shell, which aside from its insulating and distance-increasing effects, could also hinder a proper access to the pores. The use of engineered HRP-variants could circumvent this problem as has initially been shown for gold electrodes [103]. Our group recently immobilized His_6-tagged dgHRP on a mesoporous TCO electrode support and investigated its spectroelectrochemical as well as electrocatalytic properties [38]. Here, antimony tin oxide (ATO) was employed due to its previously discovered binding affinity for His_6-tags [18,104]. A direct electronic communication of the heme center with the electrode surface was demonstrated by spectroelectrochemical measurements as well as electrocatalytic reduction of hydrogen peroxide in absence of a mediator. The larger potential window of ATO in comparison to SnO_2 enabled the determination of the reduction onset potential. The latter was with +439 mV vs. Ag/AgCl high enough to confirm the formation of Compounds I and II and to enable hydrogen peroxide determination in aerobic conditions without interference of oxygen (Figure 4a). While the linear concentration range was comparable to that of HRP on PLL-modified mesoporous SnO_2, the sensitivity was significantly lower, which can be attributed at least in part to the 400 mV higher working potential of our system where the Fenton-type reaction is avoided (Figure 4b, Table 1).

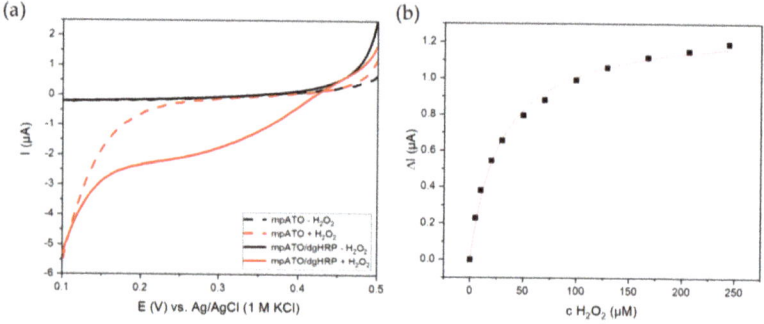

Figure 4. Electrocatalytic reduction of hydrogen peroxide by dgHRP adsorbed on mpATO. (**a**) Linear sweep voltammograms of bare (dashed lines) and dgHRP-modified (solid lines) mpATO before (black) and after (red) addition of 2 mM hydrogen peroxide in air-saturated 100 mM phosphate buffer, pH 7.4. Scan rate 2 mV/s, stirring speed 500 rpm. (**b**) Concentration dependent current increase of a dgHRP modified mpATO upon hydrogen peroxide addition obtained from amperometric measurements at 0.2 V vs. Ag/AgCl. Data were fit to the Michaelis–Menten equation. Adapted from [38].

While the vast majority of peroxidase-based electrochemical biosensors were constructed for the determination of hydrogen peroxide, only a few reports on phenol detection by peroxidases on semiconductors have been published. Rosatto et al. exploited the comparatively low conductivity of silica gels for suppression of the direct reduction of hydrogen peroxide by HRP on a carbon paste electrode and thereby increased the biosensor's sensitivity for various phenolic substrates [4]. Dai et al. on the other hand, coupled the reaction of HRP with that of tyrosinase [105]. Co-immobilization of both enzymes on mesoporous silica yielded a biosensor that exhibited a higher sensitivity for phenol than the respective monoenzyme systems and that was also applied for detection of catechol and p-cresol. Interestingly, the addition of hydrogen peroxide was not required in this system as it was generated in situ via the reduction of dissolved oxygen. Hydrogen peroxide is also produced via oxygen reduction by irradiated TiO_2 and functioned as oxidant for the soybean peroxidase catalyzed oxidation of 2,4,6-trichlorophenol by a TiO_2-soybean peroxidase composite material [27]. Here, the enzyme and the TCO material were entrapped in an UV-cured acrylic polymer matrix coated on glass. The authors reported that in presence of the enzyme less toxic intermediates are formed during degradation than by TiO_2 alone, making the system more attractive for bioremediation applications. Kamada et al. reported an increased UV-tolerance of HRP intercalated into semiconducting titanate layers [106], a circumstance they then exploited for the photoswitched oxidation of Amplex Ultrared initiated by direct oxidation of bound HRP to Compound I upon UV-irradiation of Fe-doped titanate [107]. Subsequent oxidation of the substrate by the reaction intermediate was followed by formation of the fluorescent product. The authors demonstrated a precise control of the enzymatic activity by irradiation without the need for external or in situ produced hydrogen peroxide. Later, the group further modified this approach by immobilizing HRP on a layer of platinum doped hematite on gold or platinum supports, as shown in Figure 5. The more narrow band gap of hematite compared to titanate enabled the initiation of the enzymatic reaction by visible light irradiation [108]. Though the authors did not report on the construction of a biosensor, this approach can avoid photodeactivation of peroxidases, making it an attractive starting point for the development of various applications.

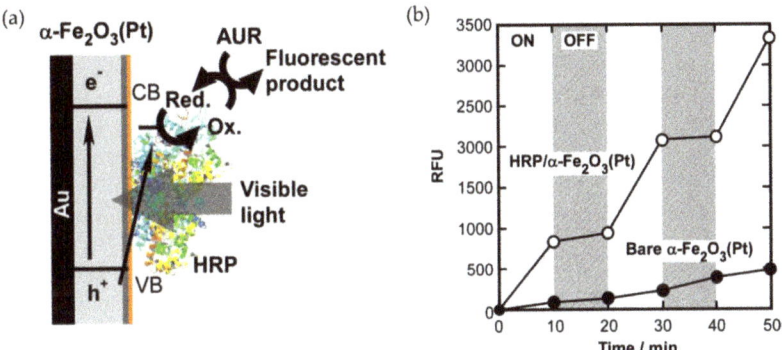

Figure 5. Light-driven conversion of Amplex Ultrared (AUR) by HRP on Pt-doped hematite thin films. (**a**) Schematic illustration of photoinduced enzymatic reaction by HRP adsorbed on Pt-doped α-Fe$_2$O$_3$ thin film. AUR is catalytically oxidized to a fluorescent product by the HRP bound to the film under visible light illumination. (**b**) Photoswitching behaviors of catalytic oxidation of 0.1 mM AUR by bare or HRP-adsorbed α-Fe$_2$O$_3$(Pt) under intermittent blue light irradiation (2 mW/cm^2). Reprinted with permission from Kamada et al., 2012 [108]. Copyright (2012) American Chemical Society.

In 1990, Wollenberger et al., reported for the first time the one-step fabrication of a hydrogen peroxide biosensor based on entrapment of HRP in a PPy matrix during electropolymerization on pyrolytic graphite and platinum [109]. Here as well, a bienzyme approach for glucose determination was established, employing a laminated GOx membrane on top of the electrode. Despite a decrease in sensitivity for H$_2$O$_2$ compared to directly adsorbed HRP, a significant increase in long-term stability was observed. The HRP/PPy system was shortly after transferred to tin oxide by Tatsuma et al. in 1992 who measured with 10 nM a four magnitudes lower detection limit for hydrogen peroxide [110]. In both cases, electrocatalytic reduction was observed at potentials much more positive than the Fe$^{2+/3+}$ transition. However, the question remained, if PPy served as conducting matrix facilitating electron transfer between HRP Compounds I and II and the electrode or if this process was mediated by pyrrole dimers entrapped in the polymer matrix. Both hypotheses were found to be eligible as Tatsuma et al. demonstrated the usability of PPy as conductive material as well as the ability of pyrrole oligomers to function as mediators [110]. Later, the group also extended their system to glucose detection by incorporation of GOx during electropolymerization [111] while Yoshida et al. coupled the HRP reaction with that of glutamate oxidase for fabrication of a mediator-free glutamate sensor [112]. Further optimizations of the HRP/PPy system with respect to hydrogen peroxide detection were reported. For example, coating the HRP/PPy electrode with a film incorporating catalase as "substrate purging catalyst" increased the upper limit for hydrogen peroxide detection by two orders of magnitude [113]. Razola et al. further deposited a thin PPy layer between electrode surface (platinum or glassy carbon) and enzyme layer in order to prevent denaturation of HRP and surface blocking [114]. Indeed, the obtained sensor detected hydrogen peroxide at +150 mV vs. Ag/AgCl in a lower concentration range than the initially reported system [109], but could not reach the sensitivity of HRP/PPy on SnO$_2$ [110]. Intriguingly, Razola et al. excluded the possibility of electron transfer mediation by pyrrole or its oligomers in their system and declared a direct electrocatalytic reduction of hydrogen peroxide by HRP.

As alternative to entrapment of peroxidases in electropolymerized films, various approaches based on either grafting of enzyme/polymer mixtures on surfaces or adsorption of the catalyst on pre-formed polymer films have been published (Figure 6). Both procedures have the advantage that less enzyme amounts are required than for batch polymerization. But, in contrast to pyrrole, for example, polymers used for mixing with enzymes need to be soluble in aqueous solution in order to ensure that the catalyst retains its activity. Again, Tatsuma et al. were among the first to report an electrochemical biosensor based on various peroxidases mixed with the water-soluble polymer poly(3-(3'-thienyl)propanesulfonic acid [115]. The mixture was cast on SnO_2 and electrocatalytic hydrogen peroxide reduction was observed at potentials up to 1 V vs. Ag/AgCl. By removal of thiophene monomers and oligomers after the chemical polymerization, the authors excluded participation of these species as mediators in the electron transfer and concluded that the reaction intermediates of HRP, a microbial peroxidase and lactoperoxidase received electrons directly from the polymer matrix. Moreover, using this hydrophilic polymer enabled stable measurements in the organic solvent acetonitrile [116]. For several systems based on adsorption of HRP on polymer films, aniline was the monomer of choice. In 1997, Yang et al. deposited polyaniline on platinum foil or glassy carbon via electropolymerization and subsequently adsorbed the positively charged enzyme during reduction of the polyaniline film at -0.5 V vs. SCE [117]. The thus constructed sensor detected hydrogen peroxide at moderate potential without the need for a mediator (Table 1). Hua et al. used composites of polyaniline and multiwalled carbon nanotubes for immobilizing HRP on gold and obtained a biosensor detecting hydrogen peroxide with a high sensitivity, though at negative potentials (Table 1) [118]. Bartlett et al. on the other hand,Bartlett et al. exploited the direct electrochemical communication between HRP and polyaniline for the fabrication of an enzyme switch, a so-called microelectrochemical enzyme transistor [119]. Here, HRP was adsorbed to electrodeposited polyaniline on dual carbon microband electrodes. Compounds I and II formed upon reaction of HRP with hydrogen peroxide oxidize the polyaniline matrix leading to a switch from its conducting to an insulating form which is then reversed by the potentiostat. The potential change of the polymer and the drain current served as measuring parameters for the concentration-dependent detection of hydrogen peroxide. Furthermore, cross-conjugated polymer networks from self-assembled nanoparticles were created by electropolymerization serving as conductive surface for HRP immobilization and the thus created biosensors were reported to be highly sensitive [120,121].

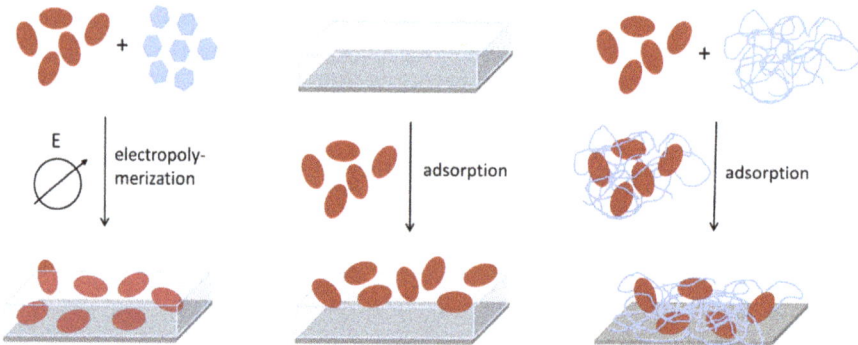

Figure 6. Schematic illustration of different ways of combining biomolecules and organic semiconductors for the fabrication of sensors: entrapment during electropolymerization or co-polymerization, adsorption to pre-polymerized films or adsorption of enzyme/polymer mixtures on the surface.

Some more recently published articles on peroxidase-based electrochemical biosensors reported on the combined use of organic semiconductors and innovative technology from areas like nanotechnology and sensor printing. Li et al., for instance, transferred the HRP/PPy system with an incorporated mediator coating to screen-printed carbon paste electrodes and thereby fabricated a disposable hydrogen peroxide biosensor for low sample volumes of down to 1 µl [122]. Qian et al. reported on the generation of microporous PPy films for HRP immobilization [123]. They adsorbed SiO_2 spheres of defined size on gold surfaces and electrogenerated a PPy film on top, followed by etching of the silica template, thus generating homogeneous PPy films with a pore diameter of 180 nm. Subsequently, a mixture of HRP and chitosan was electrochemically co-deposited in the pores and hydrogen peroxide could be detected in absence of a mediator. Further approaches for preparation of micro- or nanostructured polymer films for HRP-based biosensors involved the generation of oxygen microbubbles during PPy electrogeneration on stainless steel [124], as well as the use of a nanoparticulate polyaniline derivative for electrodeposition [125]. Zhu et al., on the other hand, deposited an HRP/PPy layer on top of a layer of single wall carbon nanotubes and reported a 50 times increase in sensitivity for hydrogen peroxide compared to a system employing graphite powder (Table 1) [126]. The authors used this setup in combination with GOx for sensitive determination of glucose in serum samples. In 2007, Setti and co-workers presented an approach for combining organic electronics with enzyme immobilization by thermal inkjet technology [127]. They printed an organic conductive ink made of poly(ethylenedioxythiophene) (PEDOT) on an ITO-covered glass and subsequently printed an HRP-ink on top. Although the biosensor had to be covered in a cellulose acetate membrane in order to avoid leaching of the layers and a mediator was required for a sensitive hydrogen peroxide detection, this technology is a promising step towards completely printed peroxidase-based biosensors that can be also exploited for the fabrication of sensor arrays on various materials.

3.3. Heme-Peptide Complexes

Several heme proteins including hemoglobin and myoglobin were shown to exhibit a pseudo-peroxidase activity when immobilized on electrode surfaces where usually reduction of the heme iron initiates a Fenton-type reduction of hydrogen peroxide. Respective studies have also been conducted with gold nanoparticles on ITO [128] or TCO nanomaterials like ZrO_2 nanoparticles, TiO_2 nanotubes and nanosheets [129–131]. Only a few reports exist on the electrochemical properties of the initially hemoglobin-derived mimochromes, none of them analyzing the electrocatalysis of peroxidase-like reactions [61,132]. However, in 2014, Vitale et al. employed mesoporous ITO electrodes for immobilization and spectroelectrochemical analysis of Fe^{III}- and Co^{III}-mimochrome VI [133]. The authors observed a direct electrochemical communication of heme and electrode, thus paving the way for a potential application of engineered heme-peptide complexes as catalysts in mediator-free electrochemical biosensors.

Microperoxidases, on the other hand, were extensively used as catalysts in electrochemical biosensors, some of which also involved semiconducting electrode materials. In 1991, Tatsuma et al. seamlessly followed up on their work with HRP and were the first to immobilize a microperoxidase on a TCO for sensing of hydrogen peroxide [134]. Again, APTES-modified SnO_2 served as electrode support onto which MP-9 was covalently attached via glutaraldehyde. The authors reported an almost ten times higher surface coverage of MP-9 compared to HRP and in contrast to the natural enzyme, electrocatalytic reduction of hydrogen peroxide by MP-9 was already observed at potentials of +300 mV vs. Ag/AgCl in absence of a mediator. Both effects were attributed to the significant size reduction of the heme catalyst. On the other hand, the sensitivity of the MP-9 modified electrodes was with 9×10^{-4} A cm^{-2} M^{-1}, almost 50 times lower than that of HRP on SnO_2 (Table 1). The linear range was shifted to higher concentrations (>1 µM) making this sensor attractive for the analysis of different samples. Using the same system, the authors also designed a biosensor for imidazole and derivates based on the inhibiting effect of these compounds on the direct reduction of hydrogen peroxide by MP-9 [135]. Astuti et al. too extended their spectroelectrochemical studies of HRP on poly-lysine modified mesoporous SnO_2 to microperoxidases [136]. The authors reported a 30 times higher surface coverage of the heme-peptide than obtained with the natural enzyme with more than 90% of the molecules being electroactive. Here, the sensitivity of the MP-11 electrode towards hydrogen peroxide reduction was almost four times higher than that of HRP on SnO_2 (Table 1). However, in contrast to HRP and the MP-9 system developed by Tatsuma et al., Astuti and co-workers proposed a Fenton-like reaction mechanism. Later, our group pursued a similar approach when we adsorbed MP-11 in mesoporous ATO electrodes modified with the positively charged binding promotor polydiallyldiammonium chloride [137]. Although we as well observed an almost ten times increase in surface coverage, the sensitivity was seven times lower compared to dgHRP on mesoporous ATO (Table 1), which at least in part can be attributed to formation of six-coordinated low-spin MP-11 as verified by resonance Raman spectra. Still, we observed formation of high-potential reaction intermediates demonstrated via hydrogen peroxide reduction at potentials below +450 mV vs. Ag/AgCl. A comparable behavior was observed for MP-11 in a three-dimensional layer-by-layer assembly of the heme-peptide and gold nanoparticles on APTES-modified ITO [138]. Tian et al. as well combined both materials and fabricated microstructured silica on an ITO surface with gold nanoparticles electrodeposited inside the cavities in order to immobilize MP-11 via covalent coupling to a mercaptobenzoic acid SAM [139]. The SiO_2 cavities enhanced the electron transfer as well as the sensitivity for hydrogen peroxide reduction though in this system, the latter was observed only at negative potentials. Renault and co-workers performed extensive spectroelectrochemical analyses of MP-11 in porous ITO and TiO_2 with respect to the MP-11-catalyzed electrocatalytic reduction of molecular oxygen and used it as a model compound for thorough investigations of electron and charge transfer processes in porous TCO electrodes [67,140–142]. Much less attention on the other hand has been directed to combinations of microperoxidases and organic semiconductors for biosensor construction. Korri-Youssoufi et al. transferred the well-established HRP/PPy-system to MP-8 and reported a calculated limit of detection for hydrogen peroxide of 3.7 nM [143].

Additionally, microperoxidase/semiconductor systems have also been used for the electrochemical detection of other compounds than hydrogen peroxide, including glucose via combination with glucose oxidase [144] or nitric oxide [145]. Recently, Ioannidis et al. presented an approach for the detection of the antimalarial drug artemisinin where MP-11 was adsorbed to a film of surfactant-modified mesoporous SnO_2 on ITO [146]. Ferrous MP-11 catalyzes the cleavage of an organic peroxide within the artemisinin molecule, which leads to heme re-oxidation, thus invoking an enhanced electrocatalytic reduction current proportional to the concentration of the drug in solution (Figure 7).

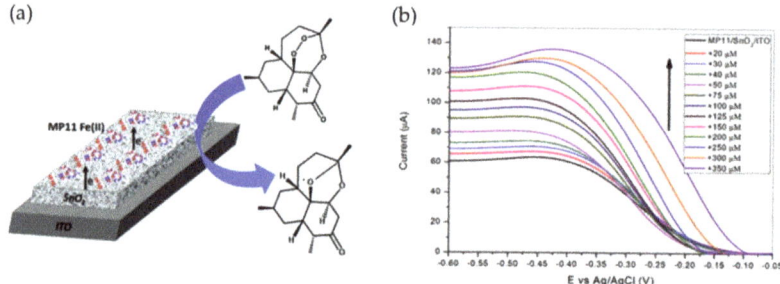

Figure 7. Artemisinin (ART) sensing by MP-11 on mesoporous SnO_2. (a) Schematic representation of MP-11 immobilized on didodecyldimethylammonium bromide (DDAB) modified SnO_2 film electrodes causing the electrocatalytic reduction of ART. (b) DPVs of the sensor after the addition of increasing ART concentrations in 10 mM NaH_2PO_4 pH 7 buffer at a scan rate 0.1 V s^{-1}. Reprinted from Ioannidis et al., 2019 [146]. Published by the Royal Society of Chemistry.

3.4. Hemin and Other Fe-Porphyrins

Due to the aforementioned inherent peroxidase activity of the heme cofactor, hemin and other types of Fe-porphyrins have been widely used as catalysts for the electrochemical determination of the typical peroxidase substrates hydrogen peroxide [44,147,148] or phenolic compounds [149] as well as other substances like superoxide [150], and peroxynitrite [151]. In addition, the MP-11-based artemisinin sensor mentioned in Section 3.3 was initially developed using hemin as the peroxide reducing catalyst adsorbed on TiO_2-modified silica [152]. The here achieved sensitivity for the target was almost two orders of magnitude higher compared to the MP-11-system. Other TCO-based electrochemical sensors with hemin involved its immobilization on SiO$_2$-modified iron oxide particles [153] or mesoporous SnO$_2$ on ITO grafted on a polyethylene terephthalate support, creating a flexible sensor for hydrogen peroxide [154]. The latter approach was further modified by the same group around Topoglidis using Metglas ribbons as support, thus enabling electrochemical as well as magnetoelastic sensing of hydrogen peroxide [155]. Unfortunately, the peroxide determination had to be performed at negative potentials in these systems, although some previous studies on gold and carbon have demonstrated that the direct electrocatalytic reduction of hydrogen peroxide by immobilized hemin can proceed at high potentials similar to heme peroxidases [44,148,156].

Combinations of TCOs with Fe-porphyrins have mainly been used for photocatalytic approaches. Amadelli et al. performed solution studies on the photooxidation of cyclohexane and cyclohexene by an Fe-porphyrin covalently linked to APTES-modified TiO$_2$ [157]. The authors proposed that the ferrous porphyrin formed upon illumination reduces oxygen to reactive oxygen species like superoxide which then are involved in the oxidation of the organic substrates. A few years later, they presented an interesting modification to their system by employing a silane-modified Fe-porphyrin for direct covalent coupling to a TiO$_2$ film on glass [158]. Spectroscopic analyses showed reduction of the iron center which could be reversed by oxygenation. Furthermore, they could demonstrate that the photooxidation of cyclohexane proceeded more efficiently and selectively in presence of the porphyrin than in its absence. The group of Meyer reported multiple studies on the photoinduced reaction of hemin on nanocrystalline TiO$_2$ with organohalide pollutants like chloroform [159–162]. The authors spectroscopically followed the light-induced reduction of the iron center and reported that the ferrous catalyst was stable in the dark for days. In 2015, Gu et al. reported on the fabrication of a photoelectrochemical sensor for hydrogen peroxide employing hemin adsorbed on nanoporous NiO modified ITO [163]. Substrate determination was performed via photocurrent generation at -0.05 V vs. Ag/AgCl and the sensor was also successfully tested in real samples like milk or pharmaceutical eye drops.

The higher stability of Fe-porphyrins in organic solvents and in presence of various conducting salts compared to polypeptide catalysts offers more possibilities for its electrodeposition on a transducer surface. The catalyst can either be incorporated into a polymer matrix or it can be employed as a monomer itself. Peteu et al. entrapped hemin in a PEDOT film electropolymerized on carbon fiber electrodes and investigated its use as sensor for peroxynitrite [164]. Later, the authors employed the direct polymerization of hemin at oxidizing potentials up to 1.3 V vs. Ag/AgCl, presumably proceeding via its inherent vinyl groups, but observed a 50 times lower sensitivity for peroxynitrite [151]. Most studies on directly polymerized heme catalysts though involve Fe-porphyrins that have been chemically modified e.g., with dimethyl ester, phenyl, pyrrole, or thiophene groups with the resulting films being used for the detection of superoxide, nitrite, and nitric oxide [45,46,48,165]. However, studies about the conversion of classical peroxidase substrates by these kinds of electrocatalytic polymers are rare. Schäferling et al. observed a concentration-dependent influence of 2,4,5-trichlorophenol on the cyclic voltammograms of an Fe-porphyrin-substituted bithiophene polymer in dichloromethane but did not pursue this approach for the fabrication of a sensor [47]. Recently, our group analyzed the electrocatalytic reduction of hydrogen peroxide by thienylated Fe-porphyrins co-polymerized with EDOT on glassy carbon electrodes [166]. For the first time, we also implemented a Hangman porphyrin as catalyst and were able to demonstrate its superior catalytic behavior in terms of reduction onset and sensitivity for hydrogen peroxide at high potentials as shown in Figure 8 and Table 1.

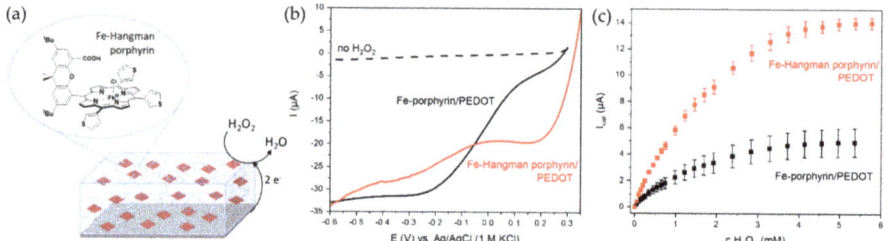

Figure 8. Hydrogen peroxide sensor based on electropolymerized Fe-porphyrins operating at moderate potentials in aqueous solution. (**a**) Schematic illustration of a co-polymer film of thienylated Fe-Hangman porphyrin and poly(ethylenedioxythiophene) (PEDOT) deposited on glassy carbon. (**b**) Linear sweep voltammograms of the porphyrin/PEDOT films before and after addition of 5 mM hydrogen peroxide at a scan rate of 2 mV/s. (**c**) Amperometric response of the porphyrin/PEDOT films to increasing hydrogen peroxide concentrations at 0.2 V vs. Ag/AgCl. Measurements were performed in 100 mM phosphate buffer, pH 7, at a stirring speed of 300 rpm.

Table 1. Performance of selected sensors for hydrogen peroxide based on natural peroxidases or peroxidase mimics immobilized on semiconductors.

Electrode Setup	E_{appl} (V)	Measuring Conditions	LR (µM)	Sensitivity (mA M^{-1} cm^{-2})	Reference
A: HRP					
HRP/APTES/SnO$_2$	0.15	pH 5.9, Med.	0.01–1	50	[14,134]
HRP/PLL/mpSnO$_2$	−0.2	pH 8	1–20	1070	[102]
dgHRP-His$_6$/mpATO	0.2	pH 7.4	5–20	73	[38]
Nafion/HRP/ZnO/ITO	−0.2	pH 7.4	500–9000	7.45	[90]
HRP/Chi-AOB/GC	−0.11	pH 7	1–121	1.44	[91]
HRP-ZnO-chitosan/GC	−0.2 [1]	pH 7, Med.	10–1800	n.d.	[89]
HRP/APTMS/npTiO$_2$	<−0.3 *[1]	pH 7	100–1,500	2864 *	[99]
Nafion/HRP-TiO2/Gr/Au	−0.3 [1]	pH 7, Med.	< 400	1090	[100]
HRP/SnO$_2$/GC	−0.3 [1]	pH 6	10–250	≈215 *	[84]
TiO$_2$/HRP/GC	−0.15 [1]	pH 7, Med.	80–560	488	[87]
HRP in PPy/pyrographite	0.01 [1]	pH 7	50–1750	0.024 *	[109]
HRP in PPy/SnO$_2$	0.15	pH 6.4/7.4	0.01–10	n.d.	[110]
HRP in PPy/SWCNT/Au	−0.1	pH 6.8	0.5–1000	430	[126]
HRP in PPy/SPCP	−0.3	pH 7, Med.	100–2000	33.2	[122]
HRP/PANI/Pt	0.2 [1]	pH 6.8	1–8 *	n.d.	[117]
HRP/PANI/MWCNT/Au	−0.35	pH 7	86–10,000	194.9	[118]
HRP+polythiophene/SnO$_2$	0.4	-	0.05–0.5	n.d.	[115]
HRP/PEDOT-PSS/ITO	−0.1 [1]	pH 6.5, Med.	<1000	0.54	[127]
B: Microperoxidases					
MP-9/APTES/SnO$_2$	0.15	pH 7.4	> 1	0.9	[134]
MP-11/PLL/mpSnO$_2$	−0.2	pH 8	0.05–30	4300	[136]
MP-11/PDADMAC/mpATO	0	pH 8	10–750	10.6	[137]
[MP-11/PEI]$_2$/ITO	0 [1]	pH 6.3	25–125	2.14	[144]
[MP-11/AuNP]$_5$/ITO	0 [1]	pH 7.3	100–1000 *	92	[138]
MP-11/npSiO$_2$-Au/ITO	−0.3	pH 7	2–600	1075 *	[139]
MP-8 in Ppy/GC	−0.1 [1]	pH 7.4	1–9 *	-	[143]
C: Fe-porphyrins					
Fe$_3$O$_4$-SiO$_2$-Hemin/GC	−0.4 [1]	pH 7	1–160	1662 *	[153]
Hemin/SnO$_2$/ITO-PET	−0.3	pH 7	1.5–90	n.d.	[154]
Hemin/SnO$_2$-metglas	−0.4	pH 7	2–90	3191 *	[155]
Hemin/npNiO/ITO	−0.05	pH 7	0.5–500	n.d.	[163]
Fe-porphyrin-PEDOT/GC	0.2	pH 7	50–550	35.2	[166]
Fe-Hangman-PEDOT/GC	0.2	pH 7	50–1000	86.6	[166]

Note: Potentials refer to Ag/AgCl, those marked with [1] refer to SCE. * Values have been estimated by the authors of this review. LR-linear range, Med.–Mediator, n.d. – not determined.

4. Summary and Conclusions

The coupling of enzymes with transducer surfaces plays a key role in the fabrication of biosensors. Expanding the spectrum of both, the electrode material as well as the biocatalyst, also extends the range of opportunities for adapting the system to the specific requirements of its application. Table 1 summarizes parameters of some of the afore-mentioned systems designed for hydrogen peroxide sensing based on the natural enzyme HRP, microperoxidases as well as hemin and other Fe-porphyrins coupled to semiconductor materials.

On average, the systems did not outperform those employing classic electrode materials. On the contrary, most sensors operated at higher concentration ranges with sensitivities considerably lower than e.g., HRP on graphite [167–169] or dgHRP on polycrystalline gold [103]. High charging currents as well as a pronounced direct reduction of hydrogen peroxide by the semiconductor matrix impede the measurements. Furthermore, it must be considered that the three-dimensional surface architecture of many TCO- and polymer-based systems is often not considered for current density calculations, thus resulting in higher sensitivities when the geometrical area is used instead. For several systems employing heme peroxidase mimics, a better sensitivity towards reduction of hydrogen peroxide compared to HRP was obtained as shown in Table 1. However, only a few operate at similarly high potentials, thus truly mimicking the enzymatic reaction via Compounds I and II. Furthermore, the mimics often exhibit higher apparent K_m values, reflecting in part a lower affinity to the substrate, resulting in higher linear ranges of the sensors. Though semiconductor materials might not be a lucrative alternative in terms of sensor performance, they in return offer new possibilities for the fabrication as well as read-out of peroxidase-based sensors. Transparent materials enable more facile spectroscopic analyses of the immobilized catalyst in the visible range of the spectrum, and thus an easier control of its active state. Immobilization on TCOs further enables the photoinduced initiation of the peroxidase reaction cycle either via direct oxidation of the catalyst or via in situ generation of hydrogen peroxide by TCOs as well as a photo-enhanced detection of hydrogen peroxide.

Especially organic semiconductors bear a high potential for novel ways of biosensor construction. Conducting polymers together with enzymes form bioink formulations and various printing techniques enable their deposition in microstructured format on materials that can be transparent, flexible, or magnetoelastic, thus further extending their field of application [170]. Here, the replacement of natural enzymes by engineered nature-derived or fully synthetic mimics also offers new ways for the facile integration of the catalysts which has been already accomplished for various peroxidase mimics. In this context, the de novo design of peroxidase mimics with integrated functional groups using techniques like solid-phase peptide synthesis or complex organic chemistry is especially attractive. This way, catalysts optimized with respect to their immobilization and/or reactivity can be created as has been aspired e.g., for fully synthetic microperoxidases, the four-helix bundle heme protein MP-3 [171] or Hangman porphyrins. All these heme peroxidase mimics have a great potential for future application in sensors which can just be expanded by their combination with semiconductor materials in organic smart devices.

Author Contributions: Conceptualization, B.N. and U.W.; writing—original draft preparation, B.N.; writing—review and editing, U.W.; supervision, U.W.; funding acquisition, U.W. All authors have read and agreed to the published version of the manuscript.

Funding: This research was funded by the Deutsche Forschungsgemeinschaft (DFG, German Research Foundation) under Germany´s Excellence Strategy–EXC 2008–390540038–UniSysCat.

Acknowledgments: UW thanks Vitaly Grigorenko for the HRP expression vector.

Conflicts of Interest: The authors declare no conflict of interest.

References

1. Hermanson, G.T. *Bioconjugate Techniques*, 2nd ed.; Academic Press: Cambridge, MA, USA, 2008; ISBN 9780123705013.
2. Ruzgas, T.; Csöregi, E.; Emnéus, J.; Gorton, L.; Marko-Varga, G. Peroxidase-modified electrodes: Fundamentals and application. *Anal. Chim. Acta* **1996**, *330*, 123–138. [CrossRef]
3. Ortega, F.; Domínguez, E.; Burestedt, E.; Emnéus, J.; Gorton, L.; Marko-Varga, G. Phenol oxidase-based biosensors as selective detection units in column liquid chromatography for the determination of phenolic compounds. *J. Chromatogr. A* **1994**, *675*, 65–78. [CrossRef]
4. Rosatto, S.S.; Kubota, L.T.; De Oliveira Neto, G. Biosensor for phenol based on the direct electron transfer blocking of peroxidase immobilising on silica-titanium. *Anal. Chim. Acta* **1999**, *390*, 65–72. [CrossRef]
5. Watt, B.E.; Proudfoot, A.T.; Vale, J.A. Hydrogen peroxide poisoning. *Toxicol. Rev.* **2004**, *23*, 51–57. [CrossRef] [PubMed]
6. Chen, W.; Cai, S.; Ren, Q.Q.; Wen, W.; Zhao, Y.D. Recent advances in electrochemical sensing for hydrogen peroxide: A review. *Analyst* **2012**, *137*, 49–58. [CrossRef] [PubMed]
7. Ekanayake, E.M.I.M.; Preethichandra, D.M.G.; Kaneto, K. Bi-functional amperometric biosensor for low concentration hydrogen peroxide measurements using polypyrrole immobilizing matrix. *Sens. Actuators B Chem.* **2008**, *132*, 166–171. [CrossRef]
8. Veal, E.A.; Day, A.M.; Morgan, B.A. Hydrogen peroxide sensing and signaling. *Mol. Cell* **2007**, *26*, 1–14. [CrossRef]
9. Halliwell, B.; Clement, M.V.; Long, L.H. Hydrogen peroxide in the human body. *FEBS Lett.* **2000**, *486*, 10–13. [CrossRef]
10. Schuhmann, W. Amperometric enzyme biosensors based on optimised electron-transfer pathways and non-manual immobilisation procedures. *Rev. Mol. Biotechnol.* **2002**, *82*, 425–441. [CrossRef]
11. Presnova, G.V.; Rybcova, M.Y.; Egorov, A.M. Electrochemical biosensors based on horseradish peroxidase. *Russ. J. Gen. Chem.* **2008**, *78*, 2482–2488. [CrossRef]
12. Wollenberger, U.; Spricigo, R.; Leimkühler, S.; Schröder, K. Protein electrodes with direct electrochemical communication. In *Advances in Biochemical Engineering/Biotechnology*; Springer: Berlin/Heidelberg, Germany, 2007; Volume 109, pp. 19–64. ISBN 9783540752004.
13. Foulds, N.C.; Lowe, C.R. Enzyme entrapment in electrically conducting polymers. Immobilisation of glucose oxidase in polypyrrole and its application in amperometric glucose sensors. *J. Chem. Soc. Faraday Trans. 1 Phys. Chem. Condens. Phases* **1986**, *82*, 1259–1264. [CrossRef]
14. Tatsuma, T.; Watanabe, T.; Okawa, Y. Enzyme monolayer- and bilayer-modified tin oxide electrodes for the determination of hydrogen peroxide and glucose. *Anal. Chem.* **1989**, *61*, 2352–2355. [CrossRef]
15. Ramanavicius, A.; Habermuller, K.; Csöregi, E.; Laurinavicius, V.; Schuhmann, W. Polypyrrole-entrapped quinohemoprotein alcohol dehydrogenase. Evidence for direct electron transfer via conducting-polymer chains. *Anal. Chem.* **1999**, *71*, 3581–3586. [CrossRef]
16. Trashin, S.A.; Haltrich, D.; Ludwig, R.; Gorton, L.; Karyakin, A.A. Improvement of direct bioelectrocatalysis by cellobiose dehydrogenase on screen printed graphite electrodes using polyaniline modification. *Bioelectrochemistry* **2009**, *76*, 87–92. [CrossRef] [PubMed]
17. Yoshioka, K.; Kato, D.; Kamata, T.; Niwa, O. Cytochrome P450 modified polycrystalline indium tin oxide film as a drug metabolizing electrochemical biosensor with a simple configuration. *Anal. Chem.* **2013**, *85*, 9996–9999. [CrossRef] [PubMed]
18. Frasca, S.; Molero Milan, A.; Guiet, A.; Goebel, C.; Pérez-Caballero, F.; Stiba, K.; Leimkühler, S.; Fischer, A.; Wollenberger, U. Bioelectrocatalysis at mesoporous antimony doped tin oxide electrodes—Electrochemical characterization and direct enzyme communication. *Electrochim. Acta* **2013**, *110*, 172–180. [CrossRef]
19. Kazmerski, L.L. Photovoltaics: A review of cell and module technologies. *Renew. Sustain. Energy Rev.* **1997**, *1*, 71–170. [CrossRef]
20. Grätzel, M. Photoelectrochemical cells. *Nature* **2001**, *414*, 338–344. [CrossRef]
21. Wang, F.; Liu, X.; Willner, I. Integration of photoswitchable proteins, photosynthetic reaction centers and semiconductor/biomolecule hybrids with electrode supports for optobioelectronic applications. *Adv. Mater.* **2013**, *25*, 349–377. [CrossRef]
22. Dunford, H.B. *Heme Peroxidases*; John Wiley and Sons: New York, NY, USA, 1999; ISBN 0471242446.

23. Battistuzzi, G.; Bellei, M.; Bortolotti, C.A.; Sola, M. Redox properties of heme peroxidases. *Arch. Biochem. Biophys.* **2010**, *500*, 21–36. [CrossRef]
24. Zámocký, M.; Hofbauer, S.; Schaffner, I.; Gasselhuber, B.; Nicolussi, A.; Soudi, M.; Pirker, K.F.; Furtmüller, P.G.; Obinger, C. Independent evolution of four heme peroxidase superfamilies. *Arch. Biochem. Biophys.* **2015**, *574*, 108–119. [CrossRef] [PubMed]
25. Wang, B.; Li, B.; Wang, Z.; Xu, Q.; Wang, Q.; Dong, S. Sol-gel thin-film immobilized soybean peroxidase biosensor for the amperometric determination of hydrogen peroxide in acid medium. *Anal. Chem.* **1999**, *71*, 1935–1939. [CrossRef] [PubMed]
26. Wang, B.; Li, B.; Cheng, G.; Dong, S. Acid-stable amperometric soybean peroxidase biosensor based on a self-gelatinizable grafting copolymer of polyvinyl alcohol and 4-vinylpyridine. *Electroanalysis* **2001**, *13*, 555–558. [CrossRef]
27. Calza, P.; Avetta, P.; Rubulotta, G.; Sangermano, M.; Laurenti, E. TiO2-soybean peroxidase composite materials as a new photocatalytic system. *Chem. Eng. J.* **2014**, *239*, 87–92. [CrossRef]
28. Munteanu, F.D.; Lindgren, A.; Emnéus, J.; Gorton, L.; Ruzgas, T.; Csöregi, E.; Ciucu, A.; Van Huystee, R.B.; Gazaryan, I.G.; Lagrimini, L.M. Bioelectrochemical monitoring of phenols and aromatic amines in flow injection using novel plant peroxidases. *Anal. Chem.* **1998**, *70*, 2596–2600. [CrossRef] [PubMed]
29. Bollella, P.; Medici, L.; Tessema, M.; Poloznikov, A.A.; Hushpulian, D.M.; Tishkov, V.I.; Andreu, R.; Leech, D.; Megersa, N.; Marcaccio, M.; et al. Highly sensitive, stable and selective hydrogen peroxide amperometric biosensors based on peroxidases from different sources wired by Os-polymer: A comparative study. *Solid State Ion.* **2018**, *314*, 178–186. [CrossRef]
30. Gazaryan, I.G.; Gorton, L.; Ruzgas, T.; Csoregi, E.; Schuhmann, W.; Lagrimini, L.M.; Khushpul'yan, D.M.; Tishkov, V.I. Tobacco peroxidase as a new reagent for amperometric biosensors. *J. Anal. Chem.* **2005**, *60*, 629–638. [CrossRef]
31. Lindgren, A.; Ruzgas, T.; Gorton, L.; Csöregi, E.; Bautista Ardila, G.; Sakharov, I.Y.; Gazaryan, I.G. Biosensors based on novel peroxidases with improved properties in direct and mediated electron transfer. *Biosens. Bioelectron.* **2000**, *15*, 491–497. [CrossRef]
32. Veitch, N.C. Horseradish peroxidase: A modern view of a classic enzyme. *Phytochemistry* **2004**, *65*, 249–259. [CrossRef]
33. Gajhede, M.; Schuller, D.J.; Henriksen, A.; Smith, A.T.; Poulos, T.L. Crystal structure of horseradish peroxidase c at 2.15 å resolution. *Nat. Struct. Biol.* **1997**, *4*, 1032–1038. [CrossRef]
34. Rodríguez-López, J.N.; Lowe, D.J.; Hernandez-Ruiz, J.; Hiner, A.N.P.; Garcia-Canovas, F.; Thorneley, R.N.F. Mechanism of Reaction of Hydrogen Peroxide with Horseradish Peroxidase: Identification of Intermediates in the Catalytic Cycle. *J. Am. Chem. Soc.* **2001**, *123*, 11838–11847. [CrossRef] [PubMed]
35. Dunford, B.H. How do enzymes work? Effect of electron circuits on transition state acid dissociation constants. *J. Biol. Inorg. Chem.* **2001**, *6*, 819–822. [CrossRef] [PubMed]
36. Hayashi, Y.; Yamazaki, I. The oxidation-reduction potentials of compound I/compound II and compound II/ferric couples of horseradish peroxidases A2 and C. *J. Biol. Chem.* **1979**, *254*, 9101–9106. [PubMed]
37. Humphrey, W.; Dalke, A.; Schulten, K. VMD: Visual molecular dynamics. *J. Mol. Graph.* **1996**, *14*, 33–38. [CrossRef]
38. Neumann, B. Bioelectrocatalytic Activity of Surface-Confined Heme Catalysts: From Natural Enzymes to Synthetic Analogs. Ph.D. Thesis, University Potsdam, Potsdam, Brandenburg, Germany, 2019.
39. Robinson, S.R.; Dang, T.N.; Dringen, R.; Bishop, G.M. Hemin toxicity: A preventable source of brain damage following hemorrhagic stroke. *Redox Rep.* **2009**, *14*, 228–235. [CrossRef]
40. Sadeghi, S.J.; Gilardi, G.; Cass, A.E.G. Mediated electrochemistry of peroxidases - Effects of variations in protein and mediator structures. *Biosens. Bioelectron.* **1997**, *12*, 1191–1198. [CrossRef]
41. Ferapontova, E.E.; Grigorenko, V.G.; Egorov, A.M.; Börchers, T.; Ruzgas, T.; Gorton, L. Direct electron transfer in the system gold electrode-recombinant horseradish peroxidases. *J. Electroanal. Chem.* **2001**, *509*, 19–26. [CrossRef]
42. Ferapontova, E.; Schmengler, K.; Börchers, T.; Ruzgas, T.; Gorton, L. Effect of cysteine mutations on direct electron transfer of horseradish peroxidase on gold. *Biosens. Bioelectron.* **2002**, *17*, 953–963. [CrossRef]
43. Presnova, G.; Grigorenko, V.; Egorov, A.; Ruzgas, T.; Lindgren, A.; Gorton, L.; Börchers, T. Direct heterogeneous electron transfer of recombinant horseradish peroxidases on gold. *Faraday Discuss.* **2000**, *116*, 281–289. [CrossRef]

44. Lötzbeyer, T.; Schuhmann, W.; Schmidt, H.-L. Minizymes. A new strategy for the development of reagentless amperometric biosensors based on direct electron-transfer processes. *Bioelectrochem. Bioenerg.* **1997**, *42*, 1–6. [CrossRef]
45. Younathan, J.N.; Wood, K.S.; Meyer, T.J. Electrocatalytic reduction of nitrite and nitrosyl by iron(III) protoporphyrin IX dimethyl ester immobilized in an electropolymerized film. *Inorg. Chem.* **1992**, *31*, 3280–3285. [CrossRef]
46. Bedioui, F.; Trevin, S.; Albin, V.; Guadalupe, M.; Villegas, G. Design and characterization of chemically modified electrodes with iron (III) porphyrinic-based polymers: Study of their reactivity toward nitrites and nitric oxide in aqueous solution. *Anal. Chim. Acta* **1997**, *341*, 177–185. [CrossRef]
47. Schäferling, M.; Bäuerle, P. Porphyrin-functionalized oligo- and polythiophenes. *J. Mater. Chem.* **2004**, *14*, 1132–1141. [CrossRef]
48. Yuasa, M.; Oyaizu, K.; Yamaguchi, A.; Ishikawa, M.; Eguchi, K.; Kobayashi, T.; Toyoda, Y.; Tsutsui, S. Electrochemical sensor for superoxide anion radical using polymeric iron porphyrin complexes containing axial 1-methylimidazole ligand as cytochrome c mimics. *Polym. Adv. Technol.* **2005**, *16*, 287–292. [CrossRef]
49. Chng, L.L.; Chang, C.J.; Nocera, D.G. Catalytic O-O activation chemistry mediated by iron hangman porphyrins with a wide range of proton-donating abilities. *Org. Lett.* **2003**, *5*, 2421–2424. [CrossRef]
50. Di Costanzo, L.; Geremia, S.; Randaccio, L.; Nastri, F.; Maglio, O.; Lombardi, A.; Pavone, V. Miniaturized heme proteins: Crystal structure of Co(III)-mimochrome IV. *J. Biol. Inorg. Chem.* **2004**, *9*, 1017–1027. [CrossRef]
51. Kosman, J.; Juskowiak, B. Peroxidase-mimicking DNAzymes for biosensing applications: A review. *Anal. Chim. Acta* **2011**, *707*, 7–17. [CrossRef]
52. Ricoux, R.; Sauriat-Dorizon, H.; Girgenti, E.; Blanchard, D.; Mahy, J.P. Hemoabzymes: Towards new biocatalysts for selective oxidations. *J. Immunol. Methods* **2002**, *269*, 39–57. [CrossRef]
53. Atamna, H.; Boyle, K. Amyloid-beta peptide binds with heme to form a peroxidase: relationship to the cytopathologies of Alzheimer's disease. *Proc. Natl. Acad. Sci. USA* **2006**, *103*, 3381–3386. [CrossRef]
54. Neumann, B.; Yarman, A.; Wollenberger, U.; Scheller, F. Characterization of the enhanced peroxidatic activity of amyloid β peptide-hemin complexes towards neurotransmitters. *Anal. Bioanal. Chem.* **2014**, *406*, 3359–3364. [CrossRef]
55. Marques, H.M. Insights into porphyrin chemistry provided by the microperoxidases, the haempeptides derived from cytochrome c. *Dalton Trans.* **2007**, *9226*, 4371–4385. [CrossRef]
56. Tanabe, J.; Nakano, K.; Hirata, R.; Himeno, T.; Ishimatsu, R.; Imato, T.; Okabe, H.; Matsuda, N. Totally synthetic microperoxidase-11. *R. Soc. Open Sci.* **2018**, *5*, 1–10. [CrossRef] [PubMed]
57. Yarman, A.; Badalyan, A.; Gajovic-Eichelmann, N.; Wollenberger, U.; Scheller, F.W. Enzyme electrode for aromatic compounds exploiting the catalytic activities of microperoxidase-11. *Biosens. Bioelectron.* **2011**, *30*, 320–323. [CrossRef] [PubMed]
58. Osman, A.M.; Koerts, J.; Boersma, M.G.; Boeren, S.; Veeger, C.; Rietjens, I.M. Microperoxidase/H_2O_2-catalyzed aromatic hydroxylation proceeds by a cytochrome-P-450-type oxygen-transfer reaction mechanism. *Eur. J. Biochem.* **1996**, *240*, 232–238. [CrossRef] [PubMed]
59. Nastri, F.; Lombardi, A.; Morelli, G.; Maglio, O.; D'Auria, G.; Pedone, C.; Pavone, V. Hemoprotein models based on a covalent helix-heme-helix sandwich: 1. design, synthesis, and characterization. *Chem. A Eur. J.* **1997**, *3*, 340–349. [CrossRef]
60. Nastri, F.; Lista, L.; Ringhieri, P.; Vitale, R.; Faiella, M.; Andreozzi, C.; Travascio, P.; Maglio, O.; Lombardi, A.; Pavone, V. A heme-peptide metalloenzyme mimetic with natural peroxidase-like activity. *Chemistry* **2011**, *17*, 4444–4453. [CrossRef]
61. Caserta, G.; Chino, M.; Firpo, V.; Zambrano, G.; Leone, L.; D'Alonzo, D.; Nastri, F.; Maglio, O.; Pavone, V.; Lombardi, A. Enhancement of peroxidase activity in artificial mimochrome VI catalysts through rational design. *ChemBioChem* **2018**, *19*, 1823–1826. [CrossRef]
62. Vitale, R.; Lista, L.; Cerrone, C.; Caserta, G.; Chino, M.; Maglio, O.; Nastri, F.; Pavone, V.; Lombardi, A. An artificial heme-enzyme with enhanced catalytic activity: Evolution, functional screening and structural characterization. *Org. Biomol. Chem.* **2015**, *13*, 4859–4868. [CrossRef]
63. Bratov, A.; Abramova, N.; Ipatov, A. Recent trends in potentiometric sensor arrays-A review. *Anal. Chim. Acta* **2010**, *678*, 149–159. [CrossRef]
64. Fortunato, E.; Barquinha, P.; Martins, R. Oxide semiconductor thin-film transistors: A review of recent advances. *Adv. Mater.* **2012**, *24*, 2945–2986. [CrossRef]

65. Kong, F.T.; Dai, S.Y.; Wang, K.J. Review of recent progress in dye-sensitized solar cells. *Adv. Optoelectron.* **2007**, *2007*, 75384. [CrossRef]
66. Stadler, A. Transparent conducting oxides—An up-to-date overview. *Materials* **2012**, *5*, 661–683. [CrossRef] [PubMed]
67. Renault, C.; Andrieux, C.P.; Tucker, R.T.; Brett, M.J.; Balland, V.; Limoges, B. Unraveling the mechanism of catalytic reduction of O2 by microperoxidase-11 adsorbed within a transparent 3D-nanoporous ITO film. *J. Am. Chem. Soc.* **2012**, *134*, 6834–6845. [CrossRef]
68. Bachmeier, A.; Armstrong, F. Solar-driven proton and carbon dioxide reduction to fuels—Lessons from metalloenzymes. *Curr. Opin. Chem. Biol.* **2015**, *25*, 141–151. [CrossRef]
69. Zucca, P.; Sanjust, E. Inorganic materials as supports for covalent enzyme immobilization: Methods and mechanisms. *Molecules* **2014**, *19*, 14139–14194. [CrossRef]
70. Thissen, P.; Valtiner, M.; Grundmeier, G. Stability of phosphonic acid self-assembled monolayers on amorphous and single-crystalline aluminum oxide surfaces in aqueous solution. *Langmuir* **2010**, *26*, 156–164. [CrossRef]
71. Chen, X.; Chockalingam, M.; Liu, G.; Luais, E.; Gui, A.L.; Gooding, J.J. A molecule with dual functionality 4-aminophenylmethylphosphonic acid: A comparison between layers formed on indium tin oxide by in situ generation of an aryl diazonium salt or by self-assembly of the phosphonic acid. *Electroanalysis* **2011**, *23*, 2633–2642. [CrossRef]
72. Pinson, J.; Podvorica, F. Attachment of organic layers to conductive or semiconductive surfaces by reduction of diazonium salts. *Chem. Soc. Rev.* **2005**, *34*, 429–439. [CrossRef]
73. Hench, L.L.; West, J.K. The sol-gel process. *Chem. Rev.* **1990**, *90*, 33–72. [CrossRef]
74. Wang, J. Sol-gel materials for electrochemical biosensors. *Anal. Chim. Acta* **1999**, *399*, 21–27. [CrossRef]
75. Cosnier, S. Biomolecule immobilization on electrode surfaces by entrapment or attachment to electrochemically polymerized films. A review. *Biosens. Bioelectron.* **1999**, *14*, 443–456. [CrossRef]
76. Livache, T.; Roget, A.; Dejean, E.; Barthet, C.; Bidan, G.; Teoule, R. Preparation of a DNA matrix via an electrqchemically directed copolymerization of pyrrole and oligonucleotides bearing a pyrrole group. *Nucleic Acids Res.* **1994**, *22*, 2915–2921. [CrossRef] [PubMed]
77. Wolowacz, S.E.; Yon Hin, B.F.Y.; Lowe, C.R. Covalent electropolymerization of glucose oxidase in polypyrrole. *Anal. Chem.* **1992**, *64*, 1541–1545. [CrossRef]
78. Wu, S.; Lin, J.; Chan, S.I. Oxidation of dibenzothiophene catalyzed by heme-containing enzymes encapsulated in sol-gel glass—A new form of biocatalysts. *Appl. Biochem. Biotechnol.* **1994**, *47*, 11–20. [CrossRef]
79. Li, J.; Wang, K.M.; Yang, X.H.; Xiao, D. Sol-gel horseradish peroxidase biosensor for the chemiluminescent flow determination of hydrogen peroxide. *Anal. Commun.* **1999**, *36*, 195–197. [CrossRef]
80. Kumar, A.; Malhotra, R.; Malhotra, B.D.; Grover, S.K. Co-immobilization of cholesterol oxidase and horseradish peroxidase in a sol-gel film. *Anal. Chim. Acta* **2000**, *414*, 43–50. [CrossRef]
81. Lloyd, C.R.; Eyring, E.M. Protecting heme enzyme peroxidase activity from H2O2 inactivation by sol-gel encapsulation. *Langmuir* **2000**, *16*, 9092–9094. [CrossRef]
82. Xu, X.; Zhao, J.; Jiang, D.; Kong, J.; Liu, B.; Deng, J. TiO2 sol-gel derived amperometric biosensor for H_2O_2 on the electropolymerized phenazine methosulfate modified electrode. *Anal. Bioanal. Chem.* **2002**, *374*, 1261–1266. [CrossRef] [PubMed]
83. Jia, N.; Zhou, Q.; Liu, L.; Yan, M.; Jiang, Z. Direct electrochemistry and electrocatalysis of horseradish peroxidase immobilized in sol-gel-derived tin oxide/gelatin composite films. *J. Electroanal Chem.* **2005**, *580*, 213–221. [CrossRef]
84. Jia, N.Q.; Xu, J.; Sun, M.H.; Jiang, Z.Y. A mediatorless hydrogen peroxide biosensor based on horseradish peroxidase immobilized in tin oxide sol-gel film. *Anal. Lett.* **2005**, *38*, 1237–1248. [CrossRef]
85. Lia, J.; Tana, S.N.; Geb, H. Silica sol-gel immobilized amperometric biosensor for hydrogen peroxide. *Anal. Chim. Acta* **1996**, *335*, 137–145. [CrossRef]
86. Li, J.; Tan, S.N.; Oh, J.T. Silica sol-gel immobilized amperometric enzyme electrode for peroxide determination in the organic phase. *J. Electroanal. Chem.* **1998**, *448*, 69–77. [CrossRef]
87. Yu, J.; Ju, H. Preparation of porous titania sol-gel matrix for immobilization of horseradish peroxidase by a vapor deposition method. *Anal. Chem.* **2002**, *74*, 3579–3583. [CrossRef] [PubMed]

88. Chen, H.; Dong, S. Direct electrochemistry and electrocatalysis of horseradish peroxidase immobilized in sol-gel-derived ceramic-carbon nanotube nanocomposite film. *Biosens. Bioelectron.* **2007**, *22*, 1811–1815. [CrossRef] [PubMed]
89. Liu, Y.L.; Yang, Y.H.; Yang, H.F.; Liu, Z.M.; Shen, G.L.; Yu, R.Q. Nanosized flower-like ZnO synthesized by a simple hydrothermal method and applied as matrix for horseradish peroxidase immobilization for electro-biosensing. *J. Inorg. Biochem.* **2005**, *99*, 2046–2053. [CrossRef]
90. Zhang, W.; Guo, C.; Chang, Y.; Wu, F.; Ding, S. Immobilization of horseradish peroxidase on zinc oxide nanorods grown directly on electrodes for hydrogen peroxide sensing. *Monatshefte für Chemie Chem. Mon.* **2014**, *145*, 107–112. [CrossRef]
91. Lu, X.; Wen, Z.; Li, J. Hydroxyl-containing antimony oxide bromide nanorods combined with chitosan for biosensors. *Biomaterials* **2006**, *27*, 5740–5747. [CrossRef]
92. Gong, J.-M.; Lin, X.-Q. Direct electrochemistry of horseradish peroxidase embedded in nano-Fe_3O_4 matrix on paraffin impregnated graphite electrode and its electrochemical catalysis for H2O2. *Chin. J. Chem.* **2010**, *21*, 761–766. [CrossRef]
93. Chen, W.; Weng, W.; Yin, C.; Niu, X.; Li, G.; Xie, H.; Liu, J.; Sun, W. Fabrication of an electrochemical biosensor based on Nafion/horseradish peroxidase/Co_3O_4 NP/CILE and its electrocatalysis. *Int. J. Electrochem. Sci.* **2018**, *13*, 4741–4752.
94. Zhang, Y.; He, P.; Hu, N. Horseradish peroxidase immobilized in TiO2 nanoparticle films on pyrolytic graphite electrodes: Direct electrochemistry and bioelectrocatalysis. *Electrochim. Acta* **2004**, *49*, 1981–1988. [CrossRef]
95. Wu, F.; Xu, J.; Tian, Y.; Hu, Z.; Wang, L.; Xian, Y.; Jin, L. Direct electrochemistry of horseradish peroxidase on TiO2 nanotube arrays via seeded-growth synthesis. *Biosens. Bioelectron.* **2008**, *24*, 198–203. [CrossRef]
96. Liu, S.; Chen, A. Coadsorption of horseradish peroxidase with thionine on TiO_2 nanotubes for biosensing. *Langmuir* **2005**, *21*, 8409–8413. [CrossRef]
97. Xiao, P.; Garcia, B.B.; Guo, Q.; Liu, D.; Cao, G. TiO2 nanotube arrays fabricated by anodization in different electrolytes for biosensing. *Electrochem. commun.* **2007**, *9*, 2441–2447. [CrossRef]
98. Kafi, A.K.M.; Wu, G.; Chen, A. A novel hydrogen peroxide biosensor based on the immobilization of horseradish peroxidase onto au-modified titanium dioxide nanotube arrays. *Biosens. Bioelectron.* **2008**, *24*, 566–571. [CrossRef] [PubMed]
99. Deva Kumar, E.T.; Ganesh, V. Immobilization of horseradish peroxidase enzyme on nanoporous titanium dioxide electrodes and its structural and electrochemical characterizations. *Appl. Biochem. Biotechnol.* **2014**, *174*, 1043–1058. [CrossRef] [PubMed]
100. Rahemi, V.; Trashin, S.; Meynen, V.; De Wael, K. An adhesive conducting electrode material based on commercial mesoporous titanium dioxide as a support for Horseradish peroxidase for bioelectrochemical applications. *Talanta* **2016**, *146*, 689–693. [CrossRef] [PubMed]
101. Wang, L.; Wang, E. A novel hydrogen peroxide sensor based on horseradish peroxidase immobilized on colloidal Au modified ITO electrode. *Electrochem. Commun.* **2004**, *6*, 225–229. [CrossRef]
102. Astuti, Y.; Topoglidis, E.; Cass, A.G.; Durrant, J.R. Direct spectroelectrochemistry of peroxidases immobilised on mesoporous metal oxide electrodes: Towards reagentless hydrogen peroxide sensing. *Anal. Chim. Acta* **2009**, *648*, 2–6. [CrossRef] [PubMed]
103. Ferapontova, E.E.; Grigorenko, V.G.; Egorov, A.M.; Börchers, T.; Ruzgas, T.; Gorton, L. Mediatorless biosensor for H2O2 based on recombinant forms of horseradish peroxidase directly adsorbed on polycrystalline gold. *Biosens. Bioelectron.* **2001**, *16*, 147–157. [CrossRef]
104. Jetzschmann, K.J.; Yarman, A.; Rustam, L.; Kielb, P.; Urlacher, V.B.; Fischer, A.; Weidinger, I.M.; Wollenberger, U.; Scheller, F.W. Molecular LEGO by domain-imprinting of cytochrome P450 BM3. *Colloids Surf. B Biointerfaces* **2018**, *164*, 240–246. [CrossRef]
105. Dai, Z.; Xu, X.; Wu, L.; Ju, H. Detection of trace phenol based on mesoporous silica derived tyrosinase-peroxidase biosensor. *Electroanalysis* **2005**, *17*, 1571–1577. [CrossRef]
106. Kamada, K.; Tsukahara, S.; Soh, N. Enhanced ultraviolet light tolerance of peroxidase intercalated into titanate layers. *J. Phys. Chem. C* **2011**, *115*, 13232–13235. [CrossRef]
107. Kamada, K.; Nakamura, T.; Tsukahara, S. Photoswitching of enzyme activity of horseradish peroxidase intercalated into semiconducting layers. *Chem. Mater.* **2011**, *23*, 2968–2972. [CrossRef]

108. Kamada, K.; Moriyasu, A.; Soh, N. Visible-light-driven enzymatic reaction of peroxidase adsorbed on doped hematite thin films. *J. Phys. Chem. C* **2012**, *116*, 20694–20699. [CrossRef]
109. Wollenberger, U.; Bogdanovskaya, V.; Scheller, F.; Bobrin, S.; Tarasevich, M. Enzyme electrodes using bioelectrocatalytic reduction of hydrogen peroxide. *Anal. Lett.* **1990**, *23*, 1795–1808. [CrossRef]
110. Tatsuma, T.; Gondaira, M.; Watanabe, T. Peroxidase-incorporated polypyrrole membrane electrodes. *Anal. Chem.* **1992**, *64*, 1183–1187. [CrossRef]
111. Tatsuma, T.; Watanabe, T.; Watanabe, T. Electrochemical characterization of polypyrrole bienzyme electrodes with glucose oxidase and peroxidase. *J. Electroanal Chem.* **1993**, *356*, 245–253. [CrossRef]
112. Yoshida, S.; Kanno, H.; Watanabe, T. Glutamate sensors carrying glutamate oxidase/peroxidase bienzyme system on tin oxide electrode. *Anal. Sci.* **1995**, *11*, 251–256. [CrossRef]
113. Tatsuma, T.; Watanabe, T.; Tatsuma, S.; Watanabe, T. Substrate-purging enzyme electrodes. peroxidase/catalase electrodes for H_2O_2 with an improved upper sensing limit. *Anal. Chem.* **1994**, *66*, 290–294. [CrossRef]
114. Razola, S.S.; Ruiz, B.L.; Diez, N.M.; Mark, H.B.; Kauffmann, J.M. Hydrogen peroxide sensitive amperometric biosensor based on horseradish peroxidase entrapped in a polypyrrole electrode. *Biosens. Bioelectron.* **2002**, *17*, 921–928. [CrossRef]
115. Tatsuma, T.; Ariyama, K.; Oyama, N. Electron Transfer from a Polythiophene Derivative to Compounds I and II of Peroxidases. *Anal. Chem.* **1995**, *67*, 283–287. [CrossRef]
116. Tatsuma, T.; Ariyama, K.; Oyama, N. Peroxidase-incorporated hydrophilic polythiophene electrode for the determination of hydrogen peroxide in acetonitrile. *Anal. Chim. Acta* **1996**, *318*, 297–301. [CrossRef]
117. Yang, Y.; Mu, S. Bioelectrochemical responses of the polyaniline horseradish peroxidase electrodes. *J. Electroanal. Chem.* **1997**, *432*, 71–78. [CrossRef]
118. Hua, M.Y.; Lin, Y.C.; Tsai, R.Y.; Chen, H.C.; Liu, Y.C. A hydrogen peroxide sensor based on a horseradish peroxidase/polyaniline/ carboxy-functionalized multiwalled carbon nanotube modified gold electrode. *Electrochim. Acta* **2011**, *56*, 9488–9495. [CrossRef]
119. Bartlett, P.N.; Birkin, P.R.; Wang, J.H.; Palmisano, F.; De Benedetto, G. An enzyme switch employing direct electrochemical communication between horseradish peroxidase and a poly(aniline) film. *Anal. Chem.* **1998**, *70*, 3685–3694. [CrossRef] [PubMed]
120. Zhang, R.; Jiang, C.; Fan, X.; Yang, R.; Sun, Y.; Zhang, C. A gold electrode modified with a nanoparticulate film composed of a conducting copolymer for ultrasensitive voltammetric sensing of hydrogen peroxide. *Microchim. Acta* **2018**, *185*, 1–9. [CrossRef] [PubMed]
121. Zhang, R.; Wang, L.; Zhang, C.; Yang, R.; Sun, X.; Song, B.; Wong, C.P.; Xu, Y. An electrochemical biosensor based on conductive colloid particles self-assembled from poly(3-thiophenecarboxylic acid) and chitosan. *J. Appl. Polym. Sci.* **2018**, *135*, 1–7. [CrossRef]
122. Li, G.; Wang, Y.; Xu, H. A hydrogen peroxide sensor prepared by electropolymerization of pyrrole based on screen-printed carbon paste electrodes. *Sensors* **2007**, *7*, 239–250. [CrossRef]
123. Qian, J.; Peng, L.; Wollenberger, U.; Scheller, F.W.; Liu, S. Analytical & three-dimensionally microporous polypyrrole film as an efficient matrix for enzyme immobilization. *Anal. Bioanal. Electrochem.* **2011**, *3*, 233–248.
124. Kathuroju, P.K.; Jampana, N. Growth of polypyrrole-horseradish peroxidase microstructures for H_2O_2 biosensor. *IEEE Trans. Instrum. Meas.* **2012**, *61*, 2339–2345. [CrossRef]
125. Morrin, A.; Ngamna, O.; Killard, A.J.; Moulton, S.E.; Smyth, M.R.; Wallace, G.G. An amperometric enzyme biosensor fabricated from polyailine nanoparticles. *Electroanalysis* **2005**, *17*, 423–430. [CrossRef]
126. Zhu, L.; Yang, R.; Zhai, J.; Tian, C. Bienzymatic glucose biosensor based on co-immobilization of peroxidase and glucose oxidase on a carbon nanotubes electrode. *Biosens. Bioelectron.* **2007**, *23*, 528–535. [CrossRef] [PubMed]
127. Setti, L.; Fraleoni-Morgera, A.; Mencarelli, I.; Filippini, A.; Ballarin, B.; Di Biase, M. An HRP-based amperometric biosensor fabricated by thermal inkjet printing. *Sens. Actuators B Chem.* **2007**, *126*, 252–257. [CrossRef]
128. Zhang, J.; Oyama, M. A hydrogen peroxide sensor based on the peroxidase activity of hemoglobin immobilized on gold nanoparticles-modified ITO electrode. *Electrochim. Acta* **2004**, *50*, 85–90. [CrossRef]
129. Liu, S.; Dai, Z.; Chen, H.; Ju, H. Immobilization of hemoglobin on zirconium dioxide nanoparticles for preparation of a novel hydrogen peroxide biosensor. *Biosens. Bioelectron.* **2004**, *19*, 963–969. [CrossRef]

130. Zheng, W.; Zheng, Y.F.; Jin, K.W.; Wang, N. Direct electrochemistry and electrocatalysis of hemoglobin immobilized in TiO2 nanotube films. *Talanta* **2008**, *74*, 1414–1419. [CrossRef]
131. Zhang, L.; Zhang, Q.; Li, J. Layered titanate nanosheets intercalated with myoglobin for direct electrochemistry. *Adv. Funct. Mater.* **2007**, *17*, 1958–1965. [CrossRef]
132. Ranieri, A.; Monari, S.; Sola, M.; Borsari, M.; Battistuzzi, G.; Ringhieri, P.; Nastri, F.; Pavone, V.; Lombardi, A. Redox and electrocatalytic properties of mimochrome VI, a synthetic heme peptide adsorbed on gold. *Langmuir* **2010**, *26*, 17831–17835. [CrossRef]
133. Vitale, R.; Lista, L.; Lau-Truong, S.; Tucker, R.T.; Brett, M.J.; Limoges, B.; Pavone, V.; Lombardi, A.; Balland, V. Spectroelectrochemistry of Fe(III)- and Co(III)-mimochrome VI artificial enzymes immobilized on mesoporous ITO electrodes. *Chem. Commun.* **2014**, *50*, 1894–1896. [CrossRef]
134. Tatsuma, T.; Watanabe, T. Peroxidase model electrodes: Heme peptide modified electrodes as reagentless sensors for hydrogen peroxide. *Anal. Chem.* **1991**, *63*, 1580–1585. [CrossRef]
135. Tatsuma, T.; Watanabe, T. Peroxidase model electrodes: Sensing of imidazole derivatives with heme peptide-modified electrodes. *Anal. Chem.* **1992**, *64*, 143–147. [CrossRef]
136. Astuti, Y.; Topoglidis, E.; Durrant, J.R. Use of microperoxidase-11 to functionalize tin dioxide electrodes for the optical and electrochemical sensing of hydrogen peroxide. *Anal. Chim. Acta* **2011**, *686*, 126–132. [CrossRef]
137. Neumann, B.; Kielb, P.; Rustam, L.; Fischer, A.; Weidinger, I.M.; Wollenberger, U. Bioelectrocatalytic reduction of hydrogen peroxide by microperoxidase-11 immobilized on mesoporous antimony-doped tin oxide. *ChemElectroChem* **2017**, *4*, 913–919. [CrossRef]
138. Patolsky, F.; Gabriel, T.; Willner, I. Controlled electrocatalysis by microperoxidase-11 and Au-nanoparticle superstructures on conductive supports. *J. Electroanal. Chem.* **1999**, *479*, 69–73. [CrossRef]
139. Tian, S.; Zhou, Q.; Gu, Z.; Gu, X.; Zhao, L.; Li, Y.; Zheng, J. Hydrogen peroxide biosensor based on microperoxidase-11 immobilized in a silica cavity array electrode. *Talanta* **2013**, *107*, 324–331. [CrossRef] [PubMed]
140. Renault, C.; Harris, K.D.; Brett, M.J.; Balland, V.; Limoges, B. Time-resolved UV-visible spectroelectrochemistry using transparent 3D-mesoporous nanocrystalline ITO electrodes. *Chem. Commun.* **2011**, *47*, 1863–1865. [CrossRef] [PubMed]
141. Renault, C.; Balland, V.; Limoges, B.; Costentin, C. Chronoabsorptometry to investigate conduction-band-mediated electron transfer in mesoporous TiO_2 thin films. *J. Phys. Chem. C* **2015**, *119*, 14929–14937. [CrossRef]
142. Renault, C.; Nicole, L.; Sanchez, C.; Costentin, C.; Balland, V.; Limoges, B. Unraveling the charge transfer/electron transport in mesoporous semiconductive TiO_2 films by voltabsorptometry. *Phys. Chem. Chem. Phys.* **2015**, *17*, 10592–10607. [CrossRef] [PubMed]
143. Korri-Youssoufi, H.; Desbenoit, N.; Ricoux, R.; Mahy, J.P.; Lecomte, S. Elaboration of a new hydrogen peroxide biosensor using microperoxidase 8 (MP8) immobilized on a polypyrrole coated electrode. *Mater. Sci. Eng. C* **2008**, *28*, 855–860. [CrossRef]
144. Graça, J.S.; De Oliveira, R.F.; De Moraes, M.L.; Ferreira, M. Amperometric glucose biosensor based on layer-by-layer films of microperoxidase-11 and liposome-encapsulated glucose oxidase. *Bioelectrochemistry* **2014**, *96*, 37–42. [CrossRef] [PubMed]
145. Abdelwahab, A.A.; Koh, W.C.A.; Noh, H.B.; Shim, Y.B. A selective nitric oxide nanocomposite biosensor based on direct electron transfer of microperoxidase: Removal of interferences by co-immobilized enzymes. *Biosens. Bioelectron.* **2010**, *26*, 1080–1086. [CrossRef]
146. Ioannidis, L.A.; Nikolaou, P.; Panagiotopoulos, A.; Vassi, A.; Topoglidis, E. Microperoxidase-11 modified mesoporous SnO2 film electrodes for the detection of antimalarial drug artemisinin. *Anal. Methods* **2019**, *11*, 3117–3125. [CrossRef]
147. Shigehara, K.; Anson, F.C. Electrocatalytic activity of three iron porphyrins in the reductions of dioxygen and hydrogen peroxide at graphite electrodes. *J. Phys. Chem.* **1982**, *86*, 2776–2783. [CrossRef]
148. Brusova, Z.; Magner, E. Kinetics of oxidation of hydrogen peroxide at hemin-modified electrodes in nonaqueous solvents. *Bioelectrochemistry* **2009**, *76*, 63–69. [CrossRef] [PubMed]
149. Yarman, A.; Neumann, B.; Bosserdt, M.; Gajovic-Eichelmann, N.; Scheller, F.W. Peroxide-Dependent Analyte Conversion by the Heme Prosthetic Group, the Heme Peptide "Microperoxidase-11" and Cytochrome c on Chitosan Capped Gold Nanoparticles Modified Electrodes. *Biosensors* **2012**, *2*, 189–204. [CrossRef]

150. Chen, J.; Wollenberger, U.; Lisdat, F.; Ge, B.; Scheller, F.W. Superoxide sensor based on hemin modified electrode. *Sens. Actuators B Chem.* **2000**, *70*, 115–120. [CrossRef]
151. Peteu, S.F.; Bose, T.; Bayachou, M. Polymerized hemin as an electrocatalytic platform for peroxynitrite's oxidation and detection. *Anal. Chim. Acta* **2013**, *780*, 81–88. [CrossRef]
152. Reys, J.R.M.; Lima, P.R.; Cioletti, A.G.; Ribeiro, A.S.; De Abreu, F.C.; Goulart, M.O.F.; Kubota, L.T. An amperometric sensor based on hemin adsorbed on silica gel modified with titanium oxide for electrocatalytic reduction and quantification of artemisinin. *Talanta* **2008**, *77*, 909–914. [CrossRef]
153. Feng, J.-J.; Li, Z.-H.; Li, Y.-F.; Wang, A.-J.; Zhang, P.-P. Electrochemical determination of dioxygen and hydrogen peroxide using Fe_3O_4@SiO_2@hemin microparticles. *Microchim. Acta* **2011**, *176*, 201–208. [CrossRef]
154. Panagiotopoulos, A.; Gkouma, A.; Vassi, A.; Johnson, C.J.; Cass, A.E.G.; Topoglidis, E. Hemin modified SnO_2 films on ITO-PET with enhanced activity for electrochemical sensing. *Electroanalysis* **2018**, *30*, 1956–1964. [CrossRef]
155. Samourgkanidis, G.; Nikolaou, P.; Gkovosdis-Louvaris, A.; Sakellis, E.; Blana, I.M.; Topoglidis, E. Hemin-modified SnO_2/metglas electrodes for the simultaneous electrochemical and magnetoelastic sensing of H2O2. *Coatings* **2018**, *8*, 1–21. [CrossRef]
156. Chen, J.; Zhao, L.; Bai, H.; Shi, G. Electrochemical detection of dioxygen and hydrogen peroxide by hemin immobilized on chemically converted graphene. *J. Electroanal. Chem.* **2011**, *657*, 34–38. [CrossRef]
157. Amadelli, R.; Bregola, M.; Polo, E.; Carassiti, V.; Maldotti, A. Photooxidation of hydrocarbons on porphyrin-modified titanium dioxide powders. *J. Chem. Soc. Chem. Commun.* **1992**, 1355–1357. [CrossRef]
158. Molinari, A.; Amadelli, R.; Antolini, L.; Maldotti, A.; Battioni, P.; Mansuy, D. Phororedox and photocatalytic processes on Fe (III)—Porphyrin surface modified nanocrystalline TiO2. *J. Mol. Catal. A Chem.* **2000**, *158*, 521–531. [CrossRef]
159. Obare, S.O.; Ito, T.; Balfour, M.H.; Meyer, G.J. Ferrous hemin oxidation by organic halides at nanocrystalline TiO2 interfaces. *Nano Lett.* **2003**, *3*, 1151–1153. [CrossRef]
160. Obare, S.O.; Ito, T.; Meyer, G.J. Controlling reduction potentials of semiconductor-supported molecular catalysts for environmental remediation of organohalide pollutants. *Environ. Sci. Technol.* **2005**, *39*, 6266–6272. [CrossRef]
161. Obare, S.O.; Ito, T.; Meyer, G.J. Multi-electron transfer from heme-functionalized nanocrystalline TiO2 to organohalide pollutants. *J. Am. Chem. Soc.* **2006**, *128*, 712–713. [CrossRef]
162. Stromberg, J.R.; Wnuk, J.D.; Pinlac, R.A.F.; Meyer, G.J. Multielectron transfer at heme-functionalized nanocrystalline TiO2: Reductive dechlorination of DDT and CCl 4 forms stable carbene compounds. *Nano Lett.* **2006**, *6*, 1284–1286. [CrossRef]
163. Gu, T.T.; Wu, X.M.; Dong, Y.M.; Wang, G.L. Novel photoelectrochemical hydrogen peroxide sensor based on hemin sensitized nanoporous NiO based photocathode. *J. Electroanal. Chem.* **2015**, *759*, 27–31. [CrossRef]
164. Peteu, S.; Peiris, P.; Gebremichael, E.; Bayachou, M. Nanostructured poly (3,4-ethylenedioxythiophene)-metalloporphyrin films: Improved catalytic detection of peroxynitrite. *Biosens. Bioelectron.* **2010**, *25*, 1914–1921. [CrossRef]
165. Matsuoka, R.; Kobayashi, C.; Nakagawa, A.; Aoyagi, S.; Aikawa, T.; Kondo, T.; Kasai, S.; Yuasa, M. A reactive oxygen/nitrogen species sensor fabricated from an electrode modified with a polymerized iron porphyrin and a polymer electrolyte membrane. *Anal. Sci.* **2017**, *33*, 911–915. [CrossRef] [PubMed]
166. Neumann, B.; Götz, R.; Wrzolek, P.; Scheller, F.W.; Weidinger, I.M.; Schwalbe, M.; Wollenberger, U. Enhancement of the electrocatalytic activity of thienyl-substituted iron porphyrin electropolymers by a hangman effect. *ChemCatChem* **2018**, *10*, 4353–4361. [CrossRef]
167. Kulys, J.; Bilitewski, U.; Schmid, R.D. The kinetics of simultaneous conversion of hydrogen peroxide and aromatic compounds at peroxidase electrodes. *J. Electroanal. Chem. Interfacial Electrochem.* **1991**, *26*, 277–286. [CrossRef]
168. Csöregi, E.; Jönsson-Pettersson, G.; Gorton, L. Mediatorless electrocatalytic reduction of hydrogen peroxide at graphite electrodes chemically modified with peroxidases. *J. Biotechnol.* **1993**, *30*, 315–337. [CrossRef]
169. Domínguez Sánchez, P.; Miranda Ordieres, A.J.; Costa García, A.; Blanco Tuñón, P. Peroxidase–ferrocene modified carbon paste electrode as an amperometric sensor for the hydrogen peroxide assay. *Electroanalysis* **1991**, *3*, 281–285. [CrossRef]

170. Elkington, D.; Wasson, M.; Belcher, W.; Dastoor, P.C.; Zhou, X. Printable organic thin film transistors for glucose detection incorporating inkjet-printing of the enzyme recognition element. *Appl. Phys. Lett.* **2015**, *106*, 263301. [CrossRef]
171. Faiella, M.; Maglio, O.; Nastri, F.; Lombardi, A.; Lista, L.; Hagen, W.R.; Pavone, V. De novo design, synthesis and characterisation of MP3, a new catalytic four-helix bundle hemeprotein. *Chem. A Eur. J.* **2012**, *18*, 15960–15971. [CrossRef] [PubMed]

© 2020 by the authors. Licensee MDPI, Basel, Switzerland. This article is an open access article distributed under the terms and conditions of the Creative Commons Attribution (CC BY) license (http://creativecommons.org/licenses/by/4.0/).

Review

Beyond Sensitive and Selective Electrochemical Biosensors: Towards Continuous, Real-Time, Antibiofouling and Calibration-Free Devices

Susana Campuzano *, María Pedrero, Maria Gamella, Verónica Serafín, Paloma Yáñez-Sedeño and José Manuel Pingarrón *

Departamento de Química Analítica, Facultad de CC. Químicas, Universidad Complutense de Madrid, E-28040 Madrid, Spain; mpedrero@quim.ucm.es (M.P.); mariagam@quim.ucm.es (M.G.); veronicaserafin@quim.ucm.es (V.S.); yseo@quim.ucm.es (P.Y.-S.)
* Correspondence: susanacr@quim.ucm.es (S.C.); pingarro@quim.ucm.es (J.M.P.)

Received: 31 May 2020; Accepted: 13 June 2020; Published: 16 June 2020

Abstract: Nowadays, electrochemical biosensors are reliable analytical tools to determine a broad range of molecular analytes because of their simplicity, affordable cost, and compatibility with multiplexed and point-of-care strategies. There is an increasing demand to improve their sensitivity and selectivity, but also to provide electrochemical biosensors with important attributes such as near real-time and continuous monitoring in complex or denaturing media, or in vivo with minimal intervention to make them even more attractive and suitable for getting into the real world. Modification of biosensors surfaces with antibiofouling reagents, smart coupling with nanomaterials, and the advances experienced by folded-based biosensors have endowed bioelectroanalytical platforms with one or more of such attributes. With this background in mind, this review aims to give an updated and general overview of these technologies as well as to discuss the remarkable achievements arising from the development of electrochemical biosensors free of reagents, washing, or calibration steps, and/or with antifouling properties and the ability to perform continuous, real-time, and even in vivo operation in nearly autonomous way. The challenges to be faced and the next features that these devices may offer to continue impacting in fields closely related with essential aspects of people's safety and health are also commented upon.

Keywords: electrochemical biosensors; real-time; continuous operation; reagentless; reusable; calibration-free; antibiofouling

1. Introduction

The availability of technologies for tracking the levels of specific molecules in real time in food production lines or in the living body would revolutionize various applications involved in aspects of people's life safety and physical health, such as clinical diagnosis, food analysis, or environment monitoring [1,2].

However, other than for glucose, point-of-care molecular testing (POCT) is largely restricted to lateral-flow dipstick tests, a technology that is often hard to adapt to multiplexed detections and quantitative measurements [3]. Even so, this technology is making great strides and some methods do allow analyte quantification [4,5].

Motivated by overcoming these limitations and with the aim of improving the adaptation to the real world, electrochemical biosensors, i.e., those incorporating a biological molecule (enzyme, antibody, oligonucleotide, peptide, etc.) as recognition element, pursue features other than high sensitivity and selectivity. These aimed characteristics imply near real-time and continuous monitoring "at home" of molecular targets directly in complex or denaturing media, even under flowing conditions after minimal intervention. Successful meeting of these challenges requires the development of electrochemical devices with antibiofouling properties that are reagentless, single-step, no-wash, and calibration-free.

The great advances in recent years regarding the modification of electrode surfaces with nanomaterials and antibiofouling reagents as well as the irruption of folding-based biosensors have led to the development of electroanalytical bioplataforms filling many of these attributes. They allow the continuous, real-time, in situ measurement of specific molecules directly in flowing complex food samples [6], biological fluids [3,7–9], or even in the bodies of awake [10,11] and free moving individuals.

With this background in mind, this review aims to give an updated overview of the main technologies and recent advances involving electrochemical biosensors free of reagents, washing, or calibration steps, and/or with antifouling properties useful for performing the determination of relevant molecular targets in untreated complex samples or after minimal pre-treatment both in vitro and in vivo. As far as we know, this is the first review that critically and jointly discusses the achievement of these outstanding attributes in electrochemical biosensing.

2. Continuous, Real-Time Electrochemical Biosensors: Towards Antibiofouling, Reagentless, No-Wash, Single-Step, Reusable, and Calibration-Free Devices

Electrochemical biosensors exhibit distinct advantages compared to other biosensors, such as a lack of the high complexity of the sensor setup and the high cost. They are also robust, easy to miniaturize or multiplex, involve low-cost and portable instrumentation, and provide low detection limits even when small sample volumes are available. Furthermore, they can be used to analyze turbid fluids with optically absorbing or fluorescing compounds.

Electrochemical biosensors have evolved as attractive tools to perform single or multiplexed determinations of molecular targets in a simple, affordable, and decentralized way. Although they are constantly seeking higher sensitivity and specificity, in addition, they search other particularly challenging attributes essential to make the reality of their marketing and implementation in the real world to come true.

Currently, scientists are more and more aware that the single-handed pursuit of the sensitivity and accuracy cannot meet the demands of many in situ or POCT circumstances, especially in the fields of clinical diagnosis, food analysis, and environment monitoring. Increasing attention is focused on simplifying their operation and reducing detection time by developing no-wash electrochemical sensors, which make them more suitable for application in the in situ and POCT circumstances [1].

In addition, the ability to monitor specific molecules in real-time would greatly enhance the understanding of diseases as well as their early detection, monitoring, and treatment, thus helping to achieve the promise of personalized medicine. Moreover, triggering of timely countermeasure actions in the food safety and environmental fields would be expected. However, to achieve these goals, sensors must: (i) provide relevant selectivity, precision, and sensitivity; (ii) operate continuously with no sample preparation, batch processing (such as washing steps), or addition of exogenous reagents; and (iii) be insensitive to biofouling such as the detrimental accumulation of proteins and blood cells on the sensor surface [12].

Furthermore, the development of quantitative single-step and calibration-free biosensors is particularly relevant for sensors deployed in vivo to minimize the variability of their fabrication and baseline drift. The willingness to control test conditions and perform calibrations is not only inconvenient but impossible in these operating conditions [13].

The electrode surface modification with antibiofouling reagents, the rapid growth of nanoscience and nanotechnology, and the great advances experienced by folded-based biosensors and in the detection strategies have imparted upon electrochemical biosensors these highly pursued attributes. The remarkable characteristics achieved by electrochemical biosensors beyond sensitivity and selectivity are discussed below based on representative examples selected from recent literature and summarized in Table 1 to provide an overall picture. They are grouped in the following sections according to the more remarkable attribute exhibited. However, it is worth noting that many of these biosensors have additional features with great practical relevance. The important advances provided by wearable devices for real-time electrochemical biosensing have been extensively reviewed in the last few years [14–19] and are not discussed in this article.

Table 1. Representative electrochemical (bio)sensors exhibiting remarkable sensing attributes beyond sensitivity and selectivity.

Electrode	Sensor Fundamentals	Transduction Technique	Attribute (used Approach)	Additional Features	Molecular Target Tested	L.R./LOD	Sample	Ref.
16× Au electrode arrays prepared by photolithography	Sandwich hybridization assay at arrays modified with SHCP/HDT+MCH	Chrono-amperometry (TMB/H_2O_2)	Antibiofouling (thiolated ternary monolayer)	—	Synthetic target DNA (characteristic region of *E. coli* 16S rRNA)	LOD: 7 pM and 17 pM in spiked undiluted human serum and urine	Raw undiluted human serum and urine	[20]
Au/SPEs	Sandwich hybridization assay at arrays modified with SHCP/HDT+MCH	Chrono-amperometry (TMB/H_2O_2)	Antibiofouling (thiolated ternary monolayer)	—	Synthetic target DNA and *E. coli* 16S rRNA	LOD: 25 pM and 100 pM in spiked undiluted human serum and urine and 16S rRNA *E. coli* corresponding to 3000 CFU mL^{-1} in raw cell lysate samples	Untreated raw serum, urine, and crude bacterial lysate solutions	[21]
AuE	E-AB: Aptamer dually modified with a thiol and a redox reporter + PC-terminated SAM	SWV (MB)	Antibiofouling (PC-terminated SAM)	Continuous operation label-free	Kanamycin, doxorubicin	—	Flowing whole blood, both in vitro and in vivo (sensors placed in the jugular veins of live rats)	[12]
GOx-PB-graphite SPEs	Electrode modified with Eudragit® L100	CV ([Fe(CN)$_6$]$^{4-/3-}$) and Chrono-amperometry (PB/H_2O_2)	Antibiofouling (pH-sensitive transient polymer coating)	Continuous operation	Glucose	—	Raw undiluted blood and saliva	[22]
Edible carbon paste GOx biosensors	Electrodes coated with Eudragit® E PO (pH < 5.0) or Eudragit®L100 (pH > 6.0)	Chrono-amperometry (PB/H_2O_2)	Antibiofouling (pH-sensitive transient polymer coating)	Continuous operation Biocatalytic activity preservation at media with denaturing pH values	Glucose	L.R.: 2–10 mM	GI fluids	[23]

Table 1. Cont.

Electrode	Sensor Fundamentals	Transduction Technique	Attribute (used Approach)	Additional Features	Molecular Target Tested	L.R./LOD	Sample	Ref.
PEDOT-citrate film-modified GCE	Covalent immobilization using EDC/NHS chemistry of a peptide with anchoring, antifouling, and recognizing capabilities onto GCE/PEDOT-citrate	DPV ([Fe(CN)$_6$]$^{4-/3-}$)	Antibiofouling (multifunctional peptide)	—	APN, HepG2 cells	L.R.: 1 ng mL^{-1}–15 µg mL^{-1} (APN) and 50–5 × 10^5 cells mL^{-1} (HepG2 cells) LOD: 0.4 ng mL^{-1} (APN) and 20 cells mL^{-1} (HepG2 cells)	Human urine	[24]
Au disk	E-DNA: DNA probe dually modified with a thiol and a redox reporter + MCH SAM	SWV (MB)	Continuous and real-time operation (Folded-biosensor)	Reagentless and single-step	Melamine	LOD: 150 µM (~19 ppm) in buffered solutions and 20 µM (~2.5 ppm) in whole milk	Flowing undiluted whole milk	[6]
Au	E-DNA: TDNs with two functional DNA/aptamer strands, one of them modified with MB	SWV (MB)	Continuous and real-time operation (Folded-biosensor)	Reagentless and single-step Antibiofouling Reusability	Target DNA, ATP	LOD: 300 fM (target DNA), 5 nM (ATP)	Flowing whole blood	[8]
AuE	E-DNA: nucleic acid "scaffold" attached on one end to an electrode and presenting both a redox reporter and a specific epitope on the other	SWV (MB)	Reagentless and single-step (Folded-biosensor)	—	Three types of HIV-diagnostic antibodies	—	Human serum	[25]

Table 1. Cont.

Electrode	Sensor Fundamentals	Transduction Technique	Attribute (used Approach)	Additional Features	Molecular Target Tested	L.R./LOD	Sample	Ref.
Microfabricated Au onto MECAS chip	E-AB: Aptamer dually modified with a thiol and a redox reporter + MCH SAM	ACV (MB)	Continuous and real-time operation (Folded-biosensor)	Reagentless and single-step Reusability	Cocaine	—	Flowing undiluted blood serum	[9]
100 nm Au layer sputtered on glass slides	E-AB: Hairpin structure aptamer dually modified with a thiol and a redox reporter (MB or AQ) + MCH SAM	SWV (MB, AQ)	Continuous operation (Folded-biosensor)	Antibiofouling Reagentless and single-step	IFN-γ + TNF-α	LOD: 6.35 ng mL^{-1} (IFN-γ), 5.46 ng mL^{-1} (TNF-α)	Integrated into microfluidic devices to dynamically monitoring of cytokine release from immune cells (2.5 h)	[26]
Au wire	E-AB: Aptamer dually modified with a thiol and a redox reporter + MCH SAM	SWV (MB)	Continuous and real-time and in vivo operation (Folded-biosensor)	Reagentless and single-step	Doxorubicin, Kanamycin, Gentamycin, and Tobramycin	—	Bloodstream awake, ambulatory rats	[10]
Au disk, Au-plated SPCEs	E-PB: Peptide dually modified with a thiol and a redox reporter + MCH SAM	ACV, CV (MB)	Real-time operation (Folded-biosensor)	Reagentless and single-step	Pb^{2+}	LOD: 5 μM (ACV)	Diluted tap water, saliva, and urine samples	[27]
Au disk	E-ION: T-rich ssDNA dually modified with thiol and redox reporter + Hg^{2+} + MCH SAM	ACV (MB)	Real-time operation (Folded-biosensor)	Reagentless and single-step Reusable	GSH (displaces Hg^{2+} by chelation)	LOD: 5 nM	50% synthetic human saliva	[28]
AuE	E-AB: Aptamer dually modified with a thiol and a redox reporter + MCH SAM	SWV (MB)	Calibration-free ("dual-frequency")	Continuous and real-time operation Reagentless and single-step	Cocaine, doxorubicin	—	Continuous measurement in flowing, undiluted whole blood	[7]

Table 1. Cont.

Electrode	Sensor Fundamentals	Transduction Technique	Attribute (used Approach)	Additional Features	Molecular Target Tested	L.R/LOD	Sample	Ref.
Au-SPE	E-AB: Aptamer dually modified with a thiol and a redox reporter + MCH	SWV (MB)	Calibration-free ("dual-frequency")	Reagentless and single-step	Phenylalanine	L.R.: 90 nM–7 µM	Whole blood (diluted 1000-fold to match the sensor's dynamic range)	[3]
AuE	E-AB: Aptamer modified with a thiol and two different redox reporters + PC-terminated SAM	SWV (MB and AQ)	Calibration-free, ("dual reporter") and in vivo operation	Continuous operation Antibiofouling Reagentless and single-step	Cocaine, ATP, kanamycin	—	Flowing whole blood, both in vitro and in vivo (sensors placed in the jugular veins of live rats)	[2]

Abbreviations: ACV: alternating current voltammetry; APN: aminopeptidase N; AQ: anthraquinone; ATP: anti-adenosine triphosphate; CFU: colony forming unit; CV: cyclic voltammetry; DPV: differential pulse voltammetry; E-AB: electrochemical aptamer-based; E. coli: Escherichia coli; EDC: carbodiimide; E-DNA: electrochemical DNA-based biosensor; E-ION: electrochemical for ion determination; E-PB: electrochemical peptide-based biosensor; GCE: glassy carbon electrode; GI: gastrointestinal fluids; GOx: glucose oxidase; GSH: glutathione; HDT: 1,6-hexanedithiol; HIV: Human Immunodeficiency Virus; INF-γ: interferon-γ; LOD: limit of detection; L.R.: linear range; MB: methylene blue; MCH: 6-mercapto-1-hexanol; MECAS: Microfluidic Electrochemical Aptamer-based Sensor; NHS: succinimide; PB: Prussian Blue; PC: phosphatidylcholine; PEDOT: poly (3,4-ethylenedioxythiophene); SAM: self-assembled monolayer; SHCP: thiolated capture probe; SPCEs: screen-printed carbon electrodes; SPEs: screen-printed electrodes; SWV: square wave voltammetry; T: thymine; TDNs: tetrahedral DNA nanostructures; TMB: tetramethylbenzidine; TNF-α: tumor necrosis factor-α.

2.1. Electrochemical Biosensors with Antibiofouling Properties

The impressive opportunities and capabilities that electrochemical biosensors offer for the monitoring of a wide variety of molecules in situ in complex or biological fluids over a prolonged period of time are limited by the gradual passivation of the (bio)sensing surface due to the nonspecific accumulation of macromolecules present in such matrices. These biofouling issues reduce the direct contact of the target analyte with the electrode surface, hampering the electron transfer, and may severely affect the sensitivity, reproducibility, stability, and overall reliability of the resulting (bio)sensors [29]. Therefore, the development of biosensors with antibiofouling properties able to keep their performance after direct/prolonged incubation in complex and protein-rich media has encouraged the utmost interest. In order to do this, the sensing interfaces are modified with several kinds of antifouling materials, such as poly(ethylene glycol) (PEG) and oligo(ethyleneglycol) (OEG). Although these are considered as the "gold standard" materials, they show various defects such as their oxidative damage, poor water-solubility, and low protein resistance at high temperature. Peptides have come up as possible alternative antifouling materials in electrochemical affinity biosensors [30–32], providing additional advantages of biocompatibility, tunable and simple structure and synthesis [24].

Among all the strategies currently available to develop antibiofouling surfaces, the modification of electrode substrates with different biomaterials, including monolayers, transient polymeric coatings, or multifunctional peptides, is particularly attractive and promising. These strategies have been recently reviewed [29], and, therefore, this section discusses only remarkable features of a limited number of selected methods.

A wide variety of monolayers that exhibit biofouling properties have been reported in the last decade [12,20,21,33–51].

In this context, binary monolayers involving thiolated nucleic acid capture probes (SHCP) and MCH self-assembled onto gold electrodes display unspecific background contributions, due to incomplete backfilling, and irreproducibility, which is attributed to the presence of surface defects and heterogeneity in the distribution of SHCP strands. This unspecific background negatively affects the performance of binary monolayers in complex biofluids and the storage stability of the resulting modified surfaces due to the displacement of SHCP by MCH [52].

Recent research has shown that a judicious design of thiolated surface chemistry involving binary or ternary mixed monolayers, prepared by co-assembling or sequential assembly (noted by "/" or "+", respectively, in their nomenclature) of the components, or the use of tetrahedral DNA nanostructures (TDNs), has led to electrochemical nucleic acid biosensors with substantially better antibiofouling properties and analytical characteristics as compared to conventional SHCP+MCH binary monolayers. This is the case, for example, of binary layers prepared by bringing SHCP into p-aminothiophenol (p-ATP) monolayers previously subjected to potential cycling at acidic pH [49], or layers prepared by attaching amino-functionalized capture probes (NH_2-CP) to p-mercaptobenzoic acid (p-MBA) SAM-modified electrodes (Figure 1a) [48].

Pioneering work by Dharuman's group described simultaneous control of probe orientation and surface passivation by ternary mixed monolayers prepared by sequential immobilization of thiolated ss-DNA probes, MCH, and mercaptopropionic acid (MPA) as diluents, also achieving higher hybridization and discrimination efficiencies due to the distance among anchored ssDNA probes. Moreover, MPA was demonstrated to be more effective than MCH in reducing unspecific adsorptions, due to the generated hydrogen bonds between MPA and MCH, and by placing the DNA strands perpendicularly to the electrode surface [53].

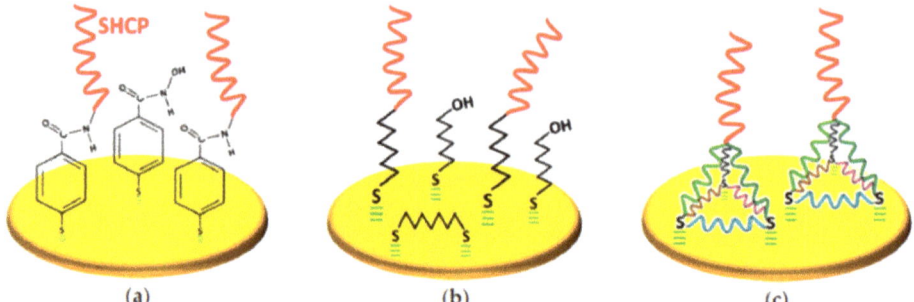

Figure 1. Schematic illustration of three different antifouling thiolated monolayers: (**a**) layers prepared by attaching NH$_2$-CP to p-MBA SAM-modified electrodes, (**b**) ternary SHCP/HDT+MCH layers, and (**c**) layers prepared with TDNs. Reprinted from [29] with permission.

Wang's group were pioneers in preparing ternary DNA SAM-interfaces by co-immobilizing a short (cyclic or linear) dithiol with the SHCP, followed by assembling MCH. The ternary interfaces provided significantly larger signal-to-blank ratios (~100-fold improvement) than the common binary SHCP+MCH SAMs [20,21,36,54]. Results obtained with different lineal dithiols (dithiothreitol, DTT, 1,3-propanedithiol, PDT, 1,6-hexanedithiol, HDT, and 1,9-nonanedithiol, NDT) showed that the SAMs formed with PDT and HDT exhibited better analytical performance due to the preferential lying-down orientation adopted by these linear dithiols. The dithiol lying down configuration minimized nonspecific adsorptions while maintaining SHCP favorable orientation and good permeability of small signaling molecules when compared to the compact surface coverage obtained with the ternary DTT SAM, which resulted in higher signals. The smaller signals observed at interfaces modified with NDT were attributed to the lower amount of attached SHCP [20].

The SHCP/HDT+MCH layers (Figure 1b), assembled onto photolithography-prepared Au electrode arrays [20] or gold screen-printed electrodes [21], exhibited better storage stability than the binary SHCP+MCH layers, and excellent resistance to fouling after 24 h incubation in undiluted human serum and urine. The as-prepared biosensors allowed direct measurement of target nucleic acid concentrations at pM levels in these raw liquid biopsies.

TDNs are assembled on a gold surface through the reproducible immobilization of four especially designed ss-DNA strands, which constitute the six edges of a DNA tetrahedron. Thiol linkers are used at the ends of three component strands, and a linear sequence at the fourth vertex at the top of the bound tetrahedron is left pendant to anchor bio-probes (Figure 1c) [55,56]. This simple, rapid, and high yield one-step process leads to rigid, stable, and reproducible scaffolds adequate for anchoring recognition probes in an upright orientation, spatially segregated, and far away from electrode surfaces, in a solution-phase-like environment ensuring optimal hybridization without requiring a subsequent backfilling step [57]. These tetrahedral DNA nanostructured scaffolds exhibited higher stability and affinity and are less-susceptible to non-specific adsorptions than those fabricated with single point-tethered oligonucleotides [58].

Cell membrane-mimicking phosphatidylcholine (PC)-terminated monolayers are also an attractive option to prepare antibiofouling electrochemical biosensors. PC head-groups mimic the fouling resistance of eukaryotic cellular membranes by strongly binding water to produce a hydration layer that forms a barrier against protein or cell non-specific adsorption. These PC-terminated SAMs can be relatively short, thus supporting rapid electron transfer [12].

The use of biocompatible pH-sensitive transient polymer coatings has been recently exploited by Wang´s group to develop electrochemical biosensors that exhibit antibiofouling properties and good performance after prolonged incubation in complex biological fluids or media with denaturing pH values [22,23]. The method is based on the use of commercial biocompatible polymers sensitive to pH and of controlled dissolution, which are deposited on the (bio)sensor surface temporarily protecting it from undesirable adsorption processes during its prolonged incubation in the biological fluid of interest. The dissolution of these temporary methacrylate-based coatings can be controlled by varying the density and/or thickness of the deposited polymer layer, which allows leaving the "intact" (bio)sensing surface exposed only at the desired moment (Figure 2).

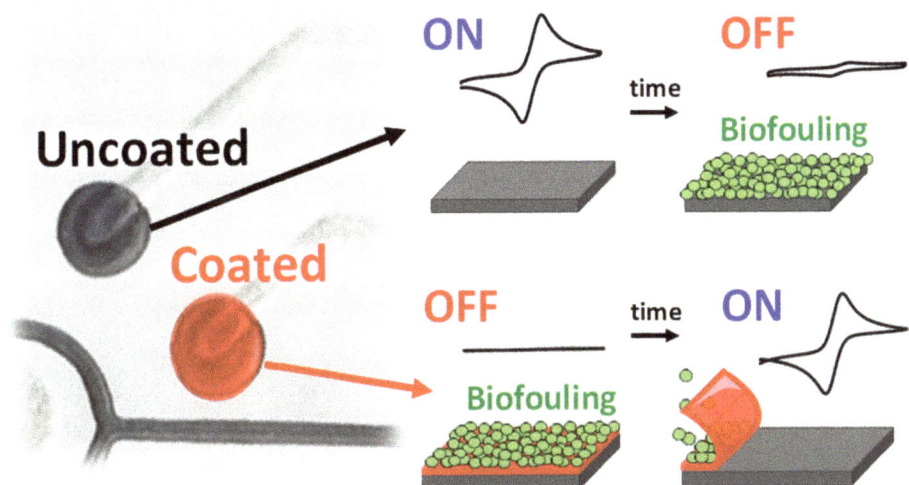

Figure 2. Schematic display of the methodology involving the use of biocompatible pH-sensitive transient polymer coatings to impart electrochemical biosensor antibiofouling properties. Figure drawn based on [22].

This strategy was applied to a four working electrodes-multisensing platform coated with different layers of Eudragit L100 polymer (which dissolves at pH ≥ 6). The sensing platform exhibited excellent operational characteristics in terms of reproducibility and controlled coating dissolution/tunable sequential actuation (0, 2, 4, or 6 h) of the individual electrodes. Monitoring was carried out by cyclic voltammetry with the $[Fe(CN)_6]^{4-/3-}$ redox system. The practical usefulness of this antibiofouling strategy was demonstrated with glucose enzyme biosensors, allowing the sequential enzymatic actuation every 2 h (0, 2, 4, or 6 h) and the direct glucose monitoring in untreated blood and saliva samples over prolonged periods (2 h) without compromising the sensitivity of the biosensors [22]. The excellent antifouling properties imparted by the temporary coatings allowed coated biosensors to maintain 100% of the initial response after 2 h of incubation in these complex biological fluids in comparison with the 65–70% lost observed for the unmodified biosensors.

The unique advantages imparted by pH-responsive protective coatings were also exploited to ensure enzyme activity in media of denaturing pH values to develop edible electrochemical biosensors (based on carbon-paste prepared from olive oil, activated charcoal, and glucose oxidase) with remarkable prolonged resistance to extreme acidic conditions for glucose sensing in gastrointestinal fluids [23]. The active surface of the edible biosensors was modified with commercial polymers such as Eudragit E PO or L-100, which are dissolved at pHs ≤ 5.0 or ≥ 6.0, respectively, for dissolution in gastric fluid (pH between 1.5–5) or intestinal fluid (pH 6.5). The combination of edible olive oil-carbon pastes and transitory coatings preserved the catalytic activity of biomolecules in strongly acidic gastrointestinal fluids and protected the active surface of the biosensor from nonspecific adsorptions, allowing the dissolution/tuning activation of the biosensor selectively in gastric or intestinal fluids at previously fixed times.

Some peptide sequences have shown good antifouling performances. However, complicated chemical reactions or self-assembling on metal surfaces like Au are usually employed for their immobilization at different surfaces, and they need additional reagents to block the peptide-modified interface [24]. To overcome these disadvantages, Song et al. [24] have recently proposed the preparation of an antifouling biosensor for the determination of both aminopeptidase N (APN) and human hepatocellular carcinoma cells (HepG2 cells). The preparation of the biosensor involved a GCE modified with electrodeposited poly (3,4-ethylenedioxythiophene) (PEDOT)-citrate film and the use of a multifunctional peptide with anchoring, antifouling, and recognizing abilities (Figure 3). In the designed three-in-one peptide, one end is a unique anchoring part rich in amine groups to allow its covalent immobilization using carbodiimide/succinimide (EDC/NHS) chemistry on GCE/PEDOT-citrate. The other end is the recognized part for target molecules, while the middle and the anchoring sides are designed to be antifouling. The as-devised biosensor showed, using DPV in the presence of $[Fe(CN)_6]^{4-/3-}$, antifouling properties after incubation in different charged protein solutions and human serum, as well as high sensitivity for the determination of the target analytes with detection limits of 0.4 ng mL^{-1} and 20 cells mL^{-1} for APN and HepG2 cells, respectively, in human urine.

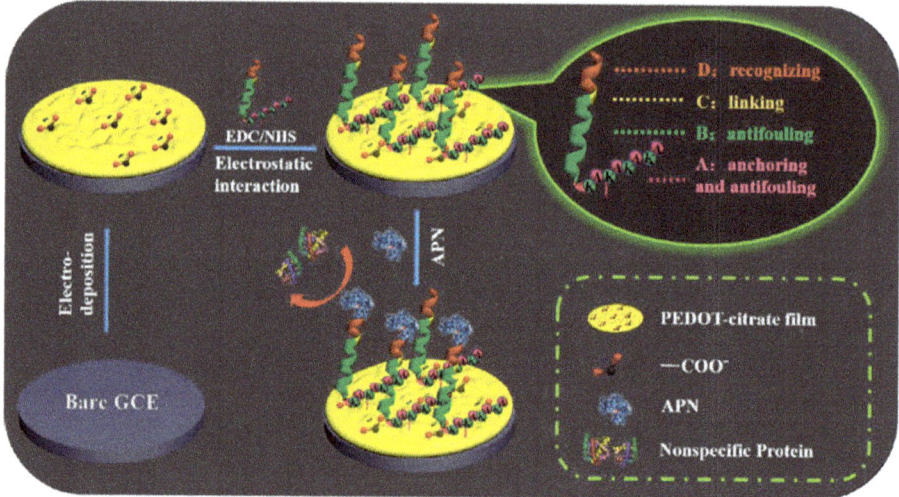

Figure 3. Electrochemical biosensor with antibiofouling properties for the determination of APN and HepG2 cells involving the immobilization of a multifunctional peptide onto a GCE modified with a PEDOT-citrate film. Reprinted from [24] with permission.

It is also important to note at this point that the electrochemical switch-based biosensors have demonstrated to be less prone to fouling because of their transduction mechanism (this issue is discussed in more detail in the next section) [59,60].

2.2. Reagentless, No-Wash, Single-Step, Near Real-Time, and Reusable Electrochemical Biosensors

In the field of reagentless, real-time, and continuous monitoring, folding-based electrochemical biosensors have a fundamental role [59–62]. Recently, we have comprehensively reviewed the main features of this particular type of electrochemical biosensors [60]. Therefore, this section deals only with the relevant aspects within the context of this review article and addresses some methods that have emerged very recently.

Switch-based electrochemical biosensors use biomolecular switches, DNAs, aptamers, or peptides that reversibly change between at least two conformations in response to the specific binding of a wide range of molecular targets. The switches are modified with a linker for their immobilization on an interrogating electrode and at least one redox-active reporter [59,60]. The enzyme-free conformation-linked signal transduction mechanism relies on the target binding induction of a change in the conformation of the probe, which alters the efficiency with which the redox reporter transfers electrons to the electrode. This produces an easily measured signal, using common electrochemical techniques, which makes these biosensors rapid (often reaching effective equilibrium in seconds), drastically simple, and quite insensitive to nonspecific adsorption and response variability [63].

They can be classified as electrochemical DNA (E-DNA) [6,8,25] aptamer (E-AB) [2,3,7,9,10,26], and peptide (E-PB) [27] biosensors, and electrochemical biosensors for ion determination (E-ION) [28]. They have targeted the single or simultaneous determination [10,25,64–66] of a great number of significant analytes (DNAs, polymerase chain reaction amplification products, proteins, hormones, autoantibodies, drugs, toxins, adulterants, explosives, ions, and other biologically relevant molecules), and provide LODs as low as aM-fM for target DNAs [8,67,68], and pM for autoantibodies [69] and proteins [70], with compliance with threshold values and current regulations.

An illustrative example is shown in Figure 4, where the scheme of a recently reported E-DNA sensor for the multidetermination of three HIV-diagnostic antibodies in human serum [25] is displayed. The comparison of the biosensor performance with those of gold standard serological techniques shows that this strategy merged the quantitation and multiplexing of ELISAs with the convenience and speed of dipsticks.

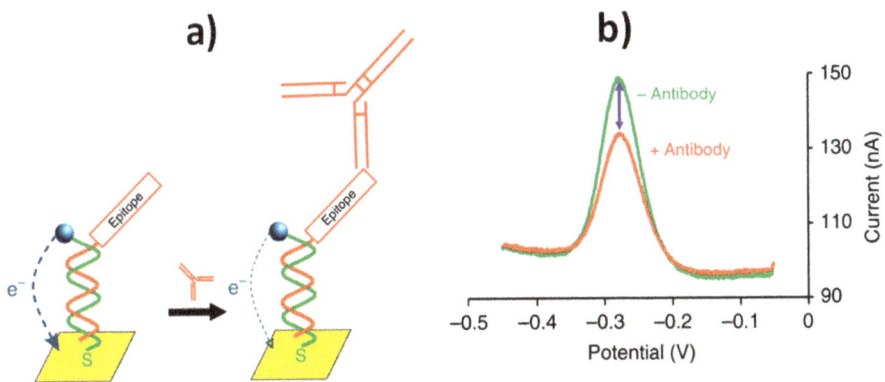

Figure 4. E-DNA sensor developed for the multiplexed determination of HIV-diagnostic antibodies using a nucleic acid "scaffold" anchored on one end to an electrode and presenting both a redox reporter and a specific epitope on the other. When the targeted antibody is not present, the DNA scaffold efficiently transfers electrons to the gold electrode, the electron transfer being reduced due to steric hindrance upon antibody binding (**a**); the square wave voltammetric signals obtained in the absence and in the presence of the targeted antibody (**b**). Reprinted and adapted from [25] with permission.

Other interesting characteristics exhibited by this type of electrochemical biosensors include near real-time response, wash-free, reagentless, and single-step operation, reusability, and autonomous and selective enough read-outs. They were applied in multicomponent and protein-rich samples (blood serum, saliva, urine, seawater, soil suspensions, and foodstuffs such beer and milk). In addition, they have been operated in continuous mode in flowing undiluted samples (milk, blood, and secretions released by immune cells [26]) or even in situ within the living body (Figure 5) [2,3,10–12].

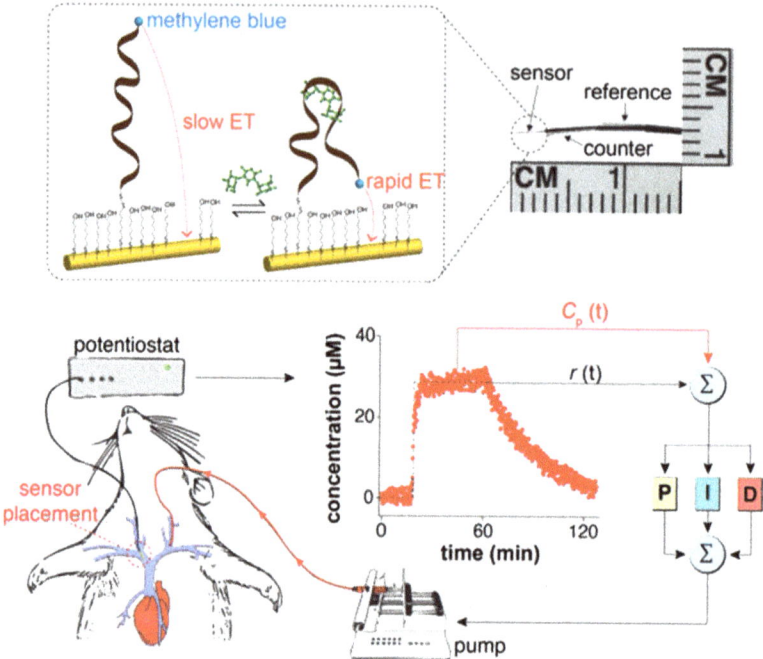

Figure 5. E-AB sensor implanted in the jugular of rats to monitor drug levels in vivo. Reprinted and adapted from [11] with permission.

Other remarkable appealing characteristics and challenging achievements of these biosensors include their stability after storage for more than one week in room-temperature blood serum [6,61,71], their capability to respond to ups and downs in analyte concentration within seconds or minutes in a reversible way even in flowing complex samples without invoking reagents (which may contaminate the sample/product stream, or batch processing) [61,72], and their integration into microfluidic systems (Figure 6) [9,26].

It is important to mention that the great advances in nanotechnology and nanomaterials have allowed the development of other no-wash electrochemical biosensors apart from the folded-based ones [1,73,74]. However, they display some limitations that hamper their real-world application, such as signal drifting due to the change in environmental conditions, and electrode surface passivation and contamination after exposure to samples.

Indeed, although they are little more than a decade old, the rational and relentless research on switch-based electrochemical biosensors, mainly by Plaxco's group, has imparted these biosensors with additional attributes. As it is discussed in the following section, they have proved to be able to operate in flowing highly complex samples with the required accuracy and without the need for calibration.

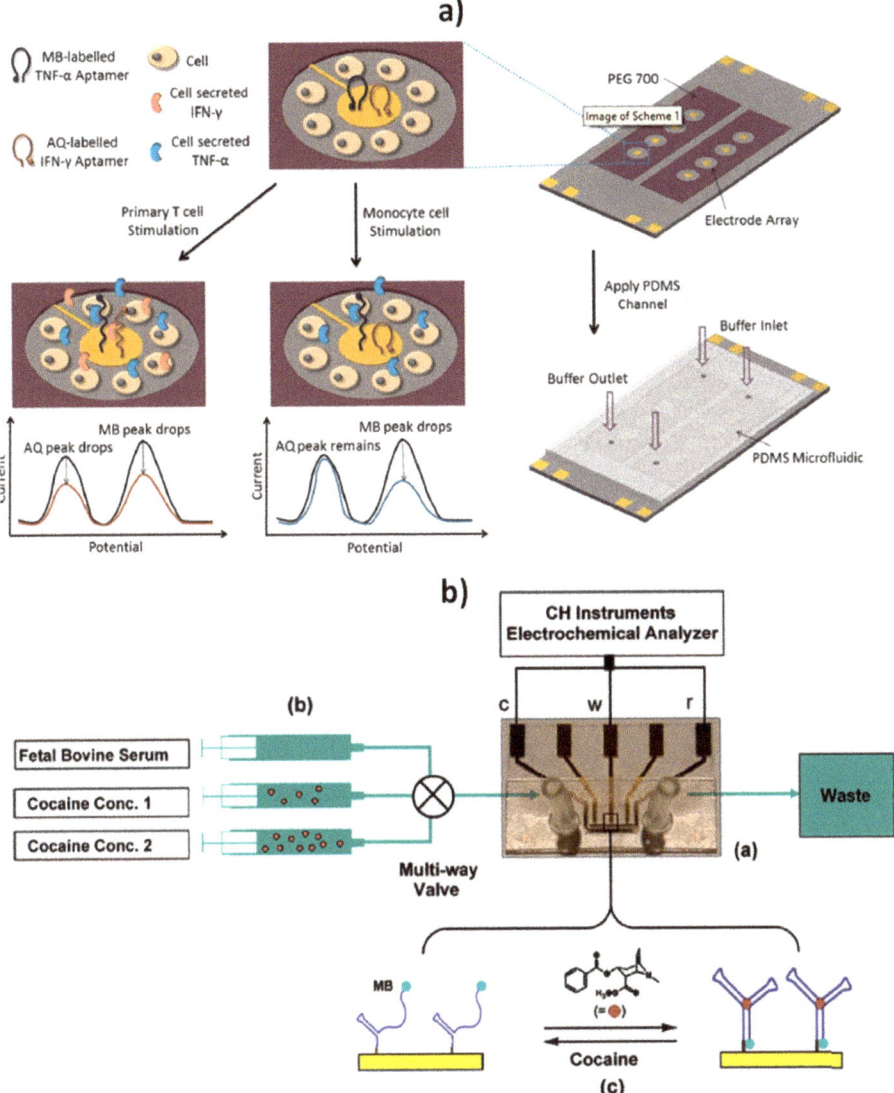

Figure 6. Schematic display of the microfluidic sensing platform using E-AB sensors for the dynamic monitoring of interferon-γ (IFN-γ) and tumor necrosis factor-α (TNF-α) from immune cells (**a**), and the experimental setup for the real time detection of cocaine in continuously flowing, undiluted blood serum using a E-AB sensor constructed onto a microfabricated MECAS chip (**b**). Reprinted and adapted from (**a**) [26] and (**b**) [9] with permission.

2.3. Calibration-Free Biosensors

To achieve acceptable accuracy, electrochemical biosensors require calibration to correct for sensor-to-sensor fabrication variation and sensor drift. This requirement of calibration or recalibration several times a day (commercial continuous glucose monitors) has proven to be one of the significant hurdles limiting the widespread use of biosensors, decreasing convenience and increasing sensor complexity and cost and opportunities for errors, leading in turn to inappropriate clinical action [2,7].

The overcoming of this limitation is even more important and more complex for continuous, in vivo monitoring due to the drift invariably seen in sensors operating in situ within the body over many days. Under these challenging operation conditions, in-factory calibration has proven to be insufficient to assure clinical accuracy and reagent-using on-device autocalibration is impractical [7].

Calibration-free operating E-AB biosensors have been reported using two different ratiometric strategies: (1) the "dual-reporter" approach initially proposed by Ellington and coworkers [75] and (2) the "dual-frequency" operating method developed by Plaxco´s group [7]. Both strategies use as readout unit-less ratiometric values that are largely independent of sensor-to-sensor fabrication variation (attributed in this particular kind of biosensors to variations in electrode surface area and aptamer packing density) and sensor degradation. They overcome the irreproducibility of electrochemical DNA sensors and obviate the need for calibrating each individual sensor without scarifying sensitivity or selectivity. Importantly, both approaches are easily adaptable to nearly any electrochemical system that undergoes a change in its electron transfer kinetics in response to a target binding, and they may be employed in situ in the living body where calibrations are particularly difficult from a practical point of view [7].

The "dual-reporter" strategy uses the ratio of the signal output provided by two different reporters (named as sensing and reference reporters), which are interrogated independently at non-overlapping redox potentials (Figure 7a) [2,75]. Conversely, the "dual-frequency" operating mode uses as output the ratio of peak currents collected at responsive and non- or low responsive square-wave frequencies (Figure 7b) [3,7].

Plaxco's group demonstrated the potential of the "dual-frequency" interrogation strategy to develop calibration-free E-AB biosensors, exhibiting good accuracy and precision for the continuous measurement of two drugs in flowing whole blood over the course of hours, despite the significant drift observed in the absolute current recorded with the sensor [7]. The same group recently exploited this strategy to construct a calibration-free E-AB sensor for determining phenylalanine levels in blood compatible with POCT applications [3]. This biosensor can be deployed on screen-printed electrodes and allows the rapid (<10 min) determination of clinically relevant phenylalanine levels with an accuracy of ±20%, and specificity when challenged to in finger-prick-scale volumes of diluted unprocessed blood, thus offering the possibility to perform at-home measurements as an advance to personalized medicine.

Plaxco et al. also reported an E-AB biosensor combining the "dual reporter" approach [75] with drift-eliminating surface passivation using a phosphatidylcholine monolayer (Figure 7a) [12]. The biosensor was employed to perform calibration-free in vivo measurements of ATP or kanamycin using sensors placed in situ in the jugular veins of live rats over multi-hour measurements [2]. The "sensing" reporter (methylene blue, MB), was placed on the probe distal terminus and, therefore, the produced current was strongly modified by the conformational change associated with target recognition. The second, "reference," reporter (anthraquinone, AQ) was placed in a position that responded differently to the presence of the target and served as an internal reference to correct the sensor-to-sensor variability. The use of a PC-terminated SAM (a phosphatidylcholine alkanethiolate derivative from 2-methacryloyloxyethyl phosphorylcholine) was largely responsible of the good biosensor functioning in vivo, allowing the elimination of the often severe-baseline drift observed in biosensors placed in the living body for long periods. It should be noted that the use of PC-terminated SAMs as backfilling agent demonstrated to be advantageous compared to the hydroxyl-terminated ones (the traditionally employed 6-mercapto-1-hexanol, MCH), lowering the baseline drift from around 70% to a 10% after 12 h in flowing whole blood in vitro or in situ in the veins of live animals. The method achieved precision in the micromolar range over many hours without invoking physical barriers (membranes or fluid sheaths to prevent cells from approaching the sensor surface, that increase the sensor bulk and slow sensor response times) or active drift-correction algorithms that require the collection of additional data at each time point, thereby degrading the time resolution.

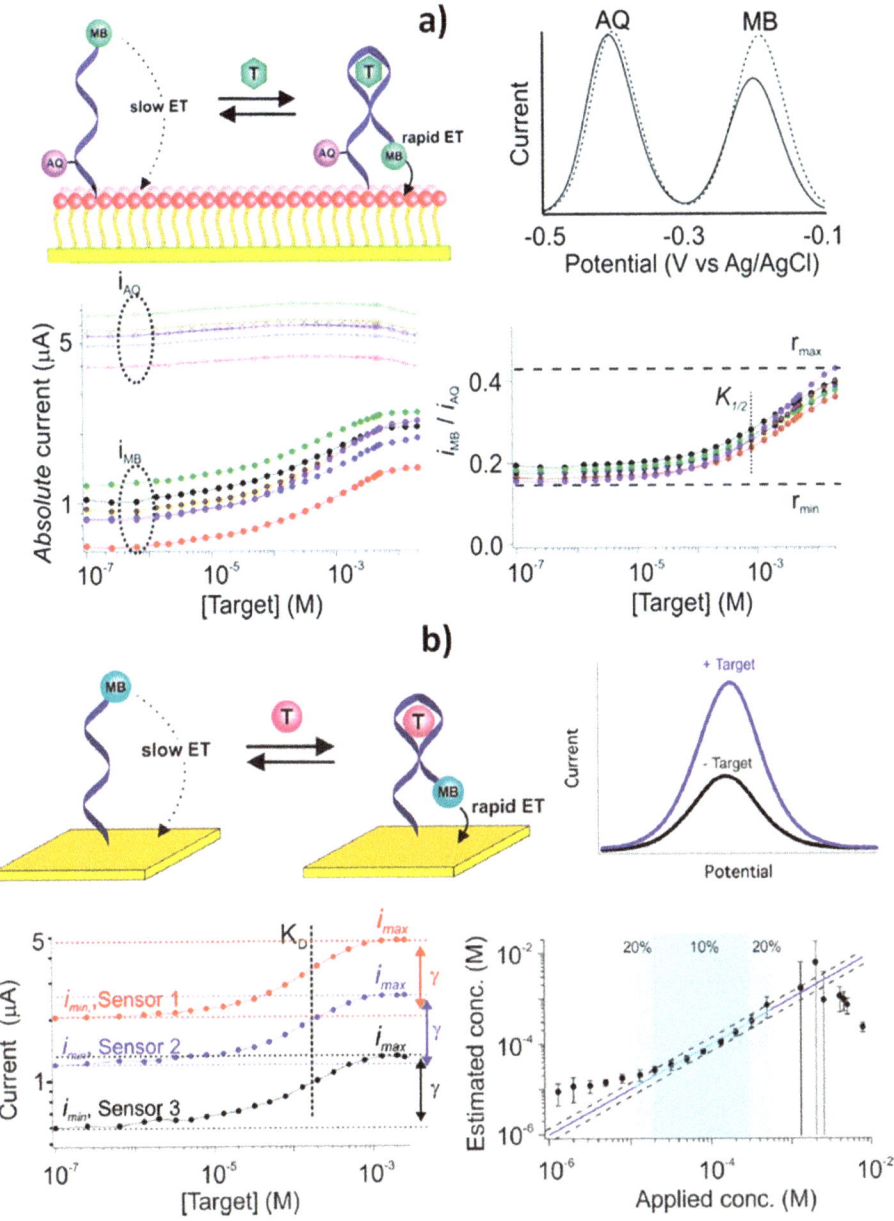

Figure 7. Schematic display of calibration-free E-AB biosensors based on "dual-reporter" (a) and "dual-frequency" (b) strategies. Reprinted and adapted from [2] (a) and [7] (b) with permission.

3. Opportunities, Impact, Challenges, and Future Insights

With the aim to enhance the market adaptation and acceptance of electrochemical biosensors, significant progress has been made recently in the development of bioplatforms able to support continuous, real-time measurements of molecular biomarkers in unprocessed and/or flowing samples involving reagentless and single-step processes that are quantitative, easily multiplexed, and user-friendly for deployment at the POCT.

These advances have come hand in hand with rational modification of electrode surfaces with antibiofouling reagents (monolayers, transient polymeric coatings, or multifunctional peptides) or nanomaterials, and with relentless research into switch-based electrochemical biosensors and their coupling to ratiometric detection techniques.

According to the methods summarized in Table 1, one can deduce that there are electrochemical biosensors:

i. Able to achieve high sensitivity and selectivity when defied punctually in multicomponent and protein-rich samples or continuously in flowing undiluted samples.
ii. Capable of responding to ups and downs in analyte concentration within seconds or minutes in a reversible way and without batch processing or addition of exogenous reagents.
iii. Insensitive to biofouling and stable after storage for more than one week in room-temperature blood serum.
iv. In ingestible formats coupled to transitory commercial polymer coatings with remarkable prolonged resistance to complex media of denaturing pH values such as gastrointestinal fluids.
v. Integrated into microfluidic systems to monitor cell secretions.
vi. Deployed on screen-printed electrodes to provide rapid and accurate determinations when challenged to in finger-prick-scale sample volumes, suitable for application in POCT circumstances.
vii. Able to minimize the variability of the sensors fabrication and baseline drift and provide the required accuracy when operating continuously in vivo without the need for calibration, invoking physical barriers or using active drift-correction algorithms, thus surpassing main limitations of the commercial continuous glucose monitors.

We postulate these outstanding features beyond sensitivity and selectivity allow us to envision molecular detection moving away from complex, multi-step, benchtop assays towards direct, single-step devices (such as the home glucose monitor). These attributes will boost translational progress of electrochemical biosensors beyond the well-controlled laboratory benchtop into areas such as clinical diagnostics and field-portable devices and gain ground on other cumbersome methodologies to provide unprecedented quality control, safety monitoring, and clinical diagnosis even in resource limited areas. These impressive developments are decisive to deploy in vivo determinations, where the tuning of assay conditions is not so much inconvenient as it is impossible.

These biosensors, competitive to gold standard molecular detection techniques as referred to clinical sensitivity and selectivity, are deemed to merge the quantitation and multiplexing of ELISAs with the convenience of use by nonspecialized user and speed of dipsticks, thus significantly improving current molecular detection. The detection limits achieved with ELISAs also come at a significant cost in terms of time, workflow, and equipment overhead that renders them not suitable for application at the POCT.

Compared to lateral-flow assays, which until recently have only given a binary output ("yes/no") and request the user to read the test at an exact time, advanced electrochemical biosensors do not require a fixed readout time window, match them in terms of ease of use and surpass them in terms of the clinically relevant information they provide. Electrochemical bioplatforms provide quantitative and objective readout, thus giving a more precise picture of the tackled problem and allow the possibility to establish reliable clinical cut-off values. Moreover, unlike the dipsticks, which per design and per ease of use cannot integrate more than a few test lines, electrochemical biosensors are easily multiplexed to increase clinical sensitivity. However, there are many reasons why, so far, quantitative multiplexed biosensors are not popular and widespread in the market as lateral flow systems. Firstly, the interest in multiplexed determination to improve the reliability of the diagnostics is relatively recent. Secondly, multiplexing bioplatforms can be designed through two different approaches: using multielectrode arrays where each immunoreagent is attached to each electrode, or by means of barcode configurations involving a unique electrode platform and different electroactive labels with dissimilar electrochemical properties for each analyte. While multi-electrode arrays require complex and independent n-channel electrochemical workstations, the barcode approach makes use of distinct electroactive labels capable of generating appropriate and distinguishable signals at different potentials in a single amperometric or voltammetric scan. Unfortunately, the number of different electroactive labels with dissimilar electrochemical properties is quite limited. Therefore, the development of low-cost, custom designed, and field-portable multiplexed potentiostats and the use of novel nanomaterials and/or electroactive probes producing independent electrochemical signals is required. In addition, multiplexed quantitative bioplatforms should be adopted by the individual user in a clinical environment. This, in turn, requires both the identification and clinical validation of new and reliable signatures of biomarkers for each particular application and what is more, laboratory personnel gains familiarity with the new methodologies, and medical personnel themselves expand their knowledge to implement the technology and produce trustworthy results interpretation. Moreover, since the identified biomarkers panel could comprise biomarkers with high differences in the threshold levels, additional efforts should be focused on developing electrochemical biosensing strategies suitable to simultaneously determine biomarkers at very different concentration levels and both of genetic or protein nature.

Despite the great strides made, other research efforts should also be devoted to couple electrochemical biosensing devices with nanozymes and implementation on paper-based substrates. These advances would impart unprecedented opportunities upon electrochemical biosensors in terms of cost, stability, and eco-friendliness, and would allow us to face them with a large number of samples, multi-determination of several analytes, and continuous quantitative analysis for long times, by different users and in different environments beyond the well-controlled laboratory benchtop. Furthermore, although advanced sensors are envisioned to be a part of many more emerging technologies such as wearable devices, significant development stemming from multidisciplinary efforts in material sciences, electrochemistry, biophysics, biology, and pharmacology will be needed. Funding of innovative R&D and productive collaboration between universities, research centers, companies, and end users should also be enhanced to face up to the continuous evolving market demands.

Other attributes worth pursuing include tuning the concentration range (matching the specificity window of the receptor with the expected range of target concentrations in a given application) and the selectivity (minimizing cross-reactivity with close structural analogues of the targeted molecule) of biomolecular receptors through the rational adaptation of two strategies employed by nature, structure-switching and allosteric control, and expanding the variety of analytes to be detected by exploring other type of receptors in switch-based electrochemical biosensors.

In summary, although there are still several bottlenecks to overcome and we must be aware that there are no yet many commercially available electrochemical biosensors, the intense work performed and in due course to push the outstanding and unique opportunities they provide brings us closer and closer to this point. This will aid with having devices designed on demand, which will end up offering any attribute we can imagine while requiring less and less attention to do their job. The great promise they offer to largely simplify and speed up molecular detection makes electrochemical biosensors particularly appealing to support a broad range of applications while considerably improving our quality and way of life.

We hope this review will help newcomers to the field catch up with the current state of the art technology and will stimulate new researchers to join those with long experience in this field to continue to work on bringing out their full potential.

Author Contributions: Writing—review and editing, S.C., M.G., V.S., M.P., P.Y.-S. and J.M.P.; funding acquisition, S.C. and P.Y.-S. All authors have read and agreed to the published version of the manuscript.

Funding: The financial support of the PID2019-103899RB-I00 (Ministerio de Ciencia e Innovación) and RTI2018-096135-B-I00 (Ministerio de Ciencia, Innovación y Universidades) Research Projects and the TRANSNANOAVANSENS-CM Program from the Comunidad de Madrid (Grant S2018/NMT-4349) are gratefully acknowledged.

Conflicts of Interest: The authors declare no conflict of interest.

References

1. Zhang, Y.; Chen, X. Nanotechnology and nanomaterials-based no-wash electrochemical biosensors: From design to application. *Nanoscale* **2019**, *11*, 19105–19118. [CrossRef]
2. Li, H.; Li, S.; Dai, J.; Li, C.; Zhu, M.; Li, H.; Lou, X.; Xia, F.; Plaxco, K.W. High frequency, calibration-free molecular measurements in situ in the living body. *Chem. Sci.* **2019**, *10*, 10843–10848. [CrossRef]
3. Idili, A.; Parolo, C.; Ortega, G.; Plaxco, K.W. Calibration-free measurement of phenylalanine levels in the blood using an electrochemical aptamer-based sensor suitable for point-of-care applications. *ACS Sens.* **2019**, *4*, 3227–3233. [CrossRef]
4. Duan, D.; Fan, K.; Zhang, D.; Tan, S.; Liang, M.; Liu, Y.; Zhang, J.; Zhang, P.; Liu, W.; Qiu, X.; et al. Nanozyme-strip for rapid local diagnosis of Ebola. *Biosens. Bioelectron.* **2015**, *74*, 134–141. [CrossRef]
5. Qiu, W.; Baryeh, K.; Takalkar, S.; Chen, W.; Liu, G. Carbon nanotube-based lateral flow immunoassay for ultrasensitive detection of proteins: Application to the determination of IgG. *Microchim. Acta* **2019**, *186*, 436. [CrossRef]
6. Li, H.; Somerson, J.; Xia, F.; Plaxco, K.W. Electrochemical DNA-based sensors for molecular quality control: Continuous, real-time melamine detection in flowing whole milk. *Anal. Chem.* **2018**, *90*, 10641–10645. [CrossRef]
7. Li, H.; Dauphin-Ducharme, P.; Ortega, G.; Plaxco, K.W. Calibration-free electrochemical biosensors supporting accurate molecular measurements directly in undiluted whole blood. *J. Am. Chem. Soc.* **2017**, *139*, 11207–11213. [CrossRef]
8. Li, C.; Hu, X.; Lu, J.; Mao, X.; Xiang, Y.; Shu, Y.; Li, G. Design of DNA nanostructure-based interfacial probes for the electrochemical detection of nucleic acids directly in whole blood. *Chem. Sci.* **2018**, *9*, 979–984. [CrossRef]
9. Swensen, J.S.; Xiao, Y.; Ferguson, B.S.; Lubin, A.A.; Lai, R.Y.; Heeger, A.J.; Plaxco, K.W.; Soh, H.T. Continuous, real-time monitoring of cocaine in undiluted blood serum via a microfluidic, electrochemical aptamer-based sensor. *J. Am. Chem. Soc.* **2009**, *131*, 4262–4266. [CrossRef]
10. Arroyo-Currás, N.; Somerson, J.; Vieira, P.A.; Ploense, K.L.; Kippine, T.E.; Plaxco, K.W. Real-time measurement of small molecules directly in awake, ambulatory animals. *Proc. Natl. Acad. Sci. USA* **2017**, *114*, 645–650. [CrossRef]
11. Arroyo-Currás, N.; Ortega, G.; Copp, D.A.; Ploense, K.L.; Plaxco, Z.A.; Kippin, T.E.; Hespanha, J.P.; Plaxco, K.W. High-precision control of plasma drug levels using feedback-controlled dosing. *ACS Pharmacol. Transl. Sci.* **2018**, *1*, 110–118. [CrossRef]
12. Li, H.; Dauphin-Ducharme, P.; Arroyo-Currás, N.; Tran, C.H.; Vieira, P.A.; Li, S.; Shin, C.; Somerson, J.; Kippin, T.E.; Plaxco, K.W. A biomimetic phosphatidylcholine-terminated monolayer greatly improves the in vivo performance of electrochemical aptamer-based sensors. *Angew. Chem. Int. Ed.* **2017**, *56*, 1–5.
13. Bissonnette, S.; del Grosso, E.; Simon, A.J.; Plaxco, K.W.; Ricci, F.; Vallée-Bélisle, A. Optimizing the specificity window of biomolecular receptors using structure-switching and allostery. *ACS Sens.* **2020**, in press. [CrossRef]
14. Kim, J.; Kumar, R.; Bandodkar, A.J.; Wang, J. Advanced materials for printed wearable electrochemical devices: A review. *Adv. Electron. Mater.* **2017**, *3*, 1600260. [CrossRef]
15. Hubble, L.J.; Wang, J. Sensing at Your Fingertips: Glove-based wearable chemical sensors. *Electroanalysis* **2019**, *31*, 428–436. [CrossRef]
16. Khan, S.; Ali, S.; Bermak, A. Recent developments in printing flexible and wearable sensing electronics for healthcare applications. *Sensors* **2019**, *19*, 1230. [CrossRef]

17. Tu, J.; Torrente-Rodríguez, R.M.; Wang, M.; Gao, W. The era of digital health: A review of portable and wearable affinity biosensors. *Adv. Funct. Mater.* **2019**, 1906713. [CrossRef]
18. Kim, J.; Campbell, A.S.; Esteban-Fernández de Ávila, B.; Wang, J. Wearable biosensors for healthcare monitoring. *Nat. Biotechnol.* **2019**, *37*, 389–406. [CrossRef]
19. Sempionatto, J.R.; Jeerapan, I.; Krishnan, S.; Wang, J. Wearable chemical sensors: Emerging systems for on-body analytical chemistry. *Anal. Chem.* **2020**, *92*, 378–396. [CrossRef]
20. Campuzano, S.; Kuralay, F.; Lobo-Castañón, M.J.; Bartošik, M.; Vyavahare, K.; Paleček, E.; Haake, D.A.; Wang, J. Ternary monolayers as DNA recognition interfaces for direct and sensitive electrochemical detection in untreated clinical samples. *Biosens. Bioelectron.* **2011**, *26*, 3577–3583. [CrossRef]
21. Kuralay, F.; Campuzano, S.; Haake, D.A.; Wang, J. Highly sensitive disposable nucleic acid biosensors for direct bioelectronic detection in raw biological samples. *Talanta* **2011**, *85*, 1330–1337. [CrossRef] [PubMed]
22. Ruiz-Valdepeñas Montiel, V.; Sempionatto, J.R.; Esteban-Fernández de Ávila, B.; Whitworth, A.; Campuzano, S.; Pingarrón, J.M.; Wang, J. Delayed sensor activation based on transient coatings: Biofouling protection in complex biofluids. *J. Am. Chem. Soc.* **2018**, *140*, 14050–14053. [CrossRef] [PubMed]
23. Ruiz-Valdepeñas Montiel, V.; Sempionatto, J.R.; Campuzano, S.; Pingarrón, J.M.; Esteban-Fernández de Ávila, B.; Wang, J. Direct electrochemical biosensing in gastrointestinal fluids. *Anal. Bioanal. Chem.* **2019**, *411*, 4597–4604. [CrossRef] [PubMed]
24. Song, Z.; Chen, M.; Ding, C.; Luo, X. Designed three-in-one peptides with anchoring, antifouling and recognizing capabilities for highly sensitive and low fouling electrochemical sensing in complex biological media. *Anal. Chem.* **2020**, *92*, 5795–5802. [CrossRef]
25. Parolo, C.; Greenwood, A.S.; Ogden, N.E.; Kang, D.; Hawes, C.; Ortega, G.; Arroyo-Currás, N.; Plaxco, K.W. E-DNA scaffold sensors and the reagentless, single step, measurement of HIV-diagnostic antibodies in human serum. *Microsyst. Nanoeng.* **2020**, *6*, 13. [CrossRef]
26. Liu, Y.; Liu, Y.; Matharu, Z.; Rahimian, A.; Revzin, A. Detecting multiple cell-secreted cytokines from the same aptamer functionalized electrode. *Biosens. Bioelectron.* **2015**, *64*, 43–50. [CrossRef] [PubMed]
27. Zhad, H.R.L.Z.; Lai, R.Y. Application of calcium-binding motif of E-cadherin for electrochemical detection of Pb(II). *Anal. Chem.* **2018**, *90*, 6519–6525. [CrossRef]
28. Zhad, H.R.L.Z.; Lai, R.Y. A Hg(II)-mediated "signal-on" electrochemical glutathione sensor. *Chem. Commun.* **2014**, *50*, 8385–8387. [CrossRef]
29. Campuzano, S.; Pedrero, M.; Yáñez-Sedeño, P.; Pingarrón, J.M. Antifouling (bio)materials for electrochemical (bio)sensing. *Int. J. Mol. Sci.* **2019**, *20*, 423. [CrossRef]
30. Cui, M.; Wang, Y.; Jiao, M.; Jayachandran, S.; Wu, Y.; Fan, X.; Luo, X. Mixed self-assembled aptamer and newly designed zwitterionic peptide as antifouling biosensing interface for electrochemical detection of alpha-fetoprotein. *ACS Sens.* **2017**, *2*, 490–494. [CrossRef]
31. Wang, G.; Han, R.; Sua, X.; Lia, Y.; Xua, G.; Luo, X. Zwitterionic peptide anchored to conducting polymer PEDOT for the development of antifouling and ultrasensitive electrochemical DNA sensor. *Biosens. Bioelectron.* **2017**, *92*, 396–401. [CrossRef]
32. Wang, Y.; Cui, M.; Jiao, M.; Luo, X. Antifouling and ultrasensitive biosensing interface based on self-assembled peptide and aptamer on macroporous gold for electrochemical detection of immunoglobulin E in serum. *Anal. Bioanal. Chem.* **2018**, *410*, 5871–5878. [CrossRef] [PubMed]
33. Henry, O.Y.F.; Acero Sanchez, J.L.; O'Sullivan, C.K. Bipodal PEGylated alkanethiol for the enhanced electrochemical detection of genetic markers involved in breast cancer. *Biosens. Bioelectron.* **2010**, *26*, 1500–1506. [CrossRef] [PubMed]
34. Dharuman, V.; Chang, B.-Y.; Park, S.-M.; Hahn, J.H. Ternary mixed monolayers for simultaneous DNA orientation control and surface passivation for label free DNA hybridization electrochemical sensing. *Biosens. Bioelectron.* **2010**, *25*, 2129–2134. [CrossRef] [PubMed]
35. Dharuman, V.; Vijayaraj, K.; Radhakrishnan, S.; Dinakaran, T.; Narayanan, J.S.; Bhuvana, M.; Wilson, J. Sensitive label-free electrochemical DNA hybridization detection in the presence of 11-mercaptoundecanoic acid on the thiolated single strand DNA and mercaptohexanol binary mixed monolayer surface. *Electrochim. Acta* **2011**, *56*, 8147–8155. [CrossRef]

36. Campuzano, S.; Kuralay, F.; Wang, J. Ternary monolayer interfaces for ultrasensitive and direct bioelectronics detection of nucleic acids in complex matrices. *Electroanalysis* **2012**, *24*, 483–493. [CrossRef]
37. Blaszykowski, C.; Sheikh, S.; Thompson, M. Surface chemistry to minimize fouling from blood-based fluids. *Chem. Soc. Rev.* **2012**, *41*, 5599–5612. [CrossRef]
38. Josephs, E.A.; Ye, T. A Single-Molecule View of conformational switching of DNA tethered to a gold electrode. *J. Am. Chem. Soc.* **2012**, *134*, 10021–10030. [CrossRef]
39. Halford, C.; Gonzalez, R.; Campuzano, S.; Hu, B.; Babbitt, J.T.; Liu, J.; Wang, J.; Churchill, B.M.; Haake, D.A. Rapid antimicrobial susceptibility testing by sensitive detection of precursor rRNA using a novel electrochemical biosensing platform. *Antimicrob. Agents Chemother.* **2013**, *57*, 936–943. [CrossRef]
40. Goda, T.; Tabata, M.; Sanjoh, M.; Uchimura, M.; Iwasaki, Y.; Miyahara, Y. Thiolated 2-methacryloyloxyethyl phosphorylcholine for an antifouling biosensor platform. *Chem. Commun.* **2013**, *49*, 8683–8685. [CrossRef]
41. McQuistan, A.; Zaitouna, A.J.; Echeverria, E.; Lai, R.Y. Use of thiolated oligonucleotides as anti-fouling diluents in electrochemical peptide-based sensors. *Chem. Commun.* **2014**, *50*, 4690–4692. [CrossRef] [PubMed]
42. Jolly, P.; Formisano, N.; Tkáč, J.; Kasák, P.; Frost, C.G.; Estrela, P. Label-free impedimetric aptasensor with antifouling surface chemistry: A prostate specific antigen case study. *Sens. Actuators B Chem.* **2015**, *209*, 306–312. [CrossRef]
43. González-Fernández, E.; Avlonitis, N.; Murray, A.F.; Mount, A.R.; Bradley, M. Methylene blue not ferrocene: Optimal reporters for electrochemical detection of protease activity. *Biosens. Bioelectron.* **2016**, *84*, 82–88. [CrossRef] [PubMed]
44. Hu, Y.; Liang, B.; Fang, L.; Ma, G.; Yang, G.; Zhu, Q.; Chen, S.; Ye, X. Antifouling Zwitterionic Coating via Electrochemically Mediated Atom Transfer Radical Polymerization on Enzyme-Based Glucose Sensors for Long-Time Stability in 37 °C Serum. *Langmuir* **2016**, *32*, 11763–11770. [CrossRef]
45. Gao, J.; Jeffries, L.; Mach, K.E.; Craft, D.W.; Thomas, N.J.; Gau, V.; Liao, J.C.; Wong, P.K. A multiplex electrochemical biosensor for bloodstream infection diagnosis. *SLAS Technol.* **2017**, *22*, 466–474. [CrossRef]
46. Altobelli, E.; Mohan, R.; Mach, K.E.; Sin, M.L.Y.; Anikst, V.; Buscarini, M.; Wong, P.K.; Gau, V.; Banaei, N.; Liao, J.C. Integrated biosensor assay for rapid uropathogen identification and phenotypic antimicrobial susceptibility testing. *Eur. Urol. Focus* **2017**, *3*, 293–299. [CrossRef]
47. Ding, S.; Mosher, C.; Lee, X.Y.; Das, S.R.; Cargill, A.A.; Tang, X.; Chen, B.; McLamore, E.S.; Gomes, C.; Hostetter, J.M.; et al. Rapid and label-free detection of interferon gamma via an electrochemical aptasensor comprising a ternary surface monolayer on a gold interdigitated electrode array. *ACS Sens.* **2017**, *2*, 210–217. [CrossRef]
48. Miranda-Castro, R.; Sánchez-Salcedo, R.; Suárez-Álvarez, B.; de-los-Santos-Álvarez, N.; Miranda-Ordieres, A.J.; Lobo-Castañón, M.J. Thioaromatic DNA monolayers for target-amplification-free electrochemical sensing of environmental pathogenic bacteria. *Biosens. Bioelectron.* **2017**, *92*, 162–170. [CrossRef]
49. Miranda-Castro, R.; de-los-Santos-Álvarez, N.; Lobo-Castañón, M.J. Understanding the factors affecting the analytical performance of sandwich-hybridization genosensors on gold electrodes. *Electroanalysis* **2018**, *30*, 1229–1240. [CrossRef]
50. González-Fernández, E.; Staderini, M.; Yussof, A.; Scholefield, E.; Murray, A.F.; Mount, A.R.; Bradley, M. Electrochemical sensing of human neutrophil elastase and polymorphonuclear neutrophil activity. *Biosens. Bioelectron.* **2018**, *119*, 209–214. [CrossRef]
51. González-Fernández, E.; Staderini, M.; Avlonitis, N.; Murray, A.F.; Mount, A.R.; Bradley, M. Effect of spacer length on the performance of peptide-based electrochemical biosensors for protease detection. *Sens. Actuators B Chem.* **2018**, *255*, 3040–3046. [CrossRef]
52. Boozer, C.; Chen, S.F.; Jiang, S.Y. Controlling DNA orientation on mixed ssDNA/OEG SAMs. *Langmuir* **2006**, *22*, 4694–4698. [CrossRef] [PubMed]
53. Dharuman, V.; Hahn, J.H. Label-free electrochemical DNA hybridization discrimination effects at the binary and ternary mixed monolayers of single stranded DNA/diluent/s in presence of cationic intercalators. *Biosens. Bioelectron.* **2008**, *23*, 1250–1258. [CrossRef] [PubMed]
54. Wu, J.; Campuzano, S.; Halford, C.; Haake, D.A.; Wang, J. Ternary surface monolayers for ultrasensitive (zeptomole) amperometric detection of nucleic acid hybridization without signal amplification. *Anal. Chem.* **2010**, *82*, 8830–8837. [CrossRef]

55. Goodman, R.P.; Berry, R.M.; Turberfield, A.J. The single-step synthesis of a DNA tetrahedron. *Chem. Commun.* **2004**, *12*, 1372–1373. [CrossRef]
56. Pei, H.; Lu, N.; Wen, Y.; Song, S.; Liu, Y.; Yan, H.; Fan, C. A DNA Nanostructure-based biomolecular probe carrier platform for electrochemical biosensing. *Adv. Mater.* **2010**, *22*, 4754–4758. [CrossRef]
57. Lin, M.; Song, P.; Zhou, G.; Zuo, X.; Aldalbahi, A.; Lou, X.; Shi, J.; Fan, C. Electrochemical detection of nucleic acids, proteins, small molecules and cells using a DNA-nanostructure-based universal biosensing platform. *Nat. Protoc.* **2016**, *11*, 1244–1263. [CrossRef]
58. Dong, S.; Zhao, R.; Zhu, J.; Lu, X.; Li, Y.; Qiu, S.; Ji, L.; Jiao, X.; Song, S.; Fan, C.; et al. Electrochemical DNA biosensor based on a tetrahedral nanostructure probe for the detection of Avian Influenza A (H7N9) virus. *ACS Appl. Mater. Interfaces* **2015**, *5*, 8834–8842. [CrossRef]
59. Lubin, A.A.; Plaxco, K.W. Folding-based electrochemical biosensors: The case for responsive nucleic acid architectures. *Acc. Chem. Res.* **2010**, *43*, 496–505. [CrossRef]
60. Campuzano, S.; Yáñez-Sedeño, P.; Pingarrón, J.M. Reagentless and reusable electrochemical affinity biosensors for near real-time and/or continuous operation. Advances and prospects. *Curr. Opin. Electrochem.* **2019**, *16*, 35–41. [CrossRef]
61. Plaxco, K.W.; Soh, H.T. Switch-based biosensors: A new approach towards real-time, in vivo molecular detection. *Trends Biotechnol.* **2011**, *29*, 1–5. [CrossRef] [PubMed]
62. Arroyo-Currás, N.; Dauphin-Ducharme, P.; Scida, K.; Chávez, J.L. From the beaker to the body: Translational challenges for electrochemical, aptamer-based sensors. *Anal. Methods* **2020**, *12*, 1288–1310.
63. Ranallo, S.; Rossetti, M.; Plaxco, K.W.; Vallée-Bélisle, A.; Ricci, F. A modular, DNA-based beacon for single-step fluorescence detection of antibodies and other proteins. *Angew. Chem. Int. Ed.* **2015**, *54*, 13214–13218. [CrossRef] [PubMed]
64. White, R.J.; Kallewaard, H.M.; Hsieh, W.; Patterson, A.S.; Kasehagen, J.B.; Cash, K.J.; Uzawa, T.; Soh, H.T.; Plaxco, K.W. Wash-free, electrochemical platform for the quantitative, multiplexed detection of specific antibodies. *Anal. Chem.* **2012**, *84*, 1098–1103. [CrossRef]
65. Ferguson, B.S.; Hoggarth, D.A.; Maliniak, D.; Ploense, K.; White, R.J.; Woodward, N.; Hsieh, K.; Bonham, A.J.; Eisenstein, M.; Kippin, T.E.; et al. Real-time, aptamer-based tracking of circulating therapeutic agents in living animals. *Sci. Transl. Med.* **2013**, *5*, 213ra165. [CrossRef]
66. Kang, D.; Sun, S.; Kurnik, M.; Morales, D.; Dahlquist, F.W.; Plaxco, K.W. New architecture for reagentless, protein-based electrochemical biosensors. *J. Am. Chem. Soc.* **2017**, *139*, 12113–12116. [CrossRef]
67. Xiao, Y.; Lubin, A.A.; Baker, B.R.; Plaxco, K.W.; Heeger, A.J. Single-step electronic detection of femtomolar DNA by target-induced strand displacement in an electrode-bound duplex. *Proc. Natl. Acad. Sci. USA* **2006**, *103*, 16677–16680. [CrossRef] [PubMed]
68. Wang, Q.; Gao, F.; Ni, J.; Liao, X.; Zhang, X.; Lin, Z. Facile construction of a highly sensitive DNA biosensor by in-situ assembly of electro-active tags on hairpin-structured probe fragment. *Sci. Rep.* **2016**, *6*, 22441. [CrossRef] [PubMed]
69. Zaitouna, A.J.; Lai, R.Y. An electrochemical peptide-based Ara h 2 antibody sensor fabricated on a nickel(II)-nitriloacetic acid self-assembled monolayer using a His-tagged peptide. *Anal. Chim. Acta* **2014**, *828*, 85–91. [CrossRef]
70. Mayer, M.D.; Lai, R.Y. Effects of redox label location on the performance of an electrochemical aptamer-based tumor necrosis factor-alpha sensor. *Talanta* **2018**, *189*, 585–591. [CrossRef]
71. Lai, R.Y.; Seferos, D.S.; Heeger, A.J.; Bazan, G.C.; Plaxco, K.W. Comparison of the signaling and stability of electrochemical DNA sensors fabricated from 6- or 11-carbon self-assembled monolayers. *Langmuir* **2016**, *22*, 10796–10800. [CrossRef]
72. Baker, B.R.; Lai, R.Y.; Wood, M.S.; Doctor, E.H.; Heeger, A.J.; Plaxco, K.W. An electronic, aptamer-based small-molecule sensor for the rapid, label-free detection of cocaine in adulterated samples and biological Fluids. *J. Am. Chem. Soc.* **2006**, *128*, 3138–3139. [CrossRef] [PubMed]
73. Parate, K.; Karunakaran, C.; Claussen, J.C. Electrochemical cotinine sensing with a molecularly imprinted polymer on a graphene-platinum nanoparticle modified carbon electrode towards cigarette smoke exposure monitoring. *Sens. Actuators B Chem.* **2019**, *287*, 165–172. [CrossRef]

74. Wang, Z.; Ma, B.; Shen, C.; Cheong, L.Z. Direct, selective and ultrasensitive electrochemical biosensing of methyl parathion in vegetables using Burkholderia cepacia lipase@MOF nanofibersbased biosensor. *Talanta* **2019**, *197*, 356–362. [CrossRef] [PubMed]
75. Du, Y.; Lim, B.J.; Li, B.; Jiang, Y.S.; Sessler, J.L.; Ellington, A.D. Reagentless, ratiometric electrochemical DNA sensors with improved robustness and reproducibility. *Anal. Chem.* **2014**, *86*, 8010–8016. [CrossRef] [PubMed]

© 2020 by the authors. Licensee MDPI, Basel, Switzerland. This article is an open access article distributed under the terms and conditions of the Creative Commons Attribution (CC BY) license (http://creativecommons.org/licenses/by/4.0/).

Review

Biosensors—Recent Advances and Future Challenges in Electrode Materials

Fernando Otero and Edmond Magner *

Department of Chemical Sciences and Bernal Institute, University of Limerick, V94 T9PX Limerick, Ireland; fernando.otero.diez@ul.ie
* Correspondence: edmond.magner@ul.ie

Received: 29 May 2020; Accepted: 19 June 2020; Published: 23 June 2020

Abstract: Electrochemical biosensors benefit from the simplicity, sensitivity, and rapid response of electroanalytical devices coupled with the selectivity of biorecognition molecules. The implementation of electrochemical biosensors in a clinical analysis can provide a sensitive and rapid response for the analysis of biomarkers, with the most successful being glucose sensors for diabetes patients. This review summarizes recent work on the use of structured materials such as nanoporous metals, graphene, carbon nanotubes, and ordered mesoporous carbon for biosensing applications. We also describe the use of additive manufacturing (AM) and review recent progress and challenges for the use of AM in biosensing applications.

Keywords: electrochemical biosensors; glucose biosensors; nanoporous metals; nanoporous gold; graphene; carbon nanotube; ordered mesoporous carbon; additive manufacturing

1. Introduction

In comparison with other methods of detection such as optical, spectroscopic, and chromatographic, electrochemical sensors possess advantages such as simplicity, rapid response times, and high sensitivity [1]. Electrochemical sensors can be easily adapted for the detection of a wide range of analytes and can be incorporated into robust, portable, low cost, minituarized devices that can be tailored for particular applications [2]. Taking advantage of these attributes and the incorporation of highly specific biological recognition elements (enzymes, nucleic acids, cells, tissues, and so on), electrochemical biosensors are capable of selectively detecting a broad range of target analytes. As defined by IUPAC, electrochemical biosensors are "self-contained integrated devices, which are capable of providing specific quantitative or semi-quantitative analytical information using a biological recognition element (biochemical receptor), which is retained in direct spatial contact with an electrochemical transduction element" [3]. Bioelectrochemical sensors are used in environmental monitoring, healthcare, and biological analysis, among others. Depending on the recognition process, biosensors can be subdivided into two main categories: affinity and biocatalytic sensors. Affinity sensors operate via selective binding between the analyte and the biological component (i.e., antibody and nucleic acid) [4]. In contrast, biocatalytic devices incorporate enzymes, whole cells, or tissue slices that recognize the target analyte, and subsequently produce an electroactive species [5].

The first biosensor was described by Clark and Lyons in 1962 [6]. This biosensor was composed of an oxygen electrode, an inner oxygen semipermeable membrane, and a thin layer of glucose oxidase (GOx, EC 1.1.3.4) entrapped by a dialysis membrane. The decrease in the level of oxygen was proportional to the concentration of glucose resulting from the enzyme catalysed oxidation of β-D-glucose to β-D-glucono-δ-lactone [7]. Since this pioneering work, extensive efforts have been made to develop electrochemical biosensors for a wide range of analytes. Bioelectrochemical sensing devices have been effectively transferred from the laboratory to the point-of-care (POC) with global sales growing from less than $5 million per annum [8] thirty years ago to over $18 billion in 2018. Although commercial systems are available for a range of small molecules (lactose, uric acid, cholesterol, lactate, ketone, and so on), the market is dominated by glucose sensors, with approximately 90% of the market associated with glucose monitoring for diabetes [9]. Diabetes mellitus is one of the leading causes of death and disability in the world [10]. It is a metabolic disorder that causes insulin deficiency and hyperglycemia, resulting in blood glucose concentration deviating from the normal range of 3.9–6.2 mM [11]. According to the International Diabetes Federation (IDF), the number of diabetic patients increased from 151 million in 2000 to 415 in 2015. The IDF also predicted that the number of diabetic patients would increase to 642 million in 2040, with diabetes becoming the seventh-leading cause of mortality [12]. Commercial home use blood glucose sensors generally detect glucose in the concentration range of 1.1–33.3 mM with test times of less than 30 s [4]. GOx is widely employed as the recognition element in glucose biosensors owing its relatively low cost, high selectivity, and stability [13]. First and second generation sensors rely on the immobilization of the enzyme onto an electrode surface. As the redox active centers of GOx are at least 13–18 Å from the electrode surface, mediators are employed to shuttle electrons between the electrode surface and enzyme's active site [14]. The direct oxidation of GOx occurs in third generation sensors, where the enzyme is specifically wired to minimize the distance between the active site of the enzyme and the electrode surface. Despite the considerable progress that has been made, the majority of commercial glucose sensors are based on second generation glucose sensors. The vast majority of commercial devices utilise blood samples obtained via a finger prick. The development of glucose biosensors based on the detection of glucose in fluids such as tears [15], saliva [16], and sweat [17] has been described. Such systems face challenges, in particular the poor correlation between glucose levels in blood and in other fluids and also significantly lower concentrations of glucose in fluids such as tears. Individually optimized designs must be developed [18] for commercially viable sensors, where challenges such as low cost, ease of manufacture, robustness, and portability are additional factors for consideration [19].

In contrast, detection of larger biomolecules such as nucleic acids and proteins faces significant additional challenges that include electrode fouling, non-specific adsorption of biological components at the electrode surface, lack of sensitivity in the appropriate concentration range, and in particular at low concentrations (femtomolar to attomolar) [8]. Commercial systems for the detection of larger biomolecules are dominated by pregnancy tests that rely on the detection of human chorionic gonadotrophin (hCG), a glycoprotein hormone secreted during pregnancy [20].

In this review, we describe recent advances on the use of materials as supports in electrochemical biosensors, and in particular the use of materials such as nanoporous metals, graphene, carbon nanotubes, and mesoporous carbon. Examples of the detection of clinically relevant molecules are provided, with a focus on the detection of glucose. An overview of invasive and non-invasive glucose monitoring with case studies is given. In addition, we discuss the use of additive manufacturing for electrochemical sensing applications.

2. Electrode Materials

Owing to their intrinsic conductivity, biocompatibility, and ease of manufacture, high surface area materials such as nanoporous gold, graphene, carbon nanotubes, and mesoporous carbon have been used for the preparation of electrodes for bioelectrochemical applications.

2.1. Nanoporous Metal Electrodes

Nanoporous metals are 3D bicontinuous structures with tuneable pore diameters and lengths that possess large surface areas, mechanical resistance, and high conductivity [21,22]. Although nanoporous electrodes have been prepared using a range of metals such as copper, silver, and palladium, the majority of research has focused on nanoporous gold (NPG) owing to its ease of manufacture, chemical stability, and biocompatibility [23]. NPG is a 3D nanostructured material with pore sizes that can be tuned over the range 5 nm to greater than 2 µm [14]. The morphology of NPG is generally characterised using atomic force microscope (AFM) and scanning or transmission electron microscopy (SEM/TEM). The electrochemically addressable accessible surface area is evaluated by measuring the roughness factor calculated from the charge associated with reduction of gold oxide in 0.5 M H_2SO_4 solution and applying a conversion factor of 390 µC/cm^2 [24]. NPG electrodes possess good electrical conductivity, catalytic activity, high surface-to-volume ratio, permeability, chemical, and thermal and mechanical stability [25,26], as well as properties of interest for a range of applications including biocatalysis [27], nucleic acid sensors [28], enzymatic sensors [29], non-enzymatic sensors [30], immunosensors [31], supercapacitors [32], enzymatic fuel cells [33], and so on.

Different methods have been studied for the controlled manufacture of nanoporous gold [33]. For example, using anodization methods, the 3D structure is generally formed by the anodization of gold in oxalic acid at different applied potentials, which enables the formation of specific nanoporous structures [34]. Recently, a NPG microelectrode was fabricated via electrochemical anodization-reduction steps in 0.5 M H_2SO_4, exhibiting pore sizes in the range of 30–50 nm [35]. Although anodization of gold avoids the use of corrosive chemicals, the pore diameters are typically ca. 20 nm in size [36], making it potentially difficult to achieve high loading of biomolecules. Another route entails using hydrogen bubbles formed via the electrochemical reduction of H$^+$ as the template [37]. Gabriella Sanzo et al. synthesized a gold nanocorals porous structure with an electroactive area 500 times higher than a gold screen printed electrode that was used as the base substrate [38]. The nanocoral electrode was modified with glucose oxidase for the development of an enzymatic biosensor based on the detection of H_2O_2. The nanocoral electrode showed a sensitivity of 48.3 µA/mMcm^2, two times higher than that of the bare gold electrode. The hydrogen template produces materials with pore sizes in the micrometer region. In order to overcome the limitation on pore size, other template routes can be used. The hard template route usually involves two steps: assembly of monodisperse spheres, then electrodeposition of the metal followed by removal of the hard template, where the diameter and thickness of the porous structure are controlled in the range of 100–2000 nm [39]. The spatial arrangement and size of the pores can be controlled using colloidal crystals as a template. For instance, Szamocki et al. fabricated macroporous gold electrodes of different sizes for the electrochemical oxidation of glucose with glucose dehydrogenase (GDH), with an enhanced electrochemical response by more than one order of magnitude compared with planar gold electrodes [39]. Gamero et al. immobilised lactate oxidase (LOx) on NPG with a pore size of 500 nm, with a linear response observed up to a concentration of 1.3 mM [40].

An alternative approach relies on chemical dealloying of the less noble metal of an alloy, which can be prepared by sputtering a gold-metal alloy onto a support or by using commercially available gold alloys, for example "white gold". During the dealloying process, atoms of the less noble metal are detached from the surface and subsequently dissolved under the etchant conditions, forming nanoporous structures. Different alloy systems including Au-Zn [41], Au-Ni [42], Au-Al [43], Au-Si [44], and Au-Ag [45], have been used for the formation of nanoporous gold by dealloying the least noble metal component. Au-Ag is the most commonly used owing to the ease of removal of silver, which is generally removed under corrosive conditions (usually 70% nitric acid). In a systematic study, different alloy compositions $Ag_{70}Au_{30}$, $Ag_{50}Au_{50}$, and $Ag_{35}Au_{65}$ were prepared (Figure 1 A–D) [46]. The silver content in the alloy Ag35/Au65 was too low to enable nanoporous structures to be formed. A homogeneous distribution of nanopores was formed using the $Ag_{70}Au_{30}$ alloy. The thickness and composition of the layer were controlled by the sputtering conditions, while the pore sizes were controlled by factors such as the time period and the temperature of the process. For instance, by varying the temperature and time of dealloying of a 100 nm thick $Ag_{70}Au_{30}$ alloy, the pore size of the dealloyed sheets ranged from 4 to 78 nm, with a maximum surface area 44 times greater than the geometric area [46]. NPG prepared using this approach exhibits a controllable pore size range from 5 to 700 nm [47], a range sufficiently large to accommodate biomolecules. As with planar gold electrodes, the surfaces of NPG can subsequently be modified. For example, carboxylic acid terminated diazonium salts were covalently attached onto NPG and the immobilization of fructose dehydrogenase (FDH) was subsequently accomplished via crosslinking with CMC [48]. The sensor showed a linear range of 0.05–0.3 mM, with a sensitivity of 3.7 µA/cm² mM and a limit of detection (LOD) of 1.2 µM with a fast response of less than 5 s. The linear range encompasses that observed in juices and the sensor displayed excellent selectivity.

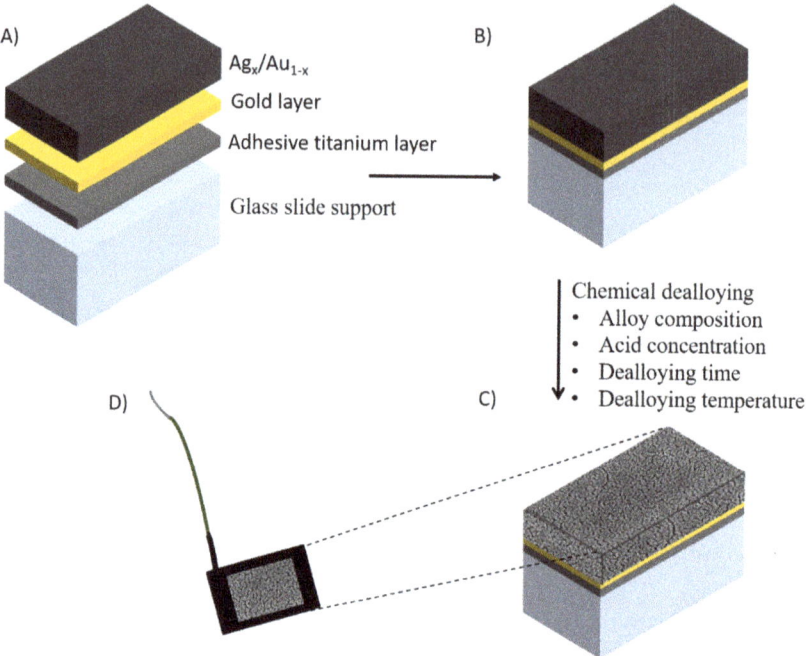

Figure 1. Schematic representation of the manufacture of nanoporous gold (NPG) electrodes with (**A**) different layers and thicknesses, (**B**) sputtered glass sheet prior to etching, (**C**) formation of nanopores after etching, and (**D**) the completed NPG electrode. Adapted from [46].

Wearable sensors have the potential to play a major role in the development of continuous monitoring for glucose and other biomarkers in different fluids such as tears, saliva, interstitial fluids, and sweat. Flexible NPG was prepared using an electrochemical dealloying approach. NPG electrodes were further modified with lactate oxidase and bilirubin oxidase to develop a lactate/O_2 enzymatic fuel cell, which was successfully tested in artificial lachrymal fluids [33].

Matharu et al. described the fabrication of NPG with different pore sizes via dealloying of a 600 nm thick Au-Ag alloy to investigate DNA hybridization in the presence of biofouling species [49]. The thiolated capture probe DNA and its target DNA were used to investigate hybridization using methylene blue as intercalator. In the absence of biofouling conditions, the accessibility of target DNA increased with larger pores, resulting in higher signal suppression with maximum values of ~70% of that for a pore size of about 45 nm. However, in the presence of biofouling conditions, electrodes with average pore sizes of 25–30 nm maximized the accessibility of target DNA as the pores were sufficiently small to block the entrance of biofouling molecules. In contrast, larger pores were susceptible to electrode blockage by biofouling, decreasing the biosensor performance.

Owing to the expensive nature of gold, electrodes have been manufactured using lower cost non-noble metals such as Cu, Ni, Ti, or Fe [23]. However, the reproducible preparation of nanoporous structures from such alloys needs to be addressed [50].

NPG is the most widely used metal support used for biosensing, with reviews on the preparation and application of nanoporous gold published recently [14,51]. The high surface area per volume, biocompatibility, and the ability to prepare flexible electrodes make NPG an attractive material for use with biological systems. However, the high cost of gold and the complexity of the manufacturing process currently limit the applicability of NPG to research applications [43].

2.2. Carbon Based Materials

2.2.1. Graphene

Graphene is a flat sheet of two dimensional layer sp^2 bonded carbon that is one atom in thickness [52]. The carbon atoms are sp^2 hybridized, with out of plane π bonds that are responsible for the high electrical conductivity of the graphene. Graphene is of interest owing to its high specific surface area; electrical conductivity; and thermal, optical, and mechanical properties [53]. These remarkable properties have potential applicability in electrochemical biosensors [54]. In comparison with more traditional carbon materials, graphene has a large theoretical surface area (2630 m^2/g) [55], higher electrical conductivity (200 S/m) [56], and good mechanical strength (1.0 TPA) [57].

Graphene was first prepared via mechanical exfoliation of highly oriented pyrolytic graphite [52]. Other methods such as the exfoliation and cleavage of graphite, chemical vapor deposition, plasma enhanced chemical vapor deposition, solution based reduction of graphene oxide (GO), and so on have been reported [54]. Each of these strategies produces graphene material with different characteristics. These methods focus on the production of large areas of single layers of graphene at low cost and high scale. The primary obstacle to achieving single or a small number of graphene layers is to overcome the strong, interlayer, van der Waal's forces. To date, the most common approach to graphite exfoliation is the use of strong oxidizing agents that yield GO, a non-conductive hydrophilic carbon material, in a process known as Hummers method [58]. GO produced via this route can be reduced or used for the immobilization of biomolecules. GO can be also obtained using an improved version of Hummers' method, with a material that contains fewer defects in the basal plane [59]. Liu et al. reported a glucose biosensor obtained via covalent immobilization between the carboxyl and amine groups of GO and GOx, respectively [60]. A nanocomposite film based on chitosan-ferrocene GO (positively charged) was used to immobilise negatively-charged GOx [61]. The biosensor showed a linear response to glucose in the concentration rage of 0.02 to 6.78 mM, with a sensitivity of 10 µA/mMcm^2 and an LOD of 7.6 µM. Using thermal, electrochemical, or chemical reduction processes to eliminate oxygen-functional groups (ketone, epoxy, carboxyl, and so on) results in graphene with properties that include excellent electrical conductivity, large surface area, and ease of functionalization. Furthermore, residual functional oxygen groups are available for the immobilization of biomolecules [62].

However, owing to the lack of oxygen functional groups to anchor biomolecules, it is necessary to functionalise graphene. Fenzt et al. anchored 1-pyrenebutyric acid onto graphene and an aptamer against the coagulation factor thrombin was subsequently covalently attached [63]. The biosensor displayed a limit of detection of 1 and 5 pM in buffer and serum, respectively. Lee et al. developed a patch-based strip-type disposable sweat glucose sensor and microneedle-based point-of-care therapy [64]. In addition to the detection of glucose, the wearable device consisted of stretchable sensors for humidity, pH, and temperature. A mixture of graphene, GOx, and chitosan was drop cast onto a gold working electrode, followed by Nafion®(Chemours, US) and subsequently glutaraldehyde to cross-link the enzyme layer. The patch was reusable and could be reattached several times. The response for the detection of glucose was corrected via simultaneous measurement of pH. When tested on human subjects, pre- and post-prandial glucose levels correlated with those obtained using a commercial glucose kit. Multiplexed biosensors aim to detect several target biomarkers by integrating a series of sensors on a chip [65]. Such systems are of assistance for the correct diagnosis/treatment of specific diseases. For instance, it was recently shown that lactate is the most important carrier for cancer cells and diabetic patients are prone to accumulate lactate in their tissue [66], and thus a multiplexed biosensor that can be used to discriminate between diseases would be very beneficial.

Owing to the lack of functional groups to anchor biomolecules, pristine graphene has not been extensively used as a biosensor. Functional doping of graphene with heteroatoms such as N, S, B, P, and F is an excellent pathway to enhance electron transfer processes [63]. Among them, nitrogen-doped graphene (NG) offers better electrochemical activity owing to the positive charge density in carbon atoms adjacent to N dopants, enhancing the conductivity of the material [67]. A multilayer biosensor containing GOx, nitrogen-doped graphene, chitosan, and poly(styrene sulfonate) was constructed layer-by-layer [68]. The presence of NG decreased the charge transfer resistance of the assembly, increased the interfacial capacitance, and provided a film matrix with significant charge separation. The biosensor operated at a low potential of −0.2 V versus Ag/AgCl and exhibited a short linear range between 0.2 and 1.8 mM. Nevertheless, the selective doping of N in specific sites is still a challenge and further research needs to be performed for the development of more reproducible methods of preparation. Reviews on the synthesis, characterization, and applications of NG have been published recently [69,70].

To avoid the loss of electrochemical active area and irreversible π–π stacking aggregation, graphene is generally combined with different nanomaterials (e.g., gold nanoparticles, polyaniline, carbon nanotubes, chitosan, Nafion, methylene green, and so on) to enhance the sensitivity of detection [71]. Recently, a graphene thionine gold nanoparticles (AuNP) composite material was used as a paper-based electrochemical immunosensor for the detection of the cancer antigen 125, a biomarker related to ovarian cancer [72]. An impedimetric HIV-1 biosensor based on graphene-Nafion composite was reported. The decrease in electron transfer resistance was proportional to the concentration of HIV-1 gene over the concentration range 1.0×10^{-13} to 1.0×10^{-10} M and displayed a limit of detection of 2.3×10^{-14} M [73]. A third-generation glucose biosensor was fabricated using a graphene/polyethyleneimine/gold nanoparticle for the immobilization of GOx using glutaraldehyde as a crosslinker. The biosensor displayer a linear response to the concentration of glucose over the range of 1–100 µM with a sensitivity of 93 µA/mMcm^2 [74]. An enzymatic amperometric sensor based on a graphene/PANI/AuNPs modified glassy carbon electrode was reported [75]. The adsorption of GOx facilitated direct electron transfer between the modified electrode and enzymes. Although adsorbed enzyme molecules retained their activity, the leakage of enzymes is a major drawback, a drawback that can be overcome by encapsulation, otherwise covalent binding of the enzymes may be required. Conductive polymers such as polyaniline, polythiophene, polyacetylene, and polypyrrole have been extensively used for the entrapment of biomolecules. The thickness of the polymer film, and thus the barrier to diffusion, could be controlled by tuning the deposition parameters [76]. Such polymer provides high conductivity, biocompatibility, and high stability. For example, a glucose biosensor based on GOx immobilized onto 3,4-ethylenedioxythiophene microspheres modified with platinum nanoparticles retained 97% of its sensitivity after 12 days of storage at room temperature [77]. Owing to its high surface area, high conductivity, and ease of functionalization, graphene has been used extensively as a platform for the construction of a wide range of biosensors references [78], and it holds promise in the development of biosensors for minimally invasive continuous monitoring in, for example, interstitial fluids. The main source of graphene is graphite, which is inexpensive and readily available. However, issues with the degree of biocompatibility of graphene have yet to be fully resolved. Additional challenges include the development of robust biosensors that can function in a range of operating conditions and the preparation of mechanically robust single-sheet graphene electrodes.

2.2.2. Carbon Nanotubes

Carbon nanotubes (CNTs) are one dimensional (1D) carbon tubes prepared by rolling a graphite sheet of variable length and diameter. CNTs are light and possess a large surface area, excellent conductivity, and good mechanical strength, together with chemical and thermal stability. Thanks to these properties, CNTs can be used as transducer or nanocarrier in biosensors [79]. It has a theoretical surface area of 1315 m^2/g, 50% of that of a single graphene sheet [80]. CNTs can be divided into two main groups: single-walled nanotubes (SWNTs) and multi-walled nanotubes (MWNTs). SWNT is a single layer nanomaterial formed by rolling a graphene sheet into a seamless molecular cylinder with diameter and length ranging between 0.75–3 nm and 1–50 µm, respectively. MWCNT is composed of at least two layers of graphite sheets, separated by approximately 0.42 nm, with a diameter ranging from 2 to more than 100 nm [81].

Different routes have been developed for the manufacture of CNTs. The main method is chemical vapor deposition (CVD), which is based on the decomposition of a carbon source gas at 600–1000 °C producing CNTs. CNTs can be grown directly on the substrate at large scale and low cost. In spite of the simplicity of the process, the use of a metal catalyst such as Co, Fe, Cu, Cr, and so on is required, which can subsequently be incorporated into defects [82]. Another approach that uses metals as catalysts is based on laser ablation, as reported by Smalley and co-workers in 1995 [83]. Carbon atoms from graphite and metal catalysts atoms are irradiated using high energy laser beam. This method results in high purity materials with few defects, but is expensive with high levels of energy consumption and is not practical for large-scale production. Another high-cost approach is the arc discharge method [84], where CNTs are deposited onto a graphite cathode under the action of a current in a vacuum reactor.

Although the structural integrity of enzymes is preserved via non-covalent functionalization, the interaction between enzymes and CNTs is weak, resulting in leakage of the enzyme. This limitation can be overcome by functionalizing the CNT surface or using nanoparticles or polymers for enzyme immobilization [85]. Paolo et al. electrochemically grafted 2-aminoantrhracene diazonium salt onto SWNCT-based electrodes that were further incubated in a solution of FDH [86]. The biosensor displayed a linear range from 0.05 to 5 mM, a sensitivity of 47 µA/mMcm^2, an LOD of 0.9 µM, and great stability (90% of retained signal after 60 days). A Pt electrode was modified with a rGO/CNT/AuNPs composite for the detection of lactate. At a potential of 0.2 V, the sensor had a wide linear range of 0.05 to 100 mM with high sensitivity (35.3 µA/mMcm^2) and a low LOD (2.3 µM) [87]. A wearable glucose biosensor was prepared by immobilizing GOx onto SWCNTs with Nafion®(Chemours, US), which could detect glucose with a response time of less than 5 s [88]. The response to glucose was transmitted to a smartphone using a wireless connection and a linear response to glucose over the range 0.05 to 1 mM was observed.

However, it is important to remark that toxic effects, mainly owing to the presence of metallic impurities, can occur with CNTs. Further studies require the creation of biocompatible CNT-based electrodes that can be addressed by adding dialysis bags [89] or by coating with biocompatible polymers (e.g., chitosan, collagen, Nafion®(Chemours, US), and so on) [90]. CNTs have been also successfully tested for a wide range of biomolecules such as DNA [91], immunosensors [92], proteins [93], and other biological molecules. Graphene and CNTs possess high thermal, mechanical, and electronic properties and both materials can be produced on a large scale. However, the synthesis of CNTs is a high cost process and usually involves the use of metal nanoparticles, which can be toxic, limiting its potential use.

2.2.3. Ordered Mesoporous Carbon

Ordered mesoporous carbon (OMC) is a flexible material that provides interconnected channels for the diffusion of electroactive compounds in electrochemical systems. OMC possess high specific surface area; large and open porous structure; high conductivity; and excellent chemical, thermal, and mechanical stability [94]. OMC can be synthesized via catalytic activation of carbon precursors, carbonization of the blends of one thermosetting precursors and one thermally unstable polymer, and carbonization of organic aerogels. Nevertheless, the resulting mesopores have abundant micropores and a wide pore distribution [95].

A more reliable pore size distribution with a symmetric ordering can be obtained through a template method, which can be subdivided into two categories: hard and soft-templating. In hard-templating, the pore size is controlled using a mesoporous silica template that controls the pore size. Other templates such as nickel oxide can also be used. The overall process involves the impregnation of the pores of the template with a carbon source (e.g., sucrose, ethylene, furfuryl alcohol), followed by polymerization and carbonization upon pyrolysis. Dissolution of the silica in HF or alkaline solution results in a mesoporous carbon replica. The largest used mesoporous silica are the hexagonal SBA-15 and cubic MCM-48 materials, leading to CMK-3 and CMK-1 respectively [94]. The structure of CMK-1 is dependent on the carbon precursor used. CMK-3 exhibited a highly ordered hexagonal packed mesopores interconnected owing to the presence of micropores. The process for the formation of OMC via hard-templating is schematically represented in Figure 2A. The pore arrangement of CMK-1 carbon replicas resulted in a more accessible structure owing to the more favorable rate of diffusion of reactant molecules during catalytic processes [96]. CKM-1 modified with an ionic liquid showed a good electrocatalytic response for the direct oxidation of dsDNA with a detection limit of 1.2 µg/mL [97]. A CMK-3 was used for the construction of alcohol and glucose biosensors, based on alcohol dehydrogenase and glucose oxidase [98]. To date, the majority of OMC biosensors rely on the use of CMK-1 and CMK-3 [94]. Other mesoporous ordered silica have been used as a hard template for the synthesis of OMC. For instance, a 1D-carbon nanotube array, designated as CMK-5, was synthesized when the channels of SBA-15 were partially filled [99]. The covalent immobilization of GOx was performed using a 4-nitrophenyl functionalized CMK-5, exhibiting a linear response over the range of 1–14 mM [100]. The electrochemical response of the sensor was reduced by 6% after one month of storage.

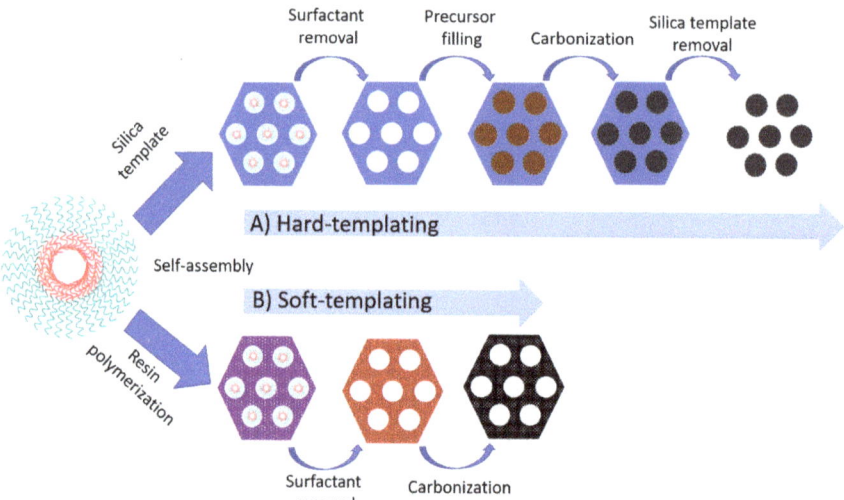

Figure 2. Two typical methods for the preparation of ordered mesoporous carbon materials: (**A**) the nanocasting strategy from mesoporous silica hard templates and (**B**) the direct synthesis from block copolymer soft templates. Adapted from [95].

The hard-template route requires the use of expensive reagents for the impregnation of the template and toxic reagents such as HF for the selective removal of the silica template. Besides, it is a time-consuming and multi-step complex process and, consequently, the manufacture of OMC at high scale is not suitable [101]. Efforts have been made for the development of OMC at cost-effective approaches with controllable pore size. Via soft-templating, OMC is obtained via self-assembling of supramolecular aggregates of carbon precursors (thermosetting agents such as phenolic resin or resorcinol-formaldehyde mixtures) and an amphiphilic copolymer surfactant (F127, CTAB, P123, and so on) as a template (Figure 2B) [102]. The carbon precursor was polymerized to form a highly cross-linked composite, followed by the template removal and carbonization. The direct process is simple, cost-effective, and suitable for large-scale industrial applications [103]. Using F127 as a soft-template, a GOx-based biosensor displayed a linear concentration range from 5 to 100 mg/mL [104].

More recent developments are focusing on the preparation of OMC with large mesopores and graphite walls. A hierarchically porous partially graphitic carbon membrane with three-dimensionally networked nanotunnels was used as a monolithic electrode matrix for the construction of a glucose biosensor [105]. The nanotunnels (~40–80 nm in diameter) are composed of partially graphitic carbon with ordered mesoporous (~6.5 nm in diameter). The carbon material was subsequently modified with polydopamine and decorated with AuNPs for the immobilization of GOx. The biosensor displays an LOD of 4.8 pM, which is four orders of magnitude lower than conventional nanostructured enzymatic glucose sensors.

High surface volume and ordered mesoporous make OMC an interesting material for biosensing, although they still suffer from a number of limitations. The removal of silica or polymer requires the use of HF, NaOH, or high temperature [106]. OMC materials are usually powdered materials and the use of a binder is required in order to improve the mechanical stability that can be tackle with the development of continuous OMC directly attached onto the electrode surface. Finally the majority of studies to date rely on the use of CMK-3 or CMK-1, materials that are not suitable for production on a large scale. Future studies require the development of alternative OMC materials that could provide the same advantages of graphene or carbon nanotubes.

3. 3D-Printing Technology

Additive manufacturing (AM) or three-dimensional printing is an emerging eco-friendly technology that holds promise to revolutionize the fabrication process. AM is based on layer-by-layer deposition of materials onto a substrate capable to manufacture geometrically complex objects in a one-step digitally controlled process [107]. 3D-printed devices are manufactured in a highly flexible manner with fast process times, generating minimum waste while offering precise replication and reducing constraints of creativity. In contrast, conventional technologies require complex, expensive machinery and tools (drilling, milling, and so on) [108]. The specific applications and requirements (material, composition, transparency, and so on) of the printed device define the most suitable 3D-printer technology. To date, various 3D-printing processes have been examined, including fused deposition modeling (FDM), inkjet and polyjet printing, and selective laser melting (SLM) [109], for the development of biosensors. A summary of the processes, printable materials, build volume, advantages, limitations, and printers, as well as cost and printing effectiveness, has been recently described [110,111]. The main applications of 3D-printing technology in the development of electrochemical biosensors are based on the development of fluidic platforms, electroactive, and catalytic surfaces and the manufacture of structures that include 3D-electrodes, flow channels, and auxiliary structures such as microneedles [112].

In comparison with traditional techniques for the manufacture of thin or thick electrodes (focused ion beam milling, electron-beam lithography, photolithography, and screen-printing), AM holds promise in overcoming issues such as high equipment and process costs. Screen-printing is a commonly used approach in the preparation of electrodes. In contrast to screen-printing, AM minimizes the consumption of materials to be printed, and thus reduces waste. AM also allows for the formation of small sized electrodes and the deposition of biomolecules with high spatial resolution. To date, the majority of 3D-printing methods use stainless steel owing to its cost-effectiveness and its passivated surface. In order to make 3D-printed stainless steel suitable for electrochemical sensing, the steel needs to be coated with another metal (Au, Bi, IrO2, Pt, and so on) [109]. For instance, Ambrosini et al. used selective laser melting for the manufacture of 3D-printed stainless steel electrodes, which were then modified with three different catalysts via electrodeposition [113]. A similar approach was also reported by Pumera and co-workers on the use of 3D-printed helical-shaped stainless steel electrodes that were subsequently electroplated with gold [114]. Gold-plated 3D-printed electrodes were utilized as a platform for DNA hybridization with different target DNA sequences (Figure 3). Upon hybridization with complementary DNA, the biosensor displayed a linear response over the range of 1–1000 nM. The selectivity of the sensor was examined using a non-complementary DNA sequence, resulting in a similar electrochemical response to the probe DNA owing to ineffective levels of hybridization. DNA biosensors require the selective discrimination of single-nucleotide polymorphism [115], and thus further investigations of 3D-printed biosensors are needed.

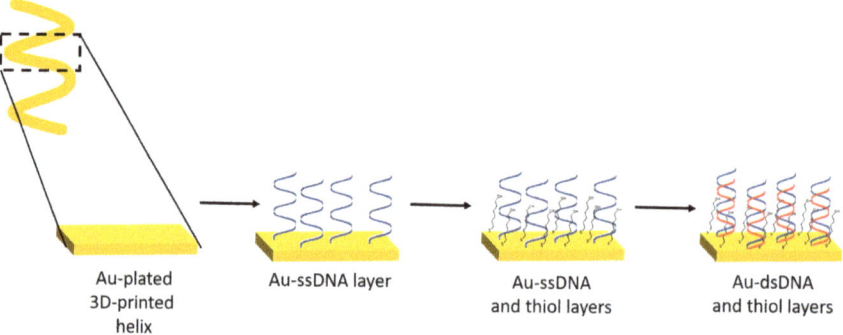

Figure 3. Schematic representation of the preparation of DNA biosensor. The thiolated DNA was covalently immobilized onto a gold-plated 3D-printed helix electrode. The modified electrode was then incubated with a DNA target, and the electrode was then exposed to methylene blue. Adapted from [114].

3D-printed metal electrodes are expensive and offer a limited electrochemical potential window that can restrict applications in biosensing. Carbon-based materials are more attractive materials owing to reduced costs. Carbon nanotubes, graphene, and carbon black are commonly used for the development of 3D-printed electrodes. However, 3D-printed carbon electrodes suffer from poor electrochemical performance as the carbon material is combined with polymeric binders, often in the presence of surfactants [116]. The presence of high levels of binder leads to printing issues owing to the high viscosity and tackiness of the ink, whereas low concentrations of binder may result in film cracking [117]. Different methods have been used to improve the electrochemical performance by removing the protective polymer of the top layer and exposing the carbon materials to solvents such as dimethylformamide or by electrochemical activation. Using both methods can enhance the electrochemical performance [118]. Katseli et al. described a carbon black/PLA electrode modified with GOx and Nafion®(Chemours, US) [119]. The glucose biosensor relied on the detection of H_2O_2 and exhibited a linear response over the range of 2–28 mM. A 3D-printed graphene/PLA was modified with AuNPs and horseradish peroxidase for the electrochemical detection of hydrogen peroxide at 0.0 V versus Ag/AgCl via direct electron transfer [120]. The biosensor was used to detect H_2O_2 in human serum and had a stable response after 7 days of incubation. A 3D-printed graphene/PLA was treated with DMF before the immobilization of GOx by crosslinking with glutaraldehyde [121]. The biosensor relied on the use of ferrocene-carboxylic acid as mediator for the indirect detection of H_2O_2 generated from the enzymatic reaction and was utilised for the detection of nitrite and uric acid in human saliva and urine respectively. A review on 3D-printed electrochemical sensors has recently been published [122].

4. Conclusions and Future Perspective

Electrochemical biosensors can be readily incorporated into miniaturized, portable devices. Although biorecognition elements provide reliability and good analytical performance, they can suffer from disadvantages such as high cost, short lifetime, and low levels of stability. The preparation of more stable biorecognition elements using a range of genetic engineering approaches to overcome these limitations is a major focus of current research. The development of structured materials with properties tailored to the effective and selective immobilization of the biorecognition elements will be needed for each particular system. The point of care detection of small molecules such as glucose, lactate, cholesterol, and so on has been successfully demonstrated. However, the detection of large molecules such as proteins and nucleic acids suffers from issues such as electrode fouling and non-specific adsorption of biomolecules. Resolution of these issues would enable more widespread use of electrochemical techniques. In the analysis of large biomolecules, low limits of detection are required, levels that can be enhanced with the use of nanomaterials such as nanoporous gold, graphene, or carbon nanotubes. The pore size of nanoporous gold can be tailored in the range 5 to 700 nm, a size range sufficient to accommodate large amounts of biomolecules. Nanoporous gold possesses advantages such as high surface area, good conductivity, and biocompatibility that make it an attractive material for biosensing. However, the complexity of the manufacturing process currently limits the applicability of NPG to research applications. In comparison, the manufacture of carbon nanomaterials can be performed at a relatively low cost. Graphene has enhanced sensitivity for a wide range of biomolecules when compared with other carbon materials such as carbon nanotubes or ordered mesoporous carbon. The unique properties of graphene (high conductivity, high surface area, excellent mechanical properties, ease functionalization, and scalability) and the low cost of manufacture make it an attractive material for the manufacture of biosensors. The use of graphene in non-invasive biosensors can open up new applications in wearable sensors and personalized health. However, such devices face challenges such as improved comfort and analytical performance.

Additive manufacturing has the ability to produce geometrically complex devices in a digitally controlled process. Additive manufacturing methods have been used to prepare a range of structures and electrodes. However, the exploitation of additive manufacturing is still at an early stage and further, detailed investigations are required. For example, the immobilization of enzymes deep within pores and channels may give rise to issues with substrate transport. The use of 3D-printed customized microfluidic devices can potentially overcome such transport limitations. To date, the electrochemical performance of 3D-printed electrodes shows diminished performance when compared with electrodes manufactured using more established methods. The high cost of consumables and instrumentation needed to prepare 3D-printed metal electrodes, and the difficulty in manufacturing porous structures, will possibly limit applications in biosensing. 3D-printed carbon electrodes hold more promise owing to their low cost, ease of fabrication, and suitability for large-scale production. To date, only a relatively small number of biosensors based on 3D-printed electrodes has been reported. Further research is required to produce 3D-printed electrodes at a large scale and with the performance required for clinically relevant analytes.

Author Contributions: This work has been carried out between all authors. F.O. and E.M. prepared the initial outline. F.O. wrote the original draft. E.M. critically reviewed the paper. E.M. contributed to supervision and project management. All authors have read and agreed to the published version of the manuscript.

Funding: This research was funded by the European Union's Horizon 2020 research and innovation programme, M3DLoC (Additive Manufacturing of 3D Microfluidics MEMs for Lab-on-a-Chip applications) under grant agreement no. 760662.

Conflicts of Interest: The authors declare no conflict of interest.

References

1. Zhu, C.; Yang, G.; Li, H.; Du, D.; Lin, Y. Electrochemical sensors and biosensors based on nanomaterials and nanostructures. *Anal. Chem.* **2014**, *87*, 230–249. [CrossRef]
2. Kimmel, D.W.; LeBlanc, G.; Meschievitz, M.E.; Cliffel, D.E. Electrochemical sensors and biosensors. *Anal. Chem.* **2012**, *84*, 685–707. [CrossRef]
3. Thevenot, D.R.; Toth, K.; Durst, R.A.; Wilson, G.S. Electrochemical biosensors: Recommended definitions and classification. *Pure Appl. Chem.* **1999**, *71*, 2333–2348. [CrossRef]

4. Ronkainen, N.J.; Halsall, H.B.; Heineman, W.R. Electrochemical biosensors. *Chem. Soc. Rev.* **2010**, *39*, 1747–1763. [CrossRef]
5. Turner, A.P. Biosensors: Sense and sensibility. *Chem. Soc. Rev.* **2013**, *42*, 3184–3196. [CrossRef]
6. Clark Jr, L.C.; Lyons, C. Electrode systems for continuous monitoring in cardiovascular surgery. *Ann. N. Y. Acad. Sci.* **1962**, *102*, 29–45. [CrossRef]
7. Magner, E. Trends in electrochemical biosensors. *Analyst* **1998**, *123*, 1967–1970. [CrossRef]
8. Labib, M.; Sargent, E.H.; Kelley, S.O. Electrochemical methods for the analysis of clinically relevant biomolecules. *Chem. Rev.* **2016**, *116*, 9001–9090. [CrossRef]
9. Wang, J. Electrochemical glucose biosensors. *Chem. Rev.* **2008**, *108*, 814–825. [CrossRef]
10. Balakumar, P.; Maung-U, K.; Jagadeesh, G. Prevalence and prevention of cardiovascular disease and diabetes mellitus. *Pharmacol. Res.* **2016**, *113*, 600–609. [CrossRef]
11. Chen, C.; Xie, Q.; Yang, D.; Xiao, H.; Fu, Y.; Tan, Y.; Yao, S. Recent advances in electrochemical glucose biosensors: A review. *RSC Adv.* **2013**, *3*, 4473–4491. [CrossRef]
12. Hwang, D.-W.; Lee, S.; Seo, M.; Chung, T.D. Recent advances in electrochemical non-enzymatic glucose sensors–a review. *Anal. Chim. Acta* **2018**, *1033*, 1–34. [CrossRef]
13. Heller, A.; Feldman, B. Electrochemical glucose sensors and their applications in diabetes management. *Chem. Rev.* **2008**, *108*, 2482–2505. [CrossRef]
14. Xiao, X.; Si, P.; Magner, E. An overview of dealloyed nanoporous gold in bioelectrochemistry. *Bioelectrochemistry* **2016**, *109*, 117–126. [CrossRef]
15. Iguchi, S.; Kudo, H.; Saito, T.; Ogawa, M.; Saito, H.; Otsuka, K.; Funakubo, A.; Mitsubayashi, K. A flexible and wearable biosensor for tear glucose measurement. *Biomed. Microdevices* **2007**, *9*, 603–609. [CrossRef]
16. Arakawa, T.; Kuroki, Y.; Nitta, H.; Chouhan, P.; Toma, K.; Sawada, S.-i.; Takeuchi, S.; Sekita, T.; Akiyoshi, K.; Minakuchi, S. Mouthguard biosensor with telemetry system for monitoring of saliva glucose: A novel cavitas sensor. *Biosen. Bioelectron.* **2016**, *84*, 106–111. [CrossRef]
17. Lipani, L.; Dupont, B.G.; Doungmene, F.; Marken, F.; Tyrrell, R.M.; Guy, R.H.; Ilie, A. Non-invasive, transdermal, path-selective and specific glucose monitoring via a graphene-based platform. *Nat. Nanotechnol.* **2018**, *13*, 504–511. [CrossRef]
18. Lee, H.; Hong, Y.J.; Baik, S.; Hyeon, T.; Kim, D.H. Enzyme-based glucose sensor: From invasive to wearable device. *Adv. Healthc. Mater.* **2018**, *7*, 1701150. [CrossRef]
19. Bahadır, E.B.; Sezgintürk, M.K. Applications of commercial biosensors in clinical, food, environmental, and biothreat/biowarfare analyses. *Anal. Biochem.* **2015**, *478*, 107–120. [CrossRef]
20. Bahadır, E.B.; Sezgintürk, M.K. Applications of electrochemical immunosensors for early clinical diagnostics. *Talanta* **2015**, *132*, 162–174. [CrossRef]
21. Sattayasamitsathit, S.; O'Mahony, A.M.; Xiao, X.; Brozik, S.M.; Washburn, C.M.; Wheeler, D.R.; Gao, W.; Minteer, S.; Cha, J.; Burckel, D.B. Highly ordered tailored three-dimensional hierarchical nano/microporous gold–carbon architectures. *J. Mater. Chem.* **2012**, *22*, 11950–11956. [CrossRef]
22. Juarez, T.; Biener, J.; Weissmüller, J.; Hodge, A.M. Nanoporous metals with structural hierarchy: A review. *Adv. Eng. Mater.* **2017**, *19*, 1700389. [CrossRef]
23. Zhang, J.; Li, C.M. Nanoporous metals: Fabrication strategies and advanced electrochemical applications in catalysis, sensing and energy systems. *Chem. Soc. Rev.* **2012**, *41*, 7016–7031. [CrossRef]
24. Trasatti, S.; Petrii, O. Real surface area measurements in electrochemistry. *Pure Appl. Chem.* **1991**, *63*, 711–734. [CrossRef]
25. Seker, E.; Reed, M.L.; Begley, M.R. Nanoporous gold: Fabrication, characterization, and applications. *Materials* **2009**, *2*, 2188–2215. [CrossRef]
26. Wittstock, A.; Wichmann, A.; Baumer, M. Nanoporous gold as a platform for a building block catalyst. *ACS Catal.* **2012**, *2*, 2199–2215. [CrossRef]
27. Qiu, H.; Xue, L.; Ji, G.; Zhou, G.; Huang, X.; Qu, Y.; Gao, P. Enzyme-modified nanoporous gold-based electrochemical biosensors. *Biosen. Bioelectron.* **2009**, *24*, 3014–3018. [CrossRef]
28. Hu, K.; Lan, D.; Li, X.; Zhang, S. Electrochemical DNA biosensor based on nanoporous gold electrode and multifunctional encoded DNA–Au bio bar codes. *Anal. Chem.* **2008**, *80*, 9124–9130. [CrossRef]
29. Xiao, X.; Ulstrup, J.; Li, H.; Zhang, J.; Si, P. Nanoporous gold assembly of glucose oxidase for electrochemical biosensing. *Electrochim. Acta* **2014**, *130*, 559–567. [CrossRef]

30. Meng, F.; Yan, X.; Liu, J.; Gu, J.; Zou, Z. Nanoporous gold as non-enzymatic sensor for hydrogen peroxide. *Electrochim. Acta* **2011**, *56*, 4657–4662. [CrossRef]
31. Sun, X.; Ma, Z. Electrochemical immunosensor based on nanoporpus gold loading thionine for carcinoembryonic antigen. *Anal. Chim. Acta* **2013**, *780*, 95–100. [CrossRef]
32. Xiao, X.; Conghaile, P.Ó.; Leech, D.; Ludwig, R.; Magner, E. An oxygen-independent and membrane-less glucose biobattery/supercapacitor hybrid device. *Biosen. Bioelectron.* **2017**, *98*, 421–427. [CrossRef]
33. Xiao, X.; Siepenkoetter, T.; Conghaile, P.O.; Leech, D.n.; Magner, E. Nanoporous gold-based biofuel cells on contact lenses. *ACS Appl. Mater. Interfaces* **2018**, *10*, 7107–7116. [CrossRef]
34. Xu, J.; Kou, T.; Zhang, Z. Anodization strategy to fabricate nanoporous gold for high-sensitivity detection of p-nitrophenol. *CrystEngComm* **2013**, *15*, 7856–7862. [CrossRef]
35. Sáenz, H.S.C.; Hernández-Saravia, L.P.; Selva, J.S.; Sukeri, A.; Espinoza-Montero, P.J.; Bertotti, M. Electrochemical dopamine sensor using a nanoporous gold microelectrode: A proof-of-concept study for the detection of dopamine release by scanning electrochemical microscopy. *Microchim. Acta* **2018**, *185*, 367. [CrossRef]
36. Nishio, K.; Masuda, H. Anodization of gold in oxalate solution to form a nanoporous black film. *Angew. Chem. Int. Ed.* **2011**, *50*, 1603–1607. [CrossRef]
37. Cherevko, S.; Chung, C.-H. Impact of key deposition parameters on the morphology of silver foams prepared by dynamic hydrogen template deposition. *Electrochim. Acta* **2010**, *55*, 6383–6390. [CrossRef]
38. Sanzó, G.; Taurino, I.; Antiochia, R.; Gorton, L.; Favero, G.; Mazzei, F.; De Micheli, G.; Carrara, S. Bubble electrodeposition of gold porous nanocorals for the enzymatic and non-enzymatic detection of glucose. *Bioelectrochemistry* **2016**, *112*, 125–131. [CrossRef]
39. Szamocki, R.; Reculusa, S.; Ravaine, S.; Bartlett, P.N.; Kuhn, A.; Hempelmann, R. Tailored mesostructuring and biofunctionalization of gold for increased electroactivity. *Angew. Chem. Int. Ed.* **2006**, *45*, 1317–1321. [CrossRef]
40. Gamero, M.; Sosna, M.; Pariente, F.; Lorenzo, E.; Bartlett, P.; Alonso, C. Influence of macroporous gold support and its functionalization on lactate oxidase-based biosensors response. *Talanta* **2012**, *94*, 328–334. [CrossRef]
41. Huang, J.F.; Sun, I.W. Fabrication and surface functionalization of nanoporous gold by electrochemical alloying/dealloying of Au–Zn in an ionic liquid, and the self-assembly of L-Cysteine monolayers. *Adv. Funct. Mater.* **2005**, *15*, 989–994. [CrossRef]
42. Rouya, E.; Reed, M.L.; Kelly, R.G.; Bart-Smith, H.; Begley, M.; Zangari, G. Synthesis of nanoporous gold structures via dealloying of electroplated Au-Ni alloy films. *ECS Trans.* **2007**, *6*, 41–50. [CrossRef]
43. Zhang, Q.; Wang, X.; Qi, Z.; Wang, Y.; Zhang, Z. A benign route to fabricate nanoporous gold through electrochemical dealloying of Al–Au alloys in a neutral solution. *Electrochim. Acta* **2009**, *54*, 6190–6198. [CrossRef]
44. Gupta, G.; Thorp, J.; Mara, N.; Dattelbaum, A.; Misra, A.; Picraux, S. Morphology and porosity of nanoporous Au thin films formed by dealloying of $AuSi_{1-x}$. *J. Appl. Phys.* **2012**, *112*, 094320. [CrossRef]
45. Snyder, J.; Asanithi, P.; Dalton, A.B.; Erlebacher, J. Stabilized nanoporous metals by dealloying ternary alloy precursors. *Adv. Mater.* **2008**, *20*, 4883–4886. [CrossRef]
46. Siepenkoetter, T.; Salaj-Kosla, U.; Xiao, X.; Belochapkine, S.; Magner, E. Nanoporous gold electrodes with tuneable pore sizes for bioelectrochemical applications. *Electroanalysis* **2016**, *28*, 2415–2423. [CrossRef]
47. Qian, L.; Yan, X.; Fujita, T.; Inoue, A.; Chen, M. Surface enhanced Raman scattering of nanoporous gold: Smaller pore sizes stronger enhancements. *Appl. Phys. Lett.* **2007**, *90*, 153120. [CrossRef]
48. Siepenkoetter, T.; Salaj-Kosla, U.; Magner, E. The immobilization of fructose dehydrogenase on nanoporous gold electrodes for the detection of fructose. *ChemElectroChem* **2017**, *4*, 905–912. [CrossRef]
49. Matharu, Z.; Daggumati, P.; Wang, L.; Dorofeeva, T.S.; Li, Z.; Seker, E. Nanoporous-gold-based electrode morphology libraries for investigating structure–property relationships in nucleic acid based electrochemical biosensors. *ACS Appl. Mater. Interfaces* **2017**, *9*, 12959–12966. [CrossRef]
50. Qiu, H.-J.; Li, X.; Xu, H.-T.; Zhang, H.-J.; Wang, Y. Nanoporous metal as a platform for electrochemical and optical sensing. *J. Mater. Chem. C* **2014**, *2*, 9788–9799. [CrossRef]
51. Kim, S.H. Nanoporous gold: Preparation and applications to catalysis and sensors. *Curr. Appl. Phys.* **2018**, *18*, 810–818. [CrossRef]

52. Novoselov, K.S.; Geim, A.K.; Morozov, S.V.; Jiang, D.; Zhang, Y.; Dubonos, S.V.; Grigorieva, I.V.; Firsov, A.A. Electric field effect in atomically thin carbon films. *Science* **2004**, *306*, 666–669. [CrossRef]
53. Geim, A.K.; Novoselov, K.S. The rise of graphene. In *Nanoscience and Technology: A Collection of Reviews from Nature Journals*; World Scientific: Singapore, 2010; pp. 11–19.
54. Kuila, T.; Bose, S.; Khanra, P.; Mishra, A.K.; Kim, N.H.; Lee, J.H. Recent advances in graphene-based biosensors. *Biosen. Bioelectron.* **2011**, *26*, 4637–4648. [CrossRef]
55. Stoller, M.D.; Park, S.; Zhu, Y.; An, J.; Ruoff, R.S. Graphene-based ultracapacitors. *Nano Lett.* **2008**, *8*, 3498–3502. [CrossRef]
56. Stankovich, S.; Dikin, D.A.; Piner, R.D.; Kohlhaas, K.A.; Kleinhammes, A.; Jia, Y.; Wu, Y.; Nguyen, S.T.; Ruoff, R.S. Synthesis of graphene-based nanosheets via chemical reduction of exfoliated graphite oxide. *Carbon* **2007**, *45*, 1558–1565. [CrossRef]
57. Lee, C.; Wei, X.; Kysar, J.W.; Hone, J. Measurement of the elastic properties and intrinsic strength of monolayer graphene. *Science* **2008**, *321*, 385–388. [CrossRef]
58. Hummers, W.S., Jr.; Offeman, R.E. Preparation of graphitic oxide. *J. Am. Chem. Soc.* **1958**, *80*, 1339. [CrossRef]
59. Marcano, D.C.; Kosynkin, D.V.; Berlin, J.M.; Sinitskii, A.; Sun, Z.; Slesarev, A.; Alemany, L.B.; Lu, W.; Tour, J.M. Improved synthesis of graphene oxide. *ACS Nano* **2010**, *4*, 4806–4814. [CrossRef]
60. Liu, Y.; Yu, D.; Zeng, C.; Miao, Z.; Dai, L. Biocompatible graphene oxide-based glucose biosensors. *Langmuir* **2010**, *26*, 6158–6160. [CrossRef]
61. Qiu, J.-D.; Huang, J.; Liang, R.-P. Nanocomposite film based on graphene oxide for high performance flexible glucose biosensor. *Sens. Actuators B Chem.* **2011**, *160*, 287–294. [CrossRef]
62. Quast, T.; Mariani, F.; Scavetta, E.; Schuhmann, W.; Andronescu, C. Reduced-Graphene-Oxide-Based Needle-Type Field-Effect Transistor for Dopamine Sensing. *ChemElectroChem* **2020**, *7*, 1922–1927. [CrossRef]
63. Wang, H.; Maiyalagan, T.; Wang, X. Review on recent progress in nitrogen-doped graphene: Synthesis, characterization, and its potential applications. *ACS Catal.* **2012**, *2*, 781–794. [CrossRef]
64. Lee, H.; Song, C.; Hong, Y.S.; Kim, M.S.; Cho, H.R.; Kang, T.; Shin, K.; Choi, S.H.; Hyeon, T.; Kim, D.-H. Wearable/disposable sweat-based glucose monitoring device with multistage transdermal drug delivery module. *Sci. Adv.* **2017**, *3*, e1601314. [CrossRef]
65. Li, L.; Pan, L.; Ma, Z.; Yan, K.; Cheng, W.; Shi, Y.; Yu, G. All inkjet-printed amperometric multiplexed biosensors based on nanostructured conductive hydrogel electrodes. *Nano Lett.* **2018**, *18*, 3322–3327. [CrossRef]
66. Adeva-Andany, M.; López-Ojén, M.; Funcasta-Calderón, R.; Ameneiros-Rodríguez, E.; Donapetry-García, C.; Vila-Altesor, M.; Rodríguez-Seijas, J. Comprehensive review on lactate metabolism in human health. *Mitochondrion* **2014**, *17*, 76–100. [CrossRef]
67. Rahsepar, M.; Foroughi, F.; Kim, H. A new enzyme-free biosensor based on nitrogen-doped graphene with high sensing performance for electrochemical detection of glucose at biological pH value. *Sens. Actuators B Chem.* **2019**, *282*, 322–330. [CrossRef]
68. Barsan, M.M.; David, M.; Florescu, M.; Țugulea, L.; Brett, C.M. A new self-assembled layer-by-layer glucose biosensor based on chitosan biopolymer entrapped enzyme with nitrogen doped graphene. *Bioelectrochemistry* **2014**, *99*, 46–52. [CrossRef]
69. Kaur, M.; Kaur, M.; Sharma, V.K. Nitrogen-doped graphene and graphene quantum dots: A review on synthesis and applications in energy, sensors and environment. *Adv. Colloid Interface Sci.* **2018**, *259*, 44–64. [CrossRef]
70. Yadav, R.; Dixit, C. Synthesis, characterization and prospective applications of nitrogen-doped graphene: A short review. *J. Sci. Adv. Mater. Devices* **2017**, *2*, 141–149. [CrossRef]
71. Tang, J.; Yan, X.; Engelbrekt, C.; Ulstrup, J.; Magner, E.; Xiao, X.; Zhang, J. Development of graphene-based enzymatic biofuel cells: A minireview. *Bioelectrochemistry* **2020**, *134*, 107537. [CrossRef]
72. Fan, Y.; Shi, S.; Ma, J.; Guo, Y. A paper-based electrochemical immunosensor with reduced graphene oxide/thionine/gold nanoparticles nanocomposites modification for the detection of cancer antigen 125. *Biosen. Bioelectron.* **2019**, *135*, 1–7. [CrossRef]
73. Gong, Q.; Wang, Y.; Yang, H. A sensitive impedimetric DNA biosensor for the determination of the HIV gene based on graphene-Nafion composite film. *Biosen. Bioelectron.* **2017**, *89*, 565–569. [CrossRef]

74. Rafighi, P.; Tavahodi, M.; Haghighi, B. Fabrication of a third-generation glucose biosensor using graphene-polyethyleneimine-gold nanoparticles hybrid. *Sens. and Actuators B Chem.* **2016**, *232*, 454–461. [CrossRef]
75. Xu, Q.; Gu, S.-X.; Jin, L.; Zhou, Y.-E.; Yang, Z.; Wang, W.; Hu, X. Graphene/polyaniline/gold nanoparticles nanocomposite for the direct electron transfer of glucose oxidase and glucose biosensing. *Sens. Actuators B Chem.* **2014**, *190*, 562–569. [CrossRef]
76. Cosnier, S.; Holzinger, M. Electrosynthesized polymers for biosensing. *Chem. Soc. Rev.* **2011**, *40*, 2146–2156. [CrossRef]
77. Liu, Y.; Turner, A.P.; Zhao, M.; Mak, W.C. Processable enzyme-hybrid conductive polymer composites for electrochemical biosensing. *Biosen. Bioelectron.* **2018**, *100*, 374–381. [CrossRef]
78. Chen, H.; Müller, M.B.; Gilmore, K.J.; Wallace, G.G.; Li, D. Mechanically strong, electrically conductive, and biocompatible graphene paper. *Adv. Mat.* **2008**, *20*, 3557–3561. [CrossRef]
79. Wang, Z.; Dai, Z. Carbon nanomaterial-based electrochemical biosensors: An overview. *Nanoscale* **2015**, *7*, 6420–6431. [CrossRef]
80. Nagar, B.; Balsells, M.; de la Escosura-Muñiz, A.; Gomez-Romero, P.; Merkoçi, A. Fully printed one-step biosensing device using graphene/AuNPs composite. *Biosen. Bioelectron.* **2019**, *129*, 238–244. [CrossRef]
81. Yang, N.; Chen, X.; Ren, T.; Zhang, P.; Yang, D. Carbon nanotube based biosensors. *Sens. Actuators B Chem.* **2015**, *207*, 690–715. [CrossRef]
82. ChandraKishore, S.; Pandurangan, A. Electrophoretic deposition of cobalt catalyst layer over stainless steel for the high yield synthesis of carbon nanotubes. *Appl. Surf. Sci.* **2012**, *258*, 7936–7942. [CrossRef]
83. Guo, T.; Nikolaev, P.; Thess, A.; Colbert, D.T.; Smalley, R.E. Catalytic growth of single-walled manotubes by laser vaporization. *Chem. Phys.Lett.* **1995**, *243*, 49–54. [CrossRef]
84. Iijima, S. Helical microtubules of graphitic carbon. *Nature* **1991**, *354*, 56. [CrossRef]
85. Zhu, Z. An overview of carbon nanotubes and graphene for biosensing applications. *Nano-Micro Lett.* **2017**, *9*, 25. [CrossRef]
86. Bollella, P.; Hibino, Y.; Kano, K.; Gorton, L.; Antiochia, R. Enhanced direct electron transfer of fructose dehydrogenase rationally immobilized on a 2-aminoanthracene diazonium cation grafted single-walled carbon nanotube based electrode. *ACS Catal.* **2018**, *8*, 10279–10289. [CrossRef]
87. Hashemzadeh, S.; Omidi, Y.; Rafii-Tabar, H. Amperometric lactate nanobiosensor based on reduced graphene oxide, carbon nanotube and gold nanoparticle nanocomposite. *Microchim. Acta* **2019**, *186*, 680. [CrossRef]
88. Kang, B.-C.; Park, B.-S.; Ha, T.-J. Highly sensitive wearable glucose sensor systems based on functionalized single-wall carbon nanotubes with glucose oxidase-nafion composites. *Appl. Surf. Sci.* **2019**, *470*, 13–18. [CrossRef]
89. Cinquin, P.; Gondran, C.; Giroud, F.; Mazabrard, S.; Pellissier, A.; Boucher, F.; Alcaraz, J.-P.; Gorgy, K.; Lenouvel, F.; Mathé, S. A glucose biofuel cell implanted in rats. *PLoS ONE* **2010**, *5*, e10476. [CrossRef]
90. Cosnier, S.; Gross, A.J.; Le Goff, A.; Holzinger, M. Recent advances on enzymatic glucose/oxygen and hydrogen/oxygen biofuel cells: Achievements and limitations. *J. Power Sources* **2016**, *325*, 252–263. [CrossRef]
91. Han, S.; Liu, W.; Zheng, M.; Wang, R. Label-Free and Ultrasensitive Electrochemical DNA Biosensor Based on Urchinlike Carbon Nanotube-Gold Nanoparticle Nanoclusters. *Anal. Chem.* **2020**, *92*, 4780–4787. [CrossRef]
92. Viet, N.X.; Hoan, N.X.; Takamura, Y. Development of highly sensitive electrochemical immunosensor based on single-walled carbon nanotube modified screen-printed carbon electrode. *Mater. Chem. Phys.* **2019**, *227*, 123–129. [CrossRef]
93. Janssen, J.; Lambeta, M.; White, P.; Byagowi, A. Carbon Nanotube-Based Electrochemical Biosensor for Label-Free Protein Detection. *Biosensors* **2019**, *9*, 144. [CrossRef]
94. Walcarius, A. Recent trends on electrochemical sensors based on ordered mesoporous carbon. *Sensors* **2017**, *17*, 1863. [CrossRef]
95. Ma, T.-Y.; Liu, L.; Yuan, Z.-Y. Direct synthesis of ordered mesoporous carbons. *Chem. Soc. Rev.* **2013**, *42*, 3977–4003. [CrossRef]
96. Jarczewski, S.; Drozdek, M.; Michorczyk, P.; Cuadrado-Collados, C.; Gandara-Loe, J.; Silvestre-Albero, J.; Kuśtrowski, P. Oxidative dehydrogenation of ethylbenzene over CMK-1 and CMK-3 carbon replicas with various mesopore architectures. *Microporous Mesoporous Mater.* **2018**, *271*, 262–272. [CrossRef]

97. Zhu, Z.; Li, X.; Zeng, Y.; Sun, W. Ordered mesoporous carbon modified carbon ionic liquid electrode for the electrochemical detection of double-stranded DNA. *Biosen. Bioelectron.* **2010**, *25*, 2313–2317. [CrossRef]
98. Zhou, M.; Shang, L.; Li, B.; Huang, L.; Dong, S. Highly ordered mesoporous carbons as electrode material for the construction of electrochemical dehydrogenase-and oxidase-based biosensors. *Biosen. Bioelectron.* **2008**, *24*, 442–447. [CrossRef]
99. Joo, S.H.; Choi, S.J.; Oh, I.; Kwak, J.; Liu, Z.; Terasaki, O.; Ryoo, R. Ordered nanoporous arrays of carbon supporting high dispersions of platinum nanoparticles. *Nature* **2001**, *412*, 169–172. [CrossRef]
100. Ghasemi, E.; Shams, E.; Nejad, N.F. Covalent modification of ordered mesoporous carbon with glucose oxidase for fabrication of glucose biosensor. *J. Electroanal. Chem.* **2015**, *752*, 60–67. [CrossRef]
101. Nishihara, H.; Kyotani, T. Templated nanocarbons for energy storage. *Adv. Mater.* **2012**, *24*, 4473–4498. [CrossRef]
102. Walcarius, A. Electrocatalysis, sensors and biosensors in analytical chemistry based on ordered mesoporous and macroporous carbon-modified electrodes. *TrAC Trends Anal. Chem.* **2012**, *38*, 79–97. [CrossRef]
103. Liang, C.; Hong, K.; Guiochon, G.A.; Mays, J.W.; Dai, S. Synthesis of a large-scale highly ordered porous carbon film by self-assembly of block copolymers. *Angew. Chem. Int. Ed.* **2004**, *43*, 5785–5789. [CrossRef] [PubMed]
104. Dai, M.; Maxwell, S.; Vogt, B.D.; La Belle, J.T. Mesoporous carbon amperometric glucose sensors using inexpensive, commercial methacrylate-based binders. *Anal. Chim. Acta* **2012**, *738*, 27–34. [CrossRef] [PubMed]
105. Fu, C.; Yi, D.; Deng, C.; Wang, X.; Zhang, W.; Tang, Y.; Caruso, F.; Wang, Y. A partially graphitic mesoporous carbon membrane with three-dimensionally networked nanotunnels for ultrasensitive electrochemical detection. *Chem. Mater.* **2017**, *29*, 5286–5293. [CrossRef]
106. Sanati, A.; Jalali, M.; Raeissi, K.; Karimzadeh, F.; Kharaziha, M.; Mahshid, S.S.; Mahshid, S. A review on recent advancements in electrochemical biosensing using carbonaceous nanomaterials. *Microchim. Acta* **2019**, *186*, 773. [CrossRef]
107. O'Donnell, J.; Kim, M.; Yoon, H.-S. A review on electromechanical devices fabricated by additive manufacturing. *J. Manufac. Sci. Eng.* **2017**, *139*, 010801. [CrossRef]
108. Rocha, V.G.; Garcia-Tunon, E.; Botas, C.; Markoulidis, F.; Feilden, E.; D'Elia, E.; Ni, N.; Shaffer, M.; Saiz, E. Multimaterial 3D printing of graphene-based electrodes for electrochemical energy storage using thermoresponsive inks. *ACS Appl. Mater. Interfaces* **2017**, *9*, 37136–37145. [CrossRef]
109. Hamzah, H.H.; Shafiee, S.A.; Abdalla, A.; Patel, B.A. 3D printable conductive materials for the fabrication of electrochemical sensors: A mini review. *Electrochem. Commun.* **2018**, *96*, 27–31. [CrossRef]
110. Tofail, S.A.; Koumoulos, E.P.; Bandyopadhyay, A.; Bose, S.; O'Donoghue, L.; Charitidis, C. Additive manufacturing: Scientific and technological challenges, market uptake and opportunities. *Mater. Today* **2018**, *21*, 22–37. [CrossRef]
111. Lee, J.-Y.; An, J.; Chua, C.K. Fundamentals and applications of 3D printing for novel materials. *Appl. Mater. Today* **2017**, *7*, 120–133. [CrossRef]
112. Palenzuela, C.L.M.; Pumera, M. (Bio) Analytical chemistry enabled by 3D printing: Sensors and biosensors. *TrAC Trends Anal. Chem.* **2018**, *103*, 110–118. [CrossRef]
113. Ambrosi, A.; Pumera, M. Self-Contained Polymer/Metal 3D Printed Electrochemical Platform for Tailored Water Splitting. *Adv. Funct. Mater.* **2018**, *28*, 1700655. [CrossRef]
114. Loo, A.H.; Chua, C.K.; Pumera, M. DNA biosensing with 3D printing technology. *Analyst* **2017**, *142*, 279–283. [CrossRef] [PubMed]
115. Ferapontova, E.E. Hybridization biosensors relying on electrical properties of nucleic acids. *Electroanalysis* **2017**, *29*, 6–13. [CrossRef]
116. Lawes, S.; Riese, A.; Sun, Q.; Cheng, N.; Sun, X. Printing nanostructured carbon for energy storage and conversion applications. *Carbon* **2015**, *92*, 150–176. [CrossRef]
117. Somalu, M.R.; Brandon, N.P. Rheological studies of nickel/scandia-stabilized-zirconia screen printing inks for solid oxide fuel cell anode fabrication. *J. Am. Ceram. Soc.* **2012**, *95*, 1220–1228. [CrossRef]
118. Manzanares Palenzuela, C.L.; Novotný, F.; Krupička, P.; Sofer, Z.k.; Pumera, M. 3D-printed graphene/polylactic acid electrodes promise high sensitivity in electroanalysis. *Anal. Chem.* **2018**, *90*, 5753–5757. [CrossRef]

119. Katseli, V.; Economou, A.; Kokkinos, C. Single-step fabrication of an integrated 3D-printed device for electrochemical sensing applications. *Electrochem. Commun.* **2019**, *103*, 100–103. [CrossRef]
120. Marzo, A.M.L.; Mayorga-Martinez, C.C.; Pumera, M. 3D-printed graphene direct electron transfer enzyme biosensors. *Biosen. Bioelectron.* **2020**, *151*, 111980. [CrossRef]
121. Cardoso, R.M.; Silva, P.R.; Lima, A.P.; Rocha, D.P.; Oliveira, T.C.; do Prado, T.M.; Fava, E.L.; Fatibello-Filho, O.; Richter, E.M.; Muñoz, R.A. 3D-Printed graphene/polylactic acid electrode for bioanalysis: Biosensing of glucose and simultaneous determination of uric acid and nitrite in biological fluids. *Sens. Actuators B Chem.* **2020**, *307*, 127621. [CrossRef]
122. Cardoso, R.M.; Kalinke, C.; Rocha, R.G.; dos Santos, P.L.; Rocha, D.P.; Oliveira, P.R.; Janegitz, B.C.; Bonacin, J.A.; Richter, E.M.; Munoz, R.A. Additive-manufactured (3D-printed) electrochemical sensors: A critical review. *Anal. Chim. Acta* **2020**, *1118*, 73–91. [CrossRef] [PubMed]

© 2020 by the authors. Licensee MDPI, Basel, Switzerland. This article is an open access article distributed under the terms and conditions of the Creative Commons Attribution (CC BY) license (http://creativecommons.org/licenses/by/4.0/).

Review

Biosensors Based on Advanced Sulfur-Containing Nanomaterials

Chunmei Li, Yihan Wang, Hui Jiang and Xuemei Wang *

State Key Laboratory of Bioelectronics, National Demonstration Center for Experimental Biomedical Engineering Education, School of Biological Science and Medical Engineering, Southeast University, Nanjing 210096, China; li_chunmei@foxmail.com (C.L.); yihanwangxynu@163.com (Y.W.); sungi@seu.edu.cn (H.J.)
* Correspondence: xuewang@seu.edu.cn

Received: 1 June 2020; Accepted: 17 June 2020; Published: 19 June 2020

Abstract: In recent years, sulfur-containing nanomaterials and their derivatives/composites have attracted much attention because of their important role in the field of biosensor, biolabeling, drug delivery and diagnostic imaging technology, which inspires us to compile this review. To focus on the relationships between advanced biomaterials and biosensors, this review describes the applications of various types of sulfur-containing nanomaterials in biosensors. We bring two types of sulfur-containing nanomaterials including metallic sulfide nanomaterials and sulfur-containing quantum dots, to discuss and summarize the possibility and application as biosensors based on the sulfur-containing nanomaterials. Finally, future perspective and challenges of biosensors based on sulfur-containing nanomaterials are briefly rendered.

Keywords: sulfur-containing nanomaterials; metallic sulfide nanomaterials; sulfur-containing quantum dots; biosensors

1. Introduction

As a by-product of oil refining and natural-gas purification, sulfur usually exists in the form of sulfide, sulfate or elementary substance in nature and is one of most abundant elements [1–3]. Since the discovery of sulfur, research involving sulfur has always been at the center of scientific research topics. Researchers have dedicated to exploiting the wide applications of sulfur. Until now, sulfur has been important in our daily life with a wide variety of applications, such as vulcanization of rubber, being cathode of rechargeable battery, raw material for fertilizer, insecticide, plastic and gunpowder [1,4–6]. Under the right conditions, sulfur is also well-known to form compounds with numerous other elements (e.g., lead, calcium or iron), and even form sulfur-containing nanomaterials.

A variety of sulfur-containing nanomaterials have been reported, such as metallic sulfide nanomaterials, sulfur-containing quantum dots, sulfur-containing organosilicon compounds, and lithium sulfide materials [7–10]. Sulfur-containing nanomaterials (e.g., metallic sulfide nanomaterials and sulfur-containing quantum dots) exhibit excellent properties, such as nanometric scale, water-dispersible, non-toxicity, excellent catalytic activity, conductivity, photoactivity and fascinating optical properties, and they have proven useful in many biomedical applications including imaging and sensing [7,8]. As known, metallic sulfide nanomaterials have been used as photoactive materials which can generate photocurrent excited by light in biosensing systems. Some sulfur-containing quantum dots can stably bind with biomolecules or other nanomaterials due to their functional groups (e.g., amino, carboxyl and sulfhydryl groups) as common reaction sites within biological systems. This allows their versatile roles as functional biomaterials in biosensor, biolabeling, drug delivery and diagnostic imaging technology [7,11–13]. Moreover, some sulfur-containing quantum dots (e.g., Ag_2S quantum dots), exhibit high absorption in near-infrared (NIR) region, which enables their applications in bioimaging, biolabeling, deep tissue imaging, diagnostics and photodynamic therapy [7]. In this review, we will summarize the most recent advances on the applications of biosensors fabricated based on sulfur-containing nanomaterials and their composites (Scheme 1). Since there are too many sulfur-containing nanomaterials, it is impossible to provide a comprehensive overview of all sulfur-containing nanomaterials in a mini-review. Thus, we aim to provide two categories of sulfur-containing nanomaterials, i.e., metallic sulfide nanomaterials and sulfur-containing quantum dots. Concretely speaking, the metallic sulfide nanomaterials include binary, ternary, quaternary and non-metallic/metallic hetero-sulfides. The sulfur-containing quantum dots consist of sulfur, sulfide and sulfur-doped quantum dots. Firstly, we will briefly introduce various kinds of metallic sulfide nanomaterials or sulfur-containing quantum dots and summarize their synthetic approaches, respectively. Then, we will discuss the possibility as biosensors of the two categories, respectively. We also summarize the applications of biosensors based on metallic sulfide nanomaterials or sulfur-containing quantum dots, respectively. Lastly, the future perspectives and challenges of biosensors based on metallic sulfide nanomaterials or sulfur-containing quantum dots are briefly rendered.

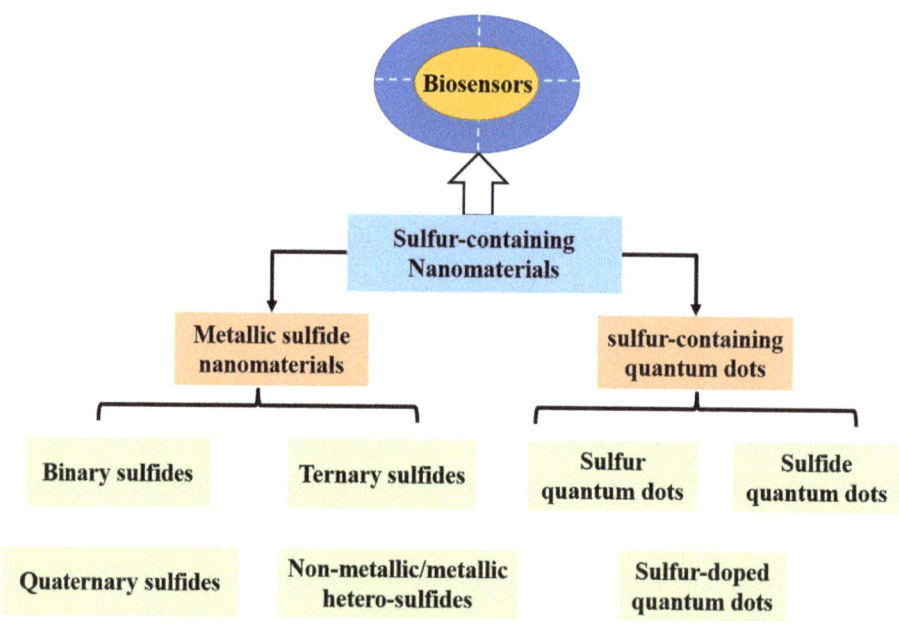

Scheme 1. A summary of sulfur-containing nanomaterials used as biosensors.

2. Metallic Sulfide Nanomaterials

2.1. Generalities

Metal sulfides contain chemical bonding of one or more sulfur atoms (S) to a metal (M) [7]. They can be broken down into four main categories: binary, ternary, quaternary, and non-metallic/metallic hetero-sulfides, which can be denoted by the chemical formulas of M_xS_n, $M_xM'_yS_n$, $M_xM'_yM''_zS_n$ and $M_xA_iB_j \ldots C_kS_n$ (A, B, C = non-metallic atoms), respectively. It should be noted that actually metal sulfide nanomaterials also include metal sulfide quantum dots, which will be illustrated in the section of "sulfur-containing quantum dots" below.

Binary sulfides. Binary sulfides (M_xS_n, e.g., MoS_2, NiS, Cu_2S, Bi_2S_3, CuS, SnS, In_2S_3 and Ag_2S) [14–18], containing one type of metal and S atom in their chemical formulas, have received substantial attention for their applications in fields of sensing [19,20], photothermal therapy [21], antibacterial and antifungal activity [22], ablation therapy [23], optoelectronics [24], photovoltaic [25,26] and magnetic device [27].

Among binary sulfides, transitional metal disulfides, such as ZnS, CuS, CdS, MoS$_2$, WS$_2$ and NiS [19,28–32], have been widely studied during the past few years as new members of 2-dimensional (2-D) family. The transitional metal disulfides are typical layered materials with sandwich-like structures, where metal atoms sandwich between two layers of S atoms by strong chemical bonds and two layers of S atoms are stacked together by weak van der Waals forces [25]. Similar to graphene, graphene oxide and other 2-D materials, transitional metal disulfides are promising biosensing materials due to their excellent properties, such as large active surface areas, and the suitable bandgaps. Large active surface areas in their sandwich-like structure can provide abundant active sites to establish particular bonds between layers and biological analytes, then target specific biomolecules, and finally promote specific reactions on the surface of 2-D transitional metal disulfides. In addition, suitable bandgaps endow transitional metal disulfides with advantageous optoelectronic properties, which can improve sensitivity in electrochemical, electrochemiluminescence (ECL) and photoluminescence (PL) biosensors.

Ternary sulfides. Ternary sulfides ($M_xM'_yS_n$, e.g., $Ni_3In_2S_2$, $Ni_3Tl_2S_2$ and $NiCo_2S_4$ [33,34]) contain two types of metals and S atoms in their chemical formulas. By changing the two types of metals, tuning the atomic ratios of metal or S atoms, researchers have validated different properties of ternary metal sulfides [35,36]. These ternary sulfides exhibit more flexible properties arising from their enhanced chemical and structural freedoms. These increased freedoms endow ternary sulfides with more suitable chemical and physical properties to satisfy a certain requirement, such as for more sophisticated biosensors. The bandgaps of some ternary sulfides vary those of binary sulfides [37,38], and the changed bandgaps make ternary sulfides more suitable for application in biosensors.

Quaternary sulfides. Quaternary sulfides contain three types of metal and S atoms in their chemical formulas, which have common composition of $M_xM'_yM''_zS_n$ where M, M', M'' = Zn, Cd, Mn, Hg, Cu, Ge, Sn, Cd, Fe, Co or Ba [39–44].

Non-metallic/metallic hetero-sulfides. Non-metallic/metallic hetero-sulfides have attracted considerable interests recently, which contain not only metal and S atoms but also other non-metallic atoms in their chemical formulas. For example, phosphor-chalcogenides [45,46] (e.g., $Pd_3(PS_4)_2$) are an emerging class of non-metallic/metallic hetero-sulfides.

Literature [7,23,25,34,37,47–52] has reported that different morphologies of metallic sulfide nanomaterials such as nanowires, nanoplates, hollow ellipsoid, nanotubes, hollow spheres, nanorods, flowerlike structures, core-shell nanoparticles, nanoribbons and complex hierarchal micro/nanostructures been synthesized. Different synthetic approaches, such as dip-coating, chemical vapor deposition (CVD), aqueous one-step wet chemistry, hydrothermal, coprecipitation, exfoliation, sputtering, solid-state reaction, ball-milling and biosynthetic methods, were used to synthesize metal sulfide nanomaterials (shown in Table 1). Even same metal sulfide nanomaterials synthesized with different methods may exhibit different properties and be applied in different areas [30]. Therefore, it is necessary to find the suitable synthetic techniques for metallic sulfide nanomaterials. For convenience, we can also buy metallic sulfides in the market for experimental research. After investigation, most of binary sulfides (e.g., WS$_2$ powders, Cu$_2$S powders, ZnS powders and CuS powders) have been available in the market, but ternary, quaternary, and non-metallic/metallic hetero-sulfides have not been available in the market.

Table 1. Synthetic methods and potential applications of metallic sulfide nanomaterials.

Metallic Sulfides	Synthetic Methods	Examples	Potential Applications	References
Binary sulfide	Dip-coating method	WS_2 nanoflakes	Sensing	[19]
	Chemical vapor deposition method	CdS nanoflakes	Sensing	[28]
	Aqueous one-step wet chemistry method	Ag_2S nanoparticles	Photothermal cancer treatment	[21]
	Hydrothermal approach	CuS flower shaped nanoparticles	Photothermal ablation cancer therapy	[23]
	Coprecipitation method	Fe_3S_4 nanoparticles	Glucose detection	[53]
	Exfoliation method	ZnS quasi-spherical nanoparticles	Antibacterial and antifungal activity	[22]
	Biosynthesis	ZnS nanoparticles	N/A *	[54]
Ternary sulfide	Exfoliation method	Ta_2NiS_5 nanosheets	N/A	[55]
	Hydrothermal approach	NiCo sulfide multistage nanowire array	Glucose detection	[56]
	Lithium intercalation assisted exfoliation	Cu_2WS_4 nanosheets	Supercapacitors	[57]
	Electrochemical Li-intercalation and exfoliation method	Ta_2NiS_5 nanosheets	Photoacoustic therapy	[58]
Quaternary sulfide	Sputtering method	Cu_2BaSnS_4 nanosheets	Solar energy storage	[37]
	Eco-friendly ball-milling method	Cu_2FeSnS_4 powder	Solar Cell Absorbers	[59]
	Solid-state reaction	$Li_4HgGe_2S_7$ diamond-like nanoparticles	Femtosecond laser systems	[60]
Non-metal/metal hetero-sulfide	Exfoliation method	$Pd_3(PS_4)_2$ nanosheets	Photocatalyst for water splitting	[46]

* N/A: Not available.

2.2. Applications in Biosensors

Metallic sulfide nanomaterials, as important and emerging materials, have arisen quickly in the area of biosensing due to their specific properties, namely, nanometric scale, water-dispersibility, large specific surface area, excellent catalytic activity, conductivity, biosafety, PL quenching abilities, photoactivity, and fascinating optical properties [7,48,61].

Catalysis: The metallic sulfide nanomaterials had excellent catalytic activity due to their high density of active sites, which can be used as modifiers in fabrication of novel biosensors [39,48]. Catalytic activities by the unsaturated sulfur commonly localize on the edge sites of metallic sulfide nanomaterials, which leads to fast heterogeneous electron transfer rate at the edge sites and enhanced catalytic activities [62].

Conductivity: The metallic sulfide nanomaterials had high electronic conductivity due to their low bandgaps, which make them be used as electrode materials in fabricating biosensors via electrochemical or ECL assays. For example, Chen et al. [20] have constructed a non-enzymatic glucose biosensor based on the high electronic conductivity of NiS nanospheres.

Biosafety: Some metallic sulfide nanomaterials, such as silver, copper and iron sulfides, were non-toxic [7]. These non-toxic metallic sulfide nanomaterials showed good biocompatibility in vitro, thus they could be used to fabricate biosensors.

PL quenching effect: Quencher is one of important component in PL (especially, fluorescence) sensing platforms for detection of biomolecules. In our previous work, we have demonstrated graphene possesses unprecedented PL quenching abilities [63]. Just like graphene, some representative 2-D metallic sulfide nanomaterials with 2-D layer structure also exhibited PL quenching abilities. These metallic sulfide nanomaterials with PL quenching abilities were suitable for constructing PL biosensors via PL quenching effect. Wang et al. [64] have used CuS nanoplates as quencher for fast, sensitive and selective detection of DNA via fluorescence quenching effect.

Photoactivity: some metallic sulfide nanomaterials were photoactive materials which can convert light illumination into electrical signals. When excited by light, electrons of metallic sulfide nanomaterials transferred from valence band to conduction band, resulting in the separation of photogenerated electrons and holes [65,66].

Fascinating optical properties: Some metallic sulfide nanomaterials, especially 0-D metal sulfide nanomaterials (namely, sulfide quantum dots), emitted high fluorescence. In comparison with organic fluorophores or alloy nanoclusters, metallic sulfide nanomaterials are superior as biomarkers due to their water-dispersibility, long lifetimes, resistance to photobleaching and biosafety [67,68]. Moreover, some metallic sulfide nanomaterials (e.g., Ag_2S quantum dots), emitted tunable fluorescence in near-infrared (NIR) region, which enabled their applications in bioimaging, biolabeling, deep tissue imaging, diagnostics and photodynamic therapy [69]. Even some metallic sulfide nanomaterials (e.g., EuS nanocrystals) have been used as ECL luminescent signal source, which endowed them with possibility of fabricating ECL biosensors [70].

Based on these specific properties of metallic sulfides that can be used to prepare biosensors, much research efforts have been devoted to developing biomolecule sensors for understanding physiological or pathological functions of biomolecules in living body or cells. In recent years, metal sulfide nanomaterials mainly have been applied to establish four types of biosensors, including electrochemical, photoelectrochemical (PEC), ECL and PL biosensors, for probing various types of biomolecules.

2.2.1. Electrochemical Biosensors

Metallic sulfide nanomaterials have been applied to establish electrochemical biosensors commonly due to their properties of conductivity, catalysis and biosafety. For example, Guo et al. [34] have developed a nonenzymatic glucose biosensor utilizing hierarchically porous $NiCo_2S_4$ nanowires due to their novel catalytic properties (Figure 1A). In synthetic processes, using electrospum graphitic nanofiber (EGF) as skeletons, $NiCo_2S_4$ nanowires were grown on the EGF toward different directions to decrease the agglomeration of $NiCo_2S_4$. In addition, the $NiCo_2S_4$ nanowires on EGF were core-shell structures with rough surface and polycrystalline nature. When applied to glucose determination, the skeletons of EGF and core-shell structures of $NiCo_2S_4$ enlarged effective surface to interact with glucose in solution and supplied more electrochemical active sites for accelerating glucose oxidation. Due to good sensing performances and biocompatibility toward glucose, an electrochemical biosensor based on $NiCo_2S_4$/EGF system was proposed with fast response (reaching a stable state within 5 s), a wide linear range (0.0005~3.571 mM, R^2 = 0.995) and low detection limit (0.167 µM, S/N = 3) via amperometric strategy.

Wang et al. [62] have also constructed a high-performance electrochemical platform for biosensing glucose and lactate in sweat based on their catalytic properties and conductivity of MoS_2 nanocrystals. The MoS_2 nanocrystals displayed enhanced catalytic activities and fast heterogeneous electron transfer rate because of unsaturated sulfur on the edge sites and stronger quantum confinement. As shown in Figure 1B, the biosensor for glucose detection was fabricated by sequentially growing MoS_2 nanocrystals and Cu submicron-buds on graphene paper (GP) via hydrothermal and electrodeposition method to form GP-MoS_2-Cu biosensor. Further coating of lactate oxidase (LOD) on the GP-MoS_2-Cu electrode, GP-MoS_2-Cu-LOD biosensor for lactate detection was obtained. Due to the electron transport property and high specific surface area of GP, enhanced catalytic activities, fast electron transfer rate and biosafety of MoS_2 nanocrystals, the electrochemical biosensor showed excellent sensing performances. For glucose, the electrochemical biosensor had a linear range of 5~1775 µM with a detection limit of 500 nM (S/N = 3). For lactate, the electrochemical biosensor had a linear range of 0.01~18.4 mM with a detection limit of 0.1 µM (S/N = 3).

In order to further enhance the catalytic performances of MoS_2 nanocrystals, Zhang et al. [71] have incorporated of a secondary metal sulfide (CoS_2) into MoS_2 nanocrystals to obtain binary metal sulfide composites (CoS_2-MoS_2). Based on CoS_2-MoS_2, a non-enzymatic electrochemical biosensor for determination of ascorbic acid, dopamine and nitrite have been proposed with linear ranges of 9.9~6582, 0.99~261.7 and 0.5~5160 µM, respectively. In addition, due to its good electrochemical activity caused by synergistic effect between CoS_2 and MoS_2, low detection limits of the electrochemical biosensor for determining ascorbic acid (3.0 µM), dopamine (0.25 µM) and nitrite (0.20 µM) have obtained, respectively.

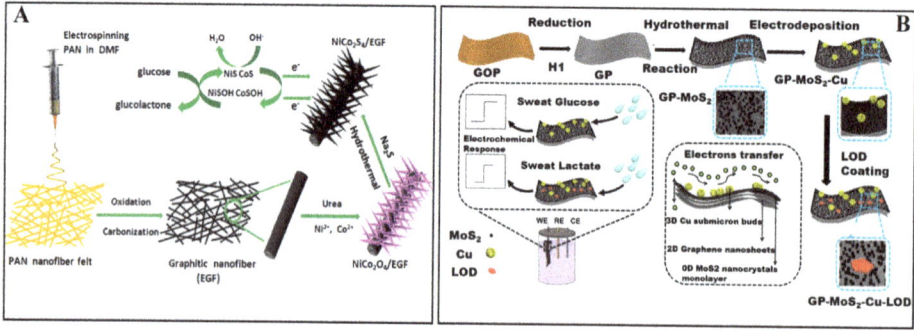

Figure 1. (**A**) The fabrication processes of porous $NiCo_2S_4$ nanowires and their applications in glucose biosensing [34]. Reproduced with permission Copyright 2019, Elsevier B.V. (**B**) The fabrication of biosensor based on MoS_2 nanocrystals for sensing glucose and lactate in sweat [62]. Reproduced with permission Copyright 2017, Elsevier B.V.

2.2.2. PEC Biosensors

Due to their superior photoactivities and conductivity, metallic sulfide nanomaterials have been used in various PEC biosensing systems to be photoactive materials or be one of other components through combining with photoactive materials [72–77].

Guo and Liu et al. [50] have fabricated a PEC biosensor for detecting a breast cancer biomarker human epidermal growth factor receptor-2 (HER2) (Figure 2A) using WS_2 nanowire on Ti mesh (WS_2 NW/TM) as photoactive material. Under visible light excitation, photo energy collected by WS_2 NW/TM electrode was higher than that of its bandgap. Accordingly, electron was transferred from valence band (VB) to conduction band (CB), and then CB electron was transferred to the surface of Ti mesh, finally the hole in VB was scavenged by H_2O_2. Based on the electron transfer process, photocurrent was generated. Moreover, to obtain a dual signal PEC amplification strategy, AuNPs modified with glucose oxidase (GOx) and HER2 specific peptide for signal amplification were utilized. The localized surface plasmon resonance (LSPR) of AuNPs generated a collective oscillation of free electrons when excited by visible light. The free electrons can transfer from Au to the CB of the WS_2 NW/TM electrode, which enhanced the photoelectric transfer efficiency and then achieved dual signal amplification. GOx modified in AuNPs catalyzed glucose to produce H_2O_2, which scavenged the hole in VB of the WS_2 NW/TM electrode.

For binding with HER2 molecules, HER2 aptamers were modified on the WS_2 NW surface via oxygen containing sulfur species of WS_2 NW. The HER2 specific peptides modified on the surface of AuNPs were also utilized to bind with HER2 molecules. When detected HER2 molecules, a sandwich type dual signal PEC amplification biosensor was established with a wide linear range (0.5~10 ng/mL) and low detection limit (0.36 ng/mL, S/N = 3).

Cui et al. [78] have also reported a PEC biosensor for determination of polynucleotide kinase (PNK) based on Bi_2S_3 nanorods as the photoactive materials (Figure 2B). The Bi_2S_3 nanorods displayed photoactive properties and generated a high photocurrent when excited by visible light. For fabricating PNK biosensor, a hybrid film consists of Bi_2S_3 nanorod and AuNPs was used to modify ITO electrode and to bind with capture probe (P1). Manganese based mimic enzymes (MnME) were modified with AuNPs to obtain MnME@AuNPs composites, which could label signal probes (P2). The capture probe on the modified electrode can specifically hybridize with the MnME@ AuNPs-labeled signal probe to form a double-stranded DNA. In the absence of PNK, MnME can catalyze H_2O_2 with 3,3-diaminobenzidine (DAB) as substrate, and generated MnME catalytic precipitations on the modified ITO electrode. The MnME catalytic precipitations were insulating barriers and blocked the interfacial electron transfer and eventually leaded to a low PEC signal. In the presence of PNK, the double-stranded DNA was phosphorylated and subsequently cleaved by lambda exonuclease to release the MnME@AuNPs from the modified electrode, leading to a high PEC signal. Based on the signal on-off PEC strategy, the PNK biosensor was proposed and exhibited high sensitivity with a detection limit of 1.27×10^{-5} U/mL.

In addition to being photoactive materials, metallic sulfide nanomaterials also have been one of other components through combining with photoactive materials in PEC biosensing systems. For example, Zhao et al. [73] have also reported a PEC biosensor for determination of prostate specific antigen (PSA) using $CdTe/TiO_2$ sensitized structures as photoactive materials and CuS nanocrystal as electronic extinguisher (Figure 2C). For fabricating the PSA biosensor, a peptide was fixed to the $CdTe/TiO_2$ electrode surface and used to immobilize a double-helix DNA (dsDNA). Then, CuS nanocrystal was efficiently immobilized on the dsDNA via doxorubicin (Dox) inserting into the dsDNA. In absence of PSA, electron donor and radiant light were consumed by CuS nanocrystals, and steric hindrance effect of insulating substances (e.g., peptides and DNA) generated, leading to a low PEC signal. In the presence of PSA, the PSA specifically cleaved the peptide, and DNA/Dox-CuS probes were released from the electrode surface, resulting in a high PEC signal. Take advance of the signal on-off PEC strategy, the PSA biosensor revealed good sensing performance with a linear range from 0.005 to 20 ng/mL and a low detection limit of 0.0015 ng/mL.

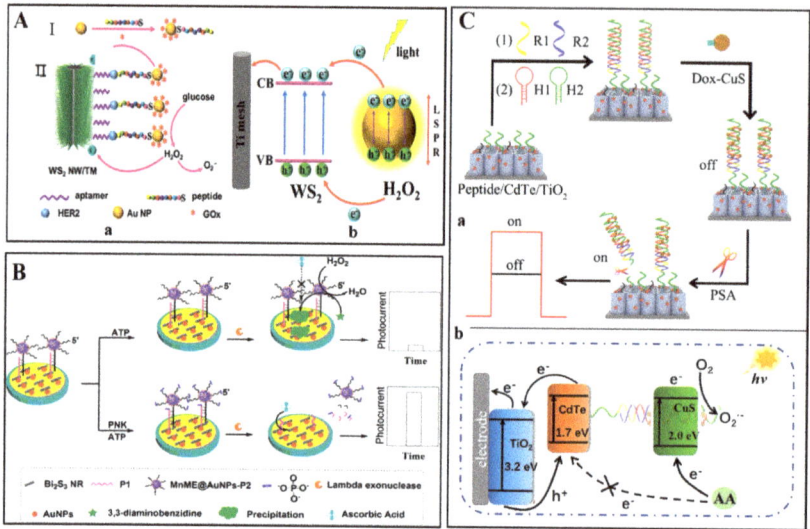

Figure 2. (**A**) (a) The process to fabricate PEC biosensor based on WS$_2$ nanowire array on Ti mesh (TM) for breast cancer biomarker HER2 detection; (b) Schematic mechanism of the PEC system [50]. Reproduced with permission Copyright 2019, Elsevier B.V. (**B**) A signal-on PEC biosensor for PNK assay with the MnME@AuNPs-P2 catalytic precipitation on Bi$_2$S$_3$ nanorod as the photoactive materials [78]. Reproduced with permission Copyright 2018, American Chemical Society. (**C**) Construction (a) and response mechanism (b) of PEC biosensor based on CuS nanocrystals [73]. Reproduced with permission Copyright 2019, Elsevier B.V.

2.2.3. ECL Biosensors

Due to their ECL properties, metallic sulfide nanomaterials have been used to establish ECL biosensors. For example, Babamiri et al. [70] have prepared an ECL biosensor for determining human immunodeficiency virus (HIV) DNA sequence utilizing EuS nanocrystals as ECL luminophore through a molecularly imprinted polymer ECL (MIP-ECL) system (Figure 3A). In the MIP-ECL system, HIV aptamer as template and o-phenylenediamine as the functional monomer were electropolymerized directly on the surfaces of the ITO electrode. After removing HIV aptamer template, the MIP modified electrode was obtained. The MIP modified electrode can bind with HIV-1 gene when immersed into different concentrations of HIV-1 gene standard solution. Then, the HIV-1 gene on the MIP modified electrode reacted with the HIV DNA strand functionalized on EuS nanocrystals by hybridization reaction. Based on the hybridization reaction between HIV-1 gene and HIV DNA strand, the MIP-ECL biosensor was proposed. Using K$_2$S$_2$O$_8$ as co-reactant, the ECL signal of the MIP-ECL biosensor significantly enhanced with increased concentrations of HIV-1 gene. Taking advantage of both MIP-ECL assays and the ECL properties of EuS nanocrystals, the HIV gene biosensor was sensitive and selective with a wide linear range (3.0 fM~0.3 nM) and low detection limit (3.0 fM).

Moreover, Zhu et al. [79] have also fabricated a sandwich-type ECL biosensor for detecting insulin based on the ECL property of zinc-doping cadmium sulfide (Au-ZnCd$_{14}$S) (Figure 3B). Au-ZnCd$_{14}$S combined nitrogen doping mesoporous carbons (Au-ZnCd$_{14}$S/NH$_2$-NMCs) acted as sensing platform and Au-Cu alloy nanocrystals were employed as labels to quench the ECL of Au-ZnCd$_{14}$S/NH$_2$NMCs. On the basis of the ECL quenching effects between ZnCd$_{14}$S and Au-Cu alloy nanocrystals, a sensitive ECL immunosensor for insulin detection was successfully constructed with a linear response range from 0.1 pg/mL to 30 ng/mL and detection limit of 0.03 pg/mL (S/N = 3). Although some metallic sulfide nanomaterials did not exhibit ECL properties, they have also been used to construct ECL biosensors via being as electrode materials based on their superior conductivity and large specific surface area. For example, Wei et al. [80] have reported an ECL biosensor for detection of amlodipine besylate (AML) based on reduced graphene oxide-copper sulfide (rGO-CuS) composite coupled with capillary electrophoresis (CE) (Figure 3C). The rGO-CuS composite was synthesized based on flowerlike CuS wrapped with rGO sheet and utilized to modify electrode. Due to the presence of rGO-CuS composite, the electron transfer rate between the electroactive center of Ru (bpy)$_3^{2+}$ and the electrode was facilitated. At the present of AML, the ECL intensity of Ru (bpy)$_3^{2+}$ increased which induced the development of AML biosensor. Take advance of large specific surface area of rGO-CuS composite and powerful CE separation technique, the ECL biosensor for the detection of AML was successfully fabricated with a linear response range of 0.008 to 5.0 µg/mL and a detection limit of 2.8 ng/mL (S/N = 3).

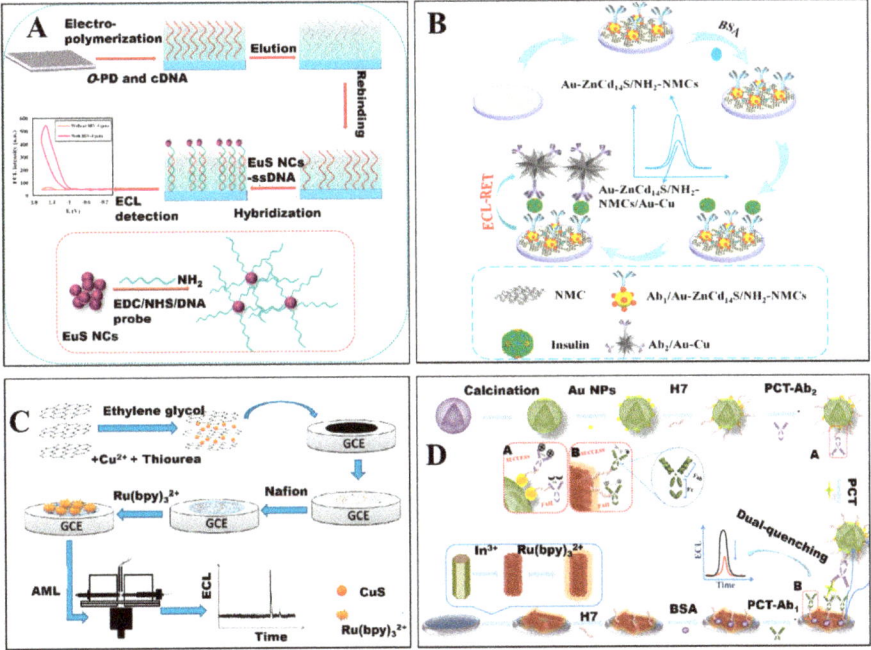

Figure 3. (**A**) Schematic diagram of the HIV gene biosensor using EuS nanocrystals as luminophore [70]. Reproduced with permission Copyright 2018, Elsevier B.V. (**B**) The fabrication of insulin biosensor based on Au-ZnCd$_{14}$S [79]. Reproduced with permission Copyright 2017, Elsevier B.V. (**C**) Schematic fabrication process of ECL sensor and CE-ECL detection [80]. Reproduced with permission Copyright 2016, Elsevier B.V. (**D**) The fabrication of immunosensor based on hollow In$_2$S$_3$ nanotubes for procalcitonin detection [52]. Reproduced with permission Copyright 2019, Elsevier B.V.

Moreover, Xue et al. [52] have also designed a procalcitonin (PCT) biosensor based on dual-quenching ECL-RET strategy utilizing hollow Ru-In$_2$S$_3$ nanocomposite as ECL acceptor and porous α-MoO$_3$-Au structure as ECL donor (Figure 3D). Specifically, Ru-In$_2$S$_3$ nanocomposite was prepared by hollow In$_2$S$_3$ nanotubes as substrate adsorbing Ru (bpy)$_3^{2+}$. For fabricating PCT biosensor, HWRGWVC heptapeptide (H7), which could provide -SH, was immobilized on the surface of nanomaterials through amide bond (with Ru-In$_2$S$_3$ nanocomposite) and Au-S bond (with α-MoO$_3$-Au structures) and used to capture antibody (Ab$_1$ and Ab$_2$). In the presence of PCT, Ru-In$_2$S$_3$ nanocomposite captured Ab$_1$ and α-MoO$_3$-Au structures captured Ab$_2$ connected together, and ECL-RET from Ru-In$_2$S$_3$ to α-MoO$_3$-Au occurred which was further confirmed by testing the overlap between ECL emission of Ru-In$_2$S$_3$ and UV-vis spectra of α-MoO$_3$-Au. Take advantage of huge specific surface area of Ru-In$_2$S$_3$

or α-MoO$_3$-Au and dual-quenching ECL-RET strategy, the ECL biosensor for detecting PCT was obtained with sensitive response, linear range from 0.0001 to 50 ng/mL and low detection limit of 12.49 fg/mL (S/N = 3).

2.2.4. PL Biosensors

Metal sulfide nanomaterials also have been used to establish PL biosensors due to their fascinating optical properties. However, to our best well know, metal sulfide nanomaterials used to fabricate PL biosensors mainly were 0-D metal sulfide nanomaterials (namely, sulfide quantum dots). Thus, PL biosensors based on metal sulfide nanomaterials will be illustrated in the section of "sulfur-containing quantum dots" below.

As described above, biosensors based on metal sulfide nanomaterials have been used for detection of various analytes, including glucose, dopamine, proteins, DNA, etc. Moreover, these biosensors displayed good sensing performance toward analytes detection. In addition, these biosensors also showed other outstanding advantages, including simple of preparation, low cost and good selectivity, stability, and great promising practical applications in clinical diagnosis, as shown in Table 2.

Table 2. Properties of biosensors based on metal sulfide nanomaterials listed above.

Biosensors	Analytes	Linear Range with Detection Limit (S/N = 3)	Practical Application	References
Electrochemical biosensor based on NiCo$_2$S$_4$	glucose	0.0005~3.571 mM (R^2 = 0.995) with a detection limit 0.167 µM	glucose determination in human blood serum sample, recoveries 98.23~100.61% with RSDs of 3.53~5.12%	[34]
Electrochemical biosensor based on MoS$_2$	glucose	5~1775 µM (R^2 = 0.998) with a detection limit of 500 nM	glucose determination in sweat, recoveries (N/A *)	[62]
	lactate	0.01~18.4 mM (R^2 = 0.996) with a detection limit of 0.1 µM	lactate determination in sweat, recoveries (N/A)	
Electrochemical biosensor CoS$_2$-MoS$_2$	ascorbic acid (AA)	9.9~6582 µM (R^2 = 0.997) with a detection limit of 3.0 µM	AA determination in urine sample, recoveries 96.5%~102.7% with RSD within 3%	[71]
	dopamine (DA)	0.99~261.7 µM (R^2 = 0.996) with a detection limit of 0.25 µM	DA determination in urine sample, recoveries 96.5%~102.7% with RSD within 3%	
	nitrite	0.5~5160 µM (R^2 = 0.997) with a detection limit of 0.20 µM	nitrite determination in urine sample, recoveries 96.5%~102.7% with RSD within 3%	
PEC biosensor based on WS$_2$ NW	HER2 molecules	0.5~10 ng/mL (R^2 = 0.998) with a detection limit of 0.36 ng/mL	HER2 determination in serum sample, recoveries 108.2%, 98.6% and 101.3% with RSD 1.5%, 2.3% and 3.2%	[50]
PEC biosensor based on Bi$_2$S$_3$ nanorods	polynucleotide kinase (PNK)	0.0005~10 U/mL (R^2 = 0.995) with a detection limit of 1.27 × 10^{-5} U/mL	PNK activity in HEK293T cells, intra-assay with a RSD of 6.27% and interassay with a RSD of 5.52%	[78]
PEC biosensor based on CuS nanocrystal	prostate specific antigen (PSA)	0.005~20 ng/mL (R^2 = 0.991) with a detection limit of 0.0015 ng/mL	PSA determination in human serum sample, recoveries (N/A)	[73]

Table 2. Cont.

Biosensors	Analytes	Linear Range with Detection Limit (S/N = 3)	Practical Application	References
ECL biosensor based on EuS nanocrystals	HIV-1 gene	3.0 fM~0.3 nM (R^2 = 0.996) with a low detection limit of 3.0 fM	HIV-1 gene determination in serum samples, recoveries 95.00~101.2% with RSDs of 1.78~4.2%	[70]
ECL biosensor based on $ZnCd_{14}S$	insulin	0.1 pg/mL~30 ng/mL (R^2 = 0.996) with a detection limit of 0.03 pg/mL	insulin determination in human serum samples, recoveries 98.5~103.1% with RSDs of 2.1%~3.7%	[79]
ECL biosensor based on rGO-CuS composite	amlodipine besylate (AML)	0.008~5.0 µg/mL (R^2 = 0.998) with a detection limit of 2.8 ng/mL	AML determination in plasma samples, recoveries 95.42%~98.50% with RSDs of 3.2% to 4.5%.	[80]
ECL biosensor based on $Ru-In_2S_3$ nanocomposite	procalcitonin	0.0001~50 ng/mL (R^2 = 0.996) with a low detection limit of 12.49 fg/mL	procalcitonin determiantion in human serum, recoveries 95.2%~96.8% with RSD under 3.6%	[52]

* N/A: Not available.

3. Sulfur-Containing Quantum Dots

3.1. Generalities

Sulfur-containing quantum dots are quantum dots containing central sulfur-containing nanodots and surface functional groups (e.g., carboxyl groups or amino groups), and possess fascinating photophysical properties, small size (typically below 10 nm), good biocompatibility, and chemical inertness. They can be broken down into three main categories: sulfur quantum dots, sulfide quantum dots and sulfur-doped quantum dots.

Sulfur quantum dots. Sulfur quantum dots are pure elemental quantum dots, mainly including S central nanodots and surface functional groups [81,82].

As a new class of quantum dots, sulfur quantum dots were firstly synthesized by Li's group through phase interfacial reactions in 2014 [83]. Since then, researchers have eagerly pursued synthetic approaches of sulfur quantum dots due to their excellent aqueous dispersibility, small size, excellent photostability, low toxicity, narrow size distribution and ultrahigh photostability [84]. To date, sulfur quantum dots have not been available in the market. Generally, sulfur quantum dots were synthesized by hydrothermal methods based on "top-down" synthetic approaches. Literature has reported detailed synthetic approaches including phase interfacial reaction [81,83], "assemble-fission" approach [82,85], H_2O_2-assisted "top-down" approach [86] and oxygen accelerated scalable approach [84]. Synthetic details for each approach are described as follows:

For phase interfacial reaction, CdS quantum dots or ZnS quantum dots were diluted by n-hexane and then sonicated to form a homogeneous solution. HNO_3 aqueous solution was mixed with CdS quantum dots or ZnS quantum dots solution with a slowly stirring at room temperature. The resulting white mixture was separated by a funnel, and sulfur quantum dots were synthesized as a white suspension in hexane.

For "assemble-fission" approach, sulfur quantum dots were synthesized by simply treating sublimated sulfur powders with alkali using polyethylene glycol-400 as passivation agents.

For H_2O_2-assisted "top-down" approach, sulfur quantum dots were synthesized by dissolved bulk sulfur powder into small particles in an alkaline environment in the presence of polyethylene glycol, followed by the H_2O_2-assisted etching of polysulfide species.

For oxygen accelerated scalable approach, sulfur quantum dots were synthesized by dissolved bulk sulfur powder into small particles in an alkaline environment in the presence of polyethylene glycol to form polysulfide species (S_x^{2-}), followed by oxidation of S_x^{2-} to zero-valent sulfur under a pure O_2 atmosphere sulfide quantum dots. Sulfide quantum dots commonly included central sulfide nanodots, especially metal sulfide, and surface functional groups. Much research efforts have been devoted to synthesize sulfide quantum dots, such as ZnS, CdS, PbS, Ag_2S, SnS_2, In_2S_3 and AgInZnS quantum dots [87–92].

Sulfide quantum dots were commonly synthesized by simple aqueous method and used various stabilizing agents or sulfides to maintain metal atoms in order to assembly into nanodots. The stabilizing agents or sulfides were surface ligands and S sources, and the stabilizing agents included cysteamine, mercaptoacetic acid, l-cysteine, N-acetyl-l-cysteine, bovine serum albumin [67,69,87,93–95]. To date, various of sulfide quantum dots, such as PbS quantum dots, Ag_2S quantum dots and CdSeS/ZnS quantum dots, have been available in the market.

Sulfur-doped quantum dots. Sulfur-doped quantum dots are obtained by doping S atoms into other quantum dots, such as silicon, carbon, phosphorus and graphene quantum dots [82]. Among sulfur-doped quantum dots, sulfur-doped carbon or graphene quantum dots were the most widely studied in the recent years [96–101]. This review will focus on sulfur-doped carbon or graphene quantum dots.

The approaches used to synthesize sulfur-doped carbon or graphene quantum dots can be divided into two categories: "top-down" approaches and "bottom-up" approaches. The "top-down" approaches included hydrothermal, solvothermal, ultrasound, chemical exfoliation, microwave-assisted exfoliation methods, and so on [102–106]. Due to their superiority such as time-saving and easy to operation, the "top-down" approaches have attracted much excitement for synthesizing sulfur-doped carbon or graphene quantum dots. The "bottom-up" approaches used to synthesize sulfur-doped carbon or graphene quantum dots can be controlled by "step-by-step" chemical reactions through various precursors [107,108]. To date, carbon or graphene quantum dots have been available in the market, but sulfur-doped carbon or graphene quantum dots haven't been available in the market yet.

3.2. Applications in Biosensors

Sulfur-containing quantum dots are considered to be suitable alternative nanomaterials in biosensing applications [67,109–112]. Their stable photoelectric properties made sulfur-containing quantum dots be adapted as excellent probes in biosensors via various strategies, such as electrochemical, PEC, PL and ECL strategy [113–115]. Soluble sulfur-containing quantum dots can react with biomolecules, thus biosensors for detection biomolecules could be established through specific physiochemical reactions between them [81,116,117]. Functionalization of sulfur-containing quantum dots (especially, sulfide quantum dots) with different stabilizing agents to form surface groups can enhance their hydrophilicity and interaction ability with other biomolecules [115,118–120]. The low toxicity of sulfur-containing quantum dots made them suitable to be used for sensing in cells or living bodies [83,121–123].

3.2.1. Biosensors Based on Sulfur Quantum Dots

As emerging quantum dots, sulfur quantum dots have been paid much attention due to their possessions of inexpensive S atoms and unique physicochemical properties [82–84]. Literature has demonstrated that sulfur quantum dots were applied in the field of sensor [81,85,124]. For example, sulfur quantum dots have been used for sensing metal ions or detecting drug [81,125]. Very recently, sulfur quantum dots also have gradually been applied to living cells imaging [124]. However, applications of sulfur quantum dots in biosensing or bio-medical diagnosis field were still far from satisfactory. To this end, there is an urgent need of efficient approaches to exploit biosensing applications of sulfur quantum dots in the next few years.

3.2.2. Biosensors Based on Sulfide Quantum Dots

Due to optical responses of sulfide quantum dots from the visible to the near infrared (NIR), sulfide quantum dots have received extensive attention in the field of biosensing [8,126]. They have been widely used as alternative probes for biomolecules via various strategies, such as electrochemical, PEC, ECL and PL strategies [94,109,126,127].

Electrochemical Biosensors

Due to excellent electrochemical activities of sulfide quantum dots and inexpensive instruments and simple operations of electrochemical methods, electrochemical biosensors based on sulfide quantum dots have attracted increased attention. Zhang et al. [128] have reported an electrochemical biosensor for detecting clenbuterol antibody based on ZnS quantum dots. Amor-Gutiérrez et al. [129] have established an electrochemical biosensor for determination bacteria based on Ag_2S quantum dots.

PEC Biosensors

PEC sensing, a branch of electrochemistry, is a newly developed technology and has attracted great interest in biosensing fields. For fabricating PEC biosensor, photoactive materials are vital because they can generate photocurrent excited by light. To our best well know, sulfide quantum dots not only have directly been photoactive materials to establish PEC sensors [89,92,130–132], but also been one of other components through combining with photoactive materials to indirectly establish PEC sensors [87,133].

Wang et al. [130] have proposed a PEC biosensor for detection of H_2S released from MCF-7 cells based on heterostructures formed by CdS quantum dots and branched TiO_2 nanorods (CdS-B-TiO_2). Herein, CdS-B-TiO_2 heterostructures in the PEC biosensors were directly as photoactive materials. In addition, due to the formation of CdS-B-TiO2 heterostructures, a significant enhancement in photocurrent was obtain, thus leading to sensitive PEC recording of the H_2S level in cellular environments.

Moreover, Deng et al. [133] have utilized CdS quantum dots as one of other components through combining with photoactive materials to indirectly establish PEC biosensors for determination of PSA. The PSA biosensor was utilized reduced graphene oxide-TiO_2 (ERGO-TiO_2) as reduced graphene oxide-TiO_2 (ERGO-TiO_2) and CdS quantum dots as a PEC signal amplifier. For preparing the PSA biosensor, ERGO-TiO_2 was utilized to immobilize capture antibody (Ab1) for PSA detection, and quinone-rich PDA nanospheres (PDANS) loaded with CdS quantum dots were used to load detection antibody (Ab2) for PSA detection. In the presence of PSA, photo-generated electron transferred between PDANS loaded with CdS quantum dots and ERGO-TiO_2. Due to the good conductivity of PDANS, ERGO and CdS quantum dots, a PSA biosensor has been proposed with a linear range from 0.02 pg/mL to 200 ng/mL with the detection limit of 6.8 fg/mL.

ECL Biosensors

Sulfide quantum dots have been widely used as alternative probes for biomolecules (e.g., dopamine, thrombin, laminin or enzyme) via ECL strategies.

Liu et al. [134] have fabricated a dopamine (DA) biosensor based on water-dispersible CdS quantum dots (CdS QDs) via ECL strategy. As shown in Figure 4A, they synthesized four sizes of CdS QDs, namely 1.8, 2.7, 3.2 and 3.7 nm. Each size of CdS QDs had various ECL performance. Under the optimized conditions, the ECL biosensor displayed excellent sensing properties with linear detection range from 8 pM to 20 nM and detection limit of 3.6 pM (S/N = 3).

In addition, Wang et al. [135] have also proposed a thrombin (TB) biosensor based on lanthanum ion-doped CdS quantum dots (CdS: La QDs) via ECL strategy (Figure 4B). The detection mechanism of the ECL biosensor was based on a distance-dependent ECL intensity enhanced or quenched system between CdS: La QDs and AuNPs. In the presence of Hg^{2+}, ECL quenching (signal off) achieved lie in

RET between the CdS: La QDs and AuNPs at a close distance. In the presence of TB, ECL enhance (signal on) achieved lie in surface plasmon resonances (SPR) between the CdS: La QDs and AuNPs at a separated distance. The "on-off-on" approach was used to detect TB, and the linear range were 1.00×10^{-16} to 1.00×10^{-6} mol/L with limit of detection (S/N = 3) of 3.00×10^{-17} mol/L.

Moreover, Wu et al. [136] have prepared a laminin (LN) biosensor based on Mn doped Ag_2S quantum dots (Ag_2S: Mn QDs) as ECL materials (Figure 4C). The optical response of Ag_2S: Mn QDs was in IR window (i.e., about at 626 nm) obtained by ECL spectrum. Based on a sandwiched ECL immunoassay, the biosensor displayed a wide linear range of 10 pg/mL to 100 ng/mL with a low detection limit of 3.2 pg/mL for LN detection.

Furthermore, Zhou et al. [137] have proposed an enzyme (namely, DNA methyltransferase (MTase)) biosensor based on CdS quantum dots (CdS QDs) as ECL materials (Figure 4D). For fabricating the MTase biosensor, double-stranded DNA containing 5'-CCGG-3' sequence was bonded to a CdS QDs modified glassy carbon electrode, then the modified electrode was incubated with M.SssI CpG MTase which catalyzed the methylation of the specific CpG dinucleotides. Subsenquently, the electrode was treated with a restriction endonuclease HpaII. The HpaII can recognize and cut off the 5'-CCGG-3' sequence, but recoginition function was blocked when the CpG site in the 5'-CCGG3' was methylated. Double-stranded DNA having been methylated can immobilized AuNPs with glucose oxidase mimicking activity. AuNPs immobilized on double-stranded DNA can catalyze the oxidation of glucose to genetate H_2O_2 which served as coreactant of CdS QDs. Thus, the ECL intensity of CdS QDs was linear correlation with the activity of M.SssI MTase.

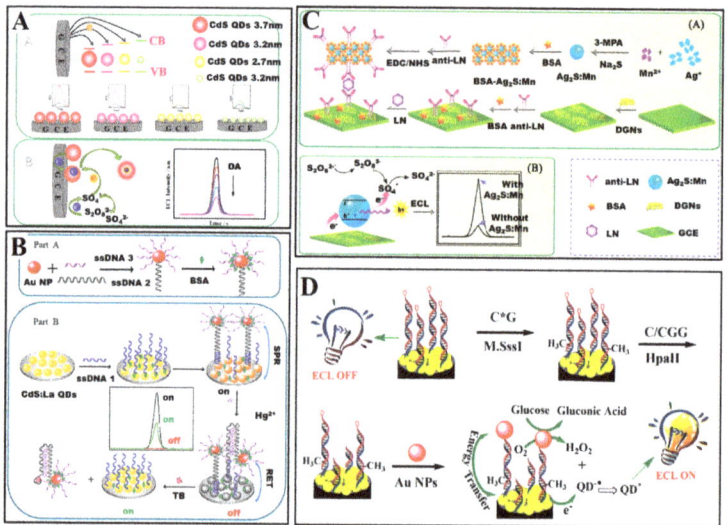

Figure 4. (A) ECL biosensors for dopamine based on CdS QDs [134]. Reproduced with permission Copyright 2016, Elsevier B.V. (B) Fabrication of the ECL-RET aptasensor for thrombin based on CdS: La quantum dots film and AuNPs [135]. Reproduced with permission Copyright 2019, Elsevier B.V. (C) Schematic diagram to show (A) the ECL immunosensor fabrication process and (B) ECL mechanism of Ag_2S: Mn QDs [136]. Reproduced with permission Copyright 2017, Elsevier B.V. (D) Schematic illustration of ECL biosensors bade on CdS QDs for detection of the MTase activity [137]. Reproduced with permission Copyright 2016, American Chemical Society.

PL Biosensors

Sulfide quantum dots, as 0-D metal sulfide nanomaterials, have also attracted great attention to establish PL biosensors due to their fascinating optical properties [138]. Until now, there have been various types of PL biosensors based on sulfide quantum dots, such as phosphorescence biosensors, fluorescence biosensors.

Phosphorescence biosensors were proposed based on sulfide quantum dots with phosphorescence emission. For example, Gong and Fan [139] have proposed a phosphorescence biosensor for detection of DNA based on riboflavin-modulated Mn doped ZnS quantum dots. Due to the longer average life of phosphorescence emitted by Mn doped ZnS quantum dots, the DNA biosensor allowed appropriate delay time and avoided any scattering light.

Because most of sulfide quantum dots have fluorescent properties, fluorescence biosensors based on sulfide quantum dots were the most common biosensors. For example, Liu et al. [140] have prepared a fluorescence biosensor for detecting alkaline phosphatase based on l-cysteine-capped CdS QDs. Moreover, Du et al. [67] have synthesized VS_2 quantum dots, and constructed a glutathione biosensor based on the efficient fluorescence RET from VS_2 quantum dots to MnO_2 nanosheets and fast redox reaction between MnO_2 and glutathione. Adegoke et al. [141] have utilized CdZnSeS/ZnSeS quantum dots to fabricate a fluorescence biosensor for determining influenza virus RNA. Rong et al. [142] have synthesized novel Eu^{3+} ion-functionalized fluorescent MoS_2 quantum dots for biosensing Guanosine 3′-diphosphate-5′-diphophate (ppGpp) (Figure 5A). Literature has also reported fluorescence biosensors for determining other biomolecules, such as thrombin, glutathione S-transferase enzyme, bilirubin [91,94,143,144] via PL strategies.

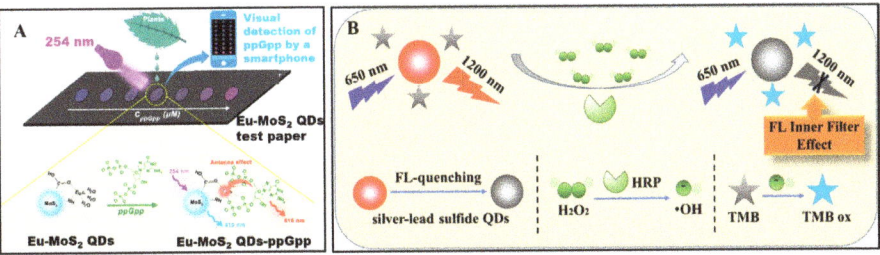

Figure 5. (**A**) Schematic illustration for ppGpp detection using Eu-MoS$_2$ QDs test paper [142]. Reproduced with permission Copyright 2020, Elsevier B.V. (**B**) Illustration of H$_2$O$_2$ detection based on NIR-II fluorescence Pb-doped Ag$_2$S quantum dots [69]. Reproduced with permission Copyright 2019, Elsevier B.V.

NIR fluorescence biosensors were emerging types of fluorescence biosensors and have also attracted great attention. The NIR fluorescence biosensors were fabricated based on fluorescence materials with emissions in NIR window (750 to 1000 nm) which can achieve higher imaging depth without complications from tissue autofluorescence. Sulfide quantum dots, such as Ag_2S quantum dots, displayed fluorescence emission in NIR window, therefore they have been used to establish biosensors. Moreover, Ding et al. [123] have fabricated a NIR biosensor for detecting F^- in living cells based on NIR emitting Ag_2S quantum dots. The fluorescence intensity of Ag_2S quantum dots enhanced when various rare earth ions were added. In the presence of F^-, F^- coordinated with rare earth ions leaded to fluorescence quenching of Ag_2S quantum dots. Based on the on-off fluorescence findings, a label-free NIR fluorescence biosensor for F^- in living has been proposed. Moreover, Shu et al. [69] have ameliorated Ag_2S quantum dots by doping Pb ions to synthesized Pb-doped Ag_2S quantum dots. The Pb-doped Ag_2S quantum dots emitted fluorescence in NIR-II window (950 to 1200 nm). Based on the NIR-II emitting Pb-doped Ag_2S quantum dots, a biosensor for H_2O_2 have been proposed (Figure 5B).

3.2.3. Biosensors Based on Sulfur-Doped Carbon or Graphene Quantum Dots

Carbon or graphene quantum dots have attracted intensive interest and have also been used to determine biomolecules due to their fascinating properties [145–148]. Doping heteroatoms (e.g., nitrogen, sulfur, phosphorus/or metal atoms) in carbon or graphene quantum dots is an effective way to tune their properties [107,112,149,150]. As the third most abundant element in fossil fuels, S and its derived material have attracted a lot of interest. Sulfur doped carbon or graphene quantum dots, as one of derived sulfur-containing nanomaterials, have also attracted intense interest and been widely used to established biosensors [151–154]. However, sulfur doped carbon or graphene quantum dots still belonged to carbon or graphene quantum dots. Since too much literature has reported biosensors based on carbon or graphene quantum dots [155–158], we won't explore them in this review. As described above, biosensors based on sulfur-containing quantum dots have been used for detection of various analytes, including antibody, dopamine, proteins, DNA, RNA, glutathione, bacteria, F^- in living cells, etc. These biosensors displayed good sensing performance toward analytes detection. In addition, these biosensors also showed other outstanding advantages, including simple of preparation, low cost and good selectivity, stability, and great promising practical applications in clinical diagnosis, as shown in Table 3.

Table 3. Properties of biosensors based on sulfur-containing quantum dots listed above.

Biosensors	Analytes	Linear Range with Detection Limit (S/N = 3)	Practical Application	References
Electrochemical biosensor based on ZnS quantum dots	clenbuterol antibody	0.01~10 ng/mL (R^2 = 0.991) with a detection limit of 5.5 pg/mL	clenbuterol antibody in pig urine, recoveries 96.39%~103% with RSDs of 0.09%~0.27%	[128]
Electrochemical biosensor based on Ag_2S quantum dots	bacteria	10^{-1}~10^3 bacteria/mL (R^2 = 0.993) with a detection limit of 1 bacteria/mL	bacterica in human serumm, recoveries (N/A*)	[129]
PEC biosensor based on CdS quantum dots	H_2S	1.0 nM~5 mM (R^2 = 0.991) with a detection limit of 29 ng/mL	H_2S released from MCF-7 cells, recoveries (N/A)	[130]
PEC biosensor based on CdS quantum dots	prostate specific antigen (PAS)	0.02 pg/mL~200 ng/mL (R^2 = 0.997) with a detection limit of 6.8 fg/mL.	PAS in human serum samples, recoveries 96.2%~110.0% with RSDs less than 9.7%	[133]
ECL biosensor based on CdS quantum dots	dopamine	8 pM~20 nM (R^2 = 0.998) with a detection limit of 3.6 pM	dopamine in human urine and serum samples, recoveries 95.4%~102.6% with RSDs of 0.34%~5.14%	[134]
ECL biosensor based on lanthanum ion-doped CdS quantum dots	thrombin	1.00×10^{-16}~1.00×10^{-6} mol/L (R^2 = 0.996) with limit of detection of 3.00×10^{-17} mol/L	thrombin in human serum, recoveries 98.0%~100.1%	[135]
ECL biosensor based on Mn doped Ag_2S quantum dots	laminin	10 pg/mL~100 ng/mL (R^2 = 0.993) with a low detection limit of 3.2 pg/mL	laminin in human serum, recoveries 96.08%~105.56%	[136]
PL biosensor based on Mn doped ZnS quantum dots	DNA	15 µg/L~40 mg/L (R^2 = 0.998) with a detection limit of 15 µg/mL	DNA in urine samples, recoveries 97%~103%	[139]
PL biosensor based on l-cysteine-capped CdS quantum dots	alkaline phosphatase (ALP)	1~10 nM (R^2 = 0.999) with a detection limit of 96 pM	ALP in human serums, recoveries 98.58%~106.60% with RSDs of 1.59%~9.50%	[140]

Table 3. Cont.

Biosensors	Analytes	Linear Range with Detection Limit (S/N = 3)	Practical Application	References
PL biosensor based on VS$_2$ quantum dots	glutathione	0~500 μM (R^2 = 0.996) with a detection limit of 0.31 μM	glutathione detection in human serum samples, recoveries 101.0%~109.0% with RSDs of 0.7%~2.7%	[67]
PL biosensor based on CdZnSeS/ZnSeS quantum dots	influenza virus RNA	detection limit of 5.2 copies/mL	N/A	[141]
PL biosensor based on MoS$_2$ quantum dots	ppGpp	25~250 μM (R^2 = 0.997) with a detection limit of 23.8 μM	ppGpp in plants, recoveries 100.0%~138.0% with RSDs below 1.4%	[142]
PL biosensor based on NIR emitting Ag$_2$S quantum dots	F$^-$	5~260 μM (R^2 = 0.9978) with a detection limit of 1.5 μM	F$^-$ in living cells	[123]
PL biosensor based on the NIR-II emitting Pb-doped Ag$_2$S quantum dots	H$_2$O$_2$	40~800 μM with a detection limit of 5 μM	H$_2$O$_2$ analysis in disinfectant	[69]

* N/A: Not available.

4. Brief Comparison between Biosensors Based on Sulfur-Containing Nanomaterials and Others

All in all, biosensors based on sulfur-containing nanomaterials have been used for detection of various biomolecules in the last five years, including glucose, dopamine, proteins, DNA, RNA, etc. The biosensors based on sulfur-containing nanomaterials displayed enhanced selectivity, lower sensitivity, faster response time, and low detection limit in comparison to biosensors based on other nanomaterials. Taking glucose as analytes, Table 4 displays brief comparison between biosensors based on sulfur-containing nanomaterials and other biosensors. Data listed in Table 4 illustrates that sulfur-containing nanomaterials are promising materials to established biosensors and can be widely used in biomedical field.

Table 4. Brief comparison between biosensors based on sulfur-containing nanomaterials and others.

Biosensors	Linear Range	Limit of Detection (LOD) (S/N = 3)	References
biosensor based on morphous Co$_x$S$_y$ nanosheets	0.2~1380 μM	0.079 μM	[159]
biosensor based on VS$_2$ nanoparticles	0.5 μM~3.0 mM,	0.224 μM	[160]
biosensor based on flowerlike NiCo$_2$S$_4$	0.5 μM~6 mM	50 nM	[51]
biosensor based on Ag$_2$S quantum dots	0.1 mM~12.2 mM	0.324 μM	[161]
biosensor based on ZnS:Ni/ZnS Quantums Dots	0.1~100 μM	35 nM	[162]
biosensor based on TiO$_2$-SnS$_2$ nanocomposite	0.008~1.13 mM; 1.13~5.53 mM	1.8 μM	[163]
biosensor based on bienzyme and carbon nanotubes incorporated into an Os-complex thin film	0.05~1.5 mM	3 μM	[164]
biosensor based on Fe$_3$O$_4$/PPy@ZIF-8 nanocomposite	1 μM~2 mM	0.333 μM	[165]
biosensor based on silver nanowires and chitosan-glucose oxidase film	10 μM~0.8 mM	2.83 μM	[166]

5. Conclusions and Outlooks

In summary, this paper provides a brief overview of recent researches on the applications of sulfur-containing nanomaterials, including metallic sulfide nanomaterials and sulfur-containing quantum dots in biosensors. The sulfur-containing nanomaterials have excellent properties, such as nanometric scale, water-dispersibility, excellent catalytic activity, conductivity, biosafety, photoactivity, and fascinating optical properties, and have been proven useful in various biosensing applications via electrochemical, PEC, ECL and PL strategies. Though many achievements have been obtained for biosensors based on sulfur-containing nanomaterials, there are still significant challenges that need to be solved.

(1) As an emerging quantum dots, sulfur quantum dots possess excellent optical properties and biocompatibility which make them possible to prepare biosensors. However, researches on biosensing applications of sulfur quantum dots are still inadequate. Therefore, it is urgent to further exploit the biosensing applications of sulfur quantum dots in the next few days. Taking advantage of optical properties of sulfur quantum dots and PL-based technologies (e.g., fluorescence detection technologies), PL probes for detecting various biomolecules based on sulfur quantum dots can be established. (2) Real-time biosensing in vivo or intracellular based on sulfur-containing nanomaterials remains a challenge because typical analytical measurements only capture a single-time-point in samples. Biosensing in vivo or intracellular are a new class of detecting technologies that can be established by means of a number of sophisticated analysis platforms providing an in vivo read-out of the spatial, temporal, and quantitative information of biomolecules. Therefore, the vast majority of analytes detected by biosensors based on sulfur-containing nanomaterials are limited to exist in vitro or extracellular. In order to fabricate vivo or intracellular biosensors based on sulfur-containing nanomaterials, sophisticated analysis platforms providing real-time information of biomolecules should be tried to fabricate by utilizing various of technologies and methods (such as, confocal fluorescence microscopic techniques and Raman spectroscopy methods).

(3) With the development of materials science and nanotechnology, wide variety of nanomaterials have emerged in our life. Some nanomaterials, such as metal organic frameworks and gold nanoclusters, are easy to synthesize without complicated operations. Considering time cost and experimental safety, more and more researchers have dedicated to exploiting these nanomaterials. In view of the abundant storage and the pressure on the environment of S elements, more and more sulfur-containing nanomaterials should be synthesized and used to construct biosensors. Therefore, green synthetic methods (e.g., coprecipitation and hydrothermal methods) should be exploited and utilized to synthesize sulfur-containing nanomaterials.

Author Contributions: Data collection, research design, writing (original draft preparation), software, and validation were contributed by C.L. and Y.W. Supervision, research design, writing (review and editing), and funding acquisition are the contribution of C.L., H.J. and X.W. All authors have read and agreed to the published version of the manuscript.

Funding: This work was supported by National Key Research and Development Program of China (2017YFA0205300), the National Natural Science Foundation of China (21675023 and 91753106), the Postgraduate Research & Practice Innovation Program of Jiangsu Province (KYCX19_0109).

Conflicts of Interest: The authors declare no conflict of interest.

References

1. Lim, J.; Pyun, J.; Char, K. Recent Approaches for the Direct Use of Elemental Sulfur in the Synthesis and Processing of Advanced Materials. *Angew. Chem. Int. Ed.* **2015**, *54*, 3249–3258. [CrossRef] [PubMed]
2. Boyd, D.A. Sulfur and its role in modern materials science. *Angew. Chem. Int. Ed.* **2016**, *55*, 15486–15502. [CrossRef]
3. Saleh, T.A. Characterization, determination and elimination technologies for sulfur from petroleum: Toward cleaner fuel and a safe environment. *Trends Environ. Anal. Chem.* **2020**, *25*, e00080. [CrossRef]

4. Fang, R.; Xu, J.; Wang, D.-W. Covalent fixing of sulfur in Metal-Sulfur batteries. *Energy Environ. Sci.* **2020**, *13*, 432–471. [CrossRef]
5. Mann, M.; Kruger, J.E.; Andari, F.; McErlean, J.; Gascooke, J.R.; Smith, J.A.; Worthington, M.J.H.; McKinley, C.C.C.; Campbell, J.A.; Lewis, D.A.; et al. Sulfur polymer composites as Controlled-Release fertilisers. *Org. Biomol. Chem.* **2019**, *17*, 1929–1936. [CrossRef] [PubMed]
6. Nguyen, T.B. Recent advances in organic reactions involving elemental sulfur. *Adv. Synth. Catal.* **2017**, *359*, 1066–1130. [CrossRef]
7. Argueta-Figueroa, L.; Martinez-Alvarez, O.; Santos-Cruz, J.; Garcia-Contreras, R.; Acosta-Torres, L.S.; de la Fuente-Hernandez, J.; Arenas-Arrocena, M.C. Nanomaterials made of Non-toxic metallic sulfides: A systematic review of their potential biomedical applications. *Mater. Sci. Eng. C Biomim. Supramol. Syst.* **2017**, *76*, 1305–1315. [CrossRef] [PubMed]
8. Xu, G.; Zeng, S.; Zhang, B.; Swihart, M.T.; Yong, K.T.; Prasad, P.N. New generation Cadmium-Free quantum dots for biophotonics and nanomedicine. *Chem. Rev.* **2016**, *116*, 12234–12327. [CrossRef] [PubMed]
9. Vlasova, N.; Sorokin, M.; Oborina, E. Carbofunctional Sulfur-Containing Organosilicon Compounds: Synthesis and Application Fields. *Russ. J. Appl. Chem.* **2016**, *89*, 1031–1042. [CrossRef]
10. Lee, S.K.; Lee, Y.J.; Sun, Y.K. Nanostructured lithium sulfide materials for Lithium-Sulfur batteries. *J. Power Sources* **2016**, *323*, 174–188. [CrossRef]
11. Sheng, J.; Wang, L.; Han, Y.; Chen, W.; Liu, H.; Zhang, M.; Deng, L.; Liu, Y.N. Dual roles of protein as a template and a sulfur provider: A general approach to metal sulfides for efficient photothermal therapy of cancer. *Small* **2018**, *14*. [CrossRef] [PubMed]
12. Xie, Z.; Wang, D.; Fan, T.; Xing, C.; Li, Z.; Tao, W.; Liu, L.; Bao, S.; Fan, D.; Zhang, H. Black phosphorus analogue tin sulfide nanosheets: Synthesis and application as Near-Infrared photothermal agents and drug delivery platforms for cancer therapy. *J. Mater. Chem. B* **2018**, *6*, 4747–4755. [CrossRef] [PubMed]
13. Heckert, B.; Banerjee, T.; Sulthana, S.; Naz, S.; Alnasser, R.; Thompson, D.; Normand, G.; Grimm, J.; Perez, J.M.; Santra, S. Design and synthesis of new Sulfur-Containing hyperbranched polymer and theranostic nanomaterials for bimodal imaging and treatment of cancer. *ACS Macro Lett.* **2017**, *6*, 235–240. [CrossRef] [PubMed]
14. Clark, R.M.; Kotsakidis, J.C.; Weber, B.; Berean, K.J.; Carey, B.J.; Field, M.R.; Khan, H.; Ou, J.Z.; Ahmed, T.; Harrison, C.J.; et al. Exfoliation of Quasi-Stratified Bi_2S_3 crystals into Micron-Scale ultrathin corrugated nanosheets. *Chem. Mat.* **2016**, *28*, 8942–8950. [CrossRef]
15. Cao, M.; Wang, H.; Kannan, P.; Ji, S.; Wang, X.; Zhao, Q.; Linkov, V.; Wang, R. Highly efficient Non-Enzymatic glucose sensor based on Cu_xS hollow nanospheres. *Appl. Surf. Sci.* **2019**, *492*, 407–416. [CrossRef]
16. Han, G.; Popuri, S.R.; Greer, H.F.; Zhang, R.; Ferre-Llin, L.; Bos, J.G.; Zhou, W.; Reece, M.J.; Paul, D.J.; Knox, A.R.; et al. Topotactic Anion-Exchange in thermoelectric nanostructured layered tin chalcogenides with reduced selenium content. *Chem. Sci.* **2018**, *9*, 3828–3836. [CrossRef]
17. Jannat, A.; Haque, F.; Xu, K.; Zhou, C.; Zhang, B.Y.; Syed, N.; Mohiuddin, M.; Messalea, K.A.; Li, X.; Gras, S.L.; et al. Exciton-Driven chemical sensors based on Excitation-Dependent photoluminescent Two-Dimensional SnS. *ACS Appl. Mater. Interfaces* **2019**, *11*, 42462–42468. [CrossRef]
18. Kalantar-zadeh, K.; Ou, J.Z. Biosensors based on Two-Dimensional MoS_2. *ACS Sens.* **2015**, *1*, 5–16. [CrossRef]
19. Li, X.; Li, X.; Li, Z.; Wang, J.; Zhang, J. WS_2 nanoflakes based selective ammonia sensors at room temperature. *Sens. Actuators B Chem.* **2017**, *240*, 273–277. [CrossRef]
20. Kubendhiran, S.; Sakthivel, R.; Chen, S.-M.; Mutharani, B. Functionalized-Carbon black as a conductive matrix for nickel sulfide nanospheres and its application to Non-Enzymatic glucose sensor. *J. Electrochem. Soc.* **2018**, *165*, B96–B102. [CrossRef]
21. Munaro, J.; Dolcet, P.; Nappini, S.; Magnano, E.; Dengo, N.; Lucchini, G.; Speghini, A.; Gross, S. the role of the synthetic pathways on properties of Ag_2S nanoparticles for photothermal applications. *Appl. Surf. Sci.* **2020**, *514*, 145856. [CrossRef]
22. Labiadh, H.; Lahbib, K.; Hidouri, S.; Touil, S.; Chaabane, T.B. Insight of ZnS nanoparticles contribution in different biological uses. *Asian Pac. J. Trop. Med.* **2016**, *9*, 757–762. [CrossRef] [PubMed]
23. Tian, Q.; Tang, M.; Sun, Y.; Zou, R.; Chen, Z.; Zhu, M.; Yang, S.; Wang, J.; Wang, J.; Hu, J. Hydrophilic Flower-Like CuS superstructures as an efficient 980 nm Laser-Driven photothermal agent for ablation of cancer cells. *Adv. Mater.* **2011**, *23*, 3542–3547. [CrossRef] [PubMed]

24. Xia, C.; Li, J. Recent advances in optoelectronic properties and applications of Two-Dimensional metal chalcogenides. *J. Semicond.* **2016**, *37*, 051001. [CrossRef]
25. Zhang, Y.; Zhang, L.; Lv, T.; Chu, P.K.; Huo, K. Two-Dimensional transition metal chalcogenides for alkali metal ions storage. *ChemSusChem* **2020**, *13*, 1114–1154. [CrossRef]
26. Matthews, P.D.; McNaughter, P.D.; Lewis, D.J.; O'Brien, P. Shining a light on transition metal chalcogenides for sustainable photovoltaics. *Chem. Sci.* **2017**, *8*, 4177–4187. [CrossRef]
27. Baranov, N.V.; Selezneva, N.V.; Kazantsev, V.A. Magnetism and superconductivity of transition metal chalcogenides. *Phys. Met. Met.* **2019**, *119*, 1301–1304. [CrossRef]
28. Li, H.-Y.; Yoon, J.-W.; Lee, C.-S.; Lim, K.; Yoon, J.-W.; Lee, J.-H. Visible light assisted NO_2 sensing at room temperature by CdS nanoflake array. *Sens. Actuators B Chem.* **2018**, *255*, 2963–2970. [CrossRef]
29. Du, Y.; Yin, Z.; Zhu, J.; Huang, X.; Wu, X.J.; Zeng, Z.; Yan, Q.; Zhang, H. A general method for the Large-Scale synthesis of uniform ultrathin metal sulphide nanocrystals. *Nat. Commun.* **2012**, *3*, 1177. [CrossRef]
30. Goel, S.; Chen, F.; Cai, W. Synthesis and biomedical applications of copper sulfide nanoparticles: From sensors to theranostics. *Small* **2014**, *10*, 631–645. [CrossRef]
31. Joo, J.; Na, H.B.; Yu, T.; Yu, J.H.; Kim, Y.W.; Wu, F.; Zhang, J.Z.; Hyeon, T. Generalized and facile synthesis of semiconducting metal sulfide nanocrystals. *J. Am. Chem. Soc.* **2003**, *125*, 11100–11105. [CrossRef] [PubMed]
32. Cui, J.; Wang, L.; Yu, X. A simple and generalized Heat-Up method for the synthesis of metal sulfide nanocrystals. *New J. Chem.* **2019**, *43*, 16007–16011. [CrossRef]
33. Kuznetsov, A.N.; Stroganova, E.A.; Serov, A.A.; Kirdyankin, D.I.; Novotortsev, V.M. New Quasi-2D Nickel-Gallium mixed chalcogenides based on the Cu_3Au-Type extended fragments. *J. Alloys Compd.* **2017**, *696*, 413–422. [CrossRef]
34. Guo, Q.; Wu, T.; Liu, L.; He, Y.; Liu, D.; You, T. Hierarchically porous $NiCo_2S_4$ nanowires anchored on flexible electrospun graphitic nanofiber for High-Performance glucose biosensing. *J. Alloys Compd.* **2020**, *819*, 153376. [CrossRef]
35. Behera, C.; Samal, R.; Rout, C.S.; Dhaka, R.S.; Sahoo, G.; Samal, S.L. Synthesis of $CuSbS_2$ nanoplates and $CuSbS_2$-Cu_3SbS_4 nanocomposite: Effect of sulfur source on different phase formation. *Inorg. Chem.* **2019**, *58*, 15291–15302. [CrossRef] [PubMed]
36. Zou, Y.; Gu, Y.; Hui, B.; Yang, X.; Liu, H.; Chen, S.; Cai, R.; Sun, J.; Zhang, X.; Yang, D. Nitrogen and sulfur vacancies in carbon shell to tune charge distribution of $Co_6Ni_3S_8$ core and boost sodium storage. *Adv. Energy Mater.* **2020**, *10*, 1904147. [CrossRef]
37. Stroyuk, O.; Raevskaya, A.; Gaponik, N. Solar Light harvesting with multinary metal chalcogenide nanocrystals. *Chem. Soc. Rev.* **2018**, *47*, 5354–5422. [CrossRef]
38. Zheng, L.; Teng, F.; Ye, X.; Zheng, H.; Fang, X. Photo/Electrochemical applications of metal sulfide/TiO_2 heterostructures. *Adv. Energy Mater.* **2019**, *10*, 1902355. [CrossRef]
39. Zhou, B.; Song, J.; Xie, C.; Chen, C.; Qian, Q.; Han, B. Mo–Bi–Cd ternary metal chalcogenides: Highly efficient photocatalyst for CO_2 reduction to formic acid under visible light. *ACS Sustain. Chem. Eng.* **2018**, *6*, 5754–5759. [CrossRef]
40. Liu, Y.; Song, X.; Guo, Y.; Zhong, Y.; Li, Y.; Sun, Y.; Ji, M.; You, Z.; An, Y. Mild solvothermal syntheses and characterizations of two layered sulfides $Ba_2Cu_2Cd_2S_5$ and $Ba_3Cu_4Hg_4S_9$. *J. Alloys Compd.* **2020**, *829*, 154586. [CrossRef]
41. Wang, T.; Huo, T.; Wang, H.; Wang, C. Quaternary chalcogenides: Promising thermoelectric material and recent progress. *Sci. China-Mater.* **2019**, *63*, 8–15. [CrossRef]
42. Tiwari, K.J.; Prem Kumar, D.S.; Mallik, R.C.; Malar, P. Ball mill synthesis of bulk quaternary $Cu_2ZnSnSe_4$ and thermoelectric studies. *J. Electron. Mater.* **2016**, *46*, 30–39. [CrossRef]
43. Song, Q.; Qiu, P.; Chen, H.; Zhao, K.; Ren, D.; Shi, X.; Chen, L. Improved thermoelectric performance in nonstoichiometric $Cu_{2+\delta}Mn_{1-\delta}SnSe_4$ quaternary diamondlike compounds. *ACS Appl. Mater. Interfaces* **2018**, *10*, 10123–10131. [CrossRef] [PubMed]
44. Song, Q.; Qiu, P.; Hao, F.; Zhao, K.; Zhang, T.; Ren, D.; Shi, X.; Chen, L. Quaternary pseudocubic $Cu_2TMSnSe_4$ (TM = Mn, Fe, Co) chalcopyrite thermoelectric materials. *Adv. Electron. Mater.* **2016**, *2*, 1600312. [CrossRef]
45. Kempt, R.; Kuc, A.; Heine, T. Two-Dimensional Noble-Metal chalcogenides and phosphochalcogenides. *Angew. Chem. Int. Ed.* **2020**, *59*, 9242–9254. [CrossRef]
46. Tang, C.; Zhang, C.; Matta, S.K.; Jiao, Y.; Ostrikov, K.; Liao, T.; Kou, L.; Du, A. Predicting New Two-Dimensional $Pd_3(PS_4)_2$ as an Efficient Photocatalyst for Water Splitting. *J. Phys. Chem. C* **2018**, *122*, 21927–21932. [CrossRef]

47. Nie, L.; Zhang, Q. Recent progress in crystalline metal chalcogenides as efficient photocatalysts for organic pollutant degradation. *Inorg. Chem. Front.* **2017**, *4*, 1953–1962. [CrossRef]
48. Cheng, C.; Kong, D.; Wei, C.; Du, W.; Zhao, J.; Feng, Y.; Duan, Q. Self-Template synthesis of hollow ellipsoid Ni-Mn sulfides for supercapacitors, electrocatalytic oxidation of glucose and water treatment. *Dalton Trans.* **2017**, *46*, 5406–5413. [CrossRef]
49. Radhakrishnan, S.; Kim, H.-Y.; Kim, B.-S. A novel CuS microflower superstructure based sensitive and selective nonenzymatic glucose detection. *Sens. Actuators B Chem.* **2016**, *233*, 93–99. [CrossRef]
50. Guo, X.; Liu, S.; Yang, M.; Du, H.; Qu, F. Dual Signal amplification photoelectrochemical biosensor for highly sensitive human epidermal growth factor Receptor-2 Detection. *Biosens. Bioelectron.* **2019**, *139*, 111312. [CrossRef]
51. Babu, K.J.; Raj Kumar, T.; Yoo, D.J.; Phang, S.-M.; Gnana Kumar, G. Electrodeposited nickel cobalt sulfide flowerlike architectures on disposable cellulose filter paper for Enzyme-Free glucose sensor applications. *ACS Sustain. Chem. Eng.* **2018**, *6*, 16982–16989. [CrossRef]
52. Xue, J.; Yang, L.; Jia, Y.; Zhang, Y.; Wu, D.; Ma, H.; Hu, L.; Wei, Q.; Ju, H. Dual-Quenching electrochemiluminescence resonance energy transfer system from Ru-In$_2$S$_3$ to alpha-MoO$_3$-Au based on protect of protein bioactivity for procalcitonin detection. *Biosens. Bioelectron.* **2019**, *142*, 111524. [CrossRef] [PubMed]
53. Simeonidis, K.; Liébana-Viñas, S.; Wiedwald, U.; Ma, Z.; Li, Z.A.; Spasova, M.; Patsia, O.; Myrovali, E.; Makridis, A.; Sakellari, D.; et al. A Versatile Large-Scale and green process for synthesizing magnetic nanoparticles with tunable magnetic hyperthermia features. *RSC Adv.* **2016**, *6*, 53107–53117. [CrossRef]
54. Vena, M.P.; Jobbagy, M.; Bilmes, S.A. Microorganism mediated biosynthesis of metal chalcogenides; a powerful tool to transform toxic effluents into functional nanomaterials. *Sci. Total. Environ.* **2016**, *565*, 804–810. [CrossRef]
55. Tan, C.; Lai, Z.; Zhang, H. Ultrathin Two-Dimensional multinary layered metal chalcogenide nanomaterials. *Adv. Mater.* **2017**, *29*, 1701392. [CrossRef]
56. Zhang, Y.; Ma, Y.; Li, Y.; Zhu, W.; Wei, Z.; Sun, J.; Li, T.; Wang, J. Ambient Self-Derivation of Nickel-Cobalt hydroxysulfide multistage nanoarray for High-Performance electrochemical glucose sensing. *Appl. Surf. Sci.* **2020**, *505*, 144636. [CrossRef]
57. Hu, X.; Shao, W.; Hang, X.; Zhang, X.; Zhu, W.; Xie, Y. Superior electrical conductivity in hydrogenated layered ternary chalcogenide nanosheets for flexible All-Solid-State supercapacitors. *Angew. Chem. Int. Ed.* **2016**, *55*, 5733–5738. [CrossRef]
58. Zhu, H.; Lai, Z.; Fang, Y.; Zhen, X.; Tan, C.; Qi, X.; Ding, D.; Chen, P.; Zhang, H.; Pu, K. Ternary chalcogenide nanosheets with ultrahigh photothermal conversion efficiency for photoacoustic theranostics. *Small* **2017**, *13*, 1604139. [CrossRef]
59. Baláž, P.; Hegedüs, M.; Achimovičová, M.; Baláž, M.; Tešinský, M.; Dutková, E.; Kaňuchová, M.; Briančin, J. Semi-industrial green mechanochemical syntheses of solar cell absorbers based on quaternary sulfides. *ACS Sustain. Chem. Eng.* **2018**, *6*, 2132–2141. [CrossRef]
60. Wu, K.; Yang, Z.; Pan, S. The first quaternary Diamond-Like semiconductor with 10-Membered LiS$_4$ rings exhibiting excellent nonlinear optical performances. *Chem. Commun.* **2017**, *53*, 3010–3013. [CrossRef]
61. Kubendhiran, S.; Thirumalraj, B.; Chen, S.M.; Karuppiah, C. Electrochemical Co-Preparation of cobalt sulfide/reduced graphene oxide composite for electrocatalytic activity and determination of H$_2$O$_2$ in biological samples. *J. Colloid Interface Sci.* **2018**, *509*, 153–162. [CrossRef]
62. Wang, Z.; Dong, S.; Gui, M.; Asif, M.; Wang, W.; Wang, F.; Liu, H. Graphene paper supported MoS$_2$ nanocrystals monolayer with Cu Submicron-Buds: High-Performance flexible platform for sensing in sweat. *Anal. Biochem.* **2018**, *543*, 82–89. [CrossRef] [PubMed]
63. Li, C.; Zhang, J.; Jiang, H.; Wang, X.; Liu, J. orthogonal adsorption of carbon dots and DNA on nanoceria. *Langmuir* **2020**, *36*, 2474–2481. [CrossRef] [PubMed]
64. Wang, D.; Li, Q.; Xing, Z.; Yang, X. Copper sulfide nanoplates as nanosensors for fast, sensitive and selective detection of DNA. *Talanta* **2018**, *178*, 905–909. [CrossRef] [PubMed]
65. Okoth, O.K.; Yan, K.; Liu, Y.; Zhang, J. Graphene-Doped Bi$_2$S$_3$ nanorods as Visible-Light photoelectrochemical aptasensing platform for sulfadimethoxine detection. *Biosens. Bioelectron.* **2016**, *86*, 636–642. [CrossRef] [PubMed]

66. Foo, C.Y.; Lim, H.N.; Pandikumar, A.; Huang, N.M.; Ng, Y.H. Utilization of reduced graphene oxide/cadmium Sulfide-Modified carbon cloth for Visible-Light-Prompt Photoelectrochemical Sensor For copper (II) ions. *J. Hazard. Mater.* **2016**, *304*, 400–408. [CrossRef] [PubMed]
67. Du, C.; Shang, A.; Shang, M.; Ma, X.; Song, W. Water-Soluble VS_2 quantum dots with unusual fluorescence for biosensing. *Sens. Actuators B Chem.* **2018**, *255*, 926–934. [CrossRef]
68. Zheng, Y.; Jiang, H.; Wang, X. Multiple strategies for controlled synthesis of atomically precise alloy nanoclusters. *Acta Phys. Chim. Sin.* **2018**, *34*, 740–754. [CrossRef]
69. Shu, Y.; Yan, J.; Lu, Q.; Ji, Z.; Jin, D.; Xu, Q.; Hu, X. Pb Ions Enhanced Fluorescence of Ag_2S QDs with Tunable Emission in the NIR-II Window: Facile One Pot Synthesis and Their Application in NIR-II Fluorescent Biosensing. *Sens. Actuators B Chem.* **2020**, *307*, 127593. [CrossRef]
70. Babamiri, B.; Salimi, A.; Hallaj, R. A Molecularly imprinted electrochemiluminescence sensor for ultrasensitive HIV-1 gene detection using eus nanocrystals as luminophore. *Biosens. Bioelectron.* **2018**, *117*, 332–339. [CrossRef]
71. Zhang, Y.; Wen, F.; Huang, Z.; Tan, J.; Zhou, Z.; Yuan, K.; Wang, H. Nitrogen doped lignocellulose/binary metal sulfide modified electrode: Preparation and application for Non-Enzymatic ascorbic acid, dopamine and nitrite sensing. *J. Electroanal. Chem.* **2017**, *806*, 150–157. [CrossRef]
72. Huang, D.; Wang, L.; Zhan, Y.; Zou, L.; Ye, B. Photoelectrochemical biosensor for CEA detection based on SnS_2-GR with multiple quenching effects of Au@CuS-GR. *Biosens. Bioelectron.* **2019**, *140*, 111358. [CrossRef] [PubMed]
73. Zhao, J.; Wang, S.; Zhang, S.; Zhao, P.; Wang, J.; Yan, M.; Ge, S.; Yu, J. Peptide Cleavage-Mediated photoelectrochemical signal on-off via CuS electronic extinguisher for PSA detection. *Biosens. Bioelectron.* **2020**, *150*, 111958. [CrossRef] [PubMed]
74. Shang, M.; Qi, H.; Du, C.; Huang, H.; Wu, S.; Zhang, J.; Song, W. One-Step electrodeposition of High-Quality amorphous molybdenum Sulfide/RGO photoanode for Visible-Light sensitive photoelectrochemical biosensing. *Sens. Actuators B Chem.* **2018**, *266*, 71–79. [CrossRef]
75. Zhang, K.; Lv, S.; Zhou, Q.; Tang, D. CoOOH Nanosheets-Coated g-C_3N_4/$CuInS_2$ nanohybrids for photoelectrochemical biosensor of carcinoembryonic antigen coupling hybridization chain reaction with etching reaction. *Sens. Actuators B Chem.* **2020**, *307*, 127631. [CrossRef]
76. Wang, J.; Long, J.; Liu, Z.; Wu, W.; Hu, C. Label-Free and High-Throughput biosensing of multiple tumor markers on a single Light-Addressable photoelectrochemical sensor. *Biosens. Bioelectron.* **2017**, *91*, 53–59. [CrossRef] [PubMed]
77. Wang, B.; Cao, J.T.; Dong, Y.X.; Liu, F.R.; Fu, X.L.; Ren, S.W.; Ma, S.H.; Liu, Y.M. An in situ electron donor consumption strategy for photoelectrochemical biosensing of proteins based on ternary Bi_2S_3/Ag_2S/TiO_2 NT arrays. *Chem. Commun.* **2018**, *54*, 806–809. [CrossRef] [PubMed]
78. Cui, L.; Hu, J.; Wang, M.; Diao, X.K.; Li, C.C.; Zhang, C.Y. Mimic Peroxidase- and Bi_2S_3 Nanorod-Based photoelectrochemical biosensor for Signal-On detection of polynucleotide kinase. *Anal. Chem.* **2018**, *90*, 11478–11485. [CrossRef]
79. Zhu, W.; Wang, C.; Li, X.; Khan, M.S.; Sun, X.; Ma, H.; Fan, D.; Wei, Q. Zinc-Doping enhanced cadmium sulfide electrochemiluminescence behavior based on Au-Cu alloy nanocrystals quenching for insulin detection. *Biosens. Bioelectron.* **2017**, *97*, 115–121. [CrossRef]
80. Wei, Y.; Wang, H.; Sun, S.; Tang, L.; Cao, Y.; Deng, B. An Ultrasensitive electrochemiluminescence sensor based on reduced graphene Oxide-Copper sulfide composite coupled with capillary electrophoresis for determination of amlodipine besylate in Mice Plasma. *Biosens. Bioelectron.* **2016**, *86*, 714–719. [CrossRef]
81. Chen, D.; Li, S.; Zheng, F. Water Soluble Sulphur Quantum Dots for Selective Ag^+ Sensing based on the Ion Aggregation-Induced Photoluminescence Enhancement. *Anal. Methods* **2016**, *8*, 632–636. [CrossRef]
82. Shen, L.; Wang, H.; Liu, S.; Bai, Z.; Zhang, S.; Zhang, X.; Zhang, C. Assembling of sulfur quantum dots in fission of sublimed sulfur. *J. Am. Chem. Soc.* **2018**, *140*, 7878–7884. [CrossRef] [PubMed]
83. Li, S.; Chen, D.; Zheng, F.; Zhou, H.; Jiang, S.; Wu, Y. Water-Soluble and lowly toxic sulphur quantum dots. *Adv. Funct. Mater.* **2014**, *24*, 7133–7138. [CrossRef]
84. Song, Y.; Tan, J.; Wang, G.; Gao, P.; Lei, J.; Zhou, L. Oxygen accelerated scalable synthesis of highly fluorescent sulfur quantum dots. *Chem. Sci.* **2020**, *11*, 772–777. [CrossRef]
85. Wang, S.; Bao, X.; Gao, B.; Li, M. A novel sulfur quantum dot for the detection of cobalt ions and norfloxacin as a fluorescent "switch". *Dalton Trans.* **2019**, *48*, 8288–8296. [CrossRef] [PubMed]

86. Wang, H.; Wang, Z.; Xiong, Y.; Kershaw, S.V.; Li, T.; Wang, Y.; Zhai, Y.; Rogach, A.L. Hydrogen peroxide assisted synthesis of highly luminescent sulfur quantum dots. *Angew. Chem. Int. Ed.* **2019**, *58*, 7040–7044. [CrossRef]
87. Wang, Y.; Wang, P.; Wu, Y.; Di, J. A cathodic "Signal-On" photoelectrochemical sensor for Hg^{2+} detection based on Ion-Exchange with ZnS quantum dots. *Sens. Actuators B Chem.* **2018**, *254*, 910–915. [CrossRef]
88. Saleviter, S.; Fen, Y.W.; Omar, N.A.S.; Daniyal, W.M.E.M.M.; Abdullah, J.; Zaid, M.H.M. Structural and optical studies of cadmium sulfide quantum Dot-Graphene Oxide-Chitosan nanocomposite thin film as a novel SPR spectroscopy active layer. *J. Nanomater.* **2018**, *2018*, 4324072. [CrossRef]
89. Wang, P.; Cao, L.; Wu, Y.; Di, J. A cathodic photoelectrochemical sensor for chromium(VI) based on the use of PbS quantum dot semiconductors on an ito electrode. *Microchim. Acta* **2018**, *185*, 356. [CrossRef]
90. Li, Y.; Tang, L.; Li, R.; Xiang, J.; Teng, K.S.; Lau, S.P. SnS_2 quantum dots: Facile synthesis, properties, and applications in ultraviolet photodetector. *Chin. Phys. B* **2019**, *28*, 037801. [CrossRef]
91. Aydemir, D.; Hashemkhani, M.; Acar, H.Y.; Ulusu, N.N. In vitro interaction of glutathione S-Transferase-pi enzyme with Glutathione-Coated silver sulfide quantum dots: A novel method for biodetection of glutathione S-Transferase enzyme. *Chem. Biol. Drug Des.* **2019**, *94*, 2094–2102. [CrossRef] [PubMed]
92. Liu, Y.; Du, L.; Gu, K.; Zhang, M. Effect of Tm dopant on luminescence, photoelectric properties and electronic structure of In_2S_3 quantum dots. *J. Lumin.* **2020**, *217*, 116775. [CrossRef]
93. Kang, D.; Bharath Kumar, M.; Son, C.; Park, H.; Park, J. Simple synthesis method and characterizations of Aggregation-Free cysteamine capped PbS quantum dot. *Appl. Sci.* **2019**, *9*, 4661. [CrossRef]
94. Abha, K.; Nebu, J.; Anjali Devi, J.S.; Aparna, R.S.; Anjana, R.R.; Aswathy, A.O.; George, S. Photoluminescence sensing of bilirubin in human serum using L-cysteine tailored manganese doped zinc sulphide quantum dots. *Sens. Actuators B Chem.* **2019**, *282*, 300–308. [CrossRef]
95. Na, W.; Liu, X.; Hu, T.; Su, X. Highly sensitive fluorescent determination of sulfide using BSA-Capped CdS quantum dots. *New J. Chem.* **2016**, *40*, 1872–1877. [CrossRef]
96. Liu, Z.; Xiao, J.; Wu, X.; Lin, L.; Weng, S.; Chen, M.; Cai, X.; Lin, X. Switch-On fluorescent strategy based on N and S co-Doped graphene quantum dots (N-S/GQDs) for monitoring pyrophosphate ions in synovial fluid of arthritis patients. *Sens. Actuators B Chem.* **2016**, *229*, 217–224. [CrossRef]
97. Chen, C.; Zhao, D.; Hu, T.; Sun, J.; Yang, X. Highly fluorescent nitrogen and sulfur co-Doped graphene quantum dots for an inner filter Effect-based cyanide sensor. *Sens. Actuators B Chem.* **2017**, *241*, 779–788. [CrossRef]
98. Yao, D.; Liang, A.; Jiang, Z. A fluorometric clenbuterol immunoassay using sulfur and nitrogen doped carbon quantum dots. *Microchim. Acta* **2019**, *186*, 323. [CrossRef]
99. Guo, Z.; Luo, J.; Zhu, Z.; Sun, Z.; Zhang, X.; Wu, Z.-c.; Mo, F.; Guan, A. A facile synthesis of High-Efficient N,S co-Doped carbon dots for temperature sensing application. *Dyes Pigment.* **2020**, *173*, 107952. [CrossRef]
100. Sharma, V.; Kaur, N.; Tiwari, P.; Saini, A.K.; Mobin, S.M. Multifunctional fluorescent "Off-On-Off" nanosensor for Au^{3+} and S^{2-} employing N-S co-Doped Carbon-Dots. *Carbon* **2018**, *139*, 393–403. [CrossRef]
101. Peng, J.; Zhao, Z.; Zheng, M.; Su, B.; Chen, X.; Chen, X. Electrochemical synthesis of phosphorus and sulfur co-Doped graphene quantum dots as efficient electrochemiluminescent immunomarkers for monitoring okadaic acid. *Sens. Actuators B Chem.* **2020**, *304*, 127383. [CrossRef]
102. Bian, S.; Shen, C.; Qian, Y.; Liu, J.; Xi, F.; Dong, X. Facile synthesis of Sulfur-Doped graphene quantum dots as fluorescent sensing probes for Ag^+ ions detection. *Sens. Actuators B Chem.* **2017**, *242*, 231–237. [CrossRef]
103. Fan, T.; Zhang, G.; Jian, L.; Murtaza, I.; Meng, H.; Liu, Y.; Min, Y. Facile synthesis of Defect-Rich nitrogen and Sulfur co-doped graphene quantum dots as Metal-Free electrocatalyst for the oxygen reduction reaction. *J. Alloys Compd.* **2019**, *792*, 844–850. [CrossRef]
104. Xu, Q.; Pu, P.; Zhao, J.; Dong, C.; Gao, C.; Chen, Y.; Chen, J.; Liu, Y.; Zhou, H. Preparation of highly photoluminescent Sulfur-Doped carbon dots for Fe(iii) detection. *J. Mater. Chem. A* **2015**, *3*, 542–546. [CrossRef]
105. Amjadi, M.; Manzoori, J.L.; Hallaj, T.; Azizi, N. Sulfur and Nitrogen co-Doped carbon quantum dots as the chemiluminescence probe for detection of Cu^{2+} Ions. *J. Lumin.* **2017**, *182*, 246–251. [CrossRef]
106. Haque, E.; Kim, J.; Malgras, V.; Reddy, K.R.; Ward, A.C.; You, J.; Bando, Y.; Hossain, M.S.A.; Yamauchi, Y. Recent advances in graphene quantum dots: Synthesis, properties, and applications. *Small Methods* **2018**, *2*, 1800050. [CrossRef]

107. Wang, W.; Xu, S.; Li, N.; Huang, Z.; Su, B.; Chen, X. Sulfur and phosphorus co-Doped graphene quantum dots for fluorescent monitoring of nitrite in pickles. *Spectrochim. Acta Part A Mol. Biomol. Spectrosc.* **2019**, *221*, 117211. [CrossRef]
108. Zhu, S.; Zhao, X.; Song, Y.; Lu, S.; Yang, B. Beyond Bottom-up carbon nanodots: Citric-Acid derived organic molecules. *Nano Today* **2016**, *11*, 128–132. [CrossRef]
109. Ganiga, M.; Cyriac, J. An ascorbic acid sensor based on cadmium sulphide quantum dots. *Anal. Bioanal. Chem.* **2016**, *408*, 3699–3706. [CrossRef]
110. Kulchat, S.; Boonta, W.; Todee, A.; Sianglam, P.; Ngeontae, W. A fluorescent sensor based on thioglycolic acid capped cadmium sulfide quantum dots for the determination of dopamine. *Spectrochim. Acta Part A Mol. Biomol. Spectrosc.* **2018**, *196*, 7–15. [CrossRef]
111. Ngamdee, K.; Kulchat, S.; Tuntulani, T.; Ngeontae, W. Fluorescence sensor based on d-Penicillamine capped cadmium sulfide quantum dots for the detection of cysteamine. *J. Lumin.* **2017**, *187*, 260–268. [CrossRef]
112. Miao, X.; Yan, X.; Qu, D.; Li, D.; Tao, F.F.; Sun, Z. Red emissive sulfur, nitrogen codoped carbon dots and their application in ion detection and theraonostics. *ACS Appl. Mater. Interfaces* **2017**, *9*, 18549–18556. [CrossRef]
113. Omar, N.A.S.; Fen, Y.W.; Abdullah, J.; Zaid, M.H.M.; Daniyal, W.M.E.M.M.; Mahdi, M.A. Sensitive surface plasmon resonance performance of cadmium sulfide quantum Dots-Amine functionalized graphene oxide based thin film towards dengue virus E-protein. *Opt. Laser Technol.* **2019**, *114*, 204–208. [CrossRef]
114. Omar, N.A.S.; Fen, Y.W.; Abdullah, J.; Anas, N.A.A.; Ramdzan, N.S.M.; Mahdi, M.A. Optical and structural properties of cadmium sulphide quantum dots based thin films as potential sensing material for dengue Virus E-protein. *Results Phys.* **2018**, *11*, 734–739. [CrossRef]
115. Han, X.-L.; Li, Q.; Hao, H.; Liu, C.; Li, R.; Yu, F.; Lei, J.; Jiang, Q.; Liu, Y.; Hu, J. Facile One-Step synthesis of quaternary aginzns quantum dots and their applications for causing bioeffects and detecting Cu^{2+}. *RSC Adv.* **2020**, *10*, 9172–9181. [CrossRef]
116. Haldar, D.; Dinda, D.; Saha, S.K. High selectivity in water soluble MoS_2 quantum dots for sensing nitro explosives. *J. Mater. Chem. C* **2016**, *4*, 6321–6326. [CrossRef]
117. Grinyte, R.; Barroso, J.; Saa, L.; Pavlov, V. Modulating the growth of Cysteine-Capped cadmium sulfide quantum dots with enzymatically produced hydrogen peroxide. *Nano Res.* **2017**, *10*, 1932–1941. [CrossRef]
118. Malik, R.; Pinnaka, A.K.; Kaur, M.; Kumar, V.; Tikoo, K.; Singh, S.; Kaushik, A.; Singhal, S. Water-Soluble Glutathione-CdS QDs with exceptional antimicrobial properties synthesized via green route for fluorescence sensing of fluoroquinolones. *J. Chem. Technol. Biotechnol.* **2019**, *94*, 1082–1090. [CrossRef]
119. Bhardwaj, H.; Singh, C.; Pandey, M.k.; Sumana, G. Star shaped zinc sulphide quantum dots Self-Assembled monolayers: Preparation and applications in food toxin detection. *Sens. Actuators B Chem.* **2016**, *231*, 624–633. [CrossRef]
120. Ngamdee, K.; Ngeontae, W. Circular dichroism glucose biosensor based on chiral cadmium sulfide quantum dots. *Sens. Actuators B Chem.* **2018**, *274*, 402–411. [CrossRef]
121. Jacob, J.M.; Rajan, R.; Tom, T.C.; Kumar, V.S.; Kurup, G.G.; Shanmuganathan, R.; Pugazhendhi, A. Biogenic design of ZnS quantum dots—insights into their In-Vitro cytotoxicity, photocatalysis and biosensing properties. *Ceram. Int.* **2019**, *45*, 24193–24201. [CrossRef]
122. Nathiya, D.; Gurunathan, K.; Wilson, J. Size controllable, pH triggered reduction of bovine serum albumin and its adsorption behavior with SnO_2/SnS_2 quantum dots for biosensing application. *Talanta* **2020**, *210*, 120671. [CrossRef] [PubMed]
123. Ding, C.; Cao, X.; Zhang, C.; He, T.; Hua, N.; Xian, Y. Rare earth ions enhanced near infrared fluorescence of Ag_2S quantum dots for the detection of fluoride ions in living cells. *Nanoscale* **2017**, *9*, 14031–14038. [CrossRef] [PubMed]
124. Qiao, G.; Liu, L.; Hao, X.; Zheng, J.; Liu, W.; Gao, J.; Zhang, C.C.; Wang, Q. Signal transduction from small particles: Sulfur nanodots featuring mercury sensing, cell entry mechanism and in vitro tracking performance. *Chem. Eng. J.* **2020**, *382*, 122907. [CrossRef]
125. Zhao, J.; Fan, Z. Using Zinc Ion-Enhanced fluorescence of sulfur quantum dots to improve the detection of the Zinc(II)-Binding antifungal drug clioquinol. *Microchim. Acta* **2019**, *187*, 3. [CrossRef]
126. Bansal, A.K.; Antolini, F.; Zhang, S.; Stroea, L.; Ortolani, L.; Lanzi, M.; Serra, E.; Allard, S.; Scherf, U.; Samuel, I.D.W. Highly luminescent colloidal CdS quantum dots with efficient Near-Infrared electroluminescence in Light-Emitting diodes. *J. Phys. Chem. C* **2016**, *120*, 1871–1880. [CrossRef]

127. Elakkiya, V.; Menon, M.P.; Nataraj, D.; Biji, P.; Selvakumar, R. Optical detection of CA 15.3 breast cancer antigen using CdS quantum dot. *IET Nanobiotechnol.* **2017**, *11*, 268–276. [CrossRef]
128. Zhang, Z.; Duan, F.; He, L.; Peng, D.; Yan, F.; Wang, M.; Zong, W.; Jia, C. electrochemical clenbuterol immunosensor based on a gold electrode modified with zinc sulfide quantum dots and polyaniline. *Microchim. Acta* **2016**, *183*, 1089–1097. [CrossRef]
129. Amor-Gutierrez, O.; Iglesias-Mayor, A.; Llano-Suarez, P.; Costa-Fernandez, J.M.; Soldado, A.; Podadera, A.; Parra, F.; Costa-Garcia, A.; de la Escosura-Muniz, A. Electrochemical quantification of Ag_2S quantum dots: Evaluation of different surface coating ligands for bacteria determination. *Microchim. Acta* **2020**, *187*, 169. [CrossRef]
130. Wang, Y.; Ge, S.; Zhang, L.; Yu, J.; Yan, M.; Huang, J. Visible photoelectrochemical sensing platform by in situ generated cds quantum dots decorated Branched-TiO_2 nanorods equipped with prussian blue electrochromic display. *Biosens. Bioelectron.* **2017**, *89*, 859–865. [CrossRef]
131. Raevskaya, A.; Rozovik, O.; Novikova, A.; Selyshchev, O.; Stroyuk, O.; Dzhagan, V.; Goryacheva, I.; Gaponik, N.; Zahn, D.R.T.; Eychmüller, A. Luminescence and photoelectrochemical properties of Size-Selected aqueous Copper-Doped Ag-In-S quantum dots. *RSC Adv.* **2018**, *8*, 7550–7557. [CrossRef]
132. Mo, F.; Han, Q.; Song, J.; Wu, J.; Ran, P.; Fu, Y. An ultrasensitive "On-Off-On" photoelectrochemical thrombin aptasensor based on perylene tetracarboxylic acid/gold nanoparticles/cadmium sulfide quantum dots amplified matrix. *J. Electrochem. Soc.* **2018**, *165*, B679–B685. [CrossRef]
133. Deng, K.; Wang, H.; Xiao, J.; Li, C.; Zhang, S.; Huang, H. Polydopamine nanospheres loaded with L-cysteine-coated cadmium sulfide quantum dots as photoelectrochemical signal amplifier for PSA detection. *Anal. Chim. Acta* **2019**, *1090*, 143–150. [CrossRef] [PubMed]
134. Liu, J.-X.; Ding, S.-N. Multicolor electrochemiluminescence of cadmium sulfide quantum dots to detect dopamine. *J. Electroanal. Chem.* **2016**, *781*, 395–400. [CrossRef]
135. Wang, C.; Chen, M.; Wu, J.; Mo, F.; Fu, Y. Multi-functional electrochemiluminescence aptasensor based on resonance energy transfer between Au nanoparticles and lanthanum Ion-doped cadmium sulfide quantum dots. *Anal. Chim. Acta* **2019**, *1086*, 66–74. [CrossRef]
136. Wu, F.-F.; Zhou, Y.; Wang, J.-X.; Zhuo, Y.; Yuan, R.; Chai, Y.-Q. A novel electrochemiluminescence immunosensor based on Mn doped Ag_2S quantum dots probe for laminin detection. *Sens. Actuators B Chem.* **2017**, *243*, 1067–1074. [CrossRef]
137. Zhou, H.; Han, T.; Wei, Q.; Zhang, S. Efficient enhancement of electrochemiluminescence from cadmium sulfide quantum dots by glucose oxidase mimicking gold nanoparticles for highly sensitive assay of methyltransferase activity. *Anal. Chem.* **2016**, *88*, 2976–2983. [CrossRef]
138. Mishra, H.; Umrao, S.; Singh, J.; Srivastava, R.K.; Ali, R.; Misra, A.; Srivastava, A. pH dependent optical switching and fluorescence modulation of molybdenum sulfide quantum dots. *Adv. Opt. Mater.* **2017**, *5*, 1601021. [CrossRef]
139. Gong, Y.; Fan, Z. Room-Temperature phosphorescence Turn-on detection of DNA based on Riboflavin-Modulated manganese doped Zinc sulfide quantum dots. *J. Fluoresc.* **2016**, *26*, 385–393. [CrossRef]
140. Liu, Y.; Dong, P.; Jiang, Q.; Wang, F.; Pang, D.-W.; Liu, X. Assembly-Enhanced fluorescence from metal nanoclusters and quantum dots for highly sensitive biosensing. *Sens. Actuators B Chem.* **2019**, *279*, 334–341. [CrossRef]
141. Adegoke, O.; Seo, M.W.; Kato, T.; Kawahito, S.; Park, E.Y. Gradient band gap engineered alloyed quaternary/ternary CdZnSeS/ZnSeS quantum dots: An ultrasensitive fluorescence reporter in a conjugated molecular beacon system for the biosensing of influenza virus RNA. *J. Mater. Chem. B* **2016**, *4*, 1489–1498. [CrossRef] [PubMed]
142. Rong, M.; Ye, J.; Chen, B.; Wen, Y.; Deng, X.; Liu, Z.-Q. Ratiometric fluorescence detection of stringent ppGpp using Eu-MoS_2 QDs test paper. *Sens. Actuators B Chem.* **2020**, *309*, 127807. [CrossRef]
143. Saa, L.; Diez-Buitrago, B.; Briz, N.; Pavlov, V. CdS quantum dots generated In-Situ for fluorometric determination of thrombin activity. *Microchim. Acta* **2019**, *186*, 657. [CrossRef] [PubMed]
144. Chowdhury, A.D.; Takemura, K.; Khorish, I.M.; Nasrin, F.; Ngwe Tun, M.M.; Morita, K.; Park, E.Y. The detection and identification of dengue virus serotypes with quantum dot and AuNP regulated localized surface plasmon resonance. *Nanoscale Adv.* **2020**, *2*, 699–709. [CrossRef]

145. Liu, M.L.; Chen, B.B.; Li, C.M.; Huang, C.Z. Carbon dots: Synthesis, formation mechanism, fluorescence origin and sensing applications. *Green Chem.* **2019**, *21*, 449–471. [CrossRef]
146. Yan, Y.; Gong, J.; Chen, J.; Zeng, Z.; Huang, W.; Pu, K.; Liu, J.; Chen, P. Recent advances on graphene quantum dots: From chemistry and physics to applications. *Adv. Mater.* **2019**, *31*, e1808283. [CrossRef]
147. Li, C.; Wang, Y.; Jiang, H.; Wang, X. Review-Intracellular sensors based on carbonaceous nanomaterials: A review. *J. Electrochem. Soc.* **2020**, *167*, 037540. [CrossRef]
148. Hui, B.; Zhang, K.; Xia, Y.; Zhou, C. Natural Multi-Channeled wood frameworks for electrocatalytic hydrogen evolution. *Electrochim. Acta* **2020**, *330*, 135274. [CrossRef]
149. Li, C.; Zheng, Y.; Ding, H.; Jiang, H.; Wang, X. Chromium(III)-Doped carbon dots: Fluorometric detection of p-Nitrophenol via inner filter effect quenching. *Microchim. Acta* **2019**, *186*, 384. [CrossRef]
150. Li, C.; Qin, Z.; Wang, M.; Liu, W.; Jiang, H.; Wang, X. Manganese oxide doped carbon dots for Temperature-Responsive biosensing and target bioimaging. *Anal. Chim. Acta* **2020**, *1104*, 125–131. [CrossRef]
151. Tran, H.L.; Doong, R.-a. Sustainable fabrication of green luminescent Sulfur-Doped graphene quantum dots for rapid visual detection of hemoglobin. *Anal. Methods* **2019**, *11*, 4421–4430. [CrossRef]
152. Luo, X.; Zhang, W.; Han, Y.; Chen, X.; Zhu, L.; Tang, W.; Wang, J.; Yue, T.; Li, Z. N,S co-Doped carbon dots based fluorescent "On-Off-On" sensor for determination of ascorbic acid in common fruits. *Food Chem.* **2018**, *258*, 214–221. [CrossRef] [PubMed]
153. Zhang, Q.; Liu, Y.; Nie, Y.; Liu, Y.; Ma, Q. Wavelength-Dependent surface plasmon coupling electrochemiluminescence biosensor based on Sulfur-Doped carbon nitride quantum dots for K-RAS gene detection. *Anal. Chem.* **2019**, *91*, 13780–13786. [CrossRef] [PubMed]
154. Fan, D.; Bao, C.; Khan, M.S.; Wang, C.; Zhang, Y.; Liu, Q.; Zhang, X.; Wei, Q. A novel Label-Free photoelectrochemical sensor based on N,S-GQDs and CdS co-Sensitized hierarchical Zn_2SnO_4 cube for detection of cardiac troponin I. *Biosens. Bioelectron.* **2018**, *106*, 14–20. [CrossRef]
155. Feng, H.; Qian, Z.S. Functional carbon quantum dots: A versatile platform for chemosensing and biosensing. *Chem. Rec.* **2018**, *18*, 491–505. [CrossRef] [PubMed]
156. Campuzano, S.; Yanez-Sedeno, P.; Pingarron, J.M. Carbon dots and graphene quantum dots in electrochemical biosensing. *Nanomaterials* **2019**, *9*, 634. [CrossRef]
157. Suvarnaphaet, P.; Pechprasarn, S. Graphene-Based materials for biosensors: A review. *Sensors* **2017**, *17*, 2161. [CrossRef]
158. Karimzadeh, A.; Hasanzadeh, M.; Shadjou, N.; de la Guardia, M. Optical bio(sensing) using nitrogen doped graphene quantum dots: Recent advances and future challenges. *Trac Trends Anal. Chem.* **2018**, *108*, 110–121. [CrossRef]
159. Meng, A.L.; Sheng, L.Y.; Zhao, K.; Li, Z.J. A controllable Honeycomb-like amorphous cobalt sulfide architecture directly grown on the reduced graphene Oxide-poly(3,4-ethylenedioxythiophene) composite through electrodeposition for Non-enzyme glucose sensing. *J. Mat. Chem. B* **2017**, *5*, 8934–8943. [CrossRef]
160. Sarkar, A.; Ghosh, A.B.; Saha, N.; Bhadu, G.R.; Adhikary, B. Newly designed amperometric biosensor for hydrogen peroxide and glucose based on vanadium sulfide nanoparticles. *ACS Appl. Nano Mater.* **2018**, *1*, 1339–1347. [CrossRef]
161. Zhang, X.R.; Liu, M.S.; Liu, H.X.; Zhang, S.S. Low-Toxic Ag_2S quantum dots for photoelectrochemical detection glucose and cancer cells. *Biosens. Bioelectron.* **2014**, *56*, 307–312. [CrossRef] [PubMed]
162. Wang, Y.; Qu, J.H.; Li, S.F.; Qu, J.Y. Catechol biosensor based on ZnS:Ni/ZnS quantums dots and laccase modified glassy carbon electrode. *J. Nanosci. Nanotechnol.* **2016**, *16*, 8302–8307. [CrossRef]
163. Yao, P.; Yu, S.H.; Shen, H.F.; Yang, J.; Min, L.F.; Yang, Z.J.; Zhu, X.S. A TiO_2-SnS_2 nanocomposite as a novel matrix for the development of an enzymatic electrochemical glucose biosensor. *New J. Chem.* **2019**, *43*, 16748–16752. [CrossRef]
164. Liu, J.; Sun, S.H.; Shang, H.; Lai, J.H.; Zhang, L.L. Electrochemical biosensor based on bienzyme and carbon nanotubes incorporated into an os-complex thin film for continuous glucose detection in Human Saliva. *Electroanalysis* **2016**, *28*, 2016–2021. [CrossRef]

165. Hou, C.; Zhao, D.Y.; Wang, Y.; Zhang, S.F.; Li, S.Y. Preparation of magnetic Fe_3O_4/PPy@ZIF-8 nanocomposite for glucose oxidase immobilization and used as glucose electrochemical biosensor. *J. Electroanal. Chem.* **2018**, *822*, 50–56. [CrossRef]
166. Wang, L.S.; Gao, X.; Jin, L.Y.; Wu, Q.; Chen, Z.C.; Lin, X.F. amperometric glucose biosensor based on silver nanowires and glucose oxidase. *Sens. Actuators B Chem.* **2013**, *176*, 9–14. [CrossRef]

© 2020 by the authors. Licensee MDPI, Basel, Switzerland. This article is an open access article distributed under the terms and conditions of the Creative Commons Attribution (CC BY) license (http://creativecommons.org/licenses/by/4.0/).

Review

MXenes-Based Bioanalytical Sensors: Design, Characterization, and Applications

Reem Khan and Silvana Andreescu *

Department of Chemistry and Biomolecular Science, Clarkson University, Potsdam, New York, NY 13676, USA; rekhan@clarkson.edu
* Correspondence: eandrees@clarkson.edu

Received: 23 August 2020; Accepted: 18 September 2020; Published: 22 September 2020

Abstract: MXenes are recently developed 2D layered nanomaterials that provide unique capabilities for bioanalytical applications. These include high metallic conductivity, large surface area, hydrophilicity, high ion transport properties, low diffusion barrier, biocompatibility, and ease of surface functionalization. MXenes are composed of transition metal carbides, nitrides, or carbonitrides and have a general formula $M_{n+1}X_n$, where M is an early transition metal while X is carbon and/or nitrogen. Due to their unique features, MXenes have attracted significant attention in fields such as clean energy production, electronics, fuel cells, supercapacitors, and catalysis. Their composition and layered structure make MXenes attractive for biosensing applications. The high conductivity allows these materials to be used in the design of electrochemical biosensors and the multilayered configuration makes them an efficient immobilization matrix for the retention of activity of the immobilized biomolecules. These properties are applicable to many biosensing systems and applications. This review describes the progress made on the use and application of MXenes in the development of electrochemical and optical biosensors and highlights future needs and opportunities in this field. In particular, opportunities for developing wearable sensors and systems with integrated biomolecule recognition are highlighted.

Keywords: MXenes; 2D nanomaterials; biosensors; wearables

1. Introduction

Since the discovery of graphene, two-dimensional (2D) nanomaterials have gained significant attention due to their high surface area, electrical conductivity, functionalized surfaces, and mechanical properties. Prime examples include materials such as silicene, germanene, boron nitride, and molybdenum sulfide. 2D materials offer exceptional physical, chemical, and structural properties that make them useful in a wide variety of applications such as sensors, energy storage and conversion, optoelectronics, and catalysis. MXene is an emerging class of 2D transition metal carbide, nitride, or carbonitride added to the 2D nanomaterial group initially developed by Gogotsi and coworkers in 2011 [1]. MXenes have general formula $M_{n+1}X_n$ where M is an early transition metal while X is carbon and/or nitrogen, synthesized by the selective etching of MAX phases [2]. MAX phases are layered ternary carbides and nitrides with a general formula $M_{n+1}AX_n$, where A represents elements from the group 13 and 14 of the periodic table. The term MAX phases were established in the late 1990s, but most of these phases were discovered 40 years ago. A renewed interest in the application of MAX phases started in 1996 due to their unusual combination of chemical, physical, electric, and mechanical properties resulting from their layered structure supported by a strong mixed metallic-covalent *M X* bond and a relatively weak *M A* bond [3].

To date, 70 different MAX phases have been reported, while more than 20 members of the MXenes family have been synthesized, and dozens more are predicted, making it one of the fastest-growing 2D material families. Typically, 2D materials MXene are synthesized by removing the A layer from their parent MAX (remaining MX-) phases and are structurally similar with graphene (hence named -ene). Additionally, the MXene family comes in three atomic structures, ranging from M_2X to M_3X_2 and M_4X_3, yielding tunability, and opportunity to discover and mold materials based on application requirements. Recently, the MXene family has been expanded to include double transition metal MXenes with formula $M'_2M''C_2$ and $M'_2M''_2C_3$.

The most successful synthesis route of Ti_3C2 involves the wet chemical etching in hydrofluoric (HF) acid or HF containing/forming etchants. The wet chemical route is based on three simple steps: (1) Etching, (2) delamination, and (3) intercalation. Due to the use of HF as an etchant, MXene layers are primarily terminated with F and OH/=O functional groups, abbreviated as T_x to give a general formula of $M_{n+1}X_nT_x$, n = 1–3. The etching is simply done by immersing the MAX phase in HF. Alhabeb et al. studied the effect of different concentrations of HF on the morphology of the resulting MXene. The results of the study have shown that Al can be etched by using HF concentration as low as five weight percent; however, for low HF concentrations, the MXene did show the characteristic accordion-like morphology that was observed in the case of 30% HF. After completing the etching process, the powder was washed thoroughly with deionized water by successive centrifugation cycles at 3500 rpm. After each cycle, the supernatant was decanted and replaced with fresh deionized water. The washing process continued until the pH of the supernatant has reached 6 to 7. Delamination is the second crucial step in the synthesis protocol that enhances the accessible surface area of the nanomaterial. In order to delaminate MXene nanosheets, the van der Waals forces between the adjacent MX sheets have to be broken. This barrier is relatively strong for MXenes (with their ~2.2 Å interlayer distance) compared to that of graphene (3.35 Å). A well-established method for the delamination of MXenes is to increase the interlayer distance through intercalation, achieved by inserting external elements (ions or molecules) in between the layers of the laminated material. Large, organic molecules, such as dimethyl sulfoxide (DMSO), isopropylamine, tetraalkylammonium hydroxides (TBAOH), were some of the first intercalants used for expanding the interlayer spacing of Ti_3C_2Tx MXenes synthesized with HF [4]. Figure 1 shows a general synthesis scheme for producing MXenes (left) and examples of SEM micrographs (right) displaying the 'accordion'-like morphology that is typical of various MXenes compositions. For description of detailed synthesis methodologies, readers are referred to other literature, specifically discussing the synthesis protocol of MXenes [5,6]. Most applications of these materials to date are in energy conversion, catalysis, and electronics with emerging areas in structural, biomedical, and environmental fields [7–9]. The biosensing and analytical measurement field remains relatively unexplored and provides unique opportunities for future development.

Several prior works summarized the application of MXenes in chemical sensors [10], particularly gas sensors [11] and their environment-related applications [10–13]. Here we focus on the properties and suitability of MXenes as an immobilization matrix, signal transducer, and amplifier of biomolecular recognition for the design of biological sensors. The first half of the review provides an overview of the physical and chemical properties of MXenes that are of interest for the development of bioanalytical sensors. The second half discusses specific examples of MXenes-based biosensors with enzyme, antibodies, DNA, and aptamer recognition, and highlights recent developments on the use of MXenes as supporting material for wearable devices.

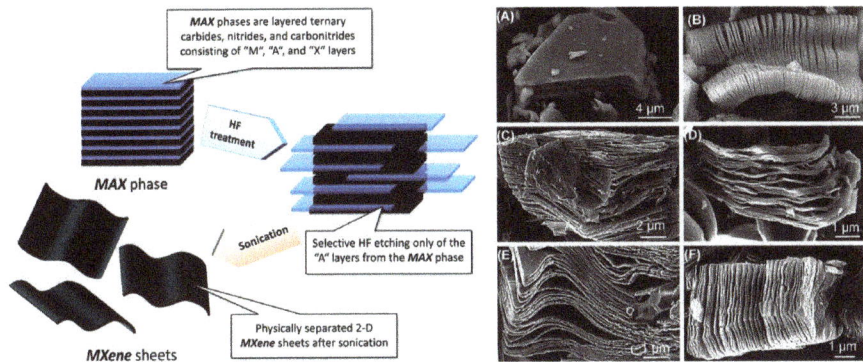

Figure 1. (Left) Schematic for the exfoliation process of MAX phases and formation of MXenes. (right) SEM micrographs for (**A**) Ti_3AlC_2 MAX phases, (**B**) Ti_3AlC_2 after HF treatment, (**C**) Ti_2AlC after HF treatment, (**D**) Ta_4AlC_3 after HF treatment, (**E**) TiNbAlC after HF treatment, and (**F**) Ti_3AlCN after HF treatment. Reprint with permission [2]. Copyright © 2012, American Chemical Society.

2. Properties of MXenes for Bioanalytical Sensing

2D nanomaterials such as graphene, MoS_2, boron nitride nanosheets, either alone or in nanocomposite form, have demonstrated superior properties such as high surface area platforms for the construction of different types of chemical and biological sensors. When used in sensing design, these materials improved the performance of sensors in terms of sensitivity, selectivity, and analyte binding. However, limitations such as high hydrophobicity, low biocompatibility, low electrical conductivity, and difficulty in surface functionalization limit their incorporation in sensor design at a commercial scale. MXenes complement the emerging class of 2D materials and provide advantages for sensors design in terms of hydrophilicity, electrical conductivity, biocompatibility, and above all, ease of functionalization due to the presence of surface functional groups. Furthermore, MXene has high intercalation capacity, which is missing in other 2D nanomaterials. A comparison of the fundamental characteristics of different nanomaterials is given in Table 1.

Table 1. Comparison of fundamental properties of various nanomaterials.

Nanomaterials	Surface Area ($m^2\,g^{-1}$)	Conductivity (S/cm)	Band Gap (eV)	Biocompatibility	References
Graphene	450	2700	0	Biocompatible	[14–16]
h-Boron nitride	150–550	insulator	5.9	Dependent on size and shape	[17,18]
SWCNT	600	10^2–10^6	0.042	Unclear/under debate	[19–21]
MWCNT	122	10^3–10^5	1.82	Unclear/under debate	[19,20,22]
MoS_2	8.6	10^{-4}	1.89	Biocompatible	[23–25]
δ-MnO_2	257.5	10^{-5} to 10^{-6}	1.33	Biocompatible	[15,26,27]
MXene (Ti_3C_2)	93.6	2410	0.1	Biocompatible	[28–30]

The high electrical conductivity, thermal stability, hydrophilic nature, large interlayer spacing, high surface area, and easily tunable structure make this new family of 2D nanomaterials extremely attractive for a wide variety of applications. Because MXenes have a relatively low number of atomic layers, a single stack of MXene layers is typically less than 1 nm thickness, while the lateral dimension is in the nm to µm range, depending on the composition, synthesis, and processing steps. Moreover, by selecting double metal MXenes, it is possible to tune the valence states and relativistic spin-orbit coupling [31]. Furthermore, the MXenes have a ceramic-like nature that is responsible for their high chemical and mechanical stability. However, unlike traditional ceramics, MXenes have inherent conductivity and a significantly larger surface area imparted by the composition and layered structure of their molecular sheets made of carbides and nitrides of transition metals such as Ti. In addition, the MXene surface can be functionalized with different functional groups that provides numerous opportunities for surface state engineering, greatly extending their properties and applications. The ease of functionalization is a particularly important feature of using MXenes in the design of bioanalytical sensors that require materials with abundant functional groups at their surface. This high surface functionality, in addition to their layered structure, provides the ability to bind, protect, and retain the activity of biomolecules for specific targeting and recognition properties that are essential for bioanalytical sensing. The versatility and compositional variety make MXene an extraordinarily diverse and appealing class of materials for applications [31].

In the last 10 years, there has been growing interest in exploring the use of MXenes in the design of supercapacitors [32,33], transparent conductors [34,35], field-effect transistors [36], Li-ion batteries [37,38], electromagnetic interface shielders [39], catalysts [40], hybrid nanocomposites [41], fillers in polymeric composites [42], dual-responsive surfaces, purifiers, [43,44] suitable substrates for dyes [45], photocatalysts for hydrogen production [31], methane storage [46], and as well as photothermal conversion [47]. Applications in electronic [48,49], magnetic [50,51], optical [1,52,53], thermoelectric [54,55], and sensing devices [5,56] are being explored while new utilizations such as reaction media to facilitate catalytic and photocatalytic processes, hydrogen storage [57,58], and nanoscale superconductivity [59] are beginning to be investigated. Some of the MXenes are predicted to be topological insulators with large band gaps involving only d orbitals [60,61]. This review focuses on the applications of MXenes in the field of biological sensors, where MXenes have been used as an electrode material exploiting their high catalytic, high surface area, charge transport properties, and biocompatibility with biological matrices. The different classes of biosensing platforms and the applications covered in this review are summarized in Figure 2.

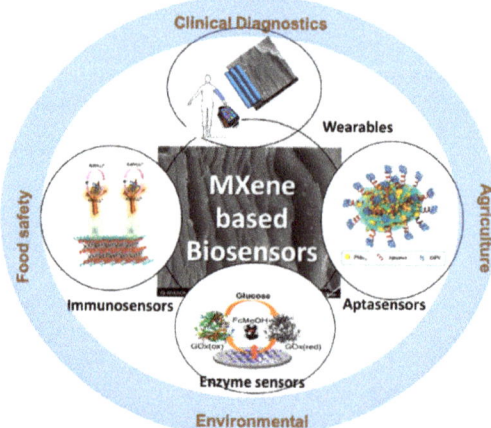

Figure 2. Summary of different classes of biosensing platforms based on MXenes and their applications. Immunosensors (reprinted with permission from [62]) Aptasensors (reprinted with permission from [63]), Enzyme sensors (reprinted with permission from [64]).

3. Classes of Bioanalytical Sensors Based on MXenes

3.1. Enzyme Sensors

Enzymes, as a bio-recognition element, offer distinct advantages such as substrate specificity and high efficiency under mild conditions. Enzyme based electrochemical biosensing devices have been developed extensively in the last few decades [1]. Despite progress, enzymes can lose their bioactivity when directly immobilized onto electrode surfaces. Moreover, due to the deeply rooted location of redox-active centers in enzymes, the direct electron transfer (DET) between these biomolecules and the electrode surface is generally difficult and has been the subject of extensive investigations as reviewed in a recent work [65]. Therefore, the use of nanomaterials has been found beneficial to facilitate the electron transfer along with promoting retention of the bioactivity of immobilized enzymes. Among different types of nanomaterials, 2D nanomaterials proved to be effective for improving the DET from the enzyme to the transducer surface. Because MXene have a high surface to volume ratio and are biocompatible, they provide a highly suitable matrix for the fabrication of enzymatic biosensors. They can be used as a supporting platform for the immobilization of enzymes, promote diffusion, and accelerate the electrode kinetics, thus improving DET transfer. MXenes are also expected to enhance sensitivity and lower the detection limit of sensing devices. The most explored application of MXenes reported in the literature is for glucose sensing using glucose oxidase (GOx) immobilized within stacked layers of MXenes. For example, Chia et al. reported a Ti_3C_2 MXene, produced via HF etching and subsequent delamination with tetrabutylammonium hydroxide (TBAOH), as a transducer platform for the development of an electrochemical glucose biosensor with chronoamperometric detection. The biosensor exhibited high selectivity and good electrocatalytic activity toward the detection of glucose, with a linear range spanning from 50 to 750 µM and a limit of detection (LOD) of 23.0 µM (Figure 3) [64].

Other works explored hybrid configurations that combine the layered structure of MXenes with the properties of other materials to add additional functions. Gu et al. constructed a porous MXene-graphene (MG) nanocomposite-based glucose biosensor [66]. The 3D porous nanostructure was prepared using a mixing-drying process in which the size of the pores was controlled by tuning the content of Ti_3C_2 and graphene. The synthetic methodology and the preparation process of the sensor are displayed in Figure 4. The porous structure provided more open structures to embed GOx within the internal pores, favoring retention of the GOx activity. The biosensor exhibited good electrocatalytic properties towards glucose biosensing for the detection of glucose in human sera.

Other improvements have been achieved by decorating the surface of MXenes with metal oxide nanoparticles (NPs), further increasing the surface area and conductivity of the electrodes and maximizing the enzyme loading. A composite of Nafion-Au NPs-MXene was reported as electrode material for the immobilization of GOx and subsequent detection of glucose [67]. The synergistic effects of Au NPs and MXene sheets resulted in unique electrocatalytic properties, which enabled the detection of glucose in the µM to mM range. Similarly, $Ti_3C_2T_x$ nanosheets were modified with β-hydroxybutyrate dehydrogenase and then used as a biosensor for amperometric sensing of β-hydroxybutyrate, a biomarker for diabetic ketoacidosis. The developed biosensor best operated at a potential of −0.35 V (vs. Ag/AgCl), displayed a linear range between (0.36 to 17.9 mM), a sensitivity of 0.480 µA mM^{-1} cm^{-2}, and a LOD of 45 µM. Later, the biosensor was successfully applied to the determination of β-hydroxybutyrate in (spiked) real serum samples [68].

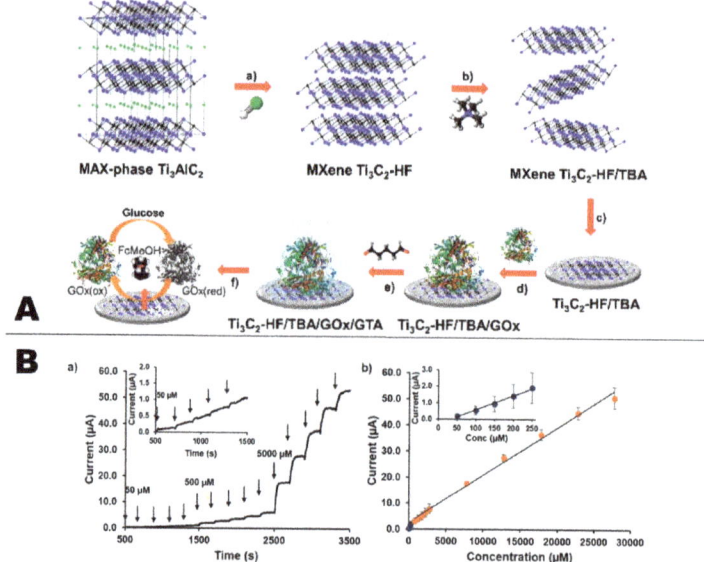

Figure 3. (**A**) Schematic illustration of sensor fabrication; (a) exfoliation of Ti$_3$AlC$_2$ via etching with HF; (b) delamination with TBAOH; (c) modifying the glassy carbon electrode with MXene; (d) loading of glucose oxidase (GOx), and; (e) cross-linking glutaraldehyde (GTA) with GOx; (f) glucose detection mechanism of the proposed biosensing system. (**B**) Chronoamperometry data (a) and calibration plot (b) for Ti$_3$C$_2$–HF/TBA-based electrochemical glucose biosensor conducted using FcMeOH (2 mM) in pH 7.2 PBS (electrolyte) and 0.15 V potential (reproduced with permission from reference [64]).

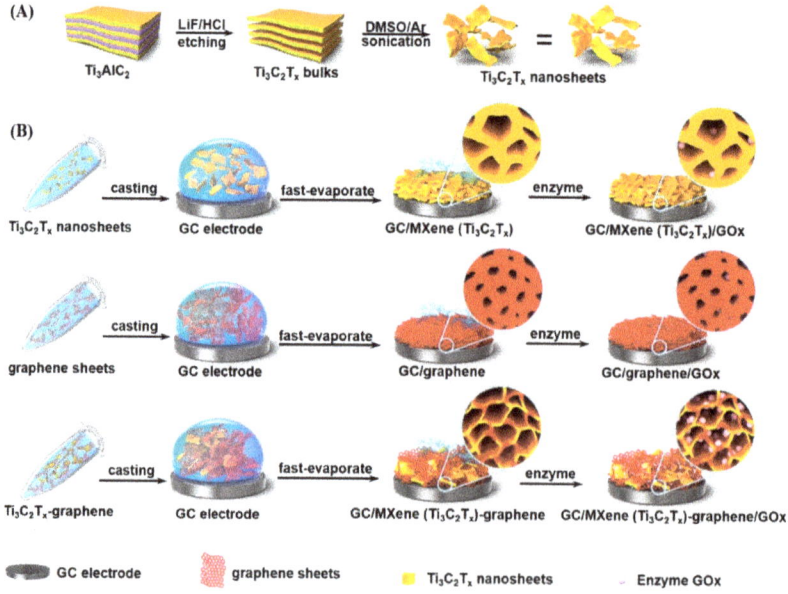

Figure 4. Preparation of (**A**) Ti$_3$C$_2$T$_x$ MNS; (**B**) pure Ti$_3$C$_2$T$_x$ film, pure graphene film, and MG hybrid film for enzyme immobilization (reproduced with permission from reference [66]).

Wang et al. reported a mediator free biosensor for the detection of H2O2 by immobilizing hemoglobin (Hb) on an MXene modified electrode [69]. The use of MXene lowered the LOD of the sensor to the nM range. A similar type of Hb-MXene based biosensor was reported for the electrochemical detection of nitrite using cyclic voltammetry [70]. Kai et al. modified the MXene surface with horseradish peroxidase (HRP) to fabricate an electrochemical biosensor for the detection of H_2O_2. The HRP enzyme was immobilized on an MXene/chitosan/GCE electrode, and the film demonstrated good electrocatalytic activity toward the reduction of H_2O_2. The amperometric biosensor displayed a wide linear range from 5 to 1650 µmol. L^{-1}, LOD of 0.74 µmol. L^{-1} and good operation stability. The biosensor was successfully applied for the sensing of H_2O_2 traces in food samples [71].

MXene composites were also used to develop biosensors used for the ultra-sensitive detection of phenol in real water samples. These sensors are based on the use of the Tyrosinase (Tyr) enzyme entrapped within a chitosan composite that increased the adhesion of MXene-Tyr onto the GCE. The mechanism for phenol detection involves the oxidation of phenol by Tyr into the corresponding o-quinone. Afterward, the electrochemical reduction of the o-quinone producing polyhydric phenol is measured electrochemically. A noticeable increase in the current vs. time was observed with increasing the concentration of phenol. The sensitivity of Tyr-MXene-Chi/GC (414.4 mA M^{-1}) was about 1.5 times higher than that of Tyr-Chi/GC (290.8 mA M^{-1}). The LOD was 12 nmol L^{-1} while the linear range was from 5.0×10^{-8} to 15.5×10^{-6} mol L^{-1} [4].

Other sensor configurations reported the use of MXene as a transducer for the detection of organophosphorus pesticides (OPs). Zhou and his coworkers reported an acetylcholinesterase (AChE) based sensor for the detection of malathion using Ti_3C_2 nanosheets and chitosan as an immobilization matrix. The electrochemical behavior of the AChE/CS-Ti_3C_2/GCE biosensor was studied by cyclic voltammetry and electrochemical impedance spectroscopy (EIS). The biosensor displayed excellent performance against malathion with a LOD as low as 0.3×10^{-14} M [72]. In another report, the same analyte, i.e., malathion was detected by AChE inhibition, using Ag modified MXene as a transducer. The modification of MXene with Ag NPs amplified the electrochemical signal; a LOD of 3.27×10^{-15} M was reported in this case [73].

Another configuration was adopted to design a biosensor for OPs by combining Ti_3C_2 MXene nanosheets with metal-organic frameworks (MOFs). The combined material, a MOF-derived MnO_2/Mn_3O_4, Ti_3C_2 MXene/Au NPs composite, was used as an electrochemical biosensing platform with immobilized AChE enzyme, constructed as shown in Figure 5. The vertically aligned, highly ordered nanosheets of Mn-MOF derived 3D MnO_2/Mn_3O_4 combined with MXene/Au NPs yielded a synergistic amplification effect, providing enlarged specific surface area and good environmental biocompatibility. Using this biosensor, the detection of methamidophos was achieved over a broad concentration range (10^{-12}–10^{-6} M). TheLOD (1.34×10^{-13} M) of the biosensor far exceeds the maximum residue limits (MRLs) for methamidophos (0.01 mg/kg) established by European Union [74]. Applicability of several of these enzyme-based MXene biosensors has been tested for food quality analysis. A Ti_3C_2 based double enzyme biosensor was reported for the detection of inosine monophosphate (IMP), which imparts flavor to the meat and can be used as an indicator to assess meat quality [75]. The developed biosensor was based on a nanohybrid structure consisting of Ti_3C_2 MXene as a highly conductive and stable base material, modified with two enzymes (5'-nucleotidase and xanthine oxidase) and a core-shell bimetallic nanoflower composed of an Au core and a Pt shell. In this configuration, the MXene provided a stable biocompatible microenvironment for enzyme loading, while the double-enzyme was used to hydrolyze IMP to produce H_2O_2, which can be easily determined electrochemically through the bimetallic nanoflowers. Therefore, the content of IMP was indirectly quantified by monitoring the current change due to H_2O_2 production. The LOD achieved by this double enzyme biosensor was 2.73 ng mL^{-1} with a correlation coefficient of 0.9964 and a linear range between 0.04 and 17 g L^{-1}.

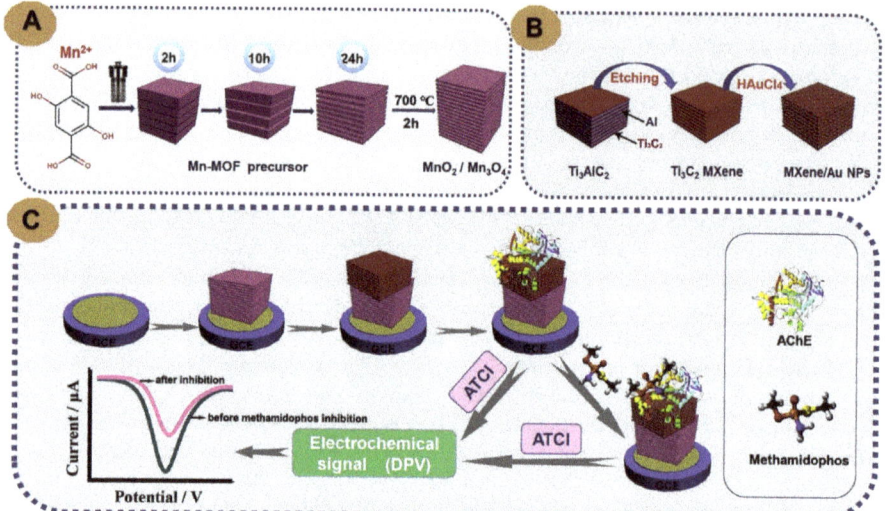

Figure 5. Illustration of the formation of MnO_2/Mn_3O_4 composite (**A**), MXene/Au NPs (**B**), the fabrication process of AChE-Chit/MXene/Au NPs/MnO_2/Mn_3O_4/GCE biosensor for methamidophos assay (**C**) (reproduced with permission from reference [74]).

3.2. MXenes Based Electrochemical Immunosensors

Immunosensors combining immunochemical reactions based on the antibody (Ab)-antigen (Ag) recognition with an appropriate electrochemical transducer provide significant advantages for bioanalysis, including cost-effectiveness, low reagent and sample volume, portability, high-specificity and sensitivity, and high-throughput analysis [76]. Immunosensors have applications in various fields of analysis, such as clinical medicine [77,78], environmental pollutant evaluation [79–81], food inspection [82,83], and pathogenic microorganism detection [84,85]. The incorporation of MXenes in the fabrication of immunosensors provide opportunities for increasing bioactivity of immobilized Abs on electrode surfaces, increasing surface area, and improving detection sensitivity. Kumar et al. fabricated an immunosensor for the detection of carcinoembryonic antigen (CEA) [62], an important cancer biomarker found in the liver, breast, lung, colorectal, ovarian, and pancreatic cancer patients. To fabricate the sensor, amino-functionalized Ti_3C_2 MXene was used to chemically immobilize -COOH terminated-CEA. The electrochemical detection mechanism of the sensors is shown in Figure 6. The MXenes were synthesized using a layer delamination method that reduces the surface defects in the MXene and produces MXene nanoflakes with higher electrical conductivity. The biosensor showed a linear detection range of 0.0001 to 2000 ng mL^{-1} with a sensitivity of 37.9 µA ng^{-1} mL cm^{-2} and a stability up to 7 days. Recovery studies for the quantification of CEA in serum samples indicated promising results.

Another MXene based CEA biosensor has been reported using a sandwich-type immunoassay format in which Ti_3C_2 was first functionalized with amino silane (APTES) for covalent immobilization of monoclonal anti-CEA antibodies (Ab_1) with surface plasmon resonance (SPR) detection. A MXene-hollow Au NPs (HGNPs) nanohybrid was synthesized and further decorated with staphylococcal protein A (SPA) to immobilize the polyclonal anti-CEA detection antibody (Ab2) and serve as signal enhancers. The capture of CEA resulted in the formation of an Ab2-conjugated SPA/HGNPs/N-Ti3C2-MXene sandwiched nanocomplex on the SPR chip and the generation of a response signal. This SPR immunoassay had a reported LOD of 0.15 fM and a linear range of 0.001 to 1000 pM.

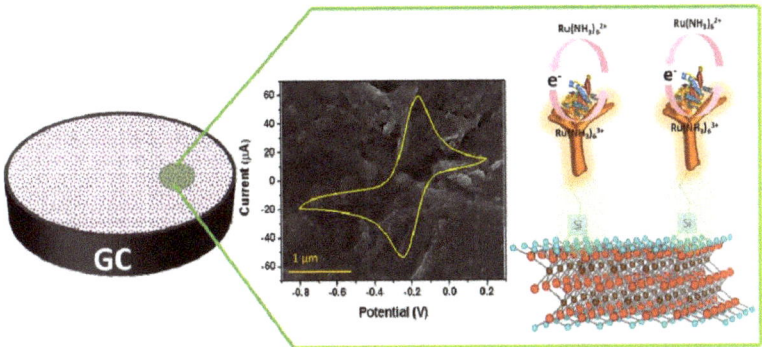

Figure 6. Schematic of the electrochemical carcinoembryonic antigen (CEA) detection mechanism (reproduced with permission from reference [62]).

In other configurations, Chen et al. developed a Ti_3C_2 MXene-based interdigitated capacitance immunosensor for the detection of prostate-specific antigen (PSA), an important biomarker for used to screen prostate cancer. The normal levels of PSA in the serum of healthy males are less than 4.0 ng mL^{-1}, whereas rising levels are associated with prostate cancer. The biosensor utilized an MXene modified interdigitated micro comb electrode to immobilize anti-PSA capture Ab, whereas Au NPs functionalized with horseradish peroxidase (HRP) and detection antibody were used as the signal-transducer tags. To further increase the sensitivity of the assay, a tyramine signal amplification and enzymatic biocatalytic precipitation were coupled with the main immunochemical reaction. Under optimum conditions, the change of the immunosensor in the capacitance increased with the increasing target PSA concentrations from 0.1 ng mL^{-1} to 50 ng mL^{-1} at a detection limit of 0.031 ng mL^{-1}. The characteristics of the developed system are shown in Figure 7. Despite these merits, one of the limitations of this immunosensor is the complex fabrication procedure and long reaction time as compared to commercial enzyme-linked immunosorbent assays [86].

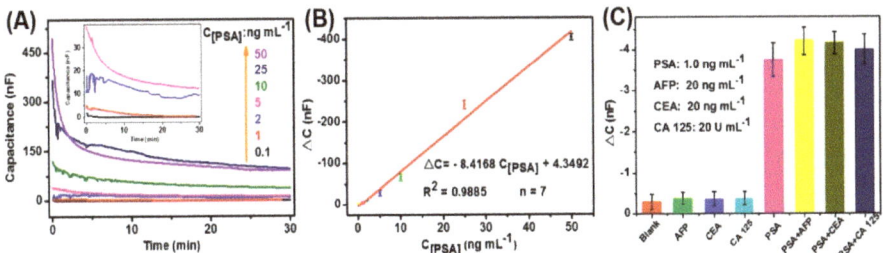

Figure 7. Analytical performance of the developed immunoassay (**A**) capacitance responses of Ti_3C_2 MXene-based interdigitated immunosensor toward target PSA standards; (**B**) the corresponding calibration plots; and (**C**) the specificity of capacitance immunosensor against target PSA and non-targets including AFP, CEA, and CA 125 (reproduced with permission from reference [86]).

3.3. DNA/Aptamer Based Biosensor

Biosensors with nucleic acid detection using either optical or electrical output to monitor the hybridization event have been widely reported as bioanalytical platforms for many applications [87]. Recently, MXenes have been explored as an electrode material to monitor the hybridization event and enhance detection sensitivity. Zheng et al. reported a DNA/Pd/Pt nanocomposite for the electrochemical detection of dopamine (DA) [88], an important catecholamine neurotransmitter in living organisms that plays an essential role in the function of human renal, metabolism, cardiovascular, and central nervous systems [89]. Abnormal DA levels may indicate neurological disorders and a variety of acute and chronic diseases such as Schizophrenia, Parkinson, and Alzheimer [90]. In the developed biosensor, MXene nanosheets act as a matrix for the loading of the DNA and Pd/Pt. First, the DNA was adsorbed on the MXene surface through π-π stacking interaction between the nucleotide bases and MXene nanosheets, then the Pd and Pt NPs were synthesized in-situ in the presence of DNA/MXene nanocomposite. The results revealed that the presence of DNA prevents the restacking of Ti_3C_2 nanosheets and facilitates the even growth of PdNPs and Pd/Pt NPs. Moreover, the deposition of Pd/Pt NPs onto Ti_3C_2 nanosheets enhanced the electrocatalytic activity of the nanocomposites towards DA. In the final step, a GCE electrode was modified with the DNA/MXene/NPs composite to create the DA biosensor. The amperometric biosensor exhibited DA detection capabilities in concentration range of 0.2 to 1000 µM with a LOD of 30 nM (S/N = 3) and high selectivity against uric acid, ascorbic acid, and glucose [88].

The detection of specific DNA sequences is significant, not only in clinical diagnostics but also in the environment and food analysis fields. Wang and his coworkers developed a nanobiosensor for gliotoxin [84], one of the most toxic mycotoxins produced by Aspergillus fumigatus, which poses a serious threat to humans and animals health. The biosensor was prepared by modifying MXene nanosheets with a tetrahedral DNA nanostructure (TDN), which acts as the main sensing element. The stated benefits of incorporating MXenes in the sensor design included increasing the sensitivity and providing an ample surface area for immobilizing a much greater amount of TDN onto the electrode. Moreover, the titanium element on the surface of MXene nanosheets offers a facile method for assembly via strong chelation interaction between the titanium and phosphate groups of TDNs, thus eliminating the need for complex and costly chemical modification of TDNs, which is generally required for the immobilization of TDNs onto the electrode. The amperometric response of the optimized biosensor responded to gliotoxin concentrations increasing from 5 pM to 10 nM, with an LOD of 5 pM. Zhoe et al. reported an impedimetric aptasensor for electrochemical detection of osteopontin (OPN), a cancer biomarker that is also responsible for tumor growth and progression in human cervical cancer. The aptasensor was based on Ti_3C_2 MXene-phosphomolybdic acid (PMo_{12})-polypyrrole (PPy) nanohybrid used as an immobilization matrix for the anti-OPN aptamer. The fabrication procedure of the developed aptasensor is represented in Figure 8. The PPy@Ti_3C_2/PMo_{12}-based aptasensor exhibited high selectivity and stability along with an extremely low detection limit of 0.98 fg mL^{-1}, and applicability in human serum samples [63].

Apart from its use as transducer surface in electrochemical biosensors, MXenes have also been used as bioimmobilization material or quencher in optical DNA assays. For example, Peng et al. designed an "off-on" fluorescent biosensor for the detection of Human papillomavirus (HPV) infection [91]. HPV is a human pathogen known to induce cervical cancer, the second most common cancer in women [92]. The assay was designed based on a fluorescence quenching mechanism, which has already been used in various fluorescence-based biosensors [93]. A fluorescent dye (FAM) labeled ssDNA was used as a fluorescent probe while Ti_3C_2 MXene was used as a nanoquencher. In the presence of Ti_3C_2, the fluorescence of the FAM tagged ssDNA probe (P) was completely quenched while after interacting with its complementary DNA target (T), a P/T hybridized dsDNA structure is formed, which resulted in the recovery of the fluorescence signal. Furthermore, Exonuclease III (Exo III) was used to improve the sensitivity by enhancing fluorescence. Interestingly, when Exo III was introduced in the hybridization process, the 3' end of the newly formed dsDNA was recognized by the Exo III, and then the hydrolysis of the P/T complex was initiated. This cycling process facilitated by the Exo III enhanced the fluorescence of P/T lowering the limit for HPV-18 detection to 100 pM [91].

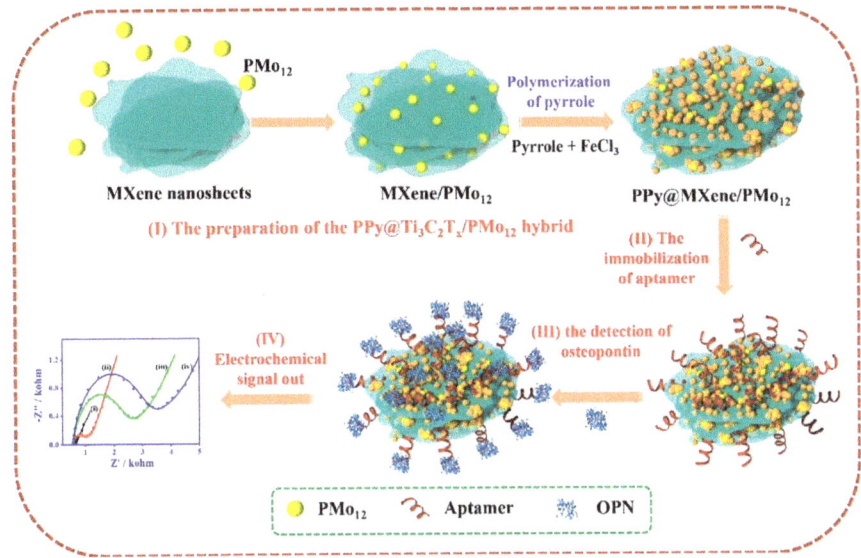

Figure 8. Schematic diagram of the aptasensor fabrication based on PPy@Ti$_3$C$_2$/PMo$_{12}$ for the OPN detection. (reproduced with permission from reference [63]).

Utilizing MXenes as a nanoquencher, Zhang et al. designed a fluorescence resonance energy transfer (FRET) bioassay for the detection of exosomes (Figure 9). This platform is based on a Cy3 dye-labeled CD63 aptamer (Cy3-CD63 aptamer)/Ti$_3$C$_2$ MXenes nanocomplex in which the fluorescence of the dye was quenched by the MXene nanosheets, while the addition of exosomes in the reaction mixture immediately recovered the fluorescence. The turn-on fluorescence phenomenon was ascribed to the release of the Cy3-CD63 aptamer from the surface of MXenes because of the relative strong specific recognition between the aptamer and the CD63 protein on exosomes. The assay achieved a LOD 1.4×10^3 particles mL^{-1}, which is reported as 1000× lower than the conventional ELISA based assay [94].

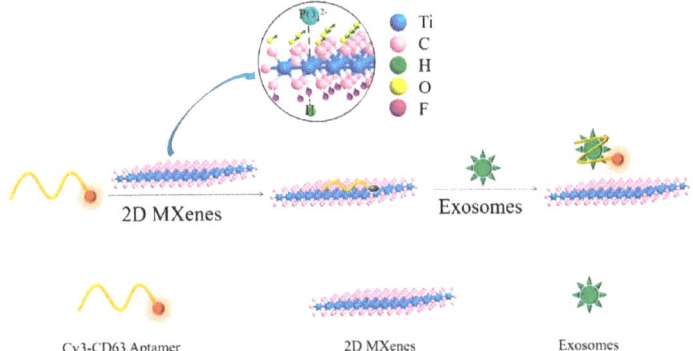

Figure 9. Schematic illustration of the developed fluorescence resonance energy (FRET)-based aptasensor (reprinted with permission from reference [94]).

3.4. MXene for Next Generation Wearable Biosensors

Wearable electronics attracted considerable attention since the commercialization of smartphones and other portable health monitoring devices due to their ability to provide substantial information regarding the health of an individual. Early research in this field focused on the development of physical sensors that could monitor the heart rate, oxygen level, movement such as steps and calories burned, etc., but in the last years, the focus broadened to tackle challenges in health care applications such as monitoring of disease biomarkers. The high specificity, portability, and low power consumption of wearable biosensing devices hold promise for such applications. These include biosensor platforms for noninvasive analysis of biofluids, such as interstitial fluid (ISF), sweat, saliva, or tears. These fluids have been targeted mostly because of the advantage of noninvasive sampling, minimizing the risk of infection during the sampling procedure and user-friendly operation.

Key characteristics when designing wearable devices are the flexibility, mechanical properties, and conductivity of the material used as a sensing platform. Challenges include the integration and scalability of the bioelectronics components in order to achieve the maximum performance in an accurate and sensitive manner and ensure manufacturability. Large surface area nanomaterials have been explored as materials for flexible sensors to increase the effective contact area and impart favorable electrical and mechanical properties. MXenes represent a new wave of materials for wearable sensors and biosensors. The main applications explored thus far include pressure and strain sensors [95], piezoresistive sensors [96], and chemical sensors for detection of subtle pressure, simultaneous monitoring of human activities, quantitative illustration of pressure distribution, human-computer interaction, and electronic skin. The incorporation of MXenes in conjunction with biomolecules for wearable applications is relatively unexplored but rapidly emerging.

Cheng et al. used MXene to fabricate a piezoresistive sensor with spinous microstructures inspired by the human skin, which can be considered a sensitive biological sensor. The sensor was prepared using a simple abrasive paper stencil printing method. The spinous microstructures effectively increased contact area of the conductive channels, consequently improving the performance of the pressure sensor: Low-pressure detection limit (4.4 Pa), fast response time (<130 ms), high sensitivity (151.4 kPa^{-1}), and excellent stability over 10,000 cycles [97]. Guo et al. presented a similar kind of flexible, highly sensitive, and degradable pressure sensor. A porous MXene-impregnated tissue paper was sandwiched between an interdigitated electrode-coated PLA thin sheet and a biodegradable polylactic acid (PLA) thin sheet. The flexible sensor exhibited high sensitivity (10.2 Pa), fast response (11 ms), low power consumption (10–8 W), biodegradability, and excellent reproducibility over 10,000 cycles [98]. Li and Du reported an MXene-Ag nanocomposite based sensitive strain sensor. The sensor is based on 1D Ag nanowires, 0D Ag NPs and 2D MXene nanosheets. The Ag NPs act as a bridge between the MXene sheets and the Ag nanowires. This design helped to increase the elasticity and conductivity of the sensor. This fabric-based strain sensor was composed of elastic textile material (double-covered yarn) that was doped and blended with Ag/MXene nanocomposite, creating a wearable clothing material. The as-prepared sensor showed highly sensitive and stretchable performance with a high gauge factor at an exceptionally large strain (350%). In addition to wearable textiles, the developed sensor has potential applications in biomedical and fire safety applications [99]. Other configurations use MXene in conjunction with hydrogels to produce wearable sensors [100,101].

Apart from these physical sensors, Lei et al. have developed an MXene-based wearable biosensor for sweat analysis. The device was fabricated by using 2D MXene (Ti$_3$C$_2$T$_x$) nanosheets as material to develop an oxygen-rich enzyme biosensor for H$_2$O$_2$ detection based on a Ti$_3$C$_2$T$_x$/PB (Prussian Blue) composite and enzyme. Due to the high conductivity and excellent electrochemical activity of exfoliated MXene, Ti$_3$C$_2$T$_x$/PB composites showed better electrochemical performance compared to carbon nanotubes/PB and graphene/PB composites toward H$_2$O$_2$ detection. The sensing device contains a versatile replaceable sensor component, which can be inserted and changed with customized sensors prepared to track different analytes, i.e., glucose, lactate, or pH value (Figure 10). The substrate of the device was composed of superhydrophobic carbon fiber that was used to create a tri-phase interface and protect the connector from sweat corrosion. The sensing performance of the as-prepared device was evaluated using artificial sweat. Electrochemical results showed sensitivities of 11.4 µA mm^{-1} cm^{-2} for lactate 35.3 µA mm^{-1} cm^{-2} for glucose. Furthermore, the sensor was tested on human subjects and used for sweat analysis. The obtained results showed simultaneous measurements of glucose and lactate with high sensitivity and repeatability [102].

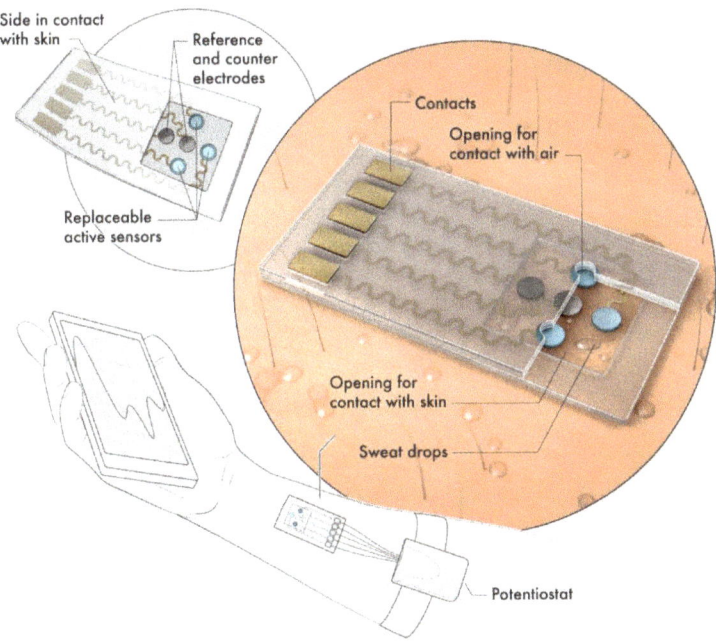

Figure 10. Schematic diagram of wearable sweat sensor (reprinted with permission from reference [102]).

4. Conclusions and Future Prospects

In summary, MXenes are an emerging class of 2D nanomaterials mainly composed of transition metal carbides, nitrides, or carbonitrides. MXenes have gained substantial attention in fields such as batteries and supercapacitors, and their application in chemical and biological sensors is growing. This paper provided an overview of the MXenes properties and their incorporation in the development of enzyme, antibodies, and aptamer-based bioanalytical sensors. As compared to the previously discovered 2D nanomaterials, MXenes possess some superior characteristics such as hydrophilicity, ease of functionalization, high electron transport capability, and vast compositional variety, high surface area due to their "accordion-like" morphology, and biocompatibility, which makes them attractive for biosensing applications. The surface terminal functional groups such as -O, -OH, -F are amenable to functionalization with biomolecules and enable the development of innovative bioanalytical platforms. A growing area is to incorporate MXene in wearable technologies and devices that have the required stability, flexibility, and targeting properties to be interfaced with the human body. This area is expected to grow in the future with new developments in processing, integration, and manufacturing.

Several challenges need to be addressed in order to utilize the full potential of this novel class of materials. (1) Current synthetic methods to create MXenes involve the use of HF as an etchant. The method is not environmentally-friendly and it cannot be easily scaled up, thus hindering production at a commercial scale. Other mild etchants such as LiF/HCl, molten $ZnCl_2$ have also been reported, but the final yield is very low. Significant progress is needed to advance the sustainable and scalable synthesis of MXenes. Moreover, a large number of MXenes have been predicted theoretically, but only a few have been actually produced by exfoliating MAX phases. Thus, it is important to shift the synthesis procedure from theoretical prediction to wet chemical labs. (2) The HF based etching method produces MXenes with randomly functionalized surfaces and it is difficult to obtain MXenes with specific and uniformly distributed surface termination. (3) The presence of fluorine on the surface hinders the use of MXenes in biomedical applications or for the immobilization of biomolecules. Therefore, new fluorine-free synthesis routes with high yield need to be developed, or these groups should be removed from the surface. (4) The top-down synthesis gives less control over tuning the morphology and surface modification of the MXenes. Hence, a bottom-up synthesis approach of MXenes should be developed as an alternative. (5) The long-time sonication for delamination produces MXenes with high surface defects that could possibly alter the properties of MXenes. Additionally, organic solvents are being used for intercalation and delamination, which is also not a very environmentally friendly approach. (6) Lastly, the oxidation of MXenes in oxygenated solvents is a major obstacle. Other approaches are needed to make it possible to use MXenes in the anodic potential window without compromising the inherent physicochemical properties of MXenes. Looking ahead, more effective synthesis and surface modification strategies are required to advance application of MXene in the development of practical bioanalytical sensors and other applications. Scalability in synthesis and processing as well as integration and more effective transduction, are needed to ensure future growth and implementation of these materials in wearable sensing applications.

Author Contributions: R.K. contributed with literature review, writing, original draft preparation and selection of the relevant scientific literature. S.A. contributed with review and editing, supervision, project administration and funding acquisition. All authors have read and agreed to the published version of the manuscript.

Funding: This work was funded by the National Science Foundation under Grant No. 1561491.

Acknowledgments: This work was funded by the National Science Foundation under Grant No. 1561491. Any opinions, findings, and conclusions or recommendations expressed in this material are those of the author(s) and do not necessarily reflect the views of the National Science Foundation. Reem Khan gratefully acknowledges the Higher Education Commission (HEC) of Pakistan and the US-Pakistan Knowledge Corridor Program for support of her studies at Clarkson University.

Conflicts of Interest: The authors declare no conflict of interest.

References

1. Hantanasirisakul, K.; Gogotsi, Y. Electronic and Optical Properties of 2D Transition Metal Carbides and Nitrides (MXenes). *Adv. Mater.* **2018**, *30*, 1804779. [CrossRef] [PubMed]
2. Naguib, M.; Mashtalir, O.; Carle, J.; Presser, V.; Lu, J.; Hultman, L.; Gogotsi, Y.; Barsoum, M.W. Two-Dimensional Transition Metal Carbides. *ACS Nano* **2012**, *6*, 1322–1331. [CrossRef] [PubMed]
3. Sun, Z.M. Progress in research and development on MAX phases: A family of layered ternary compounds. *Int. Mater. Rev.* **2011**, *56*, 143–166. [CrossRef]
4. Mashtalir, O.; Naguib, M.; Mochalin, V.; Dall'Agnese, Y.; Heon, M.; Barsoum, M.W.; Gogotsi, Y. Intercalation and delamination of layered carbides and carbonitrides. *Nat. Commun.* **2013**, *4*, 1716. [CrossRef] [PubMed]
5. Zhu, J.; Ha, E.; Zhao, G.; Zhou, Y.; Huang, D.; Yue, G.; Hua, L.; Sun, N.; Wang, Y.; Lee, L.Y.S.; et al. Recent advance in MXenes: A promising 2D material for catalysis, sensor and chemical adsorption. *Co-Ord. Chem. Rev.* **2017**, *352*, 306–327. [CrossRef]
6. Alhabeb, M.; Maleski, K.; Anasori, B.; Lelyukh, P.; Clark, L.; Sin, S.; Gogotsi, Y. Guidelines for synthesis and processing of two-dimensional titanium carbide ($Ti_3C_2T_x$ MXene). *Chem. Mater.* **2017**, *29*, 7633–7644. [CrossRef]
7. Huang, K.; Li, Z.; Lin, J.; Han, G.; Huang, P. Two-dimensional transition metal carbides and nitrides (MXenes) for biomedical applications. *Chem. Soc. Rev.* **2018**, *47*, 5109–5124. [CrossRef]
8. Soleymaniha, M.; Shahbazi, M.-A.; Rafieerad, A.R.; Maleki, A.; Amiri, A. Promoting Role of MXene Nanosheets in Biomedical Sciences: Therapeutic and Biosensing Innovations. *Adv. Health Mater.* **2018**, *8*, 1801137. [CrossRef]

9. Gogotsi, Y.; Anasori, B. The Rise of MXenes. *ACS Nano* **2019**, *13*, 8491–8494. [CrossRef]
10. Sinha, A.; Dhanjai; Zhao, H.; Huang, Y.; Lu, X.; Chen, J.; Jain, R. MXene: An emerging material for sensing and biosensing. *TrAC Trends Anal. Chem.* **2018**, *105*, 424–435. [CrossRef]
11. Zhang, Y.; Wang, L.; Zhang, N.; Zhou, Z. Adsorptive environmental applications of MXene nanomaterials: A review. *RSC Adv.* **2018**, *8*, 19895–19905. [CrossRef]
12. Rasool, K.; Pandey, R.P.; Rasheed, P.A.; Buczek, S.; Gogotsi, Y.; Mahmoud, K.A. Water treatment and environmental remediation applications of two-dimensional metal carbides (MXenes). *Mater. Today* **2019**, *30*, 80–102. [CrossRef]
13. Sinha, A.; Dhanjai; Mugo, S.M.; Chen, J.; Lokesh, K.S. *Handbook of Nanomaterials in Analytical Chemistry*; MXene-based sensors and biosensors: Next-generation detection platforms; Elsevier: Amsterdam, The Netherlands, 2020; pp. 361–372.
14. Chang, W.-C.; Cheng, S.-C.; Chiang, W.-H.; Liao, J.-L.; Ho, R.-M.; Hsiao, T.-C.; Tsai, D.-H. Quantifying Surface Area of Nanosheet Graphene Oxide Colloid Using a Gas-Phase Electrostatic Approach. *Anal. Chem.* **2017**, *89*, 12217–12222. [CrossRef] [PubMed]
15. Chen, J.; Meng, H.; Tian, Y.; Yang, R.; Du, D.; Li, Z.; Qu, L.; Lin, Y. Recent advances in functionalized MnO_2 nanosheets for biosensing and biomedicine applications. *Nanoscale Horiz.* **2018**, *4*, 321–338. [CrossRef]
16. Mohamed, H. Post-Synthetic Immobilization of Ni Ions in Porous-Organic Polymer-Graphene Composite for the Non-Noble Metal Electrocatalytic Water Oxidation. *Eur. Soc. J. Catal.* **2017**, *9*, 2894.
17. Kim, J.; Han, J.; Seo, M.; Kang, S.; Kim, D.; Ihm, J. High-surface area ceramic-derived boron-nitride and its hydrogen uptake properties. *J. Mater. Chem. A* **2013**, *1*, 1014–1017. [CrossRef]
18. Mateti, S.; Wong, C.S.; Liu, Z.; Yang, W.; Li, Y.; Chen, Y. Biocompatibility of boron nitride nanosheets. *Nano Res.* **2017**, *11*, 334–342. [CrossRef]
19. Birch, M.E.; Ruda-Eberenz, T.A.; Chai, M.; Andrews, R.; Hatfield, R.L. Properties that influence the specific surface areas of carbon nanotubes and nanofibers. *Ann. Occup. Hyg.* **2013**, *57*, 1148–1166. [CrossRef]
20. Ma, P.-C.; Siddiqui, N.A.; Marom, G.; Kim, J. Dispersion and functionalization of carbon nanotubes for polymer-based nanocomposites: A review. *Compos. Part A Appl. Sci. Manuf.* **2010**, *41*, 1345–1367. [CrossRef]
21. Ouyang, M.; Huang, J.-L.; Cheung, C.L.; Lieber, C.M. Energy Gaps in "Metallic" Single-Walled Carbon Nanotubes. *Science* **2001**, *292*, 702–705. [CrossRef]
22. Lakshmi, A.; Gracelin, D.L.; Vigneshwari, M.; Karpagavinayagam, P.; Veeraputhiran, V.; Vedhi, C. Microwave Synthesis and Characterization of Multiwalled Carbon Nanotubes (MWCNT) and Metal Oxide Doped MWCNT. *J. Nanosci. Technol.* **2015**, *1*, 19–22.
23. El Beqqali, O.; Zorkani, I.; Rogemond, F.; Chermette, H.; Ben Chaabane, R.; Gamoudi, M.; Guillaud, G. Electrical properties of molybdenum disulfide MoS_2. Experimental study and density functional calculation results. *Synth. Met.* **1997**, *90*, 165–172. [CrossRef]
24. Sobanska, Z.; Zapor, L.; Szparaga, M.; Stępnik, M. Biological effects of molybdenum compounds in nanosized forms under in vitro and in vivo conditions. *Int. J. Occup. Med. Environ. Health* **2020**, *33*, 1–19. [CrossRef] [PubMed]
25. Ikram, M.; Liu, L.; Liu, Y.; Ma, L.; Lv, H.; Ullah, M.; He, L.; Wu, H.; Wang, R.; Shi, K. Fabrication and characterization of a high-surface area MoS_2@WS_2 heterojunction for the ultra-sensitive NO_2 detection at room temperature. *J. Mater. Chem. A* **2019**, *7*, 14602–14612. [CrossRef]
26. Mahmood, N.; De Castro, I.A.; Pramoda, K.; Khoshmanesh, K.; Bhargava, S.K.; Kalantar-Zadeh, K. Atomically thin two-dimensional metal oxide nanosheets and their heterostructures for energy storage. *Energy Storage Mater.* **2019**, *16*, 455–480. [CrossRef]
27. Jia, Z.; Wang, J.; Wang, Y.; Li, B.; Wang, B.; Qi, T.; Wang, X. Interfacial Synthesis of δ-MnO_2 Nano-sheets with a Large Surface Area and Their Application in Electrochemical Capacitors. *J. Mater. Sci. Technol.* **2016**, *32*, 147–152. [CrossRef]
28. Ren, C.E.; Zhao, M.-Q.; Makaryan, T.; Halim, J.; Boota, M.; Kota, S.; Anasori, B.; Barsoum, M.W.; Gogotsi, Y. Porous Two-Dimensional Transition Metal Carbide (MXene) Flakes for High-Performance Li-Ion Storage. *ChemElectroChem* **2016**, *3*, 689–693. [CrossRef]
29. Wang, H.; Wu, Y.; Zhang, J.; Li, G.; Huang, H.; Zhang, X.; Jiang, Q. Enhancement of the electrical properties of MXene Ti_3C_2 nanosheets by post-treatments of alkalization and calcination. *Mater. Lett.* **2015**, *160*, 537–540. [CrossRef]

30. Naguib, M.; Kurtoglu, M.; Presser, V.; Lu, J.; Niu, J.; Heon, M.; Hultman, L.; Gogotsi, G.A.; Barsoum, M.W. Two-Dimensional Nanocrystals Produced by Exfoliation of Ti$_3$AlC$_2$. *Adv. Mater.* **2011**, *23*, 4248–4253. [CrossRef]

31. Khazaei, M.; Ranjbar, A.; Arai, M.; Sasaki, T.; Yunoki, S. Electronic properties and applications of MXenes: A theoretical review. *J. Mater. Chem. C* **2017**, *5*, 2488–2503. [CrossRef]

32. Lukatskaya, M.R.; Mashtalir, O.; Ren, C.E.; Dall'Agnese, Y.; Rozier, P.; Taberna, P.L.; Naguib, M.; Simon, P.; Barsoum, M.W.; Gogotsi, Y. Cation Intercalation and High Volumetric Capacitance of Two-Dimensional Titanium Carbide. *Science* **2013**, *341*, 1502–1505. [CrossRef] [PubMed]

33. Ghidiu, M.; Lukatskaya, M.R.; Zhao, M.-Q.; Gogotsi, Y.; Barsoum, M.W. Conductive two-dimensional titanium carbide 'clay' with high volumetric capacitance. *Nature* **2014**, *516*, 78–81. [CrossRef] [PubMed]

34. Halim, J.; Lukatskaya, M.R.; Cook, K.M.; Lu, J.; Smith, C.R.; Näslund, L.-Å.; May, S.J.; Hultman, L.; Gogotsi, Y.; Eklund, P.; et al. Transparent Conductive Two-Dimensional Titanium Carbide Epitaxial Thin Films. *Chem. Mater.* **2014**, *26*, 2374–2381. [CrossRef] [PubMed]

35. Mariano, M.; Mashtalir, O.; Antonio, F.Q.; Ryu, W.-H.; Deng, B.; Xia, F.; Gogotsi, Y.; Taylor, A.D. Solution-processed titanium carbide MXene films examined as highly transparent conductors. *Nanoscale* **2016**, *8*, 16371–16378. [CrossRef] [PubMed]

36. Lai, S.; Jeon, J.; Jang, S.K.; Xu, J.; Choi, Y.J.; Park, J.-H.; Hwang, E.; Lee, S. Surface group modification and carrier transport properties of layered transition metal carbides (Ti$_2$CT$_x$, T: –OH, –F and –O). *Nanoscale* **2015**, *7*, 19390–19396. [CrossRef] [PubMed]

37. Naguib, M.; Halim, J.; Lu, J.; Cook, K.M.; Hultman, L.; Gogotsi, Y.; Barsoum, M.W. New Two-Dimensional Niobium and Vanadium Carbides as Promising Materials for Li-Ion Batteries. *J. Am. Chem. Soc.* **2013**, *135*, 15966–15969. [CrossRef]

38. Xie, Y.; Dall'Agnese, Y.; Naguib, M.; Gogotsi, Y.; Barsoum, M.W.; Zhuang, H.L.; Kent, P.R.C. Prediction and Characterization of MXene Nanosheet Anodes for Non-Lithium-Ion Batteries. *ACS Nano* **2014**, *8*, 9606–9615. [CrossRef]

39. Shahzad, F.; Alhabeb, M.; Hatter, C.B.; Anasori, B.; Hong, S.M.; Koo, C.M.; Gogotsi, Y. Electromagnetic interference shielding with 2D transition metal carbides (MXenes). *Science* **2016**, *353*, 1137–1140. [CrossRef]

40. Fan, G.; Li, X.; Ma, Y.; Zhang, Y.; Wu, J.; Xu, B.; Sun, T.; Gao, D.-J.; Bi, J. Magnetic, recyclable Pt$_y$Co$_{1-y}$/Ti$_3$C$_2$X$_2$ (X = O, F) catalyst: A facile synthesis and enhanced catalytic activity for hydrogen generation from the hydrolysis of ammonia borane. *New J. Chem.* **2017**, *41*, 2793–2799. [CrossRef]

41. Xue, M.; Wang, Z.; Yuan, F.; Zhang, X.; Wei, W.; Tang, H.; Li, C. Preparation of TiO$_2$/Ti$_3$C$_2$T$_x$ hybrid nanocomposites and their tribological properties as base oil lubricant additives. *RSC Adv.* **2017**, *7*, 4312–4319. [CrossRef]

42. Zhang, X.; Xu, J.; Wang, H.; Zhang, J.; Yan, H.; Pan, B.; Zhou, J.; Xie, Y. Ultrathin nanosheets of MAX phases with enhanced thermal and mechanical properties in polymeric compositions: Ti$_3$Si$_{0.75}$Al$_{0.25}$C$_2$. *Angew. Chem. Int. Ed.* **2013**, *52*, 4361–4365. [CrossRef] [PubMed]

43. Peng, Q.; Guo, J.; Zhang, Q.; Xiang, J.; Liu, B.; Zhou, A.; Liu, R.; Tian, Y. Unique Lead Adsorption Behavior of Activated Hydroxyl Group in Two-Dimensional Titanium Carbide. *J. Am. Chem. Soc.* **2014**, *136*, 4113–4116. [CrossRef] [PubMed]

44. Guo, J.; Peng, Q.; Fu, H.; Zou, G.; Zhang, Q. Heavy-Metal Adsorption Behavior of Two-Dimensional Alkalization-Intercalated MXene by First-Principles Calculations. *J. Phys. Chem. C* **2015**, *119*, 20923–20930. [CrossRef]

45. Mashtalir, O.; Cook, K.M.; Mochalin, V.; Crowe, M.; Barsoum, M.W.; Gogotsi, Y. Dye adsorption and decomposition on two-dimensional titanium carbide in aqueous media. *J. Mater. Chem. A* **2014**, *2*, 14334–14338. [CrossRef]

46. Liu, F.; Zhou, A.; Chen, J.; Zhang, H.; Cao, J.; Wang, L.; Hu, Q. Preparation and methane adsorption of two-dimensional carbide Ti$_2$C. *Adsorption* **2016**, *22*, 915–922. [CrossRef]

47. Lin, H.; Wang, X.; Yu, L.; Chen, Y.; Cui, X. Two-Dimensional Ultrathin MXene Ceramic Nanosheets for Photothermal Conversion. *Nano Lett.* **2016**, *17*, 384–391. [CrossRef]

48. Khazaei, M.; Arai, M.; Sasaki, T.; Chung, C.-Y.; Venkataramanan, N.S.; Estili, M.; Sakka, Y.; Kawazoe, Y. Novel Electronic and Magnetic Properties of Two-Dimensional Transition Metal Carbides and Nitrides. *Adv. Funct. Mater.* **2012**, *23*, 2185–2192. [CrossRef]

49. Khazaei, M.; Ranjbar, A.; Ghorbani-Asl, M.; Arai, M.; Sasaki, T.; Liang, Y.; Yunoki, S. Nearly free electron states in MXenes. *Phys. Rev. B* **2016**, *93*, 205125. [CrossRef]
50. Si, C.; Zhou, J.; Sun, Z. Half-Metallic Ferromagnetism and Surface Functionalization-Induced Metal–Insulator Transition in Graphene-like Two-Dimensional Cr2C Crystals. *ACS Appl. Mater. Interfaces* **2015**, *7*, 17510–17515. [CrossRef]
51. Gao, G.; Ding, G.; Li, J.; Yao, K.; Wu, M.; Qian, M. Monolayer MXenes: Promising half-metals and spin gapless semiconductors. *Nanoscale* **2016**, *8*, 8986–8994. [CrossRef]
52. Berdiyorov, G.R. Optical properties of functionalized $Ti_3C_2T_2$ (T = F, O, OH) MXene: First-principles calculations. *AIP Adv.* **2016**, *6*, 55105. [CrossRef]
53. Zhang, H.; Yang, G.; Zuo, X.; Tang, H.; Yang, Q.; Li, G. Computational studies on the structural, electronic and optical properties of graphene-like MXenes (M_2CT_2, M = Ti, Zr, Hf; T = O, F, OH) and their potential applications as visible-light driven photocatalysts. *J. Mater. Chem. A* **2016**, *4*, 12913–12920. [CrossRef]
54. Khazaei, M.; Arai, M.; Sasaki, T.; Estili, M.; Sakka, Y. Two-dimensional molybdenum carbides: Potential thermoelectric materials of the MXene family. *Phys. Chem. Chem. Phys.* **2014**, *16*, 7841–7849. [CrossRef] [PubMed]
55. Kim, H.; Anasori, B.; Gogotsi, Y.; Alshareef, H.N. Thermoelectric Properties of Two-Dimensional Molybdenum-Based MXenes. *Chem. Mater.* **2017**, *29*, 6472–6479. [CrossRef]
56. Yu, X.-F.; Li, Y.; Cheng, J.-B.; Liu, Z.-B.; Li, Q.-Z.; Li, W.-Z.; Yang, X.; Xiao, B. Monolayer Ti_2CO_2: A Promising Candidate for NH3 Sensor or Capturer with High Sensitivity and Selectivity. *ACS Appl. Mater. Interfaces* **2015**, *7*, 13707–13713. [CrossRef]
57. Yadav, A.; Dashora, A.; Patel, N.; Miotello, A.; Press, M.; Kothari, D. Study of 2D MXene Cr2C material for hydrogen storage using density functional theory. *Appl. Surf. Sci.* **2016**, *389*, 88–95. [CrossRef]
58. Hu, Q.; Sun, D.; Wu, Q.; Wang, H.; Wang, L.; Liu, B.; Zhou, A.; He, J. MXene: A New Family of Promising Hydrogen Storage Medium. *J. Phys. Chem. A* **2013**, *117*, 14253–14260. [CrossRef]
59. Lei, J.; Kutana, A.; Yakobson, B.I. Predicting stable phase monolayer Mo2C (MXene), a superconductor with chemically-tunable critical temperature. *J. Mater. Chem. C* **2017**, *5*, 3438–3444. [CrossRef]
60. Si, C.; Jin, K.-H.; Zhou, J.; Sun, Z.; Liu, F. Large-Gap Quantum Spin Hall State in MXenes: D-Band Topological Order in a Triangular Lattice. *Nano Lett.* **2016**, *16*, 6584–6591. [CrossRef]
61. Liang, Y.; Khazaei, M.; Ranjbar, A.; Arai, M.; Yunoki, S.; Kawazoe, Y.; Weng, H.; Fang, Z. Theoretical prediction of two-dimensional functionalized MXene nitrides as topological insulators. *Phys. Rev. B* **2017**, *96*, 195414. [CrossRef]
62. Kumar, S.; Lei, Y.; Alshareef, N.H.; Quevedo-Lopez, M.; Salama, K.N. Biofunctionalized two-dimensional Ti_3C_2 MXenes for ultrasensitive detection of cancer biomarker. *Biosens. Bioelectron.* **2018**, *121*, 243–249. [CrossRef] [PubMed]
63. Zhou, S.; Gu, C.; Li, Z.; Yang, L.; He, L.; Wang, M.; Huang, X.; Zhou, N.; Zhang, Z. $Ti_3C_2T_x$ MXene and polyoxometalate nanohybrid embedded with polypyrrole: Ultra-sensitive platform for the detection of osteopontin. *Appl. Surf. Sci.* **2019**, *498*, 143889. [CrossRef]
64. Chia, H.L.; Mayorga-Martinez, C.C.; Antonatos, N.; Sofer, Z.; Gonzalez-Julian, J.J.; Webster, R.D.; Pumera, M. MXene Titanium Carbide-based Biosensor: Strong Dependence of Exfoliation Method on Performance. *Anal. Chem.* **2020**, *92*, 2452–2459. [CrossRef]
65. Bollella, P.; Katz, E. Enzyme-Based Biosensors: Tackling Electron Transfer Issues. *Sensors* **2020**, *20*, 3517. [CrossRef] [PubMed]
66. Gu, H.; Xing, Y.; Xiong, P.; Tang, H.; Li, C.; Chen, S.; Zeng, R.; Han, K.; Shi, G. Three-Dimensional Porous $Ti_3C_2T_x$ MXene–Graphene Hybrid Films for Glucose Biosensing. *ACS Appl. Nano Mater.* **2019**, *2*, 6537–6545. [CrossRef]
67. Rakhi, R.B.; Nayak, P.; Xia, C.; Alshareef, H.N.; Nayuk, P. Novel amperometric glucose biosensor based on MXene nanocomposite. *Sci. Rep.* **2016**, *6*, 36422. [CrossRef]
68. Koyappayil, A.; Chavan, S.G.; Mohammadniaei, M.; Go, A.; Hwang, S.Y.; Lee, M.-H. β-Hydroxybutyrate dehydrogenase decorated MXene nanosheets for the amperometric determination of β-hydroxybutyrate. *Microchim. Acta* **2020**, *187*, 277. [CrossRef]
69. Wang, F.; Yang, C.; Duan, M.; Tang, Y.; Zhu, J. TiO_2 nanoparticle modified organ-like Ti_3C_2 MXene nanocomposite encapsulating hemoglobin for a mediator-free biosensor with excellent performances. *Biosens. Bioelectron.* **2015**, *74*, 1022–1028. [CrossRef]

70. Liu, H.; Duan, C.; Yang, C.; Shen, W.; Wang, F.; Zhu, Z. A novel nitrite biosensor based on the direct electrochemistry of hemoglobin immobilized on MXene-Ti$_3$C$_2$. *Sens. Actuators B Chem.* **2015**, *218*, 60–66. [CrossRef]
71. Ma, B.-K.; Li, M.; Cheong, L.-Z.; Weng, X.-C.; Shen, C.; Huang, Q. Enzyme-MXene Nanosheets: Fabrication and Application in Electrochemical Detection of H$_2$O$_2$. *J. Inorg. Mater.* **2020**, *35*, 131–138.
72. Zhou, L.; Zhang, X.; Ma, L.; Gao, J.; Jiang, Y. Acetylcholinesterase/chitosan-transition metal carbides nanocomposites-based biosensor for the organophosphate pesticides detection. *Biochem. Eng. J.* **2017**, *128*, 243–249. [CrossRef]
73. Jiang, Y.; Zhang, X.; Pei, L.; Yue, S.; Ma, L.; Zhou, L.; Huang, Z.; He, Y.; Gao, J. Silver nanoparticles modified two-dimensional transition metal carbides as nanocarriers to fabricate acetycholinesterase-based electrochemical biosensor. *Chem. Eng. J.* **2018**, *339*, 547–556. [CrossRef]
74. Song, D.; Jiang, X.; Li, Y.; Lu, X.; Luan, S.; Wnag, Y.; Li, Y.; Gao, F. Metal–Organic frameworks-derived MnO$_2$/Mn$_3$O$_4$ microcuboids with hierarchically ordered nanosheets and Ti$_3$C$_2$ MXene/Au NPs composites for electrochemical pesticide detection. *J. Hazard. Mater.* **2019**, *373*, 367–376. [CrossRef] [PubMed]
75. Wang, G.; Sun, J.; Yao, Y.; An, X.; Zhang, H.; Chu, G.; Jiang, S.; Guo, Y.; Sun, X.; Liu, Y. Detection of Inosine Monophosphate (IMP) in Meat Using Double-Enzyme Sensor. *Food Anal. Methods* **2019**, *13*, 420–432. [CrossRef]
76. Fang, L.; Liao, X.; Jia, B.; Shi, L.; Kang, L.; Zhou, L.; Kong, W. Recent progress in immunosensors for pesticides. *Biosens. Bioelectron.* **2020**, *164*, 112255. [CrossRef]
77. Cheng, N.; Song, Y.; Shi, Q.; Du, D.; Liu, D.; Luo, Y.; Xu, W.; Lin, Y. Au@Pd Nanopopcorn and Aptamer Nanoflower Assisted Lateral Flow Strip for Thermal Detection of Exosomes. *Anal. Chem.* **2019**, *91*, 13986–13993. [CrossRef]
78. Fan, Y.; Shi, S.; Ma, J.; Guo, Y. A paper-based electrochemical immunosensor with reduced graphene oxide/thionine/gold nanoparticles nanocomposites modification for the detection of cancer antigen 125. *Biosens. Bioelectron.* **2019**, *135*, 1–7. [CrossRef]
79. Cheng, N.; Song, Y.; Zeinhom, M.M.A.; Chang, Y.-C.; Sheng, L.; Li, H.; Du, D.; Li, L.; Zhu, M.-J.; Luo, Y.; et al. Nanozyme-Mediated Dual Immunoassay Integrated with Smartphone for Use in Simultaneous Detection of Pathogens. *ACS Appl. Mater. Interfaces* **2017**, *9*, 40671–40680. [CrossRef]
80. Cheng, N.; Xu, Y.; Huang, K.; Chen, Y.; Yang, Z.; Luo, Y.; Xu, W. One-step competitive lateral flow biosensor running on an independent quantification system for smart phones based in-situ detection of trace Hg(II) in tap water. *Food Chem.* **2017**, *214*, 169–175. [CrossRef]
81. Pan, G.; Zhao, G.; Wei, M.; Wang, Y.; Zhao, B. Design of nanogold electrochemical immunosensor for detection of four phenolic estrogens. *Chem. Phys. Lett.* **2019**, *732*, 136657. [CrossRef]
82. Angulo-Ibáñez, A.; Eletxigerra, U.; Lasheras, X.; Campuzano, S.; Merino, S. Electrochemical tropomyosin allergen immunosensor for complex food matrix analysis. *Anal. Chim. Acta* **2019**, *1079*, 94–102. [CrossRef] [PubMed]
83. Kunene, K.; Weber, M.; Sabela, M.; Voiry, D.; Kanchi, S.; Bisetty, K.; Bechelany, M. Highly-efficient electrochemical label-free immunosensor for the detection of ochratoxin A in coffee samples. *Sens. Actuators B Chem.* **2020**, *305*, 127438. [CrossRef]
84. Wang, H.; Li, H.; Huang, Y.; Xiong, M.; Wang, F.; Li, C. A label-free electrochemical biosensor for highly sensitive detection of gliotoxin based on DNA nanostructure/MXene nanocomplexes. *Biosens. Bioelectron.* **2019**, *142*, 111531. [CrossRef] [PubMed]
85. Xu, Y.; Wei, Y.; Cheng, N.; Huang, K.; Wang, W.; Zhang, L.; Xu, W.; Luo, Y. Nucleic Acid Biosensor Synthesis of an All-in-One Universal Blocking Linker Recombinase Polymerase Amplification with a Peptide Nucleic Acid-Based Lateral Flow Device for Ultrasensitive Detection of Food Pathogens. *Anal. Chem.* **2017**, *90*, 708–715. [CrossRef] [PubMed]
86. Chen, J.; Tong, P.; Huang, L.; Yu, Z.; Tang, D. Ti$_3$C$_2$ MXene nanosheet-based capacitance immunoassay with tyramine-enzyme repeats to detect prostate-specific antigen on interdigitated micro-comb electrode. *Electrochim. Acta* **2019**, *319*, 375–381. [CrossRef]
87. Khan, R.; Ben Aissa, S.; Sherazi, T.A.; Catanante, G.; Hayat, A.; Marty, J.-L. Development of an Impedimetric Aptasensor for Label Free Detection of Patulin in Apple Juice. *Molecules* **2019**, *24*, 1017. [CrossRef]
88. Zheng, J.; Wang, B.; Ding, A.; Weng, B.; Chen, J. Synthesis of MXene/DNA/Pd/Pt nanocomposite for sensitive detection of dopamine. *J. Electroanal. Chem.* **2018**, *816*, 189–194. [CrossRef]

89. Yan, X.; Gu, Y.; Li, C.; Tang, L.; Zheng, B.; Li, Y.; Zhang, Z.; Yang, M. Synergetic catalysis based on the proline tailed metalloporphyrin with graphene sheet as efficient mimetic enzyme for ultrasensitive electrochemical detection of dopamine. *Biosens. Bioelectron.* **2016**, *77*, 1032–1038. [CrossRef]
90. Salamon, J.; Sathishkumar, Y.; Ramachandran, K.; Lee, Y.S.; Yoo, D.; Kim, A.R.; Kumar, G.G. One-pot synthesis of magnetite nanorods/graphene composites and its catalytic activity toward electrochemical detection of dopamine. *Biosens. Bioelectron.* **2015**, *64*, 269–276. [CrossRef]
91. Peng, X.; Zhang, Y.; Lu, D.; Guo, Y.-J.; Guo, S. Ultrathin Ti_3C_2 nanosheets based "off-on" fluorescent nanoprobe for rapid and sensitive detection of HPV infection. *Sens. Actuators B Chem.* **2019**, *286*, 222–229. [CrossRef]
92. Bray, F.; Ferlay, J.; Soerjomataram, I.; Siegel, R.L.; Torre, L.A.; Jemal, A. Global cancer statistics 2018: GLOBOCAN estimates of incidence and mortality worldwide for 36 cancers in 185 countries. *CA Cancer J. Clin.* **2018**, *68*, 394–424. [CrossRef] [PubMed]
93. Khan, R.; Sherazi, T.A.; Catanante, G.; Rasheed, S.; Marty, J.-L.; Hayat, A. Switchable fluorescence sensor toward PAT via CA-MWCNTs quenched aptamer-tagged carboxyfluorescein. *Food Chem.* **2020**, *312*, 126048. [CrossRef] [PubMed]
94. Zhang, Q.; Wang, F.; Zhang, H.; Zhang, Y.; Liu, M.; Liu, Y. Universal Ti_3C_2 MXenes Based Self-Standard Ratiometric Fluorescence Resonance Energy Transfer Platform for Highly Sensitive Detection of Exosomes. *Anal. Chem.* **2018**, *90*, 12737–12744. [CrossRef] [PubMed]
95. Shi, X.; Wang, H.; Xie, X.; Xue, Q.; Zhang, J.; Kang, S.; Wang, C.; Liang, J.; Chen, Y. Bioinspired Ultrasensitive and Stretchable MXene-Based Strain Sensor via Nacre-Mimetic Microscale "Brick-and-Mortar" Architecture. *ACS Nano* **2018**, *13*, 649–659. [CrossRef] [PubMed]
96. Song, D.; Li, X.; Li, X.-P.; Jia, X.; Min, P.; Yu, Z.-Z. Hollow-structured MXene-PDMS composites as flexible, wearable and highly bendable sensors with wide working range. *J. Colloid Interface Sci.* **2019**, *555*, 751–758. [CrossRef] [PubMed]
97. Cheng, Y.; Ma, Y.; Li, L.; Zhu, M.; Yue, Y.; Liu, W.; Wang, L.; Jia, S.; Li, C.; Qi, T.; et al. Bioinspired Microspines for a High-Performance Spray $Ti_3C_2T_x$ MXene-Based Piezoresistive Sensor. *ACS Nano* **2020**, *14*, 2145–2155. [CrossRef] [PubMed]
98. Guo, Y.; Zhong, M.; Fang, Z.; Wan, P.; Yu, G. A Wearable Transient Pressure Sensor Made with MXene Nanosheets for Sensitive Broad-Range Human–Machine Interfacing. *Nano Lett.* **2019**, *19*, 1143–1150. [CrossRef]
99. Li, H.; Du, Z. Preparation of a Highly Sensitive and Stretchable Strain Sensor of MXene/Silver Nanocomposite-Based Yarn and Wearable Applications. *ACS Appl. Mater. Interfaces* **2019**, *11*, 45930–45938. [CrossRef]
100. Liao, H.; Guo, X.; Wan, P.; Yu, G. Conductive MXene Nanocomposite Organohydrogel for Flexible, Healable, Low-Temperature Tolerant Strain Sensors. *Adv. Funct. Mater.* **2019**, *29*, 1904507. [CrossRef]
101. Chen, Z.; Hu, Y.; Zhuo, H.; Liu, L.; Jing, S.; Zhong, L.; Peng, X.; Sun, R.-C. Compressible, Elastic, and Pressure-Sensitive Carbon Aerogels Derived from 2D Titanium Carbide Nanosheets and Bacterial Cellulose for Wearable Sensors. *Chem. Mater.* **2019**, *31*, 3301–3312. [CrossRef]
102. Lei, Y.; Zhao, W.; Zhang, Y.-Z.; Jiang, Q.; He, J.; Baeumner, A.J.; Wolfbeis, O.S.; Wang, Z.L.; Salama, K.N.; Alshareef, H.N. A MXene-Based Wearable Biosensor System for High-Performance In Vitro Perspiration Analysis. *Small* **2019**, *15*, e1901190. [CrossRef] [PubMed]

© 2020 by the authors. Licensee MDPI, Basel, Switzerland. This article is an open access article distributed under the terms and conditions of the Creative Commons Attribution (CC BY) license (http://creativecommons.org/licenses/by/4.0/).

Review

Synthesis, Catalytic Properties and Application in Biosensorics of Nanozymes and Electronanocatalysts: A Review

Nataliya Stasyuk [1], Oleh Smutok [1,2], Olha Demkiv [1,3], Tetiana Prokopiv [1], Galina Gayda [1], Marina Nisnevitch [4] and Mykhailo Gonchar [1,2,*]

1. Institute of Cell Biology, National Academy of Sciences of Ukraine, 79005 Lviv, Ukraine; stasukne@nas.gov.ua (N.S.); smutok@cellbiol.lviv.ua (O.S.); demkivo@nas.gov.ua (O.D.); t.prokopiv@nas.gov.ua (T.P.); galina.gayda@nas.gov.ua (G.G.)
2. Department of Biology and Chemistry, Drohobych Ivan Franko State Pedagogical University, 82100 Drohobych, Ukraine
3. Faculty of Veterinary Hygiene, Ecology and Law, Stepan Gzhytskyi National University of Veterinary Medicine and Biotechnologies, 79000 Lviv, Ukraine
4. Department of Chemical Engineering, Ariel University, Kyriat-ha-Mada, Ariel 4070000, Israel; marinan@ariel.ac.il
* Correspondence: gonchar@cellbiol.lviv.ua; Tel.: +380-32-2612144; Fax: +380-32-2612148

Received: 16 July 2020; Accepted: 7 August 2020; Published: 12 August 2020

Abstract: The current review is devoted to nanozymes, i.e., nanostructured artificial enzymes which mimic the catalytic properties of natural enzymes. Use of the term "nanozyme" in the literature as indicating an enzyme is not always justified. For example, it is used inappropriately for nanomaterials bound with electrodes that possess catalytic activity only when applying an electric potential. If the enzyme-like activity of such a material is not proven in solution (without applying the potential), such a catalyst should be named an "electronanocatalyst", not a nanozyme. This paper presents a review of the classification of the nanozymes, their advantages vs. natural enzymes, and potential practical applications. Special attention is paid to nanozyme synthesis methods (hydrothermal and solvothermal, chemical reduction, sol-gel method, co-precipitation, polymerization/polycondensation, electrochemical deposition). The catalytic performance of nanozymes is characterized, a critical point of view on catalytic parameters of nanozymes described in scientific papers is presented and typical mistakes are analyzed. The central part of the review relates to characterization of nanozymes which mimic natural enzymes with analytical importance ("nanoperoxidase", "nanooxidases", "nanolaccase") and their use in the construction of electro-chemical (bio)sensors ("nanosensors").

Keywords: nanoparticle; nanocomposite; nanozyme; synthesis; catalytic properties; nano-peroxidase; nanooxidase; nanolaccase; electronanocatalyst; amperometric (bio)sensors

1. Introduction: Definition of Nanozymes, Classification, Advantages vs. Natural Enzymes, and Potential Practical Applications

Enzymes are biological catalysts that play a key role in biological processes. They have long been an indispensable tool in many chemical and biotechnological processes that are widely used in food processing, industry, agriculture and medicine. Natural enzymes (except for ribozymes) are proteins and are responsible for almost all biochemical reactions in living organisms. They are characterized by high selectivity and extremely high catalytic activity (Table 1). However, natural enzymes tend to have limited chemical and biological stability as well as a high cost due to complicated technologies employed for their isolation and purification from biological sources. Although modern molecular technologies (gene cloning, genetic and protein engineering, etc.) significantly facilitate these procedures, obtaining highly purified enzymes in commercial quantities is still a key challenge for practical enzymology.

Table 1. Natural enzymes—biocatalysts of protein origin (except for ribozymes).

Advantages	Drawbacks
An extremely high rate of enzymatic reactions: spontaneous reactions can run for millions of years, while enzymatic ones run for milliseconds. Examples of particularly active enzymes: -Catalase ($2H_2O_2 \rightarrow 2H_2O + O_2$) 1 molecule of the enzyme catalyzes the decay of 5 mln S per 1 min; -Carbanhydrase ($CO_2 + H_2O \Leftrightarrow H_2CO_3 \Leftrightarrow H^+ + HCO_3^-$): 36 mln turnovers per 1 min. High selectivity	Physicochemical instability to action of environmental (chemical and physical) factors. Biological instability (susceptibility to degradation by proteases). High costs of isolation and purification.

Artificial substitutes for enzymes were invented toward the end of the twentieth century [1,2]. Artificial enzymes include cyclodextrins with catalytic activity, abzymes (antibodies with catalytic activity), synzymes (synthetic enzymes) and aptamers (DNAzymes and RNAzymes). They are usually more stable than natural enzymes but are inferior in their catalytic activity and preparation costs due to complicated synthesis technologies. As a rule, artificial enzymes are unfortunately not as substrate specific as natural enzymes.

Nanozymes (NZs) are the newest class of functional nanomaterials [3–6] that have enzyme-like activity. They possess increased stability and greater availability due to their simpler preparation technologies. Described nanoscale materials include catalysts with different reaction specificities. They are mainly oxidoreductases: peroxidase [7], haloperoxidase, catalase, glucose oxidase, sulfite oxidase, superoxide dismutase (SOD), laccase, monoxygenase, CO oxidase, ferritin ferrooxidase [8], different hydrolases (phosphatase, phosphotriesterase, chymotrypsin, carbonic anhydrase), as well as proteases, endonucleases, DNA-ases, NO synthases, etc. [9–11].

The term "nanozyme" was defined by Wei and Wang in 2013 [3], although the first exciting discovery of Fe_3O_4-based ferromagnetic nanoparticles (NPs) with peroxidase(PO)-like catalytic activity was made in 2007 by Gao [12], as cited by Huang [6]. In our opinion, Prussian Blue is another nanomaterial candidate for which catalytic activity toward hydrogen peroxide was proven for the first time (see, for example, a review by Karyakin in 2001 [13]).

The most important advantage of nanozymes is their size/composition-dependent activity. This allows the design of materials with a broad range of catalytic activity simply by varying shape, structure, and composition. NZs also have unique properties compared to other artificial enzymes, including large surface areas which significantly facilitate their further modification and bioconjugation. The ability of nanomaterials to self-assemble is also a very important characteristic for biology and medicine, due to easier incorporation of biological components into the nanomaterial's structure (Table 2).

Table 2. Advantages of nanozymes.

(1)	Availability and low preparation costs.
(2)	Physicochemical and biological stability.
(3)	High surface area.
(4)	Self-assembling activity.
(5)	Size/composition-dependent activity. Broad possibility for modification and regulation of activity.
(6)	Compatibility with biological elements.

There is no official classification of NZs. Huang et al. (2019) [6] proposed dividing NZs into two categories: (1) oxidoreductases (oxidases, peroxidase, catalase, superoxide dismutase, and nitrate reductase); (2) hydrolases (nucleases, esterases, phosphatases, proteases, and silicatein). Such a classification could easily be expanded upon the discovery of novel NZs with other catalytic activities, similarly to other natural enzymes allocated to Enzyme Committee (EC) classes.

Additional problems arise when attempting to classify NZs according to the chemical structure of the catalytic nanocomposite (see Table 1 from [6]). Such classification will be rather cumbersome, due to the broad chemical diversity of NZs. For example, the list of known NZs includes: NPs of noble and transient metals, their hybrid forms, carbon NPs (graphene, carbon nanotubes, fullerene), metal oxides, metal sulfides and tellurides, carbon nitride, quantum dots, 2D-nanomaterials with confined single metal and nonmetal atoms [14], Prussian Blue NPs, polypirrole, hemin micelles and many nanomaterials that have been functionalized or modified by organic ligands [3,5].

In many cases, the term "NZ" or "enzyme-mimicking activity" is used without a sufficiently strong reason. On our opinion, these terms can be used properly only for materials whose enzyme-mimicking activity has been proven in solution. In most papers, the catalytic activity was revealed only after applying a potential to an electrode modified by the tested nanocomposite. In this case, the catalytic nanomaterial should be named an "electrocatalyst" ("electronanocatalyst"), but not, strictly speaking, a "nanozyme" or "enzyme mimic" (Figure 1). We therefore propose naming a catalytic nanomaterial as a "NZ" (nanooxidase, nanoperoxidase, nanocatalase, nanolaccase, and so on) only when the nanomaterial exhibits the enzyme-like catalytic activity in solution/suspension, without applying an electrochemical potential.

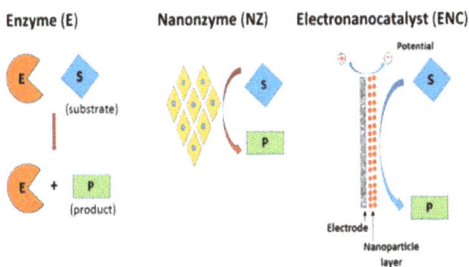

Figure 1. Principal scheme of catalytic action of enzyme, nanozyme and electronanocatalyst.

Although in most cases the catalytic efficiency of NZs is lower compared to natural analogues, it was reported that some NZs can compete with the enzymes. For example, the highest values for any NZ reported to date belong to $MnFe_2O_4$ with a nanooctahedron morphology, whose turnover number (k_{cat}) and catalytic efficiency (k_{cat}/K_M) in the oxidation of 3,3′,5,5′-tetramethylbenzidine (TMB) are 8.34×10^4 s^{-1} and 2.21×10^9 M^{-1}·s^{-1}, respectively [15,16].

It is worth noting that in many (or even most) publications, the catalytic activity is presented only by a V_{max} value, without indicating the catalyst concentration. This is problematic, since the V_{max} depends on the catalyst (enzyme) concentration ($V_{max} = [E] \cdot k_{cat}$). In order to compare the intrinsic kinetic parameters of different nanocatalysts with each other or with natural enzymes, a catalytic constant (a turnover number, k_{cat}) must be used, or at least the V_{max} value, but in conjunction with the concentration at which the maximal reaction rate of the nanocatalyst was determined. Valuable recommendations on this problem and standardization of experimental protocols for measurement of enzyme-mimicking activity are presented by Jiang et al. [17].

As a consequence of their catalytic activity, NZs can be applied for practical use in scientific research, biotechnology and food industries, agriculture, degradation of environmental pollutants (wastewater treatment, degradation of chemical warfare agents), clinical diagnostics and pharmacology [6,17–19]. In addition to their use as diagnostic tools, nanozymes are promising catalytic components of therapeutic drugs [6,8–11,20,21]. Due to the biocompatibility and magnetic properties of some NZs, they can be used for targeted treatment of malignancies. NZs have the potential for serving as antioxidants in the treatment of autoimmune, Alzheimer's and Parkinson's diseases, and for application as antibacterial agents. Using NZs in the construction of biosensors [18,19,22–24] and biofuel cells is very promising, where high catalytic activity, chemical and biological stability, nanoscale size of catalytic elements, and more cost-effective preparation are the most important challenges.

In the future, we can expect a sophisticated design of NZs in silico which will allow them to compete with natural enzymes in catalytic efficiency and selectivity. NZs sensitive to regulation by low-molecular effectors (activators and inhibitors) and able to catalyze cascade reactions will be created, and their catalytic behavior will be adapted to different environmental conditions. Large scale production will also be developed, based on cost-effective physicochemical methods and "green synthesis" approaches.

To date, over 300 types of nanomaterials were found to possess intrinsic enzyme-like activity. A large variety of NZs (or ENCs) simultaneously exhibit dual- or multienzyme mimetic activity [10]. Several reviews that summarize the new data concerning synthesis, characterization, bioanalytical and medical application of NZs are published every year, and the number of these reports is rising in a geometrical progression [10,23,24]. Thus, "nanozymology", as a new field of science connecting nanotechnology and enzymology, has great potential for further development and for many practical future applications [10,17].

The current paper is devoted to reviewing the recent advantages and current challenges of using NZs (or electronanocatalysts) as catalytic and recognition elements in biosensors. The catalytic performance of NZs is characterized, a critical point of view on catalytic parameters of NZs described in scientific papers is presented, and typical mistakes are analyzed. The central part of the review relates to characterization of NZs which mimic natural enzymes with analytical importance ("nanoperoxidase", "nanooxidases", "nanolaccase") and their use in the construction of electro-chemical (bio)sensors ("nanosensors").

2. Methods for the Synthesis of Catalytic Nanomaterials

Catalytic nanomaterials are known to possess unique properties compared with natural enzymes [25]. It was shown that activities of NZs are greatly dependent on the chemical structure, particle size, shape and surface morphology, which could be affected by charges, coatings, dopings, loadings and external fields. The morphology of synthesized NZs can be controlled due to rapid development of nanotechnology techniques [16,26].

This section summarizes different methods for preparation of nanomaterials possessing catalytic, mainly electrocatalytic, properties, in particular hydrothermal and solvothermal methods, co-precipitation and sol-gel methods, etc. Their applications for construction of biosensors are also described. Figure 2 presents the principal scheme illustrating methods of nanocatalyst synthesis. The kinetic parameters for different types of nanocatalysts are summarized in Tables 3–11. Table 3 relates to the effect of synthetic methods on the structural and functional properties of ENCs.

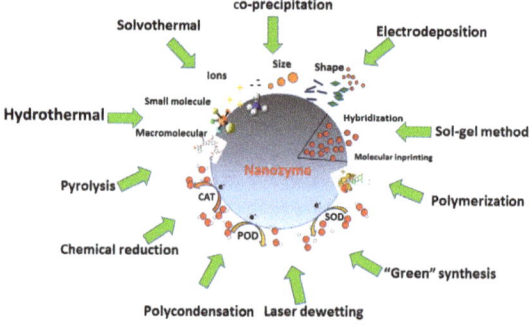

Figure 2. Principal scheme of the methods for nanocatalyst synthesis.

2.1. Hydrothermal and Solvothermal Methods

The main characteristics of the obtained NZs can be changed in accordance with the method chosen for NZ synthesis. The most promising techniques for synthesizing nanomaterials are the hydrothermal and solvothermal methods. Low-cost nanocrystals with well-controlled dimensions can be obtained using the proposed approaches [26].

A series of spinel-type $Co_xNi_{1-x}Fe_2O_4$ (x = 0, 0.2, 0.4, 0.5, 0.6, 0.8, 1.0) nanocatalysts were synthesized using the solvothermal method, where ethylene glycol served as the solvent. The obtained NZs were employed as enzyme mimics for the detection of H_2O_2. The $Co_{0.5}Ni_{0.5}Fe_2O_4$/CPE exhibited a wider linear range and a higher sensitivity compared with H_2O_2-selective enzymatic biosensors based on spinel-type ferrites, indicating a promising future for $Co_xNi_{1-x}Fe_2O_4$ as enzyme mimics for the construction of chemosensors [27]. The surface morphology of the synthesized $Co_xNi_{1-x}Fe_2O_4$ ENCs was studied and a diameter of 70 to 130 nm was observed. The small size of the nanocatalysts greatly increases the effective surface area for electrocatalytically active sites, and thus improves the physicochemical properties of the amperometric biosensor [27] (Table 3).

Solid spherical crystals of CuO (CuO-ENCs) having an average size of 45 nm were prepared by thermal decomposition of a precursor-Cu-based metal-organic gel (Cu-MOG). It was shown that the obtained CuO-ENCs possessed high electrocatalytic activity in glucose oxidation and exhibited PO-like catalytic ability. The CuO-ENC can also be used as a biomimetic for detection of cholesterol [28]. CuO-ENCs have high Brunauer-Emmett-Teller (BET) surface area of 20.16 $m^2 \cdot g^{-1}$, high pore volume of 0.11 $cm^3 \cdot g^{-1}$ and the average pore diameter of ~2.5 nm [29]. In another example, two kinds of carbon-based nanocatalysts with a size of 100–150 nm were synthesized using a combination of two methods, a thermal method and a solid-state reaction, from the zeolitic imidazolate framework-8 (ZIF-8) [30,31]. The carbon cubic nanomaterial (labelled HCC) with the hollow structure was obtained by chemically etching ZIF-8 with tannic acid, followed by a calcination process. The carbon cubic nanomaterial with the porous surface (labelled PCC) was obtained by direct pyrolysis. The two types of synthesized NZs, HCC and PCC, possess a BET specific surface area of 356 and 756 $m^2 \cdot g^{-1}$, respectively [30].

Electrochemical detection of glucose and fructose based on gold nanoparticles (AuNPs) deposited onto graphene paper has recently been proposed. These nanostructures were synthesized via two approaches: thermal and laser dewetting processes [32]. Gold nanostructures obtained by both methods exhibited major differences in their particle morphology. Both types of AuNPs were examined by their ability to oxidize glucose and fructose. The corresponding analytical characteristics of the constructed chemosensors are presented in Table 3.

It was shown that the BET of PCC was higher than that of HCC, indicating that PCC had more specific area for analyte adhesion (Table 3) [30].

The gold/cobalt (Au/Co) bimetallic nanocomposite-decorated hollow nanoporous carbon framework (Au/Co@HNCF) was synthesized as an ENC by thermal pyrolysis at 900 °C of the Au (III)-etching zeolitic imidazolate framework-67 (ZIF-67) [33]. An ultrasensitive electrochemical biosensor was developed to identify low levels of uric acid in human serum. Scanning electron microscopy images showed that the Au (III)-etching ZIF-67 has a dodecahedron shape (Table 3). The Au/Co@HNCF biosensor (7.88 cm^2) exhibited a significantly higher electrochemical active surface area than the bare glassy carbon electrode (GCE, 0.0155 cm^2), indicating the existence of abundant active sites on the Au/Co@HNCF modified layers.

2.2. Chemical Reduction

Chemical reduction is the most frequently used method, due to its rapidity and simplicity. This technique enables producing NPs in which the morphology and particle size distribution are controlled by changing the molar concentration of the reactants, the reductant type and the reaction temperature [34]. The critical factor in achieving high chemical reduction is choosing the appropriate reductants. Reduction of metal salts requires reactivity of the reduction agent to the redox potential of the metal. The obtained particles are small if the reaction rate during the synthesis process is too fast [35]. However, if the reaction rate is too slow, particle aggregation may occur [36]. The synthesis of hollow copper sulfide nanocubes (h-CuS NCs) was conducted via the chemical reduction method [37]. The average particle size was 50 nm and the specific surface area was 57.84 m$^2 \cdot$g^{-1}, which is larger than that of general solid CuS (34.76 m$^2 \cdot$g^{-1}). h-CuS NCs exhibited promising signs for high enzyme-mimicking catalytic activity (Table 3).

The chemical reduction method has been used for the synthesis of a peroxidase (PO)-like NZ based AuNPs in combination with *Pseudomonas aeruginosa*–specific aptamer [38]. The average particle size was ~20 ± 3 nm. The proposed bioelectrode structure can be used for detection of other bacterial pathogens in water or biological fluids.

ENC-carbon supported Pt-NiNPs stabilized with oleylamine (Pt-Ni) were synthesized by chemical reduction of metal salts [39]. The synthesized ENCs were near-spherical in shape, with a mean diameter of 8.4 nm. The ENCs exhibited an active electrochemical surface area of 2.3 m$^2 \cdot$g^{-1}, which is lower than that of the etched particles without surfactant (Table 3).

Metal-organic frameworks (MOFs) are composed of metal ions as nodes and organic ligands as linkers and have diversified and tailorable structures. MOFs have high surface area and porosity, exposed active sites and good biocompatibility, and for these reasons they attact wide attention as a blooming alternative material. A multifunctional artificial enzyme was synthesized through a combination of the chemical reduction method and electrodeposition technique by modifying PtNPs on the metalloporphyrin MOF [40]. In the Pt@PMOF(Fe) complex, PMOF(Fe) could prevent the aggregation of PtNPs, leading to high stability of the Pt NPs. Furthermore, PtNPs exhibited catalase-and PO-like activities. In another example, oxidase-like nanosheets were prepared [41] using the chemical reduction method modified by Hummers et al. [42], in which histidine enantiomers served as both a reducing agent and a protecting ligand. Compared to the previously developed strategies, this synthesis method is very rapid and mild and does not require heating, pressuring, and special media. It is a mild synthesis strategy, where only two reactants of HAuCl$_4$ and the His enantiomers are involved, without additional catalysts or templates. The oxidase-like and electrocatalytic activity of His@AuNCs/RGO was evaluated for the determination of nitrite (Table 3). The developed method was used for the detection of nitrites in real samples [41].

The green synthesis procedure was recently proposed as an important method for producing inorganic NZs. It is well known that living organisms can produce substances that act as reducing agents [6,11,16]. For example, Han et al. proposed a new environmentally friendly technology for synthesis of MoO$_3$ NZs in which green algae (*Enteromorpha prolifera*, EP) were used as the reducing agent [43]. The obtained MoO$_3$ NZs revealed excellent PO-like activity and were used, together with GOx, for colorimetric detection of glucose in human serum. In another work, Han and co-authors [44] used bovine serum albumin (BSA) as a biotemplate for Co$_3$O$_4$ NZs synthesis. The obtained NZs had a spherical morphology with an average diameter of 60 nm and exhibited catalase-like activities.

2.3. Sol-gel Method

In this method, a gel-like network containing both a liquid phase and a solid phase is formed. The crystallinity, morphology and magnetic properties of the NZs can be controlled by choosing an appropriate complexing agent, concentration and type of chemical additives, and temperature conditions [28].

The synthesis of PtNPs polyaniline (PAni) hydrogel heterostructures was performed via the sol-gel method [45]. Phytic acid was used as a complexing agent. The PtNPs loaded into the hydrogel matrix were found to act as active catalysts for the oxidation of H$_2$O$_2$. The obtained PtNP/PAni hydrogel had a 3D hierarchical structure consisting of connected PAni nanofibers with diameters of approximately 100 nm (Table 3). The porous structure of the PAni hydrogel allows immobilization of concentrated enzyme solutions. Since water-soluble molecules can penetrate through the hydrogel, the PtNPs preserve their ability to catalyze glucose oxidation.

2.4. Co-Precipitation

Co-precipitation is a fast method for the synthesis of different types of nanocatalysts. Co-precipitation is an excellent choice when higher purity and better stoichiometric control are required [46]. Xuan Cai et al. [47] prepared two kinds of nanomaterials by applying the co-precipitation synthesis method: carbon spheres (Mn-MPSA-HCS) and hollow carbon cubic materials (Mn-MPSA-HCC) with a size of 100 to 200 nm, respectively, for $O_2^{\bullet-}$ sensing (Table 3). Physicochemical characterization of the obtained nanocatalyst demonstrates that the $Mn_x(PO_4)_y$ monolayer was homogeneously dispersed on the surface of the carbon structures without visible size or morphologies, thus providing numerous active sites for reaction of analytes. The proposed method can be adapted as a universal strategy for fabricating transition metal phosphates with all kinds of shapes and sizes for different applications and is particularly promising for biosensing. In another example, Dashtestani et al. [48] used a combination of two methods for nanocomposite synthesis: chemical reduction of $HAuCl_4$ and co-precipitation of the obtained gold nanoparticles (GNPs) with the copper(II) complex of cysteine (GNP/Cu-Cys). The combination of GNPs and Cu-Cys complex increased the electrochemical signal toward $O_2^{\bullet}$ (Table 3).

The sensitive electrochemical biosensor based on dual aptamers was proposed for detection of cardiac troponin I (cTnI). The biosensor included DNA nanotetrahedron (NTH) capture probes and multifunctional hybrid nanoprobes. First, the NTH-based Tro4 aptamer probes were immobilized on a screen-printed gold electrode (SPGE) surface through the Au–S bond. Then, the hybrid nanoprobes were prepared using magnetic Fe_3O_4 NPs as nanocarriers for immobilization of a cTnI-specific Tro6 aptamer, horseradish peroxidase (PO), PO-mimicking Au@Pt nanozymes and G-quadruplex/hemin DNAzyme [49]. The constructed sensor exhibited a wide linear concentration range (10 pg·mL^{-1} to 100 ng·mL^{-1}) and a low LOD (7.5 pg·mL^{-1}) for cTnI (Table 3).

2.5. Polymerization and Polycondensation

NZs can be obtained either by using insoluble polymers or by cross-linking of a soluble polymer [50]. Mesoporous SiO_2–L-lysine hybrid nanodisks were synthesized by Han et al. [51] via hydrolyte polycondensation of tetraethylorthosilicate in the presence of CTMB and L-lysine. The prepared hybrid nanodisks have a high surface area (570 m^2·g^{-1}) and ordered mesopores with a size of about 2.9 nm. The obtained hybrid nanodisks possessed excellent biocompatibility to L-lysine. An electrochemical biosensor for superoxide anions ($O_2^{\bullet-}$) was constructed based on this result.

In another example, Santhosh et al. [52] synthesized composite core-shell nanofibers consisting of gold NPs on poly(methylmethacrylate) (PMMA) by the combination of an electrospinning technique and in situ polymerization of aniline. The average diameter of the PMMA fibers was 400–500 nm. The surface of the fibers was fairly smooth and randomly oriented. The proposed core-shell fibers were used for electrochemical detection of the superoxide anion ($O_2^{\bullet-}$).

2.6. Electrochemical Deposition

Electrochemical deposition is a low-cost method for obtaining metal nanocatalysts. However, it is usually used less often than chemical reduction methods [53]. The process is simple and includes an immersion of a conductive surface into a solution containing ions of the material to be deposited and application of a voltage across the solid/electrolyte interface. In the course of this procedure a charge transfer reaction causes the film deposition [54]. The disadvantage of this method is in impossibility of controlling the morphology. However, it has certain advantages: (1) short synthesis time; (2) absence of chemical reductants or oxidants; (3) absence of undesired byproducts [55]. Electrodeposition is applied using different electrochemical techniques: cyclic voltammetry, potential step deposition method and double-pulse deposition [56]. The particle size can be controlled by adjusting the current density or applied potential and the electrolysis time [57]. Electrodeposition is used for synthesis of nanostructural materials with and without templates. Templates utilized for electrodeposition include porous membranes with a 1D-channel, liquid crystal materials and surfactants [58].

Table 3. Preparation and properties of nanocatalysts and corresponding electrochemical biosensors.

Mimetic Enzyme	Enzyme-Like Activity	Preparation Method	Pore Diameter, nm	Shape	Particle Size, nm	BET, $m^2 \cdot g^{-1}$	Potential, V	Detection	Linear Range, µM	LOD, µM	References
Au/Co@HNCF	Urease	Thermal	1.2	Dodeca hedron	300–400	7.88	+0.3	Uric acid	0.1–2500	0.023	[10]
Co$_{0.5}$Ni$_{0.5}$Fe$_2$O$_4$	PO	Solvothermal		Sphere	70–130		+0.5	H_2O_2	0.01–1000	0.001	[25]
SOD/mesoporous SiO$_2$-(L)-lysine	SOD	Hydrolyte polycondensation	3	disk	40–70	570	0.05		3.11–177	800	[27]
SOD/PMMA/PANI-Au	SOD	Electrospinning/polymerization		fiber	400–500		0.3		0.5–2.4	300	[28]
CuO-NZs	GOx	Thermal	2.5	Sphere	~45	20.16	+0.55	Glucose	5–600	0.59	[28]
GOx/PtNP/PAni/Pt	PO	Sol-gel method		Sphere/fiber	2/100		+0.56	Cholesterol	1–15	0.43	[28]
HCC/SPCE		Solid statereaction		Hollow cubic	100–150		−0.5	Glucose	10–8000	0.7	[43]
PCC/SPCE	SOD	Thermal		hexahedral				Superoxide-anion O_2^-	0–192	0.207	[30]
									192–912	0.140	
									0–112		
									112–1152		
Graphene paper/AuNPs	GOx	Laser		Sphere	10–150		+0.17	Glucose	20–8000	2.5	
		Thermal		Plate	200–400		+0.5	Fructose	40–4000	≥20	[32]
							+0.19	Glucose	15–8000	2.5	
							+0.4	Fructose	5–4000		
PCC/SPCE	Ox	Thermal		Sphere	20 ± 3			Glucose	10–$1 \cdot 10^4$		[37]
	SOD						+0.4	P. aeruginosa	60.0–6.0×10^7 CFU/mL	60.0 CFU/mL	[38]
AgNP/NCF/GCE	CAT, PO	Thermal/Electrodeposition	~2	sphere/fiber	15–20/5–8 µm	2.3	−0.5	Oxygen reduction	$6.96 \cdot 10^{-11}$–72	$1.66 \cdot 10^{-4}$	[38]
	Ox			Sphere	8.4		0.04–1.0				[39]
Pt@PMOF(Fe)		Chemical reduction/Electrodeposition		Ellipsoid	300		−0.45	Glucose	100–10,000	6	[40]
His@AuNCs/RGO-GCE		Chemical reduction		Sheet	1.72		+0.8	Nitrite	1.0–7000	0.5	[41]
GOx/PtNP/PAni/Pt		Sol-gel method		Sphere/fiber	2/100		+0.56	Glucose	10–8000	0.7	[43]
Mn-MPSA-HCC	GOx	Co-precipitation		Hollow cubic	100–200		+0.75		0–1260	0.001	[47]
Mn-MPSA-HCS				Hollow sphere							
AuNPs/Cu-Cys		Chemical reduction/Co-precipitation		sphere	77		0.25		3.1–326	2.8	[48]
Fe$_3$O$_4$/Au@Pt-HRP-DNAzyme-Tnő	PO	Co-precipitation		Sphere	158		−0.25		$4.2 \cdot 10^{-13}$–$4.2 \cdot 10^{-9}$	$3.1 \cdot 10^{-15}$	[49]
GCE/MWCNTs-Av/RuNPs/biot-GOx	PO	Drop-coating/Electrodeposition		Tubes/sphere	1000		−0.05	Glucose	20–1230	3.3	[59]
nPtRu/AO		Electrodeposition		Sphere	5–150		−0.1	Ethanol	25–200	2.5	[60]
nPtRu/AMO								Methylamine	20–600	3	

Gallay et al. [59] prepared a hybrid nanocomposite using the electrochemical electrodeposition of RuNPs on the surface of avidin-functionalized multiwalled carbon nanotubes by applying a potential of −0.600 V for 15 s in a 50 ppm ruthenium solution under stirring. The nanohybrid-electrochemical interfaces had excellent PO-like properties toward H_2O_2 (Table 3).

A simple method for the electrochemical detection of methylamine and ethanol in real samples of food and alcoholic beverages using PtRu NZ as artificial PO was reported recently [60]. PtRuNPs were synthesized on the surface of a graphite electrode (nPtRu/GE) by electrodeposition of 1 mM $RuCl_3$ and a H_2PtCl_6-containing solution using the method of cyclic voltammetry in the range of −1000 to +1000 mV with a scan rate of 50 mV·min^{-1} during 10 cycles. The resultant nPtRu/GE turned dark brown-golden due to the formation of nPtRu. Morphological properties of the PtRu-film were studied. It was shown that the average thickness of the deposited layer was 60 nm and it was proven that the obtained layer was nanosized.

Electrosynthesis of AgNP/NCF/GCE was performed by a combination of thermal reduction and electrodeposition methods through thermal synthesis of an electroconductive nitrogen-doped cotton carbon fiber composite (NCF) followed by electrodeposition of AgNPs onto NCF [61]. The developed AgNP/NCF/GCE electrode exhibited outstanding performance toward $O_2^{•-}$, with a wide linear range and a super-low detection limit.

As mentioned, the applied nanocatalyst synthesis methods can influence their structural and physical properties [25,62–64]. The effect of different synthesis methods on structural properties of ENCs with electrocatalytic interfaces are presented in Table 3.

3. Catalytic Performance of Nanozymes

Development of new nanocatalysts, including NZs, with higher catalytic activity expands their applications in bioanalytics. As a rule, the activity of NZs is lower than that of enzymes and their catalytic performance strongly depends on: (1) the method of synthesis (pH, reaction time and temperature) [25]; the composition of the nanomaterial [16,65]; the shape, size, dispersity and final morphology [66]; the mass ratio of the components in the nanocomposite [67]; and surface functionalization [40,68–70].

The catalytic performance of NZs, as well as their efficiency, is quantitatively estimated by kinetic properties: K_M, V_{max} (at a defined nanocatalyst concentration), k_{cat}, k_{cat}/K_M ratio, IC_{50} and morphological characteristics: specific BET surface areas, pore volume, pore diameter, crystallite size.

The enzyme mimics follow Michaelis–Menten kinetics, which is similar to natural enzymes. It has been found that the activity of NZs is strongly dependent on the pH, temperature and substrate concentration. The rate of the reaction is generally determined by the reaction extent as a function of time. It is known that K_M and k_{cat} are two key parameters for quantifying the catalytic ability of an enzyme, where K_M characterizes the affinity of the enzyme to its substrate. The main manifestation is that low K_M values reflect high affinity of the enzyme for the substrate. V_{max} describes the reactivity of the enzyme (at a fixed concentration) when saturated with the substrate. K_M and k_{cat} are therefore among the important reference standards for judging the superiority of an enzyme vs. a NZ [27]. Different substrates (chromogenic, fluorogenic or chemiluminescent) are used for experimental detection of the enzyme-like activity of NZs. It was shown that His@AuNCs/reduced graphene oxide (His@AuNCs/RGO) exhibits phenol oxidase mimic activity and possesses a low K_M value (0.031 mM) with a V_{max} value of 6.55×10^{-8} M·s^{-1} at a nanocatalyst concentration of 8.28 mg·L^{-1}. Of the other four synthesized types of AuNCs, the best catalytic performance for oxidizing TMB was shown for the His@AuNCs/RGO variant. It was assumed that His@AuNCs react via enzyme-like catalytic centers, when substrates and active sites of His@AuNCs are located on the RGO sheets. The "co-interaction" between the His@AuNCs and RGO increases the speed of electron transfer from TMB to oxygen, leading to enhanced catalytic activity [41].

In another example, the catalytic steady-state kinetics for $CoFe_2O_4$ CF300 was estimated using TMB and H_2O_2 as substrates [71]. It was shown that the PO-like reaction followed a Michaelis-Menten kinetics toward both substrates. The K_M^{app} for TMB of the CF300 is 0.387 mM, which is practically the same as reported for PO (0.434 mM). The K_M for H_2O_2 of the CF300 is 8.89 mM, which is two-fold higher than for PO (3.7 mM), showing that CF300 has a lower affinity for H_2O_2.

Although in most cases the catalytic efficiency of NZs is lower compared to natural analogues, it was reported that some NZs can compete with the enzymes. For example, the highest values for any NZ reported to date belong to $MnFe_2O_4$ with a nanooctahedron morphology (see Introduction) [14,15].

Three-dimensional nanomaterials, containing confined single atoms, possess increased catalytic activity due to mutual effects of unique geometric and electronic structures of the matrix and intrinsic catalytic activity of confined atoms [65]. Shackery and coworkers [72] recently prepared a glucose oxidase-like catalyst based on cobalt hydroxide [$Co(OH)_2$] nanorods on a 3D-graphene network using the chemical deposition method. The obtained nanocomposite was used for glucose detection. It was demonstrated earlier that the redox reaction of $Co(OH)_2$ at the graphene surface is a diffusion-controlled electrochemical process [73]. Oxidation of $Co(OH)_2$ to CoOOH is reflected by the peak at ~0.445 V (vs. Ag/AgCl), and the reverse process corresponds to the cathodic peak at around ~−0.08 V (vs. Ag/AgCl). A possible electrochemical reaction is as follows:

$$Co(OH)_2 + OH^- \underset{\text{Oxidoreduction}}{\longleftrightarrow} CoOOH + H_2O + e^-$$

The constructed amperometric sensor demonstrated a high sensitivity for glucose (36,900 $A \cdot M^{-1} \cdot M^{-2}$) and a very low LOD (16 nM).

In another example, it was shown that octahedral cuprous oxide (Cu_2O) in combination with carbon quantum dots (CQDs) could be an efficient electrocatalyst for the detection of glucose [74]. The CQDs/octahedral Cu_2O showed higher electrocatalysis for glucose oxidation and H_2O_2 reduction than the octahedral Cu_2O. The cyclic voltammograms showed an oxidation peak at +0.6 V reflecting the conversion of Cu(II) to Cu(III) and showing high electrocatalytic ability for oxidation of glucose by the CQDs/octahedral Cu_2O [75]. Equations (1)–(5) of the electrocatalytic oxidation reaction are presented in Figure 3.

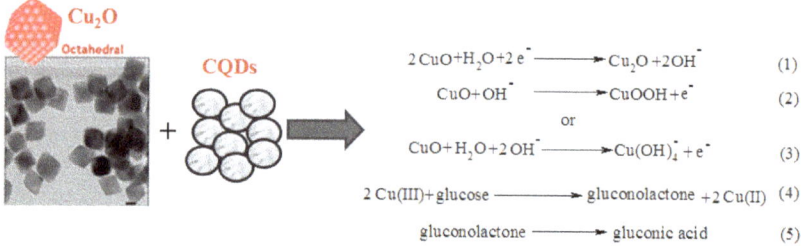

Figure 3. Carbon quantum dots/octahedral Cu_2O nanocomposites for non-enzymatic glucose assay.

Pirmohamed et al. [76] showed that nanoceria (CeNPs) is a redox active catalyst which possesses two oxidation states (Ce^{+3}/Ce^{+4}) and contains transportable lattice oxygen located on its surface which facilitates the interchangeable conversion of Ce^{+4} and Ce^{+3}. Hayat et al. [77] found that the dual oxidation state and high mobility of surface oxygen are responsible for the oxidase-like activity of CeNPs toward phenolic compounds. It was shown that CeNPs exhibits the lowest K_M^{app} value: 0.25 µM for dopamine and 180 µM for catechol, among the other nanomaterials (Table 10). For comparison, the K_M^{app} values of the optimized enzymatic reaction using tyrosinase under the same reaction conditions were 0.3 µM and 200 µM for dopamine and catechol, respectively.

Another quantitative characteristic that evaluates the effectiveness of NZs is the half maximal inhibitory concentration (IC_{50}), reflecting the ability of a substance to inhibit certain biological or biochemical functions. A nanocomposite based on gold NPs and the copper(II) complex of cysteine

(GNPs/Cu-Cys) was synthesized by Dashtestani and colleagues [48]. The SOD mimetic activity of a Cu-Cys and GNPs/Cu-Cys nanocomposite was determined using the pyrogallol autoxidation inhibition assay. The half-maximal inhibitory concentration of the nanocomposite was 0.3 µg·mL^{-1}, which is 3-fold higher than that of the native enzyme. However, the authors did not indicate how the concentrations of both catalysts were normalized. The IC$_{50}$ values of the compounds indicate that all individual components related to the GNPs/Cu-Cys nanocomposite exhibited SOD mimic activity.

The enzyme-like activity can be controlled by the shape, size, crystallinity and final morphology of the NZs [78–80]. Liu et al. [79] showed that different nanocrystalline shapes had different PO-like activities. The activities of the NZs in descending order according to shape, can be presented in the following order: cluster spheres > triangular plates > octahedral. The activity depends on the exposure of catalytically active iron atoms or crystal planes. Two kinds of nanomaterials for $O_2^{\bullet-}$ sensing were prepared: hollow carbon spheres and hollow carbon cubic nanomaterials with a size of 100 to 200 nm, respectively [47]. The electrochemical sensor prepared using the hollow carbon sphere NZ exhibited an extremely low detection limit of 1.25 nM and was successfully employed in the dynamic monitoring of $O_2^{\bullet-}$ released from HeLa cells (Table 3). In another example, a detailed study for investigating the effects of particle size and morphology on the PO-like catalytic activity of magnetic cobalt ferrite ($CoFe_2O_4$) was performed [81]. $CoFe_2O_4$ NZs having different shapes (near corner-grown cubic, near cubic, polyhedron and star-like) were synthesized when varying the amounts of iron and cobalt acetylacetonates precursors and changing the reaction temperature. To increase the suspensibility of NPs in water solution, the obtained $CoFe_2O_4$ were modified with PEG-3,4-dihydroxybenzylamine. The catalytic activity was structure dependent (in descending order according to a shape): spherical > near corner-grown cubic) > star like > near cubic (> nanopolyhedrons. The kinetic studies of the obtained $CoFe_2O_4$ showed that star-like shaped $CoFe_2O_4$NPs with 4 nm-sized had the highest affinity for TMB and H_2O_2 compared to the other NPs obtained in the study [81].

The affinity of the NZs to substrates can be changed by a procedure of surface functionalization [82]. Ling et al. recently synthesized a new artificial nanocatalyst based on metalloporphyrinic metal organic frameworks (PMOF(Fe)) and PtNPs [83]. It was shown that Pt@PMOF(Fe)NPs showed high activity toward the oxidation of o-phenylenediamine (OPD) as a chromogenic substrate and H_2O_2 as the oxidant. This mechanism can be summarized in the following reaction:

$$\text{o-phenylenediamine} \xrightarrow[H_2O_2]{PMOF(Fe)/Pt@PMOF(Fe)} \text{2,3-diaminophenazine}$$

It was found that when Pt@PMOF(Fe) was mixed with H_2O_2 and OPD, the Fe center provided an electron to PtNPs forming high-valence Fe, due to the synergistic effects between PMOF and PtNPs, and the compound, which included Fe(IV) = O and a porphyrin π cation radical, generated via the reaction between Fe(III) and H_2O_2. OPD then was easily oxidized. Thus, PtNPs enhanced the PO-like activity of Pt@PMOF(Fe) [40].

The mass ratio (Ni-MOFs to Fe-MOFs) is one of the crucial factors influencing the catalytic activity of nanocatalysts. The catalytic performance of Fe-Ni-n (where n is the mass ratios) was explored by optimizing the mass ratios (Ni-MOFs@Fe-MOFs). Due to differences in the morphology and specific surface area of Ni-MOFs and Fe-MOFs, the optimal load of Ni-MOF nanosheets on octahedral Fe-MOFs was realized with the increase in n. When n = 2, the nanosheets and octahedron achieve relatively sufficient mixing. However, the n value is too small for the nanosheets to adequately cover every surface of the octahedrons. On the other hand, if the n is too large, there would be extra nanosheets that are not coated on the octahedrons [67].

It was shown that the activity of nanocatalysts is strongly dependent on their composition. PtRuNP alloy with a suitable degree of alloying (e.g., $Pt_{90}Ru_{10}$) can mimic oxidase (ferroxidase), PO,

SOD and catalase. These enzyme-like activities correspond to the redox enzymes: oxidase and PO which catalyze oxygen and H_2O_2 reduction, respectively, and SOD and catalase which are responsible for disproportionation of superoxide and H_2O_2 decomposition, respectively. It was found that varying the composition of PtRu alloys affects their ability to facilitate electron transfer, since these reactions involve a transfer of electrons. Such dependence on the alloy composition was observed for all four enzyme-like reactions [66].

A fast, effective and low-cost method for production of $ZnFe_2O_4$ nanoparticle-decorated ZnO nanofibers using co-electrospinning and the sequence annealing process has recently been proposed [83]. It was shown that the presence of $ZnFe_2O_4$ NPs on the surface of ZnO nanofibers and formation of heterostructures significantly improved the PO-like activity compared with $ZnFe_2O_4$ NPs. Furthermore, the smaller-sized NPs had a higher activity. The obtained results suggest that the surface-to-volume ratio, the composition and the size of the NZs play a critical role in their catalytic activities [84].

4. Nanozymes as Peroxidase Mimetics

Peroxidases (PO, E.C. 1.11.1.7) are oxidoreductases from a variety of sources, including plants, animals and microbes, that contain an iron-porphyrin derivative (heme) in their active site and catalyze the oxidation of diverse organic compounds using H_2O_2 as the electron acceptor. Some of the most popular PO substrates with the highest chromogenic ability are 3,3′,5,5′-tetramethylbenzidine (TMB), 2,2′-azinobis[3-ethylbenzothiazoline-6-sulfonic acid] (ABTS) and o-phenylenediamine (OPD).

In the catalytic process, the Fe of the heme of natural PO provides H_2O_2 dissociation by changing the Fe(III) to the Fe(IV) valence state in an intermediate with high oxidative activity. The enzyme's oxidative activity enables realization of the catalytic cycle. Therefore, if a nanomaterial is able to cause a similar electron transfer, it can be called a NZ with PO-like activity ("nanoperoxidase"). Due to the high reduction potential of POs, they are very promising for use as bioelectrocatalysts in biosensorics, fuel cell technology as well as environmental biotechnology. Despite the benefits of POs, their wide usage is still restricted, due to their fast inactivation in the presence of H_2O_2 under native physiological conditions. Their low thermal and environmental stability reduces the possibilities for their practical application. Screening the highly stable synthetic nanomaterials with PO-like activity for practical application in the different fields of modern technologies thus seems very promising.

In 2007, ferromagnetic NPs became the first reported nanomaterials with PO-like enzymatic activity [12]. Since then, the number of such materials has been growing constantly. These include different material types, such as Nafion-cytochrome c [85], supramolecular complexes of hydrogel [86], nanocomplexes of lanthanides [87–92], Co_3O_4@CeO_2 hybrid microspheres [93], transition metal oxides and their composites [71,94–99], as well as noble metals: Au [100], Fe/Au [101], Au/Pt [102], Pt [103,104], Ru [105], Pt/Ru [60], Pd, Pd@Pt [106], Pd/Pt [107], Pd@γ-Fe_2O_3 [108], etc. In the last decade, special interest in bionanotechnology has focused on the use of NZs based on carbon materials such as fullerenes [109], Prussian blue [110], TiO_2 [111] or Fe_2O_3/Pt-modified [112] multi-walled carbon nanotubes, hemin-graphene hybrid nanosheets [113], Pt/Ru/3D graphene foam [114], graphene oxide [115], Au/Pt/Au-graphene oxide nanosheets [116], Pd-magnetic graphene nanosheets [117], hemin-graphene-Au hybrid [118], IrO_2 [119], Cu-Ag [120] or Fe_2O_3-modified graphene oxide [121] (Figure 4), Co-modified magnetic carbon [122], Co_3O_4 graphene composite [123], carbon nanofibers [124], graphene quantum dots [125–127], graphene nanotubes nickel-/nitrogen-doped and functionalyzed by PtNPs [128], carbon fiber-supported ultrathin CuAl layered double hydroxide nanosheets [129], Fe^{3+}-doped mesoporous carbon nanospheres [130] (Figure 5), and many others.

The kinetic parameters of different artificial POs toward H_2O_2 are presented in Table 4.

Table 4. Comparison of the kinetic parameters of different types of artificial peroxidases and natural enzyme toward H_2O_2 in solution.

Catalyst	Concentration	K_M^{app}, mM	V_{max}, µM·s^{-1}	k_{cat}, s^{-1}	Reference
Fe_3O_4 NPs	11.4×10^{-13} M	154.0	0.098	8.58×10^4	[12]
HRP	2.5×10^{-11} M	3.7	0.087	3.48×10^3	[12]
$CoFe_2O_4$	20 µg·mL^{-1}	8.89	0.019		[71]
CeO_2-MMT	300 µg·mL^{-1}	3.4	0.010		[88]
CeO_2 NPs	300 µg·mL^{-1}	3.18	0.009		[88]
H/WS2-NSs	3.2 µg·mL^{-1}	0.926	0.028		[89]
CeO_2 NPs	40 µg·mL^{-1}	0.28	0.009		[90]
BNNS@CuS	30 µg·mL^{-1}	25	0.125		[92]
$Co_3O_4@CeO_2$	50 µg·mL^{-1}	7.09	0.430		[93]
$Fe_3O_4@Cu@Cu_2O$	50 µg·mL^{-1}	2.3	0.119		[95]
H_2TCPP-γ-Fe_2O_3	18.5 µg·mL^{-1}	21.1	1.3×10^{-3}		[96]
γ-Fe_2O_3 NPs	100 µg·mL^{-1}	157.2	0.013		[97]
$Fe_3O_4@C$ YSNs	20 µg·mL^{-1}	0.035	0.033		[99]
CuO	100 µg·mL^{-1}	440	0.161		[98]
Zn-CuO	100 µg·mL^{-1}	71	0.003		[98]
H_2TCPP-CeO_2	40 µg·mL^{-1}	0.25	0.013		[90]
Pt/CeO_2 NPs	10 µg·mL^{-1}	0.21	0.085		[91]
Pt NCs	1×10^{-4} M	3.07	0.182	1.8×10^{-5}	[104]
Ru NPs	10 µg·mL^{-1}	2.2	0.580		[105]
Pd NPs	5.06×10^{-12} M	4.4	0.065	1.3×10^4	[106]
Pd@Pt NPs	1.9×10^{-12} M	2.23	0.050	2.5×10^4	[106]
Pd@γ-Fe_2O_3	1.35×10^{-6} M	0.25	0.128	9.4×10^{-2}	[108]
$C_{60}[C(COOH)_2]_2$	2×10^{-5} M	~50	0.003	1.6×10^{-4}	[109]
Fe_2O_3/Pt/CNTs	10 µg·mL^{-1}	~0.1	6×10^{-5}		[112]
GO-COOH	40 µg·mL^{-1}	3.99	0.039		[115]
H-rGO-Au	0.5 µg·mL^{-1}	3.1	0.121		[118]
IrO_2/GO	2.4 µg·mL^{-1}	5.19	~0.300		[119]
Cu-Ag/rGO	5 µg·mL^{-1}	8.63	0.070		[120]
GO-Fe_2O_3	~1.25×10^{-1} M	305.0	0.101	8.1×10^{-7}	[121]
MC	10 µg·mL^{-1}	0.74	0.028		[122]
CNFs	5 µg·mL^{-1}	3.0	0.390	7.8×10^{-2} mmole·g^{-1}·s^{-1}	[124]
CQDs	15 µg·mL^{-1}	26.77	0.306		[125]
CQDs		0.49	0.026		[126]
GQDs/CuO	70 µg·mL^{-1}	0.098	0.032		[127]
CF@CuAl-LDH	50 µg·mL^{-1}	0.59	0.003		[129]
Fe^{3+}-MCNs	25 µg·mL^{-1}	161.0	0.007		[130]

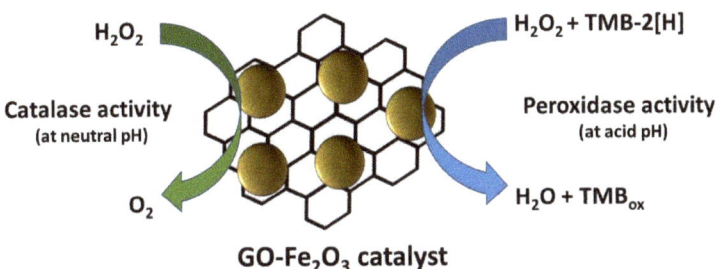

Figure 4. Scheme of catalytic action of graphene oxide-based Fe_2O_3 enzyme-like mimetic of peroxidase and catalase [121] (modified).

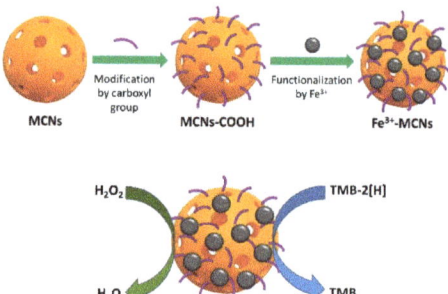

Figure 5. Schematic representation of the synthesis of Fe^{3+}-decorated mesoporous carbon nanospheres (Fe^{3+}-MCNs) and their peroxidase-mimicking catalytic activity [130] (modified).

H_2O_2 molecules are known to participate in numerous biological processes. Its analysis is routine for all newly synthesized PO mimics. Moreover, many natural enzymes (oxidases) produce H_2O_2 as a byproduct of their enzymatic reaction, so that detection of the target substrate can be performed bymeasuring generation of H_2O_2. PO-like NZs can be combined with oxidases of protein nature expanding the detection range of analytes using electrochemical biosensing. In the last decade, numerous hybrid amperometric biosensors based on PO mimetics coupled with natural enzymes, mostly oxidases (e.g., glucose oxidase, alcohol oxidase, and lactate oxidase), have been described. Because of the high pull of researches in this field, we singled out only recently described NZ-based sensors for H_2O_2 detection, without describing the hybrid biosensors. All analyzed PO-sensitive sensors belong to electrochemical ones with voltammetric or amperometric detection. Their main operational characteristics are presented in Table 5.

Table 5. The main operational characteristics of the recently described electrochemical sensors based on H_2O_2-sensitive nanozymes.

Catalyst/Electrode Type	Working Potential, V	Linearity, mM	LOD, µM	Sensitivity, $A \cdot M^{-1} \cdot m^{-2}$	Reference
Fe_3O_4/3D GNCs//GCE	−0.2	0.0008–0.33	0.08	2742	[131]
α-MnO_2//GCE	−0.4	0.0002–0.1	0.08	5.5	[132]
AuNBP/MWCNTs//GCE	−0.5	0.005–47.3	1.50	1706	[133]
Fer/rGO-Pt//GCE	+0.1	0.0075–4.27	~0.38	3400	[134]
rGO/Pt-Ag//GCE	−0.05	0.005–1.5	0.04	6996	[135]
GBR//GCE	+0.9	0.1–10.0	48.0		[136]
CMC@Pd/Al-LDH//GCE	−0.38	0.001–0.12	0.30	163	[137]
Pt/PANI/MXene//SPCE	+0.3	0.001–7.0	1.00		[138]
CuOx/NiOy//GCE	−0.35	0.0003–9.0	0.09	2711	[139]
GDCh-NiO//RDE	+0.13	0.00001–0.0039	0.0015	1072	[140]
Pt-Pd/MoS_2//GCE	−0.35	0.01–0.08	3.4	764	[141]
N-CNFht//GCE	−0.4	0.01–0.71	0.62	3570	[142]
Cu_2O/PANI/rGOn//GCE	−0.2	0.0008–12.78	0.3	394	[143]
WCC//GCE	−0.4	0.05–1.0	0.006	67	[144]

Prussian blue (PB) is one of the most effective PO mimetics. PB or iron(III) hexacyanoferrate(II) is a member of a well-documented family of synthetic coordination compounds with an extensive 300-year history. It was produced commercially in the past and used as a pigment for paints, lacquers, printing inks and laundry dyes [145,146].

PB and its analogues (PBAs) are cheap, easy to synthesize, environmentally friendly and are prospect for wide applications in different fields, including basic research and industrial purposes [10,147–151] as well as in medicine [24,152–158].

Despite their multifunctionality, the composition of PBAs is quite complicated and tightly depends on a method of synthesis and storage conditions. [146,149–151,159,160]. Insoluble PB can be described by the formula $Fe_4[Fe(CN)_6]_3$. $KFe[Fe(CN)_6]$ corresponds to a colloidal solution of PB [149,151]. The general formula of hexacyanoferrate (HCF) is $M_k[Fe(CN)_6]\cdot xH_2O$, where M is a transition metal (Figure 6). PBAs are usually obtained via various techniques, including chemical [146,149,159–162] and alternative biological methods [163].

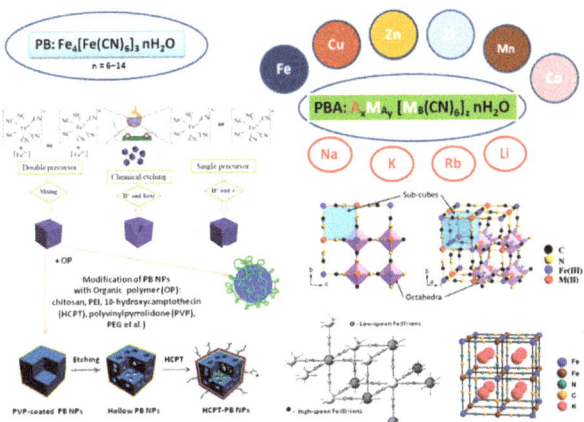

Figure 6. Hexacyanoferrates of transition metals: composition, structure and modification.

PBAs have gained extensive worldwide attention in the last two decades. The charge transfer through two transition metal ions in the complex-compound salt, as well as the nanosize of HCF particles, are the reasons for PBAs' redox activity and super-magnetic properties. Due to these remarkable properties, PBAs are widely applied as NZs in biosensors [146,147,150,164–168] and alternative energy sources [149,160,165,168].

The enzyme-like properties of PBA in solution are difficult to analyze because of the ability of PB and PBAs to simultaneously mimic the activities of several ROS-enzymes (PO, SOD and catalase). Numerous papers (see Table 6) described the ability of PBA-NZ to react with typical ROS-enzyme substrates, such as TMB, ABTS and NADH [169].

Table 6. The main kinetic characteristics of the PBA-based PO-like nanozymes in solution.

PBA	Concentration	Chromo-Gene	K_M^{app}, mM	V_{max}, $\mu M \cdot min^{-1}$	k_{cat}, $nmol \cdot \mu g^{-1} \cdot min^{-1}$	LR, µM	Reference
PB/γ-Fe$_2$O$_3$	20 µg·mL^{-1}	TMB	323.6	1.17	0.059		[12]
VOxBG hydrogel	5 µg·mL^{-1}	TMB	20	0.045	0.009		[161]
PO		TMB	3.7	0.009			[161]
PB-MIL-101 (Fe)	200 µg·mL^{-1}	TMB	0.058	1.32	0.0066	2.4–100	[166]
MoS xNi-Fe		TMB					[169]
PB, soluble form	6 µg·mL^{-1}	TMB	14.7	0.012	0.002		[170]
PS@Au@PB	300 µg·mL^{-1}	TMB	0.17	0.38	0.001		[171]
Au@HMPB (40 °C)	40 µg·mL^{-1}	TMB	88.72	2.50	0.063		[172]
PB	0.2 µM	ABTS	0.028				[173]
	0.74 nM	TMB	11.984	43.2	58.38×10^3 min^{-1}		
PB-Ferritin	0.74 nM	ABTS	0.537	0.36	0.49×10^3 min^{-1}		[174]
PB/Fe$_2$O$_3$	0.31 nM	TMB	323.6	70.2	226.5×10^3 min^{-1}		
MWCNTs-PB			1.33	6.6	0	1–1500	[175]

PB and PBAs are used successfully in optical biosensors as a result of their PO-like properties [12,16,169–175]. The first communications concerning electrochemical reduction of H_2O_2 on PB-modified glassy carbon electrode were done by Itaya and colleagues [164,165]. Karyakin and colleagues published many reports over the last 25 years on the use of PB as an artificial PO in amperometric biosensors [13,150,162,176–181]. In 2000, this group named PB an "artificial PO" [177]. At the same time, a large number of other scientific groups, especially from China, worked hard on this problem [148,154,156–159,166,168–174].

PBAs demonstrated intrinsic PO-like activity when coupled with carbon, graphene, natural polysaccharides or synthetic polymers. This property was successfully used in electrochemical (bio)sensors (Figure 7).

Table 7 presents selected examples of PBA application as amperometric chemo-sensors on H_2O_2 and the analytical characteristics of the developed sensors. The main peculiarities of these catalysts are high stability, sensitivity and selectivity to H_2O_2 in extra-wide linear ranges. PBAs are more stable in neutral and alkaline solutions compared to PB [181], and may be used in a physiologically compatible medium containing biorecognition elements. Some PBAs show good selectivity against easily oxidizable interfering species, for example organic acids, although the electrocatalytic activity in H_2O_2 reduction is similar to that of PB [146,147,149].

Table 7. Amperometric H_2O_2-sensitive sensors based on PBA as a PO-like nanozyme.

Electrode	Working Potential, V	Nanozyme	Sensitivity, $A \cdot M^{-1} \cdot m^{-2}$	LOD, µM	Linear Range, µM	Reference
GCE	−0.05	PB/BG	2850		4–83,000	[167]
		AuNPs-PB/BG	11243		9.2–8100	
GCE	0.65	MnPBA	1472	3	3–8610	[168]
GCE	−0.3	MoS xNi-Fe PBA			0.1–2500	[169]
GCE	0.18	PB	10,000/20,000	1	1–5000	[176]
GCE	0.05	PB	6000	0.1	0.1–100	[177]
GCE/	0.0	Ni-FePBA	18,000		up to 100	[178]
DBD	−0.05	PB	2100			
DBD	−0.05	Ni-FePBA	1500	0.5	0.5–1000	[179]
GE	−0.05	PB/NZ	4500			
Planar screen-printed		Ni-PB	3500		0.1–1000	[180]
Carbon planar screen-printed	0.0	PB/film PB/NZ	6500		up to 500	[181]
			8500		up to 500	
GCE	0.0	PNAANI-PB	5073	0.07	1–1000	[182]
Graphene		CoPBA		0.007	5–1200	[183]
Graphene nanocomposite		CoPBA		0.1	0.6–380	[184]
Graphite-string	0.05		6413		30–1000	[185]
Graphite paste		Cu-FePBA	2030	0.2	0.5–1000	[186]
		Ni-FePBA	1130	2	2–1000	
Nanoporous gold film		PB	7080	0.22	1–17,000	[187]
GCE		Ni-FePBA	0.192 A/M	1		[188]
GCE	0.33	Ni-Fe PBA-HNCs	361.3	0.291	0.1–20,000	[189]
GE		Thionine-NiPBA		0.557	1.67–1110	[190]
CNE		PB	500,000		10–3000	[191]

Electrochemical biosensors have long been used as an efficient way for quantitative detection of different analytes (biomarkers) of interest. PBA-based NZs, being PO-mimetics, may comprise a promising platform for the construction of biosensors that can be applied in clinical diagnostics, theranostics, for control of therapy, cell/tissue growth and proliferation [5–11,146–148,152,170,192].

The main drawback of the many H_2O_2-sensitive NZ-based electrochemical sensors is the application of rather high or low working potentials. As a result, they suffer from non-selectivity. It is known that H_2O_2 is prone to direct auto-oxidation on electroactive surfaces (e.g., Pt) at an operational potential above +0.4 V or auto-reduction at −0.4 V or less vs. Ag/AgCl. Moreover, the real samples consisted mostly of organic compounds which are easily co-oxidized/co-reduced at the above-mentioned potentials (e.g., ascorbic or uric acids, neurotransmitters, pigments, drugs, and even glucose), consequently resulting in overestimation of the target analytes. There are only a few possibilities for decreasing this interfering impact, for example by restricting the access of the potentially interfering compounds by perm-selective membranes. The weak point of this approach is the additional diffusion limitations, resulting in deterioration of the sensor's operational parameters. A much more successful strategy is the screening of new NZ types that work at operating potentials close to zero (0) V vs. Ag/AgCl [193].

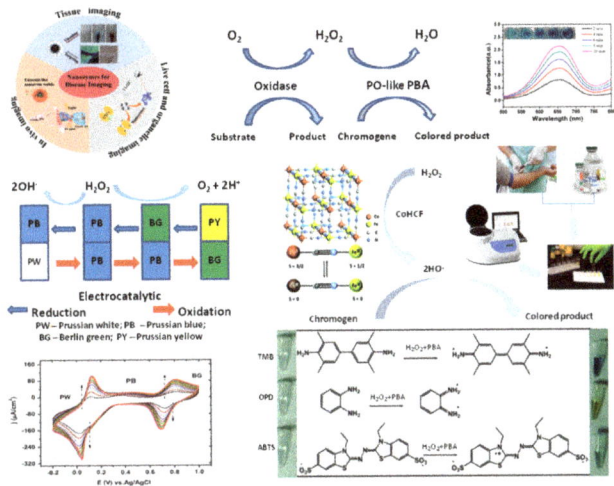

Figure 7. Peroxidase-like catalytic activity of Prussian blue analogues (PBAs) as a platform for the development of amperometric and colorimetric (bio)sensors.

5. Nanooxidases

Natural oxidases, which belong to the EC class 1.1.3, catalyze the oxidation of many substrates that contain the CH-OH group (electron donors). They use molecular oxygen as the electron acceptor, and form hydrogen peroxide as a byproduct [194]. Generally, a specific oxidase name is given according to the target oxidized substrate.

For example, glucose oxidase (GOx), alcohol oxidase (AOx), lactate oxidase (LOx), cholesterol oxidase (COx), etc. are the specific oxidases that catalyze the oxidation of glucose, ethanol, lactate, and cholesterol, respectively:

$$\text{Glucose} + O_2 \rightarrow \text{Gluconic acid} + H_2O_2 \text{ (GOx)}$$

$$\text{Alcohol} + O_2 \rightarrow \text{Aldehyde} + H_2O_2 \text{ (AOx)}$$

$$\text{Lactate} + O_2 \rightarrow \text{Pyruvate} + H_2O_2 \text{ (LOx)}$$

$$\text{Cholesterol} + O_2 \rightarrow \text{Cholestenone} + H_2O_2 \text{ (COx)}$$

Due to the generation of H_2O_2, natural oxidases and oxidase-like NZs can efficiently oxidize the colorless substrates (in the presence of PO) into corresponding colored products, which make them promising tools for analysis of a number of biological molecules. Many forms of nanomaterials that exhibit nonenzymatic oxidase-like catalytic activities have been reported in recent decades: ferrous metals, their oxides or bimetallic/alloys [195–198], nanocomplexes of lanthanides [199–201], transition metals [105,202,203], as well as noble metals and their combinations: (IrPd)/Au [204], PtNPs [205,206], Au@Pt [207], Au/Pt/Ag [208], Au-Pd NPs [209], Au/TiO$_2$ [210] and many others [211]. Although the metallic GOx-like NPs yielded >99% gluconic acid, these materials suffered from high adsorption of reaction products resulting in their oxidation and inactivation [212]. The most extensively studied nanooxidases used for selective oxidation of glucose to gluconic acid are AuNPs [4,210,213–222] (see

Table 8). Gold NPs are much more resistant to O_2 compared with Pt and PdNPs and their reaction products have a lower affinity to adsorption onto the Au surface, in addition to being more active and selective under mild conditions [212]. Their main drawbacks are strong dependence of catalytic activity on a type of Au surface and nanoparticle size affected by of sintering [223]. Carbon-based materials and metal/carbon composites have also demonstrated catalytic oxidation of glucose in the presence of O_2 [224–227] as an alternative to metallic NPs.

Table 8. Comparison of the kinetic parameters of different types of artificial GOx and natural enzyme toward glucose in solution.

Catalyst	Concentration	K_M^{app}, mM	V_{max}, µM·s^{-1}	k_{cat}, s^{-1}	Reference
Au/MCM-41	0.025 mg·mL^{-1}	55.2	18.0	14.2	[212]
AuNPs	34 × 10^{-9} M	6.97	0.63	18.52	[213]
GOx	34 × 10^{-9} M	5.0	0.69	9.7	[213]
AuNPs	0.05 mg·mL^{-1}	0.41	0.10		[221]
AuNPs-MIP	0.05 mg·mL^{-1}	0.18	0.42	3.76 × 10^{-7} mmole·g^{-1}·s^{-1}	[221]
AuNPs-PFOP	0.05 mg·mL^{-1}	0.09	0.58	5.21 × 10^{-7} mmole·g^{-1}·s^{-1}	[221]
AuNP	2 × 10^{-9} M			4.5 × 10^7	[222]
β-CD@AuNPs		9.60	1.80 × 10^{-2}		[228]
EMSN-AuNPs	562.0 mg·mL^{-1}	26.2	0.53		[229]
MnO$_2$NFs	0.02 mg·mL^{-1}	21	0.43		[230]

Unfortunately, the catalytic parameters were not determined for many of the above-mentioned nanooxidases. It is therefore not possible to compare the efficiency of GOx-mimicking nanocomposites.

Ortega-Liebana and coauthors [212] (Figure 8) described an Au-silica nanohybrid Au-MCM-41 with a determined K_M^{app} of ~55 mM, which is significantly higher compared with the values reported for free AuNZ (K_M^{app} ~7 mM) [213], polymer-coated gold-based mimics (K_M ~0.4 mM) [221], gold-supported mimicking systems (K_M^{app} ~27 mM) [229], as well as the natural GOx (K_M^{app} ~5 mM) [213]. Au-MCM-41 NZ showed a slightly lower affinity towards glucose as a substrate. Nevertheless, the corresponding catalytic constant k_{cat}, a true catalytic parameter for comparison determined as the V_{max}/concentration of catalyst ratio, is close to the value reported for natural GOx (k_{cat} ~14.2 s^{-1} vs. k_{cat} ~9.7 s^{-1}, respectively) and similar to freestanding Au NPs (k_{cat} ~18.5 s^{-1}) as reported by Luo and coauthors [213]. The Au-MCM-41 NZ showed a good response, possibly due to a good homogeneous distribution of active sites of Au in the mesoporous carrier, which improved their availability [212].

Some of the described NZs are characterized by double enzyme-like properties. They are named "tandem NZs" (nanomaterials with tandem enzyme-like characteristics). Ma and coauthors [222] investigated catalytic properties of single AuNPs and Ag-Au hybrid NPs as possible GOx and PO mimetics. The electrochemical experiments demonstrated that a high turnover of NZs was obtained from individual catalytic elements compared to results from ensemble-averaged measurements as a classic approach. The authors concluded that the unique increasing catalytic activity of single NZ supports is due to the high accessible surface area of monodispersed NPs and high activities of carbon-supported NP during single particle collisions on a carbon ultra-microelectrode. It was proposed as a new method for accurate characterization of NZs' catalytic activities that opens further prospects for the design of highly efficient catalytic nanomaterials. Kou and coauthors [228] described the synthesis of β-CD@AuNPs that are characterized by simultaneous GOx-like and PO-like activities (Table 8). Han and coauthors [230] desribed the synthesis of 2D MnO$_2$-based NPs with dual enzyme activities in a similar pH range. Moreover, a one-pot nonenzymatic approach was proposed for the colorimetric analysis of glucose, where the oxidation of glucose and the colorimetric detection of H$_2$O$_2$ are conducted simultaneously as a result of the single NZ (MnO$_2$ NPs) catalysis. This method is characterised by a high sensitivity, low LOD and a short time of analysis, because of the proximity effect and in situ reaction [230]. However, the weak point of the most widely described oxidase mimics is their nonselective oxidation of a number of substrates, contrary to natural enzymes. Improving the selectivity of oxidase-like NZs is therefore a great challenge that needs to be solved before their successful application in analytical technologies.

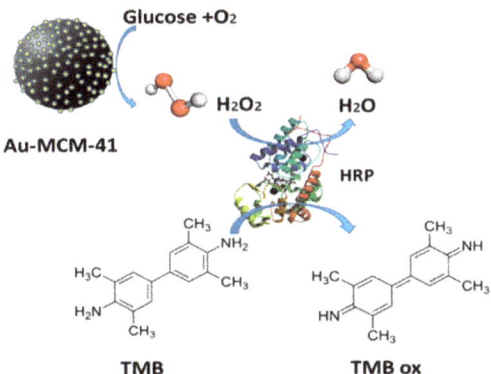

Figure 8. Determination of H_2O_2 as a product generated in the reaction of glucose oxidation [212] (modified).

Summarizing the above-mentioned data on nanocomposites with GOx-like activity, it is worth emphasizing that many of them (see Table 8) have a very low activity that does not enable defining them as NZs. There is an enormous difference in catalytic efficiency of the described GOx-mimetics: k_{cat} values differ more than 107-fold. Although according to BRENDA data this parameter is also very variable for natural GOs-from 0.005 to 2300 s^{-1} [231], it is obvious that improving the catalytic efficiency of synthetic NZs is a very important challenge. Nevertheless, the most effective nanooxidases (including those presented in Table 9) can be a good basis for the creation of non-enzymatic sensors for glucose analysis [72,222–227].

Table 9. Comparison of the main operational properties of glucose-sensitive amperometric sensors based on GOx-like nanozymes.

Electrode Material	Potential, V	Linearity, mM	LOD, µM	Sensitivity, $A \cdot M^{-1} \cdot m^{-2}$	Selectivity	Reference
3DG/Co(OH)$_2$	+0.6	0.1–10	0.016	36900	AA, UA, fructose, lactose, urea	[72]
Au/MWCNTs	+0.15	0.01–36.0	3.0	1012	DA, UA, AA, fructose, saccharose, maltose, Ca^{2+}, Cl$^-$	[133]
PtNi-ERGO	−0.35	0.01–35	10	204.2	AA, UA, urea, fructose	[224]
CuO NW/CF	+0.35	0.001–18.8	0.3	22174	AA, UA, DA, lactose, sucrose, maltose	[225]
Octahedral Cu$_2$O	+0.6	0.3–4.1	128	2410	AA, UA, DA, NaCl	[225]
CQDs/octahedral Cu$_2$O	+0.6	0.02–4.3	8.4	2980	AA, UA, DA, NaCl	[225]
CuNWs/rGO	+0.58	0.01–11	0.2	16250	AA, UA, DA, AP, fructose, sucrose	[226]
AKCN	+0.7		0.8		fructose, lactose, maltose	[227]
β-CD@AuNPs	−0.05				Cu^{2+}, Al^{3+}, UA, AA, guanine, guanosine	[228]
Ni-Pd/Si-MCP	−0.1			0.081 A·M^{-1}	AA	[232]
Pd-Pt core-shell NCs	−0.05	0.3–6.8	41.1	1700		[233]
Pt NPs	−0.05	0.3–5.2	91.8	457		[233]
Cu$_2$O/GNs	+0.1	0.3–3.3	3.3	2850		[234]
Cu$_2$O nanocubes	+0.1		5.9	2000		[234]
ITO/PbS/SiO$_2$/AuNPs	−0.2	0.001–1	0.46		AA, UA, L-Cys, lactose, maltose, sucrose	[235]

Many sensors for glucose detection are based on its electrochemical oxidation directly on a nanocatalyst-covered electrode have been reported (Table 9). A silicon-based amperometric non-enzymatic sensor for the glucose determination (nEGS) was described by Miao and coauthors [232]. Ni–Pd NPs adhered onto a supporter comprising the 3D ordered silicon microchannel plate (MCP) were proposed as sensing materials and were used as an electrode. The 3D structure provided ample space allowed a fast mass transport of ions/gas through the electrolyte/electrode interface, thus causing fast electrochemical reactions. The Ni-Pd/Si-MCP nanocomposite electrode showed strong electrocatalysis of glucose under alkaline conditions. The nanocomposite was characterized by a good selectivity even in the presence of high concentrations of interfering agents, excellent storage stability and reproducibility. Ye and coauthors [233] synthesized and employed heterostructured Pd-Pt core-shell nanocubic materials (NCs) as nEGSs, due to their electrocatalytic activity in glucose oxidation. These core-shell NCs with a large surface area show remarkable GOx catalytic activity and can potentially be applicable as nEGSs. Gao and coauthors [224] developed a PtNi alloy NP-graphene composite and found that the PtNi-ERGO nanocomposite-based nEGS possessed many merits in terms of high selectivity, superior resistance to poisoning, low LOD, rapid response, excellent reproducibility and stability, which outmatches the performance of any other reported Pt-based nEGSs. The nEGS retained 93.2 and 90.5% of its initial sensitivity at 10 and 50 days postpreparation, respectively. The combination of these unique characteristics has enabled the application of this new type of nanoelectrocatalyst-loaded electrodes for analysis of real human samples. Li and coauthors [225] proposed utilizing 3D porous copper foam (CF) as an electroconductive base and a precursor for a growth of CuO nanowires (NWs) in situ used for the construction of electrochemical nEGSs. CF has a high surface area due to its unique 3D porous structure, resulting in good sensitivity for glucose detection. The CuO NWs/CF-based nEGSs are characterized by good selectivity, reproducibility, repeatability and stability (Table 9).

The CuO NWs/CF based nEGSs have also been employed for glucose assay in human serum and saliva (which indicated that CuO NWs/CF are promising for noninvasive glucose detection). Li and coauthors [225] designed an electrochemical nEGS based on a novel nanostructured electrocatalyst of carbon quantum dots (CQDs)/octahedral cuprous oxide (Cu_2O) nanocomposites [74]. Compared to octahedral Cu_2O, the CQDs/octahedral Cu_2O exhibited preference for electrocatalysis over glucose oxidation and H_2O_2 reduction. The experimental results demonstrated that nEGSs have a good potential for practical determination of glucose in real samples of biological liquids [75,236]. Ju and coauthors [226] synthesized a nanocomposite consisting of 1D CuNWs and 2D reduced graphene oxide nanosheets (CuNWs/rGO) and constructed amperometric nEGSs. Contrary to the CuNWs, the CuNWs/rGO hybrids exhibit a higher current response relative to their auto background current, indicating a stronger electrocatalytic capacity toward the oxidation of glucose. The sensor is characterized by very high sensitivity (16,250 $A·M^{-1}·m^{-2}$) and low LOD (0.2 µM). Shackery and coauthors [72] described the porous, conducting, chemically stable structure of $Co(OH)_2$/3DG. The unique $Co(OH)_2$ NRs electrode morphology displays a unique high sensitivity—36,900 $A·M^{-1}·m^{-2}$ with sufficient selectivity (Table 9).

The bifunctional cascade catalysis was successfully tested for the real-time colorimetric glucose detection with 0.8 µM LOD for 30 s. Alkalized graphitic carbon nitride (AKCN) exhibited perfect photoactivity for H_2O_2 generation at neutral pH-conditions, which are typical for natural GOx. The photocatalytic GOx-like activity of AKCN was successfully demonstrated by an in situ photoproduction of H_2O_2 which was proportional to the rate of glucose. The production of H_2O_2 exhibited a wide linear range proportional to glucose concentration (up to 0.1 M). The production of CO_2 from the photocatalytic oxidation of glucose was negligible (2 µM), compared with that of H_2O_2. This indicates that the photocatalytic mineralization of glucose is inhibited at the applied conditions, and glucose is selectively phototransformed into gluconic acid on AKCN.

In addition to amperometric sensors, recent publications described several promising photoelectrochemical nEGSs using oxidase-like NZs. Zhang and coauthors [237] reported a synthesis of metal-free oxidase mimicking NZ based on modified graphitic carbon nitride (Figure 9). The H_2O_2 is generated as a result of coupled photocatalytic oxidation of glucose and O_2 reduction under visible light irradiation with about 100% apparent quantum efficiency. The generated in situ H_2O_2 serves for oxidation of a chromogenic substrate on the same catalyst in a dark to complete the nonenzymatic glucose detection.

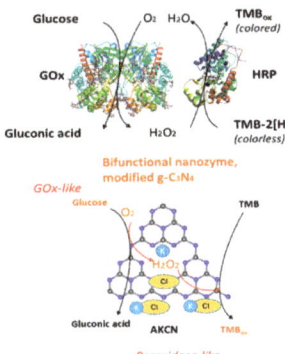

Figure 9. Comparison of two glucose detection systems: based on the glucose oxidase (GOx) combined with horseradish peroxidase (HRP) (above); based on synthetic NZ aerobic photocatalytic oxidation of glucose combined with in situ H_2O_2-production on AKCN (alkalized GCN—graphitic carbon nitride) (below) [237] (modified).

Cao and coauthors [235] recently reported the development of a photoelectrochemical glucose sensor using complicated ternary layered NPs of ITO/PbS/SiO$_2$/AuNPs (ITO-indium tin oxide). Thioglycolic acid-capped PbS quantum dots that are highly sensitive to oxygen were employed as a photoelectrochemical active probe. The AuNPs were used as the GOx-like NZ for aerobic catalytic glucose oxidation. The catalysis promoted oxygen consuming, resulting in a decrease in the cathodic photocurrent. The insertion layer of SiO$_2$ NPs between PbS and AuNPs could efficiently reduce the base current due to its low electroconductivity, which improved the LOD. The described sensor showed high sensitivity and good selectivity. The LR toward glucose was in the frames from 1.0 µM to 1.0 mM with 0.46 µM LOD.

Thus, glucose sensors are practically the only oxidase-like NZ-based electrochemical sensors that are currently being developed and characterized. GOx-like NZs incorporated on the electrode covered layer enhanced their catalytic power due to intrinsic catalytic activity and a synergetic effect of applied potential. In our opinion, due to additional electrocatalytic activity, such NZs can be a promising alternative for natural enzymes in the construction of electrochemical sensors. They are cost-effective, possess high sensitivity, favorable stability, reproducibility, simplicity in development and avoid complex enzymatic immobilization techniques. Unfortunately, performing a multielectron oxidation reaction in the presence of easily oxidized interfering agents has some severe constraints. For example, platinum electrodes lose their activity quickly in glucose solutions through accumulation of chemisorbed intermediates which block the electrocatalyst's surface [238].

6. Laccase-Mimicking Nanozymes

Natural laccases [239–242] are members of the multi-copper oxidases which catalyze the single-electron oxidation of a wide range of organic substrates, such as polyamines, aryl diamines, ortho- and para-diphenols as well as polyphenols, with the subsequent four-electron reduction of molecular oxygen to water (Figure 10). Due to their activities, laccases can be used as "green" catalysts in water treatment and soil bioremediation [243,244]. However, the poor stability of natural laccases in complex environments, the difficulty of their recycling, and the high cost of the purified enzyme preparations severely hamper practical applications of this enzyme [12,245–248].

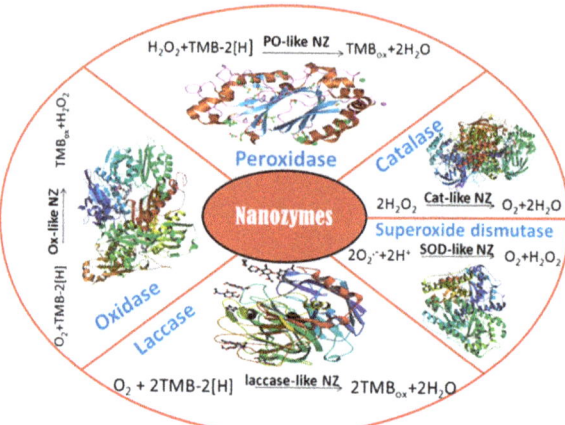

Figure 10. Scheme of reactions catalyzed by oxido-reductases-mimicking nanozymes (NZs).

Many synthetic methods for obtaining various types of laccase-like nanomaterials have been described, when most of them are based on the use of copper ions as a catalyst, because the active centers of natural laccases also contain these ions. A large number of copper-based complexes with different types of organic ligands are reported as laccase mimetics [249–253]. Ren and other authors [254–257] reported one-pot synthesis of copper-containing carbon dots as laccase mimics. Shams and coauthors [258] described the synthesis of Cu/H$_3$BTC MOF (copper ions with 1,3,5-benzene tricarboxylic acid, H$_3$BTC and metal–organic framework, MOF) possessing laccase-like activity with regard to oxidation of phenolic compounds. Cu/H$_3$BTC MOF was used for quantitative detection of epinephrine. This NZ showed excellent stability under different conditions compared with natural laccase [258] (Figure 11).

Water-soluble nucleotides have a significant potential for use as ligands for different nanostructures. Nucleotide coordinated Cu^{2+} complexes were demonstrated as having laccase-like activity [259,260]. Such coordination complexes were immobilized onto magnetic NPs forming Fe$_3$O$_4$@Cu/nucleotide NPs [260,261]. The guanosine monophosphate (GMP) based laccase mimicking Cu/GMP NZ [262] and Fe$_3$O$_4$@Cu/GMP NZ [263] demonstrated excellent laccase-like catalytic activity toward high spectra of phenolic substrates, e.g., hydroquinone, naphthol, catechol, epinephrine and o-phenylenediamine. The K_M^{app} of Cu/GMP toward 2,4-dichlorophenol was quite similar to that of natural laccase (0.59 mM vs. 0.65 mM, respectively). Although it was reported that the V_{max} of Cu/GMP was 5.4-fold higher compared with the natural enzyme, the value of the intrinsic catalytic parameter (k_{cat}) was not indicated. Cu/GMP also showed better stability over pH 3–9, temperatures of 30–90 °C, and a high ionic strength of 500 mM NaCl, as well as long-term storage for 9 days. Analysis of epinephrine with Cu/GMP was nearly 16-fold more sensitive and 2400-fold more cost-effective than using natural laccase [262]. The magnetic Fe$_3$O$_4$@Cu/GMP NZ is able to oxidize toxic o-phenylenediamine and showed higher activity and stability compared with natural laccase [263]. However, the K_M^{app} of laccase was 18-fold lower than that of the Fe$_3$O$_4$@Cu/GMP NZ, which means that laccase had a better affinity toward the substrate. On the other hand, the Vmax of Fe$_3$O$_4$@Cu/GMP was almost 4.2-fold higher than that of laccase (see our remark on the irrelevance of such a comparison). Fe$_3$O$_4$@Cu/GMP retained about 90% of its residual activity at 90 °C, with little change at pH 3–9, and showed excellent storage stability.

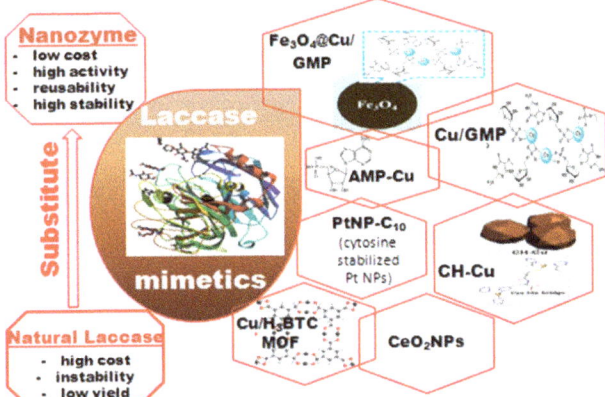

Figure 11. Laccase-mimicking nanozyme for oxidizing phenolic compounds [263–265].

Huang and coauthors [264] described a laccase-mimic NZ based on copper ions and adenosine monophosphate (AMP-CuNZ) with a 15-fold higher catalytic activity than that of natural laccase (at the same mass concentration; normalization to the molar concentration of both catalysts for calculating k_{cat} was not reported). It also has a higher V_{max} and a lower K_M^{app}. The V_{max} of AMP-Cu was 4.5-fold higher than that of natural laccase (1.30 µM·min^{-1} vs. 0.28 µM·min^{-1} at the same mass concentration of both catalysts, 0.1 mg·mL^{-1}), with a 4-fold lower K_M^{app} (0.09 mM vs. 0.36 mM). The lower K_M^{app} of AMP-Cu indicates that the simulated enzyme had a stronger affinity toward the substrate. The concentration linear range (LR) of phenolic compounds detected by AMP-CuNZ was 0.1–100 µM, and the LOD was 0.033 µM (lower than that of laccase). The AMP-Cu had good stability (over 9 days of storage) under conditions of 30–90 °C and pH > 6. AMP-Cu NZ can be used to detect a variety of phenolic compounds: phenol, hydroquinone, p-chlorophenol, resorcinol, phloroglucinol and catechol [264]. It could be predicted that due to its favorable properties, AMP-Cu NZ has a great potential for applications and can replace native laccase in biosensors.

Wang and coauthors [265] presented a new class of laccase-like NZs (denoted as CH-Cu). The electron transfer in this system was provided via the coordination of Cu^+/Cu^{2+} with a cysteine (Cys)-histidine (His) dipeptide. The CH-Cu has similar K_M^{app} values to the natural enzyme (K_M^{app} 0.42 mM and 0.41 mM, respectively) and increased V_{max} values compared to the natural enzyme (7.32 µM·min^{-1} and 6.41 µM·min^{-1}, respectively, at a catalyst concentration of 0.1 mg·mL^{-1} for CH-Cu or laccase). The k_{cat} value of CH-Cu is much higher than that of laccase (1.91 × 10^4 min^{-1} and 4.13 min^{-1}, respectively), indicating a higher turnover number and catalytic efficiency of a single CH-Cu NZ particle, which has a larger number of active sites than the natural protein molecule. A similar result was found for Fe_3O_4 NPs with PO-mimicking activity (k_{cat} = 3.1 × 10^5 s^{-1}) [17].

The catalytic activity of NPs is dependent on the copper ions content. The authors therefore measured the catalytic activity of laccase and CH-Cu NZ with the same amount of Cu atoms. The amount of copper in laccase is 0.32 wt. % and in CH-Cu it is 38.3 wt. %. As a result, the weight ratio of laccase to NZ was 120:1 for the activity test. Due to increasing the number of active sites in a single particle, CH-Cu shows good laccase-like catalytic activity. The higher efficiency of CH-Cu compared with laccase in the degradation of chlorophenols and bisphenols was also demonstrated in a batch mode. Moreover, using CH-Cu a new method for the quantitative detection of epinephrine was developed [265].

Nanomaterials that do not contain copper could also possess laccase mimicking activity. Many metal-based catalysts have been described recently, including noble metal NPs: gold [266], platinum [267], or oxides such as iron oxide [12,17], cerium(IV) oxide [268,269], manganese oxide [270], etc.

NZs placed on Pt in combination with oligonucleotides as stabilizing agents (adenine-A10, thymine-T10, cytosine-C$_{10}$, and guanine-G$_{10}$) display excellent catalytic performance in oxidation of multiple substrates, including 2,4-dichlorophenol (2,4-DCP), dopamine, p-phenylenediamine, catechol and hydroquinone [267]. This kind of Pt NZs has high stability in the range of 20–90 °C and pH 3–9, which exceeds the range of native laccase. It was shown that the laccase-like activities of PtNPs are strongly associated with particle size (2.5–5 nm). PtNPs modified with the oligonucleotides cytosine (C$_{10}$), with a particle size of 4.6 nm, showed the highest activity in the oxidation of 2,4-DCP. These NPs exhibit an apparent K_M^{app} value of 0.12 mM toward 2,4-DCP, whereas laccase exhibits a K_M^{app} of 0.40 mM. Enzyme-like kinetics was observed in the oxidation of 2,4-DCP catalyzed by DNA-stabilized PtNPs which have a three-fold better substrate affinity compared to natural laccase [267]. Nanozymes PtNP-C$_{10}$ and AMP-Cu possess lower K_M values (0.12 mM and 0.09 mM, respectively), that indicate their better affinity toward 2,4-DCP compared to natural laccases: from *Pycnoporus sanguineus* sp. CS43 (0.224 mM) and *Trametes versicolor* (0.40 mM).

The main catalytic characteristics of the synthetic laccase mimetics and natural enzyme toward different substrates are presented in Table 10.

Table 10. The main catalytic characteristics of the synthetic laccase mimetics and natural enzyme toward typical substrates.

Chemo/Biocatalyst	Substrate	K_M^{app}, mM	V_{max}, mM·min^{-1}	NP Concentration, mg·mL^{-1}	Reference
CeO$_2$NPs	Dopamine	0.25×10^{-3}			[77]
	Catechol	0.180			
Cu/H$_3$BTC MOF	Epinephrine	0.068	94×10^{-3}	0.1	[258,262]
Laccase		0.062	5.81×10^{-3}	0.1	
Cu/GMP	Epinephrine	0.59	0.83	0.1	
Laccase	2,4-Dichlorophenol	0.65	0.15	0.1	
AMP-Cu	2,4-Dichlorophenol	0.09	1.3×10^{-3}	0.1	[264]
Laccase	2,4-Dichlorophenol	0.36	0.29×10^{-3}	0.1	
CH-Cu Cys-His dipeptide	Epinephrine/2,4-dichlorophenol	0.58	2.74×10^{-2}	0.1	[265]
		0.42	7.32×10^{-3}		
Laccase	Epinephrine/2,4-Dichlorophenol	0.16	3.10×10^{-3}	0.1	
		0.41	6.41×10^{-3}		
PtNP-C$_{10}$ (cytosine stabilized Pt NPs)	2,4-Dichlorophenol	0.12	8.49×10^{-3}	0.01	[267]
Laccase from *Trametes versicolor*	2,4-Dichlorophenol	0.40	3.51×10^{-3}	0.16	
Laccase from *Pycnoporus sanguineus* sp. CS43	2,4-Dichlorophenol	0.224	2.21×10^{-3}		[271]

Since the synthetic laccase mimetics have preferential properties compared with natural enzymes, they appear to be very promising for the development of new non-enzymatic amperometric sensors for assaying phenolic compounds (Figure 12).

Garcia and coauthors [272] reported the construction of a CPS2 (copper oxide-based carbon paste) biomimetic sensor for phenol, rutin and catechol determination in natural samples, such as dried extracts of red fruits and coffee. The enhanced sensitivity towards catechol as a model substrate correlated to the amount of incorporated copper oxide, resulting in improvement of electroactive surface area and electrocatalytic ability os the sensor. The sensor can be characterized by high sensitivity and selectivity for rutin (LR of 1 to 120 μM and LOD of 0.4 μM) and catechol (LR of 10 to 600 μM) (Table 11) that is better compared to laccase-based biosensors.

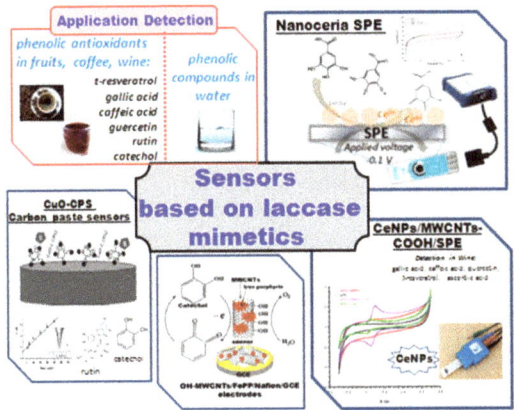

Figure 12. Sensors based on laccase-mimicking nanozymes of different chemical nature in detection of phenolic compounds [268,269,272,273].

Table 11. Analytical characteristics of the amperometric sensors based on synthetic laccase mimetics and natural enzyme toward typical substrates.

Sensor	Analyte	Linear Range, µM	LOD, µM	Reference
CeNPs	Gallic acid	2–20	1.5	[268]
	Caffeic acid	50–200	15.3	
	Quercetine	20–200	8.6	
	Ascorbic acid	0.5–20	0.4	
CeNPs/MWCNTs-COOH/SPE	Gallic acid	25–50	7	[269]
	Caffeic acid	33–100	10	
	Quercetin	25–100	8	
	t-Resveratrol	25–50	8	
	Ascorbic acid	25–100	7	
Laccase/Fc/SPE	Caffeic acid	2.0–30.0	1.6	[269]
Laccase/PAP/SWCNTs/SPE	Gallic acid	0.53–96		[269]
Laccase immobilized in polyzetidine prepolymer and SWCNTs	Gallic acid	0.53–96		[269]
CuO-C-CPS2	Rutin	1 to 120	0.4	[272]
	Catechol	10 to 600	2.0	
Laccase on CP	Catechol	20 to 700	4.5	[272]
FePP modified multi-walled carbon nanotubes (OH-MWCNTs/FePP/Nafion/GCE)	Catechol	65 to 1600	3.75	[273]

Andrei and coauthors [268] reported the construction and characterization of a disposable single-use electrochemical sensor (SPE) based on CeNPs for the detection of phenolic antioxidants. The nanoceria supported oxidation of phenolic compounds to their corresponding quinones that can be detected at the SPE surface. For analysis of reactivity of gallic acid, ascorbic acid, quercetin, caffeic acid and epicatechin the amperometric detection was performed at the working potential −0.1 V vs. Ag/AgCl. The LR for gallic acid was between 50–200 µM, with a LOD of 15.3 µM (Table 11). The electrode did not respond to glucose, tartaric acid, ethanol, citric acid and sulfur dioxide at the applied conditions. The developed CeNPs SPE electrodes are low-cost, stable and disposable, making them promising for rapid field detection of antioxidant-rich samples and for screening purposes.

Another sensor has been developed for the analysis of antioxidants in wines based on CeNPs as a biomimetic [269]. The proposed single-use electrochemical screen-printed electrode, modified by CeNPs, was applied for detection of a number of antioxidant compounds which are present in wines (Table 11). The LR of the sensor is similar, and in some cases better, than for other analogs. The sensor does not require any specific storage conditions (e.g., buffer and low temperature) as opposed to enzymatic biosensors. Furthermore, the CeNP-modified electrode has a broader LR for gallic acid and ascorbic acid [269].

Wang and coworkers [270] proposed different crystalline nanostructured manganese oxide (MnO_2) and Mn_3O_4 on an electrode surface, which showed laccase-like catalytic activity toward ABTS and 17β-estradiol (E2). The best catalytic performance of the six different crystal structures for oxidizing ABTS and E2 was shown for the γ-MnO_2 variant. ABTS oxidation by γ-MnO_2 NPs under different pH values proved high oxidation activity of γ-MnO_2 at pH values of 3–4. The findings add to the understanding of the laccase-like catalysts based on MnOx [270].

Bin and coauthors [273] developed a novel highly stable sensor for cost-effective and efficient catechol detection based on the functionalized surface of multi-walled carbon nanotubes (MWCNTs) and iron porphyrins (FePP). Under optimal condition, the LOD for this NZ is slightly higher than that of the laccase sensor with the broader LR (Table 11). The reproducibility test showed that OH-MWCNTs/FePP/Nafion/GCE has better reproducibility than other laccase sensors. Storage analysis demonstrated that the oxidation current maintained 98.3% of its initial output after storage at −4 °C for 35 days.

7. Conclusions and Potential Applications

Only a small part of the known NZs and electronanocatalysts used in the construction of electrochemical sensors is presented in this review. At present, nanozymology is progressing faster than could be described directly as a real-time review. However, we believe that the presented data will be valuable for demonstrating the preferability of a combination of novel progressive nanotechnologies with electroanalytical approaches compared to classic enzymological methods.

Enzyme-based sensors are already successful commercial products in different application fields. Using cost-effective synthetic enzyme-like nanomaterials seems to be a very promising way for sensor development. It should be noted that NZs have several limitations for their application that need to be solved. The main disadvantage of most known NZs is the lack of substrate specificity. This drawback raises concerns regarding "nanozyme" term since in some cases these nanomaterials operate like chemical or electrochemical catalysts. The second drawback is the fouling of the NZ surface due to aggressive impact or adsorption of some compounds of the real environmental or biological samples. The third is a rather low reaction rate as a result of pH-sensitivity. Many common NZs work best in an acidic pH, while the pH of biological samples is often neutral. The fourth is a limited reaction type of the current NZs that are mainly capable of mimicking oxidoreductases (oxidases, peroxidase, catalase, SOD). Known NZs are thus currently able to mimic only a small fraction of enzymes. These critical problems must be solved in order to create highly selective, stable, reliable NZs that are suitable for practical use. However, considering the high scientific interest, progress in this field will focus on the creation of novel materials with new improved catalytic as well as physic-chemical properties, maximally fitted for (bio)chemosensing devices.

Funding: This work was funded in part by the Ministry of Education and Science of Ukraine (projects Nos. 0118U000297 and 0119U100671; Ukrainian- Lithuanian R&D Project in 2020–2021: "Development of biosensors and biofuel cells based on redox enzymes"), National Academy of Sciences of Ukraine in the frame of the Scientific-Technical Program "Smart sensor devices of a new generation based on modern materials and technologies" (projects No. 13 and No. 10/3).

Acknowledgments: The authors are thankful to A. Zakalskiy for his help in literature analysis. The work was supported by the Research Authority of the Ariel University, Israel.

Conflicts of Interest: The authors declare no conflicts of interest.

Abbreviations

AA	Ascorbic acid
ABTS	2,2'-Azinobis-(3-ethylbenzthiazoline-6-sulphonate)
AKCN	Alkalized graphitic carbon nitride
AMP	Adenosine monophosphate
AOx	Alcohol oxidase
AP	Acetamidophenol

AuNPs	Gold nanoparticles
Au/Co@HNCF	Gold/cobalt nanoporous carbon framework
AuNBP/MWCNTs	Gold nanobipyramids supported by multi-walled carbon nanotubes
Fer/rGO	ferumoxytol and reduced graphene oxide
BET	Brunauer-Emmett-Teller
BG	Bucky gel consisting of carbon nanotubes and ionic liquid
biot-GOx	Biotinylated glucose oxidase
BNNS@CuS	Boron nitride nanosheet and copper sulfide nanohybrids
BSA	Bovine serum albumin
$C_{60}[C(COOH)_2]_2$	C_{60}-carboxyfullerene
CeNPs	Nanoceria particles
CF@CuAl-LDH	Carbon fiber-supported ultrathin cual layered double hydroxides (LDH) nanosheets
CF-H-Au	Carbon microfibers-hemin-gold nanoparticles
CMC@Pd/Al-LDH	Pd/Al layered double hydroxide/carboxymethyl cellulose nanocomposite;
CNE	Carbon nanoelectrodes fabricated within a quartz nanopipette and electrochemically etched
CNFs	Carbon nanofibers
CNTs	Multi-walled carbon nanotubes
COx	Cholesterol oxidase
CP	Carbon paste
CPS	Carbon paste sensors
CQDs	Carbon quantum dots
cTnI	Cardiac troponin I
CTMB	Cetyltrimethylammonium bromide
Cu-Cys	Copper(II) complex of cysteine
Cu-MOG	Cu-based metal-organic gel
CuNWs/rGO	One-dimensional copper nanowires and two-dimensional reduced graphene oxide nanosheets
DA	Dopamine
2,4-DCP	2,4-Dichlorophenol
3D GNCs	3D graphene-supported quantum dots
3DG	Three-dimensional graphene
DBD	Diamond Boron-doped
EMSN-AuNPs	Expanded mesoporous silica-encapsulated gold nanoparticles
ENC	Electronanocatalyst
ERGO	Electrochemically reduced graphene oxide
FePP	Iron porphyrins
GBR	Metal-free brominated graphene
GCE	Glassy carbon electrode
GCE/MWCNTs-Av/RuNPs/biot-GOx	Glassy carbon electrode/avidin-functionalized multi-walled carbon/nanotubes/Ru nanoparticles/biotinylated glucose oxidase
GDCh	Glucose derived sheet-like carbons
GE	Graphite electrode
GNP/Cu-Cys	Gold nanoparticles with the copper(II) complex of cysteine
GNPs	Gold nanoparticles
GNs	Graphene nanosheets
GO	Graphene oxide
GOx	Glucose oxidase
GOx/PtNP/PAni/Pt	Glucose Sensor Based on Pt Nanoparticle/Polyaniline Hydrogel
H_2TCPP	*meso*-Tetrakis(4-carboxyphenyl)-porphyrin
HCC	Carbon cubic nanomaterial
HCF	Hexacyanoferrate
h-CuS NCs	Hollow copper sulfide nanocubes
H-GNs	Hemin-graphene hybrid nanosheets
His@AuNCs	Histidine-capped gold nanoclusters

HNCs	Hollow nanocubes
H-rGO-Au	Hemin-graphene-gold nanoparticles
HRP	Horseradish peroxidase
H/WS2-NSs	Hemin-functionalized/Tungsten disulfide nanosheets
ITO	Indium tin oxide
L-Cys	L-Cysteine
LOD	Limit of detection
LOx	Lactate oxidase
LR	Linear range
MCM-41	Mobil composition of matter No. 41
MCNs	Mesoporous carbon nanospheres
MCP	Microchannel plate
MIP	Molecularly imprinted polymer
MMT	Montmorillonite
MC	Magnetic carbon
Mn-MPSA-HCC	Mn nanomaterials: hollow carbon cubic
Mn-MPSA-HCS	Mn nanomaterials: hollow carbon sphere
MOFs	Metal-Organic Frameworks
MWCNTs	Multi-walled carbon nanotubes
MWCNTs-Av	Avidin-functionalized multi-walled carbon nanotubes
NCs	Nanoclusters
NCF	Nitrogen doped cotton carbon fiber composite
N-CNFht	Nitrogen-doped carbon nanofibers
NFs	Nanoflakes
NPs	Nanoparticles
NTH	Nanotetrahedron
NW/CF	Nanowires/copper foam
NZ	Nanozyme
OPD	o-Phenylenediamine
PAni	Polyaniline
PANI/MXene	Polyaniline and $Ti_3C_2T_x$
PB	Prussian Blue
PEG-HCCs	Poly(ethylene glycolated) hydrophilic carbon clusters
PFOP	Heptadecafluoro-n-octyl bromide
PMMA	Poly(methylmethacrylate)
PNAANI	Poly(N-acetylaniline)
PO	Natural horseradish peroxidase
RDE	Rotating disk electrode
RGO	Reduced graphene oxide
rGO	Reduced graphene oxide
SOD	Superoxide dismutase
SPCE	Screen-printed carbon electrode
TMB	3,5,3′,5′-Tetramethylbenzidine
UA	Uric acid
WCC	Tungsten carbide decorated by cobalt nanoparticles
YSNs	Yolk-shell nanostructures
ZIF	Zeolitic imidazolate framework
β-CD	β-Cyclodextrin

References

1. Breslow, R.; Overman, L.E. An "Artificial Enzyme" combining a metal catalytic group and a hydrophobic binding cavity. *J. Am. Chem. Soc.* **1970**, *92*, 1075–1077. [CrossRef] [PubMed]
2. Cheng, H.; Wang, X.; Wei, H. Artificial enzymes: The next wave. In *Encyclopedia of Physical Organic Chemistry*, 1st ed.; Wang, Z., Ed.; John Wiley & Sons: Hoboken, NJ, USA, 2017; p. 64. ISBN 978-1-118-46858-6.

3. Wei, H.; Wang, E. Nanomaterials with enzyme-like characteristics (nanozymes): Next-generation artificial enzymes. *Chem. Soc. Rev.* **2013**, *42*, 6060–6093. [CrossRef] [PubMed]
4. Lin, Y.; Ren, J.; Qu, X. Nano-Gold as Artificial Enzymes: Hidden Talents. *Adv. Mater.* **2014**, *26*, 4200–4217. [CrossRef] [PubMed]
5. Wu, J.; Wang, X.; Wang, Q.; Lou, Z.; Li, S.; Zhu, Y.; Qin, L.; Wei, H. Nanomaterials with enzyme-like characteristics (nanozymes): Next-generation artificial enzymes (II). *Chem. Soc. Rev.* **2019**, *48*, 1004–1076. [CrossRef] [PubMed]
6. Huang, Y.; Ren, J.; Qu, X. Nanozymes: Classification, Catalytic Mechanisms, Activity Regulation, and Applications. *Chem. Rev.* **2019**, *119*, 4357–4412. [CrossRef]
7. Carmona-Ribeiro, A.; Prieto, T.; Nantes, I.L. Peroxidases in nanostructures. *Front. Mol. Biosci.* **2015**, *2*, 50. [CrossRef]
8. Jiang, B.; Fang, L.; Wu, K.; Yan, X.; Fan, K. Ferritins as natural and artificial nanozymes for theranostics. *Theranostics* **2020**, *10*, 687–706. [CrossRef]
9. Wang, P.; Wang, T.; Hong, J.; Yan, X.; Liang, M. Nanozymes: A New Disease Imaging Strategy. *Front. Bioeng. Biotech.* **2020**, *8*, 1–10. [CrossRef]
10. Liang, M.; Yan, X. Nanozymes: From New Concepts, Mechanisms, and Standards to Applications. *Acc. Chem. Res.* **2019**, *5*, 2190–2200. [CrossRef]
11. Jiang, D.; Ni, D.; Rosenkrans, Z.T.; Huang, P.; Yan, X.; Cai, W. Nanozyme: New horizons for responsive biomedical applications. *Chem. Soc. Rev.* **2019**, *48*, 3683–3704. [CrossRef]
12. Gao, L.Z.; Zhuang, J.; Nie, L.; Zhang, J.B.; Zhang, Y.; Gu, N.; Wang, T.H.; Feng, J.; Yang, D.L.; Perrett, S.; et al. Intrinsic peroxidase-like activity of ferromagnetic nanoparticles. *Nat. Nanotechnol.* **2007**, *2*, 577–583. [CrossRef] [PubMed]
13. Karyakin, A.A. Prussian Blue and Its Analogues: Electrochemistry and Analytical Applications. *Electroanalysis* **2001**, *13*, 813–819. [CrossRef]
14. Wang, Y.; Mao, J.; Meng, X.; Yu, L.; Deng, D.; Bao, X. Catalysis with Two-Dimensional Materials Confining Single Atoms: Concept, Design, and Applications. *Chem. Rev.* **2019**, *119*, 1806–1854. [CrossRef] [PubMed]
15. Vernekar, A.A.; Das, T.; Ghosh, S.; Mugesh, G. A Remarkably Efficient $MnFe_2O_4$-based Oxidase Nanozyme. *Chem. Asian, J.* **2016**, *11*, 72–76. [CrossRef] [PubMed]
16. Wang, H.; Wan, K.; Xinghua, S. Recent Advances in Nanozyme Research. *Adv. Mater.* **2019**, *31*, 1805368. [CrossRef] [PubMed]
17. Jiang, B.; Duan, D.; Gao, L.; Zhou, M.; Fan, K.; Tang, Y.; Xi, J.; Bi, Y.; Tong, Z.; Gao, G.F.; et al. Standardized assays for determining the catalytic activity and kinetics of peroxidase-like nanozymes. *Nat. Protoc.* **2018**, *13*, 1506–1520. [CrossRef]
18. Liu, J. Special Topic: Nanozyme-Based Analysis and Testing. *J. Anal. Test.* **2019**, *3*, 189–190. [CrossRef]
19. Golchin, J.; Golchin, K.; Alidadian, N.; Ghaderi, S.; Eslamkhah, S.; Eslamkhah, M.; Akbarzadeh, A. Nanozyme applications in biology and medicine: An overview. *Artif. Cells Nanomed. Biotechnol.* **2017**, *45*, 1–8. [CrossRef]
20. Liu, X.; Gao, Y.; Chandrawati, R.; Hosta-Rigau, L. Therapeutic applications of multifunctional nanozymes. *Nanoscale* **2019**, *11*, 21046–21060. [CrossRef]
21. Gu, Y.; Huang, Y.; Qiu, Z.; Xu, Z.; Li, D.; Chen, L.; Jiang, J.; Gao, L. Vitamin B_2 functionalized iron oxide nanozymes for mouth ulcer healing. *Sci. China Life Sci.* **2019**, *62*, 12. [CrossRef]
22. Wang, Q.; Wei, H.; Zhang, Z.; Wang, E.; Dong, S. Nanozyme: An emerging alternative to natural enzyme for biosensing and immunoassay. *Trends Anal. Chem.* **2018**, *105*, 218–224. [CrossRef]
23. Nayl, A.A.; Abd-Elhamid, A.I.; El-Moghazy, A.Y.; Hussin, M.; Abu-Saied, M.A.; El-Shanshory, A.A.; Soliman, H.M.A. The nanomaterials and recent progress in biosensing systems: A review. *Trends Environ. Anal. Chem.* **2020**, e00087. [CrossRef]
24. Mahmudunnabi, R.G.; Farhana, N.F.; Kashaninejad, Z.; Firoz, S.H.; Shim, Y.; Shiddiky, M.J.A. Nanozymes-based electrochemical biosensors for disease biomarker detection. *Analyst* **2020**, *145*, 4398–4420. [CrossRef] [PubMed]
25. Qin, L.; Hu, Y.; Wei, H. Nanozymes: Preparation and Characterization. In *Nanostructure Science and Technology*; Yan, X., Ed.; Springer: Singapore, 2020.
26. Li, J.; Wu, Q.; Wu, J. Synthesis of Nanoparticles via Solvothermal and Hydrothermal Methods. In *Handbook of Nanoparticles*; Aliofkhazraei, M., Ed.; Springer International Publishing: Cham, Switzerland, 2016; pp. 295–328.

27. Luo, L.; Zhang, Y.; Li, F.; Si, X.; Ding, Y.; Wang, T. Enzyme mimics of spinel-type $Co_xNi_{1-x}Fe_2O_4$ magnetic nanomaterial for eletroctrocatalytic oxidation of hydrogen peroxide. *Anal. Chim. Acta* **2013**, *788*, 46–51. [CrossRef]
28. Wu, Q.; He, L.; Jiang, Z.W.; Li, Y.; Zheng, M.C.; Cheng, Z.H.; Li, Y.F. CuO Nanoparticles Derived from Metal-Organic Gel with Excellent Electrocatalytic and Peroxidase-Mimicking Activities for Glucose and Cholesterol Detection. *Biosens. Bioelectron.* **2019**, *145*, 111704. [CrossRef]
29. Zheng, S.; Li, B.; Tang, Y.; Li, Q.; Xue, H.; Pang, H. Ultrathin Nanosheet-Assembled $[Ni_3(OH)_2(PTA)_2(H_2O)_4]\cdot 2H_2O$ Hierarchical Flowers for High-Performance Electrocatalysis of Glucose Oxidation Reactions. *Nanoscale* **2018**, *10*, 13270–13276. [CrossRef]
30. Li, Y.; Zhang, H.; Cai, X.; Zhao, H.; Magdassi, S.; Lan, M. Electrochemical detection of superoxide anions in HeLa cells by using two enzyme-free sensors prepared from ZIF-8-derived carbon nanomaterials. *Mikrochim. Acta* **2019**, *186*, 370. [CrossRef]
31. Wang, C.; Liu, C.; Li, J.; Sun, X.; Shen, J.; Han, W.; Wang, L. Electrospun metal–organic framework derived hierarchical carbon nanofibers with high performance for supercapacitors. *Chem. Commun.* **2017**, *53*, 1751–1754. [CrossRef]
32. Scandurra, A.; Ruffino, F.; Sanzaro, S.; Grimaldi, M.G. Laser and Thermal Dewetting of Gold Layer onto Graphene Paper for non-Enzymatic Electrochemical Detection of Glucose and Fructose. *Sens. Actuators B Chem.* **2019**, *301*, 127113. [CrossRef]
33. Wang, K.; Wu, C.; Wang, F.; Liao, M.; Jiang, G. Bimetallic nanoparticles decorated hollow nanoporous carbon framework as nanozyme biosensor for highly sensitive electrochemical sensing of uric acid. *Biosens. Bioelectron.* **2020**, *150*, 111869. [CrossRef]
34. Chou, K.S.; Ren, C.Y. Synthesis of nanosized silver particles by chemical reduction method. *Mater. Chem. Phys.* **2000**, *64*, 241–246. [CrossRef]
35. Rane, A.V.; Kanny, K.; Abitha, V.K.; Sabu, T. Chapter 5. Methods for Synthesis of Nanoparticles and Fabrication of Nanocomposites. In *Synthesis of Inorganic Nanomaterials Advances and Key Technologies Micro and Nano Technologies*, 1st ed.; Bhagyaraj, S.M., Oluwafemi, O.S., Kalarikkal, N., Sabu, T., Eds.; Woodhead Publishing Company: Sawston, UK, 2018; pp. 121–139.
36. Suriati, G.M.; Mariatti, M.; Azizan, A. Synthesis of silver nanoparticles bu chemical reduction method: Effect of reducing agent and surfactant concentration. *Int. J. Automot. Mech. Eng.* **2014**, *10*, 1920–1927. [CrossRef]
37. Zhu, J.; Peng, X.; Nie, W.; Wang, Y.; Gao, J.; Wen, W.; Wang, S. Hollow copper sulfide nanocubes as multifunctional nanozymes for colorimetric detection of dopamine and electrochemical detection of glucose. *Biosens. Bioelectron.* **2019**, *141*, 111450. [CrossRef] [PubMed]
38. Das, R.; Dhiman, A.; Kapil, A.; Bansal, V.; Sharma, T.K. Aptamer-mediated colorimetric and electrochemical detection of *Pseudomonas aeruginosa* utilizing peroxidase-mimic activity of gold NanoZyme. *Anal. Bioanal. Chem.* **2019**, *411*, 1229–1238. [CrossRef] [PubMed]
39. Benedetti, T.M.; Andronescu, C.; Cheong, S.; Wilde, P.; Wordsworth, J.; Kientz, M.; Tilley, R.D.; Schuhmann, W.; Gooding, J.J. Electrocatalytic Nanoparticles That Mimic the Three-Dimensional Geometric Architecture of Enzymes: Nanozymes. *J. Am. Chem. Soc.* **2018**, *140*, 13449–13455. [CrossRef] [PubMed]
40. Ling, P.; Cheng, S.; Chen, N.; Qian, C.; Gao, F. Nanozyme-Modified Metal–Organic Frameworks with Multienzymes Activity as Biomimetic Catalysts and Electrocatalytic Interfaces. *ACS Appl. Mater. Interfaces* **2020**, *12*, 17185–17192. [CrossRef]
41. Liu, L.; Du, J.; Liu, W.E.; Guo, Y.; Wu, G.; Qi, W.; Lu, X. Enhanced His@AuNCs Oxidase-Like Activity by Reduced Graphene Oxide and Its Application for Colorimetric and Electrochemical Detection of Nitrite. *Anal. Bioanal. Chem.* **2019**, *411*, 2189–2200. [CrossRef]
42. Hummers, W.S.; Offeman, R.E. Preparation of graphitic oxide. *J. Am. Chem. Soc.* **1958**, *80*, 1339. [CrossRef]
43. Han, R.; Lu, Y.; Mingjun, L.; Yanbo, W.; Xuan, L.; Chongyang, L.; Kun, L.; Lingxing, Z.; Aihua, L. Green tide biomass templated synthesis of molybdenum oxide nanorods supported on carbon as efficient nanozyme for sensitive glucose colorimetric assay. *Sens. Actuators B Chem.* **2019**, *296*, 126517. [CrossRef]
44. Han, L.; Zhang, H.; Li, F. Bioinspired Nanozymes with pH-Independent and Metal Ions-Controllable Activity: Field-Programmable Logic Conversion of Sole Logic Gate System. *Part. Part. Syst. Char.* **2018**, 1800207. [CrossRef]
45. Zhai, D.; Liu, B.; Shi, Y.; Pan, L.; Wang, Y.; Li, W.; Zhang, R.; Yu, G. Highly Sensitive Glucose Sensor Based on Pt Nanoparticle/Polyaniline Hydrogel Heterostructures. *ACS Nano* **2013**, *7*, 3540–3546. [CrossRef] [PubMed]

46. Chen, H.I.; Chang, H.Y. Synthesis of nanocrystalline cerium oxide particles by the precipitation method. *Ceram. Int.* **2005**, *31*, 795–802. [CrossRef]
47. Cai, X.; Wang, Z.; Zhang, H.; Li, Y.; Chen, K.; Zhao, H.; Lan, M. Carbon-mediated synthesis of shape-controllable manganese phosphate as nanozymes for modulation of superoxide anions in HeLa cells. *J. Mater. Chem. B* **2019**, *7*, 401–407. [CrossRef] [PubMed]
48. Dashtestani, F.; Ghourchian, H.; Eskandari, K.; Rafiee-Pour, H.A. A superoxide dismutase mimic nanocomposite for amperometric sensing of superoxide anions. *Microchim. Acta* **2015**, *82*, 1045–1053. [CrossRef]
49. Sun, D.; Lin, X.; Lu, J.; Wei, P.; Luo, Z.; Lu, X.; Chen, Z.; Zhang, L. DNA Nanotetrahedron-Assisted Electrochemical Aptasensor for Cardiac Troponin I Detection Based on the Co-Catalysis of Hybrid Nanozyme, Natural Enzyme and Artificial DNAzyme. *Biosens. Bioelectron.* **2019**, *42*, 111578. [CrossRef] [PubMed]
50. Wurm, F.R.; Weiss, C.K. Nanoparticles from renewable polymers. *Front. Chem.* **2014**, *2*, 49. [CrossRef]
51. Han, M.; Guo, P.; Wang, X.; Tu, W.; Bao, J.; Dai, Z. Mesoporous SiO_2–(L)-lysine hybrid nanodisks: Direct electron transfer of superoxide dismutase, sensitive detection of superoxide anions and its application in living cell monitoring. *RSC Adv.* **2013**, *3*, 20456–20463. [CrossRef]
52. Manesh, K.M.; Lee, S.H.; Uthayakumar, S.; Gopalan, A.I.; Lee, K.P. Sensitive electrochemical detection of superoxide anion using gold nanoparticles distributed poly(methyl methacrylate)–polyaniline core–shell electrospun composite electrode. *Analyst* **2011**, *136*, 1557–1561.
53. Tonelli, D.; Scavetta, E.; Gualandi, I. Electrochemical Deposition of Nanomaterials for Electrochemical Sensing. *Sensors* **2019**, *19*, 1186. [CrossRef]
54. Al-Bat'hi, S.A.M. Electrodeposition of nanostructure materials. In *Electroplating of Nanostructures*; Aliofkhazraei, M., Ed.; InTech: Rijeka, Croatia, 2015; pp. 3–25.
55. Cioffi, N.; Colaianni, L.; Ieva, E.; Pilolli, R.; Ditaranto, N.; Daniela, M.; Cotrone, S.; Buchholt, K.; Lloyd, A.; Sabbatini, L.; et al. Electrosynthesis and characterization of gold nanoparticles for electronic capacitance sensing of pollutants. *Electrochim. Acta* **2011**, *56*, 3713–3720. [CrossRef]
56. Dominguez-Dominguez, S.; Arias-Pardilla, J.; Berenguer-Murcia, A.; Morallon, E.; Cazorla-Amoros, D. Electrochemical deposition of platinum nanoparticles on different carbon supports and conducting polymers. *J. Appl. Electrochem.* **2008**, *38*, 259–268. [CrossRef]
57. Rodriguez-Sanchez, L.; Blanco, M.C.; Lopez-Quintela, M.A. Electrochemical Synthesis of Silver Nanoparticles. *J. Phys. Chem. B* **2000**, *104*, 9683–9688. [CrossRef]
58. Guangwei, S.; Lixuan, M.; Wensheng, S. Electroplating of nanostructures. *Recent Pat. Nanotech.* **2009**, *3*, 182–191.
59. Gallay, P.; Eguílaz, M.; Rivas, G. Designing Electrochemical Interfaces Based on Nanohybrids of Avidin Functionalized-Carbon Nanotubes and Ruthenium Nanoparticles as Peroxidase-Like Nanozyme With Supramolecular Recognition Properties for Site-Specific Anchoring of Biotinylated Residues. *Biosens. Bioelectron.* **2020**, *148*, 111764. [CrossRef]
60. Stasyuk, N.; Gayda, G.; Zakalskiy, A.; Zakalska, O.; Serkiz, R.; Gonchar, M. Amperometric biosensors based on oxidases and PtRu nanoparticles as artificial peroxidase. *Food Chem.* **2019**, *285*, 213–220. [CrossRef]
61. Wu, T.; Li, L.; Song, G.; Ran, M.; Lu, X.; Liu, X. An ultrasensitive electrochemical sensor based on cotton carbon fiber composites for the determination of superoxide anion release from cells. *Microchim. Acta* **2019**, *186*, 198. [CrossRef]
62. Lu, Z.; Wu, L.; Zhang, J.; Dai, W.; Mo, G.; Ye, J. Bifunctional and highly sensitive electrochemical non-enzymatic glucose and hydrogen peroxide biosensor based on $NiCo_2O_4$ nanoflowers decorated 3D nitrogen doped holey graphene hydrogel. *Mater. Sci. Eng. C Mater. Biol. Appl.* **2019**, *102*, 708–717. [CrossRef]
63. Lin, S.; Zhang, Y.; Cao, W.; Wang, X.; Qin, L.; Zhou, M.; Wei, H. Nucleobase-mediated synthesis of nitrogen-doped carbon nanozymes as efficient peroxidase mimics. *Dalton Trans.* **2019**, *48*, 1993–1999. [CrossRef]
64. Jiang, B.; Yan, L.; Zhang, J.; Zhou, M.; Shi, G.; Tian, X.; Fan, K.; Hao, C.; Yan, X. Biomineralization Synthesis of the Cobalt Nanozyme in SP94-Ferritin Nanocages for Prognostic Diagnosis of Hepatocellular Carcinoma. *ACS Appl. Mater. Interfaces* **2019**, *11*, 9747–9755. [CrossRef]
65. Zhang, D.; Shen, N.; Zhang, J.; Zhu, J.; Guo, Y.; Xu, L. A novel nanozyme based on selenopeptide-modified gold nanoparticles with a tunable glutathione peroxidase activity. *RSC Adv.* **2020**, *10*, 8685–8691. [CrossRef]

66. Liu, L.; Yan, Y.; Zhang, X.; Mao, Y.; Ren, X.; Hu, C.; He, W.; Yin, J. Regulating the Pro- and Anti-Oxidant Capability of Bimetallic Nanozymes for Detection of Fe^{2+} and Protection of *Monascus* pigments. *Nanoscale* **2020**, *12*, 3068–3075. [CrossRef] [PubMed]
67. Liu, M.; Kong, L.; Wang, X.; He, J.; Bu, X.H. Engineering Bimetal Synergistic Electrocatalysts Based on Metal–Organic Frameworks for Efficient Oxygen Evolution. *Nano Micro. Small* **2019**, *15*, 1903410. [CrossRef] [PubMed]
68. Cao-Milán, R.; He, L.D.; Shorkey, S.; Tonga, G.Y.; Wang, L.S.; Zhang, X.; Uddin, I.; Das, R.; Sulak, M.; Rotello, V.M. Modulating the Catalytic Activity of Enzyme-like Nanoparticles Through their Surface Functionalization. *Mol. Syst. Des. Eng.* **2017**, *2*, 624–628. [CrossRef] [PubMed]
69. Wang, X.; Sun, H.; Wang, C. Functionalization of GroEL nanocages with hemin for label-free colorimetric assays. *Anal. Bioanal. Chem.* **2019**, *411*, 3819–3827. [CrossRef]
70. Gao, M.; Lu, X.; Nie, G.; Chi, M.; Wang, C. Hierarchical $CNFs/MnCo_2O_{4.5}$ nanofibers as a highly active oxidase mimetic and its application in biosensing. *Nanotechnology* **2017**, *28*, 485708. [CrossRef]
71. Wu, L.; Wan, G.; Hu, N.; He, Z.; Shi, S.; Suo, Y.; Wang, K.; Xu, X.; Tang, Y.; Wang, G. Synthesis of Porous $CoFe_2O_4$ and Its Application as a Peroxidase Mimetic for Colorimetric Detection of H_2O_2 and Organic Pollutant Degradation. *Nanomaterials* **2018**, *8*, 451. [CrossRef]
72. Shackery, I.; Patil, U.; Pezeshki, A.; Shinde, N.M.; Im, S.; Jun, S.C. Enhanced Non-enzymatic amperometric sensing of glucose using $Co(OH)_2$ nanorods deposited on a three dimensional graphene network as an electrode material. *Microchim. Acta* **2016**, *183*, 2473–2479. [CrossRef]
73. Tyagi, M.; Tomar, M.; Gupta, V. Glad assisted synthesis of NiO nanorods for realization of enzymatic reagentless urea biosensor. *Biosens. Bioelectron.* **2014**, *52*, 196–201. [CrossRef]
74. Li, Y.; Zhong, Y.; Zhang, Y.; Weng, W.; Li, S. Carbon quantum dots/octahedral Cu_2O nanocomposites for non-enzymatic glucose and hydrogen peroxide amperometric sensor. *Sens. Actuators B Chem.* **2015**, *206*, 735–743. [CrossRef]
75. Wu, H.X.; Cao, W.M.; Li, Y.; Liu, G.; Wen, Y.; Yang, H.F.; Yang, S.P. In situ growth of copper nanoparticles on multiwalled carbon nanotubes and their application as non-enzymatic glucose sensor materials. *Electrochim. Acta* **2010**, *55*, 3734–3740. [CrossRef]
76. Pirmohamed, T.; Dowding, J.M.; Singh, S.; Wasserman, B.; Heckert, E.; Karakoti, A.S.; King, J.E.S.; Seal, S.; Self, W.T. Nanoceria exhibit redox state-dependent catalase mimetic activity. *Chem. Commun.* **2010**, *46*, 2736–2738. [CrossRef] [PubMed]
77. Hayat, A.; Cunningham, J.; Bulbula, G.; Andreescu, S. Evaluation of the oxidase like activity of nanoceria and its application in colorimetric assays. *Anal. Chim. Acta* **2015**, *885*, 140–147. [CrossRef] [PubMed]
78. Mumtaz, S.; Wang, L.-S.; Abdullah, M.; Hussain, S.Z.; Iqbal, Z.; Rotello, V.M.; Hussain, I. Facile method to synthesize dopamine-capped mixed ferrite nanoparticles and their peroxidase-like activity. *J. Phys. D Appl. Phys.* **2017**, *50*, 11LT02. [CrossRef]
79. Liu, S.; Lu, F.; Xing, R.; Zhu, J.J. Structural effects of Fe_3O_4 nanocrystals on peroxidase-like activity. *Chem. Eur. J.* **2011**, *17*, 620–625. [CrossRef]
80. Rauf, S.; Ali, N.; Tayyab, Z.; Shah, M.Y.; Yang, C.P.; Hu, J.F.; Kong, W.; Huang, Q.A.; Hayat, A.; Muhammad, M. Ionic liquid coated zerovalent manganese nanoparticles with stabilized and enhanced peroxidase-like catalytic activity for colorimetric detection of hydrogen peroxide. *Mater. Res. Express* **2020**, *7*, 035018. [CrossRef]
81. Zhang, K.; Zuo, W.; Wang, Z.; Liu, J.; Li, T.; Wang, B.; Yang, Z. A simple route to $CoFe_2O_4$ nanoparticles with shape and size control and their tunable peroxidase-like activity. *RSC Adv.* **2015**, *5*, 10632–10640. [CrossRef]
82. Liu, B.; Liu, J. Surface modification of nanozymes. *Nano Res.* **2017**, *10*, 1125–1148. [CrossRef]
83. Zhao, M.; Huang, J.; Zhou, Y.; Pan, X.; He, H.; Ye, Z.; Pan, X. Controlled synthesis of spinel $ZnFe_2O_4$ decorated ZnO heterostructures as peroxidase mimetics for enhanced colorimetric biosensing. *Chem. Commun.* **2013**, *49*, 7656–7658. [CrossRef]
84. Cheng, X.; Huang, L.; Yang, X.; Elzatahry, A.A.; Alghamdi, A.; Deng, Y. Rational design of a stable peroxidase mimic for colorimetric detection of H_2O_2 and glucose: A synergistic CeO_2/Zeolite Y nanocomposite. *J. Colloid Interface Sci.* **2019**, *535*, 425–435. [CrossRef]
85. Hong, J.; Wang, W.; Huang, K.; Yang, W.; Zhao, Y.; Xiao, B.; Gao, Y.; Moosavi-Movahedi, Z.; Ahmadian, S.; Bohlooli, M.; et al. A self-assembled nano-cluster complex based on cytochrome c and nafion: An efficient nanostructured peroxidise. *Biochem. Eng. J.* **2012**, *65*, 16–22. [CrossRef]

86. Rui, Q.; Liangliang, S.; Aoting, Q.; Ruolin, W.; Yingli, A.; Linqi, S. Artificial Peroxidase/Oxidase Multiple Enzyme System Based on Supramolecular Hydrogel and Its Application as a Biocatalyst for Cascade Reactions. *ACS Appl. Mater. Interfaces* **2015**, *7*, 16694–16705. [CrossRef]
87. Zeng, H.H.; Qiu, W.B.; Zhang, L.; Liang, R.P.; Qiu, J.D. Lanthanide Coordination Polymer Nanoparticles as an Excellent Artificial Peroxidase for Hydrogen Peroxide Detection. *Anal. Chem.* **2016**, *88*, 6342–6348. [CrossRef] [PubMed]
88. Sun, L.; Ding, Y.; Jiang, Y.; Liu, Q. Montmorillonite-loaded ceria nanocomposites with superior peroxidase-like activity for rapid colorimetric detection of H_2O_2. *Sens. Actuators B Chem.* **2017**, *239*, 848–856. [CrossRef]
89. Chen, Q.; Chen, J.; Gao, C.; Zhang, M.; Chen, J.; Qiu, H. Hemin-functionalized WS2 nanosheets as highly active peroxidase mimetics for label-free colorimetric detection of H_2O_2 and glucose. *Analyst* **2015**, *140*, 2857–2863. [CrossRef]
90. Liu, Q.; Yang, Y.; Lv, X.; Ding, Y.; Zhang, Y.; Jing, J.; Xu, C. Onestep synthesis of uniform nanoparticles of porphyrin functionalized ceria with promising peroxidase mimetics for H_2O_2 and glucose colorimetric detection. *Sens. Actuators B Chem.* **2017**, *240*, 726–734. [CrossRef]
91. Li, Z.; Yang, X.; Yang, Y.; Tan, Y.; He, Y.; Liu, M.; Liu, X.; Yuan, Q. Peroxidase-mimicking nanozyme with enhanced activity and high stability based on metal-support interaction. *Chem. Eur. J.* **2018**, *24*, 409–415. [CrossRef]
92. Zhang, Y.; Wang, Y.N.; Sun, X.T.; Chen, L.; Xu, Z.R. Boron nitride nanosheet/CuS nanocomposites as mimetic peroxidase for sensitive colorimetric detection of cholesterol. *Sens. Actuators B Chem.* **2017**, *246*, 118–126. [CrossRef]
93. Jampaiah, D.; Reddy, T.S.; Coyle, V.E.; Nafady, A.; Bhargava, S.K. $Co_3O_4@CeO_2$ hybrid flower-like microspheres: A strong synergistic peroxidase-mimicking artificial enzyme with high sensitivity for glucose detection. *J. Mater. Chem. B* **2017**, *5*, 720–773. [CrossRef]
94. Vallabani, N.V.; Karakoti, A.S.; Singh, S. ATP-mediated intrinsic peroxidase-like activity of Fe_3O_4-based nanozyme: One step detection of blood glucose at physiological pH. *Colloids Surf. B Biointerfaces* **2017**, *153*, 52–60. [CrossRef]
95. Wang, Z.H.; Chen, M.; Shu, J.X.; Li, Y. One-step solvothermal synthesis of $Fe_3O_4@Cu@Cu_2O$ nanocomposite as magnetically recyclable mimetic peroxidase. *J. Alloys Compd.* **2016**, *682*, 432–440. [CrossRef]
96. Liu, Q.Y.; Zhang, L.Y.; Li, H.; Jia, Q.; Jiang, Y.; Yang, Y.; Zhu, R. One-pot synthesis of porphyrin functionalized gamma-Fe_2O_3 nanocomposites as peroxidase mimics for H_2O_2 and glucose detection. *Mat. Sci. Eng. C Mater.* **2015**, *55*, 193–200. [CrossRef] [PubMed]
97. Roy, A.; Sahoo, R.; Ray, C.; Dutta, S.; Pa, T. Soft template induced phase selective synthesis of Fe_2O_3 nanomagnets: One step towards peroxidase-mimic activity allowing colorimetric sensing of thioglycolic acid. *RSC Adv.* **2016**, *6*, 32308–32318. [CrossRef]
98. Nagvenkar, A.P.; Gedanken, A. Cu0.89Zn0.11O, A New Peroxidase-Mimicking Nanozyme with High Sensitivity for Glucose and Antioxidant Detection. *ACS Appl. Mater. Interfaces* **2016**, *8*, 22301–22308. [CrossRef] [PubMed]
99. Lu, N.; Zhang, M.; Ding, L.; Zheng, J.; Zeng, C.; Wen, Y.; Liu, G.; Aldalbahi, A.; Shi, J.; Song, S.; et al. Yolk-shell nanostructured $Fe_3O_4@C$ magnetic nanoparticles with enhanced peroxidase-like activity for label-free colorimetric detection of H_2O_2 and glucose. *Nanoscale* **2017**, *9*, 4508–4515. [CrossRef]
100. He, W.W.; Zhou, Y.T.; Wamer, W.G.; Hu, X.; Wu, X.; Zheng, Z.; Boudreau, M.D.; Yin, J.J. Intrinsic catalytic activity of Au nanoparticles with respect to hydrogen peroxide decomposition and superoxide scavenging. *Biomaterials* **2013**, *34*, 765–773. [CrossRef]
101. Bustami, Y.; Murray, M.Y.; William, A.A. Manipulation of Fe/Au Peroxidase-Like Activity for Development of a Nanocatalytic-Based Assay. *J. Eng. Sci.* **2017**, *13*, 29–52. [CrossRef]
102. He, W.; Han, X.; Jia, H.; Cai, J.; Zhou, Y.; Zheng, Z. AuPt Alloy Nanostructures with Tunable Composition and Enzyme-like Activities for Colorimetric Detection of Bisulfide. *Sci. Rep.* **2017**, *7*, e40103. [CrossRef]
103. He, S.B.; Chen, R.T.; Wu, Y.Y.; Wu, G.W.; Peng, H.P.; Liu, A.L.; Deng, H.H.; Xia, X.H.; Chen, W. Improved enzymatic assay for hydrogen peroxide and glucose by exploiting the enzyme-mimicking properties of BSA-coated platinum nanoparticles. *Mikrochim. Acta* **2019**, *186*, e778. [CrossRef]
104. Jin, L.; Meng, Z.; Zhang, Y.; Cai, S.; Zhang, Z.; Li, C.; Shang, L.; Shen, Y. Ultrasmall Pt Nanoclusters as Robust Peroxidase Mimics for Colorimetric Detection of Glucose in Human Serum. *ACS Appl. Mater. Interfaces* **2017**, *9*, 10027–10033. [CrossRef]

105. Cao, G.; Jiang, X.; Zhang, H.; Croleyb, T.R.; Yin, J. Mimicking horseradish peroxidase and oxidase using ruthenium nanomaterials. *RSC Adv.* **2017**, *7*, 52210–52217. [CrossRef]
106. Wei, J.; Chen, X.; Shi, S.; Moa, S.; Zheng, N. An investigation of the mimetic enzyme activity of two-dimensional Pd-based nanostructures. *Nanoscale* **2015**, *7*, 19018–19026. [CrossRef] [PubMed]
107. Shkotova, L.; Bohush, A.; Voloshina, I.; Smutok, O.; Dzyadevych, S. Amperometric biosensor modified with platinum and palladium nanoparticles for detection of lactate concentrations in wine. *SN Appl. Sci.* **2019**, *1*, 306. [CrossRef]
108. Kluenker, M.; Tahir, M.N.; Ragg, R.; Korschelt, K.; Simon, P.; Gorelik, T.E.; Barton, B.; Shylin, S.I.; Panthöfer, M.; Herzberger, J.; et al. Pd@Fe$_2$O$_3$ Superparticles with Enhanced Peroxidase Activity by Solution Phase Epitaxial Growth. *Chem. Mater.* **2017**, *29*, 1134–1146. [CrossRef]
109. Li, R.; Zhen, M.; Guan, M.; Chen, D.; Zhang, G.; Ge, J.; Gong, P.; Wang, C.; Shu, C. A novel glucose colorimetric sensor based on intrinsic peroxidase-like activity of C$_{60}$-carboxyfullerenes. *Biosens. Bioelectron.* **2013**, *47*, 502–507. [CrossRef]
110. Sajjadi, S.; Keihan, A.H.; Norouzi, P.; Habibi, M.M.; Eskandari, K.; Shirazi, N.H. Fabrication of an Amperometric Glucose Biosensor Based on a Prussian Blue/Carbon Nanotube/Ionic Liquid Modified Glassy Carbon Electrode. *J. Appl. Biotechnol. Rep.* **2017**, *4*, 603–608.
111. Guerrero, L.A.; Fernández, L.; González, G.; Montero-Jiménez, M.; Uribe, R.; Díaz Barrios, A.; Espinoza-Montero, P.J. Peroxide Electrochemical Sensor and Biosensor Based on Nanocomposite of TiO$_2$ Nanoparticle/Multi-Walled Carbon Nanotube Modified Glassy Carbon Electrode. *Nanomaterials* **2020**, *10*, 64. [CrossRef]
112. Chen, Y.; Yuchi, Q.; Li, T.; Yang, G.; Miao, J.; Huang, C.; Liu, J.; Li, A.; Qin, Y.; Zhang, L. Precise engineering of ultra-thin Fe$_2$O$_3$ decorated Pt-based nanozymes via atomic layer deposition to switch off undesired activity for enhanced sensing performance. *Sens. Actuators B Chem.* **2020**, *305*, 127436. [CrossRef]
113. Guo, Y.; Deng, L.; Li, J.; Guo, S.; Wang, E.; Dong, S. Hemin-graphene hybrid nanosheets with intrinsic peroxidase-like activity for label-free colorimetric detection of single-nucleotide polymorphism. *ACS Nano* **2011**, *5*, 1282–1290. [CrossRef]
114. Kung, C.C.; Lin, P.Y.; Buse, F.J.; Xue, Y.; Yu, X.; Dai, L.; Liu, C.C. Preparation and characterization of three-dimensional graphene foam supported platinum-ruthenium bimetallic nanocatalysts for hydrogen peroxide based electrochemical biosensors. *Biosens. Bioelectron.* **2014**, *52*, 1–7. [CrossRef]
115. Song, Y.; Qu, K.; Zhao, C.; Ren, J.; Qu, X. Graphene oxide: Intrinsic peroxidase catalytic activity and its application to glucose detection. *Adv. Mater.* **2010**, *22*, 2206–2210. [CrossRef]
116. Li, X.-R.; Xu, M.; Chen, H.; Xu, J. Bimetallic Au@Pt@Au core–shell nanoparticles on graphene oxide nanosheets for high-performance H$_2$O$_2$ bi-directional sensing. *J. Mater. Chem. B* **2015**, *3*, 4355–4362. [CrossRef] [PubMed]
117. Li, S.; Li, H.; Chen, F.; Liu, J.; Zhang, H.; Yang, Z.; Wang, B. Strong coupled palladium nanoparticles decorated on magnetic graphene nanosheets as enhanced peroxidase mimetics for colorimetric detection of H$_2$O$_2$. *Dye. Pigment.* **2016**, *125*, 64–71. [CrossRef]
118. Lv, X.; Weng, J. Ternary composite of hemin, gold nanoparticles and graphene for highly efficient decomposition of hydrogen peroxide. *Sci. Rep.* **2013**, *3*, e3285. [CrossRef] [PubMed]
119. Sun, H.; Liu, X.; Wang, X.; Han, Q.; Qi, C.; Li, Y.; Wang, C.; Chen, Y.; Yang, R. Colorimetric determination of ascorbic acid using a polyallylamine-stabilized IrO$_2$/graphene oxide nanozyme as a peroxidase mimic. *Microchim. Acta* **2020**, *187*, 110. [CrossRef]
120. Darabdhara, G.; Sharma, B.; Das, M.R.; Boukherroub, R.; Szunerits, S. Cu-Ag bimetallic nanoparticles on reduced graphene oxidenanosheets as peroxidase mimic for glucose and ascorbic aciddetection. *Sens. Actuators B Chem.* **2017**, *238*, 842–851. [CrossRef]
121. Song, L.N.; Huang, C.; Zhang, W.; Ma, M.; Chen, Z.; Gu, N.; Zhang, Y. Graphene oxide-based Fe$_2$O$_3$ hybrid enzyme mimetic with enhanced peroxidase and catalase-like activities. *Colloid Surf. A* **2016**, *506*, 747–755. [CrossRef]
122. Dong, W.; Zhuang, Y.; Li, S.; Zhang, X.; Chai, H.; Huang, Y. High peroxidase-like activity of metallic cobalt nanoparticles encapsulated in metal–organic frameworks derived carbon for biosensing. *Sens. Actuators B Chem.* **2018**, *255*, 2050–2057. [CrossRef]
123. Karuppiah, C.; Palanisamy, S.; Chen, S.; Veeramani, V.; Periakaruppan, P. A novel enzymatic glucose biosensor and sensitive non-enzymatic hydrogen peroxide sensor based on graphene and cobalt oxide nanoparticles composite modified glassy carbon electrode. *Sens. Actuators B Chem.* **2014**, *196*, 450–456. [CrossRef]

124. Bahreini, M.; Movahedi, M.; Peyvandi, M.; Nematollahi, F.; Tehrani, H.S. Thermodynamics and kinetic analysis of carbon nanofibers as nanozymes. *Nanotechnol. Sci. Appl.* **2019**, *12*, 3–10. [CrossRef]
125. Shi, W.; Wang, Q.; Long, Y.; Cheng, Z.; Chen, S.; Zheng, H.; Huang, Y. Carbon nanodots as peroxidase mimetics and their applications to glucose detection. *Chem. Commun.* **2011**, *47*, 6695–6697. [CrossRef]
126. Zhang, Y.; Wu, C.; Zhou, X.; Wu, X.; Yang, Y.; Wu, H.; Guo, S.; Zhang, J. Graphene quantum dots/gold electrode and its application in living cell H_2O_2 detection. *Nanoscale* **2013**, *5*, 1816–1819. [CrossRef]
127. Zhang, L.; Hai, X.; Xia, C.; Chen, X.W.; Wang, J.H. Growth of CuO nanoneedles on graphene quantum dots as peroxidase mimics for sensitive colorimetric detection of hydrogen peroxide and glucose. *Sens. Actuators B Chem.* **2017**, *248*, 374–384. [CrossRef]
128. Fakhri, N.; Salehnia, F.; Beigi, S.M.; Aghabalazadeh, S.; Hosseini, M.; Ganjali, M.R. Enhanced peroxidase-like activity of platinum nanoparticles decorated on nickel- and nitrogen-doped graphene nanotubes: Colorimetric detection of glucose. *Microchim. Acta* **2019**, *186*, e385. [CrossRef] [PubMed]
129. Wu, L.; Wan, G.; Shi, S.; He, Z.; Xu, X.; Tang, Y.; Hao, C.; Wang, G. Atomic layer deposition-assisted growth of CuAl LDH on carbon fiber as a peroxidase mimic for colorimetric determination of H_2O_2 and glucose. *New J. Chem.* **2019**, *15*, 5826–5832. [CrossRef]
130. Sang, Y.; Huang, Y.; Li, W.; Ren, J.; Qu, X. Bioinspired Design of Fe^{3+}-doped mesoporous carbon nanospheres for enhanced nanozyme activity. *Chemistry* **2018**, *24*, 7259–7263. [CrossRef] [PubMed]
131. Zhao, Y.; Huo, D.; Bao, J.; Yang, M.; Chen, M.; Hou, J.; Fa, H.; Hou, C. Biosensor based on 3D graphene-supported Fe_3O_4 quantum dots as biomimetic enzyme for in situ detection of H_2O_2 released from living cells. *Sens. Actuators B Chem.* **2017**, *244*, 1037–1044. [CrossRef]
132. Song, H.; Zhao, H.; Zhang, X.; Xu, Y.; Cheng, X.; Gao, S.; Huo, L. Ahollow urchin-like α-mno2 as an electrochemical sensor for hydrogen peroxide and dopamine with high selectivity and sensitivity. *Microchim. Acta* **2019**, *186*, 210–221. [CrossRef]
133. Mei, H.; Wang, X.; Zeng, T.; Huang, L.; Wang, Q.; Ru, D.; Huang, T.; Tian, F.; Wu, H.; Gao, J. A nanocomposite consisting of gold nanobipyramids and multiwalled carbon nanotubes for amperometric nonenzymatic sensing of glucose and hydrogen peroxide. *Microchim. Acta* **2019**, *186*, 235–242. [CrossRef]
134. Zhang, Y.; Duan, Y.; Shao, Z.; Chen, C.; Yang, M.; Lu, G.; Xu, W.; Liao, X. Amperometric hydrogen peroxide sensor using a glassy carbon electrode modified with a nanocomposite prepared from ferumoxytol and reduced graphene oxide decorated with platinum nanoparticles. *Microchim. Acta* **2019**, *186*, 386. [CrossRef]
135. Zhang, C.; Zhang, Y.; Du, X.; Chen, Y.; Dong, W.; Han, B.; Chen, Q. Facile fabrication of Pt-Ag bimetallic nanoparticles decorated reduced graphene oxide for highly sensitive non-enzymatic hydrogen peroxide sensing. *Talanta* **2016**, *159*, 280–286. [CrossRef]
136. Singh, S.; Singh, M.; Mitra, K.; Singh, R.; Kumar, S.; Gupta, S.; Tiwari, I.; Ray, B. Electrochemical sensing of hydrogen peroxide using brominated graphene as mimetic catalase. *Electrochim. Acta* **2017**, *258*, 1435–1444. [CrossRef]
137. Fazli, G.; Bahabadi, S.E.; Adlnasab, L.; Ahmar, H. A glassy carbon electrode modified with a nanocomposite prepared from Pd/Al layered double hydroxide and carboxymethyl cellulose for voltammetric sensing of hydrogen peroxide. *Mikrochim. Acta* **2019**, *186*, 821. [CrossRef] [PubMed]
138. Neampet, S.; Ruecha, N.; Qin, J.; Wonsawat, W.; Chailapakul, O.; Rodthongkum, N. A nanocomposite prepared from platinum particles, polyaniline and a Ti_3C_2 MXene for amperometric sensing of hydrogen peroxide and lactate. *Mikrochim. Acta* **2019**, *186*, 752. [CrossRef] [PubMed]
139. Long, L.; Liu, X.; Chen, L.; Li, D.; Jia, J. A hollow CuO_x/NiO_y nanocomposite for amperometric and non-enzymatic sensing of glucose and hydrogen peroxide. *Microchim. Acta* **2019**, *186*, 74–84. [CrossRef] [PubMed]
140. Sivakumar, M.; Veeramani, V.; Chen, S.M.; Madhu, R.; Liu, S.B. Porous carbon-NiO nanocomposites for amperometric detection of hydrazine and hydrogen peroxide. *Microchim. Acta* **2019**, *186*, 59–66. [CrossRef]
141. Sha, R.; Vishnu, N.; Badhulika, S. Bimetallic Pt-Pd nanostructures supported on MoS_2 as an ultra-high performance electrocatalyst for methanol oxidation and nonenzymatic determination of hydrogen peroxide. *Microchim. Acta* **2018**, *185*, 399–409. [CrossRef]
142. Lyu, Y.P.; Wu, Y.S.; Wang, T.P.; Lee, C.L.; Chung, M.Y.; Lo, C.T. Hydrothermal and plasma nitrided electrospun carbon nanofibers for amperometric sensing of hydrogen peroxide. *Microchim. Acta* **2018**, *185*, 371–377. [CrossRef]

143. Liu, J.; Yang, C.; Shang, Y.; Zhang, P.; Liu, J.; Zheng, J. Preparation of a nanocomposite material consisting of cuprous oxide, polyaniline and reduced graphene oxide, and its application to the electrochemical determination of hydrogen peroxide. *Microchim. Acta* **2018**, *185*, 172–179. [CrossRef]
144. Annalakshmi, M.; Balasubramanian, P.; Chen, S.M.; Chen, T.W. Enzyme-free electrocatalytic sensing of hydrogen peroxide using a glassy carbon electrode modified with cobalt nanoparticle-decorated tungsten carbide. *Mikrochimica. Acta* **2019**, *186*, 265. [CrossRef]
145. Davidson, D. The Prussian blue paradox. *J. Chem. Educ.* **1937**, *14*, 238–241. [CrossRef]
146. Guari, Y.; Larionova, J. (Eds.) *Prussian Blue-Type Nanoparticles and Nanocomposites: Synthesis, Devices and Applications*; Jenny Stanford Publishing: Singapore, 2019; p. 314. ISBN 97898148000510.
147. Matos-Peralta, Y.; Antuch, M. Review—Prussian Blue and Its Analogs as Appealing Materials for Electrochemical Sensing and Biosensing. *J. Electrochem. Soc.* **2020**, *167*, 037510. [CrossRef]
148. Cinti, S.; Basso, M.; Moscone, D.; Arduini, F. A paper-based nanomodified electrochemical biosensor for ethanol detection in beers. *Anal. Chim. Acta* **2017**, *960*, 123–130. [CrossRef] [PubMed]
149. Ojwang, D.O. Prussian Blue Analogue Copper Hexacyanoferrate: Synthesis, Structure Characterization and Its Applications as Battery Electrode and CO_2 Adsorbent. Ph.D. Thesis, Stockholm University, Stockholm, Sweden, 13 October 2017. Available online: http://www.diva-portal.org/smash/record.jsf?pid=diva2%3A1136799&dswid=8693 (accessed on 7 May 2020).
150. Komkova, M.A.; Andreev, E.A.; Ibragimova, O.A.; Karyakin, A.A. Prussian Blue based flow-through (bio)sensors in power generation mode: New horizons for electrochemical analyzers. *Sens. Actuators B Chem.* **2019**, *292*, 284–288. [CrossRef]
151. Ivanov, V.D. Four decades of electrochemical investigation of Prussian blue. *Ionics* **2020**, *26*, 531–547. [CrossRef]
152. Chen, W.; Gao, G.; Jin, Y.; Deng, C. A facile biosensor for $A\beta_{40}O$ based on fluorescence quenching of prussian blue nanoparticles. *Talanta* **2020**, *216*, 120390. [CrossRef]
153. Meng, X.; Gao, L.; Fan, K.; Yan, X. Nanozyme-Based Tumor Theranostics. In *Nanostructure Science and Technology*; Yan, X., Ed.; Springer: Singapore, 2020; pp. 425–457. ISBN 978-981-15-1489-0.
154. He, L.; Li, Z.; Guo, C.; Hu, B.; Wang, M.; Zhang, Z.; Du, M. Bifunctional bioplatform based on NiCo Prussian blue analogue: Label-free impedimetric aptasensor for the early detection of carcino-embryonic antigen and living cancer cells. *Sens. Actuators B Chem.* **2019**, *298*. [CrossRef]
155. Tabrizi, M.A.; Shamsipur, M.; Saber, R.; Sarkar, S.; Zolfaghari, N. An ultrasensitive sandwich-type electrochemical immunosensor for the determination of SKBR-3 breast cancer cell using rGOTPA/FeHCF-labeld Anti-HCT as a signal tag. *Sens. Actuators B Chem.* **2017**, *243*, 823–830. [CrossRef]
156. Wang, M.; Hu, B.; Ji, H.; Song, Y.; Liu, J.; Peng, D.; He, L.; Zhang, Z. Aptasensor based on hierarchical core-shell nanocomposites of zirconium hexacyanoferrate nanoparticles and mesoporous mFe_3O_4@mC: Electrochemical quantitation of epithelial tumor marker mucin-1. *ACS Omega* **2017**, *2*, 6809–6818. [CrossRef]
157. Gao, Z.; Li, Y.; Zhang, C.; Zhang, S.; Jia, Y.; Dong, Y. An enzyme-free immunosensor for sensitive determination of procalcitonin using NiFe PBA nanocubes@TB as the sensing matrix. *Anal. Chim. Acta* **2020**, *1097*, 169–175. [CrossRef]
158. Jia, Q.; Li, Z.; Guo, C.; Huang, X.; Kang, M.; Song, Y.; He, L.; Zhou, N.; Wang, M.; Zhang, Z.; et al. PEGMA-modified bimetallic NiCo Prussian blue analogue doped with Tb (III) ions: Efficiently pH-responsive and controlled release system for anticancer drug. *Chem. Eng.* **2020**, *389*. [CrossRef]
159. Qin, Z.; Chen, B.; Huang, X.; Mao, Y.; Li, Y.; Yang, F.; Gu, N. Magnetic internal heating-induced high performance Prussian blue nanoparticle preparation and excellent catalytic activity. *Dalton. Trans.* **2019**, *48*, 17169–17173. [CrossRef] [PubMed]
160. Nwamba, O.C.; Echeverria, E.; McIlroy, D.N.; Shreeve, J.M.; Aston, D.R. Electrochemical stability and capacitance of in-situ synthesized Prussian blue on thermally-activated graphite. *SN. Appl. Sci.* **2019**, *1*, 731–746. [CrossRef]
161. Sahar, S.; Zeb, A.; Ling, C.; Raja, A.; Wang, G.; Ullah, N.; Lin, X.M.; Xu, A.-W. A Hybrid VOx Incorporated Hexacyanoferrate Nanostructured Hydrogel as a Multienzyme Mimetic via Cascade Reactions. *ACS Nano* **2020**, *14*, 3017–3031. [CrossRef] [PubMed]
162. Komkova, M.A.; Karyakina, E.E.; Karyakin, A.A. Catalytically Synthesized Prussian Blue Nanoparticles Defeating Natural Enzyme Peroxidase. *J. Am. Chem. Soc.* **2018**, *140*, 11302–11307. [CrossRef] [PubMed]

163. Koshiyama, T.; Tanaka, M.; Honjo, M.; Fukunaga, Y.; Okamura, T.; Ohba, M. Direct Synthesis of Prussian Blue Nanoparticles in Liposomes Incorporating Natural Ion Channels for Cs+ Adsorption and Particle Size Control. *Langmuir* **2018**, *34*, 1666–1672. [CrossRef]
164. Itaya, K.; Shoji, N.; Uchida, I. Catalysis of the reduction of molecular oxygen to water at Prussian blue modified electrodes. *J. Am. Chem. Soc.* **1984**, *106*, 3423–3429. [CrossRef]
165. Itaya, K.; Uchida, I.; Neff, V.D. Electrochemistry of polynuclear transition metal cyanides: Prussian blue and its analogues. *Acc. Chem. Res.* **1986**, *19*, 162–168. [CrossRef]
166. Cui, F.; Deng, Q.; Sun, L. Prussian blue modified metal–organic framework MIL-101(Fe) with intrinsic peroxidase-like catalytic activity as a colorimetric biosensing platform. *RSC Adv.* **2015**, *5*, 98215–98221. [CrossRef]
167. Keihan, A.H.; Karimi, R.R.; Sajjadi, S. Wide dynamic range and ultrasensitive detection of hydrogen peroxide based on beneficial role of gold nanoparticles on the electrochemical properties of prussian blue. *J. Electroanal. Chem.* **2020**, *862*, 114001. [CrossRef]
168. Pang, H.; Zhang, Y.; Cheng, T.; Lai, W.Y.; Huang, W. Uniform manganese hexacyanoferrate hydrate nanocubes featuring superior performance for low-cost supercapacitors and nonenzymatic electrochemical sensors. *Nanoscale* **2015**, *7*, 16012–16019. [CrossRef]
169. Wang, C.; Ren, G.; Yuan, B.; Zhang, W.; Lu, M.; Liu, J.; Li, K.; Lin, Y. Enhancing Enzyme-like Activities of Prussian Blue Analog Nanocages by Molybdenum Doping: Toward Cytoprotecting and Online Optical Hydrogen Sulfide Monitoring. *Anal. Chem.* **2020**, *92*, 7822–7830. [CrossRef] [PubMed]
170. Zhang, W.; Hu, S.; Yin, J.J.; He, W.; Ma, M.; Gu, N.; Zhang, Y. Prussian Blue Nanoparticles as Multienzyme Mimetics and Reactive Oxygen Species Scavengers. *J. Am. Chem. Soc.* **2016**, *138*, 5860–5865. [CrossRef] [PubMed]
171. Zhang, X.Z.; Zhou, Y.; Zhang, W.; Zhang, Y.; Gu, N. Polystyrene@Au@Prussian Blue Nanocomposites with Enzyme-Like Activity and Their Application in Glucose Detection. *Colloids Surf. A* **2016**, *490*, 291–299. [CrossRef]
172. Zhou, D.; Zeng, K.; Yang, M. Gold Nanoparticle-Loaded Hollow Prussian Blue Nanoparticles with Peroxidase-Like Activity for Colorimetric Determination of L-Lactic Acid. *Microchim. Acta* **2019**, *186*, 121. [CrossRef] [PubMed]
173. Zhang, W.; Ma, D.; Du, J. Prussian blue nanoparticles as peroxidase mimetics for sensitive colorimetric detection of hydrogen peroxide and glucose. *Talanta* **2014**, *120*, 362–367. [CrossRef]
174. Zhang, W.; Zhang, Y.; Chen, Y.; Li, S.; Gu, N.; Hu, S.; Sun, Y.; Chen, X.; Li, Q. Prussian Blue Modified Ferritin as Peroxidase Mimetics and Its Applications in Biological Detection. *J. Nanosci. Nanotechnol.* **2013**, *13*, 60–67. [CrossRef]
175. Wang, T.; Fu, Y.; Chai, L.; Chao, L.; Bu, L.; Meng, Y.; Chen, C.; Ma, M.; Xie, Q.; Yao, S. Filling Carbon Nanotubes with Prussian Blue Nanoparticles of High Peroxidase-Like Catalytic Activity for Colorimetric Chemo- And Biosensing. *Chem. Eur. J.* **2014**, *20*, 2623–2630. [CrossRef]
176. Karyakin, A.A.; Gitelmacher, O.V.; Karyakina, E.E. Prussian Blue-Based First-Generation Biosensor. A Sensitive Amperometric Electrode for Glucose. *Anal. Chem.* **1995**, *67*, 2419–2423. [CrossRef]
177. Karyakin, A.A.; Karyakina, E.E.; Gorton, L. Amperometric Biosensor for Glutamate Using Prussian Blue-Based "Artificial Peroxidase" as a Transducer for Hydrogen Peroxide. *Anal. Chem.* **2000**, *72*, 1720–1723. [CrossRef]
178. Karpova, E.V.; Karyakina, E.E.; Karyakin, A.A. Communication—Accessing Stability of Oxidase-Based Biosensors via Stabilizing the Advanced H_2O_2 Transducer. *J. Electrochem. Soc.* **2017**, *164*, B3056–B3058. [CrossRef]
179. Komkova, M.A.; Pasquarelli, A.; Andreev, E.A.; Galushin, A.A.; Karyakin, A.A. Prussian Blue modified boron-doped diamond interfaces for advanced H_2O_2 electrochemical sensors. *Electrochim. Acta* **2020**, *339*. [CrossRef]
180. Vokhmyanina, D.V.; Andreeva, K.D.; Komkova, M.A.; Karyakina, E.E.; Karyakin, A.A. Artificial peroxidase" nanozyme—Enzyme based lactate biosensor. *Talanta* **2020**, *208*, 120393. [CrossRef] [PubMed]
181. Sitnikova, N.A.; Borisova, A.V.; Komkova, M.A.; Karyakin, A.A. Superstable advanced hydrogen peroxide transducer based on transition metal hexacyanoferrates. *Anal. Chem.* **2011**, *83*, 2359–2363. [CrossRef]
182. Zhou, L.; Wu, S.; Xu, H.; Zhao, Q.; Zhang, Z.; Yao, Y. Preparation of poly (N-acetylaniline)–Prussian blue hybrid composite film and its application to hydrogen peroxide sensing. *Anal. Methods* **2014**, *6*, 8003–8010. [CrossRef]

183. Zhao, H.-C.; Zhang, P.; Li, S.-H.; Luo, H.-X. Cobalt hexacyanoferrate-modified graphene platform electrode and its electrochemical sensing toward hydrogen peroxide. *Chin. J. Anal. Chem.* **2017**, *45*, 830–836. [CrossRef]
184. Yang, S.; Li, G.A.; Wang, G.; Zhao, J.; Hu, M.; Qu, L. A novel nonenzymatic H_2O_2 sensor based on cobalt hexacyanoferrate nanoparticles and graphene composite modified electrode. *Sens. Actuators B Chem.* **2015**, *208*, 593. [CrossRef]
185. Lee, S.H.; Chung, J.-H.; Park, H.-K.; Lee, G.-J. A Simple and Facile Glucose Biosensor Based on Prussian Blue Modified Graphite String. *J. Sens.* **2016**, *2016*, 1859292. [CrossRef]
186. Pandey, P.C.; Panday, D.; Pandey, A.K. Polyethylenimine mediated synthesis of copper-iron and nickel-iron hexacyanoferrate nanoparticles and their electroanalytical applications. *J. Electroanal. Chem.* **2016**, *780*, 90–102. [CrossRef]
187. Huang, J.; Fang, X.; Liu, X.; Lu, S.; Li, S.; Yang, Z.; Feng, X. High-Linearity Hydrogen Peroxide Sensor Based on Nanoporous Gold Electrode. *J. Electrochem. Soc.* **2019**, *166*, B814. [CrossRef]
188. Kumar, A.V.; Harish, S.; Joseph, J.; Phani, K.L. Nix–Fe(1−x)Fe(CN)6 hybrid thin films electrodeposited on glassy carbon: Effect of tuning of redox potentials on the electrocatalysis of hydrogen peroxide. *J. Electroanal. Chem.* **2011**, *659*, 128–133. [CrossRef]
189. Niu, Q.; Bao, C.; Cao, X.; Liu, C.; Wang, H.; Lu, W. Ni–Fe PBA hollow nanocubes as efficient electrode materials for highly sensitive detection of guanine and hydrogen peroxide in human whole saliva. *Biosens. Bioelectron.* **2019**, *141*, 111445. [CrossRef] [PubMed]
190. Sangeetha, N.S.; Narayanan, S.S. A novel bimediator amperometric sensor for electrocatalytic oxidation of gallic acid and reduction of hydrogen peroxide. *Anal. Chim. Acta* **2014**, *828*, 34–45. [CrossRef] [PubMed]
191. Clausmeyer, J.; Actis, P.; Córdoba, A.L.; Korchev, Y.; Schuhmann, W. Nanosensors for the detection of hydrogen peroxide. *Electrochem. Commun.* **2014**, *40*, 28–30. [CrossRef]
192. Ju, H.; Zhang, X.; Wang, J. NanoBiosensing: Principles, Development and Application, Biological and Medical Physics, Biomedical Engineering Series; Springer: New York, NY, USA, 2016; p. 586. ISBN 978-1441996213.
193. Dimcheva, N. Nanostructures of noble metals as functional materials in biosensors. *Curr. Opin. Electrochem.* **2020**, *19*, 35–41. [CrossRef]
194. Singh, S. Nanomaterials Exhibiting Enzyme-Like Properties (Nanozymes): Current Advances and Future Perspectives. *Front. Chem.* **2019**, *7*, 46. [CrossRef]
195. Fan, J.; Yin, J.J.; Ning, B.; Wu, X.; Hu, Y.; Ferrari, M.; Anderson, G.J.; Wei, J.; Zhao, Y.; Nie, G. Direct evidence for catalase and peroxidase activities of ferritin-platinum nanoparticles. *Biomater* **2011**, *32*, 1611–1618. [CrossRef]
196. Shah, K.; Bhagat, S.; Varade, D.; Singh, S. Novel synthesis of polyoxyethylene cholesteryl ether coated Fe-Pt nanoalloys: A multifunctional and cytocompatible bimetallic alloy exhibiting intrinsic chemical catalysis and biological enzyme-like activities. *Colloids Surf. A Physicochem. Eng. Asp.* **2018**, *553*, 50–57. [CrossRef]
197. Su, L.; Feng, J.; Zhou, X.; Ren, C.; Li, H.; Chen, X. Colorimetric detection of urine glucose based $ZnFe_2O_4$ magnetic nanoparticles. *Anal. Chem.* **2012**, *84*, 5753–5758. [CrossRef]
198. Zhang, X.; He, S.; Chen, Z.; Huang, Y. $CoFe_2O_4$ Nanoparticles as Oxidase Mimic-Mediated Chemiluminescence of Aqueous Luminol for Sulfite in White Wines. *J. Agric. Food Chem.* **2013**, *61*, 840–847. [CrossRef]
199. Dalapati, R.; Sakthivel, B.; Ghosalya, M.K.; Dhakshinamoorthy, A.; Biswas, S. A cerium-based metal–organic framework having inherent oxidase-like activity applicable for colorimetric sensing of biothiols and aerobic oxidation of thiols. *Cryst. Eng. Comm.* **2017**, *19*, 5915–5925. [CrossRef]
200. Estevez, A.Y.; Stadler, B.; Erlichman, J.S. In vitro analysis of catalase-, oxidase- and SOD-mimetic activity of commercially available and custom synthesized cerium oxide nanoparticles and assessment of neuroprotective effects in a hippocampal brain slice model of ischemia. *FASEB J.* **2017**, *31*, 693–695. [CrossRef]
201. Song, Y.; Zhao, M.; Li, H.; Wang, X.; Cheng, Y.; Ding, L.; Chen, S. Facile preparation of urchin-like $NiCo_2O_4$ microspheres as oxidase mimetic for colormetric assay of hydroquinone. *Sens. Actuators B Chem.* **2018**, *255*, 1927–1936. [CrossRef]
202. Yan, X.; Song, Y.; Wu, X.; Zhu, C.; Su, X.; Du, D.; Lin, Y. Oxidase-mimicking activity of ultrathin MnO_2 nanosheets in colorimetric assay of acetylcholinesterase activity. *Nanoscale* **2017**, *9*, 2317–2323. [CrossRef] [PubMed]
203. Cui, M.; Zhao, Y.; Wang, C.; Song, Q. The oxidase-like activity of iridium nanoparticles, and their application to colorimetric determination of dissolved oxygen. *Microchim. Acta* **2017**, *184*, 3113–3119. [CrossRef]
204. Zhang, H.; Lu, L.; Kawashima, K.; Okumura, M.; Haruta, M.; Toshima, N. Synthesis and Catalytic Activity of Crown Jewel-Structured (IrPd)/Au Trimetallic Nanoclusters. *Adv. Mater.* **2015**, *27*, 1383–1388. [CrossRef]

205. Deng, H.-H.; Lin, X.-L.; Liu, Y.-H.; Li, K.-L.; Zhuang, Q.-Q.; Peng, H.-P.; Chen, W. Chitosan-stabilized platinum nanoparticles as effective oxidase mimics for colorimetric detection of acid phosphatase. *Nanoscale* **2017**, *9*, 10292–10300. [CrossRef]
206. Jin, X.; Zhao, M.; Shen, J.; Yan, W.; He, L.; Thapa, P.S.; Ren, S.; Subramaniam, B.; Chaudhari, R.V. Exceptional performance of bimetallic Pt_1Cu_3/TiO_2 nanocatalysts for oxidation of gluconic acid and glucose with O_2 to glucaric acid. *J. Catal.* **2015**, *330*, 323–329. [CrossRef]
207. He, W.; Liu, Y.; Yuan, J.; Yin, J.J.; Wu, X.; Hu, X.; Zhang, K.; Liu, J.; Chen, C.; Ji, Y.; et al. Au@Pt nanostructures as oxidase and peroxidase mimetics for use in immunoassays. *Biomater* **2011**, *32*, 1139–1147. [CrossRef]
208. Zhang, H.; Toshima, N. Preparation of novel Au/Pt/Ag trimetallic nanoparticles and their high catalytic activity for aerobic glucose oxidation. *Appl. Catal. A Gen.* **2011**, *400*, 9–13. [CrossRef]
209. Khawaji, M.; Zhang, Y.; Loh, M.; Graça, I.; Ware, E.; Chadwick, D. Composition dependent selectivity of bimetallic Au-Pd NPs immobilised on titanate nanotubes in catalytic oxidation of glucose. *Appl. Catal. B Environ.* **2019**, *256*, 117799. [CrossRef]
210. Guo, S.; Fang, Q.; Li, Z.; Zhang, J.; Zhang, J.; Li, G. Efficient base-free direct oxidation of glucose to gluconic acid over TiO_2-supported gold clusters. *Nanoscale* **2019**, *11*, 1326–1334. [CrossRef] [PubMed]
211. Xie, J.; Zhang, X.; Wang, H.; Zheng, H.; Huang, Y. Analytical and environmental applications of nanoparticles as enzyme mimetics. *TrAC Trends Anal. Chem.* **2012**, *39*, 114–129. [CrossRef]
212. Ortega-Liebana, M.C.; Bonet-Aleta, J.; Hueso, J.L.; Santamaria, J. Gold-Based Nanoparticles on Amino-Functionalized Mesoporous Silica Supports as Nanozymes for Glucose Oxidation. *Catalysts* **2020**, *10*, 333. [CrossRef]
213. Luo, W.J.; Zhu, C.F.; Su, S.; Li, D.; He, Y.; Huang, Q.; Fan, C.H. Self-Catalyzed, Self-Limiting Growth of Glucose Oxidase-Mimicking Gold Nanoparticles. *ACS Nano* **2010**, *4*, 7451–74584. [CrossRef] [PubMed]
214. Kusema, B.T.; Murzin, D.Y. Catalytic oxidation of rare sugars over gold catalysts. *Catal. Sci. Technol.* **2013**, *3*, 297–307. [CrossRef]
215. Benkó, T.; Beck, A.; Geszti, O.; Katona, R.; Tungler, A.; Frey, K.; Guczi, L.; Schay, Z. Selective oxidation of glucose versus CO oxidation over supported gold catalysts. *Appl. Catal. A Gen.* **2010**, *388*, 31–36. [CrossRef]
216. Cao, X.; Wang, N. A novel non-enzymatic glucose sensor modified with Fe_2O_3 nanowire arrays. *Analyst* **2011**, *136*, 4241–4246. [CrossRef]
217. Megías-Sayago, C.; Ivanova, S.; López-Cartes, C.; Centeno, M.A.; Odriozola, J.A. Gold catalysts screening in base-free aerobic oxidation of glucose to gluconic acid. *Catal. Today* **2017**, *279*, 148–154. [CrossRef]
218. Megías-Sayago, C.; Bobadilla, L.F.; Ivanova, S.; Penkova, A.; Centeno, M.A.; Odriozola, J.A. Gold catalyst recycling study in base-free glucose oxidation reaction. *Catal. Today* **2018**, *301*, 72–77. [CrossRef]
219. Okatsu, H.; Kinoshita, N.; Akita, T.; Ishida, T.; Haruta, M. Deposition of gold nanoparticles on carbons for aerobic glucose oxidation. *Appl. Catal. A Gen.* **2009**, *369*, 8–14. [CrossRef]
220. Lang, N.J.; Liu, B.; Liu, J. Characterization of glucose oxidation by gold nanoparticles using nanoceria. *J. Colloid Interface Sci.* **2014**, *428*, 78–83. [CrossRef] [PubMed]
221. Fan, L.; Lou, D.D.; Wu, H.A.; Zhang, X.Z.; Zhu, Y.F.; Gu, N.; Zhang, Y. A Novel AuNP-Based Glucose Oxidase Mimic with Enhanced Activity and Selectivity Constructed by Molecular Imprinting and O_2-Containing Nanoemulsion Embedding. *Adv. Mater. Interfaces* **2018**, *5*, 1801070. [CrossRef]
222. Ma, W.; Hafez, M.E.; Ma, H.; Long, Y.-T. Unveiling the Intrinsic Catalytic Activities of Single Gold Nanoparticle-based Enzyme Mimetics. *Angew. Chem.* **2019**, *58*, 6327–6332. [CrossRef]
223. Cao, Y.; Liu, X.; Iqbal, S.; Miedziak, P.J.; Edwards, J.K.; Armstrong, R.D.; Morgan, D.J.; Wang, J.; Hutchings, G.J. Base-free oxidation of glucose to gluconic acid using supported gold catalysts. *Catal. Sci. Technol.* **2016**, *6*, 107–117. [CrossRef]
224. Gao, H.; Xiao, F.; Ching, C.B.; Duan, H. One-Step Electrochemical Synthesis of PtNi Nanoparticle-Graphene Nanocomposites for Nonenzymatic Amperometric Glucose Detection. *ACS Appl. Mater. Interfaces* **2011**, *3*, 3049–3057. [CrossRef]
225. Li, Z.; Chen, Y.; Xin, Y.; Zhang, Z. Sensitive electrochemical nonenzymatic glucose sensing based on anodized CuO nanowires on three-dimensional porous copper foam. *Sci. Rep.* **2015**, *5*, 16115. [CrossRef]
226. Ju, L.; Wu, G.; Lu, B.; Li, X.; Wu, H.; Liu, A. Non-enzymatic Amperometric Glucose Sensor Based on Copper Nanowires Decorated Reduced Graphene Oxide. *Electroanalysis* **2016**, *28*, 2543–2551. [CrossRef]
227. Zhang, X.; Shi, H.; Chi, Q.; Liu, X.; Chen, L. Cellulose-supported Pd nanoparticles: Effective for the selective oxidation of glucose into gluconic acid. *Polym. Bull.* **2019**, *77*, 1003–1014. [CrossRef]

228. Kou, B.; Yuan, Y.; Yuan, R.; Chai, Y. Electrochemical biomolecule detection based on the regeneration of high-efficiency cascade catalysis for bifunctional nanozymes. *Chem. Commun.* **2020**, *56*, 2276–2279. [CrossRef]
229. Lin, Y.; Li, Z.; Chen, Z.; Ren, J.; Qu, X. Mesoporous silica-encapsulated gold nanoparticles as artificial enzymes for self-activated cascade catalysis. *Biomater* **2013**, *34*, 2600–2610. [CrossRef]
230. Han, L.; Zhang, H.; Chen, D.; Li, F. Protein-Directed Metal Oxide Nanoflakes with Tandem Enzyme-Like Characteristics: Colorimetric Glucose Sensing Based on One-Pot Enzyme-Free Cascade Catalysis. *Adv. Funct. Mater.* **2018**, *28*, e1800018. [CrossRef]
231. Glucose Oxidase BRENDA. Available online: https://www.brenda-enzymes.org/all_enzymes.php?ecno=1.1.3.4&table=Turnover_%20Number#TABSingh (accessed on 5 May 2020).
232. Miao, F.; Tao, B.; Sun, L.; Liu, T.; You, J.; Wang, L.; Chu, P.K. Amperometric glucose sensor based on 3D ordered nickel–palladium nanomaterial supported by silicon MCP array. *Sens. Actuators Chem.* **2009**, *141*, 338–342. [CrossRef]
233. Ye, J.-S.; Hong, B.-D.; Wu, Y.-S.; Chen, H.-R.; Lee, C.-L. Heterostructured palladium-platinum core-shell nanocubes for use in a nonenzymatic amperometric glucose sensor. *Microchim. Acta* **2016**, *183*, 3311–3320. [CrossRef]
234. Liu, M.; Liu, R.; Chen, W. Graphene wrapped Cu_2O nanocubes: Non-enzymatic electrochemical sensors for the detection of glucose and hydrogen peroxide with enhanced stability. *Biosens. Bioelectron.* **2013**, *45*, 206–212. [CrossRef] [PubMed]
235. Cao, L.; Wang, P.; Chen, L.; Wu, Y.; Di, J. A photoelectrochemical glucose sensor based on gold nanoparticles as a mimic enzyme of glucose oxidase. *RSC Adv.* **2019**, *9*, 15307–15313. [CrossRef]
236. Espro, C.; Marini, S.; Giusi, D.; Ampelli, C.; Neri, G. Non-enzymatic screen printed sensor based on Cu_2O nanocubes for glucose determination in bio-fermentation processes. *J. Electroanal. Chem.* **2020**, *873*, 114354. [CrossRef]
237. Zhang, P.; Sun, D.; Cho, A.; Weon, S.; Lee, S.; Lee, J.; Choi, W. Modified carbon nitride nanozyme as bifunctional glucose oxidase-peroxidase for metal-free bioinspired cascade photocatalysis. *Nat. Commun.* **2019**, *10*, 940. [CrossRef]
238. Beden, B.; Largeaud, F.; Kokoh, K.B.; Lamy, C. Fourier transform infrared reflectance spectroscopic investigation of the electrocatalytic oxidation of d-glucose: Identification of reactive intermediates and reaction products. *Electrochim. Acta* **1996**, *41*, 701–709. [CrossRef]
239. Chandra, R.; Chowdhary, P. Properties of bacterial laccases and their application in bioremediation of industrial wastes. *Environ. Sci. Process. Impacts* **2015**, *17*, 326–342. [CrossRef]
240. Fathali, Z.; Rezaei, S.; Faramarzi, M.A.; Habibi-Rezaei, M. Catalytic phenol removal using entrapped cross-linked laccase aggregates. *Int. J. Biol. Macromol.* **2019**, *122*, 359–366. [CrossRef]
241. Galli, C.; Madzak, C.; Vadalà, R.; Jolivalt, C.; Gentili, P. Concerted electron/proton transfer mechanism in the oxidation of phenols by laccase. *Chembiochem* **2013**, *14*, 2500–2505. [CrossRef] [PubMed]
242. Jones, S.M.; Solomon, E.I. Electron transfer and reaction mechanism of laccases. *Cell. Mol. Life Sci.* **2015**, *72*, 869–883. [CrossRef] [PubMed]
243. Vernekar, M.; Lele, S.S. Laccase: Properties and Applications. *BioResources* **2009**, *4*, 1694–1717.
244. Yesilada, O.; Birhanli, E.; Geckil, H. Bioremediation and Decolorization of Textile Dyes by White Rot Fungi and Laccase Enzymes. In *Mycoremediation and Environmental Sustainability*. Fungal Biology; Prasad, R., Ed.; Springer: Berlin, Germany, 2018; pp. 121–153.
245. Shekher, R.; Sehgal, S.; Kamthania, M.; Kumar, A. Laccase: Microbial sources, production, purification, and potential biotechnological applications. *Enzym. Res.* **2011**, *2011*, 217861. [CrossRef]
246. Gao, L.; Yan, X. Nanozymes: An emerging field bridging nanotechnology and biology. *Sci. China Life Sci.* **2016**, *59*, 400–402. [CrossRef]
247. He, X.; Tan, L.; Chen, D.; Wu, X.; Ren, X.; Zhang, Y.; Meng, X.; Tang, F. Fe_3O_4-Au@mesoporous SiO_2 microspheres: An ideal artificial enzymatic cascade system. *Chem. Commun.* **2013**, *49*, 4643–4645. [CrossRef]
248. Zhou, Y.B.; Liu, B.W.; Yang, R.H.; Liu, J.W. Filling in the gaps between nanozymes and enzymes: Challenges and opportunities. *Bioconjug. Chem.* **2017**, *28*, 2903–2909. [CrossRef]
249. Tanaka, Y.; Hoshino, W.; Shimizu, S.; Youfu, K.; Aratani, N.; Maruyama, N.; Fujita, S.; Osuka, A. Thermal splitting of bis-Cu(II) octaphyrin (1.1.1.1.1.1.1.1) into two Cu(II) porphyrins. *J. Am. Chem. Soc.* **2004**, *126*, 3046–3047. [CrossRef]

250. Zhou, L.; Powell, D.; Nicholas, K.M. Tripodal bis(imidazole) thioether copper(I) complexes: Mimics of the Cu(b) site of hydroxylase enzymes. *Inorg. Chem.* **2006**, *45*, 3840–3842. [CrossRef]
251. Rivas, M.V.; De Leo, L.P.; Hamer, M.; Carballo, R.; Williams, F.J. Self-assembled monolayers of disulfide Cu porphyrins on Au surfaces: Adsorption induced reduction and demetalation. *Langmuir* **2011**, *27*, 10714–10721. [CrossRef]
252. Uhlmann, C.; Swart, I.; Repp, J. Controlling the orbital sequence in individual Cu-phthalocyanine molecules. *Nano Lett.* **2013**, *13*, 777–780. [CrossRef] [PubMed]
253. Ge, S.; Li, D.; Huang, S.; Chen, W.; Tu, G.; Yu, S.; He, Q.; Gong, J.; Li, Y.; Hong, C.; et al. A magnetically recyclable Fe_3O_4@C@TNCuPc composite catalyst for chromogenic identification of phenolic pollutants. *J. Mol. Catal. A Chem.* **2015**, *410*, 193–201. [CrossRef]
254. Ren, X.; Liu, J.; Ren, J.; Tang, F.; Meng, X. One-pot synthesis of active copper-containing carbon dots with laccase-like activities. *Nanoscale* **2015**, *7*, 19641–19646. [CrossRef] [PubMed]
255. Meng, X.Q.; Fan, K.L. Application of nanozymes in disease diagnosis. *Prog. Biochem. Biophys.* **2018**, *45*, 218–236. [CrossRef]
256. Tang, Y.; Qiu, Z.Y.; Xu, Z.B.; Gao, L.Z. Antibacterial mechanism and applications of nanozymes. *Prog. Biochem. Biophys.* **2018**, *45*, 118–128. [CrossRef]
257. Lv, Y.; Ma, M.; Huang, Y.; Xia, Y. Carbon Dot Nanozymes: How to Be Close to Natural Enzymes. *Chem. Eur. J.* **2019**, *25*, 954. [CrossRef]
258. Shams, S.; Ahmad, W.; Memon, A.H.; Wei, Y.; Yuan, Q.; Liang, H. Facile synthesis of laccase mimic Cu/H 3 BTC MOF for efficient dye degradation and detection of phenolic pollutants. *RSC Adv.* **2019**, *9*, 40845–40854. [CrossRef]
259. Zhou, P.; Shi, R.; Yao, J.-F.; Sheng, C.-F.; Li, H. Supramolecular self-assembly of nucleotide–metal coordination complexes: From simple molecules to nanomaterials. *Coordin. Chem. Rev.* **2015**, *292*, 107–143. [CrossRef]
260. Liang, H.; Zhang, Z.; Yuan, Q.; Liu, J. Self-healing metal-coordinated hydrogels using nucleotide ligands. *Chem. Commun.* **2015**, *51*, 15196–15199. [CrossRef]
261. Frey, N.A.; Peng, S.; Cheng, K.; Sun, S. Magnetic nanoparticles: Synthesis, functionalization, and applications in bioimaging and magnetic energy storage. *Chem. Soc. Rev.* **2009**, *38*, 2532–2542. [CrossRef]
262. Liang, H.; Lin, F.; Zhang, Z.; Liu, B.; Jiang, S.; Yuan, Q.; Liu, J. Multicopper laccase mimicking nanozymes with nucleotides as ligands. *ACS Appl. Mater. Inter.* **2017**, *9*, 1352–1360. [CrossRef] [PubMed]
263. Zhang, S.; Lin, F.; Yuan, Q.; Liu, J.; Li, Y.; Liang, H. Robust magnetic laccase-mimicking nanozyme for oxidizing o-phenylenediamine and removing phenolic pollutants. *J. Environ. Sci. (China)* **2020**, *88*, 103–111. [CrossRef] [PubMed]
264. Huang, H.; Lei, L.; Bai, J.; Zhang, L.; Song, D.; Zhao, J.; Li, J.; Li, Y. Efficient elimination and detection of phenolic compounds in juice using laccase mimicking nanozymes. *Chin. J. Chem. Eng.* **2020**. [CrossRef]
265. Wang, J.; Huang, R.; Qi, W.; Su, R.; Binks, B.P.; He, Z. Construction of a bioinspired laccase-mimicking nanozyme for the degradation and detection of phenolic pollutants. *Appl. Catal. B Environ.* **2019**, *254*, 452–462. [CrossRef]
266. Manea, F.; Houillon, F.B.; Pasquato, L.; Scrimin, P. Nanozymes: Gold-nanoparticle-based transphosphorylation catalysts. *Angew. Chem. Int. Ed. Engl.* **2004**, *43*, 6165–6169. [CrossRef] [PubMed]
267. Wang, Y.; He, C.; Li, W.; Zhang, J.; Fu, Y. Catalytic Performance of Oligonucleotide-Templated Pt Nanozyme Evaluated by Laccase Substrates. *Catal. Lett.* **2017**, *147*, 2144–2152. [CrossRef]
268. Andrei, V.; Sharpe, E.; Vasilescu, A.; Andreescu, S. A single use electrochemical sensor based on biomimetic nanoceria for the detection of wine antioxidants. *Talanta* **2016**, *156*, 112–118. [CrossRef] [PubMed]
269. Tortolini, C.; Bollella, P.; Zumpano, R.; Favero, G.; Mazzei, F.; Antiochia, R. Metal Oxide Nanoparticle Based Electrochemical Sensor for Total Antioxidant Capacity (TAC) Detection in Wine Samples. *Biosensors* **2018**, *8*, 108. [CrossRef]
270. Wang, X.; Liu, J.; Qu, R.; Wang, Z.; Huang, Q. The laccase-like reactivity of manganese oxide nanomaterials for pollutant conversion: Rate analysis and cyclic voltammetry. *Sci. Rep.* **2017**, *7*, 7756. [CrossRef] [PubMed]
271. Rodríguez-Delgado, M.; Orona-Navar, C.; García-Morales, R.; Hernandez-Luna, C.; Parra, R.; Mahlknecht, J.; Ornelas-Soto, N. Biotransformation kinetics of pharmaceutical and micropollutants in groundwaters by a laccase cocktail from *Pycnoporus sanguineus* CS43 fungi. *Int. Biodeterior. Biodegrad.* **2016**, *108*, 34–41.

272. Garcia, L.F.; Souza, A.R.; Lobón, G.S.; dos Santos, W.T.P.; Alecrim, M.F.; Santiago, M.F.; de Sotomayor, R.L.Á.; Gil, E.S. Efficient Enzyme-Free Biomimetic Sensors for Natural Phenol Detection. *Molecules* **2016**, *21*, 1060. [CrossRef] [PubMed]
273. Bin, Z.; Yanhong, C.; Jiaojiao, X. Biomimetic oxidase sensor based on functionalized surface of carbon nanotubes and iron prophyrins for catechol detection. *Bioprocess Biosyst. Eng.* **2018**, *42*. [CrossRef] [PubMed]

© 2020 by the authors. Licensee MDPI, Basel, Switzerland. This article is an open access article distributed under the terms and conditions of the Creative Commons Attribution (CC BY) license (http://creativecommons.org/licenses/by/4.0/).

Review

3D-Printed Immunosensor Arrays for Cancer Diagnostics

Mohamed Sharafeldin [1], Karteek Kadimisetty [2], Ketki S. Bhalerao [1], Tianqi Chen [1] and James F. Rusling [1,3,4,*]

1. Department of Chemistry, University of Connecticut, Storrs, CT 06269, USA; mohamed.sharafeldin@uconn.edu (M.S.); ketki.bhalerao@uconn.edu (K.S.B.); tianqi.chen@uconn.edu (T.C.)
2. LifeSensors Inc., 271 Great Valley Parkway, Suite 100, Malvern, PA 19355, USA; karteek.kadimisetty@gmail.com
3. Department of Surgery and Neag Cancer Center, UConn Health, Farmington, CT 06032, USA
4. School of Chemistry, National University of Ireland at Galway, Galway H91 TK33, Ireland
* Correspondence: james.rusling@uconn.edu

Received: 20 July 2020; Accepted: 6 August 2020; Published: 12 August 2020

Abstract: Detecting cancer at an early stage of disease progression promises better treatment outcomes and longer lifespans for cancer survivors. Research has been directed towards the development of accessible and highly sensitive cancer diagnostic tools, many of which rely on protein biomarkers and biomarker panels which are overexpressed in body fluids and associated with different types of cancer. Protein biomarker detection for point-of-care (POC) use requires the development of sensitive, noninvasive liquid biopsy cancer diagnostics that overcome the limitations and low sensitivities associated with current dependence upon imaging and invasive biopsies. Among many endeavors to produce user-friendly, semi-automated, and sensitive protein biomarker sensors, 3D printing is rapidly becoming an important contemporary tool for achieving these goals. Supported by the widely available selection of affordable desktop 3D printers and diverse printing options, 3D printing is becoming a standard tool for developing low-cost immunosensors that can also be used to make final commercial products. In the last few years, 3D printing platforms have been used to produce complex sensor devices with high resolution, tailored towards researchers' and clinicians' needs and limited only by their imagination. Unlike traditional subtractive manufacturing, 3D printing, also known as *additive manufacturing*, has drastically reduced the time of sensor and sensor array development while offering excellent sensitivity at a fraction of the cost of conventional technologies such as photolithography. In this review, we offer a comprehensive description of 3D printing techniques commonly used to develop immunosensors, arrays, and microfluidic arrays. In addition, recent applications utilizing 3D printing in immunosensors integrated with different signal transduction strategies are described. These applications include electrochemical, chemiluminescent (CL), and electrochemiluminescent (ECL) 3D-printed immunosensors. Finally, we discuss current challenges and limitations associated with available 3D printing technology and future directions of this field.

Keywords: 3D printing; POC; microfluidics; immunosensor; cancer; biomarkers

1. Introduction

Cancer is one of the leading causes of death worldwide. Globally, it was responsible for approximately 9.6 million deaths in 2018 [1]. A major contributing factor to the high mortality is late diagnosis due to the unavailability of modern diagnostic tools in low income countries and their limited accessibility or application in developed countries. Currently, cancer diagnosis rely on techniques such as magnetic resonance imaging (MRI), computed tomography (CT), endoscopy, mammography and pathological examination of tissue biopsies [2–4]. Because the tumor needs to be located first with these techniques, in the majority of cancer cases, cancers will only be found as patients start to show symptoms, where treatment options become limited and health is already in jeopardy [5]. Providing early diagnosis and effective screening for different cancers are major challenges to improve life expectancy and treatment outcomes [6].

The crucial need for effective cancer screening and accessible diagnostic tools has driven research endeavors utilizing cancer biomarkers in liquid biopsy samples like blood, urine, and saliva. Analyzing cancer markers in liquid biopsy samples overcome hurdles associated with solid tumor biopsy as it provides a rapid, precise, and non-invasive assay strategy [7,8], and does not require a tumor to be located. Protein biomarkers provide an opportunity to assess risk of cancer development and to detect cancer at very early stage where treatment interventions are most effective [9]. Sensors utilizing ligand-binding assay formats for candidate cancer protein biomarkers have drawn a remarkable interest in the last two decades indicated by increased number of publications as seen in Figure 1.

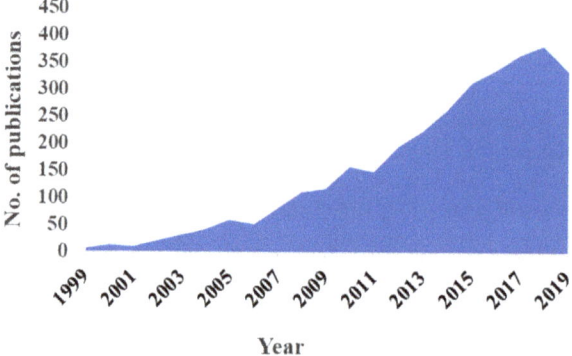

Figure 1. Number of publications per year focusing on protein biomarker cancer diagnostics from 1999–2019. Results generated using web of science® report generation tool for "Cancer Protein sensors" on 9 May 2020.

Several analytical strategies have been adapted for development of ultrasensitive detection of protein biomarkers associated with different types of cancer. Immunoassay format is the most commonly used technique for analysis due to the inherent specificity associated with the use of antibodies as molecular-recognition agents [10,11]. Immunoassay formats have been integrated with several detection strategies in order to develop cancer diagnostics including colorimetric [12], fluorescence [13], electrochemical [14], chemiluminescence [15], electrochemiluminescence [16], and plasmon resonance sensors [17].

The vast development of sensor assembly techniques encompassed a great leap in the progress of immunoassay-based cancer biomarker diagnostics. Several immunoassay-based diagnostic tools have been recently commercialized with promises of unprecedented sensitivities including electrochemiluminescence-based Meso Scale Discovery (MSD) platform and single molecule array technology (Simoa® technology) by Quanterix® (MA, USA) [18,19]. Although these techniques provided an excellent opportunity for early diagnosis and understanding cancer biology, they are limited to centralized laboratories as they require expensive bulky instrumentation and trained operators. With advanced manufacturing techniques, sensors developed acquired better automation, higher sensitivities, far-reaching accessibility, and multiplexing capabilities [20,21]. These developments promise the realization of point-of-care (POC) testing for cancer screening, detection, and staging. Among various approaches utilized for fulfilling these POC testing requirements, additive manufacturing furnished a launchpad for innovative yet easy cancer biomarker sensor manufacturing tool [22].

Additive manufacturing, also known as 3D printing, is making rapid inroads in manufacturing, and advanced fabrications that are quickly moving into production [23]. 3D printing has been utilized in development and fabrication of sensors for detection of glucose [24], drugs [25], trace elements [26], neurotransmitters [27], nucleic acids [28], and proteins [29]. The vast scope and innovative nature of 3D printing in development of biosensors is backed by the versatility of options provided by the immense progress in the design and production of desktop 3D printers. These 3D printers now offer access to hundreds of printable substrate materials that can be used to make products with spectrum of properties including transparency, electrical conductivity, elasticity, chemical and thermal resistivity. This has allowed the design and fabrication of previously hard to rapidly fabricate sensors and sensor arrays at very low cost.

The process of 3D printing is rather simple, a computer software often available at no charge to academics, is used to create the initial design. The initial computer aided design (CAD) file is then sliced into printable layers using slicing software specific for each desktop 3D printer. The 3D printer then physically prints layers on top of each other to form the final product [30]. Recently, a more advanced printing technique, tomographic volumetric 3D printing, has been used to print the whole design in one step eliminating the need for slicing and layer-by-layer printing which in turn drastically reduces printing time [31].

In this review, we provide a summary of different 3D printing techniques currently utilized in desktop 3D printers. In addition, we describe, as examples, the application of the 3D printing technology for development and fabrication of electrochemical, chemiluminescence, and electrochemiluminescence sensors for cancer biomarker proteins. We also discuss the design of complex hybrid sensors that can be achieved with 3D printing. Finally, we give a brief account on limitations associated with current 3D printing technologies and possible future impact of 3D printing.

2. 3D Printing Technologies

Several strategies have been adapted in the production of desktop 3D printers depending on the principle of printing and nature of printable substrate. Printable substrates can be divided into polymerizable materials, thermoplastics, and curable inks. Based on the nature of the substrate material different printing technologies have been developed. In this section, we will describe common 3D printing technologies, principles, limitations, and applications.

2.1. Fused Deposition Modeling (FDM)

FDM is one of the first 3D printing techniques utilized in desktop 3D printers commercially available at a large scale. This method utilizes a thermoplastic material which is melted through a heated printing nozzle and extruded onto solid printing platform. The printing nozzle head moves in X, Y, and Z directions in order to extrude layers on top of each other. Once extruded out of the printing head, thermoplastics tend to restore their solid nature before being heated in the printing nozzle (Figure 2A). FDM offers a low-cost 3D printing technology with easily changeable materials and minimal waste [32]. It also allows printing of multi-component parts simultaneously with printers equipped with double or triple printing nozzles [33]. The most common substrate materials utilized in FDM printers are acrylonitrile butadiene styrene (ABS) and polylactic acid (PLA) [34]. Interestingly, FDM was successfully utilized in printing of carbonaceous conductive substrates like carbon black, graphene, or carbon nanotubes mixed with thermoplastic materials [35]. Printing conductive materials paved the way for 3D printing of electronic components [36], integrated electrochemical sensors [37], and batteries [38]. FDM suffers from several drawbacks associated with low printing resolution (~5 µm), relatively high energy requirements, hazardous vapors, and adhesion problems with multi-materials printing [39].

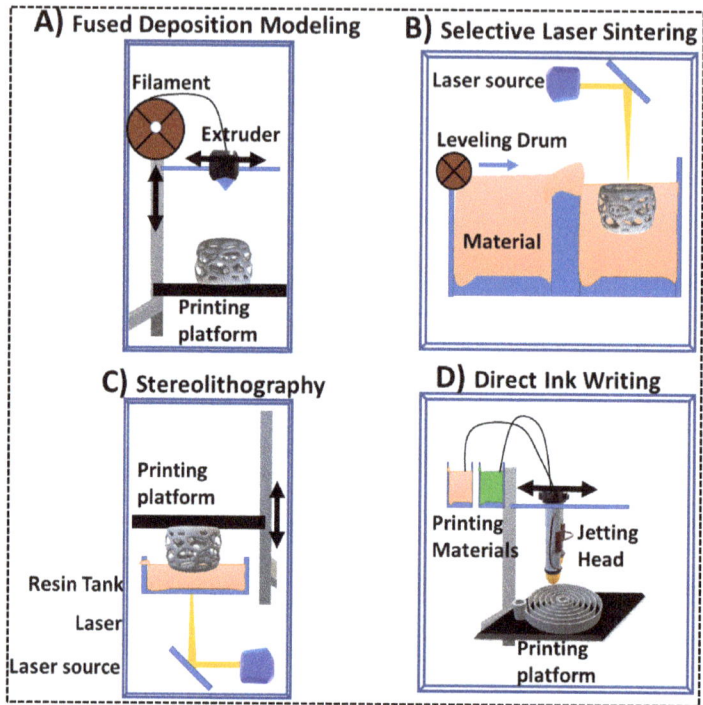

Figure 2. Schematic illustration of 3D printing techniques commonly utilized in prototyping and production of cancer immunosensors. Arrows indicates direction of movement.

2.2. Photopolymerization/Stereolithography (SLA)

Photopolymerization and digital light processing utilizes a selective light-aided curing process of special polymerizable liquid where polymerization is induced by light of specific wavelength. A moving laser beam hitting a resin-filled vat at a programmed pattern controlled by the printer software induces polymerization at that specific point. As laser beam moves, it cures a pre-designed layer point-by-point onto a solid printing platform that was immersed into the resin tank at a very close proximity to the tank bottom. After each layer the printing platform moves in the Y direction for next layer to be printed allow printing of successive layers on top of each other (Figure 2C). Digital light processing utilizes a similar system while curing one layer at each projection which allow faster printing compared to point-by-point curing [40]. Polymerizable liquids usually consist of monomers and oligomers of epoxides and acrylates mixed with photo-initiators. Light focused onto a single point or a projected layer activates crosslinking of the monomers and/or oligomers in the liquid mixture into a solid polymer. The moving printing platform allows the liquid mixture to fill the small gap between the tank bottom and the printing surface for the next layer to be printed. SLA is used to print materials with a spectrum of different properties including transparent, flexible, heat resistant, castable, and biocompatible pieces [41,42]. SLA also offers very good print resolution (~0.2 µm) at a relatively affordable cost [43]. A new technology utilizing two-photon polymerization was recently introduced to achieve nanometer resolution SLA printing [44]. Although, SLA can be used to produce smooth, high resolution complex architectures, it is still limited to printing a single polymer, by the need for internal supports, and requires post printing cleaning and processing.

2.3. Direct Ink Writing (DIW)/Material Jetting

Direct inkjet printing is utilized to deposit materials from inkjet print head onto a build platform or substrate. It depends on the on-demand delivery of adjustable amounts of printable materials onto the printing platform drop-by-drop in a predetermined pattern for layer-by-layer printing. Actuation of material jetting from inkjet head is either thermally or piezoelectrically induced. Thermal jetting requires a heating element that produces a localized heat enough to increase the vapor pressure inside the printing head, leading to ejection of small volume of material. While piezoelectric jetting utilizes a piezoelectric element, which upon application of electric current, generates a mechanical movement enough to eject the ink. An inkjet printing head moving in X, Y and Z directions guides the deposition of a viscous liquid, hydrogel or dispersion onto the printing platform in a desired pattern (Figure 2D) [45]. DIW can be also used to bond powder particles together with the aid of an adhesive polymer [46]. Due to the mechanical ejection mechanism, DIW is time consuming that may take up to a few days for a single print and usually requires post-printing drying. It is utilized for high resolution printing of electronic circuits [47], smooth flexible materials [48], cells and biomaterials [49]. Printing of biomaterials is possible with DIW due to its room temperature piezoelectric printing capabilities and ease of loading into biomimetic printable dispersions.

2.4. Selective Laser Sintering (SLS)

Similar to photopolymerization, SLS utilizes a high energy CO_2 laser beam to melt powder beads into metallic, plastic or ceramic layers. The laser beam scans through the powder bead to print one layer, then printing platform moves down allowing addition of a fresh layer of the powder beads which is then sintered and bind to the previous layer (Figure 2B). This system allows printing complex structures without the need for internal supports, usually required in SLA, while recovered powder after printing can be reused reducing cost and waste [42]. Recently, the use of high energy electronic beam was used to replace laser beam for printing of metallic objects with improved mechanical properties [50]. The use of SLS in sensor fabrication is not common as it is limited to metal and ceramic printing and relatively high operation and maintenance cost compared to other printing techniques.

2.5. Tomographic Volumetric Additive Manufacturing

Unlike other 3D printing techniques that utilize sequential layer-by-layer printing, multi-beam 3D printing technology prints objects by irradiating transparent resin from multiple angles simultaneously (Figure 3). This results in the polymerization of the whole object at the same time, greatly reducing printing time and permitting the production of highly complex architectures [31]. Although it offers a very high throughput ($>10^5$ mm^3/hr), it suffers from low resolution (80 µm) and complex printing setup [51].

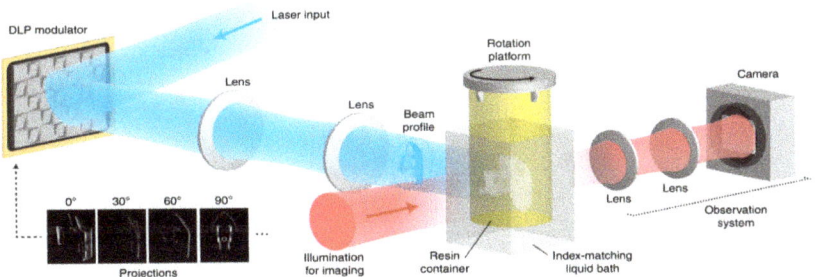

Figure 3. Schematic illustration of tomographic volumetric 3D printing. Reproduced with the permission from [31]. Copyright (2020) Springer Nature available under the terms and conditions of Creative Commons Attribution 4.0 International License.

2.6. Bioprinting

3D bioprinting encompass a spectrum of printing strategies compatible with the labile nature of cells and biomaterials. 3D bioprinting can be adapted in some of the aforementioned 3D printing technologies like direct ink writing, while most of them would have inherent limitations associated with the thermal stability and compatibility of biomaterials. Alternatively, 3D printing techniques, aimed primarily at biomaterial and cell printing, have been developed to overcome these limitations [52]. Syringe-Based extrusion bioprinting, an extrusion-based technique, extrudes a bio-ink at an optimized rate from a moving syringe onto printing platform. Bio-Ink is usually a photocurable polymer or a hydrogel loaded with cells or biomaterials. Extrusion is driven through pneumatic, mechanical, or solenoid valve activation process and the extruded bio-ink is printed layer-by-layer in a computer-aided predetermined pattern. Syringe-Based extrusion is the most widely used bioprinting technique and has been used in most of commercially available bio printers [53]. Syringe-Based extrusion has been mainly utilized in the production of cell-laden architectures for tissue engineering and drug testing [54–56]. Another common bioprinting technique is the laser-induced bioprinting, where a biomaterial-laden layer adsorbed on a donor substrate is transferred under the effect of pulsed laser to the receiving substrate. Donor substrate is usually a transparent material like glass coated with laser-absorbing layer that generate a high pressure upon exposure to pulsed laser propelling itself out of the underlying glass onto the receiving substrate [57]. Laser-Induced bioprinting has been investigated for printing of cell-laden collagen architectures and tissue models for cancer studies [58]. Bioprinting promises an easy, accessible, and cost-effective one-step fabrication platform for organs on chip for cancer studies and online high throughput drug-tissue interactions [59,60].

3. 3D Printed Electrochemical Sensors

A typical electrochemical biosensor contains two key parts: electrochemical transduction element (e.g., electrode) and biorecognition element (BRE) (e.g., antibody or enzyme). Analyte (e.g., proteins and nucleic acids) from the sample interacts with BRE and generates electroactive products, of which the electrochemical signal is then converted through the transducer and measured. Traditional electrode materials mainly fall into four groups: (1) Noble metals (gold, silver, platinum) have their excellent conductivity, electron transfer kinetics and stability. Gold electrodes are especially favored in bioassays, easy to functionalize with biomolecules and have potential window of about −0.4 to 0.7 V vs. Ag/Ag/Cl at neutral pH [61,62]. (2) Semiconductors, including organic (polymer) and inorganic (indium tin oxide ITO) semiconductors, have lower cost and larger potential window (0.0 to 1.8 V [61]) than gold, but also lower conductivity. (3) Carbon-Based electrodes (pyrolytic graphite, glassy carbon, and graphene) are easy to process, and have large potential window but may have minor stability problems in some applications. (4) Conductive polymers offer a variety of material choices, but have lower conductivity than noble metal and carbon-based electrodes. Electric signals from the electrode are measured by methods such as potentiometry, voltammetry, impedance spectroscopy, conductometry and stripping techniques [63]. Most work discussed in this section employed voltammetry such as cyclic voltammetry (CV) and square wave voltammetry (SWV), and Ag/AgCl as the reference electrode.

Electrodes can be 3D printed using fused deposition modeling (FDM) and selective laser melting (SLM). The printing materials used in FDM are conductive filaments, which are polylactic acid (PLA) or acrylonitrile butadiene styrene (ABS) filament mixed with conductive carbonaceous material such graphite, graphene, carbon nanofibers, carbon nanotubes and carbon black. Two filaments graphene/PLA (Black Magic) and carbon black/PLA (Proto-pasta) are commercially available and can be used in electrode fabrication [64]. SLM uses metal powder to print electrodes, such as iron, steel, and aluminum. 3D printing technique brings great flexibility in electrochemical sensor design. Different electrode geometries can be printed, and the influence on sensor performance studied. The high precision improves the quality of 3D printed sensors, but challenges still exist. FDM printed electrodes have poor conductivity because of the low amount of conductive material in the filament, and therefore need surface treatment before use. Methods such as mechanical polishing, electrochemical or chemical activation and enzyme digestion are used to partially remove the non-conductive material from the electrode surface [64]. SLM 3D printers and metal powders are expensive and require some post-print cleaning [65].

3.1. 3D Printed Chip Integrated with Traditional Electrodes

Here, we class recently reported sensors into two types. The first is traditionally fabricated electrodes integrated into a 3D printed chip. One example is electrochemiluminescent (ECL) sensors, which will be specially covered in next section. Damiati et al. [66] developed such an array for real-time immunodetection of liver cancer cell HepG2. The recombinant S-layer fusion protein (rSbpA/ZZ) was recrystallized on the surface of a screen-printed gold electrode, serving as an intermediate layer to aid the efficient capture of anti-CD133 antibody, which recognizes and binds to the CD133 protein on the surface of liver cancer cells HepG2. A 3D microfluidic chamber was printed by FDM using co-polyester polymer (dimension: $1.5 \times 1 \times 7$ mm) and assembled to the electrode with a double-sided adhesive film (Figure 4A). Cyclic voltammetry (CV) was performed and a detection range of 1×10^5–3×10^6 cells/mL was reported. Similarly, a flow system based on multiwall carbon nanotube (MWCNT) electrode was present by the same group one year later, targeting hepatic oval cells (HOCs), which is an important origin of liver stem cells in hepatocellular carcinoma [67]. The crosslinking chemistry of chitosan and glutaraldehyde was applied on the electrode to immobilize oval cell marker antibody (anti-OV6), which binds to the surface maker OV6 from HOCs (Figure 4B). Digital light processing (DLP) was used to print the flow cell, housing the modified electrode. Square wave voltammetry (SWV) was performed in the assay, and a detection range of 1×10^2–5×10^5 cells/mL was reported. In Sun et al.'s work, the built-in electronics was further expanded to a 1.5 inch $\times$ 2.5 inch printed circuit board (PCB) connected to a smartphone, both harvesting energy from the phone and communicating data to the phone for analysis and display [68]. The system was designed to track secretory leukocyte protease inhibitor (SLPI), a biomarker in cystic fibrosis. A case was 3D printed to house the electronics with a screen-printed electrode inserted, then the whole system was connected to a smartphone (Figure 4C). CV was applied and a detection limit of 1 nM was achieved.

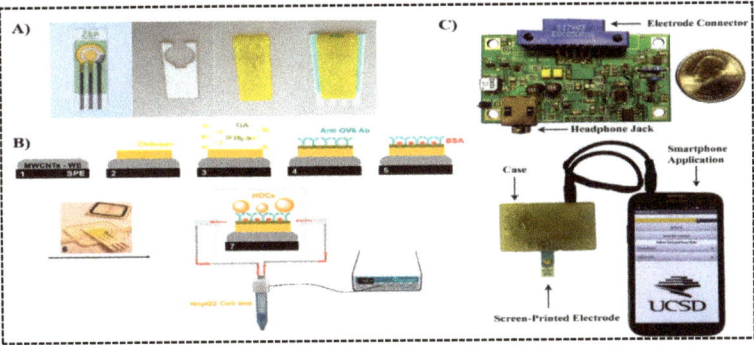

Figure 4. Pre-Fabricated electrodes integrated into 3D printed devices as electrochemical biosensors. (**A**) From left to right: the screen-printed electrode, adhesive layer, 3D printed microfluidic chamber (yellow), and the assembled device. The device was used for detection of liver cancer cell HepG2. Reproduced with permission from [66]. Copyright (2017) Elsevier. (**B**) Immunoassay procedures on a multiwall carbon nanotube (MWCNT) modified screen-printed electrode (SPE) (1–5), electrode-embedded 3D printed flow cell (6), and connected to a flow control system (7), and targeting hepatic oval cells (HOCs). Reproduced with permission from [67]. Copyright (2018) MDPI available under Creative Commons Attribution. (**C**) Printed circuit board module housed in a 3D printed case, with screen-printed electrode inserted, connected to the smart phone for powering, data communication, and display, tracking lung infection in cystic fibrosis. Reproduced with permission from [68]. Copyright (2016) Elsevier.

Progress has also been made developing lab-made electrodes, then combining with a 3D printed device. Scordo et al. [69] used wax printing and screen printing to fabricate a paper-based electrode equipped with both reference electrode (Ag/AgCl ink) and working electrode (graphite-based carbon black/prussian blue nanocomposite ink, or CB/PBNBs ink). Preloading the substrate onto the filter paper made this to be an 'all-in-paper', 'reagent-free' device. A 3D holder was printed by stereolithography (SLA) to house the electrode (Figure 5A). The assay monitored the activity of butyrylcholinesterase using amperometry detection. The sensor achieved a linear range of 1–12 IU/mL with a detection limit of 0.1 IU/mL, and was also tested in serum samples. Tang et al. designed multiple dual (ratiometric) [70,71] and single [72] aptasensors for the detection of carcinoembryonic antigen (CEA), a broad-spectrum biomarker of pancreatic carcinoma, breast cancer and gastric carcinoma, using photo-electrochemistry. The electrode was fabricated from organic or inorganic semiconductors. A 3D printed platform was used to house the whole system (Figure 5B). Photoelectric current was generated by CdTe quantum dots, harvesting light energy from the nanoparticles activated by the near-infrared light. A two working photoelectrode (WP) system was introduced in Figure 5B, where CEA aptamer 1 (A1) was immobilized on WP1, and CEA aptamer 2(A2)-gold nanoparticle conjugate with its complimentary DNA called capture DNA were immobilized on WP2. The binding of CEA to A2 released gold nanoparticles, leading to a signal change between two WPs, from WP1 > WP2 to WP1 < WP2. For detection, the constant potential was set at 0 V and the photocurrent~time curves for both electrodes were recorded. Lowest detection limit achieved in their works was 4.8 pg/mL.

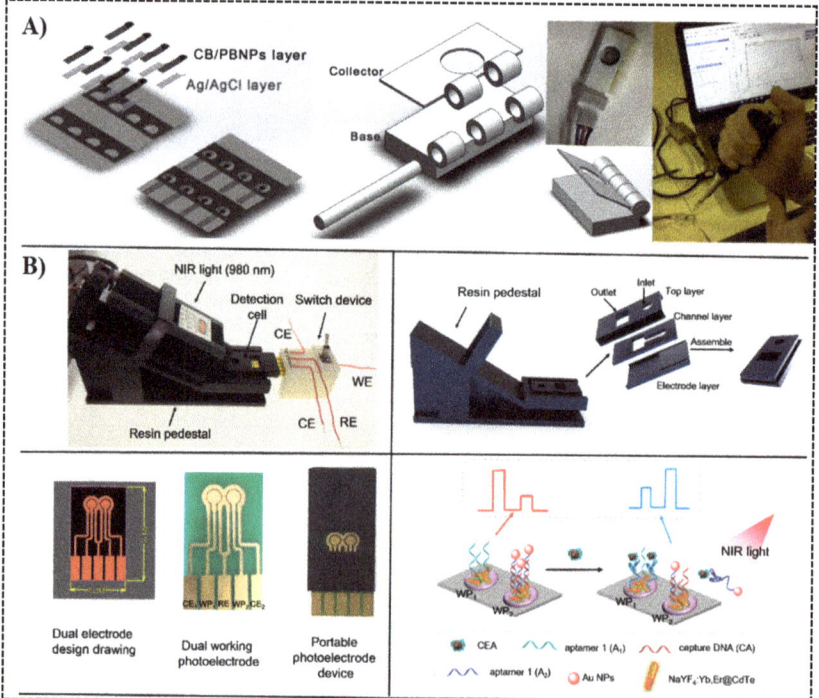

Figure 5. Self-designed electrodes integrated into 3D printed devices as electrochemical biosensors. (**A**) From left to right: wax- & screen-printed paper-based electrode, 3D printed holder for the electrode, and the connected system, used for detection of butyrylcholinesterase activity. Reproduced with permission from [69]. Copyright (2017) Elsevier. (**B**) Dual-Channel ratiometric photoelectrochemical detection of carcinoembryonic antigen (CEA) housed in a 3D printed device. Reproduced with permission from [70]. Copyright (2018) American Chemical Society.

3.2. 3D Printed Electrodes

Pumera et al. pioneered this area by introducing a helical-shaped stainless-steel electrode made by selective laser melting (SLM) printing [73]. The printed electrode had dimensions of 1.5 cm × 0.5 cm (Figure 6A). After deposition of an IrO_2 film, the steel-IrO_2 electrode gave excellent catalytic properties for oxygen generation and as a pH sensor. Moving forward onto biosensors, this group electro-plated the same SLM-printed helical-shaped steel electrode with gold to study DNA hybridization [28]. DNA recognition element SH-L probe was immobilized onto the gold electrode by thiol-gold interactions, the electrode was then blocked to cover any free surface, and incubated with DNA targets for hybridization. Finally, methylene blue solution was added, and methylene blue molecules intercalated into the double helix structure. Differential pulse voltammetry (DPV) was used to measure the reduction peak from the electroactive methylene blue and quantify the extent of DNA hybridization (Figure 6B). A detection range of 1–1000 nM was achieved.

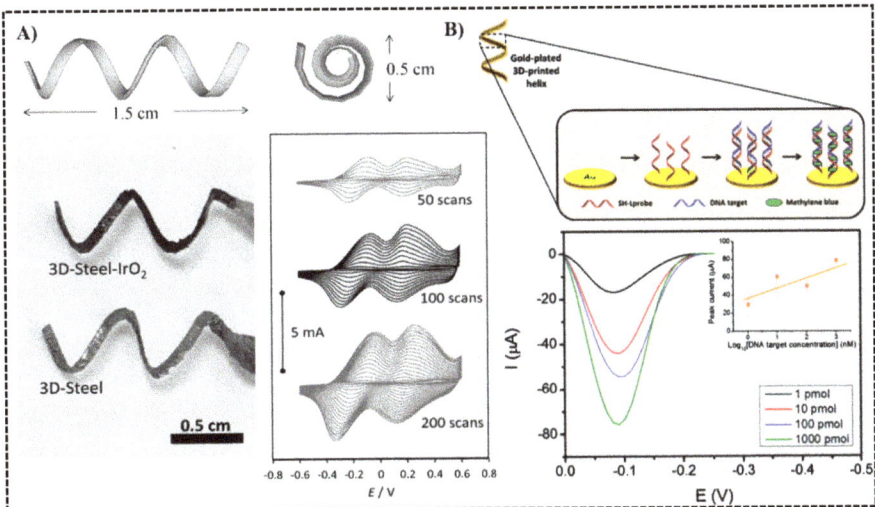

Figure 6. 3D printed electrodes as electrochemical biosensors. (**A**) Dimensions of the helical-shaped IrO2-deposited stainless-steel electrode printed by selective laser melting (SLM) and cyclic voltammograms. Reproduced with permission from [73]. Copyright (2015) Wiley. (**B**) Gold-Electroplated helical steel electrode used in measuring DNA hybridization. Differential pulse voltammograms at various DNA concentrations are shown. Reproduced with permission from [28]. Copyright (2017) Royal Society of Chemistry.

Many researches have detected metals ions and small organic molecules utilizing FDM or SLM printed electrodes, but not much has been done on the biomedical side—some analyte examples are H_2O_2, glucose, lactate, and dopamine [64], still not much on proteins and DNA. Several reasons exist: (1) Difficulty in protein/DNA immobilization onto the electrode. (2) Smaller microelectrodes compared to current printed ones are needed for immunoassays. Our group has reported measurements of various cancer biomarker proteins using an eight-electrode array fabricated by inkjet printing using gold nanoparticle ink [48,74–76]. Each printed electrode array has an overall surface area of 0.299 ± 0.015 mm^2, and the electrode contact is only 465 µm × 465 µm (~0.216 mm^2). The dimensions of reported FDM or SLM printed electrodes are a few cm or larger. (3) High quality electrodes are needed for immunoassays. Interference such as electrode fouling and non-specific binding from the sample matrix are problems for all electrochemical biosensors, but can be a bigger problem in the case of complex biological samples, which puts high requirement on non-specific binding inhibition on the printed electrodes.

4. 3D Printed ECL/CL Sensors

Electrochemiluminescent (ECL), chemiluminescent (CL), and nanoparticle-assisted assays utilizing signal amplification strategies have come into the limelight to produce immunosensors that can overcome sensitivity limitations. These immunosensors can have high sensitivity, low detection limits, low background signal, and enhanced signal transduction [77]. ECL biosensors utilize an ECL-active dye as the detection label, responsible for generating the signal via an ECL-producing pathway initiated by a complex redox reaction driven by a conductive electrode interface [78]. ECL active dye emits energy in the form of light as it transits from an excited state produced by the redox chemistry to the ground state when the proper potential is applied. Initial trends involved the development of 3D printed electrodes and channels along with ECL detection. Some of the commonly used ECL substrates include complexes of ruthenium [79,80], iridium [81], and osmium [82]. They can be used in solution, as polymerized films, as nanoparticle bead-based systems or as quantum dots [83–85]. Ruthenium poly(vinylpyridine) $[Ru(bpy)_2(PVP)_{10}]^{2+}$ (RuPVP) and ruthenium tris (2,20-bipyridyl) dichlororuthenium-(II) hexahydrate are among the most commonly used substrates [86,87].

ECL assays with 3D printed microfluidic arrays can be automated, cheap, and disposable. Our group used RuBPY silica nanoparticles to evaluate the chemical genotoxicity on DNA damage from cigarettes, electronic cigarettes and in aqueous environmental samples by ECL using a 3D printed device. Here, RuBPY was loaded into silica nanoparticles for signal amplification [88,89]. The redox reaction occurred upon applying a potential of 0.95 V vs SCE to produce ECL at 610 nm with ECL signal proportional to degree of DNA damage. The 3D printed device consisted of three sample chambers running into three detection chambers fitted with a pyrolytic graphite block bearing 10 nm deep nano-wells. The reaction took place in the nano-wells coated with RuBPY/cytochrome P450 enzyme/DNA layers for a layer by layer assembly.

We also developed 3D printed immunoarrays automated by a programmable syringe pump delivering reagents sequentially into the detection chambers. The sandwich immunoassay was carried out at the detection chamber of 10 nm deep nano-wells on a pyrolytic graphite chip, facilitating lower sample and reagent volumes. The size of the sample chamber was governed by the number of cancer biomarker proteins being detected, starting with 3 biomarkers [90] and moving up to detect 8 biomarkers simultaneously in a single array [91]. In all the devices, the detection chamber was designed to accommodate a working and reference electrode to which potential was applied to generate ECL light in the presence of the co-reactant tripropylamine (TPrA). Images of the ECL signal intensity were captured in a dark box with a CCD (charged couple device) camera (Figure 7). In an early experiment, our group used a non-transparent 3D device printed by FDM 3D printer that had a sample chamber and reagent reservoirs to facilitate gravity driven reagent delivery. The system was powered by a capacitor without a potentiostat (Figure 7B). Electrodes were screen printed and then functionalized with capture antibodies. RuBPY SiNPs coated with secondary antibodies were used to carry out the immunoassay on the surface of the electrode. 3 prostate cancer biomarkers were measured, prostate-specific antigen (PSA), prostate-specific membrane antigen (PSMA) and platelet factor 4 (PF4). Limits of detection (LODs) ranged from 300–500 fg/mL [91]. After this initial system, designs evolved to include many improvements. The 3D printed devices moved to semi-transparent devices, SLA printers were used for array printing, Krylon spray was used in order to make the devices more transparent, pyrolytic graphite blocks were replaced by thin pyrolytic graphite sheets, the cost of development decreased. Finally, panels of biomarkers were expanded for more reliable prognosis and treatment [90,92].

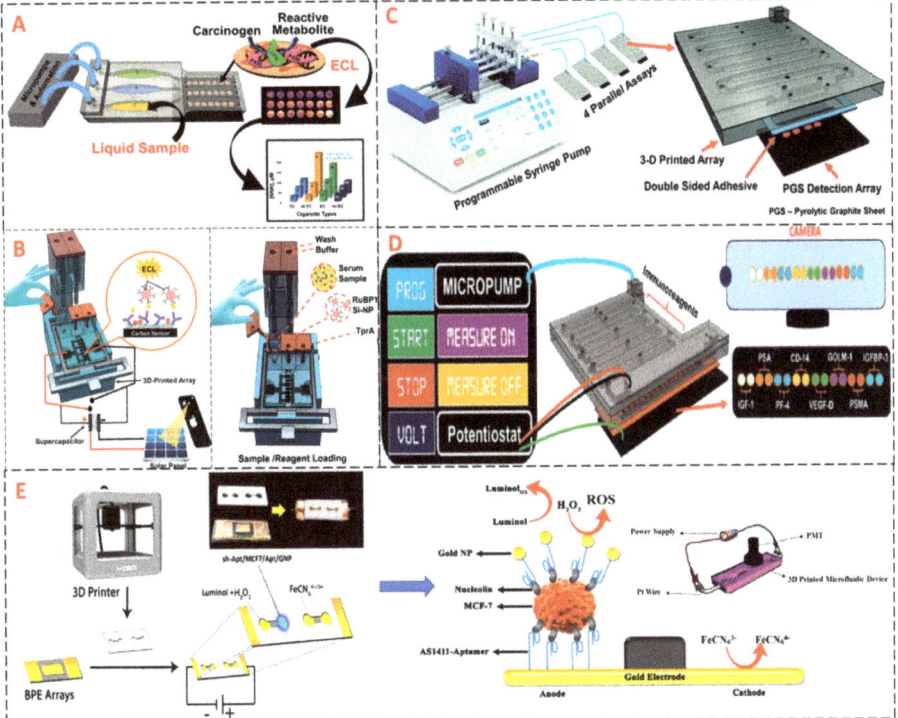

Figure 7. Schematic illustration of 3D printed biosensor arrays that employ electrochemiluminescent (ECL) detection used for cancer diagnostics. (**A**) Automated 3D-printed ECL microfluidic array used in genotoxicity screening Reprinted with permission from Copyright (2017) American Chemical Society. (**B**) Automated 3D printed supercapacitor-powered ECL Protein Immunoarray. Reproduced with permission from [90]. Copyright (2016) Elsevier. (**C**) Automated 3D printed microfluidics immunoassay detecting 4 protein samples simultaneously. Reproduced with permission from Copyright (2018) The Royal Society of Chemistry. (**D**) Automated 3D printed microfluidic array for detection of 8 cancer biomarker proteins simultaneously. Reproduced with permission from [90]. Copyright (2018) American Chemical Society. (**E**) Complete pathway to the detection of human breast cancer cells by using bipolar electrode modified by an aptamer coated inside a 3D printed microchannel by ECL. Reproduced with permission from [92] Copyright (2018) Elsevier.

In related work, Montaghi et al. developed a system for sensitive detection of breast cancer cells (MCF-7) using ECL via a functionalized bipolar electrode (BPE) mounted in a 3D printed microchannel (Figure 7E) [93]. Functionalization involved attachment of aptamer specific to nucleolin on the anode of the BPE. Gold nanoparticles were modified by a secondary aptamer. ECL was generated using luminol in the presence of hydrogen peroxide. The assay was able to detect limit of 10 breast cancer cells MCF-7.

Chemiluminescence (CL) can also be used to determine the concentration of an analyte by measurement of the luminescence intensity initiated by a chemical reaction [94]. Unlike ECL, CL has no need for electrodes and requires on light detection for operation [95,96]. The most commonly used CL substrates are luminol with peroxidases and alkaline phosphatase (ALP) for activation. The signal is generated from the reaction between the CL substrate and hydrogen peroxide in the presence of horseradish peroxidase enzyme (HRP) as the catalyst, emitting light at 425 nm when the oxidized triplet dianion decays from its excited state to the ground state. This system has wide applications to immunoassays, environmental analysis [97,98], clinical diagnosis [99,100], the food safety [101,102], and pharmaceutical analysis [103,104]. Gold nanoparticles (AuNPs) have been used to label antibodies and enhance CL signals [105]. Similarly, polymers multi-labelled with enzymes like poly-HRP has been recently investigated to enhance the CL signal increasing assay sensitivities [106]. Signals are measured in a dark box using a CCD camera same as in ECL detection. CL combined with 3D printing technology opened doors to automation and multiplexed detection of cancer biomarker proteins [107,108].

We reported the first non-polydimethylsiloxane (PDMS), transparent 3D printed device with channels, detection chamber and reagent mixer (Figure 8A) integrated with an immunoarray in 2017 to measure protein biomarkers. Proteins PF-4 and PSA were studied in this example, giving LODs of 0.5 pg/mL for both along with broad dynamic ranges [109]. We also developed an assay using CL for multiplexed ELISA in 3D printed pipette tips (Figure 8B) [29]. Both colorimetric and chemiluminescence detection methods were considered in the novel TIP ELISA approach. A smartphone was utilized to enable the electronic delivery of results proving it to be suitable for POC testing. The immunoassay was done in the side of the pipette tips. It proved to be more sensitive, faster and required less sample and reagent volumes than traditional ELISA assays. Four prostate cancer biomarkers were studied giving LODs down to 0.5 pg/mL concentration level. The results showed good correlations with ELISA cutting down the cost to less than 25% of conventional ELISA. Other 3D printed systems that have been developed that detect lactate and H_2O_2 in biological fluids and plant extracts.

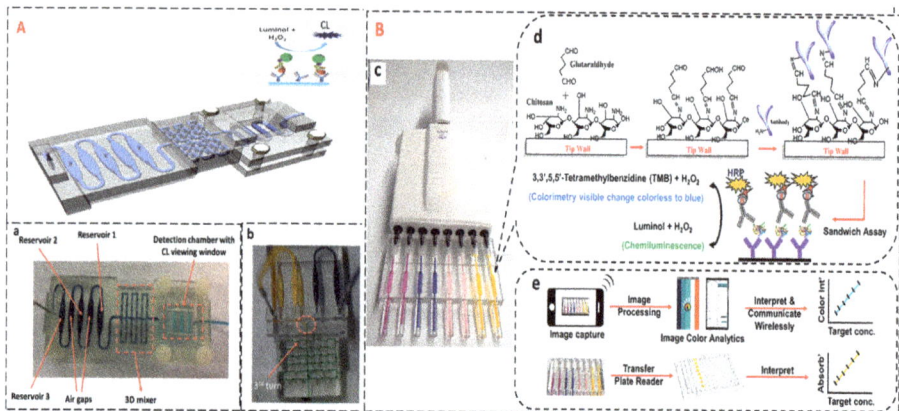

Figure 8. Schematic illustration of 3D printed biosensor arrays that employ chemiluminescent (CL) detection for cancer diagnostics. (**A**) 3D printed design of a unibody microfluidic CL array device. Inset: (**a**) Details of the unibody immunoarray showing upstream reservoir chambers separated by air chambers for air gaps to prevent intermixing, followed by a 3D mixing network of 96 turns, finally detection chamber that houses the antibody array; (**b**) The mixer highlight containing 96 turns that are 0.8 mm × 0.8 mm × 0.8 mm 90 turns. 2 different solutions pumped into the mixer at the rate of 50 µL/min mix at the third turn (indicated by the arrow). It shows excellent mixing efficiency (indicated by the difference in the colors before and after mixing). Reproduced with permission from [109]. Copyright (2017) The Royal Chemical Society. (**B**) Graphical representation of ELISA sandwich immunoassay in 3D printed pipette tips. Inset: (**c**) Fully transparent 3D printed pipette tips filled with different color food dyes attached to a multi tip pipette; (**d**) steps involved in the pre coating showing the immobilization of capture antibodies on the inner walls of the tips coated with chitosan followed by the sandwich immunoassay and the generation of the CL signal and colorimetry; (**e**) Signal capture and processing flow for both colorimetry and CL using a smartphone and a microplate reader. Reproduced with permission from [29]. Copyright (2019) American Chemical Society.

5. Hybrid 3D Printed Sensors

Hybrid sensors with capability to integrate multiple components play a crucial role in developing newer technologies and deliver better user interaction. Ability to design and fabricate such integrated hybrid sensors during the proof of concept stage provides greater advantage than using simple off the shelf commercial sensors. Such an attempt requires addressing unique challenges, and 3D printing with ability to make complex shapes and sizes, using multi-material, nano-material integration and 3 dimensional structures using conductive inks/materials will provide enhanced sensing capabilities. Most importantly, 3D printing aids in consolidating a working prototype of such hybrid sensors with less creation time and cost.

Integration of 3D printed electrodes made of materials composed of metal-based inks and conductive materials have played a key role in development of novel electrochemical and electro-optical micro devices and they offer several benefits 1. Rapid prototyping—full realization of a manufactured prototype can be improved 2. Manufacturing tailor made electrodes and seamless incorporation—ideal for designing miniaturized point-of-care devices 3. Ability to design features not possible by traditional methods like rough/smooth surfaces, multiple electrodes for multiplexing, control of sensor sizes, complex geometry, cost to prototype, robustness and chemical resistance supported my availability of novel nano composite materials. Evolution of electrochemical 3D printed sensors predominantly focused on making 3D printed housing to integrate commercially available electrodes. Thus, making a functional hybrid sensor that has all the components printed and integrated will help realize the novel technologies reach the commercial arena at a scale approachable by masses. Biomarker discovery and cancer diagnostics are particularly challenged by lack of standalone devices that are sensitive to detect ultra-low levels of the biomarker levels with multiplexing capabilities. 3D printing can facilitate integration of multiple components like reagent storage and delivery, sensing surfaces by printing bio-recognition surfaces, complementary electronics for automation and data sharing modules in a miniaturized format realizing a true point of need platform.

We summarized such recent hybrid systems here, Sebechlebska et al. demonstrated a 3D printed hybrid integrated sensor that integrates electrochemistry and UV/Vis absorption spectroscopy, dubbed as a first ever report of UV/Vis absorption spectroelectrochemical apparatus, employing 3D printed optically transparent working electrodes [110]. PLA based 3D printed electrodes made from carbon nanotubes (Figure 9A) were utilized in this study due to their higher electrical conductivity compared to PLA doped with carbon black and graphene. Functional electrode sensors require facile electron transfer at electrode/electrolyte interface, variation in electroactive probes reversibly transferring electrons at electrode surface is referred to intrinsic kinetic barrier. Majority of PLA/carbon composited have high kinetic barrier not suitable for electrochemical studies, whereas Ruthenium (III) acetylacetonate based activation process evolved a lowest ever reported kinetic barrier for a 3D printed electrode with magnitude of faradaic response like conventional carbon electrode (Figure 9A). A 3-electrode (gold wires as counter and reference electrodes with PLA/CNT electrodes as working electrodes) setup integrated in a Quartz cuvette was used to monitor electrochemical process at 3D printed electrode by in-situ UV/Vis absorption spectroscopy. PLA/CNT electrode was designed with optical window at the bottom end to accurately visualize UV-Vis spectra of electrochemically active species (Figure 9B). UV-Vis spectra obtained by cyclic voltammetry of Ru(acac)3 showed reduction and subsequent re-oxidation of the electroactive species. The rate of absorbance change in reduction step decreases with time confirms depletion of reactant at working electrode optical window. All the absorption transients suggested successful implementation of a hybrid sensor using 3D printed nanocomposite-based electrodes.

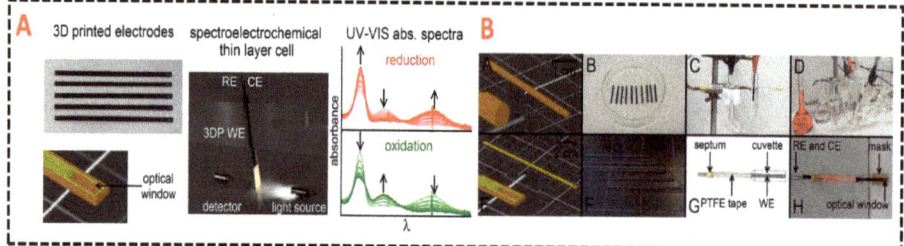

Figure 9. (**A**) Schematic illustration of 3D printed software designs and scheme of hybrid sensors arrangement with an optical window on electrode surface in the path of light beam. A complete UV-Vis spectrum with all the absorption transients shown as inset for successful description of integration. (**B**) Images of the working set up and pictures of the 3D printed electrodes and quartz housing for UV-Vis measurements. All components arrangement, working- counter—and reference electrode arrangement along with optical viewing window shown. Reproduced with permission from [110]. Copyright (2019) Elsevier.

In another attempt to integrate unique methodologies to produce a hybrid 3D printed sensors, Irudayaraj et al. integrated magnetic field/magnetic focus with lateral flow sensor to increase residence time of target-ligand interaction in turn resulting in enhanced sensitivity compared to conventional lateral flow immunoassay (LIFA) for liquid biopsy and tissue samples (Figure 10A). The proposed magnetic focus lateral flow sensor (mLFS) [111] was implemented in detection of cervical cancer biomarker valosin-containing protein (VCP) as proof of concept with exceptional detection limits of 25 fg/mL with enhanced sensitivity of 10^6 fold improvement over conventional lateral flow assays. Magnetic focus is provided by a controlling a simple magnet to manipulate magnetic probe-labelled targets with capture antibodies at the detection zones. Slower the movement the higher the interaction and higher the number of labelled probe targets at detection zone. A simple setup of 3D printed frame designed to integrate lateral flow strip with magnetic bar along with the sample application portal allowed simple operations process (Figure 10B). Thorough analysis for contribution of improved sensitivity by magnetic focus was demonstrated by surface-enhanced Raman spectroscopy (SERS) and dark field imaging of the magnetic nanoparticles along with particle image velocimetry. Signal generation on the lateral flow assay was via color change in presence of 3,3',5,5'-Tetramethylbenzidine (TMB) substrate, magnetic nanoparticle probes modified with HRP and antibody resulted in generation of color spots at detection zones upon addition of colorimetric substrate (Figure 10C).

Sarioglu et al. [112] constructed a hybrid 3D printed monolithic device for negative enrichment of circulating tumor cells from whole blood, by combining microfluidic immunoaffinity based cell capture and a membrane filter to enrich the captured circulating tumor cells for downstream applications. Microfluidic device was designed to have high surface area and increased interaction between white blood cells and functionalized surfaces with 4-32 stacked microfluidic layers and 200 µM diameter microposts (Figure 10D). The microposts serve as support of microfluidic layers as well as chemically functionalized to attach neutravidin for CTC's immunoaffinity capture. Filtration section of the platform has track etched membrane filter to facilitate minimizing cell loss from sample during enrichment and to collect capture CTCs from the device for downstream application like fluorescence microscopy or on-chip staining. Blood sample premixed with biotinylated anti-CD45 antibody to pass through Neutravidin functionalized stacked microfluidic layers. Typically, ~90% of tumor cell recovery was found with multiple cancer cell lines with ability to process clinically relevant blood volumes with at least 300 µL per microfluidic layer.

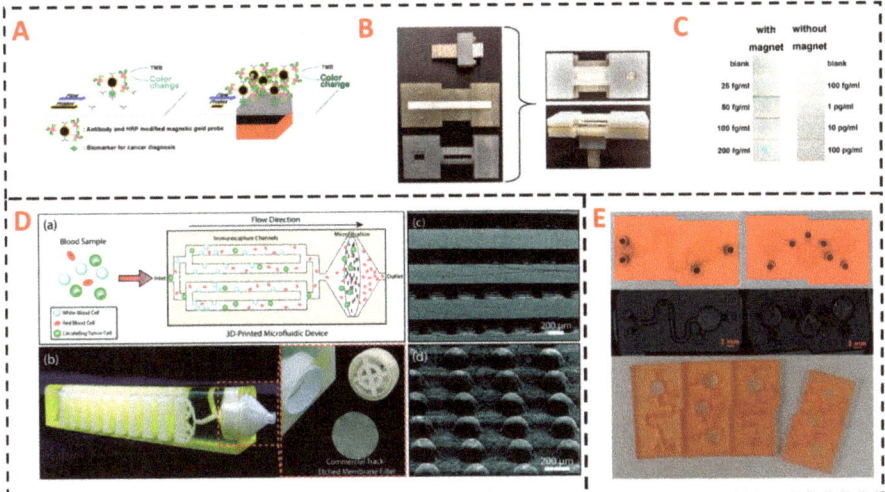

Figure 10. (**A**) Schematic representation of effect of magnetic focusing on a lateral flow assay resulting in enhanced density of magnetic probe-labelled target in the capture antibody detection zones. Effect of magnetic focus represented with and without magnets underneath lateral flow (LF) device to show improved accumulation compared to conventional lateral flow assay (LFA). (**B**) 3D printed device that acts like a frame to hold the lateral flow strip along with magnet and a sample addition zone for liquid biopsy sample aimed to detect cancer biomarkers. (**C**) Comparison of colorimetric signal as detection results with and without magnet. Reproduced with permission from [111]. Copyright (2019) American Chemical Society (**D**) Schematics of tumor cells enrichment process in a multi-layered immunocapture microfluidic layer. Microposts inside the microfluidic channels assist structural integrity and enhanced surface area to allow higher capture and enrichment efficiency. 3 µM membrane filter to retain all eluted nucleated cells for downstream applications. Reproduced with permission from [112]. Copyright (2019) The Royal Society of Chemistry (**E**) Images of 3D printed modular chips made from DLP based stereolithography based approach where monomeric resin is doped with acrylic acid to generate a platform that has intrinsic carboxylates for direct conjugation of biomolecules. Reproduced with permission from [113]. Copyright (2019) The Royal Society of Chemistry.

Frascella et al. demonstrated a hybrid sensor that possess active functional groups to conjugate biomolecules, they designed a photocurable formulation to introduction desired amount of carboxyl (-COOH) groups [113]. 3D printing with intrinsic functionalities, especially in terms of ability to easily immobilize biomolecules on polymeric surfaces is very attractive mainly because it reduces complexity of sensor manufacturing, no multiple components needed to make a functional platform and most importantly in a mass production scenario it allows reproducibility. Frascella et al. designed simple 3D printed sensors with Y-shaped mixer for single step reagent mixing assays and multiple step single chamber devices (Figure 10E) that require no chemical derivatization and showed application of colorimetry detection of cancer biomarker proteins. Three formulations of chemically modified acrylic resin bisphenol A ethoxylate diacrylate (BEDA), 1,6-hexenediol diacrylate (HDDA) and poly(ethylene glycol) diacrylate (PEGDA) were evaluated for their protein grafting ability post 3D printing and found out BEDA at 10% acrylic acid density showed highest amount of protein immobilization and subsequent detection via colorimetric signal generated by HRP labelled antibodies. Two angiogenesis biomarkers, vascular endothelial growth factor and angiopoietin-2 were detected in serum with detection limits of 11 ng/mL and 0.8 ng/mL, respectively.

Above mentioned proof-of-concept platforms highlights the need for hybrid sensors with intrinsic capabilities to evolve into true fully 3D printed diagnostic platforms. Overall, 3D printing of biosensors for diagnostics is rapidly evolving and moving in a direction to address many inherent challenges. By designing hybrid sensors with ability to integrate multiple technologies and improving biocompatibility and bio-adhering capabilities is establishing new benchmarks and a path to transform from just prototyping to commercial mass production market.

6. Challenges and Future Perspectives

Assay systems described above demonstrate that 3D printing has offered a great new low cost, asset for researchers developing biosensors. A major advantage is that sensor arrays for proteins and other analytes can be developed and fabricated on the same platform, the final optimized prototype can essentially be the final product for the clinic or hospital. That along with the speed of development and optimization, and the low cost of most of these printers, suggest a brilliant future for 3D-printed diagnostic devices measuring biomarkers of all types for cancers and other diseases. Gained protein biomarker multiplexing capabilities, increasingly supported by 3D printed immunosensors, is of great interest specially with the growing knowledge correlating different abnormalities in biomarker expression with different types of cancer. Table 1 summarize examples of protein biomarkers detection strategies and its correlated cancer.

Table 1. A summary of published biomarker-based cancer diagnostics.

Cancer	Biomarker	Sensor	Detection Range or Limit
Liver cancer	CD133	Screen-printed gold electrode integrated into a 3D printed chamber	1×10^5–3×10^6 HepG2 liver cancer cells/mL [66]
Hepatocellular carcinoma	Oval cell marker antibody (OV6)	Multiwall carbon nanotube (MWCNT) functionalized electrode integrated into a 3D printed flow cell	1×10^2–5×10^5 hepatic oval cells (HOCs)/mL [67]
Cystic fibrosis	Secretory leukocyte protease inhibitor (SLPI)	Printed circuit board with built-in screen-printed electrode integrated into a 3D printed case and connected to a smart phone for control	Limit of 1 nM [68]
Pancreatic carcinoma, breast cancer and gastric carcinoma	carcinoembryonic antigen (CEA)	Self-designed and printed photoelectrode integrated into a 3D printed platform	10.0 pg/mL–5.0 ng/mL with limit of 4.8 pg/mL [70]
Prostate cancer	Prostate-Specific antigen (PSA), prostate-specific membrane antigen (PSMA)	3D printed multiplexed ECL immunoarray with programmable syringe pump	Limits of 150 fg/mL for PSA, and 230 fg/mL for PSMA [92]
Prostate cancer	PSA, cluster of differentiation 14 (CD-14), Golgi membrane protein 1 (GOLM-1), insulin-like growth factor binding protein 3 (IGFBP-3), insulin-like growth factor 1 (IGF-1), platelet factor 4 (PF-4), vascular endothelial growth factor D(VEGF-D), PSMA	3D printed multiplexed ECL immunoarray with lab-built electronic control system	Limits of 78–110 fg /mL [90]
Prostate cancer	PSA, PSMA, PF-4	3D printed multiplexed ECL immunoarray powered by supercapacitor	Limits of 300–500 fg/mL [91]

Table 1. Cont.

Cancer	Biomarker	Sensor	Detection Range or Limit
Breast cancer	Nucleolin	Functionalized bipolar electrode (BPE) mounted in a 3D printed microchannel for ECL detection	Limit of 10 MCF-7 breast cancer cells [93]
Prostate cancer	PSA, PS-4	Unibody 3D printed multiplexed CL immunoarray	Limits of 0.5 pg/mL [109]
Prostate cancer	PSA, VEGF, IGF-1, CD-14	ELISA based 3D printed multiplexed pipette tip for CL and colorimetric detection	Limits of 5 pg/mL for PSA, 25 pg/mL for VEGF, 2.5 pg/mL for IGF-1, and 0.5 pg/mL for CD-14 [29]
Cervical cancer	Valosin-Containing protein (VCP)	Magnetic focus lateral flow immunosensor (mLFS) integrated into a 3D printed frame for colorimetric detection	Limit of 25 fg/mL [111]
Ovarian cancer, breast cancer	VEGF, angiopoietin-2 (Ang-2)	3D printed immunoarray using lab-formulated carboxyl group rich resin for colorimetric detection	Limit of 11 ng/mL for VEGF, and 0.8 ng/mL for Ang-2 [113]

One radical foreseen change that can be realized via 3D printing, is the fabrication of sensors with integrated bio-recognition elements. Most conventional sensor assembly strategies require extensive procedures to decorate the sensor interface with biomolecules to selectively capture target analytes. This is an area where 3D printing can stand out as an approach to overcome tedious interface biomolecule decoration steps. 3D printing of biomolecules is gaining great progress in this direction specially with the rapid development in 3D bioprinting where enzymes, proteins and cells can be directly integrated in the printed matrix. Some 3D printed biosensors have been printed with integrated biomolecules [114] and different printing strategies are being extensively directed towards this goal, like syringe-based or laser-induced bioprinting.

In spite of these advantages, 3D printing is still limited in aspects of multi-material printing and resolution. Developing 3D printed electrochemical sensors for cancer protein biomarkers usually necessitates multicomponent to be printed which is hindered by the very limited printing techniques capable of achieving such feature and the high variation in inter-material adhesion forces. By way of example, incorporating a conductive electrode material into a 3D printed sensor using fused deposition modeling require using conductive carbonaceous filaments that can be printed onto conventional nonconductive filaments. Although achievable [37], this process is quite complex and require extensive optimization of the printing parameters to avoid leakage and/or structure deformity. In addition, common 3D printing techniques and materials have limited compatibility with biomolecules especially with high energy required for printing processes, like high heat in FDM and high energy laser in stereolithographic 3D printing, that prevent direct printing of these biomolecules. This necessitates a post printing surface modification steps to improve the surface characters or add functionality where biorecognition moieties could be immobilized.

These drawbacks are driving research boundaries of 3D printing to address new challenges and prior limitations. Recently, 3D printing has been utilized for in situ printing of deformable sensors right onto soft tissues and organs to accommodate its movement and expansions [115]. This flexibility in the applications of 3D printing materials with different characteristics expose the power of this technique in exploring what has been previously limited to sophisticated equipment and complex fabrication facilities. Ability to integrate complex architectures in a multicomponent sensor, in one step, is a crucial progress that reduce sensor production and testing time, bring more sophisticated sensor designs, and allow the production of sensor at the point of need. 3D printing is also transforming from the method of choice for prototyping to a high scale production technique, due to progressive availability of high throughput desktop 3D printers with orders of magnitude larger printing surface compared to earlier printers. Sensors availability and affordability, granted by current 3D printing ventures, may help diagnose patients at very early stage where the disease is most responsive to treatment. 3D printing sensors with integrated biomaterials and signal readout through simple connection to portable devices like mobile phones, may shape the future of POC

Author Contributions: M.S. and J.F.R. drafted the outline. M.S., K.K., K.S.B. and T.C. wrote and contributed to the revisions of the manuscript, with the help and guidance of J.F.R. All authors have read and agreed to the published version of the manuscript.

Funding: Preparation of this article and the authors' work described therein were supported by Grant no. EB016707 from the National Institute of Biomedical Imaging and Bioengineering (NIBIB), NIH and by an Academic Plan Grant from the University of Connecticut.

Conflicts of Interest: The authors declare no conflict of interest.

References

1. Cancer. Available online: https://www.who.int/news-room/fact-sheets/detail/cancer (accessed on 9 May 2020).
2. Gollub, M.J.; Blazic, I.; Felder, S.; Knezevic, A.; Gonen, M.; Garcia-Aguilar, J.; Paty, P.P.; Smith, J.J. Value of adding dynamic contrast-enhanced MRI visual assessment to conventional MRI and clinical assessment in the diagnosis of complete tumour response to chemoradiotherapy for rectal cancer. *Eur. Radiol.* **2019**, *29*, 1104–1113. [CrossRef] [PubMed]
3. Mori, M.; Akashi-Tanaka, S.; Suzuki, S.; Daniels, M.I.; Watanabe, C.; Hirose, M.; Nakamura, S. Diagnostic accuracy of contrast-enhanced spectral mammography in comparison to conventional full-field digital mammography in a population of women with dense breasts. *Breast Cancer* **2017**, *24*, 104–110. [CrossRef] [PubMed]
4. Hisabe, T.; Tsuda, S.; Hoashi, T.; Ishihara, H.; Yamasaki, K.; Yasaka, T.; Hirai, F.; Matsui, T.; Yao, K.; Tanabe, H.; et al. Validity of conventional endoscopy using "non-extension sign" for optical diagnosis of colorectal deep submucosal invasive cancer. *Endosc. Int. Open* **2018**, *6*, 156–164. [CrossRef]
5. Danese, E.; Montagnana, M.; Lippi, G. Circulating molecular biomarkers for screening or early diagnosis of colorectal cancer: Which is ready for prime time. *Ann. Transl. Med.* **2019**, *7*, 610. [CrossRef] [PubMed]
6. Liu, B.; Rusling, J.F. Cancer diagnostics. *J. Mater. Chem. B* **2018**, *6*, 2507–2509. [CrossRef]
7. Marrugo-Ramírez, J.; Mir, M.; Samitier, J. Blood-Based Cancer Biomarkers in Liquid Biopsy: A Promising Non-Invasive Alternative to Tissue Biopsy. *Int. J. Mol. Sci.* **2018**, *19*, 2877. [CrossRef]
8. Bai, Y.; Zhao, H. Liquid biopsy in tumors: Opportunities and challenges. *Ann. Transl. Med.* **2018**, *6*, 89. [CrossRef]
9. Dasari, S.; Wudayagiri, R.; Valluru, L. Cervical cancer: Biomarkers for diagnosis and treatment. *Clin. Chim. Acta* **2015**, *445*, 7–11. [CrossRef]
10. Tian, W.; Wang, L.; Lei, H.; Sun, Y.; Xiao, Z. Antibody production and application for immunoassay development of environmental hormones: A review. *Chem. Biol. Technol. Agric.* **2018**, *5*, 5. [CrossRef]
11. Hsieh, Y.-H.P.; Rao, Q. Food Science Text Series. In *Food Analysis*; Nielsen, S.S., Ed.; Springer International Publishing: Cham, Switzerland, 2017; pp. 487–502.
12. Yin, Y.; Cao, Y.; Xu, Y.; Li, G. Colorimetric Immunoassay for Detection of Tumor Markers. *Int. J. Mol. Sci.* **2010**, *11*, 5077–5094. [CrossRef]
13. Hou, J.-Y.; Liu, T.-C.; Lin, G.-F.; Li, Z.-X.; Zou, L.-P.; Li, M.; Wu, Y.-S. Development of an immunomagnetic bead-based time-resolved fluorescence immunoassay for rapid determination of levels of carcinoembryonic antigen in human serum. *Anal. Chim. Acta* **2012**, *734*, 93–98. [CrossRef] [PubMed]
14. Dixit, C.K.; Kadimisetty, K.; Otieno, B.A.; Tang, C.; Malla, S.; Krauseb, C.E.; Rusling, J.F. Electrochemistry-based approaches to low cost, high sensitivity, automated, multiplexed protein immunoassays for cancer diagnostics. *Analyst* **2016**, *141*, 536–547. [CrossRef] [PubMed]

15. Zheng, W.; Zhou, S.; Xu, J.; Liu, Y.; Huang, P.; Liu, Y.; Chen, X. Ultrasensitive Luminescent In Vitro Detection for Tumor Markers Based on Inorganic Lanthanide Nano-Bioprobes. *Adv. Sci.* **2016**, *3*, 1600197. [CrossRef] [PubMed]
16. Sardesai, N.P.; Kadimisetty, K.; Faria, R.; Rusling, J.F. A microfluidic electrochemiluminescent device for detecting cancer biomarker proteins. *Anal. Bioanal. Chem.* **2013**, *405*, 3831–3838. [CrossRef] [PubMed]
17. Ladd, J.; Taylor, A.D.; Piliarik, M.; Homola, J.; Jiang, S. Label-free detection of cancer biomarker candidates using surface plasmon resonance imaging. *Anal. Bioanal. Chem.* **2009**, *393*, 1157–1163. [CrossRef]
18. Masucci, G.V.; Cesano, A.; Hawtin, R.; Janetzki, S.; Zhang, J.; Kirsch, I.; Dobbin, K.K.; Alvarez, J.; Robbins, P.B.; Selvan, S.R.; et al. Validation of biomarkers to predict response to immunotherapy in cancer: Volume I–pre-analytical and analytical validation. *J. Immunother. Cancer* **2016**, *4*, 76. [CrossRef]
19. Wilson, D.H.; Hanlon, D.W.; Provuncher, G.K.; Chang, L.; Song, L.; Patel, P.P.; Ferrell, E.P.; Lepor, H.; Partin, A.W.; Chan, D.W.; et al. Fifth-generation digital immunoassay for prostate-specific antigen by single molecule array technology. *Clin. Chem.* **2011**, *57*, 1712–1721. [CrossRef]
20. von Lode, P. Point-of-care immunotesting: Approaching the analytical performance of central laboratory methods. *Clin. Biochem.* **2005**, *38*, 591–606. [CrossRef]
21. Yager, P.; Edwards, T.; Fu, E.; Helton, K.; Nelson, K.; Tam, M.R.; Weigl, B.H. Microfluidic diagnostic technologies for global public health. *Nature* **2006**, *442*, 412–418. [CrossRef]
22. Xu, Y.; Wu, X.; Guo, X.; Kong, B.; Zhang, M.; Qian, X.; Mi, S.; Sun, W. The Boom in 3D-Printed Sensor Technology. *Sensors* **2017**, *17*, 1166. [CrossRef]
23. MacDonald, E.; Wicker, R. Multiprocess 3D printing for increasing component functionality. *Science* **2016**, *353*, 2093. [CrossRef] [PubMed]
24. Nesaei, S.; Song, Y.; Wang, Y.; Ruan, X.; Du, D.; Gozen, A.; Lin, Y. Micro additive manufacturing of glucose biosensors: A feasibility study. *Anal. Chim. Acta* **2018**, *1043*, 142–149. [CrossRef]
25. Liyarita, B.R.; Ambrosi, A.; Pumera, M. 3D-printed Electrodes for Sensing of Biologically Active Molecules. *Electroanalysis* **2018**, *30*, 1319–1326. [CrossRef]
26. Katseli, V.; Economou, A.; Kokkinos, C. Single-step fabrication of an integrated 3D-printed device for electrochemical sensing applications. *Electrochem. Commun.* **2019**, *103*, 100–103. [CrossRef]
27. Kalinke, C.; Neumsteir, N.V.; de Oliveira Aparecido, G.; de Barros Ferraz, T.V.; dos Santos, P.L.; Janegitz, B.C.; Bonacin, J.A. Comparison of activation processes for 3D printed PLA-graphene electrodes: Electrochemical properties and application for sensing of dopamine. *Analyst* **2020**, *145*, 1207–1218. [CrossRef] [PubMed]
28. Loo, A.H.; Chua, C.K.; Pumera, M. DNA biosensing with 3D printing technology. *Analyst* **2017**, *142*, 279–283. [CrossRef] [PubMed]
29. Sharafeldin, M.; Kadimisetty, K.; Bhalerao, K.R.; Bist, I.; Jones, A.; Chen, T.; Lee, N.H.; Rusling, J.F. Accessible Telemedicine Diagnostics with ELISA in a 3D Printed Pipette Tip. *Anal. Chem.* **2019**, *91*, 7394–7402. [CrossRef]
30. Sharafeldin, M.; Jones, A.; Rusling, J.F. 3D-Printed Biosensor Arrays for Medical Diagnostics. *Micromachines* **2018**, *9*, 394. [CrossRef]
31. Loterie, D.; Delrot, P.; Moser, C. High-resolution tomographic volumetric additive manufacturing. *Nat. Commun.* **2020**, *11*, 1–6. [CrossRef]
32. Ning, F.; Cong, W.; Qiu, J.; Wei, J.; Wang, S. Additive manufacturing of carbon fiber reinforced thermoplastic composites using fused deposition modeling. *Compos. Part B Eng.* **2015**, *80*, 369–378. [CrossRef]
33. Yin, J.; Lu, C.; Fu, J.; Huang, Y.; Zheng, Y. Interfacial bonding during multi-material fused deposition modeling (FDM) process due to inter-molecular diffusion. *Mater. Des.* **2018**, *150*, 104–112. [CrossRef]
34. Chacón, J.M.; Caminero, M.A.; García-Plaza, E.; Núñez, P.J. Additive manufacturing of PLA structures using fused deposition modelling: Effect of process parameters on mechanical properties and their optimal selection. *Mater. Des.* **2017**, *124*, 143–157. [CrossRef]
35. Hamzah, H.H.; Shafiee, S.A.; Abdalla, A.; Patel, B.A. 3D printable conductive materials for the fabrication of electrochemical sensors: A mini review. *Electrochem. Commun.* **2018**, *96*, 27–31. [CrossRef]
36. Abbasi, H.; Antunes, M.; Velasco, J.I. Recent advances in carbon-based polymer nanocomposites for electromagnetic interference shielding. *Prog. Mater. Sci.* **2019**, *103*, 319–373. [CrossRef]
37. Barragan, J.T.C.; Kubota, L.T. Minipotentiostat controlled by smartphone on a micropipette: A versatile, portable, agile and accurate tool for electroanalysis. *Electrochim. Acta* **2020**, *341*, 136048. [CrossRef]

38. Reyes, C.; Somogyi, R.; Niu, S.; Cruz, M.A.; Yang, F.; Catenacci, M.J.; Rhodes, C.P.; Wiley, B.J. Three-Dimensional Printing of a Complete Lithium Ion Battery with Fused Filament Fabrication. *ACS Appl. Energy Mater.* **2018**, *1*, 5268–5279. [CrossRef]
39. Torrado, A.R.; Shemelya, C.M.; English, J.D.; Lin, Y.; Wicker, R.B.; Roberson, D.A. Characterizing the effect of additives to ABS on the mechanical property anisotropy of specimens fabricated by material extrusion 3D printing. *Addit. Manuf.* **2015**, *6*, 16–29. [CrossRef]
40. Zhang, J.; Xiao, P. 3D printing of photopolymers. *Polym. Chem.* **2018**, *9*, 1530–1540. [CrossRef]
41. Gross, B.; Lockwood, S.Y.; Spence, D.M. Recent Advances in Analytical Chemistry by 3D Printing. *Anal. Chem.* **2017**, *89*, 57–70. [CrossRef]
42. Ambrosi, A.; Pumera, M. 3D-printing technologies for electrochemical applications. *Chem. Soc. Rev.* **2016**, *45*, 2740–2755. [CrossRef]
43. The Ultimate Guide to Stereolithography (SLA) 3D Printing (Updated for 2019). Available online: https://formlabs.com/blog/ultimate-guide-to-stereolithography-sla-3d-printing/ (accessed on 1 June 2020).
44. Barner-Kowollik, C.; Bastmeyer, M.; Blasco, E.; Delaittre, G.; Müller, P.; Richter, B.; Wegener, M. 3D Laser Micro- and Nanoprinting: Challenges for Chemistry. *Angew. Chem. Int. Ed.* **2017**, *56*, 15828–15845. [CrossRef] [PubMed]
45. Landers, R.; Mülhaupt, R. Desktop manufacturing of complex objects, prototypes and biomedical scaffolds by means of computer-assisted design combined with computer-guided 3D plotting of polymers and reactive oligomers. *Macromol. Mater. Eng.* **2000**, *282*, 17–21. [CrossRef]
46. Sachs, E.; Cima, M.; Williams, P.; Brancazio, D.; Cornie, J. Three Dimensional Printing: Rapid Tooling and Prototypes Directly from a CAD Model. *J. Eng. Ind.* **1992**, *114*, 481–488. [CrossRef]
47. Huang, L.; Huang, Y.; Liang, J.; Wan, X.; Chen, Y. Graphene-based conducting inks for direct inkjet printing of flexible conductive patterns and their applications in electric circuits and chemical sensors. *Nano Res.* **2011**, *4*, 675–684. [CrossRef]
48. Carvajal, S.; Fera, S.N.; Jones, A.L.; Baldo, T.A.; Mosa, I.M.; Rusling, J.F.; Krause, C.E. Disposable inkjet-printed electrochemical platform for detection of clinically relevant HER-2 breast cancer biomarker. *Biosens. Bioelectron.* **2018**, *104*, 158–162. [CrossRef]
49. Saunders, R.E.; Gough, J.E.; Derby, B. Delivery of human fibroblast cells by piezoelectric drop-on-demand inkjet printing. *Biomaterials* **2008**, *29*, 193–203. [CrossRef]
50. Murr, L.E.; Gaytan, S.M.; Ramirez, D.A.; Martinez, E.; Hernandez, J.; Amato, K.N.; Shindo, P.W.; Medina, F.R.; Wicker, R.B. Metal Fabrication by Additive Manufacturing Using Laser and Electron Beam Melting Technologies. *J. Mater. Sci. Technol.* **2012**, *28*, 1–14. [CrossRef]
51. Kelly, B.E.; Bhattacharya, I.; Heidari, H.; Shusteff, M.; Spadaccini, C.M.; Taylor, H.K. Volumetric additive manufacturing via tomographic reconstruction. *Science* **2019**, *363*, 1075–1079. [CrossRef]
52. Murphy, S.; Atala, A. 3D bioprinting of tissues and organs. *Nat. Biotechnol.* **2014**, *32*, 773–785. [CrossRef]
53. Ozbolat, I.T.; Hospodiuk, M. Current advances and future perspectives in extrusion-based bioprinting. *Biomaterials* **2016**, *76*, 321–343. [CrossRef]
54. Ning, L.; Zhu, N.; Mohabatpour, F.; Sarker, M.D.; Schreyer, D.J.; Chen, X. Bioprinting Schwann cell-laden scaffolds from low-viscosity hydrogel compositions. *J. Mater. Chem. B* **2019**, *7*, 4538–4551. [CrossRef]
55. Jeon, O.; Lee, Y.B.; Hinton, T.J.; Feinberg, A.W.; Alsberg, E. Cryopreserved cell-laden alginate microgel bioink for 3D bioprinting of living tissues. *Mater. Today Chem.* **2019**, *12*, 61–70. [CrossRef] [PubMed]
56. Unagolla, J.M.; Jayasuriya, A.C. Hydrogel-based 3D bioprinting: A comprehensive review on cell-laden hydrogels, bioink formulations, and future perspectives. *Appl. Mater. Today* **2020**, *18*, 100479. [CrossRef] [PubMed]
57. Guillotin, B.; Souquet, A.; Catros, S.; Duocastella, M.; Pippenger, B.; Bellance, S.; Bareille, R.; Rémy, M.; Bordenave, L.; Amédée, J.; et al. Laser assisted bioprinting of engineered tissue with high cell density and microscale organization. *Biomaterials* **2010**, *31*, 7250–7256. [CrossRef] [PubMed]
58. Hakobyan, D.; Médina, C.; Dusserre, N.; Stachowicz, M.L.; Handschin, C.; Fricain, J.C.; Guillermet-Guibert, J.; Oliveira, H. Laser-assisted 3D bioprinting of exocrine pancreas spheroid models for cancer initiation study. *Biofabrication* **2020**, *12*, 035001. [CrossRef]
59. Knowlton, S.; Yenilmez, B.; Tasoglu, S. Towards single-step biofabrication of organs on a chip via 3D printing. *Trends Biotechnol.* **2016**, *34*, 685–688. [CrossRef]

60. Han, S.; Kim, S.; Chen, Z.; Shin, H.K.; Lee, S.Y.; Moon, H.E.; Paek, S.H.; Park, S. 3D bioprinted vascularized tumour for drug testing. *Int. J. Mol. Sci.* **2020**, *21*, 2993. [CrossRef]
61. Benck, J.D.; Pinaud, B.A.; Gorlin, Y.; Jaramillo, T.F. Substrate Selection for Fundamental Studies of Electrocatalysts and Photoelectrodes: Inert Potential Windows in Acidic, Neutral, and Basic Electrolyte. *PLoS ONE* **2014**, *9*, 107942. [CrossRef]
62. Zachek, M.K.; Hermans, A.; Wightman, R.M.; McCarty, G.S. Electrochemical Dopamine Detection: Comparing Gold and Carbon Fiber Microelectrodes using Background Subtracted Fast Scan Cyclic Voltammetry. *J. Electroanal. Chem.* **2008**, *614*, 113–120. [CrossRef]
63. Wongkaew, N.; Simsek, M.; Griesche, C.; Baeumner, A.J. Functional Nanomaterials and Nanostructures Enhancing Electrochemical Biosensors and Lab-on-a-Chip Performances: Recent Progress, Applications, and Future Perspective. *Chem. Rev.* **2019**, *119*, 120–194. [CrossRef]
64. Cardoso, R.M.; Kalinke, C.; Rocha, R.G.; dos Santos, P.L.; Rocha, D.P.; Oliveira, P.R.; Janegitz, B.C.; Bonacin, J.A.; Richter, E.M.; Munoz, R.A.A. Additive-manufactured (3D-printed) electrochemical sensors: A critical review. *Anal. Chim. Acta* **2020**. [CrossRef] [PubMed]
65. Pumera, M. Three-dimensionally printed electrochemical systems for biomedical analytical applications. *Curr. Opin. Electrochem.* **2019**, *14*, 133–137. [CrossRef]
66. Damiati, S.; Küpcü, S.; Peacock, M.; Eilenberger, C.; Zamzami, M.; Qadri, I.; Choudhry, H.; Sleytr, U.B.; Schuster, B. Acoustic and hybrid 3D-printed electrochemical biosensors for the real-time immunodetection of liver cancer cells (HepG2). *Biosens. Bioelectron.* **2017**, *94*, 500–506. [CrossRef] [PubMed]
67. Damiati, S.; Peacock, M.; Leonhardt, S.; Damiati, L.; Baghdadi, M.A.; Becker, H.; Kodzius, R.; Schuster, B. Embedded Disposable Functionalized Electrochemical Biosensor with a 3D-Printed Flow Cell for Detection of Hepatic Oval Cells (HOCs). *Genes* **2018**, *9*, 89. [CrossRef]
68. Sun, A.C.; Yao, C.; Venkatesh, A.G.; Hall, D.A. An efficient power harvesting mobile phone-based electrochemicalbiosensor for point-of-care health monitoring. *Sens. Actuators B Chem.* **2016**, *235*, 126–135. [CrossRef]
69. Scordo, G.; Moscone, D.; Palleschi, G.; Arduini, F. A reagent-free paper-based sensor embedded in a 3D printing devicefor cholinesterase activity measurement in serum. *Sens. Actuators B Chem.* **2018**, *258*, 1015–1021. [CrossRef]
70. Qiu, Z.; Shu, J.; Liu, J.; Tang, D. Dual-Channel Photoelectrochemical Ratiometric Aptasensor with up-Converting Nanocrystals Using Spatial-Resolved Technique on Homemade 3D Printed Device. *Anal. Chem.* **2019**, *91*, 1260–1268. [CrossRef]
71. Zhang, K.; Lv, S.; Tang, D. Novel 3D Printed Device for Dual-Signaling Ratiometric Photoelectrochemical Readout of Biomarker Using λ-Exonuclease-Assisted Recycling Amplification. *Anal. Chem.* **2019**, *91*, 10049–10055. [CrossRef]
72. Lv, S.; Zhang, K.; Zhou, Q.; Tang, D. Plasmonic enhanced photoelectrochemical aptasensor with D-A F8BT/g-C3N4 heterojunction and AuNPs on a 3D-printed device. *Sens. Actuators B Chem.* **2020**, *310*, 127874. [CrossRef]
73. Ambrosi, A.; Moo, J.G.S.; Pumera, M. Helical 3D-Printed Metal Electrodes as Custom-Shaped 3D Platform for Electrochemical Devices. *Adv. Funct. Mater.* **2016**, *26*, 698–703. [CrossRef]
74. Jensen, G.C.; Krause, C.E.; Sotzing, G.A.; Rusling, J.F. Inkjet-printed gold nanoparticle electrochemical arrays on plastic. Application to immunodetection of a cancer biomarker protein. *Phys. Chem. Chem. Phys.* **2011**, *13*, 4888–4894. [CrossRef] [PubMed]
75. Krause, C.E.; Otieno, B.A.; Latus, A.; Faria, R.C.; Patel, V.; Gutkind, J.S.; Rusling, J.F. Rapid Microfluidic Immunoassays of Cancer Biomarker Proteins Using Disposable Inkjet-Printed Gold Nanoparticle Arrays. *ChemistryOpen* **2013**, *2*, 141–145. [CrossRef] [PubMed]
76. Otieno, B.A.; Krause, C.E.; Jones, A.L.; Kremer, R.B.; Rusling, J.F. Cancer Diagnostics via Ultrasensitive Multiplexed Detection of Parathyroid Hormone-Related Peptides with a Microfluidic Immunoarray. *Anal. Chem.* **2016**, *88*, 9269–9275. [CrossRef]
77. Forster, R.J.; Bertoncello, P.; Keyes, T.E. Electrogenerated Chemiluminescence. *Annu. Rev. Anal. Chem.* **2009**, *2*, 359–385. [CrossRef] [PubMed]
78. Leventis, N. Electrogenerated Chemiluminescence Edited by Allen, J. Bard (University of Texas at Austin). Marcel Dekker, Inc.: New York. 2004. viii + 540 pp. $165.00. ISBN 0-8247-5347-X. *J. Am. Chem. Soc.* **2005**, *127*, 2015–2016. [CrossRef]

79. Kadimisetty, K.; Malla, S.; Sardesai, N.P.; Joshi, A.A.; Faria, R.C.; Lee, N.H.; Rusling, J.F. Automated Multiplexed Ecl Immunoarrays for Cancer Biomarker Proteins. *Anal. Chem.* **2015**, *87*, 4472–4478. [CrossRef] [PubMed]
80. Hvastkovs, E.G.; So, M.; Krishnan, S.; Bajrami, B.; Tarun, M.; Jansson, I.; Schenkman, J.B.; Rusling, J.F. Electrochemiluminescent Arrays for Cytochrome P450-Activated Genotoxicity Screening. DNA Damage from Benzo[a]Pyrene Metabolites. *Anal. Chem.* **2007**, *79*, 1897–1906. [CrossRef]
81. Yang, Y.; Hu, G.B.; Liang, W.B.; Yao, L.Y.; Huang, W.; Zhang, Y.J.; Zhang, J.L.; Wang, J.M.; Yuan, R.; Xiao, D.R. An AIEgen-Based 2D Ultrathin Metal-Organic Layer as an Electrochemiluminescence Platform for Ultrasensitive Biosensing of Carcinoembryonic Antigen. *Nanoscale* **2020**, *12*, 5932–5941. [CrossRef]
82. Bist, I.; Song, B.; Mosa, I.M.; Keyes, T.E.; Martin, A.; Forster, R.J.; Rusling, J.F. Electrochemiluminescent Array to Detect Oxidative Damage in Ds-DNA Using [Os(Bpy)2(Phen-Benz-COOH)]2+/Nafion/Graphene Films. *ACS Sens.* **2016**, *1*, 272–278. [CrossRef]
83. Myung, N.; Ding, Z.; Bard, A.J. Electrogenerated Chemiluminescence of CdSe Nanocrystals. *Nano Lett.* **2002**, *2*, 1315–1319. [CrossRef]
84. Poznyak, S.K.; Talapin, D.V.; Shevchenko, E.V.; Weller, H. Quantum Dot Chemiluminescence. *Nano Lett.* **2004**, *4*, 693–698. [CrossRef]
85. Bae, Y.; Myung, N.; Bard, A.J. Electrochemistry and Electrogenerated Chemiluminescence of CdTe Nanoparticles. *Nano Lett.* **2004**, *4*, 1153–1161. [CrossRef]
86. Bist, I.; Bano, K.; Rusling, J.F. Screening Genotoxicity Chemistry with Microfluidic Electrochemiluminescent Arrays. *Sensors* **2017**, *17*, 1008. [CrossRef] [PubMed]
87. Krishnan, S.; Hvastkovs, E.G.; Bajrami, B.; Jansson, I.; Schenkman, J.B.; Rusling, J.F. Genotoxicity Screening for N-Nitroso Compounds. Electrochemical and Electrochemiluminescent Detection of Human Enzyme-Generated DNA Damage from N-Nitrosopyrrolidine. *Chem. Commun.* **2007**, *17*, 1713–1715. [CrossRef] [PubMed]
88. Sardesai, N.; Pan, S.; Rusling, J. Electrochemiluminescent Immunosensor for Detection of Protein Cancer Biomarkers Using Carbon Nanotube Forests and [Ru-(Bpy)3] 2+-Doped Silica Nanoparticles. *Chem. Commun.* **2009**, *73*, 4968–4970. [CrossRef] [PubMed]
89. Kadimisetty, K.; Malla, S.; Rusling, J.F. Automated 3-D Printed Arrays to Evaluate Genotoxic Chemistry: E-Cigarettes and Water Samples. *ACS Sens.* **2017**, *2*, 670–678. [CrossRef]
90. Kadimisetty, K.; Mosa, I.M.; Malla, S.; Satterwhite-Warden, J.E.; Kuhns, T.M.; Faria, R.C.; Lee, N.H.; Rusling, J.F. 3D-Printed Supercapacitor-Powered Electrochemiluminescent Protein Immunoarray. *Biosens. Bioelectron.* **2016**, *77*, 188–193. [CrossRef]
91. Kadimisetty, K.; Malla, S.; Bhalerao, K.S.; Mosa, I.M.; Bhakta, S.; Lee, N.H.; Rusling, J.F. Automated 3D-Printed Microfluidic Array for Rapid Nanomaterial-Enhanced Detection of Multiple Proteins. *Anal. Chem.* **2018**, *90*, 7569–7577. [CrossRef]
92. Motaghi, H.; Ziyaee, S.; Mehrgardi, M.A.; Kajani, A.A.; Bordbar, A.K. Electrochemiluminescence Detection of Human Breast Cancer Cells Using Aptamer Modified Bipolar Electrode Mounted into 3D Printed Microchannel. *Biosens. Bioelectron.* **2018**, *118*, 217–223. [CrossRef]
93. Kadimisetty, K.; Spak, A.P.; Bhalerao, K.S.; Sharafeldin, M.; Mosa, I.M.; Lee, N.H.; Rusling, J.F. Automated 4-Sample Protein Immunoassays Using 3D-Printed Microfluidics. *Anal. Methods* **2018**, *10*, 4000–4006. [CrossRef]
94. Wang, C.; Wu, J.; Zong, C.; Xu, J.; Ju, H.X. Chemiluminescent Immunoassay and Its Applications. *Chin. J. Anal. Chem.* **2012**, *40*, 3–10. [CrossRef]
95. Roda, A.; Mirasoli, M.; Dolci, L.S.; Buragina, A.; Bonvicini, F.; Simoni, P.; Guardigli, M. Portable Device Based on Chemiluminescence Lensless Imaging for Personalized Diagnostics through Multiplex Bioanalysis. *Anal. Chem.* **2011**, *83*, 3178–3185. [CrossRef] [PubMed]
96. Zhao, L.; Sun, L.; Chu, X. Chemiluminescence Immunoassay. *Trends Anal. Chem.* **2009**, *28*, 404–415. [CrossRef]
97. Yang, W.P.; Zhang, Z.J.; Hun, X. A Novel Capillary Microliter Droplet Sample Injection-Chemiluminescence Detector and Its Application to the Determination of Benzoyl Peroxide in Wheat Flour. *Talanta* **2004**, *62*, 661–666. [CrossRef] [PubMed]
98. Martín-Esteban, A.; Fernández, P.; Cámara, C. Immunosorbents: A New Tool for Pesticide Sample Handling in Environmental Analysis. *Fresenius. J. Anal. Chem.* **1997**, *357*, 927–933. [CrossRef]

99. Dai, Z.; Yan, F.; Chen, J.; Ju, H. Reagentless Amperometric Immunosensors Based on Direct Electrochemistry of Horseradish Peroxidase for Determination of Carcinoma Antigen-125. *Anal. Chem.* **2003**, *75*, 5429–5434. [CrossRef]
100. Fu, Z.; Hao, C.; Fei, X.; Ju, H. Flow-Injection Chemiluminescent Immunoassay for α-Fetoprotein Based on Epoxysilane Modified Glass Microbeads. *J. Immunol. Methods* **2006**, *312*, 61–67. [CrossRef]
101. Micheli, L.; Grecco, R.; Badea, M.; Moscone, D.; Palleschi, G. An Electrochemical Immunosensor for Aflatoxin M1 Determination in Milk Using Screen-Printed Electrodes. *Biosens. Bioelectron.* **2005**, *21*, 588–596. [CrossRef]
102. Nandakumar, R.; Nandakumar, M.P.; Mattiasson, B. Quantification of Nisin in Flow-Injection Immunoassay Systems. *Biosens. Bioelectron.* **2000**, *15*, 241–247. [CrossRef]
103. Yang, H.H.; Zhu, Q.Z.; Qu, H.Y.; Chen, X.L.; Ding, M.T.; Xu, J.G. Flow Injection Fluorescence Immunoassay for Gentamicin Using Sol-Gel-Derived Mesoporous Biomaterial. *Anal. Biochem.* **2002**, *308*, 71–76. [CrossRef]
104. Eremenko, A.V.; Bauer, C.G.; Makower, A.; Kanne, B.; Baumgarten, H.; Scheller, F.W. The Development of a Non-Competitive Immunoenzymometric Assay of Cocaine. *Anal. Chim. Acta* **1998**, *358*, 5–13. [CrossRef]
105. Huang, X.; Li, L.; Qian, H.; Dong, C.; Ren, J. A Resonance Energy Transfer between Chemiluminescent Donors and Luminescent Quantum-Dots as Acceptors (CRET). *Angew. Chem. Int. Ed.* **2006**, *45*, 5140–5143. [CrossRef] [PubMed]
106. Huang, R.-P.; Yang, W.; Yang, D.; Flowers, L.; Horowitz, I.R.; Cao, X.; Huang, R. The Promise of Cytokine Antibody Arrays in the Drug Discovery Process. *Expert Opin. Targets* **2005**, *9*, 601–615. [CrossRef] [PubMed]
107. Roda, A.; Guardigli, M.; Calabria, D.; Maddalena Calabretta, M.; Cevenini, L.; Michelini, E. A 3D-Printed Device for a Smartphone-Based Chemiluminescence Biosensor for Lactate in Oral Fluid and Sweat. *Analyst* **2014**, *139*, 6494–6501. [CrossRef] [PubMed]
108. Roda, A.; Michelini, E.; Cevenini, L.; Calabria, D.; Calabretta, M.M.; Simoni, P. Integrating Biochemiluminescence Detection on Smartphones: Mobile Chemistry Platform for Point-of-Need Analysis. *Anal. Chem.* **2014**, *86*, 7299–7304. [CrossRef] [PubMed]
109. Tang, C.K.; Vaze, A.; Rusling, J.F. Automated 3D-Printed Unibody Immunoarray for Chemiluminescence Detection of Cancer Biomarker Proteins. *Lab. Chip* **2017**, *17*, 484–489. [CrossRef]
110. Vaněčková, E.; Bouša, M.; Vivaldi, F.; Gál, M.; Rathouský, J.; Kolivoška, V.; Sebechlebská, T. UV/VIS spectroelectrochemistry with 3D printed electrodes. *J. Electroanal. Chem.* **2020**, *857*, 113760. [CrossRef]
111. Ren, W.; Mohammed, S.I.; Wereley, S.; Irudayaraj, J. Magnetic focus lateral flow sensor for detection of cervical cancer biomarkers. *Anal. Chem.* **2019**, *91*, 2876–2884. [CrossRef]
112. Chu, C.-H.; Liu, R.; Ozkaya-Ahmadov, T.; Boya, M.; Swain, B.E.; Owens, J.M.; Burentugs, E.; Bilen, M.A.; McDonald, J.F.; Sarioglu, A.F. Hybrid negative enrichment of circulating tumor cells from whole blood in a 3D-printed monolithic device. *Lab Chip* **2019**, *19*, 3427–3437. [CrossRef]
113. Chiadò, A.; Palmara, G.; Chiappone, A.; Tanzanu, C.; Pirri, C.F.; Roppolo, I.; Frascella, F. A modular 3D printed lab-on-a-chip for early cancer detection. *Lab Chip* **2020**, *20*, 665–674. [CrossRef]
114. Mandon, C.A.; Blum, L.J.; Marquette, C.A. Adding biomolecular recognition capability to 3D printed objects. *Anal. Chem.* **2016**, *88*, 10767–10772. [CrossRef] [PubMed]
115. Zhu, Z.; Park, H.S.; McAlpine, M.C. 3D printed deformable sensors. *Sci. Adv.* **2020**, *6*, 5575. [CrossRef] [PubMed]

© 2020 by the authors. Licensee MDPI, Basel, Switzerland. This article is an open access article distributed under the terms and conditions of the Creative Commons Attribution (CC BY) license (http://creativecommons.org/licenses/by/4.0/).

Review

DNA/RNA Electrochemical Biosensing Devices a Future Replacement of PCR Methods for a Fast Epidemic Containment

Manikandan Santhanam, Itay Algov and Lital Alfonta *

Departments of Life Sciences, Chemistry and Ilse Katz Institute for Nanoscale Science and Technology, PO Box 653, Ben-Gurion University of the Negev, Beer-Sheva 8410501, Israel; manikandan.mb@gmail.com (M.S.); algov@post.bgu.ac.il (I.A.)
* Correspondence: alfontal@bgu.ac.il

Received: 11 July 2020; Accepted: 17 August 2020; Published: 18 August 2020

Abstract: Pandemics require a fast and immediate response to contain potential infectious carriers. In the recent 2020 Covid-19 worldwide pandemic, authorities all around the world have failed to identify potential carriers and contain it on time. Hence, a rapid and very sensitive testing method is required. Current diagnostic tools, reverse transcription PCR (RT-PCR) and real-time PCR (qPCR), have its pitfalls for quick pandemic containment such as the requirement for specialized professionals and instrumentation. Versatile electrochemical DNA/RNA sensors are a promising technological alternative for PCR based diagnosis. In an electrochemical DNA sensor, a nucleic acid hybridization event is converted into a quantifiable electrochemical signal. A critical challenge of electrochemical DNA sensors is sensitive detection of a low copy number of DNA/RNA in samples such as is the case for early onset of a disease. Signal amplification approaches are an important tool to overcome this sensitivity issue. In this review, the authors discuss the most recent signal amplification strategies employed in the electrochemical DNA/RNA diagnosis of pathogens.

Keywords: electrochemical DNA sensor; nucleic acid sensor; signal amplification; DNA; RNA; pathogen sensing

---

1. Introduction

Rapid, specific and sensitive detection is a goal in emerging biosensor technology. Detection of pathogens using their genomes becomes a central strategy due to advancement of nucleic acid sequencing technologies [1]. Nucleic acid-based detection of pathogens provides more flexibility compared to other biomolecules, as they are present in all living organisms, while every organism or a virus encode their genes with a distinct genome and sequences. However, access to the DNA/RNA sequences in a viral pathogen is not straight forward; they are buried inside bilayers of lipids and proteins of the virion particle. Thus, DNA/RNA should be efficiently extracted from a potential sample. Additionally, the copy number of a given virion may vary, depending on the stages of the infection, virulence and the host cells' replication efficiency. In the case of RNA targets, RNA is reverse transcribed to complementary DNA using reverse transcriptase and then it is quantified.

Currently, quantitative real time-polymerase chain reaction (qPCR) is a standard method where a fluorescent signal is coupled with DNA polymerase chain reaction for quantification of DNA [2–6]. Though this method affords the detection of the presence of 1–10 copies/mL of DNA sample, PCR is still restricted to professional laboratories due to the need for specialized instrumentation [7,8]. To simplify DNA detection for point-of-care testing, other alternative approaches are being developed, namely, colorimetric [9], microfluidic platform based optical detection [10] and electrochemical methods [11]. Among these methods, electrochemical methods are ultrasensitive and well established [12]. In electrochemical DNA sensors, nucleic acid hybridization is coupled with the electrochemical reaction for selective detection of target DNA [13,14]. However, electrochemical methods may not be employed directly for the detection of a single copy of DNA sensing in biological samples. Thus, signal amplification approaches are employed to increase the sensitivity and selectivity towards the sensing of low concentration targets [15]. Additionally, in terms of the possibility to miniaturize as well as to make a quantitative measurement, electrochemical sensors is currently the most promising route for laboratory as well as for the point-of-care approaches [16,17].

Comprehensive reviews on bacterial, protozoan, viral and clinical diagnostics through a variety of electrochemical systems have been published before [18–20]. Additionally, different aspects of DNA electrochemical sensors, such as novel materials and electroanalytical methods, were reviewed in detail [21–26]. In this review, developments in signal amplification approaches for enhanced detection of DNA–DNA or DNA–RNA hybridization events using electrochemical approaches are discussed and directed to non-specialized readers. Specifically, signal amplification approaches with synthetic and/or real samples of pathogens are discussed.

2. Signal Transduction

DNA electrochemical biosensors consist of (i) a DNA recognition element where the target DNA is recognized by a probe/capture DNA(s) strand and (ii) a signal transduction part where the molecular recognition event is translated into an electrical response by an electrochemical reaction [27] (Figure 1A). Two complementary strands of DNA specifically hybridize with each other; thus detection of this hybridization event has become a central theme in biosensor studies [14,26]. The DNA hybridization is based on Watson and Crick base pairing rules (Figure 1B), i.e., specific hydrogen bond formation between two (target DNA and probe DNA) complementary single strands of DNA (ssDNA) [28]. In a sensor detection scheme, ssDNA(s) specifically hybridizes with a target DNA sequence that is being employed as a probe(s): a capture probe used to attach the target DNA to the surface of materials and/or a reporter probe labeled with signaling molecules, e.g., redox-active molecules. The hybridization reaction occurs in solution (homogeneous) or at an electrode/transducer surface (heterogeneous) [12]. The advantage of hybridization in the solution phase is that it is well controlled using known properties such as melting temperature (T_m) and ionic strength of the buffer [14,29]. In the case of hybridization on an electrode surface, the probe DNA is attached to the electrode surface in such a way that the sequence is available to target DNA in the hybridization solution. Additionally, non-specific interactions of DNA with the electrode should be avoided along with other optimization processes such as surface coverage and incubation time [30,31], these will not be discussed herein.

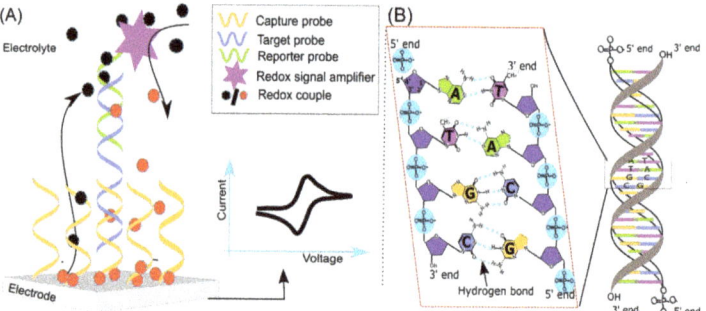

Figure 1. Nucleic acids electrochemical biosensor general principles. (**A**) A sandwich type genosensor model: A capture probe is employed to capture the target (DNA/RNA) from the solution phase to the electrode surface. The electrode bound target DNA is quantified indirectly by binding the reporter probes conjugated with a redox signal amplifier. The redox signal amplifier could be an enzyme or a

nanomaterial, which produces the redox-active molecules. The redox-active molecules undergo an oxidation/reduction reaction, which is then quantified as an electrical response (current–voltage response) using electrochemical analytical methods. The whole strategy depends solely on hybridization efficiency between the nucleic acid probes and the target molecules (RNA/DNA/PNA). In this approach, target DNA does not need any modification. (**B**) The double-helical structure of DNA and Watson and Crick base pairing in DNA. DNA consists of two strands. The two strands are held together by complementary base pairing between the bases, i.e., hydrogen bonds (A with T and G with C). Two hydrogen bonds attach A to T; three hydrogen bonds attach G to C. High temperature can denature the double-stranded DNA into single-strands. These complementary single-stranded DNAs can specifically rehybridized to form a double-stranded helix by reducing the reaction temperature.

The DNA hybridization product could be selectively and electrochemically quantified on the electrode surface. DNA adsorbed on electrode surfaces are stable below guanine oxidation potential (Table 1), this stability is advantageous for immobilization of DNA directly on the electrode surface. To make a quantitative measurement, the DNA hybridization event is coupled with electrochemical reactions, in a way that a probe-target complex increases/decreases a coupled redox reaction at the electrode surface. It can also be achieved by measuring the changes in the electrode/electrolyte interface properties due to a DNA hybridization event. In general, DNA sensors are categorized into several types based on what kind of probe DNA is being used (label-free/labeled) and how is signal transduction achieved (reagent-free or reagent-dependent). The detailed information on DNA sensor history, principles and fabrication approaches is thoroughly reviewed elsewhere [12,16,26,32,33]. In this review, approaches specific to signal amplification that involves pathogenic DNA/RNA detection is reviewed.

Table 1. Direct oxidation of DNA oxidation of guanine on different electrode supports.

Electrode	Reference Electrode	Electrolyte	Guanine Oxidation Peak (E_p) (V)	Reference
Gold	Ag/AgCl	PBS, pH 7.4	+0.7/+0.8	[34]
Nafion/Graphene	SCE	0.1 M PBS (pH 4.4)	+0.8	[35]
Glassy carbon electrode	Ag/AgCl	0.1 M PBS (pH 7.0)	+0.6	[36]
Boron doped diamond	Ag/AgCl	0.1 M acetate buffer (pH 4.5)	+0.9	[37]
Pencil graphite	Ag/AgCl	0.5 M acetate buffer and 20 mM LiClO$_4$	+0.76	[38]
DWNTs, and MWNTs	Ag/AgCl	PBS (pH 6)	+1	[39]
HOPGE	Ag/AgCl	0.1 M sodium acetate buffer (pH 7.6)	0.9	[40]

SCE—Saturated calomel electrode, PBS—Phosphate buffered saline, DWNTs—double-walled carbon nanotubes, MWNTs—multi-walled carbon nanotubes, HOPGE—Highly ordered pyrolytic graphite electrode.

3. Signal Amplification Approaches for the DNA/RNA Electrochemical Sensor

DNA hybridization is a selective process where even a single mismatch between the target and probe can be differentiated in most cases [14,29]. However, often, sensitive detection of the capture-target hybridization event is challenging. Since clinical samples may have very low copies of a pathogen in the early stages of infection, as low as 1–10 colony forming units (CFU/mL) [8]. Electrochemical methods are inherently very sensitive to detect even fM target DNA concentrations [41–47]. To overcome the limitation of instrumentation requirement for sensor deployment, signal amplification strategies have been investigated and developed to enhance the electrochemical signal (Table 2). Amplified signals are quantified using electrochemical analytical techniques such as chronoamperometry (CA), differential pulse voltammetry (DPV), square wave voltammetry (SWV), cyclic voltammetry (CV) and electrochemical impedance spectroscopy (EIS; Table 2). Signal amplification methods in combination with electrochemical analytical techniques were demonstrated for the detection of femto- and attomolar concentrations of DNA. Herein, we illustrate and discuss several signal amplification methods that were reported for pathogen detection. The methods are categorized into (i) enzyme mediated signal amplification, (ii) nanomaterials-based approaches and (iii) nucleic acid-based approaches.

Table 2. Various signal amplification strategies employed for the detection of pathogenic DNA using electrochemical analytical methods.

Pathogen	Target	Capture Probe	Reporter Probe	Electrode Modification	Amplification Strategy	Redox Signal	Limit of Detection (LOD) *	Analytical Technique	References
Ebolavirus	Biotin-ssDNA	HS-ssDNA	NA	Au	Strep-alkaline phosphatase	4-aminophenol	4.7 nM	DPV	[48]
Avian influenza A (H7N9) virus	ssDNA	SH-tetrahedral DNA	Biotin-ssDNA	Au	Strep-HRP	TMP	0.75 pM	Amperometric	[49]
Bacteria 16s RNA gene	ssDNA and genomic DNA	ssDNA (polydA SAM)	Biotin-ssDNA	Au	Strep-HRP	TMB	10 fM	Amperometric	[50]
Zika virus	ssDNA	Biotin-ssDNA (Strept-magentic beads)	Digoxigenin -ssDNA	Au	Anti- Digoxigenin coupled HRP	TMB	0.7 pM	Chronoamperometry	[51]
HIV DNA	ssDNA	SH-ssDNA	SH-ssDNA	Glucose meter	Invertase-Fe$_3$O$_4$-Au	Glucose	0.5 pM	Amperometry	[11]
Human cytomegalovirus	ssDNA(PCR product)	NA	Biotin-ssDNA	Carbon	Strep-HRP	Ophenyldimine/ 2,2'-diaminoazobenene	3.6 × 10^5 copies/mL	DPV	[52]
E. coli	gDNA	SH-ssDNA	Biotin-ssDNA	Au	Strep-HRP and redox cycling	p-aminophenyl phosphate	0.5416667	Chronoamperometric	[53]
Lactobacillus brevis	gDNA, RNA	Biotin-ssDNA	Biotin-ssDNA	Au	Strep-Lipase	Ferrocene	16 amole	CV	[54]
E. coli	ssDNA	HS-ssDNA	Biotin-ssDNA	Au	Liposome loaded with Ca^{2+}	Ca^{2+} ion-selective electrode (No redox reaction)	0.2 nM	Potentiometric method using	[55]
Dengue virus	PCR amplified target with poly (dT)	HS-ssl/Poly(dA)	Fluro-ssDNA	Au-Polyaniline/ N,S-GQDs@AuNP-dA	Nanomaterial as carrier	Methylene blue-intercalation	9.4 fM	DPV	[56]
Citrus tristeza virus	ssDNA	HS-ssDNA	NA	Au/AuNPs	Nanoparticle as carrier	[Fe(CN)$_6$]$^{3-/4-}$	100 nM	Impedance	[57]
Chikungunya virus	ssDNA	ssDNA	NA	Carbon/Fe$_3$O$_4$@Au (+ and − charge interaction to accumulate the DNA)	Nanoparticle as carrier	Methylene blue	0.1 nM	DPV and CV	[58]
Human papilloma virus	ssDNA	HS-ssDNA	NA	Nanoporous polycarbonate-AuNTs	Nanoparticles as carrier	[Fe(CN)$_6$]$^{3-/4-}$	1 fM	Impedance	[59]
Influenza and Norovirus	ssDNA	SH-ssDNA	NA	Pt-Au/Iron Oxide-CNT	Nanoparticles as carrier	NA	8.8 pM	Conductivity (the resistance change)	[60]
E. coli uropathogens	ssDNA	Biotin-ssDNA	Biotin-ssDNA	Glassy carbon	CdS quantum dots as reporter	Cd^{2+}	0.22 fM	SWV	[61]
E. coli	ssDNA	SH-ssDNA	ssDNA	AuNP-deposited on glassy carbon electrode	Nanoparticle as high amount reporter probe carrier	[Ru(NH$_3$)$_6$]$^{3+}$	1 fM	DPV	[62]
Mycobacterium tuberculosis	PCR product	SH-ssDNA	SH-ssDNA loaded AuNPs@CNT-PANI	Au	Endonuclease	Polyaniline	0.33 fM	DPV	[63]
Enterobacteriaceae	ssDNA and HAV cDNA	HS-ssDNA	biotin-ssDNA	Au	Exonuclease III and Strep-alkaline phosphatase	α-naphthyl phosphate	8.7 fM	DPV	[64]
Hepatitis B virus (HBV)	ssDNA	HS-ssDNA	ssDNA as primer	Au	CSD and RCA	Methylene blue	2.6 aM	DPV	[65]
Salmonella	gDNA	SH-ssDNA	Biotin-ssDNA	Au	DNA polymerase, T4 RNA polymerase and Strep-alkaline phosphatase	α-naphthyl phosphate	0.97 fM	DPV	[66]
Avian influenza A (H7N9) virus	ssDNA	SH-ssDNA	Molecular beacons	Au	EXPAR-HCR and G-quadruplex-hemin- (HRP like catalysis)	TMB	9.4 fM	DPV	[67]
Ebolavirus	RNA	cDNA synthesized from target RNA	Biotin-ssDNA	Carbon	Strep-glucose oxidase RCA and	H$_2$O$_2$	1 pM	Chronoamperometry	[68]
Mycobacterium tuberculosis	gDNA	Biotin-PCR product from target	Fluorescein-ssDNA	Carbon	HDA and Antifluorescein-POD Fab	TMP	0.5 aM	Chronoamperometry	[69]

DPV—Differential pulse voltammetry, CV—Cyclic Voltammetry, EIS—Electrochemical Impedance spectroscopy, CSD—Circular strand displacement, RCA—Rolling circle amplification, EXPAR—Isothermal exponential amplification, HCR—Hybridization chain reaction, HDA—Helicase dependent amplification, TMB—3,3',5,5'-tetramethylbenzidine, N,S-GQDs@AuNP—Nitrogen, sulfur codoped graphene quantum, CNT-PANI—Carbon nanotube-polyanilline, NA—Not applicable, * If limit of detection is not reported, lowest detected value is provided. ssDNA—Single stranded DNA.

3.1. Enzyme Mediated Signal Amplification

The use of an enzyme for signal amplification can aid in increasing sensitivity of a sensor in that a single recognition event that can be sensed only stoichiometrically could be transduced and recycled several times by the biocatalytic reaction mediated by an enzyme that is coupled to this recognition event. Several enzymes have been strategically conjugated with DNA hybridization complexes to amplify electrochemical signals. When an enzyme is tagged with a probe DNA, each hybridization event is coupled to an enzyme molecule. Each enzyme can produce multiple (10–1000) fold higher redox-active products. This can result in a multifold enhanced redox current at the electrode surface. Horseradish peroxidase (HRP) [70], alkaline phosphatase [64], lipase [54], invertase [11] and glucose oxidase [68] were successfully employed for signal amplification of pathogen detection studies. Different methods have been employed in signal amplification approaches to detect a low copy number of target DNA on the electrode surface. Application of magnetic beads and advancement in functionalization of nano/micro bead structures provides the ability to specifically enrich the target DNA from the background matrix components. Bioconjugates (e.g., biotin-avidin) are utilized as molecular binders with high affinity for building a network of molecular conjugations. Alzate et al. demonstrated a magnetic bead-based approach to quantify the Zika virus [51]. First, a biotinylated capture probe was immobilized on streptavidin-coated magnetic beads. Second, the target was prehybridized with the Digoxigenin (Dig)-labeled reporter probe and then added to the capture probe-coated magnetic beads to hybridize. Third, the reporter probe was recognized by an anti-Dig monoclonal antibody labeled with horseradish peroxidase (HRP). The final bead complex was magnetically attracted to the surface of a screen-printed electrode. H_2O_2 and 3,3′,5,5′-tetramethylbenzidine (TMB) were added as the HRP substrate. This strategy achieved the detection of 10 pM synthetic target ssDNA. Dong et al. reported the use of a DNA tetrahedral nanostructure-based electrochemical biosensor to detect avian influenza A (H7N9) virus [49]. The tetrahedral nanostructure was used as a biomolecule-confined surface to increase molecular recognition at the biosensing interface (Figure 2A) [71]. First, the DNA tetrahedral structure was immobilized onto a gold electrode surface via an Au-thiol bond. A single strand part of the tetrahedral DNA acted as the capture DNA to hybridize with a target ssDNA. The capture-target sequence was hybridized with a biotinylated reporter DNA sequence. Then streptavidin-horseradish peroxidase was introduced to bind the biotinylated reporter–target DNA hybrids. The reduction current for HRP oxidized TMB substrate was measured using an amperometric method. When this sensor was used for the detection of PCR products (ssDNA) amplified from cDNA isolated from positive patients, the 1.2–1200 pM range was detected. It was also shown that 1–5 cycles of asymmetric PCR generated enough target DNA for the experiment. Wang et al. reported a multiple-reporter probe approach for detection of the 16S rRNA gene of different bacteria [50] (Figure 2B). In this approach, A high-adsorption affinity of the polyA tail towards the Au surface was used to immobilize the molecular recognition complex on an Au electrode [72]. First, the target DNA was hybridized with a multiple biotinylated reporter probe. Second, the prehybridized target-reporter probe was hybridized with the capture probe immobilized at the Au electrode. The capture DNA sequence was designed to have a polyA tail. Then, an Avidin-HRP conjugate was bound to the biotinylated groups of reporter probes in the hybrid complex. The usage of multiple-reporter probes enhanced the number of HRP molecules per hybridization event [73]. The detection range was reported to be 10 fM to 1 nM of synthetic targets. The sensor was also successfully tested for specific detection of denatured genomic DNA from bacterial samples.

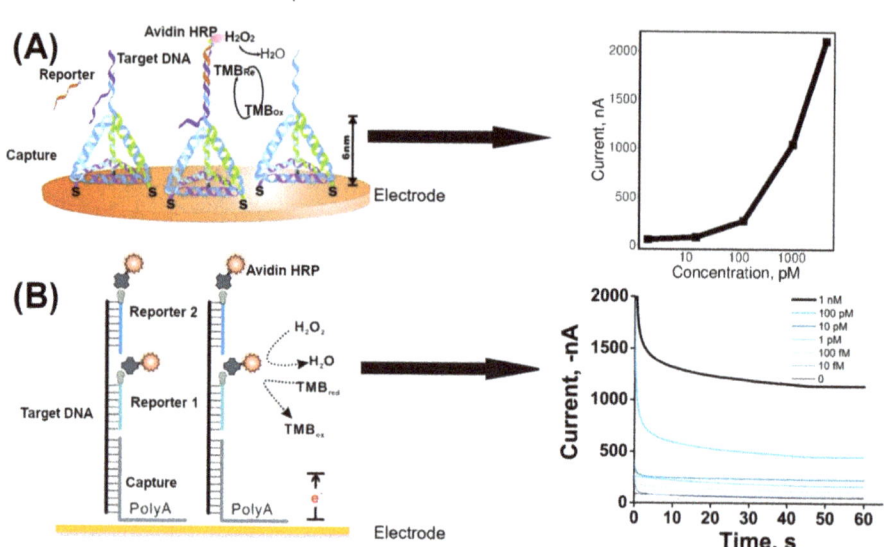

Figure 2. Schematic presentation of an HRP amplified electrochemical signal for DNA detection yth of enzyme molecules for the electrochemical signal. (**A**) DNA tetrahedral nanostructure for enhanced signal detection on gold surfaces [71]. (**B**) PolyA–gold surface interaction for immobilization of capture DNA, which was combined with multiple reporter probes and was attached to multiple HRP enzyme copies for signal amplification [50]. Adapted with permission from cited sources.

Walter et al. presented a simple approach in which signal amplification was achieved by redox cycling of p-aminophenol phosphate (p-AP) using nicotinamide adenine dinucleotide [53]. A molecular recognition complex based on sandwich-type hybridization and reporter probe was tagged with alkaline phosphatase. The electrochemically inert p-AP was converted to an electrochemically active form of p-AP by tagged alkaline phosphatase. Enzymatically generated p-AP was electro-oxidized at an Au electrode to p-quinone imine (p-QI) and in the presence of NADH, p-QI was reduced back to p-AP, which was reoxidized on the electrode. This approach overcame the drawbacks associated with the stability of p-AP. It has allowed reaching a detection limit of 1 pM of target DNA. When it was applied for the monitoring of the 16S rRNA of E. coli pathogenic bacteria it had a detection limit of 250 CFU μL^{-1}.

The signal turn-off mode system was employed for enhanced detection with enzyme-mediated signal amplification. In a signal-off mode, the current signal decreases as a function of DNA concentration. Shipovskov et al. demonstrated lipase chemistry to detect the low amount of target DNA in a signal turn-off mode (Figure 3A) [54]. They established an ester bond containing self-assembled monolayer (SAM) of 9-mercaptononyl, 4-ferrocene aminobutanoate (Fc-alkanethiol ester) on a gold surface, which exhibited high surface redox current using CVs [74]. Lipase was used to cleave the ester bond to remove ferrocene (Fc), a redox-active molecule, from a SAM layer, which resulted in a decrease in current in this system. First, a streptavidin-coated magnetic bead was decorated with biotinylated capture DNA. Then, step by step, it was allowed to react with a target DNA, biotinylated reporter probe and a streptavidin-lipase conjugate. At the end of the hybridization step, the final complex was immobilized on a magnetic bead, which was then applied on an Fc-alkanethiol ester SAM on a gold electrode. The lipase coupled to the DNA recognition complex removed the Fc from the SAM. This resulted in a decrease in the CV peak current. The lowest detected signal peak was 4 fM of the synthetic target DNA. Further, the assay could be used for the detection of down to 16 aM of denatured RNA and their cDNA copies prepared from *Lactobacillus brevis*. In a similar turn off mode, Luo et al. has employed exonuclease III for the detection of low levels of E. coli in milk samples [64]. First, a capture DNA was immobilized through its 5′-end. Second, the target DNA was hybridized with the capture probe to form a double-stranded structure, which resulted in a blunt end at the 3′-end of the capture probe. Then Exo III, an exonuclease, was introduced to catalytically remove the mononucleotides from the 3′-hydroxyl termini of DNA duplexes. The Exo III activity degraded the capture DNA strand and released the target DNA. The released target was recycled for more capture DNA degradation. After a fixed duration of treatment with the Exo III treatment, the capture DNAs that were not degraded on the sensor surface hybridized with the biotinylated reporter probe. The reporter probe was linked to streptavidin-alkaline phosphatase to produce an enzymatic electrochemical guanine signal for quantitative detection of Enterobacteriaceae bacteria. Using this approach about 40 CFU/mL of E. coli was electrochemically detected where a single strand PCR product was used as a target.

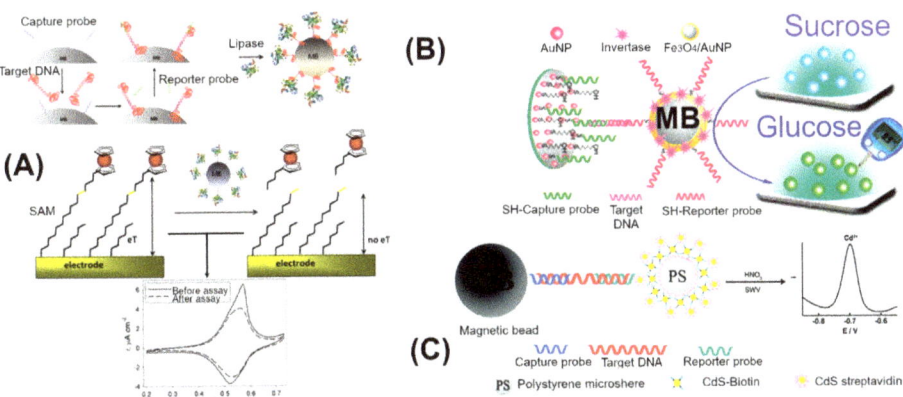

Figure 3. Schematic presentation of an electrochemical signal amplification for DNA detection. (**A**) DNA sandwich with a lipase labeled reporter probe for detection of *Lactobacillus brevis* DNA. Lipase was designed to bind with capture and target molecular recognition elements. During electrochemical analysis, lipase cleaves off the ferrocene from 9-mercaptononyl, 4-ferrocene aminobutanoate monolayer over the electrode surface. This results in the reduction of the observed current using cyclic voltammetry [74]. (**B**) Multiple invertase copies coated magnetic bead was conjugated with each capture and target molecular recognition element. The invertase was used to convert sucrose to glucose. Glucose was detected by a glucose meter. This system was reported for detection of HIV DNA [11]. (**C**) Similar to invertase, CdS coated polystyrene bead was used as a signal amplifier for the detection of urinary tract pathogens [61]. The Cds nanoparticle bound to the molecular recognition element was dissolved in the acid solution and resulting cadmium ions were quantified electrochemically. Adapted with permission from cited sources.

Screen-printed electrodes, which require a small sample volume, are widely employed in sensor development studies. However, the electrochemical response analysis for the screen-printed electrode is still limited to high-end laboratory-based instrumentation. Instead of conventional laboratory-based electrochemical techniques, the commercial glucose meter was also successfully demonstrated for the detection of pathogen's DNA. Xu et al. demonstrated multiple invertase-mediated signal amplification and the use of a glucometer as an electrochemical device for the detection of HIV DNA [11] (Figure 3B). First, a mixed layer of thiolated capture probes and 6-mercaptoethanol were self-assembled on the AuNPs via thiol–Au attachment. The capture probe coated AuNPs were applied on the glassy carbon electrode. Then target DNA was hybridized with capture probes. The reporter probe was tagged with multiple-invertase coated-Fe_3O_4/AuNPs using thiol chemistry. Hybridization of invertase coated reporter probes has led to massive quantities of invertase on the electrode surface. Glassy carbon electrode was used to characterize the loading of the probe and target DNA using [Fe(CN)$_6^{3-/4-}$] redox couple. To quantify the target DNA, sucrose was introduced onto the glassy carbon electrode surface containing a molecular recognition complex. Upon introducing sucrose, invertase converted the sucrose into glucose molecules, which were measured by a glucometer. Due to several numbers of tagged invertase per hybridization event and its high turnover number, glucose in millimolar concentrations was produced. Using this approach, about 0.5 pM to 1 nM concentration of synthetic HIV DNA was detected using a standard glucose meter sensitivity range.

3.2. Nanomaterial Enhanced Signal

The nanomaterials are used as reporter molecules and high surface area materials for high loading of probe DNA. In case of a nanomaterial as a reporter, metal-based nanoparticles were tagged with DNA for hybridization. Xiang et al. reported CdS quantum dot decorated polystyrene (PS-(CdS)$_4$) as a signal amplifier for the detection of urinary tract pathogen (Figure 3C) [61]. PS-(CdS)$_4$ was built using biotin and streptavidin functionalized PS and CdS nanoparticles. First, the biotin-capture probe was immobilized on streptavidin-magnetic beads and then incubated with the target DNA. The magnetic bead–target complex was then hybridized with a reporter probe, which was immobilized on polystyrene-CdS spheres (PS-(CdS)$_4$). The resulting complex was selectively separated by magnetic separation and was treated with nitric acid to dissolve the CdS nanoparticles. Cd ions in the solution were measured using square wave voltammetry. Using this approach, 0.5 fM to 10 pM of synthetic DNA was detected. In a similar metal nanoparticle-based approach, Zhang et al. demonstrated detection of multiple pathogens using nanoparticle-based biobarcoded electrochemical sensors [75]. Each pathogen-specific probe sequence was tagged to specific nanoparticles. The detection limit of bio-barcoded DNA sensor was 0.5 ng/mL for the insertion element (*Iel*) gene of *Salmonella enteritidis* using CdS, and 50 pg/mL for the *pagA* gene of *Bacillus anthracis* using PbS. As an alternative to toxic metal-based nanoparticles, Wang et al. reported a strategy for the detection of low levels of *E. coli* using liposome 'nanocarriers' loaded with Ca^{2+} ions [55]. Upon the successful formation of a recognition event, calcium-loaded liposomes were bound to the reporter DNA after which they were lysed by a surfactant. In this approach, sub-fmol DNA detection limit was achieved by employing Ca^{2+} ion-sensitive electrodes.

The high surface area of nanoparticles was exploited for loading of a high amount of capture DNA. Chowdhury et al. had detected Dengue virus DNA using nanocomposites of gold nanoparticles (AuNP) with nitrogen and sulfur co-doped graphene quantum dots (N,S-GQDs@AuNP) [56]. First, N,S-GQDs@AuNP were coated with a capture DNA (polydA) using Au-thiol bond formation. This led to the accumulation of a large number of single-stranded (ssDNA) capture DNA. Second, polydA was used to hybridize with a polydT tail of the target viral DNA. Finally, the complex was subjected to electrochemical quantification using methylene blue as a reporter molecule. In the presence of a target, hybridization resulted in a double-stranded DNA (dsDNA), which did not bind methylene blue effectively and resulted in low peak current. In the absence of the target, the dye binds to the capture ssDNA and gave high current in differential pulse voltammetry analysis. This signal-off mode analysis detected a synthetic target in the range of 10 fM to µM. Xu et al. employed AuNPs and exonuclease I for detection of uropathogen's DNA using [Ru(NH$_3$)$_6$]$^{3+}$ redox molecules [62]. The AuNPs were coated with multiple ssDNA that was used as the signal probe, this resulted in a greater number of DNA molecules per molecular recognition event. [Ru(NH$_3$)$_6$]$^{3+}$ bound to the excess DNA molecules electrostatically, which ultimately amplified the redox signal for every target DNA. Furthermore, exonuclease I (Exo I) treatment removed the unhybridized single-stranded capture DNA probes, which minimized the background current. The combination of signal amplification and background current reduction resulted in 1 fM detection limit. Chen et al. reported the use of redox active carbon nanotubes (CNTs) doped with polyaniline (PANI) and endonuclease mediated target recycling approach for detection of *Mycobacterium tuberculosis* [63]. In the presence of target DNA and an assistant probe that hybridized to the capture probe, a hairpin structure has opened to form a Y-shaped junction. Endonuclease recognized the sequence in the Y-shaped junction and released the assistant probe and target DNA. Released target DNA triggered the next cycle of cleavage. After hybridization between CNTs-PANI tagged reporter probe and the cleaved capture probe on the electrode surface, the electrochemical signal of CNTs-PANI was used as a readout. This strategy detected the target in a range between 1 fM to 10 nM.

3.3. Nucleic Acid Amplification and Processing Based Approaches

Though enzyme-based and nanomaterial-based signal amplification approaches reach sensitivity in the femtomolar regime, they still depend on PCR to produce sufficient target DNA. In nucleic acid-based approaches, the enzyme-mediated isothermal amplification of nucleic acids plays an important role in sample amplification and detection. Unlike PCR, which requires specialized thermal cycler instruments to mediate denaturation, annealing and subsequent extension steps, isothermal amplification could be carried out at a constant temperature to produce about a million copies of the target DNA. For isothermal amplification, in addition to DNA polymerase, ligase, nicking enzymes and helicases are employed for specific amplification of target DNA molecules. A detailed review of the method can be found in the following references [76,77]. Simple temperature control makes isothermal amplification an attractive alternative to PCR for point-of-care applications. There are several choices of methods that are available for isothermal amplification, depending on the length, secondary structures and nature of the target (RNA or DNA) [76]. The challenging aspects are electrochemical detection of specific targets amplified by the isothermal method. Cheng et al. reported a method combining circular strand displacement polymerization reaction (CSD), rolling circle amplification (RCA) and enzymatic amplification to enhance the electrochemical sensing of a target DNA (Figure 4A) [78]. First, the capture probe (SH-ssDNA with a hairpin loop structure-molecular beacon) was immobilized on a gold electrode. Second, the strand displacement (CSD) reaction was carried out by adding the target DNA, and biotinylated-primer DNA to the electrode. In the presence of the target DNA, the hairpin structure of the molecular beacon opens and parts the sequence that was hybridized with the target DNA. Another part of the capture probe sequence binds specifically with a biotinylated-primer DNA. The primer sequence was extended towards the target DNA binding region by a DNA polymerase (KF exo-), which led to the release of the target DNA. At this stage, freed target DNA binds to another capture probe to trigger another strand displacement reaction, which results in multiple biotin-tagged DNA duplexes on the electrode surface. This biotin-tagged DNA duplex anchored with another streptavidin-primer specific for rolling circle amplification. Upon addition of specific circular ssDNA templates, deoxynucleotide triphosphates and phi29 DNA polymerase, the RCA reaction produces long ssDNA molecules with tandem repeated sequences. Then alkaline phosphatase tagged reporter probe DNA was added and hybridized with repeated sequences of RCA products. Alkaline phosphatase was used as a final redox signal amplifier at the electrode surface. Using this approach, 1 fM to 100 pM of synthetic target DNA was detected. Huang et al. used a similar strand displacement and rolling circle amplification approach—without biotin and streptavidin tags—for the detection of synthetic DNA sequences specific to the hepatitis B virus [65]. The rolling circle amplification resulted in long ssDNA. They detected the final rolling circle amplification product using methylene blue and reported detection in the range of 10 aM to 0.7 fM. Yanyan et al. reported a DNA detection approach for the avian influenza virus based on isothermal exponential amplification coupled with hybridization chain reaction [67]. Catalytic G-quadruplex–hemin, HRP-mimicking DNAzymes, was tagged to the final molecular recognition complex. Electrochemical signals obtained by measuring the increase in the reduction current of oxidized 3,3',5,5'-tetramethylbenzidine sulfate, which was generated by DNAzyme in the presence of H_2O_2. This method exhibited detection limits of 9.4 fM. In a similar approach, exonuclease III mediated target DNA recycling and G-quadruplex–hemin reported for the detection of HIV gene sequence with a detection limit of 3.6 pM [79].

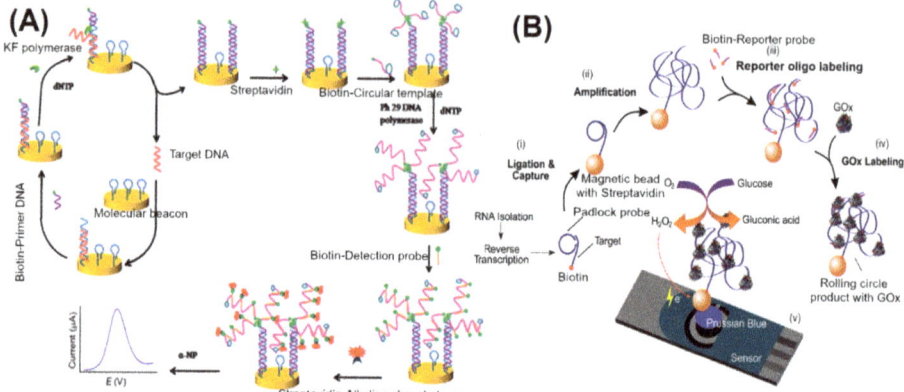

Figure 4. Enhancement of nucleic acid detection by employing polymerase and other isothermal amplification approaches on the electrode surface. (**A**) Strand displacement reaction and rolling circle amplification coupled system [78]. (**B**) Ligation and rolling circle amplification coupled system [68]. Adapted with permission from cited sources.

Ciftci et al. reported a method for tagging multiple glucose oxidase (GoX) enzymes using rolling circle amplification for the detection of *Ebolavirus* (Figure 4B) [68]. In the first step (i) biotinylated primers were used to reverse transcribe the RNA target to cDNA. Then the biotinylated cDNA target hybridized with linear pad-lock probes (PLPs). PLPs have a special sequence feature that renders the probe circular upon hybridization. These circular PLPs are ligated using the enzyme ligase; then biotinylated cDNA target-PLP complex captured on streptavidin-functionalized magnetic beads. Magnetic beads were used for the separation of target DNA from the sample. In the second step, RCA reaction was carried out to produce bulky tandem repeats of DNA coils. In the third step, RCA products were hybridized with the biotinylated-reporter probe, which was then bound to streptavidin-GoX. The final DNA recognition complex was quantified by GoX activity using chronoamperometry. The product, H_2O_2, of glucose oxidation by GoX in the presence of oxygen was electrochemically measured. In this method, 1–100 pM of synthetic target DNA was measured. An *Ebolavirus* positive clinical sample was also successfully differentiated from negative samples using this method. A similar Pad-lock-probe approach was demonstrated for the detection of *Ebolavirus* using HRP as a signal amplifier [80].

Helicase dependent amplification (HDA) is another widely employed isothermal system for amplification of target DNA [81]. Helicase is being used to unwind the double-stranded DNA instead of temperature-dependent denaturation during polymerase mediated amplification of the target gene [82]. Barreda-García et al. reported an asymmetric HDA for that resulted in a single-stranded target DNA from *Mycobacterium tuberculosis* [69]. The amplified ssDNA was selectively quantified using enzyme-mediated signal amplification assay. The system was sensitive to 0.5 aM of target DNA. In another study, HDA was also demonstrated for hybridizing the double-stranded target generated from *Salmonella* genome to a single-stranded capture DNA bound to indium-tin-oxide electrodes [83].

Yan et al. reported the detection of pathogenic DNA directly by transcription of RNA from the target DNA [66]. In this approach, hairpin structured primers were designed to open and bind to the target DNA specifically. The primer was extended using DNA polymerase (KF exo-) at 37 °C. The primer has efficiently triggered the circular primer extension reaction, i.e., the resultant dsDNA was further amplified by another primer binding and extension cycles. Additionally, the primer was designed to have a T7 RNA polymerase promoter, which served as a template for in vitro transcription of target DNA. The RNA products from the transcription reaction were directly hybridized with immobilized capture probes. The enzyme tagged signal probe was used to detect the hybridized products. With this approach, *Salmonella*'s *invA* gene from genomic DNA extract was successfully detected. Limit of detection was about 1 fM.

4. Conclusions and Future Perspectives

In nanomaterials and enzyme-mediated amplification approaches, DNA/RNA isolation and amplification of targets by PCR are commonly required for real sample electrochemical detection to increase the number of specific target DNA molecules. However, PCR can only be carried out using a specialized thermocycler. Thus, to reduce the dependence on PCR, isothermal amplification methods are explored in clinical sample target amplification in combination with an electrochemical sensor. While isothermal amplification provides the advantage of in situ amplification of target DNA at a constant temperature, more exploration of novel approaches for amplification and hybridization on-electrode surfaces is still needed to achieve practical electrochemical DNA sensors.

The requirement for single-stranded target DNA for hybridization is another constraint for electrochemical DNA sensors. To counter this issue, asymmetric PCR, thermal denaturation of target DNA followed by abrupt cooling, helicase mediated asymmetric DNA amplification, DNA to RNA transcription using RNA polymerase have been employed to yield single-stranded targets. However, novel approaches should be explored for the detection of double-stranded genomic DNA or structured single-stranded cDNA/RNA at room temperature.

Electrochemical analytical methods reached a "glass ceiling" with a limit of detection in the femtomolar concentration range. Enzymes, nanomaterials and molecular tools are successfully engineered for fM detection of DNA on an electrode surface using signal amplification approaches. With this, (i) future progress in the simplification of sample processing steps, (ii) developing approaches to detect double-stranded target DNA and (iii) making advancement in the sensitivity of instrumentation or handheld electrochemical devices will be crucial for achieving practical point-of-care electrochemical devices for pathogen detection in low titers.

In addition, for fast epidemic containment, in order to make sure that highly trained personnel will not be needed for point of care detection, it is highly needed to be able to interface such sensors with already existing simple devices such as glucometers, so it will be easy to sample (such as in non-invasive devices) and simple to read-out the signal for less trained users (e.g., airports personnel).

Author Contributions: Conceptualization, M.S. and L.A.; writing—original draft preparation, M.S. and I.A.; writing—review and editing L.A.; supervision, L.A.; project administration, L.A.; funding acquisition, All authors have read and agreed to the published version of the manuscript.

Funding: This research was funded by Israel Ministry of Science and Technology, grant number 88702. M.S. was funded by an Israeli planning and budgeting committee (PBC) post-doctoral fellowship for excellent Indian and Chinese candidates. I.A. gratefully acknowledges a Kreitman School fellowship for Biotechnology research at BGU.

Conflicts of Interest: The authors declare no conflict of interest.

References

1. Wittwer, C.T.; Makrigiorgos, G.M. Nucleic acid techniques. In *Principles and Applications of Molecular Diagnostics*; Elsevier: Amsterdam, The Netherlands, 2018; pp. 47–86. ISBN 978-0-12-816061-9.
2. Cheng, V.C.C.; Wong, S.C.; Chen, J.H.K.; Yip, C.C.Y.; Chuang, V.W.M.; Tsang, O.T.Y.; Sridhar, S.; Chan, J.F.W.; Ho, P.L.; Yuen, K.Y. Escalating infection control response to the rapidly evolving epidemiology of the coronavirus disease 2019 (COVID-19) due to SARS-CoV-2 in Hong Kong. *Infect. Control Hosp. Epidemiol.* **2020**, *41*, 493–498. [CrossRef]
3. Fritsch, A.; Schweiger, B.; Biere, B. Influenza c virus in pre-school children with respiratory infections: Retrospective analysis of data from the national influenza surveillance system in germany, 2012 to 2014. *Eurosurveillance* **2019**, *24*. [CrossRef]

4. Sonawane, A.A.; Shastri, J.; Bavdekar, S.B. Respiratory pathogens in infants diagnosed with acute lower respiratory tract infection in a tertiary care hospital of western India using multiplex Real Time PCR. *Indian J. Pediatr.* **2019**, *86*, 433–438. [CrossRef] [PubMed]
5. Deghmane, A.E.; Hong, E.; Taha, M.K. Diagnosis of meningococcal infection using internally controlled multiplex real-time PCR. In *Neisseria Meningitidis: Methods and Protocols*; Humana Press Inc.: New York, NY, USA, 2019; Volume 1969, pp. 17–31. ISBN 978-1-4939-9202-7.
6. Wlassow, M.; Poiteau, L.; Roudot-Thoraval, F.; Rosa, I.; Soulier, A.; Hézode, C.; Ortonne, V.; Pawlotsky, J.M.; Chevaliez, S. The new Xpert HCV viral load real-time PCR assay accurately quantifies hepatitis C virus RNA in serum and whole-blood specimens. *J. Clin. Virol.* **2019**, *117*, 80–84. [CrossRef] [PubMed]
7. Zhang, C.; Zheng, X.; Zhao, C.; Li, Y.; Chen, S.; Liu, G.; Wang, C.; Lv, Q.; Liu, P.; Zheng, Y.; et al. Detection of pathogenic microorganisms from bloodstream infection specimens using TaqMan array card technology. *Sci. Rep.* **2018**, *8*. [CrossRef] [PubMed]
8. Loonen, A.J.M.; Bos, M.P.; van Meerbergen, B.; Neerken, S.; Catsburg, A.; Dobbelaer, I.; Penterman, R.; Maertens, G.; van de Wiel, P.; Savelkoul, P.; et al. Comparison of pathogen DNA isolation methods from large volumes of whole blood to improve molecular diagnosis of bloodstream infections. *PLoS ONE* **2013**, *8*, e72349. [CrossRef] [PubMed]
9. Teengam, P.; Siangproh, W.; Tuantranont, A.; Vilaivan, T.; Chailapakul, O.; Henry, C.S. Multiplex paper-based colorimetric DNA sensor using pyrrolidinyl peptide nucleic acid-induced AgNPs aggregation for detecting MERS-CoV, MTB, and HPV oligonucleotides. *Anal. Chem.* **2017**, *89*, 5428–5435. [CrossRef] [PubMed]
10. Myers, F.B.; Lee, L.P. Innovations in optical microfluidic technologies for point-of-care diagnostics. *Lab. Chip* **2008**, *8*, 2015–2031. [CrossRef] [PubMed]
11. Xu, J.; Jiang, B.; Xie, J.; Xiang, Y.; Yuan, R.; Chai, Y. Sensitive point-of-care monitoring of HIV related DNA sequences with a personal glucometer. *Chem. Commun.* **2012**, *48*, 10733–10735. [CrossRef]
12. Trotter, M.; Borst, N.; Thewes, R.; von Stetten, F. Review: Electrochemical DNA sensing–Principles, commercial systems, and applications. *Biosens. Bioelectron.* **2020**, *154*, 112069. [CrossRef]
13. Liu, G.; Wan, Y.; Gau, V.; Zhang, J.; Wang, L.; Song, S.; Fan, C. An enzyme-based E-DNA sensor for sequence-specific detection of femtomolar DNA targets. *J. Am. Chem. Soc.* **2008**, *130*, 6820–6825. [CrossRef] [PubMed]
14. Zhang, D.Y.; Chen, S.X.; Yin, P. Optimizing the specificity of nucleic acid hybridization. *Nat. Chem.* **2012**, *4*, 208–214. [CrossRef] [PubMed]
15. Wang, J. Nanomaterial-Based amplified transduction of biomolecular interactions. *Small* **2005**, *1*, 1036–1043. [CrossRef] [PubMed]
16. Blair, E.O.; Corrigan, D.K. A review of microfabricated electrochemical biosensors for DNA detection. *Biosens. Bioelectron.* **2019**, *134*, 57–67. [CrossRef]
17. Hoilett, O.S.; Walker, J.F.; Balash, B.M.; Jaras, N.J.; Boppana, S.; Linnes, J.C. KickStat: A coin-sized potentiostat for high-resolution electrochemical analysis. *Sensors* **2020**, *20*, 2407. [CrossRef]
18. Chapman, J.; Power, A.; Kiran, K.; Chandra, S. Review—New twists in the plot: Recent advances in electrochemical genosensors for disease screening. *J. Electrochem. Soc.* **2017**, *164*, B665. [CrossRef]
19. Ozer, T.; Geiss, B.J.; Henry, C.S. Review—Chemical and biological sensors for viral detection. *J. Electrochem. Soc.* **2019**, *167*, 037523. [CrossRef]
20. Yáñez-Sedeño, P.; Campuzano, S.; Pingarrón, J.M. Pushing the limits of electrochemistry toward challenging applications in clinical diagnosis, prognosis, and therapeutic action. *Chem. Commun.* **2019**, *55*, 2563–2592. [CrossRef]
21. Smith, S.J.; Nemr, C.R.; Kelley, S.O. Chemistry-Driven approaches for ultrasensitive nucleic acid detection. *J. Am. Chem. Soc.* **2017**, *139*, 1020–1028. [CrossRef]
22. Bonanni, A.; del Valle, M. Use of nanomaterials for impedimetric DNA sensors: A review. *Anal. Chim. Acta* **2010**, *678*, 7–17. [CrossRef]
23. Kupis-Rozmysłowicz, J.; Antonucci, A.; Boghossian, A.A. Review—Engineering the selectivity of the DNA-SWCNT sensor. *ECS J. Solid State Sci. Technol.* **2016**, *5*, M3067. [CrossRef]
24. Rasheed, P.A.; Sandhyarani, N. Electrochemical DNA sensors based on the use of gold nanoparticles: A review on recent developments. *Microchim. Acta* **2017**, *184*, 981–1000. [CrossRef]
25. Vikrant, K.; Bhardwaj, N.; Bhardwaj, S.K.; Kim, K.-H.; Deep, A. Nanomaterials as efficient platforms for sensing DNA. *Biomaterials* **2019**, *214*, 119215. [CrossRef] [PubMed]

26. Pellitero, M.A.; Shaver, A.; Arroyo-Currás, N. Critical review—Approaches for the electrochemical interrogation of DNA-based sensors: A critical review. *J. Electrochem. Soc.* **2020**, *167*, 037529. [CrossRef]
27. Drummond, T.G.; Hill, M.G.; Barton, J.K. Electrochemical DNA sensors. *Nat. Biotechnol.* **2003**, *21*, 1192–1199. [CrossRef]
28. Leslie, A. Pray Discovery of DNA structure and function: Watson and Crick. *Nat. Educ.* **2008**, *1*, 100.
29. Inouye, M.; Ikeda, R.; Takase, M.; Tsuri, T.; Chiba, J. Single-nucleotide polymorphism detection with "wire-like" DNA probes that display quasi "on-off" digital action. *Proc. Natl. Acad. Sci. USA* **2005**, *102*, 11606–11610. [CrossRef]
30. Gong, P.; Levicky, R. DNA surface hybridization regimes. *Proc. Natl. Acad. Sci. USA* **2008**, *105*, 5301–5306. [CrossRef]
31. Levicky, R.; Horgan, A. Physicochemical perspectives on DNA microarray and biosensor technologies. *Trends Biotechnol.* **2005**, *23*, 143–149. [CrossRef]
32. Ferapontova, E.E. DNA Electrochemistry and Electrochemical Sensors for Nucleic Acids. *Annu. Rev. Anal. Chem.* **2018**. [CrossRef]
33. Paleček, E.; Bartošík, M. Electrochemistry of nucleic acids. *Chem. Rev.* **2012**, *112*, 3427–3481. [CrossRef] [PubMed]
34. Ferapontova, E.E.; Shipovskov, S.V. Electrochemically induced oxidative damage to double stranded calf thymus DNA adsorbed on gold electrodes. *Biochem. Mosc.* **2003**, *68*, 99–104. [CrossRef] [PubMed]
35. Yin, H.; Zhou, Y.; Ma, Q.; Ai, S.; Ju, P.; Zhu, L.; Lu, L. Electrochemical oxidation behavior of guanine and adenine on graphene-Nafion composite film modified glassy carbon electrode and the simultaneous determination. *Process Biochem.* **2010**, *45*, 1707–1712. [CrossRef]
36. Zhou, M.; Zhai, Y.; Dong, S. Electrochemical sensing and biosensing platform based on chemically reduced graphene oxide. *Anal. Chem.* **2009**, *81*, 5603–5613. [CrossRef] [PubMed]
37. Oliveira, S.C.B.; Oliveira-Brett, A.M. In situ DNA oxidative damage by electrochemically generated hydroxyl free radicals on a boron-doped diamond electrode. *Langmuir* **2012**, *28*, 4896–4901. [CrossRef] [PubMed]
38. Özcan, A.; Şahin, Y.; Özsöz, M.; Turan, S. Electrochemical oxidation of ds-DNA on polypyrrole nanofiber modified pencil graphite electrode. *Electroanalysis* **2007**, *19*, 2208–2216. [CrossRef]
39. Pogacean, F.; Biris, A.R.; Coros, M.; Watanabe, F.; Biris, A.S.; Clichici, S.; Filip, A.; Pruneanu, S. Electrochemical oxidation of adenine using platinum electrodes modified with carbon nanotubes. *Phys. E Low Dimens. Syst. Nanostruct.* **2014**, *59*, 181–185. [CrossRef]
40. Wu, L.; Zhou, J.; Luo, J.; Lin, Z. Oxidation and adsorption of deoxyribonucleic acid at highly ordered pyrolytic graphite electrode. *Electrochim. Acta* **2000**, *45*, 2923–2927. [CrossRef]
41. Wang, L.; Veselinovic, M.; Yang, L.; Geiss, B.J.; Dandy, D.S.; Chen, T. A sensitive DNA capacitive biosensor using interdigitated electrodes. *Biosens. Bioelectron.* **2017**, *87*, 646–653. [CrossRef]
42. Manzano, M.; Viezzi, S.; Mazerat, S.; Marks, R.S.; Vidic, J. Rapid and label-free electrochemical DNA biosensor for detecting hepatitis A virus. *Biosens. Bioelectron.* **2018**, *100*, 89–95. [CrossRef]
43. Subak, H.; Ozkan-Ariksoysal, D. Label-free electrochemical biosensor for the detection of Influenza genes and the solution of guanine-based displaying problem of DNA hybridization. *Sens. Actuators B Chem.* **2018**, *263*, 196–207. [CrossRef]
44. Deshmukh, R.; Prusty, A.K.; Roy, U.; Bhand, S. A capacitive DNA sensor for sensitive detection of: Escherichia coli O157:H7 in potable water based on the z3276 genetic marker: Fabrication and analytical performance. *Analyst* **2020**, *145*, 2267–2278. [CrossRef] [PubMed]
45. Grieshaber, D.; MacKenzie, R.; Vörös, J.; Reimhult, E. Electrochemical biosensors—Sensor principles and architectures. *Sensors* **2008**, *8*, 1400–1458. [CrossRef] [PubMed]
46. Grabowska, I.; Malecka, K.; Stachyra, A.; Góra-Sochacka, A.; Sirko, A.; Zagórski-Ostoja, W.; Radecka, H.; Radecki, J. Single electrode genosensor for simultaneous determination of sequences encoding hemagglutinin and neuraminidase of avian influenza virus type H5N1. *Anal. Chem.* **2013**, *85*, 10167–10173. [CrossRef]
47. Malecka, K.; Stachyra, A.; Góra-Sochacka, A.; Sirko, A.; Zagórski-Ostoja, W.; Radecka, H.; Radecki, J. Electrochemical genosensor based on disc and screen printed gold electrodes for detection of specific DNA and RNA sequences derived from Avian Influenza Virus H5N1. *Sens. Actuators B Chem.* **2016**, *224*, 290–297. [CrossRef]
48. Ilkhani, H.; Farhad, S. A novel electrochemical DNA biosensor for Ebola virus detection. *Anal. Biochem.* **2018**, *557*, 151–155. [CrossRef]

49. Dong, S.; Zhao, R.; Zhu, J.; Lu, X.; Li, Y.; Qiu, S.; Jia, L.; Jiao, X.; Song, S.; Fan, C.; et al. Electrochemical DNA biosensor based on a tetrahedral nanostructure probe for the detection of avian influenza A (H7N9) virus. *ACS Appl. Mater. Interfaces* **2015**, *7*, 8834–8842. [CrossRef]
50. Wang, Q.; Wen, Y.; Li, Y.; Liang, W.; Li, W.; Li, Y.; Wu, J.; Zhu, H.; Zhao, K.; Zhang, J.; et al. Ultrasensitive electrochemical biosensor of bacterial 16S rRNA gene based on polyA DNA probes. *Anal. Chem.* **2019**, *91*, 9277–9283. [CrossRef]
51. Alzate, D.; Cajigas, S.; Robledo, S.; Muskus, C.; Orozco, J. Genosensors for differential detection of Zika virus. *Talanta* **2020**, *210*, 120648. [CrossRef]
52. Azek, F.; Grossiord, C.; Joannes, M.; Limoges, B.; Brossier, P. Hybridization assay at a disposable electrochemical biosensor for the attomole detection of amplified human cytomegalovirus DNA. *Anal. Biochem.* **2000**, *284*, 107–113. [CrossRef]
53. Walter, A.; Wu, J.; Flechsig, G.-U.; Haake, D.A.; Wang, J. Redox cycling amplified electrochemical detection of DNA hybridization: Application to pathogen E. coli bacterial RNA. *Anal. Chim. Acta* **2011**, *689*, 29–33. [CrossRef] [PubMed]
54. Shipovskov, S.; Saunders, A.M.; Nielsen, J.S.; Hansen, M.H.; Gothelf, K.V.; Ferapontova, E.E. Electrochemical sandwich assay for attomole analysis of DNA and RNA from beer spoilage bacteria Lactobacillus brevis. *Biosens. Bioelectron.* **2012**, *37*, 99–106. [CrossRef] [PubMed]
55. Chumbimuni-Torres, K.Y.; Wu, J.; Clawson, C.; Galik, M.; Walter, A.; Flechsig, G.-U.; Bakker, E.; Zhang, L.; Wang, J. Amplified potentiometric transduction of DNA hybridization using ion-loaded liposomes. *Analyst* **2010**, *135*, 1618–1623. [CrossRef]
56. Chowdhury, A.D.; Ganganboina, A.B.; Nasrin, F.; Takemura, K.; Doong, R.A.; Utomo, D.I.S.; Lee, J.; Khoris, I.M.; Park, E.Y. Femtomolar detection of Dengue virus DNA with serotype identification ability. *Anal. Chem.* **2018**, *90*, 12464–12474. [CrossRef]
57. Khater, M.; de la Escosura-Muñiz, A.; Quesada-González, D.; Merkoçi, A. Electrochemical detection of plant virus using gold nanoparticle-modified electrodes. *Anal. Chim. Acta* **2019**, *1046*, 123–131. [CrossRef] [PubMed]
58. Singhal, C.; Dubey, A.; Mathur, A.; Pundir, C.S.; Narang, J. Paper based DNA biosensor for detection of chikungunya virus using gold shells coated magnetic nanocubes. *Process Biochem.* **2018**, *74*, 35–42. [CrossRef]
59. Shariati, M.; Ghorbani, M.; Sasanpour, P.; Karimizefreh, A. An ultrasensitive label free human papilloma virus DNA biosensor using gold nanotubes based on nanoporous polycarbonate in electrical alignment. *Anal. Chim. Acta* **2019**, *1048*, 31–41. [CrossRef]
60. Lee, J.; Morita, M.; Takemura, K.; Park, E.Y. A multi-functional gold/iron-oxide nanoparticle-CNT hybrid nanomaterial as virus DNA sensing platform. *Biosens. Bioelectron.* **2018**, *102*, 425–431. [CrossRef]
61. Xiang, Y.; Zhang, H.; Jiang, B.; Chai, Y.; Yuan, R. Quantum dot layer-by-layer assemblies as signal amplification labels for ultrasensitive electronic detection of uropathogens. *Anal. Chem.* **2011**, *83*, 4302–4306. [CrossRef]
62. Xu, J.; Jiang, B.; Su, J.; Xiang, Y.; Yuan, R.; Chai, Y. Background current reduction and biobarcode amplification for label-free, highly sensitive electrochemical detection of pathogenic DNA. *Chem. Commun.* **2012**, *48*, 3309–3311. [CrossRef]
63. Chen, Y.; Guo, S.; Zhao, M.; Zhang, P.; Xin, Z.; Tao, J.; Bai, L. Amperometric DNA biosensor for Mycobacterium tuberculosis detection using flower-like carbon nanotubes-polyaniline nanohybrid and enzyme-assisted signal amplification strategy. *Biosens. Bioelectron.* **2018**, *119*, 215–220. [CrossRef] [PubMed]
64. Luo, C.; Tang, H.; Cheng, W.; Yan, L.; Zhang, D.; Ju, H.; Ding, S. A sensitive electrochemical DNA biosensor for specific detection of Enterobacteriaceae bacteria by Exonuclease III-assisted signal amplification. *Biosens. Bioelectron.* **2013**, *48*, 132–137. [CrossRef] [PubMed]
65. Huang, S.; Feng, M.; Li, J.; Liu, Y.; Xiao, Q. Voltammetric determination of attomolar levels of a sequence derived from the genom of hepatitis B virus by using molecular beacon mediated circular strand displacement and rolling circle amplification. *Microchim. Acta* **2018**, *185*, 206. [CrossRef]
66. Yan, Y.; Ding, S.; Zhao, D.; Yuan, R.; Zhang, Y.; Cheng, W. Direct ultrasensitive electrochemical biosensing of pathogenic DNA using homogeneous target-initiated transcription amplification. *Sci. Rep.* **2016**, *6*, 18810. [CrossRef]
67. Yu, Y.; Chen, Z.; Jian, W.; Sun, D.; Zhang, B.; Li, X.; Yao, M. Ultrasensitive electrochemical detection of avian influenza A (H7N9) virus DNA based on isothermal exponential amplification coupled with hybridization chain reaction of DNAzyme nanowires. *Biosens. Bioelectron.* **2015**, *64*, 566–571. [CrossRef] [PubMed]

68. Ciftci, S.; Cánovas, R.; Neumann, F.; Paulraj, T.; Nilsson, M.; Crespo, G.A.; Madaboosi, N. The sweet detection of rolling circle amplification: Glucose-based electrochemical genosensor for the detection of viral nucleic acid. *Biosens. Bioelectron.* **2020**, *151*, 112002. [CrossRef]
69. Barreda-García, S.; González-Álvarez, M.J.; de-los-Santos-Álvarez, N.; Palacios-Gutiérrez, J.J.; Miranda-Ordieres, A.J.; Lobo-Castañón, M.J. Attomolar quantitation of Mycobacterium tuberculosis by asymmetric helicase-dependent isothermal DNA-amplification and electrochemical detection. *Biosens. Bioelectron.* **2015**, *68*, 122–128. [CrossRef]
70. Li, L.; Wang, L.; Xu, Q.; Xu, L.; Liang, W.; Li, Y.; Ding, M.; Aldalbahi, A.; Ge, Z.; Wang, L.; et al. Bacterial analysis using an electrochemical DNA biosensor with poly-adenine-mediated DNA self-assembly. *ACS Appl. Mater. Interfaces* **2018**, *10*, 6895–6903. [CrossRef]
71. Pei, H.; Lu, N.; Wen, Y.; Song, S.; Liu, Y.; Yan, H.; Fan, C. A DNA nanostructure-based biomolecular probe carrier platform for electrochemical biosensing. *Adv. Mater.* **2010**, *22*, 4754–4758. [CrossRef]
72. Pei, H.; Li, F.; Wan, Y.; Wei, M.; Liu, H.; Su, Y.; Chen, N.; Huang, Q.; Fan, C. Designed diblock oligonucleotide for the synthesis of spatially isolated and highly hybridizable functionalization of DNA–gold nanoparticle nanoconjugates. *J. Am. Chem. Soc.* **2012**, *134*, 11876–11879. [CrossRef]
73. Xu, L.; Liang, W.; Wen, Y.; Wang, L.; Yang, X.; Ren, S.; Jia, N.; Zuo, X.; Liu, G. An ultrasensitive electrochemical biosensor for the detection of mecA gene in methicillin-resistant Staphylococcus aureus. *Biosens. Bioelectron.* **2018**, *99*, 424–430. [CrossRef] [PubMed]
74. Ferapontova, E.E.; Hansen, M.N.; Saunders, A.M.; Shipovskov, S.; Sutherland, D.S.; Gothelf, K.V. Electrochemical DNA sandwich assay with a lipase label for attomole detection of DNA. *Chem. Commun.* **2010**, *46*, 1836–1838. [CrossRef] [PubMed]
75. Zhang, D.; Huarng, M.C.; Alocilja, E.C. A multiplex nanoparticle-based bio-barcoded DNA sensor for the simultaneous detection of multiple pathogens. *Biosens. Bioelectron.* **2010**, *26*, 1736–1742. [CrossRef]
76. Asiello, P.J.; Baeumner, A.J. Miniaturized isothermal nucleic acid amplification, a review. *Lab. Chip* **2011**, *11*, 1420–1430. [CrossRef]
77. Zhao, Y.; Chen, F.; Li, Q.; Wang, L.; Fan, C. Isothermal Amplification of Nucleic Acids. *Chem. Rev.* **2015**, *115*, 12491–12545. [CrossRef] [PubMed]
78. Cheng, W.; Zhang, W.; Yan, Y.; Shen, B.; Zhu, D.; Lei, P.; Ding, S. A novel electrochemical biosensor for ultrasensitive and specific detection of DNA based on molecular beacon mediated circular strand displacement and rolling circle amplification. *Biosens. Bioelectron.* **2014**, *62*, 274–279. [CrossRef]
79. Huang, Y.L.; Gao, Z.F.; Luo, H.Q.; Li, N.B. Sensitive detection of HIV gene by coupling exonuclease III-assisted target recycling and guanine nanowire amplification. *Sens. Actuators B Chem.* **2017**, *238*, 1017–1023. [CrossRef]
80. Carinelli, S.; Kühnemund, M.; Nilsson, M.; Pividori, M.I. Yoctomole electrochemical genosensing of Ebola virus cDNA by rolling circle and circle to circle amplification. *Biosens. Bioelectron.* **2017**, *93*, 65–71. [CrossRef] [PubMed]
81. Barreda-García, S.; Miranda-Castro, R.; de-los-Santos-Álvarez, N.; Miranda-Ordieres, A.J.; Lobo-Castañón, M.J. Helicase-dependent isothermal amplification: A novel tool in the development of molecular-based analytical systems for rapid pathogen detection. *Anal. Bioanal. Chem.* **2018**, *410*, 679–693. [CrossRef]
82. Barreda-García, S.; Miranda-Castro, R.; de-los-Santos-Álvarez, N.; Lobo-Castañón, M.J. Sequence-specific electrochemical detection of enzymatic amplification products of Salmonella genome on ITO electrodes improves pathogen detection to the single copy level. *Sens. Actuators B Chem.* **2018**, *268*, 438–445. [CrossRef]
83. Barreda-García, S.; Miranda-Castro, R.; de-los-Santos-Álvarez, N.; Miranda-Ordieres, A.J.; Lobo-Castañón, M.J. Solid-phase helicase dependent amplification and electrochemical detection of Salmonella on highly stable oligonucleotide-modified ITO electrodes. *Chem. Commun.* **2017**, *53*, 9721–9724. [CrossRef] [PubMed]

© 2020 by the authors. Licensee MDPI, Basel, Switzerland. This article is an open access article distributed under the terms and conditions of the Creative Commons Attribution (CC BY) license (http://creativecommons.org/licenses/by/4.0/).

Perspective

Restriction Endonuclease-Based Assays for DNA Detection and Isothermal Exponential Signal Amplification

Maria Smith [1], Kenneth Smith [1], Alan Olstein [2], Andrew Oleinikov [3] and Andrey Ghindilis [1,*]

1. TORCATECH, LLC, 5210 104th Street SW, Mukilteo, WA 98275, USA; mariyasmit@hotmail.com (M.S.); kencolfax@hotmail.com (K.S.)
2. Paradigm Diagnostics, Inc., 800 Transfer Rd #12, St Paul, MN 55114, USA; olstein@comcast.net
3. Department of Biomedical Science, Charles E. Schmidt College of Medicine, Florida Atlantic University, 777 Glades Road, Boca Raton, FL 33428, USA; aoleinikov@health.fau.edu
* Correspondence: andreylg@hotmail.com

Received: 16 June 2020; Accepted: 10 July 2020; Published: 11 July 2020

Abstract: Application of restriction endonuclease (REase) enzymes for specific detection of nucleic acids provides for high assay specificity, convenience and low cost. A direct restriction assay format is based on the specific enzymatic cleavage of a target–probe hybrid that is accompanied with the release of a molecular marker into the solution, enabling target quantification. This format has the detection limit in nanomolar range. The assay sensitivity is improved drastically to the attomolar level by implementation of exponential signal amplification that is based on a cascade of self-perpetuating restriction endonuclease reactions. The cascade is started by action of an amplification "trigger". The trigger is immobilized through a target-specific probe. Upon the target probe hybridization followed with specific cleavage, the trigger is released into the reaction solution. The solution is then added to the assay amplification stage, and the free trigger induces cleavage of amplification probes, thus starting the self-perpetuating cascade of REase-catalyzed events. Continuous cleavage of new amplification probes leads to the exponential release of new triggers and rapid exponential signal amplification. The proposed formats exemplify a valid isothermal alternative to qPCR with similar sensitivity achieved at a fraction of the associated costs, time and labor. Advantages and challenges of the approach are discussed.

Keywords: DNA assay; nucleic acid; isothermal; signal amplification; restriction endonuclease

---

1. Introduction

Nucleic acid assays of different formats provide a core for modern-day biotechnology and diagnostics. The critically important parameter is the assay specificity since nucleic acid target detection is usually performed in complex samples that contain DNA from different organisms. The detection specificity for most nucleic acid-based assays (qPCR, LAMP, microarrays, etc.) relies on biorecognition events of DNA strand hybridization and can be adversely affected by non-specific DNA–DNA binding. Addition of a second biorecognition event based on Class II restriction endonucleases (REases) has numerous advantages, first and foremost due to the nearly absolute specificity of these enzymes for particular double-stranded (ds) DNA recognition sites. Therefore, for the enzymatic action to take place, a hybridization event has to form a corresponding specific restriction site (usually palindromic with the total length of 4–8 bp) within the DNA double helix [1,2]. Thus, two biorecognition events are involved in signal generation, making it double-proof in terms of specificity and insensitivity to non-specific binding.

The technical principle of REase-based assays is associated with the release of an enzymatic reaction product from solid support into the liquid phase as the result of target–probe complex cleavage.

The product quantification can then be done in various ways by transferring the product-containing liquid phase into a separate reaction setup.

In addition to target–probe hybrid recognition, REases can also be used for exponential signal amplification if the initial hybrid cleavage event releases a "trigger" molecule. The trigger molecule is initially attached to the surface through an assay probe, where upon cleavage it is released into the reaction solution. The free trigger can migrate or be transferred to another surface that is modified with special "amplification" DNA probes. Specific trigger interaction with an amplification probe results in enzymatic cleavage of the probe. Each amplification probe carries additional (one or multiple) trigger molecules, thus the probe cleavage provides for the release of new triggers. This self-perpetuating cascade of cleavage events progresses exponentially until the reaction is stopped (or amplification probes are exhausted).

Several types of triggers can be used. Thus far, we have developed the following two approaches. The first is based on trigger REase enzymes that are immobilized through coupling to oligonucleotides, and can specifically cleave double-stranded oligonucleotide amplification probes. The second employs trigger oligonucleotides that hybridize to single-stranded oligonucleotide amplification probes, thus creating double-stranded REase restriction sites and subjecting them to cleavage by corresponding REase present in the solution. Both approaches are discussed in detail below, and they provide for the development of simple, low-cost, isothermal DNA hybridization assay platforms with exponential signal amplification that can achieve sensitivity similar to PCR applications.

The isothermal nucleic acid assay format is critical for the development of point of care units and field assays. One of the first isothermal assays called nucleic acid sequence-based amplification (NASBA) was introduced in 1991 by J. Compton [3]. Numerous other isothermal nucleic acid assays are reviewed in [4] including strand displacement amplification (SDA), loop-mediated amplification (LAMP), invader assay, rolling circle amplification (RCA), signal-mediated amplification of RNA technology (SMART), helicase-dependent amplification (HDA), recombinase polymerase amplification (RPA), nicking endonuclease signal amplification (NESA) and nicking endonuclease-assisted nanoparticle activation (NENNA), exonuclease-aided target recycling, junction or Y-probes, split DNAZyme and deoxyribozyme amplification strategies, template-directed chemical reactions that lead to amplified signals, non-covalent DNA catalytic reactions, hybridization chain reactions (HCR) and detection via the self-assembly of DNA probes to give supramolecular structures. However, all of them have limitations, and none are yet ready to replace PCR-based methods for the current DNA assay market. Our REase-based isothermal DNA assays are novel and promising, and the corresponding advantages and limitations are discussed in the current work. We are presenting our perspective on these novel assay formats and their potential applications.

2. Direct Restriction Assay (DRA)

We introduced DRA in 2014 [5], and the principle schematic is depicted in Figure 1. A detection probe labeled with a molecular marker is attached to a solid phase carrier (microplate, beads, resin, etc.) via streptavidin (SA)-biotin binding (Figure 1A). A single-stranded (ss) target DNA (i.e., dsDNA denatured to ssDNA, or cDNA) is added to the reaction solution and hybridizes to the probe forming dsDNA helix (Figure 1B). The probe–target hybrid carries a specific restriction site, thus the corresponding specific REase added to the reaction solution cleaves the helix. (Figure 1C). Upon cleavage, a part of the probe labeled with the molecular marker is released from the solid carrier into the reaction solution (Figure 1C). The solution is then transferred to a separate detection compartment and quantified (Figure 1D). Previously [5], we used horseradish peroxidase (HRP) as the molecular marker and quantified the signal optically by TMB (3,3',5,5'-Tetramethylbenzidine) oxidation at 655 nm. However, a variety of molecular markers can be used for this assay together with a wide range of detection techniques including fluorescent and electrochemical ones.

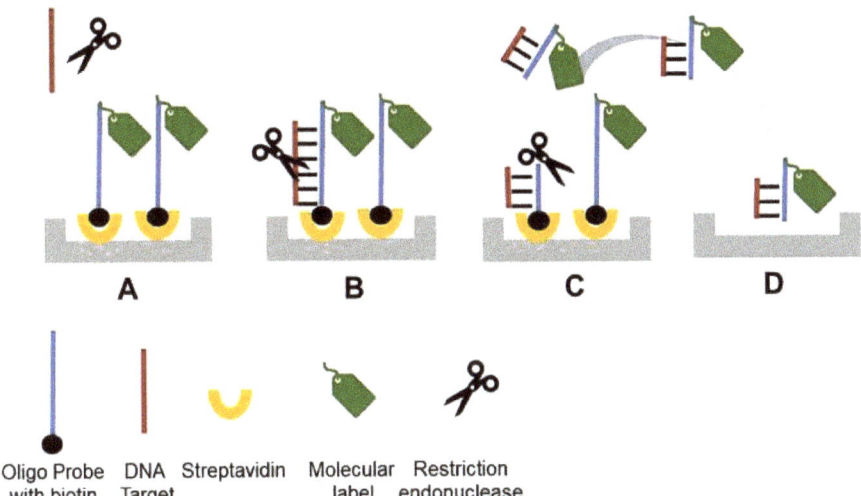

Figure 1. General schematic of the direct restriction assay (DRA). (A) A molecular marker/label is conjugated to an oligonucleotide probe that is specific for a target gene of interest and immobilized on a solid surface through biotin-SA binding. (B) Target DNA (an oligonucleotide or denatured dsDNA) is hybridized to the immobilized probe. (C) A restriction enzyme recognizes and cleaves the target–probe dsDNA hybrid, resulting in the release of the molecular marker into the reaction solution. (D) The reaction solution is transferred into a new well to quantify the molecular marker. For each target DNA molecule, one molecular marker is released, resulting in linear dependence between the assay signal and the target DNA concentration.

The developed DRA demonstrated the limit of detection of 1 nM with the dynamic range up to 30 nM [5]. The first assay was used for detection of methicillin-resistant *Staphylococcus aureus*, a bacterium with antibiotic resistance (MRSA). The assay was designed to detect a fragment of the *mecA* gene that has very high conservation (nearly 100% identity over 2 kb length) among various MRSA strains. A 40-mer probe MCA-BG (CAATTAAGTTTGCATAAGATCTATAAATATCTTCTTTATG) was designed from the *mecA* sequence commonly used for qPCR [6]. The central part of the probe had the specific recognition sequence (AGATCT) for BglII REase.

The assay was used to analyze (i) REase requirements for minimum target–probe helix sufficient for cleavage and signal generation, and (ii) the enzyme tolerance of mismatches and insertions. Our data showed a significant decrease in the assay signal when the probe–target length was reduced to 20-mer, with drastic reduction to nearly zero at the length of 16-mer. This length requirement suggested very high specificity, since on average in a random DNA sequence, a cognate 16-mer would be observed only once every 4.3 Gbp. We further analyzed the effects of mutations and showed that even a single mismatch within the restriction site eliminated the assay signal completely. In contrast, small (up to 3) target–probe mismatches and insertions (ssDNA loops) outside of the restriction site in the flanking sequences did not produce strong effects [5].

We concluded that the REase enzymatic cleavage in the process of DRA requires: (i) perfect probe–target match within a restriction site and (ii) at least 16-mer (preferably >20) of a hybridized dsDNA target–probe sequence around the restriction site. This study has been performed using BglII REase [5], with the caveat that other enzymes may be different in terms of mismatch and insertion tolerance.

The developed DRA method requires the ssDNA targets. In our previous work with dsDNA amplicons [5], heat denaturation of 95 °C was applied, followed with incubation on ice and addition to

SA-coated microplate wells carrying pre-attached biotinylated probes. Alternatively (Figure 2), the same heat denaturation can be applied to a mixture of probe and target DNA in solution. The probe used in this approach contains the biotinylated target-specific part, and an oligonucleotide tag. The tag is used for subsequent attachment of the molecular marker HRP. HRP is covalently linked to an oligonucleotide complementary to the tag and is attached through DNA–DNA hybridization (Figure 2). Thus, after the probe–target reaction solution has been denatured and cooled down, it is mixed with the tagged HRP and added to the SA-coated solid carrier (Figure 2). This leads to the quick binding of the biotinylated probe to surface SA that occurs simultaneously with the probe–target and probe–HRP tag hybridization (Figure 2). After washing to remove unbound molecules, a specific corresponding REase is added to perform enzymatic cleavage (Figure 2). The resultant cleaved HRP released into the reaction solution is then quantified colorimetrically.

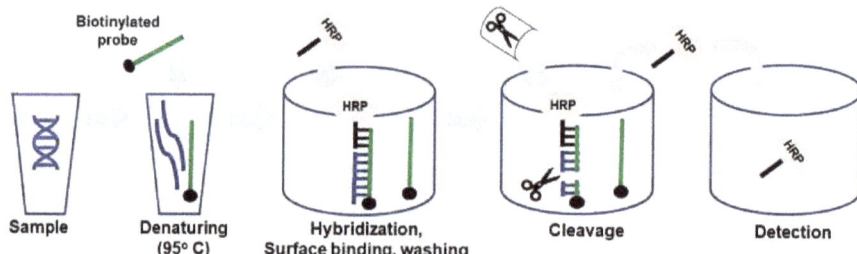

Figure 2. General schematic of the new approach to probe–target hybridization for DRA. A sample containing dsDNA targets is supplemented with a specific biotinylated probe and subjected to DNA denaturation at 95 °C followed by quick incubation on ice. The denatured probe and target mixture are supplemented with horseradish peroxidase (HRP) covalently attached to an oligonucleotide tag for hybridization to the probe. The mixture is added to the streptavidin (SA)-coated solid carrier for attachment and hybridization of the specific targets and tagged HRP to the probes. After washing to remove the unbound molecules, the specific REase is added, catalyzing enzymatic cleavage and HRP release. The free HRP is transferred to a detection cell.

Since no signal amplification is employed for the DRA platform, it provides for the nanomolar range sensitivity. The resultant practical applications are limited to analysis of amplicons for a simple and inexpensive version of semi-quantitative PCR, and to detection of precultured microbial pathogens. The former is described in [5], and the latter is currently being developed in cooperation with Paradigm Diagnostics, Inc. (http://pdx-inc.com). Paradigm Diagnostics has a technology for the detection of numerous pathogens based on culturing food industry samples in media that change color in the presence of growing microorganisms. This approach permits to detect samples with live microbes; however, the pathogen presence needs to be confirmed by an independent molecular method.

We used DRA to develop a technique to detect pre-cultured Shiga toxin-producing E.coli strains. Typically, USDA recommends qPCR testing of these strains using two genes, Eae and Stx, with a well-characterized set of corresponding primers and probes [7]. We used the qPCR probe sequences to develop DRA probes, namely Stx: CTGGATGATCTCAGTGGGCGTTCTTATGTAA and Eae: ATAGTCTCGCCAGTATTCGCCACCAATACC. The probes contain the restriction sites CTCAG and CCAGT for specific cleavage with BspCNI and BsrI REases, respectively.

The developed assay technique is based on the scheme shown in Figure 2. The SA-coated microplates were used as a solid carrier for the probes. Inoculated food samples were precultured for 5–6 h and used for total DNA extraction. The resultant DNA samples were directly used for DRA without PCR amplification. Thus, the full assay time was below 1 h including probe–target hybridization and binding to the plate (20 min) and REase cleavage (20 min). Figure 3A,B shows the results of the Eae and Stx gene detection in sample sets precultured for 5 and 6 h, respectively. In both cases, the signal obtained for inoculated samples was significantly higher than that for a negative control. Thus, the DRA technique can provide a simple, low-cost and fast alternative to PCR-based molecular detection of foodborne pathogens in precultured samples that can be carried out with minimum equipment requirements in field laboratories.

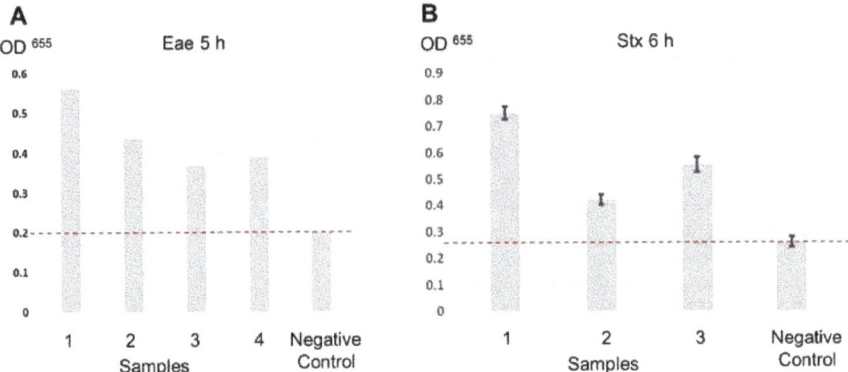

Figure 3. DRA data obtained sets of samples inoculated with Shiga toxin-producing *E.coli*. (**A**) Eae gene detection for samples precultured for 5 h. Data for Eae gene detection were obtained in singlicate. (**B**) Stx gene detection for samples precultured for 6 h. The dash lines indicate the signal level for negative control (non-inoculated samples).

3. Restriction Cascade Exponential Amplification (RCEA)

Restriction cascade exponential amplification (RCEA) has been introduced in 2015 [8]. A principle schematic of the assay is shown in Figure 4. It starts with the initial recognition stage that involves a target-specific probe modified with biotin at one end and an "amplification REase" molecule at the other end. The probe is attached to a solid carrier via SA–biotin interaction (Figure 4A). When the probe hybridizes with the corresponding target, the added free "recognition REase" cleaves the target–probe hybrid, releasing the amplification REase from the surface into the reaction solution and thus completing the first recognition stage (Figure 4B,C).

The reaction solution containing the released amplification REase is then transferred to the next amplification stage (Figure 4D). The corresponding setup contains amplification probes immobilized on a solid surface through biotin–SA interaction. The solution end of each probe is attached to the same amplification REase, as employed at the initial stage. In addition, an HRP molecule is attached to the solution probe end through complementary oligonucleotide tag hybridization (Figure 4D). The dsDNA amplification probes carry the specific restriction sites for cleavage with the attached amplification REases. However, the surface immobilization and double helix structure limit the attached REases' mobility, making them incapable to bend and cut at the restriction site.

Addition of the reaction solution from the recognition stage that contains **free** molecules of amplification REase results in cleavage of the immobilized amplification probes and release of an additional molecule of amplification REases into the reaction solution (Figure 4D,E). Thus, each cleavage event doubles the amount of free amplification REases, resulting in a cascade of cleavage reactions. In addition, each cleavage event releases immobilized HRP markers into the reaction solution (Figure 4F). The released HRP can be measured, i.e., by transferring the reaction solution to a detection cell. The described amplification setup can be common for all RCEA assays, with the target specificity determined during the initial recognition step by using the specific recognition probe and recognition REase.

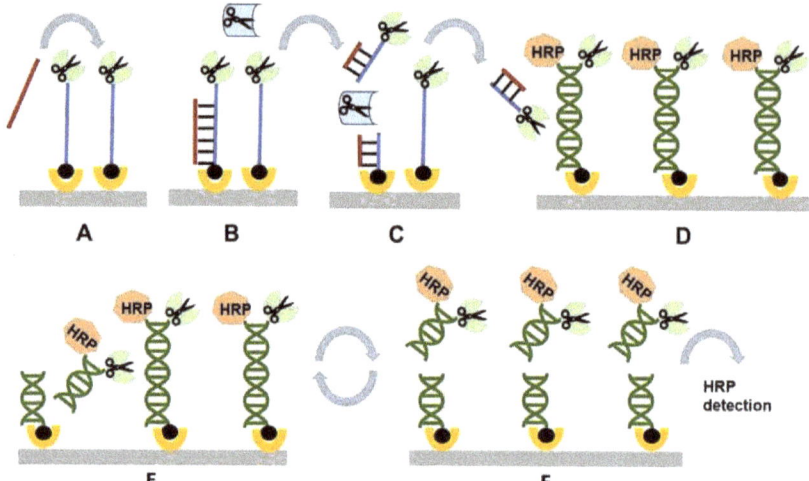

Figure 4. General schematic of the restriction cascade exponential amplification (RCEA) assay. (**A**) An oligonucleotide probe specific for a target of interest is conjugated to an REase for amplification and attached to a solid substrate using biotin. A test sample containing the target of interest is added. (**B**) The target in the test sample hybridizes to the probe. (**C**) The hybrid is specifically cleaved by a recognition REase. amplification REase is subsequently released into the solution. (**D**) The reaction solution is transferred to an amplification cell that contains an excess of immobilized amplification REase attached to the surface through an oligonucleotide linker. The linker contains the restriction site corresponding to the amplification REase, and it is double-stranded, with the second strand conjugated with HRP. All amplification REase molecules in the amplification cell are immobilized and thus incapable of cleaving their own or neighboring linkers. Addition of the free amplification REase generated in (**C**) triggers linker cleavage, releasing additional amplification REase, which in turn cleaves new linkers. (**E**) Each step of this exponential cascade of cleavage reactions doubles the amount of free amplification REase molecules in the reaction solution. (**F**) The linker cleavage releases HRP, which is quantified colorimetrically. Each initial target–probe hybridization event produces an exponentially amplified number of HRP molecules, with the value dependent on the amplification time.

Our published study [8] demonstrated highly sensitive detection of the target mecA gene related to MRSA infections. We used the same combination of recognition probe and REase: 40-mer MCA-BG and BglII, as for DRA [5]. The amplification stage was designed using two amplification REases: BamHI (restriction site GGATCC) and EcoRI (restriction site GAATTC). The most serious challenge in the RCEA assay development was associated with conjugation of REase molecules with oligonucleotide probes. All commercially available enzymes lost their enzymatic activity during standard conjugation via amino groups. Similar results were reported in the literature [9]. Successful conjugation could only be achieved by using mutant enzymes (BamHI and EcoRI) that had been engineered for ligand attachment by replacing some surface "non-essential" amino acid residues with cysteines [9].

The MRSA RCEA assay was tested using a specific target oligonucleotide complementary to the MCA-BG. As shown in Figure 5, both amplification REases, BamHI and EcoRI, demonstrated similar performance with the lower detection limit of 10 aM concentration, and the linear dynamic range (at the logarithmic scale) up to 1 nM. The plot obtained for the same target oligonucleotide without amplification using DRA is shown at the right side of the Figure 5. The data show that the RCEA assay format gained the detection limit improvement of approximately eight orders of magnitude over the DRA. The data were obtained for non-optimized assay conditions, and we could still detect as little as 200 target molecules per sample. This performance is similar to the detection limit of PCR applications and can likely be further improved by RCEA assay optimization. However, the main goal of such optimization should be the overall assay time that currently stays at about 2 h and can be significantly reduced to less than 1 h by improvement of mass transfer in the two-phase (liquid and solid) system. The improvement of mass transfer can be achieved by agitation and mixing, optimization of cell geometry, increase of surface to volume ratio, etc.

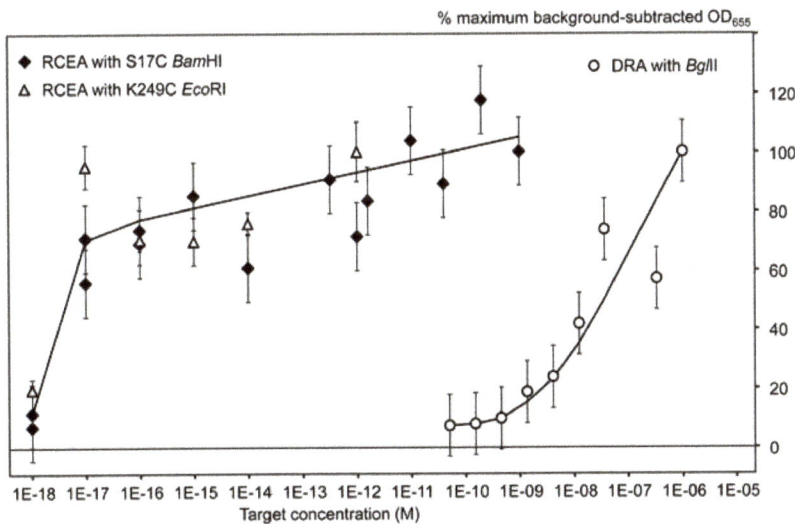

Figure 5. From [8] (Creative Commons CC-BY-NC-ND license). The RCEA limit of detection evaluated using the oligonucleotide target AMC-BG. The X-axis shows the target concentrations (M) and the Y-axis shows the background-subtracted HRP signal values (with the background calculated as the mean signal generated for zero target concentrations). For normalization and comparison of sample series, the HRP signal values were expressed as the percentages of the maximum background-subtracted OD_{655}, corresponding to each series. Open circles show the data generated using the direct restriction assay (DRA) with no amplification. The other two series were generated using the RCEA assays with the mutant S17C BamHI (closed diamonds) and K249C EcoRI (open triangles) as amplification REases. Error bars show standard deviations.

4. Tandem Oligonucleotide Repeat Cascade Amplification (TORCA)

An attractive alternative format for REase-based signal amplification employs another type of a trigger that is an unmodified oligonucleotide rather than an REase enzyme molecule. The obvious advantage is the omission of the REase conjugation step, enabling the use of standard commercially available enzymes at all stages of the assay. The first developed assay used two species of amplification trigger oligonucleotides, Tr1 and Tr2, that can start a self-perpetuating cascade of REase-catalyzed events based on trigger hybridization with each other single-stranded linker.

The assay based on this principle, tandem oligonucleotide repeat cascade amplification (TORCA), was introduced in 2019 [10]. This format employs standard REases that are suspended in the reaction solutions without immobilization. To prevent cleavage events, the restriction sites of amplification probes are kept single-stranded, and the reaction cascade is started by addition of a free trigger oligonucleotide released during the initial recognition reaction. The exponential amplification is then achieved by usage of several tandem repeats of the same trigger oligonucleotide within each probe.

The principle schematic of the TORCA assay is shown in Figure 6. It starts with the recognition step involving an oligonucleotide recognition probe specific for a target of interest that is extended with the "trigger" oligonucleotide unit Tr1 (Figure 6A). The probe is immobilized on a solid surface through biotin–SA interaction. Upon the target–probe hybridization (Figure 6B), the resultant dsDNA is cleaved with the corresponding recognition REase that is present in the reaction solution (Figure 6C). This cleavage releases the trigger Tr1 into the reaction solution at the amount proportional (ideally, equal) to the amount of the target added.

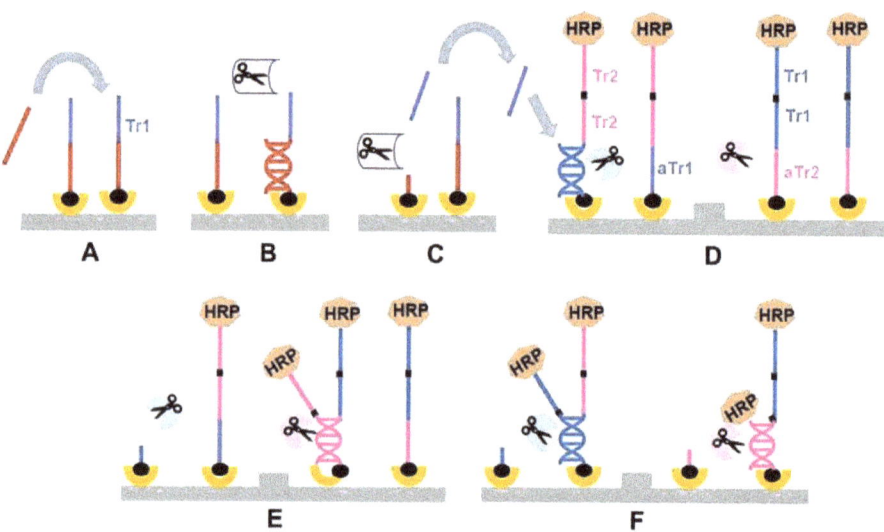

Figure 6. General schematic of tandem oligonucleotide repeat cascade amplification (TORCA). (**A**–**C**) The recognition stage: An oligonucleotide probe specific for a target of interest is extended with a "trigger" unit (Tr1) and attached to surface using biotin. A test sample containing the target of interest is added (**A**). The target in the test sample hybridizes to the probe (**B**), and the hybrid is specifically cleaved by a specific recognition REase (**C**). The Tr1 unit is subsequently released into the reaction solution. (**D**–**F**) The amplification stage: The reaction cell carries two types of amplification probes. The first contains a single unit complementary to the trigger sequence Tr1 (antisense Tr1, aTr1), and multiple identical units of a trigger sequence Tr2. The second contains multiple identical Tr1 units, and a single unit complementary to the Tr2 unit (antisense Tr2, aTr2). Both probe types are surface-attached and contain a molecular marker HRP on their solution-facing end (**D**). The reaction solution in the amplification chamber contains two common REases, specific to Tr1 and Tr2, that recognize and cleave dsDNA hybrids of Tr1-aTr1 and Tr2-aTr2, respectively. When the recognition reaction solution is transferred to the amplification cell, the free trigger Tr1 hybridizes to an aTr1 unit of the first probe leading to the probe cleavage by Tr1-REase (**D**) and release of Tr2 into the reaction solution (**E**). In turn, the released Tr2 hybridize to an aTr2 of the second probe type (**E**), causing cleavage of Tr2 and further release of additional Tr1 units. This cascade of events also results in the release of the HRP molecular marker that can be used for signal quantification (**F**).

At the next amplification stage, the recognition reaction solution is transferred to an amplification chamber that contains two types of amplification probes immobilized on a solid carrier (Figure 6D). Each probe has HRP attached to the solution end. The amplification probe AP1 consists of a sequence complementary to the trigger Tr1 (aTr1) attached to the carrier surface and **multiple** tandem repeat sequences of trigger Tr2 at the solution end. The amplification probe AP2 has a sequence complementary to Tr2 (aTr2) at the surface and multiple Tr1 sequences at the solution end (Figure 6D). The amplification chamber also contains two free amplification REases that specifically cleave the dsDNA hybrids of Tr1-aTr1 and Tr2-aTr2. Since initially all probes are present in the chamber as single-stranded, no enzymatic cleavage is observed. Addition of the recognition reaction solution containing free Tr1 results in hybridization with the complementary aTr1 part of AP1, followed by the cleavage and release of multiple Tr2 into the solution (Figure 6 D,E). In turn, the released Tr2 molecules hybridize to the immobilized complementary aTr2 within AP2, resulting in the further cleavage and release of numerous Tr1 units. Since each cleavage event is accompanied with the release of multiple trigger units and thus initiates the cleavage of the next amplification probes (Figure 6F), this process is self-perpetuating and provides for exponential accumulation of unbound HRP and thus the exponential assay signal increase over time.

Unlike RCEA, the TORCA assay format does not require REase conjugation, regular commercially available enzymes can be used at all stages. Moreover, the recognition probes do not contain the attached enzyme, thus they can be safely subjected to denaturation at high temperature. This is an important advantage to streamlining the initial recognition step: instead of separate denaturation of dsDNA targets before mixing and hybridization with recognition probes, the targets and probes can be mixed, denatured and hybridized in a single step (similar to the DRA scheme shown in Figure 2).

The main challenge for TORCA assay development is associated with prevention of physical contacts between the amplification probes AP1 and AP2. Any contact will lead to hybridization of the complementary Tr and aTr units followed by cleavage, and thus initiation of the amplification cascade without addition of a free trigger. Indeed, two types of beads, modified with either AP1 or AP2, when mixed in the presence of both amplification REases, immediately start releasing some HRP signal [10]. At the same time, if only one amplification REase is present, the HRP release does not occur. One possible solution is membrane separation, and our data showed that such physical separation of the beads with AP1 and AP2 prevents the HRP release in the presence of both amplification REases [10].

Based on these data, we designed two types of amplification chambers for the TORCA assays. The first type employs a mixture of two probe carriers without physical separation. In this chamber, addition of a trigger from the recognition step enhances the rate of HRP signal generation over a rather prominent background of the trigger-independent HRP release. The obvious disadvantage is high background values that need to be carefully measured with negative controls. The main advantage of this approach is the short assay time, approximately 15 min for the whole amplification stage [10]. The second type of amplification chamber provides for the physical separation of two different probe carriers with a membrane permeable for DNA molecules but not for carrier particles. This type is associated with a low background; however, it requires a considerably longer time for the completion of the amplification stage (over 1 h).

The two approaches have been tested in [10] using SspI (restriction site AATATT) and EcoRV (restriction site GATATC) as amplification REases. The amplification probes had seven repeats of Tr1 and Tr2. Figure 7 shows the TORCA data obtained for trigger detection using the two amplification formats: mixture of non-separated probes (Curve b) and membrane-separated probes (Curve c) [10]. They are compared to curve "a" obtained for the same trigger without amplification by using DRA.

Both TORCA formats had the same detection limit of 10 aM concentration similar to RCEA and to PCR applications. The format with the mixture of non-separated probes demonstrated a little less sensitivity as compared with the membrane-separated probes, however, it used a shorter amplification time and showed better linearity. Thus, the probe mixture assay format was chosen for further development of an assay to detect malaria *P. falciparum* parasites by using RNA as a target.

This RNA-based approach was a step toward the goal to distinguish between past and ongoing malaria infections. Such discrimination performed directly at point-of-care facilities is essential to direct drug therapy at only those patients who can benefit from it, and to conduct new drug clinical trials in malaria-endemic areas [10]. RNA stability is known to be significantly lower than stability of DNA and protein malaria markers, thus RNA detection is likely to better correlate with the presence of a live parasite as compared with stable DNA. Since REase enzymes can only cleave DNA–DNA hybrids, RNA was reverse-transcribed into cDNA.

The developed TORCA assay format was compared to PCR detection [10]. The TORCA assay sensitivity toward three different malaria RNA targets had the detection limit of about 7.5 IE per

100 µL of blood sample. This is almost two orders of magnitude better than the malaria detection limit recommended by WHO [11]. The observed linear dynamic range for the assay spanned approximately three orders of magnitude.

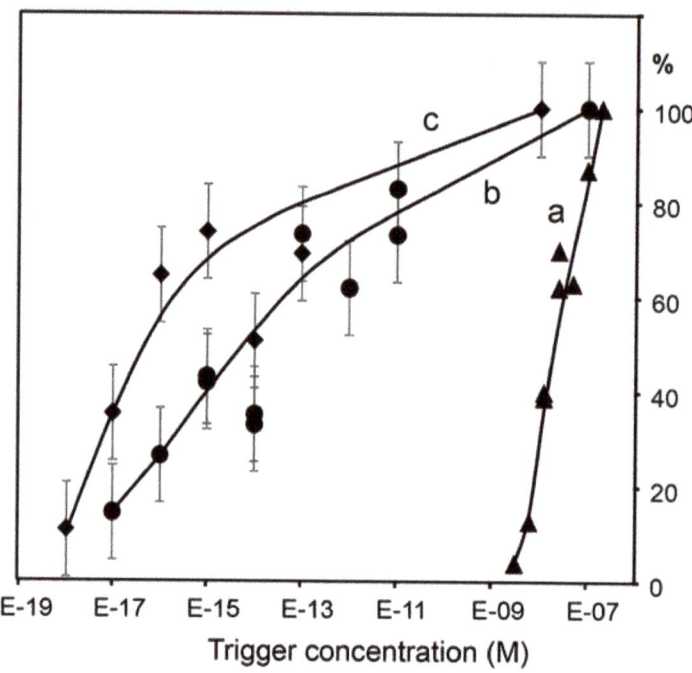

Figure 7. From [10] (Creative Commons CC-BY license). The dependence of the HRP-generated signal on the concentration of the amplification trigger added to the single (non-amplified DRA format) REase (a) or the two REases EcoRV and SspI (b-c) systems. The curves **a** and **b** are generated for mixtures of the two bead types, one modified with the amplification probe AP1-HRP and the other with AP2-HRP. The curve **c** was obtained for the same two bead types separated by a filter barrier. The X-axis shows the target concentrations (M), and the Y-axis shows the background-subtracted and normalized HRP signal values. The background was calculated as the mean signal generated for the triplicate no-trigger added negative controls. For normalization and comparison of the sample series, the HRP signal values are expressed as the percentages of the maximum background-subtracted OD_{655} corresponding to each series. Error bars show standard deviations. The data for (non-amplified DRA format) REase (a) were obtained without replicates.

Direct comparison of the TORCA assay versus common RT PCR is presented in Figure 8. Both methods successfully detected the *P. falciparum* parasite RNA targets at different times after initiation of the drug treatment of a patient [10]. The decrease of parasite RNA directly correlated with the post-treatment time, and both methods showed considerable consistency (Figure 8). The described method of distinguishing between past and ongoing infections is based on observations showing much lower stability of pathogen RNA as compared with DNA. However, for each particular infection, an independent study is needed to establish a correlation between target RNA content and disease state.

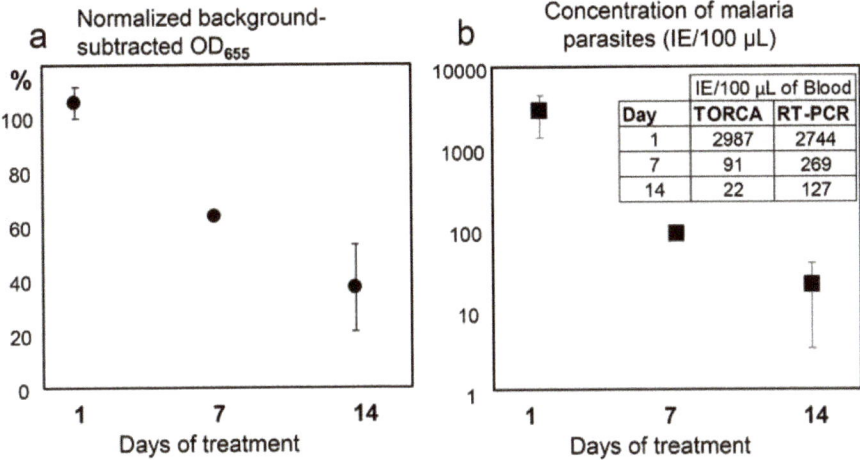

Figure 8. From [10] (Creative Commons CC-BY license). The dependence of the TORCA signal (**a**) and the calculated infected erythrocyte concentration (**b**) on the time after initiation of the patient drug treatment (X-axis, days). (**a**) The Y-axis shows the background-subtracted and normalized HRP signal values. The background was calculated as the mean signal generated for the triplicate no-trigger added negative controls. For normalization and comparison of the sample series, the HRP signal values are expressed as the percentages of the background-subtracted signal obtained for the positive control containing an equimolar mixture of targets at 10 nM concentrations. (**b**) The Y-axis shows the IE (Infected Erythrocytes) concentrations calculated using the standard calibration curve obtained separately. The inset shows mean data for IE concentrations measured using TORCA and reverse transcription-PCR methods and calculated according to corresponding calibration curves. Error bars in both graphs show standard deviations. Error bars for 7-day treatment are smaller than the marker.

An important feature of the TORCA assay is the ability to simultaneously detect multiple targets in the same chamber, generating an integrated signal to enhance assay sensitivity and to provide for mutation tolerance. For malaria detection, we successfully employed three different RNA targets, and thus three recognition probes in the same assay [10].

The TORCA optimization efforts are currently focused on finding a middle ground in the design of the amplification chamber: to combine the separation membrane to reduce backgrounds with the enhanced mass transfer to shorten the assay time. The latter can be done by using an increased membrane surface and physical agitation of the carrier particles.

An important advantage of the TORCA and RCEA formats is associated with their high tolerance of various target sizes. In contrast to PCR-based assays that normally require at least 70 bp fragments, the TORCA and RCEA minimum size requirement is 20 bp. Since the assay can be used for recognition of very short fragments, the potential applications include working with partially degraded nucleic acids in FFPE material and in liquid biopsies containing cell-free DNA and RNA from serum and plasma. Archived FFPE tissues are subjected to formalin-induced crosslinking of nucleic acids to proteins, base purination and strand breaks. As a result, the proportion of RNA fragments <200 bases is typically >50%, and can be as high as 90%, making these samples unsuitable for standard assays that require templates >150 bases. In contrast, REase-based assays do not have the 150-base size limitation [12].

In contrast to the RCEA technology, TORCA does not require enzyme engineering and complex conjugation. All required components, including custom oligonucleotides, are commercially available at low cost with a fast turn-around time, thus ensuring great flexibility towards the development of new assays towards emerging targets of interest.

5. REase Based Assays: Advantages and Limitations

All REase-based assays described above [5,8,10] have several common advantages associated with high specificity due to the "double-proof" combination of two biorecognition events. They are nearly insensitive to (i) non-specific and partially complementary DNA binding, (ii) excess foreign DNA background and (iii) various non-nucleic acid contaminants (such as proteins, PCR inhibitors, etc.).

The isothermal nature of REase-driven assays, their simplicity and flexibility provide opportunities for the development of assay cartridges with all components included. The assays can utilize various molecular markers of different nature and use colorimetric, fluorescent and electrochemical detection formats without the need of complex instrumentation, and at much lower costs as compared with other nucleic acid assays. Both fluorescent and electrochemical detection formats would allow for the direct detection of the released molecular label in the amplification chamber in real time, without the need to transfer the reaction solution to a special detection chamber. To get rid of this end-point fluid transfer, and to switch to a kinetic assay mode, one needs to place an electrode or optical detection probe into the amplification chamber and physically separate it from the solid carrier with immobilized assay reagents. This can be achieved by various engineering approaches, providing for design flexibility, and enabling the development of devices for field and point-of-care applications.

The REase-based assays provide for easy adaptation to new analytes of interest. This adaptation involves a single design step to develop a pair of a 20–30 bp recognition probes with a corresponding recognition REase. This simplicity provides for a great advantage over another isothermal assay, LAMP, that requires the use of 4–6 carefully optimized primers for each new DNA target. The current selection of commercially available REases with different restriction site sequences is expanding continuously. New England Biolabs alone offers over 285 restriction enzymes. In our experience, any sufficiently diverse 100–200 bp coding DNA sequence typically contains at least one option for a possible recognition probe with an REase restriction site. Further, in contrast to the design of recognition stage components, the same amplification stage reagents can be used for all targets of interest.

The two highly sensitive exponential amplification assay formats, RCEA and TORCA, achieve the attomolar detection limit similar to the golden standard of PCR. However, in contrast to PCR, they amplify the assay signal rather than the DNA template. The main issue of template DNA amplification is a possible sequence-dependent bias that is widely observed in PCR, when certain sequences have much higher amplification efficiency than the other [13]. The REase-based signal amplification does not employ the production of multiple new DNA copies, thus it is free of sequence-dependent bias and mis-priming issues.

The biggest challenge in the development of REase-based exponential signal amplification assays is the engineering of automated fluid transfer. Currently, reaction solutions are transferred manually, however, the operator involvement needs to be minimized for point-of-care and field applications. This will provide a competitive edge for the REase-based assays that can serve as a valid alternative to PCR due to their simplicity, low cost, high specificity and sensitivity of detection.

6. Conclusions

Our perspectives on applications of the REase-based nucleic acid assays are associated with their versatility, low cost, simplicity, specificity, isothermal nature and potential for the development of portable automated instrumentation formats. The two highly sensitive assay formats based on exponential amplification are valid alternatives to PCR-based assays. They can be performed at a fraction of the PCR cost in low-resource settings, including point-of-care laboratories, and field facilities. We believe that the described formats have high potential for biosensors development, since they can use both fluorometric and electrochemical detection. The detection of a released molecular label can be done in real time, directly in the amplification chamber during the amplification stage, with no extra fluid transfer steps, and employing a kinetic assay mode. This allows for substantial simplification of the assay setup and supporting instrumentation along with a reduction in the assay time. In our perspective, the REase-based platforms also overcome the very serious limitation of nucleic acid assays, namely the minimum target length requirement that is important for both PCR and isothermal formats such as LAMP. Thus, the novel platforms have potential for considerable expansion into new niches involving the analysis of highly fragmented nucleic acids, including liquid biopsies. However, further development of the described technologies requires optimization in terms of simplification, automation, robustness and short assay time.

Author Contributions: Conceptualization, A.G.; methodology, A.G.; M.S.; K.S.; A.O. (Andrew Oleinikov); A.O. (Alan Olstein); formal analysis, A.G.; M.S.; K.S.; A.O. (Andrew Oleinikov); A. Olstein; data curation, A.G.; M.S.; K.S.; A.O. (Andrew Oleinikov); A.O. (Alan Olstein); writing—original draft preparation, A.G.; M.S.; writing—review and editing, A.G.; M.S.; K.S.; A.O. (Andrew Oleinikov); A.O. (Alan Olstein); supervision, A.G. All authors have read and agreed to the published version of the manuscript.

Funding: Part of this research was supported by a grant from the National Institutes of Health 1R41AI129130.

Conflicts of Interest: The authors declare no conflict of interest.

References

1. Pingoud, A.; Jeltsch, A. Recognition and cleavage of DNA by Type-II restriction endonucleases. *Eur. J. Biochem.* **1997**, *246*, 1–22. [CrossRef] [PubMed]
2. Jen-Jacobson, L. Protein-DNA recognition complexes: Conservation of structure and binding energy in the transition state. *Biopolymers* **1997**, *44*, 153–180. [CrossRef]
3. Compton, J. Nucleic acid sequence-based amplification. *Nature* **1991**, *350*, 91–92. [CrossRef] [PubMed]
4. Yan, L.; Zhou, J.; Zheng, Y.; Gamson, A.S.; Roembke, B.T.; Nakayama, S.; Sintim, H.O. Isothermal amplified detection of DNA and RNA. *Mol. BioSyst.* **2014**, *10*, 970–1003. [CrossRef] [PubMed]
5. Smith, M.W.; Ghindilis, A.L.; Seoudi, I.A.; Smith, K.; Billharz, R.; Simon, H.M. A new restriction endonuclease-based method for highly-specific detection of DNA targets from methicillin-resistant *Staphylococcus aureus*. *PLoS ONE* **2014**, *9*, e97826. [CrossRef] [PubMed]
6. McDonald, R.R.; Antonishyn, N.A.; Hansen, T.; Snook, L.A.; Nagle, E.; Mulvey, M.R.; Levett, P.N.; Horsman, G.B. Development of a triplex real-time PCR Assay for detection of panton-valentine leukocidin toxin genes in clinical isolates of methicillin-resistant *Staphylococcus aureus*. *J. Clin. Microbiol.* **2005**, *43*, 6147–6149. [CrossRef] [PubMed]
7. United States Department of AgricultureFood Safety and Inspection Service, Office of Public Health Science. Laboratory GuidebookNotice of Change. 2019. Available online: https://www.fsis.usda.gov/wps/wcm/connect/1d61852b-0b71-45e9-8914-8ff95af7aaa8/mlg-5-appendix-4.pdf?MOD=AJPERES (accessed on 4 February 2019).
8. Ghindilis, A.L.; Smith, M.W.; Simon, H.M.; Seoudi, I.A.; Yazvenko, N.S.; Murray, I.A.; Fu, X.; Smith, K.; Jen-Jacobson, L.; Xu, S.-Y. Restriction Cascade Exponential Amplification (RCEA) assay with an attomolar detection limit: A novel, highly specific, isothermal alternative to qPCR. *Sci. Rep.* **2015**, *5*, 7737. [CrossRef] [PubMed]
9. Dylla-Spears, R.; Townsend, J.E.; Sohn, L.L.; Jen-Jacobson, L.; Muller, S.J. Fluorescent marker for direct detection of specific dsDNA sequences. *Anal. Chem.* **2009**, *81*, 10049–10054. [CrossRef] [PubMed]
10. Ghindilis, A.L.; Chesnokov, O.; Ngasala, B.; Smith, M.W.; Smith, K.; Mårtensson, A.; Oleinikov, A.V. Detection of sub-microscopic blood levels of *Plasmodium falciparum* using Tandem Oligonucleotide Repeat Cascade Amplification (TORCA) assay with an attomolar detection limit. *Sci. Rep.* **2019**, *9*, 2901. [CrossRef] [PubMed]
11. Malaria Policy Advisory Committee Meeting. WHO Evidence Review Group on Malaria Diagnosis in Low Transmission Settings. 2014. Available online: http://www.who.int/malaria/mpac/mpac_mar2014_diagnosis_low_transmission_settings_report.pdf (accessed on 4 February 2019).
12. Patel, P.G.; Selvarajah, S.; Guérard, K.-P.; Bartlett, J.M.S.; Lapointe, J.; Berman, D.M.; Okello, J.B.A.; Park, P.C. Reliability and performance of commercial RNA and DNA extraction kits for FFPE tissue cores. *PLoS ONE* **2017**, *12*, e0179732. [CrossRef] [PubMed]
13. Nakayama, S.; Yan, L.; Sintim, H.O. Junction probes—Sequence specific detection of nucleic acids via template enhanced hybridization processes. *J. Am. Chem. Soc.* **2008**, *130*, 12560–12571. [CrossRef] [PubMed]

 © 2020 by the authors. Licensee MDPI, Basel, Switzerland. This article is an open access article distributed under the terms and conditions of the Creative Commons Attribution (CC BY) license (http://creativecommons.org/licenses/by/4.0/).

Review

Electrochemical Immuno- and Aptamer-Based Assays for Bacteria: Pros and Cons over Traditional Detection Schemes

Rimsha Binte Jamal, Stepan Shipovskov and Elena E. Ferapontova *

Interdisciplinary Nanoscience Center (iNANO), Aarhus University Gustav Wieds Vej 14, DK-8000 Aarhus, Denmark; ribja@inano.au.dk (R.B.J.); stepan.shipovskov@gmail.com (S.S.)
* Correspondence: elena.ferapontova@inano.au.dk

Received: 31 July 2020; Accepted: 23 September 2020; Published: 28 September 2020

Abstract: Microbiological safety of the human environment and health needs advanced monitoring tools both for the specific detection of bacteria in complex biological matrices, often in the presence of excessive amounts of other bacterial species, and for bacteria quantification at a single cell level. Here, we discuss the existing electrochemical approaches for bacterial analysis that are based on the biospecific recognition of whole bacterial cells. Perspectives of such assays applications as emergency-use biosensors for quick analysis of trace levels of bacteria by minimally trained personnel are argued.

Keywords: electrochemistry; bacteria; electrochemical ELISA; electrochemical immunoassays; electrochemical aptamer-based assays

1. Introduction

Sensitive, selective and quickly responding sensors for bacterial detection are strongly required for environmental monitoring of water safety [1,2]; in diary, food and beverage industries for product quality analysis [3]; for prevention, diagnosis and antibiotic treatment of infectious diseases caused by pathogens [4]; and for combatting biocorrosion in oil and gas industries [5], biological warfare and terrorism [6]. Pathogenic bacteria cause more than 10 million deaths annually [4] and 1.5 million people pass away from diarrhea caused by microbiologically contaminated water, with >2 billion people simply lacking access to safe drinking water resources [2]. Microbiological safety requires no bacteria present in any 100 mL of drinking water [7]. Similarly low levels of pathogens should not be surpassed in dairy products (<10 CFU mL^{-1} *E. coli*, a specific indicator of the diary product spoilage, in pasteurised milk) [8], brewery products (<400 CFU L^{-1} *L. brevis* in beer) [9] and food (<100 CFU *Salmonella* gives from 0.01 to 0.56 probabilities of illness; both 17 and 36 CFU can cause illness) [10]. Less than 100 CFU mL^{-1} pathogens should be timely, within 1–2 h detected in blood [11], and unavailability of rapid sensitive tests is partially responsible for high 30–40% mortality from blood stream bacterial infections. Complex matrix effects and bacterial cross-interference further challenge the required analytical sensitivity and limits of detection (LOD).

It is thus no surprise that bacterial sensors should be uniquely robust and specific, with a minimal (if any) false signal to ensure effective protection of human health. Such sensors can be often considered as emergency-use sensors since they should provide quick analysis of trace—"alarm"—levels of bacteria, without complex sample preparation and by minimally trained people. They should be sensitive at a single cell level and trace bacteria immediately in the "alarm" spots.

The Human Microbiome Project further revealed the vital role bacteria play in human health and development of gastrointestinal diseases, several types of cancer and type 2 diabetes [12]. Poorer or uncommon gut microbiota (aka dysbiosis) can also weaken and destabilize immune responses, and its role in development of such neurologic disorders as Alzheimer's disease, through neuro-inflammation, is more and more acknowledged [13]. Antibiotic resistance is another relatively recently emerged field of bacterial sensor applications: >0.7 million people die annually from drug-resistant infections, and their number is predicted to increase to 10 million in 2050 [14]. These findings highlight the necessity of new, even more advanced analytical tools for multiplex bacterial analysis in complex biological matrix containing excessive amounts of numerous bacterial species.

Here, we overview the existing approaches for bacterial detection, placing the main focus on electrochemical immunoassays and aptamer-based assays for the whole bacterial cells, which currently represent the fastest and most straightforward way of bacterial analysis with a minimal complexity of sample preparation, and, thus, have the largest potential for practical and commercial use. We critically discuss pros and cons of such electrochemical assays compared to traditional immunoassays and their propensity to satisfy the requirements of the ideal bacterial assay's sensitivity, selectivity and response time.

2. Methodological Approaches for Bacterial Detection and Quantification

Most routine "gold standard" microbiological tests for bacterial analysis are microbiological culturing and such nucleic-acid-based approaches as the polymerase chain reaction (PCR) and fluorescence in situ hybridization (FISH) analysis of bacterial DNA and RNA (Figure 1).

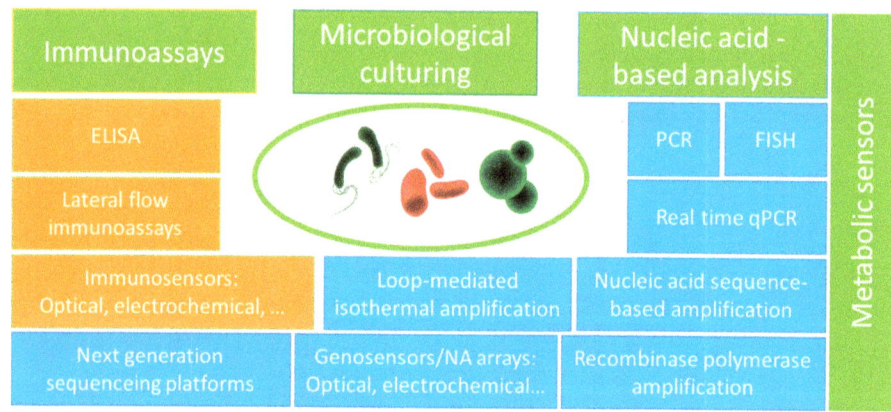

Figure 1. Existing methodologies for bacterial analysis.

Despite the established reputation in the field, some of their inherent limitations preclude their emergency-use applications. Both microbiological culturing and PCR and FISH are time consuming (from 2 to sometimes 15 h of amplification [15] and from 24–72 h to weeks of bacterial growth [16]). In addition, despite their high specificity, they are still insufficiently sensitive [16]. In the fastest PCR and FISH, from 10^3 to 10^4 CFU mL^{-1} can be detected [3,16]. Along with that, errors in PCR amplification and sequence replication and differences in the DNA extraction protocols existing for Gram-negative and Gram-positive strains may result in the wrong quantification of bacterial species [17], not to mention the probability of false-positive signals from dead cells, whose DNA can also be PCR-amplified. Amplification inhibition by the matrix components can also contribute to errors in bacteria quantification and therefore detection [18]. Real-time (quantitative) PCR (qPCR) quantifies amplified bacterial DNA in real-time and may be faster/more sensitive than traditional PCR, but requires fluorescent labels and as such a more complex optical detection equipment [19]. It may also suffer from matrix inhibitors and dead-cell produced false positives [20], though some strategies using intercalating dyes eventually allow to prevent amplification of the dead cells' DNA [21].

The recent progress made in molecular biology tools offered more promising approaches for bacterial nucleic acid (NA) analysis [22] such as:

- NA sequence-based amplification (NASBA, amplifies and detects bacterial messenger RNA, more sensitive and fast (less than 90 min) than PCR, no interference from dead cells' DNA, however, too expensive for environmental applications and suffers from errors in amplification and quantification following the amplification step) [23];
- Loop-mediated isothermal amplification (LAMP, less expensive than PCR, more sensitive and faster (1 h) DNA amplification at 60–65 °C, less sensitive to inhibitors) [24];
- Recombinase polymerase amplification (RPA, fast (<20 min) amplification of DNA/RNA at 37–42 °C; can be integrated with other, portable detection devices, however, it faces primers design difficulties and requires post-amplification purification digestion) [25].

These new approaches for NA amplification enable the qualitatively different level of analysis, particularly, when combined with a proper detection strategy (Table 1 summarizes some selected best examples). However, despite their current wide use in the clinical and molecular-biology research practice, the NA-based methods, exploiting lab-run equipment and often requiring from 24 to 72 h sample pre-enrichment, have not yet found broad applications in the in-field or alarm situations requiring fast and specific, often single-cell bacterial detection. Analytical platforms such as genosensors [26,27] and DNA microarrays [28,29], or next generation sequencing platforms [30,31] also use bacterial DNA/RNA isolation and amplification protocols as a sample preparation step, and, being more sensitive, they still rely on lab instrumentation and pre-enrichment steps. Centrifugal lab-on-chip microfluidic platforms integrating cell lysis and amplification procedures in one chip are very perspective since they do not need trained personal for handling them and can decrease the time of analysis down to 70 min [32,33]. On the other hand, due to the small, µL-volume samples they use for analysis, the LODs they show are quite high, from 10^3 to 10^4 CFU mL^{-1} (down to 10 CFU mL^{-1} when the assay time is increased, e.g., to 3.7 h) [32], making them unsuitable for many applications.

Table 1. Selected examples of nucleic acid-based optical biosensors for bacteria.

Strain	Technique	LOD [a]	Detection Range, CFU mL^{-1}	Interference Studies	Assay Time	Ref.
Campylobacter jejuni	PCR with pre-enrichment in suitable broths	3 CFU/100 mL	-	Actinomyces pyogenes, Campylobacter coli, Enterobacter cloacae, Pseudomonas aeruginosa, Salmonella saintpaul, Yersinia enterocolitica	48 h	[34]
Salmonella, Listeria monocytogene	qPCR with two fluorescently labelled primers	5 CFU/25 mL	-	B. cerus, Campylobacter, E. aerogenes, E. cowanii, Cronobacter sakazakii, E. coli, E. faecalis, S. aureus, Shigella spp, Serratia liquefaciens, S. pneumoniae	<48 h	[35]
Salmonella	q-PCR with two-step pre-filtration on filter paper	7.5 CFU/ 100 mL	-	-	3 h	[36]
Salmonella typhimurium	direct PCR with immunomagnetic preconcentration	2–3	6–6.4×10^4	-	<3 h	[37]

Table 1. Cont.

Strain	Technique	LOD [a]	Detection Range, CFU mL^{-1}	Interference Studies	Assay Time	Ref.
Salmonella	PMA-qPCR PMA was used to increase sensitivity	36 (pure culture) and 100 (raw shrimp)	36–3.6 × 10^8 (pure culture) and 100–1 × 10^8 (raw shrimp)	Vibrio parahaemolyticus, Listeria monocytogenes, E. coli O157:H7, S. aureus	1–2 h	[38]
Salmonella sp. Shigella sp. Staphylococcus aureus	Multiplex qPCR with immunomagnetic pre-concentration	2 CFU/g 6.8 CFU/g 9.6 CFU/g	-	Listeria monocytogenes, E. coli, B. cerus, Streptomyces griseus, Pseudomonus aerug, Lactobacillus plantarum, E. faecalis, Streptococcus hemolyticus, Micrococcus luteus, P. aeruginosa, Clostridium sporogenes	<8 h	[39]
Salmonella	Real-time RPA	10 CFU/g (eggs) 100 CFU/g (chicken)	-	Bacillus cereus, Campylobacter coli, E. coli O157:H7, L. casei, S. aureus, Pseudomonas aeruginosa, Vibrio vulnificus	10 min	[40]
Salmonella	LAMP	4.1	-	Listeria monocytogenes, E. coli O157:H7, S. aureus, Yersinia enterocolitica, Proteus mirabilis, Shigella Flexner, Micrococcus luteus, Bacillus cereus, Enterobacter sakazakii, Pseudomonas fluorescens	1 h	[41]
Vibrio parahaemolyticus	LAMP	530	-	Acinetobacter baumannii, Aeromonas hydrophila, Enterococcus faecalis, Haemophilus influenzae, Helicobacter pylori, Salmonella	1 h	[42]
Salmonella and Shigella	Multiplexed LAMP Simultaneous detection of two bacterial species	5 CFU/10 mL	-	S. aureus, E. coli, Bacillus cereus, Pseudomonas aeruginosa, Vibrio parahaemolyticus, Listeria monocytogenes	20 h	[43]
E. coli	NASBA	40	-	Listeria monocytogenes, Shigella sonnei, Yersinia entero Colitica, Salmonella typhimurium	40 min	[44]
Staphylococcus aureus	NASBA	1–10	-	Lactococcus lactis, Bacillus cereus, Listeria monocytogenes, Enterococcus faecalis, E. coli, Citrobacter freundii, Sal-monella, Streptococcus bovis, Klebsiella aerogenes	3–4 h	[45]
Salmonella enteritidis.	FRET with CNP for signal enhancement	150	100–3000	Salmonella typhimurium, E. coli K88	2 h	[46]
Salmonella	DNA Micro-array	2–8 CFU/g (tomato)	-	E. coli, Shigella, S. aureus, Pseudomonas aeruginosa, Citrobacter freundii, Vibrio cholera, Enterococcus fae-calis, Yersinia enterocolitica	<2 h	[47]

Table 1. Cont.

Strain	Technique	LOD [a]	Detection Range, CFU mL^{-1}	Interference Studies	Assay Time	Ref.
Salmonella	DNA Micro-array QD used in-place of fluorescent dyes	10	-	Vibrio parahaemolyticus, Vibrio fluvialis, Yersinia enterocolitica, Proteus sp.,S. aureus, Enterococcus faecalis, Campylobacter jejuni, β-hemolytic Streptococcus, Listeria monocytogenes	<2 h	[48]
Salmonella and Campylobacter	DNA Micro-array	14–57 and 11–60	-	Listeria monocytogenes, B. cereus, Cronobacter sakazakii, Citrobacter freundii, Klebsiella pneumonia, E. coli, Proteus vulgaris, Enterobacter aerogenes, Hafnia alvei, Serratia marcescens	45 min	[49]

[a] LOD: the limit of detection cited in accordance with the IUPAC definition as "the smallest amount of concentration of analyte in the sample that can be reliably distinguished from zero". CNP: Carbon nanoparticles; LAMP: Loop-mediated isothermal amplification; NASBA: Nucleic Acid Sequence Based Amplification; PMA: Propidium monoazide; RPA: Recombinase Polymerase Amplification; QD: Quantum Dots; q-PCR: quantitative Polymerase Chain Reaction.

Immunoassays for whole bacterial cells is an alternative strategy intensively used [3,16] (Figure 1). The basic principle of immediate capturing of bacteria by the immobilized highly-specific biorecognition element, such as an antibody (Ab) or an apatmer, followed by a further read-out of the binding event either label-free or through the enzyme-amplified reaction, is most attractive for the development of rapid, sensitive and specific bacterial assays. Immunological analysis of bacteria by, e.g., traditional enzyme-linked immunosorbent assays (ELISA) may be quite specific and eventually fast; however cross-interference and surface fouling in physiological matrices may obstruct the results [3]. Also, the most frequently reported LODs, from 10^4 to 10^6 CFU mL^{-1} [16], are insufficient for many emergency-use sensing applications. The required sensitivity and specificity of analysis and higher speed/lower cost can be reached by using a variety of methodologies, amplification strategies and read-out techniques, as well as bio-mimicking bioreceptors and labels (Tables 2 and 3). The existing and emerging immunoassay approaches for bacterial analysis are scrutinized in the following sections.

Table 2. Selected examples of optical and related immunoassays for whole bacterial cells.

Strain/Analytical Scheme	Technique	LOD [a], CFU mL^{-1}	Detection Range, CFU mL^{-1}	Interference Studies	Assay Time	Ref.
E. coli O157:H7 Immuno magnetic assay, separation from complex matrix	Plate counting method	16	1.6×10^1–7.2×10^7	Salmonella enteritidis, Citrobacter freundii, Listeria monocytogenes	15 min	[50]
Staphylococus aureus Aptaassay on microtiter plates	Colorimetric detection with AuNP as indicator	9	10–10^6	Vibrioparahaemolyticus, Salmonella typhimurium, Streptococcus, E. coli, Enterobacter sakazakii, Listeria monocytogenes	15 min	[51]
Pseudomonas aeruginosa Aptamer assay on MB	Fluorometric detection with magnetic separation	1	10–10^8	Listeria monocytogenes, S. aureus, Salmonella enterica, E. coli	1.5 h	[52]

Table 2. Cont.

Strain/Analytical Scheme	Technique	LOD [a], CFU mL^{-1}	Detection Range, CFU mL^{-1}	Interference Studies	Assay Time	Ref.
Vibrio cholerae O1 Sandwich immunoassay	Chromatographic	5×10^5–10^6		Shigella flexneri, Salmonella typhi, Pseudomonas aeruginosa, Proteus vulgaris, Klebsiella pneumonia, Enterobacter cloacae	15 min	[53]
E. coli O157:H7 Sandwich immunoassay	ELISA with HRP-TMB label and AuNP for signal amplification	68 (PBS)	6.8×10^2 (PBS) 6.8×10^3 (in food)	Salmonella senftenberg, Shigella sonnei, E. coli K12	3 h	[54]
Salmonella typhimurium Apta- and immunoassay on MB	Colorimetric; ELISA on MB with HRP/TMB, and AuNP for signal amplification	1×10^3	1×10^3–1×10^8	Salmonella typhi, Salmonella paratyphi, S. aureus, E. coli	3 h	[55]
E. coli ATCC 8739 Apta-ssay on AuNP	FRET	3	5–10^6	E. coli DH5a, E. coli (ATCC 25922), Bacillus subtilis; S. aureus	-	[56]
Vibrio fischeri Sandwich aptaassay on paper	Colorimetric detection with AuNP	40	40–4×10^5	Vibrio parahemolyticus, E. coli, Bacillus subtilis, Shigella sonnei, S. aureus, Salmonella choleraesuis, Listeria monocytogenes	10 min	[57]
E. coli O157:H7 Sandwich immunomagnetic assay	Fluorescence using pH sensitive fluorophore release detection labels	15	-	Streptococcus pneumoniae R6	<3 h	[58]
E. coli O157:H7, Salmonella typhimurium, Listeria monocytogens Multiplex, Sandwich immunomagnetic assay	Fluorescence	<5	-	No cross reactivity between target pathogens	2 h	[59]
Salmonella enterica Sandwich and direct immunoassays	ELISA with CNT/HRP-TMB	10^3 and 10^4	-	-	24 h (direct); 3 h (sandwich)	[60]
E. coli O157:H7, Salmonella typhimurium Sandwich immunomagnetic assay	ELISA with HRP/TMB and AuNP network for signal amplification	3–15	-	Listeria monocytogenes, Salmonella typhimurium, Salmonella enteritidis	2 h	[61]
Salmonella enterica typhi Sandwich immunoassay with pre-enrichment in BPW	dot-ELISA, with Ab-HRP conjugate and 3,3 diaminobenzidine tetrahydrochloride	10^4 before 10^2 after enrichment	-	-	4 h, 10 h with enrichment	[62]
Listeria monocytogenes, E. coli O157:H7 and Salmonella enterica Sandwich immuno-fluorescence assay	Optical fiber; multiplexed simultaneous detection	10^3	-	Cross-reactivity tested with other target pathogens	<24 h	[63]
Escherichia coli Lateral flow aptaassay on QD	Colorimetric	300–600	-	Bacillus cereus, Enterococcus faecalis, Listeria monocytogenes, Salmonella enterica	20 min	[64]

Table 2. Cont.

Strain/Analytical Scheme	Technique	LOD [a], CFU mL^{-1}	Detection Range, CFU mL^{-1}	Interference Studies	Assay Time	Ref.
Salmonella Aptamer-based lateral flow assay	Colorimetric using up-conversion of NP for detection	85	150–2000	E. coli, S. aureus, Bacillus subtilis	30 min	[65]
Salmonella typhimurium Immunoagglutination-based immunoassay	Optical Mie scattering of antigen-Ab clusters	10 inconsistent with a 15 µL sample volume	100–10^6	-	10 min (from 6 to 15 min)	[66]
E. coli O157:H7 Immunomagnetic pre-concentration	LRSP diffraction grated Au surface	50	10^3 to 10^7	E. coli K12	30 min	[67]
Escherichia coli Immunoassay in a paper-based microfluidic device	Optical Mie scattering of antigen-Ab clusters	10 inconsistent with a 3.5 µL sample volume	10 to 10^3	-	90 s	[68]
Salmonella typhimurium Sandwich immunoassay with magnetic pre-concentration	Fluorescence detection using QDNPs	10^3	10^3–10^6	E. coli	30 min	[40]
RESONANCE-FREQUENCY-BASED IMMUNOASSAYS						
Escherichia coli O157:H7 Immunoassay on Ab-modified glass b	Resonance frequency	1 (in PBS)	-	-	10 min	[69]
Salmonella enterica Aptamer-based assay on MB	Piezoelectric: QCM	100	100–4 × 10^4	E. coli	40 min	[70]
S. aureus Aptamer-based assay	Magnetoelastic resonance frequency detection	5	10–1 × 10^{11}	Listeria monocytogenes, E. coli, Enterobacter sakazakii, Streptococcus, Vibrio parahemolyticus	25–26 min	[71]
Salmonella Sandwich immunoassay	Piezoelectric: QCM using AuNP labels for mass amplification	10–20	10–10^5	Klebsiella pneumonia, Enterobacteria spp, Pseudomona spp, S. aureus	9 min	[72]
Listeria monocytogenes Sandwich immunoassay o	Resonance frequency detection on a sputtered gold/lead-zirconate-titanate surface	100	10^3–10^5	-	30 min	[73]

[a] LOD: the limit of detection cited in accordance with the IUPAC definition as "the smallest amount of concentration of analyte in the sample that can be reliably distinguished from zero". AuNPs: Gold Nanoparticles; BPW: Buffered Peptone Water; CNT: Carbon Nanotubes; ELISA: Enzyme-Linked Immunosorbent Assay; HRP: Horseradish Peroxidase; LRSP: Long Range Surface Plasmons; MB: Magnetic Beads; NP: Nanoparticles; TMB: Tetramethyl Benzidine; QCM: Quartz-Crystal Microbalance; QD: Quantum Dots; QDNPs: Quantum Dot Nanoparticles.

Table 3. Selected examples of ultrasensitive and/or specific sensors for bacterial cells based on electrochemical immunoassay approaches.

Strain/Analytical Scheme	Technique	LOD [a], CFU mL^{-1}	Detection Range, CFU mL^{-1}	Interference Studies	Assay Time	Ref.
Sulphur reducing bacteria Immunoassay on chitosan doped rGS	EIS at 10 mV vs. Ag/AgCl with ferricyanide	18	18–1.8 × 10^7	Vibrio angillarum	-	[74]
Sulphur reducing bacteria Immunoassay on AuNP-modified Ni foam	EIS at 5 mV vs. Ag/AgC with ferricyanide	21	2.1 × 10^1–2.1 × 10^7	Vibrio anguillarum, E. coli	2 h	[75]
Salmonella enterica Immunoassay on gold electrodes	EIS at 5 mV vs.Ag/AgCl with ferricyanide	100 (10 CFU in 100 µL)	100–10 × 10^4	E. coli	1.5 min (no data on incubation time)	[76]
S. aureus (protein A) Competitive magneto-immunoassay on TTF-AuSPE.	e-ELISA, HRP label; TTF mediator Amperometry at −0.15 V	1(raw milk)	1 to 10^7	E. coli, Salmonella choleraesuis	2 h	[77]
E. coli O157:H7, S. aureus Immunoassay on nano-porous alumina	EIS at 25 mV vs. Pt; no label	100	-	Both strains were used for specificity test	2 h	[78]
E. coli, Listeria monocytogenes, Campylo-bacter jejuni Sandwich immunoassay with highly dispersed carbon particles	Electrochemical detection at 105 mV; HRP as a label, TMB as a substrate	50, 10 and 50, respectively	50–10^3, 10–1500, and 50–500	-	30 min	[79]
Pseudomonas aeruginosa Aptaassay on AuNP and AuNP/SPCE	Amperometry at 0.4 V with TMB	60	60–60 × 10^7	Vibrio cholera, Listeria monocytogens, S. aureus	10 min (colorimetry)	[80]
E. coli IDE modified with anti-E. coli Ab	Impedance at 5 mV; no label, electric field perturbation	300	10^2–10^4	-	1 h	[81]
E. coli O157:H7 Immunoassay at HA modified Au electrode	EIS with ferricyanide	7	10–10^5	S. aureus, Bacillus cereus, E. coli DH5a.	-	[82]
E. coli O157:H7 Immunoassay on AuNP modified rGO paper	EIS at 5 mV vs. Ag/AgCl with ferricyanide	150	150–1.5 × 10^7	S. aureus, Listeria monocytogenes, E. coli DH5a.	-	[83]
E. coli CIP 76.24 Immunoassay on polyclonal Ab/neuroavidin/SAM/Au	EIS with no indicator, at −0.6 V in aerated solutions	10	10–10^5 and 10^3–10^7 for lysed cells	S. epidermis: Interference at ≥100 CFU mL^{-1}	1 h incubation + detection	[84]
E. coli K12, MG1655 Phage typing & assaying activity of β-D-galactosidase in cell lysates, SPCE	Amperometry at 0.22 V, oxidation of enzymatically produced p-aminophenol	1 CFU in 100 mL	1–10^9	Klebsiella pneumoniae	6–8 h	[85]
E. coli O157:H7 Immunoassay on monoclonal Ab/ITO	EIS at 0.25 V with ferricyanide as a redox indicator	10	10–10^6	S. typhimurium, E. coli K12	0.8 h incubation + wash./detect.	[86]

Table 3. Cont.

Strain/Analytical Scheme	Technique	LOD [a], CFU mL^{-1}	Detection Range, CFU mL^{-1}	Interference Studies	Assay Time	Ref.
E. coli ORN 178 Assay at carbohydrate modified SAM on Au	EIS at 5 mV vs. Ag/AgCl; with ferricyanide	100	120–2.5 × 10^3	E. coli ORN 208	<1 h	[87]
E. coli XL1-Blue; K12 Assay on non-lytic M13 phage/AuNP/GCE	EIS with ferricyanide redox indicator, at 0.15 V	14	10–10^5	Pseudomonas chlororaphis	0.5 h incubation + wash./detect.	[88]
E. coli O157:H7 Sandwich immunoassay on MB at Au IDE, detected a response to urea hydrolysis by urease	EIS at 0 V, no indicator, label: urease/AuNP/aptamer;	12	12–1.2 × 10^5	S. typhimurium, Listeria monocytogenes	ca. 2 h	[89]
E. coli K12 and DH5α Sandwich immunoassay on MBs; on nitrocellulose modified Gr	Chronocoulometry at 0.3 V; no redox indicator; label: cellulase	1 (PBS), 2 (milk)	1–4 × 10^3	E. agglomerans, S. aureus, Salmonella enteretidis, B. subtilis, P. putida	3 h	[90]
E. coli O157:H7 Sandwich immunoassay on nanoporous alumina membrane	EIS at 25 mV/Pt; no label	10	10^0–10^4	-	-	[91]
E. coli Aptaassay on ITO modified with photoelectrochemical non-metallic NM	Potentiometric detection at 0.15 V (cathodic) and −0.4 V (anodic) (ratiometric detection)	2.9	2.9–2.9 × 10^6	-	12 h	[92]
E. coli O157:H7 Immunoassay on nanoporous alumina membrane	EIS; no label	10 (PBS) 83.7 (milk)	10–10^5	S. aureus, Bacillus cereus, E. coli DH5a.	-	[93]
E. coli O157:H7 Aptaassay on a paper modified with graphene nanoplatinum composite	EIS with ferricyanide indicator at 100 mV	4	4–10^5	-	12 min	[94]
E. coli K12 Sandwich immunoassay	Amperometry at −0.35 V, HRP as a label; substrates: HQ/BQ and H$_2$O$_2$	55 (PBS) 100 (milk)	10^2–10^8	Pseudomonas putida	1 h	[95]
E. coli O157:H7 Sandwich immunoassay with PtNCs coupled to GOD	Cyclic voltammetry from −0.15 V to 0.65 V	15	32–3.2 × 10^6	Salmonella typhi, Shigella dysenteriae Shigella flexneri	30 min	[96]
E. coli O157:H7 Immunoassay on a SAM modified gold electrode	EIS with ferricyanide at 0 V vs. Ag/AgCl	2	30–3 × 10^4	Salmonella typhimurium	45 min	[97]

Table 3. *Cont.*

Strain/Analytical Scheme	Technique	LOD [a], CFU mL^{-1}	Detection Range, CFU mL^{-1}	Interference Studies	Assay Time	Ref.
E. coli Sandwich immunoassay on AuNP-structured electrode in an automated microfluidic chip	Ammperometry at –0.1 V, with an HRP label and TMB as a substrate	50	50–10^6	*Shigella, Salmonella* spp., *Salmonella typhimurium, S. aureus*	30 min	[98]
E. coli O157:H7, *S. aureus* Nano-porous alumina membrane in a PEG-modified microfluiidc chip	Electrochemical impedance	100	10^2–10^5	*E. coli* O157:H7, *S. aureus*	<1 h	[99]

[a] LOD: the limit of detection cited in accordance with the IUPAC definition as "the smallest amount of concentration of analyte in the sample that can be reliably distinguished from zero". AuNPs: Gold Nanoparticles; AuSPE: Gold Screen Printed Electrodes; BQ: Benzoquinone; EIS: Electrochemical Impedance Spectroscopy; GCE: Glassy Carbon Electrode; Gr: Spectroscopic Graphite; HRP: Horseradish Peroxidase; HA: Hyaluronic Acid; HQ: Hydroquinone; IDE: Interdigitated Electrodes; ITO: Indium Tin Oxide; GOD: Glucose Oxidase; MB: Magnetic Beads; NM: nanomaterial; PtNCs: Platinum Nanochains; PEG: Polyethylene Glycol; rGO: Reduced Graphene Oxide; rGS: Reduced Graphene Sheet; SAM: Self-Assembled Monolayers; SPE: screen printed electrodes; SPCE: Screen Printed Carbon Electrodes; TMB: 3,3′,5,5′-tetramethylbenzidin; TTF: TetraThiaFulvalene.

Standing separately are metabolic sensors based on the detection of signals associated with specific reactions related to bacterial metabolism [100,101] (Figure 1). Those approaches may be very efficient but generally are either low specific or based on very individual metabolic biomarkers of specific bacterial species, and thus they do not form general analytical platforms.

3. Immunoassays and Aptamer-Based Assays with Optical Detection

Immunoassays and aptamer-based assays exploit the specificity of a biorecognition of targeted bacterial pathogens by mono- and polyclonal Abs [102] and by their in vitro alternatives—aptamers [103], and thus rely on their availability. A variety of bacterial pathogens can be detected once Ab or aptamers are developed for a specific bacteria type. The Ab and aptamers also represent the main limitation of this approach, since the available Ab or aptamers may not be sufficiently specific or show insufficient affinities for their targets. They may be too expensive for a number of applications and, finally, they may not be available at all.

In the optical immunoassays for whole bacteria, bacterial binding to the sensor surface changes the optical properties of the sensor/reaction media, such as UV/vis absorption, fluorescence, luminescence, which is optically read out [104]. Ca. 25 and 16 CFU mL^{-1} LODs were shown for *S. typhimurium* and *E. coli* in a 1 h fluorescence assay with aptamer-modified fluorescent-magnetic multifunctional nanoprobes [105]. A 30 min fluorescence aptamer-based assay with vancomycin-Au nanoclusters and aptamer-modified Au nanoparticles (NPs) allowed to detect down to 20 CFU mL^{-1} of *S. aureus* in PBS with no interference from other species; the assay performed well in milk, juice and human serum 10- and 5-fold diluted with PBS [106]. Such assays usually rely on clinical laboratory-based equipment and are less suitable for in-field applications, though on a few occasions portable optical-fiber systems [63,107] or colorimetric lateral-flow tests [64,108] have been reported. Such label-free optical approaches as surface plasmon resonance (SPR) that follow the changes in the refractive index of the transducer bioreceptor-modified interface are generally less sensitive (Table 2), and complex matrix components can strongly interfere with analysis. They also require a bulky equipment low suitable for in-field analysis and point-of-care testing (POCT).

Among optical immunoassays, ELISA is the dominant Ab-based methodology that relies on specific binding of bacteria to Abs immobilized either to a solid support (typically polystyrene, polyvinyl or polypropylene, or in a 96 or 384 micro-well plate) or to magnetic beads [109]. Binding of bacteria by a capture Ab is followed by reactions with a secondary Ab (or an aptamer) typically labelled with such redox enzymes as horseradish peroxidase (HRP) or alkaline phosphatase (AlkP), whose enzymatic reactions with their substrates result in the optically active products. Upon addition of the corresponding enzyme substrates, the reaction mixture changes the color, being followed by the plate reader or other spectrophotometric equipment [110]. HRP and AlkP enzymes are most commonly used in the enzyme-dependent immunoassays, both due to their high turnover rates and satisfactory stability of their bio-conjugates at 4 °C (both are sensitive to freezing [111–113]). In situations of a high-level endogenous peroxidase activity, AlkP becomes a more suitable choice. Among other frequently used enzymes is β-D-galactosidase (β-gal), whose bioconjugates are much more stable and can be stored for at least one year at 4 °C [114]. However, due to its high molecular weight and low turnover number, β-gal is used less. Still, stability and sensitivity issues trigger further search of novel and advanced enzymatic labels, such as adenosine deaminase (ADA) [115], or enzyme-loaded nanostructured labels, such as HRP-loaded nano-spherical poly(acrylic acid) brushes increasing the sensitivity of conventional ELISA by 267-fold [116].

Direct ELISA, in which an antigen-coated micro-well plate is exposed directly to an enzyme-linked Ab [117] is well suited for bacterial analysis [110]. However, the reported LODs of direct ELISA (over 10^6 CFU mL^{-1} [16,118]) are insufficient for earlier discussed applications. Down to 10^3 CFU mL^{-1} LODs can be reached with a sandwich ELISA, in which the signal is generated only when a complete primary Ab(bacteria)secondary Ab sandwich is assembled, and in indirect ELISA (Table 2). Such setups are reported to be 2–5 times more sensitive than direct and other ELISA types [119]. 10^3 CFU mL^{-1} of food-borne pathogenic *Salmonella* could be detected in indirect ELISA by targeting its membrane protein, bacterial flagellin FliC [120], while whole bacterial cell sandwich ELISA allowed detecting 10^3 CFU mL^{-1} of *Brucella abortus* and *Brucella melitensis* [121] and *Bacillus cerus* [122]. Further specificity and sensitivity of analysis can be controlled by a proper choice of Abs and Ab-replacing aptamers and of aptamer-reporter labels and nanomaterial-based labels [123]. Another emerging development of ELISA is a replacement of the primary Ab by bacteriophages, which allowed specific detection of *E. coli* and *Salmonella* strains, with a LOD of 10^5 cells per well (or 10^6 cells mL^{-1}) [124]. Overall, the shown LOD and eventually the protocols requiring lab-operating equipment make traditional ELISA low suitable for sensitive emergency-use applications.

Lateral flow immunoassays (LFI) share with ELISA the basic immunological principle and offer advantage in cost and faster analysis times. Adapted for a dipstick or immunochromatographic strip operation, they can essentially simplify and accelerate the analysis of the pathogenic microorganisms, since they do not require complex equipment or special training for their handling, and thus are more suitable for POCT [125] (Table 2). Colorimetric lateral-flow tests have reported LODs varying between 100 CFU mL^{-1} (a 5 min sandwich assay for *E. coli* with an HRP label read out by a CCD camera) [108] and 10^4–10^5 CFU mL^{-1} (a 20 min LFI with up-converting phosphor NPs as reporters, 10 CFU 0.6 mg^{-1} after pre-enrichment step) [126]. Similarly, 10 CFU mL^{-1} of different bacterial species could be detected in 1 h only after from 4 to 5 h pre-enrichment steps [127]. It is clear that without pre-enrichment most of LFIs show sufficiently high LODs (from 10^2 to 10^6 CFU mL^{-1}) [128–130], which currently restricts their immediate application in "alarm" situations.

4. Electrochemical Immunoassays

Electrochemical biosensors for bacteriological analysis include: (A) genosensors for bacterial DNA or RNA; (B) bacterial metabolic sensors, and (C) biosensors for detection and quantification of the whole bacterial cells [131]. Of those, electrochemical immunoassays and aptamer-based assays for bacteria are most challenging as platform biotechnologies competing with commercially established optical immunoassays. Due to a portable and inexpensive equipment with minimal power requirements and more robust read-out techniques, electrochemical approaches often allow more rapid and accurate bacterial detection with a decreased cost of equipment/assay and easier miniaturization of the device for use in-field and at POCT sites.

In the electrochemical approach, binding of bacteria to the Ab- or aptamer-modified surfaces is detected electrochemically by means of a redox active indicator (also: a redox mediator and a redox active product of the enzymatic reaction) or label-free, through the changes in the interfacial properties of the modified electrodes (Figure 2A–C). Overall, the specificity and sensitivity of the bacterial immunoassays can be radically improved by a wise combination of bio-recognition abilities of Abs/aptamers and redox indicators with electrochemical methodologies [27,123]. In electrochemical sandwich ELISA (e-ELISA) the formation of the Ab(bacterium)Ab complex is detected electrochemically by recording the electro-enzymatic activity of the labels, such as HRP or AlkP. The signal is amplified as a result of electrochemical recycling of one of the enzyme substrates at the electrodes or due to accumulation of the product changing the electrode properties (Figure 2D–F). The electrochemical signals can be further enhanced by modifying the electrode surface and/or designing more sophisticated capturing and detection mechanisms (Table 3).

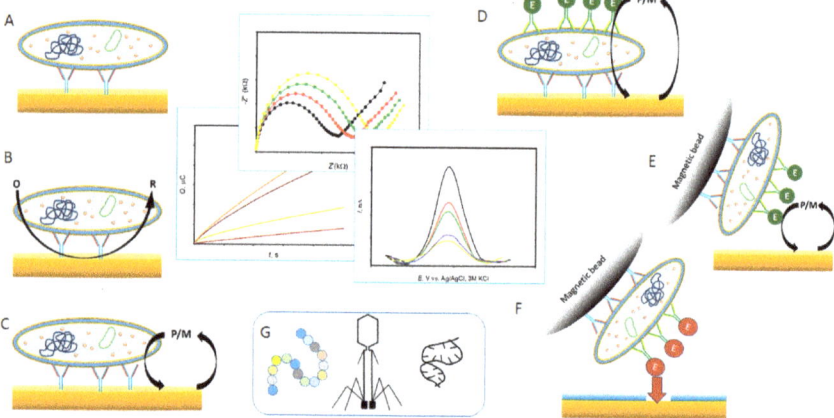

Figure 2. Schematic representation of the typical electrochemical immunosensors. (**A**) In a label-free and indicator-free immunoassay binding of a cell to the Ab-modified surface is detected impedimetrically; (**B**) In the presence of a redox indicator cell binding can also be detected voltammetrically or by chronocoulometry. In (**C**) a redox mediator of cellular metabolism is recycled at the electrode, giving rise to signals associated only with live cells. In (**D**) a bacterial cell is entrapped in an immune-sandwich formed by two Abs on the electrodes surface and labeled with a redox active enzymatic label, whose activity is electrochemically monitored through its substrate recycling at the electrode surface. (**E,F**) represent the sandwich assay adaptation to the magnetic beads-format, in (**F**) redox-inactive enzymatic labels induce changes at the electrode-solution interface that are electrochemically detected. (**G**) In (**A–F**) designs elements alternative to Ab can be used: peptides, phages and aptamers.

In contrast to optical immunoassays, electrochemical immunoassays (e-immunoassays) for bacteria may not need any labeling at all since they can rely on the interfacial changes resulting from the bacterial binding to the bioreceptor-modified surface. The microscopic, from 0.5 to 5 µm size of bacteria results in significant changes of the electrical properties of the bio-recognition interface upon bacterial binding, although surface fouling by non-specifically bound bacterial species may be a serious electroanalytical issue. The sensitivity and specificity of e-immunoassays is therewith essentially increased compared to traditional ones, the LOD being improved to just a few CFU [131,132]. The most sensitive and fastest appeared to be impedimetric biosensors and e-immunoassays on magnetic beads (MBs) enabling a few CFU mL^{-1} bacteria detection (Table 3). In addition, both the sample volume and the time of analysis can be significantly minimized, which is important for emergency applications. Detection protocols can be adapted for POCT (particularly, within the microfluidic format [98]) or portable-device in-field operation.

4.1. Electrochemical ELISA

Most intensively reported is e-ELISA exploiting HRP or AlkP bioelectrocatalytic labels, whose substrates are either electrochemically recycled at electrodes [77,133,134] or precipitate and block the redox indicator reactions at the electrode [135] (Figure 2D). Using traditional ELISA's enzymatic labels that rely on their substrate recycling at electrodes may result in a strong dependence of the e-ELISA sensitivity on the electrode surface properties. It often results in LODs close to those reported in the optical ELISA schemes. Bioelectrocatalytically-amplified ELISA on MBs is more advanced (Figure 2E). Bacteria are collected on MBs modified with bacteria-specific Ab (or aptamers), immunomagnetically separated from the original, often complex bacterial sample matrices, and finally pre-concentrated in smaller volume samples [136]. That results in sample amplification and excludes biofouling of the electrodes with matrix components. 845 CFU mL^{-1} of *S. aureus* in nasal flora samples could be specifically detected in the 4.5 h sandwich e-ELISA on MB with the AlkP label [137]. A competitive e-immunoassay with the HRP label allowed detecting 1 CFU mL^{-1} *S. aureus* in 2 h and 1.4 CFU mL^{-1} *Salmonella* in 50 min, in milk, without any sample pre-enrichment and with exceptional selectivity over other pathogens [77].

Requirements for a higher stability, simplicity and lower cost of e-immunoassays triggers application of low-cost redox inactive hydrolases as enzymatic labels in e-ELISA: lipase [138], urease [89] and cellulase [139] can also be effectively used in bacterial RNA and protein e-ELISA [9,140,141]. With hydrolase labels, products of hydrolysis of their substrates are electrochemically detected: bio-transformation of urea to ammonium carbonate increases the impedance of the system, while cellulase digestion of nitrocellulose films formed on graphite electrodes increases their electronic conductivity (Figure 2F). Such assays by itself rely not on the reactivity of the electrode, but rather on the enzymatic label reactivity with their substrates. That allowed highly sensitive and specific detection of 12 CFU mL^{-1} of *E. coli* in the buffer solution and in milk, in 2 h [89], and down to 1 CFU mL^{-1} of *E. coli* in tap water (2 CFU mL^{-1} of *E. coli* in milk) within 3 h, by assembling a hybrid aptamer (*E. coli*)Ab sandwich on MBs [90]. 100 CFU mL^{-1} of *Salmonella enteretidis*, *Enterobacter agglomerans*, *Pseudomonas putida*, *Staphylococcus aureus* and *Bacillus subtilis* did not interfere with the single *E. coli* detection when Ab was used as a capture element [90]. Both assays did not require any sample pre-treatment (e.g., cell pre-enrichment), are electrochemically label-free (no redox indicator/mediator was used), and are cost-effective due to the low cost of urease ($0.5 per mg) and cellulase (from $0.002 to $0.2 per mg) and their high storage stability. The urease assay could be well-integrated within the microfluidic format [89]. Both assays are general and can be adapted for specific detection of any other bacterial species once the corresponding Ab and aptamers are used.

Efforts are also focused on replacing enzymatic labels by different type of nanomaterial-based labels and catalysts such as quantum dots (QDs), DNAzymes and electrocatalytic nanoparticles (Table 3). For example, 3 CFU mL^{-1} of *E. coli* O157:H7 was detected in spiked milk samples by a sandwich e-ELISA assembled on the Ab-poly(*p*-aminobenzoic acid)-modified electrode and labelled with CdS QDs encapsulated in a metal organic/zeolitic imidazolate framework-8 (CdS@ZIF-8). The large load of the metal framework with CdS QDs resulted in an amplified electrochemical response detected voltammetrically [142].

The most successful example is the traditional HRP-linked sandwich ELISA integrated in the automated microfluidic electrochemical device, in which the HRP-labelled Ab was replaced by the HRP-Ab-Au NPs complex [98]. 3,3′,5,5′-Tetramethylbenzidine was an electrochemical mediator of the H_2O_2 reduction by HRP detected chronoamperometrically at −0.1 V. Sandwich assembly on the gold microelectronic chip surface then allowed down to 50 CFU mL^{-1} detection of *E. coli* in water within the overall 20 min procedure, with no interference from *Shigella*, *Salmonella* spp., *Salmonella typhimurium*, and *S. aureus*. It was possible to regenerate the Ab-modified sensor surface in a flow of 0.1 M HCl and then re-use it for another *E. coli* detections [98]. In a microfluidic assay, the fast delivery of the bacteria to the sensor surface by the microfluidic flow is equivalent to the fast capturing of bacteria on MBs.

However, due to small volumes of injected samples (e.g., 200 µL injected for 8 min at 25 µL min^{-1}) [98] and turnover limitations of redox enzymes used, a 1 CFU mL^{-1} LOD may be difficult to achieve.

4.2. Electrochemical Immunoassays (Not Enzyme-Linked)

4.2.1. Antibodies and Aptamers Based Assays

The simplest e-immunoassay strategy represents the electrode surface modification with an Ab or an aptamer, and immediate electrochemical detection of bacterial binding, either by following the interfacial changes accompanying binding impedimetrically, with or without a redox indicator, or by electrochemical monitoring of the metabolic activity of the captured cells (Figure 2A–C). The latter allows assessment of a viable cell population. *E. coli* and *N. gonorrheae* were immuno-specifically captured on the Ab-modified gold screen-printed electrodes (SPE), and 10^6 and 10^7 CFU mL^{-1} of them were quantified by voltammmetric analysis of the electroenzymatic activity of bacterial cytochrome c oxidase; synthetic substrate N,N,N',N'-tetramethyl-p-phenylene-diamine was used as a mediator [143] (Figure 3A). The assay was relatively fast (45 min binding and 1 h SPE regeneration) and inexpensive. Since the LOD was quite high, the authors suggested it for assessment of the efficiency of antibiotic treatments since it relies on the assay ability of a viable cells quantification.

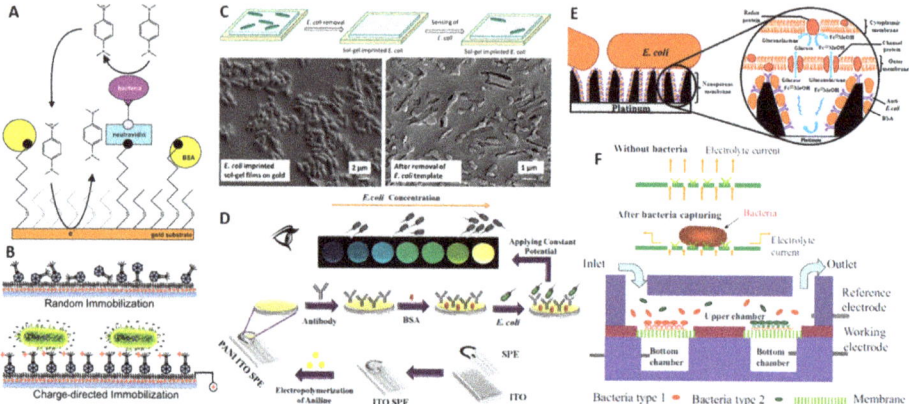

Figure 3. Selected examples of bacterial sensors based on (**A**) immune-recognition and electrochemical assessment of viable cell metabolism [143]; (**B**) Phage-based bacterial cell assay [144]; (**C**) Whole cell imprinted polymer sensor based on *E. coli* imprinting into ultrathin silica films on gold-coated glass slides [145], (**D**) Electro-chromic immunoassay [146]; and (**E**) Electrochemical nanopore immunoassay [147], and (**F**) Nanopore e-immunoassay integrated within the microfludic device [99]. Copyright (2019), (2017), (2019), and (2011) American Chemical Society and copyright (2016) and (2019) Elsevier.

The most frequently used e-immunoassays detect the total number of bacterial cells bound to the Ab or aptamer-modified electrodes by using Electrochemical Impedance Spectroscopy (EIS) either in the presence of a redox indicator, such as ferricyanide, or without it [84,148–151] (Table 3). Significant changes in the interfacial properties of the Ab-modified electrodes after binding of bacterial cells allowed 1 h detection of 10 CFU mL^{-1} of *E. coli* CIP 76.24 strain (Ab immobilization on biotinylated alkanethiol SAMs) [84] and 5.5 CFU mL^{-1} of *Listeria monocytogenes* (Ab immobilisation via protein A capable of binding of the Fc-region of Abs) [149].

10^4 CFU mL^{-1} of *S. pyogenes* was detected in human saliva in 30 min by bacteria capturing at the Ab/biotinylated Ab-modified electrodes (immobilization on NHS/EDC-activated alkanethiol SAMs or a conductive polymer via biotin-streptavidin/neutravidin linkage) [148]. 100 CFU mL^{-1} of *E. coli*

O157:H7 were detected at Ab-modified screen-printed interdigitated microelectrodes (immobilisation through the 2-dithiobis(sulfosuccinimidylpropionate)) in less than 1 h, and wheat germ agglutinin capable of binding to bacterial cell walls was used for the EIS signal enhancement [151]. Only 10 min took 600 CFU mL^{-1} *S. eneteritidis* detection at the aptamer/Au NPs-modified carbon SPE [150]. Aptamer immobilization on the Au NPs-modified electrodes was performed through the alkanethiol linker introduced in the 5′ end of the aptamer sequence.

Microfluidic formats of the impedance immunoassays further improve the assay time: 7 CFU mL^{-1} of multiple *Salmonella* serogroups were detected in 40 min in poultry and lettuce samples with no interference from *E. coli*, by bacteria trapping at the interdigitated Ab-modified Au electrodes of a microfluidic biosensor [152]. In the latter case, live and dead cells were discriminated by their different intensity of the impedance signal. However, no signal calibration data or cell growth information were given.

Generally, e-immunoassays relying on the surface-immobilized Ab and aptamers are quite specific, though, in many cases the modified electrodes also respond to other bacterial species, yet with a signal amplitude less significant than with the targeted bacteria [84,148,150]. To prevent non-specific binding of other bacteria, such antifouling strategies as electrode patterning and co-adsorption of antifouling agents were used. 100 CFU mL^{-1} of *E. coli* O157:H7 with no interference from *E. coli* K12, *Salmonella typhimurium*, or *S. aureus* were detected at the aptamer-modified three-dimensional interdigitated microelectrodes (3 µm in width and 4 µm in height) separated by insulating layers [153]. Antifouling properties of polyethylene glycol (PEG) in combination with its ability to inhibit the electrode reactions of ferri/ferrocyanide [154] were used to detect down to 10 CFU mL^{-1} of uropathogenic *E. coli* UTI89 in serum and urine at the Ab/reduced graphene oxide/polyethylenimine (PEI)-modified electrodes (Ab immobilization via formation of amide bonds with PEI and further electrode modification with pyrene-PEG) [155]. The biosensor discriminated wild type *E. coli* UTI89 from UTI89 Δfim strain.

4.2.2. Antimicrobial Peptides (AMP) Based Assays

AMP may be considered as peptide aptamer alternatives to Ab and oligonucleotide-based apatmers [156]. Though their specificity for different bacterial cell strains may be lower, they can provide the basis for broader platforms for the pathogen detection. AMP-modified micro-fabricated interdigitated electrodes allowed 12–15 min detection of down to 10^3 CFU mL^{-1} of *E. coli* and *Salmonella* [157]. AMP magainin I (GIGKFLHSAGKFGKAFVGEIMKS) bearing the net positive charge was immobilized on gold interdigitated microelectrodes via the extra Cys residue introduced in its C terminus and recognized a number of heat-killed pathogenic bacterial strains such as *E. coli* O157:H7 and *Salmonella*. AMP was concluded to primarily interact with the negatively charged phospholipids of Gram-negative bacterial membranes through the positively charged amino-acids in its N-terminal region. Also, the insignificant affinity was shown for Gram-positive (lacking the phospholipid-containing outer membrane) and non-pathogenic *E. coli* species (their cell walls miss hydrophilic O antigens essential for electrostatic and hydrogen bonding).

4.2.3. Bacteriophages Based Assays

Bacteriophages (or simply phages) are another perspective bio-recognition element alternative to Ab. Those are chemically and thermally stable viral nanoparticles capable of specific interactions with host bacteria and their infection. Phages' surface peptides display the aptamer properties towards bacterial surface proteins, and these properties can be modulated and optimized both chemically and genetically. Currently, phages are intensively explored in bacterial e-immunoassays as bio-recognition capturing probes enabling not only specific binding but also discrimination between viable and dead cells [158].

For intact bacterial sensing, phages are immobilized on electrodes by either physical adsorption, resulting in random surface orientation of phages, or directed orientation approaches (Figure 3B). Adsorption of *E. coli*-specific T4-phages on gold-nanorods-modified pencil graphite electrodes produced a biorecognition layer able of EIS detection of 10^3 CFU mL^{-1} of *E. coli* cells (no interference from *S. aureus*) [159]. The EIS responses strongly depended on the reaction time, being maximal after 25–35 min of *E. coli* binding, and then dropped down because of bacterial lysis by the phage. To improve the binding affinity of the phage SAMs, oriented T4-phage immobilization by its covalent binding either to cysteamine or already activated 3,3'-dithiodipropionic acid di(N-hydroxysuccinimide) ester was combined with the alternating electric field-modulation of the phage orientation. Such electrode polarization increased the number of phages properly oriented for *E. coli* binding, which resulted in the improved 100 CFU mL^{-1} LOD after 15 min binding reaction [160]. A very similar quantification of *Salmonella* after 50 min of bacteria binding reaction was reported for the capacitive flow system with polytyramine-modified gold electrodes modified with the covalently attached M13 phage specific for *Salmonella* spp. [161] (Figure 3B). The sensor surface could be regenerated 40 times by the alkaline treatment.

The cell-lytic properties of phages may interfere with reaching low LOD in bacterial analysis due to the fast lysis of infected bacterial cells. The sensitivity of bacterial analysis can be improved by using non-lytic phages, such as a non-lytic M13 phage that could recognize F+ pili of *E. coli* XL1-Blue and K12 strains: it was covalently attached to 3-mercaptopropionic acid-modified AuNPs via EDC/NHS chemistry [88]. That allowed increasing the time of the reaction between the immobilized phage and *E. coli* cells, and 14 CFU mL^{-1} of *E. coli* was detected by EIS. Still, the best LODs obtained with AMPs and phages do not approach those observed in the Ab- and aptamer- based e-immunoassays.

Along with that, lysis itself can be analytically useful. 10^3 CFU mL^{-1} of *E. coli* B were detected by EIS, with no interference from the K strain, at the T2 phage-modified electrodes [144]. A T2 phage specific for *E. coli* B strain was immobilized through its negatively charged head on the PEI/carbon nanotubes–modified glassy carbon electrode positively polarized. Such polarization-directed immobilization of the phage enabled selective binding of the targeted *E. coli* cells for a time sufficient for bacterial cell infection and lysis by the phage that was impedimetrically detected.

4.3. Whole Cell Imprinted Polymer Sensors as Alternative to E-Immunoassays

Cell-imprinted polymers (CIP) are biomimetic synthetic Ab alternatives for e-immunoassays. Turner's group produced a CIP sensor based on electropolymerized 3-aminophenylboronic acid (3-APBA) [162]. Polymerization of 3-APBA monomers lead to the formation of the *cis*-diol-boronic group complex within the template matrix that facilitated reversible binding and easy release of the trapped bacterial cells (upon subsequent regeneration) from the CIP. EIS responses of the CIP sensor were proportional to log 10^3–10^7 CFU mL^{-1} of *Staphylococcus epidermidis*, and the sensor did not respond to other similar shape bacteria species [162]. Impedimetric analysis with ferricyanide as a redox indicator detected less than 1 CFU mL^{-1} of uropathogenic *E. coli* UTI89 bound to ultrathin silica films prepared by sol-gel technology on gold-coated glass slides into which *E. coli* was imprinted (signal linearity from 1 to 10^4 CFU mL^{-1}), with *S. aureus* and *Pseudomonas aeruginosa* as negative controls [145] (Figure 3C). No information about the time of the assay or analyzed sample volumes versus the electrode size were reported for this very impressive assay, though. Thus, CIP can be an inexpensive replacement for the biological recognition elements if produced/shown to be sufficiently specific. Along with that, CIP may not provide the necessary specificity to discriminate between the different strains of the same bacteria, since they rely mostly on the bacterial shape and size and less on the bacterial surface peculiarities.

4.4. Electro-Optical Immunoassays

Combination of electrochemistry with an optical readout can further generate new POCT devices with improved sensitivity for bacterial analysis. Abs for *E. coli* were coupled to films of electropolymerized polyaniline (PANI) on ITO screen-printed electrodes, whose polarization changed the PANI oxidation states and generated concomitant changes in the film color different for *E. coli* bound and unbound films [146] (Figure 3D). Different electrochromic responses due to the presence of *E. coli* increasing the interfacial resistance and thus affecting the PANI oxidation states, allowed to detect down to 10^2 CFU mL^{-1} of *E. coli* by the naked eye, while 10 CFU mL^{-1} could be detected by a software.

Electrochemiluminescence (ECL) is another electro-optical approach that improves immunoassay's sensitivity by electrogenerated chemiluminescent signal amplification. Most convenient are sandwich immunoassays with a secondary Ab labelled with Ru(II) tris(bipyridine) complex emitting light after electrochemical stimulation with such co-reactant as tripropylamine [163,164] (analogues of e-ELISA in which enzyme labels are replaced by a chemiluminescent reagent). 45 CFU mL^{-1} of *Francisella tularensis* with Ab fragments as capture biomolecules could be detected in a fluidic chip with screen-printed 42 Au electrodes array within a 30 min procedure [164], and down to 2.3 CFU mL^{-1} of *E. coli* O157:H7 were detected in 2 h with the automated electroluminescent 48-well singleplex plate sensor (250 µL samples) [163]. Additional 4 h sample pre-enrichment by ultrafiltration of 10 mL samples further decreased the LOD to 0.12 CFU mL^{-1} [163]. Labelling of the secondary (reporter) Ab with graphene oxide (GO) nanosheets forming multi-complex with Ru(bpy)$_2$(phen-5-NH$_2$)$^{2+}$ allowed a 1 CFU mL^{-1} analysis of *Vibrio vulnificus* in an overall 2.2 h assay [165]. This approach was referred to as a Faraday-cage type, since the extended GO network enhanced the electron transfer exchange at the electrodes and the intermolecular ECL efficiency by extending the electrode reaction zone.

The ECL amplification can also improve the outcome of the CIP-based e-immunoassays. *E. coli* O157:H7 was imprinted into the polydopamine matrix by copolymerization, and further binding of *E. coli* was followed from the ECL signals from the next-step bound *E. coli* Abs labelled with N-doped graphene quantum nanodots (in reaction with potassium persulfate) [166]. From 10 to 10^7 CFU mL^{-1} were detected, with a LOD of 8 CFU mL^{-1}.

Another type of ECL immunoassays for bacterial cells exploits the ECL signal inhibition resulting from the electrode surface blockage with bacterial cells captured on the bioreceptor-modified surface [167] (a principle similar to some already discussed e-immunoassays [84,148–151]). Binding of *E. coli* to the aptamer-modified 3D N-doped high-surface-area graphene hydrogel was detected by following the inhibition of the ECL signal from luminol (in the presence of H_2O_2) [167]. AgBr nanoparticles were used as a catalyst for enhancing the ECL of luminol. Down to 0.5 CFU mL^{-1} were detected in spiked buffer solutions after the 40 min *E. coli* binding reaction. No information was provided on the *E. coli* strain or samples composition/volume, though. In another luminol-linked assays, the PaP1 phage specific for *Pseudomonas aeruginosa* was covalently coupled to carboxylated graphene casted on a glassy carbon electrode [168]. Down to 56 CFU mL^{-1} of *P. aeruginosa* was analyzed by following the ECL signal from luminol that decreased after binding of the bacteria [168]. *P. aeruginosa* was quantified in milk and human urine in a 30 min assay.

Despite these impressive results, the adaptation of ECL analysis of bacteria for POCT or in-field analysis seems to be not straightforward. Similarly to optical ELISA, it needs a quite complex read-out equipment. The existing commercial ECL-enabling analyzers, such as Roche cobas® 6000 analyzers, allow from 170 to 2170 test per h, but they are not yet adapted for bacterial sensing [169] and may be not suitable for direct analysis in the blood. More portable devices are nevertheless being developed [164], though their sensitivity should be further improved.

4.5. Electrochemical Immunoanalysis of Whole Cells with A Nanopore Technology

Perspective adaptations of the nanopore technology to whole cells electrochemical immunoanalysis are based either on highly specific binding of bacterial cells to Ab [147] or on less specific binding to an AMP [170] immobilized in the nanochannels of a porous alumina or silicon membrane. With this, bacterial binding results in the nanochannels being partially blocked for the ion fluxes. In the first case, *E. coli* binding blocked the nanochannels for the redox indicator reaction at 10 CFU mL^{-1}; the viability of cells was accessed with the same redox probe [147] (Figure 3E). In the second case, bacterial outer membrane liposaccharides were recognized by the AMP [170]. The binding affected the diffusivity of the redox indicator within the nanochannels and resulted in the drop of the voltammetric signal.

Though no bacterial species were analyzed in this case, the assay was claimed to be generally applicable for any Gram-negative bacteria detection [170].

Both strategies can be eventually adapted for the fast bacterial detection in easy-to-handle microfluidic devices. The Ab–modified nano-porous alumina membrane integrated within the microfluidic chip allowed the simultaneous, from 10^2 to 10^5 CFU mL^{-1}, impedimetric detection of *E. coli* 0157:H7 and *S. aureus* within ca. 40–50 min (30 min of incubation with bacteria, then washing with a buffer solution, and execution of EIS analysis) [99] (Figure 3F). To prevent non-specific bacterial adhesion, the internal surfaces of the device and the membrane were modified with PEG as an anti-biofouling agent. However, despite the simplicity of the detection scheme and the overall set-up design, the LOD of 100 CFU mL^{-1} seems to be too high for some immediate applications such as water quality analysis (according to WHO, no *E. coli* cell should be present in any 100 mL of drinking water).

5. Future Perspectives

Bacterial detection is a dynamically developing field. Emerging new methods for rapid and specific analysis of bacterial pathogens are already contributing to improving our life quality by monitoring health risk situations and decreasing the incidences of illness. For example, during the last decade, confirmed cases of *Salmonella*-caused illnesses have dropped by ca. 40%, due to significantly improved quality of food analysis, with cases reported increased six-fold [171]. Along with that, due to some inherent limitations regarding the bacterial assays' time, sensitivity and often unaffordable for some application cost, some human activity fields still challenge the bacterial sensor market.

There are complex analytical problems not solved yet in infection disease diagnostics and environmental analysis, such as both fast and specific detection and quantification of a low number of viable pathogens in clinical analysis of blood stream infections, or ultrasensitive, fast and inexpensive water quality analysis in large-volume samples in the presence of excessive amounts of other bacterial species. Fast, 1–5 min assaying of bacteria in inexpensive paper microfluidic [68] and LFI [108] devices is extremely attractive for environmental analysis, but 100–300 CFU mL^{-1} LODs reported make those immunosensors less suitable for ultrasensitive bacterial detection. In clinical analysis, from 1 to 100 CFU mL^{-1} of pathogenic species should be detected rapidly in blood samples for timely diagnosis of bloodstream infections [11], and within several hours antibiotic susceptibility testing should be performed. Despite the recent achievements, state-of-the art microfluidic systems for bacterial analysis in the blood, with their 10^3 CFU mL^{-1} LOD, are not suitable for practical applications yet [172], and current clinical analysis is still based on microbial culturing coupled with susceptibility tests; both may last for several days to result.

Considering the recent reports discussed, electrochemical immunosensors for whole bacterial cells can undoubtedly solve these problems, at low cost and by a constructive detector and instrumental design friendly to minimally trained personnel. Impedimetric, label-free 40 min e-immunoassaying of 7 CFU mL^{-1} [152] and 20 min e-ELISA of 50 CFU mL^{-1} of *E. coli* [98] in microfluidic devices are promising examples of the devices for real worlds sample analysis. The same refers to the urease-linked e-ELISA on MBs (12 CFU mL^{-1} 2 h) [89] and cellulase-linked e-ELISA on MBs (1 CFU mL^{-1} in 3 h) [90]. Both can be adapted to microfludics and electrochemical LFI, whose electrochemical adaptations are still scarce. Development of cheap electrochemical paper sensors [173,174] is another emerging biosensor trend, and their combination with electrochemical immunomagnetic and phage-based assays can deliver attractive practical solutions for cost-effective and efficient in-field/out-of-lab and POC testing systems.

However, compared to the number of excellent publications, the number of commercialized electrochemical immunosensors biosensors for bacterial pathogens is small. Development and validation of bioelectronic sensor devices capable of efficient solving the real world analytical tasks seems to be slow. Along with that, in addition to electrochemical Accu-Check (Roche Diagnostics, Basel, Switzerland) and Free Style (Abbott Diabetes Care, Chicago, IL, USA) dominating the glucose biosensor market today, Abbott Inc. introduced the e-ELISA platform iStat Systems for blood-circulating protein biomarkers of acute diseases, which becomes a breakthrough in the field of biosensors [175]. With that, e-immunoassays start to slowly crowd the optical ELISA market and may be one day will force it out with advanced electrochemical solutions addressing most urgent problems in the bacterial analysis field.

Funding: This work was supported by the European Union within the Marie Skłodowska-Curie Actions-Innovative Training Networks (ITN) project H2O2-MSCA-ITN-2018 "Break Biofilms" through the PhD fellowship to R.J.B. (grant agreement 813439).

Conflicts of Interest: The authors declare no conflict of interest.

References

1. Velusamy, V.; Arshak, K.; Korostynska, O.; Oliwa, K.; Adley, C. An overview of foodborne pathogen detection: In the perspective of biosensors. *Biotechnol. Adv.* **2010**, *28*, 232–254. [CrossRef]
2. WHO. Available online: https://www.who.int/water_sanitation_health/publications/jmp-report-2019/en/; https://www.who.int/water_sanitation_health/diseases-risks/en/ (accessed on 25 September 2020).
3. Law, J.W.-F.; Ab Mutalib, N.-S.; Chan, K.-G.; Lee, L.-H. Rapid methods for the detection of foodborne bacterial pathogens: Principles, applications, advantages and limitations. *Front. Microbiol.* **2015**, *5*, 770. [CrossRef]
4. Dye, C. After 2015: Infectious diseases in a new era of health and development. *Philos. Trans. R. Soc. B Biol. Sci.* **2014**, *369*, 20130426. [CrossRef] [PubMed]
5. Koch, G.H.; Brongers, M.P.; Thompson, N.G.; Virmani, Y.P.; Payer, J.H. Cost of corrosion in the United States. In *Handbook of Environmental Degradation of Materials*; William Andrew Inc.: New York, NY, USA, 2005; pp. 3–24.
6. Jansen, H.J.; Breeveld, F.J.; Stijnis, C.; Grobusch, M.P. Biological warfare, bioterrorism, and biocrime. *Clin. Microbiol. Infect.* **2014**, *20*, 488–496. [CrossRef] [PubMed]
7. WHO. Guidelines for drinking-water quality. In *Surveillance and Control of Community Supplies*; World Health Organization: Geneva, Switzerland, 1997; Volume 3.
8. U.S. Department of Health and Human Services; Public Health Service; Food and Drug Administration. *Grade "A" Pasteurized Milk Ordinance*; Public Health Service, Food and Drug Administration: Arden, NC, USA, 2017.
9. Shipovskov, S.; Saunders, A.M.; Nielsen, J.S.; Hansen, M.H.; Gothelf, K.V.; Ferapontova, E.E. Electrochemical sandwich assay for attomole analysis of DNA and RNA from beer spoilage bacteria *Lactobacillus brevis*. *Biosens. Bioelectron.* **2012**, *37*, 99–106. [CrossRef] [PubMed]
10. McEntire, J.; Acheson, D.; Siemens, A.; Eilert, S.; Robach, M. The Public Health Value of Reducing *Salmonella* Levels in Raw Meat and Poultry. *Food Prot. Trends* **2014**, *34*, 386–392.
11. Lee, A.; Mirrett, S.; Reller, L.B.; Weinstein, M.P. Detection of bloodstream Infections in adults: How many blood cultures are needed? *J. Clin. Microbiol.* **2007**, *45*, 3546–3548. [CrossRef]
12. Proctor, L.M.; Creasy, H.H.; Fettweis, J.M.; Lloyd-Price, J.; Mahurkar, A.; Zhou, W.; Buck, G.A.; Snyder, M.P.; Strauss, J.F.; Weinstock, G.M.; et al. The Integrative Human Microbiome Project. *Nature* **2019**, *569*, 641–648.
13. Carding, S.; Verbeke, K.; Vipond, D.T.; Corfe, B.M.; Owen, L.J. Dysbiosis of the gut microbiota in disease. *Microb. Ecol. Health Dis.* **2015**, *26*, 26191. [CrossRef]
14. WHO. *New Report Calls for Urgent Action to Avert Antimicrobial Resistance Crisis*; WHO: Geneva, Switzerland, 2019.
15. Nolan, T.; Hands, R.E.; Bustin, S.A. Quantification of mRNA using real time RT-PCR. *Nat. Protoc.* **2006**, *1*, 1559–1582. [CrossRef]
16. Wang, Y.; Salazar, J.K. Culture-independent rapid detection methods for bacterial pathogens and toxins in food matrices. *Compr. Rev. Food Sci. Food Saf.* **2016**, *15*, 183–205. [CrossRef]
17. Costea, P.I.; Zeller, G.; Sunagawa, S.; Pelletier, E.; Alberti, A.; Levenez, F.; Tramontano, M.; Driessen, M.; Hercog, R.; Jung, F.-E.; et al. Towards standards for human fecal sample processing in metagenomic studies. *Nat. Biotechnol.* **2017**, *35*, 1069. [CrossRef]
18. Ganz, K.; Gill, A. Inhibition of polymerase chain reaction for the detection of *Escherichia coli* O157:H7 and *Salmonella enterica* on walnut kernels. *Food Microbiol.* **2013**, *35*, 15–20. [CrossRef] [PubMed]
19. Zhao, X.; Lin, C.-W.; Wang, J. Advances in rapid detection methods for foodborne pathogens. *J. Microbiol. Biotechnol.* **2014**, *24*, 297–312. [CrossRef] [PubMed]

20. Margot, H.; Stephan, R.; Guarino, S.; Jagadeesan, B.; Chilton, D.; O'Mahony, E.; Iversen, C. Inclusivity, exclusivity and limit of detection of commercially available real-time PCR assays for the detection of *Salmonella*. *Int. J. Food Microbiol.* **2013**, *165*, 221–226. [CrossRef] [PubMed]
21. Nocker, A.; Cheung, C.-Y.; Camper, A.K. Comparison of propidium monoazide with ethidium monoazide for differentiation of live vs. dead bacteria by selective removal of DNA from dead cells. *J. Microbiol. Meth.* **2006**, *67*, 310–320. [CrossRef]
22. Li, J.; Macdonald, J. Advances in isothermal amplification: Novel strategies inspired by biological processes. *Biosens. Bioelectron.* **2015**, *64*, 196–211. [CrossRef]
23. Hønsvall, B.K.; Robertson, L.J. From research lab to standard environmental analysis tool: Will NASBA make the leap? *Water Res.* **2017**, *109*, 389–397. [CrossRef]
24. Li, Y.; Fan, P.; Zhou, S.; Zhang, L. Loop-mediated isothermal amplification (LAMP): A novel rapid detection platform for pathogens. *Microb. Pathog.* **2017**, *107*, 54–61. [CrossRef]
25. Lobato, I.M.; O'Sullivan, C.K. Recombinase polymerase amplification: Basics, applications and recent advances. *TrAC Tr. Anal. Chem.* **2018**, *98*, 19–35. [CrossRef]
26. Ferapontova, E. Basic concepts and recent advances in electrochemical analysis of nucleic acids. *Curr. Opin. Electrochem.* **2017**, *5*, 218–225. [CrossRef]
27. Simoska, O.; Stevenson, K.J. Electrochemical sensors for rapid diagnosis of pathogens in real time. *Analyst* **2019**, *144*, 6461–6478. [CrossRef] [PubMed]
28. Severgnini, M.; Cremonesi, P.; Consolandi, C.; De Bellis, G.; Castiglioni, B. Advances in DNA microarray technology for the detection of foodborne pathogens. *Food Bioproc. Technol.* **2011**, *4*, 936–953. [CrossRef]
29. McLoughlin, K.S. Microarrays for pathogen detection and analysis. *Brief. Funct. Gen.* **2011**, *10*, 342–353. [CrossRef] [PubMed]
30. Deurenberg, R.H.; Bathoorn, E.; Chlebowicz, M.A.; Couto, N.; Ferdous, M.; García-Cobos, S.; Kooistra-Smid, A.M.D.; Raangs, E.C.; Rosema, S.; Veloo, A.C.M.; et al. Application of next generation sequencing in clinical microbiology and infection prevention. *J. Biotechnol.* **2017**, *243*, 16–24. [CrossRef] [PubMed]
31. Vincent, A.T.; Derome, N.; Boyle, B.; Culley, A.I.; Charette, S.J. Next-generation sequencing (NGS) in the microbiological world: How to make the most of your money. *J. Microbiol. Meth.* **2017**, *138*, 60–71. [CrossRef]
32. Czilwik, G.; Messinger, T.; Strohmeier, O.; Wadle, S.; von Stetten, F.; Paust, N.; Roth, G.; Zengerle, R.; Saarinen, P.; Niittymäki, J.; et al. Rapid and fully automated bacterial pathogen detection on a centrifugal-microfluidic LabDisk using highly sensitive nested PCR with integrated sample preparation. *Lab Chip* **2015**, *15*, 3749–3759. [CrossRef]
33. Yan, H.; Zhu, Y.; Zhang, Y.; Wang, L.; Chen, J.; Lu, Y.; Xu, Y.; Xing, W. Multiplex detection of bacteria on an integrated centrifugal disk using bead-beating lysis and loop-mediated amplification. *Sci. Rep.* **2017**, *7*, 1460. [CrossRef]
34. Waage, A.S.; Vardund, T.; Lund, V.; Kapperud, G. Detection of Small Numbers of *Campylobacter jejuni* and *Campylobacter coli* Cells in Environmental Water, Sewage, and Food Samples by a Seminested PCR Assay. *Appl. Environ. Microbiol.* **1999**, *65*, 1636–1643. [CrossRef]
35. Ruiz-Rueda, O.; Soler, M.; Calvó, L.; García-Gil, J.L. Multiplex Real-time PCR for the Simultaneous Detection of *Salmonella* spp. and *Listeria monocytogenes* in Food Samples. *Food Anal. Methods* **2010**, *4*, 131–138. [CrossRef]
36. Wolffs, P.F.; Glencross, K.; Thibaudeau, R.; Griffiths, M.W. Direct quantitation and detection of salmonellae in biological samples without enrichment, using two-step filtration and real-time PCR. *Appl. Environ. Microbiol.* **2006**, *72*, 3896–3900. [CrossRef] [PubMed]
37. Vinayaka, A.C.; Ngo, T.A.; Kant, K.; Engelsmann, P.; Dave, V.P.; Shahbazi, M.-A.; Wolff, A.; Bang, D.D. Rapid detection of *Salmonella enterica* in food samples by a novel approach with combination of sample concentration and direct PCR. *Biosens. Bioelectron.* **2019**, *129*, 224–230. [CrossRef] [PubMed]
38. Xiao, L.; Zhang, Z.; Sun, X.; Pan, Y.; Zhao, Y. Development of a quantitative real-time PCR assay for viable *Salmonella* spp. without enrichment. *Food Control* **2015**, *57*, 185–189. [CrossRef]
39. Ma, K.; Deng, Y.; Bai, Y.; Xu, D.; Chen, E.; Wu, H.; Li, B.; Gao, L. Rapid and simultaneous detection of *Salmonella*, *Shigella*, and *Staphylococcus aureus* in fresh pork using a multiplex real-time PCR assay based on immunomagnetic separation. *Food Control* **2014**, *42*, 87–93. [CrossRef]
40. Kim, G.; Moon, J.-H.; Moh, C.-Y.; Lim, J.-g. A microfluidic nano-biosensor for the detection of pathogenic *Salmonella*. *Biosens. Bioelectron.* **2015**, *67*, 243–247. [CrossRef]

41. Li, J.; Zhai, L.; Bie, X.; Lu, Z.; Kong, X.; Yu, Q.; Lv, F.; Zhang, C.; Zhao, H. A novel visual loop-mediated isothermal amplification assay targeting gene62181533 for the detection of *Salmonella spp.* in foods. *Food Control* **2016**, *60*, 230–236. [CrossRef]
42. Yamazaki, W.; Ishibashi, M.; Kawahara, R.; Inoue, K. Development of a loop-mediated isothermal amplification assay for sensitive and rapid detection of *Vibrio parahaemolyticus*. *BMC Microbiol.* **2008**, *8*, 163. [CrossRef]
43. Shao, Y.; Zhu, S.; Jin, C.; Chen, F. Development of multiplex loop-mediated isothermal amplification-RFLP (mLAMP-RFLP) to detect *Salmonella* spp. and *Shigella spp.* in milk. *Int. J. Food Microbiol.* **2011**, *148*, 75–79. [CrossRef]
44. Min, J.; Baeumner, A.J. Highly sensitive and specific detection of viable *Escherichia coli* in drinking water. *Anal. Biochem.* **2002**, *303*, 186–193. [CrossRef]
45. O'Grady, J.; Lacey, K.; Glynn, B.; Smith, T.J.; Barry, T.; Maher, M. tmRNA—A novel high-copy-number RNA diagnostic target–its application for *Staphylococcus aureus* detection using real-time NASBA. *FEMS Microbiol. Lett.* **2009**, *301*, 218–223. [CrossRef]
46. Song, Y.; Li, W.; Duan, Y.; Li, Z.; Deng, L. Nicking enzyme-assisted biosensor for *Salmonella enteritidis* detection based on fluorescence resonance energy transfer. *Biosens. Bioelectron.* **2014**, *55*, 400–404. [CrossRef] [PubMed]
47. Guo, D.; Liu, B.; Liu, F.; Cao, B.; Chen, M.; Hao, X.; Feng, L.; Wang, L. Development of a DNA Microarray for Molecular Identification of All 46 *Salmonella* O Serogroups. *Appl. Environ. Microbiol.* **2013**, *79*, 3392. [CrossRef] [PubMed]
48. Huang, A.; Qiu, Z.; Jin, M.; Shen, Z.; Chen, Z.; Wang, X.; Li, J.-W. High-throughput detection of food-borne pathogenic bacteria using oligonucleotide microarray with quantum dots as fluorescent labels. *Int. J. Food Microbiol.* **2014**, *185*, 27–32. [CrossRef] [PubMed]
49. Tortajada-Genaro, L.A.; Rodrigo, A.; Hevia, E.; Mena, S.; Niñoles, R.; Maquieira, Á. Microarray on digital versatile disc for identification and genotyping of *Salmonella* and *Campylobacter* in meat products. *Anal. Bioanal. Chem.* **2015**, *407*, 7285–7294. [CrossRef] [PubMed]
50. Varshney, M.; Yang, L.; Su, X.-L.; Li, Y. Magnetic Nanoparticle-Antibody Conjugates for the Separation of *Escherichia coli* O157:H7 in Ground Beef. *J. Food Prot.* **2005**, *68*, 1804–1811. [CrossRef] [PubMed]
51. Yuan, J.; Wu, S.; Duan, N.; Ma, X.; Xia, Y.; Chen, J.; Ding, Z.; Wang, Z. A sensitive gold nanoparticle-based colorimetric aptasensor for *Staphylococcus aureus*. *Talanta* **2014**, *127*, 163–168. [CrossRef]
52. Zhong, Z.; Gao, X.; Gao, R.; Jia, L. Selective capture and sensitive fluorometric determination of *Pseudomonas aeruginosa* by using aptamer modified magnetic nanoparticles. *Microchim. Acta* **2018**, *185*, 377. [CrossRef]
53. Chaivisuthangkura, P.; Pengsuk, C.; Longyant, S.; Sithigorngul, P. Evaluation of monoclonal antibody based immunochromatographic strip test for direct detection of *Vibrio cholerae* O1 contamination in seafood samples. *J. Microbiol. Methods* **2013**, *95*, 304–311. [CrossRef]
54. Shen, Z.; Hou, N.; Jin, M.; Qiu, Z.; Wang, J.; Zhang, B.; Wang, X.; Wang, J.; Zhou, D.; Li, J. A novel enzyme-linked immunosorbent assay for detection of *Escherichia coli* O157:H7 using immunomagnetic and beacon gold nanoparticles. *Gut Pathog.* **2014**, *6*, 14. [CrossRef]
55. Wu, W.; Li, J.; Pan, D.; Li, J.; Song, S.; Rong, M.; Li, Z.; Gao, J.; Lu, J. Gold nanoparticle-based enzyme-linked antibody-aptamer sandwich assay for detection of *Salmonella* Typhimurium. *ACS Appl. Mater Interfaces* **2014**, *6*, 16974–16981. [CrossRef]
56. Jin, B.; Wang, S.; Lin, M.; Jin, Y.; Zhang, S.; Cui, X.; Gong, Y.; Li, A.; Xu, F.; Lu, T.J. Upconversion nanoparticles based FRET aptasensor for rapid and ultrasenstive bacteria detection. *Biosens. Bioelectron.* **2017**, *90*, 525–533. [CrossRef] [PubMed]
57. Shin, W.R.; Sekhon, S.S.; Rhee, S.K.; Ko, J.H.; Ahn, J.Y.; Min, J.; Kim, Y.H. Aptamer-Based Paper Strip Sensor for Detecting *Vibrio fischeri*. *ACS Comb. Sci.* **2018**, *20*, 261–268. [CrossRef] [PubMed]
58. Mouffouk, F.; da Costa, A.M.; Martins, J.; Zourob, M.; Abu-Salah, K.M.; Alrokayan, S.A. Development of a highly sensitive bacteria detection assay using fluorescent pH-responsive polymeric micelles. *Biosens. Bioelectron.* **2011**, *26*, 3517–3523. [CrossRef] [PubMed]
59. Cho, I.H.; Mauer, L.; Irudayaraj, J. In-situ fluorescent immunomagnetic multiplex detection of foodborne pathogens in very low numbers. *Biosens. Bioelectron.* **2014**, *57*, 143–148. [CrossRef] [PubMed]

60. Chunglok, W.; Wuragil, D.K.; Oaew, S.; Somasundrum, M.; Surareungchai, W. Immunoassay based on carbon nanotubes-enhanced ELISA for *Salmonella enterica serovar Typhimurium*. *Biosens. Bioelectron.* **2011**, *26*, 3584–3589. [CrossRef]
61. Cho, I.H.; Irudayaraj, J. In-situ immuno-gold nanoparticle network ELISA biosensors for pathogen detection. *Int. J. Food Microbiol.* **2013**, *164*, 70–75. [CrossRef]
62. Kumar, S.; Balakrishna, K.; Batra, H.V. Enrichment-ELISA for Detection of *Salmonella typhi* From Food and Water Samples. *Biomed. Environ. Sci.* **2008**, *21*, 137–143. [CrossRef]
63. Ohk, S.H.; Bhunia, A.K. Multiplex fiber optic biosensor for detection of *Listeria monocytogenes, Escherichia coli* O157:H7 and *Salmonella enterica* from ready-to-eat meat samples. *Food Microbiol.* **2013**, *33*, 166–171. [CrossRef]
64. Bruno, J.G. Application of DNA aptamers and quantum dots to lateral flow test strips for detection of foodborne pathogens with improved sensitivity versus colloidal gold. *Pathogens* **2014**, *3*, 341–355. [CrossRef]
65. Jin, B.; Yang, Y.; He, R.; Park, Y.I.; Lee, A.; Bai, D.; Li, F.; Lu, T.J.; Xu, F.; Lin, M. Lateral flow aptamer assay integrated smartphone-based portable device for simultaneous detection of multiple targets using upconversion nanoparticles. *Sens. Actuators B Chem.* **2018**, *276*, 48–56. [CrossRef]
66. Fronczek, C.F.; You, D.J.; Yoon, J.Y. Single-pipetting microfluidic assay device for rapid detection of *Salmonella* from poultry package. *Biosens. Bioelectron.* **2013**, *40*, 342–349. [CrossRef] [PubMed]
67. Wang, Y.; Knoll, W.; Dostalek, J. Bacterial pathogen surface plasmon resonance biosensor advanced by long range surface plasmons and magnetic nanoparticle assays. *Anal. Chem.* **2012**, *84*, 8345–8350. [CrossRef] [PubMed]
68. Park, T.S.; Yoon, J.Y. Smartphone Detection of *Escherichia coli* From Field Water Samples on Paper Microfluidics. *IEEE Sens. J.* **2015**, *15*, 1902–1907. [CrossRef]
69. Sharma, H.; Mutharasan, R. Rapid and sensitive immunodetection of *Listeria monocytogenes* in milk using a novel piezoelectric cantilever sensor. *Biosens. Bioelectron.* **2013**, *45*, 158–162. [CrossRef]
70. Ozalp, V.C.; Bayramoglu, G.; Erdem, Z.; Arica, M.Y. Pathogen detection in complex samples by quartz crystal microbalance sensor coupled to aptamer functionalized core-shell type magnetic separation. *Anal. Chim. Acta* **2015**, *853*, 533–540. [CrossRef]
71. Rahman, M.R.T.; Lou, Z.; Wang, H.; Ai, L. Aptamer immobilized magnetoelastic sensor for the determination of *Staphylococcus aureus*. *Anal. Lett.* **2015**, *48*, 2414–2422. [CrossRef]
72. Salam, F.; Uludag, Y.; Tothill, I.E. Real-time and sensitive detection of *Salmonella Typhimurium* using an automated quartz crystal microbalance (QCM) instrument with nanoparticles amplification. *Talanta* **2013**, *115*, 761–767. [CrossRef]
73. Campbell, G.A.; Mutharasan, R. A Method of measuring *Escherichia coli* O157:H7 at 1 Cell/mL in 1 liter sample using antibody functionalized piezoelectric-excited millimeter-sized cantilever sensor. *Environ. Sci. Technol.* **2007**, *41*, 1668–1674. [CrossRef]
74. Wan, Y.; Lin, Z.; Zhang, D.; Wang, Y.; Hou, B. Impedimetric immunosensor doped with reduced graphene sheets fabricated by controllable electrodeposition for the non-labelled detection of bacteria. *Biosens. Bioelectron.* **2011**, *26*, 1959–1964. [CrossRef]
75. Wan, Y.; Zhang, D.; Wang, Y.; Hou, B. A 3D-impedimetric immunosensor based on foam Ni for detection of sulfate-reducing bacteria. *Electrochem. Commun.* **2010**, *12*, 288–291. [CrossRef]
76. La Belle, J.T.; Shah, M.; Reed, J.; Nandakumar, V.; Alford, T.L.; Wilson, J.W.; Nickerson, C.A.; Joshi, L. Label-Free and Ultra-Low Level Detection of *Salmonella enterica Serovar Typhimurium* Using Electrochemical Impedance Spectroscopy. *Electroanalysis* **2009**, *21*, 2267–2271. [CrossRef]
77. Esteban-Fernández de Ávila, B.; Pedrero, M.; Campuzano, S.; Escamilla-Gómez, V.; Pingarrón, J.M. Sensitive and rapid amperometric magnetoimmunosensor for the determination of *Staphylococcus aureus*. *Anal. Bioanal. Chem.* **2012**, *403*, 917–925. [CrossRef] [PubMed]
78. Tan, F.; Leung, P.H.M.; Liu, Z.-b.; Zhang, Y.; Xiao, L.; Ye, W.; Zhang, X.; Yi, L.; Yang, M. A PDMS microfluidic impedance immunosensor for *E. coli* O157:H7 and *Staphylococcus aureus* detection via antibody-immobilized nanoporous membrane. *Sens. Actuators B Chem.* **2011**, *159*, 328–335. [CrossRef]
79. Chemburu, S.; Wilkins, E.; Abdel-Hamid, I. Detection of pathogenic bacteria in food samples using highly-dispersed carbon particles. *Biosens. Bioelectron.* **2005**, *21*, 491–499. [CrossRef] [PubMed]
80. Das, R.; Dhiman, A.; Kapil, A.; Bansal, V.; Sharma, T.K. Aptamer-mediated colorimetric and electrochemical detection of Pseudomonas aeruginosa utilizing peroxidase-mimic activity of gold NanoZyme. *Anal. Bioanal. Chem.* **2019**, *411*, 1229–1238. [CrossRef]

81. de la Rica, R.; Baldi, A.; Fernández-Sánchez, C.; Matsui, H. Selective Detection of Live Pathogens via Surface-Confined Electric Field Perturbation on Interdigitated Silicon Transducers. *Anal. Chem.* **2009**, *81*, 3830–3835. [CrossRef]
82. Joung, C.-K.; Kim, H.-N.; Im, H.-C.; Kim, H.-Y.; Oh, M.-H.; Kim, Y.-R. Ultra-sensitive detection of pathogenic microorganism using surface-engineered impedimetric immunosensor. *Sens. Actuators B Chem.* **2012**, *161*, 824–831. [CrossRef]
83. Wang, Y.; Ping, J.; Ye, Z.; Wu, J.; Ying, Y. Impedimetric immunosensor based on gold nanoparticles modified graphene paper for label-free detection of *Escherichia coli* O157:H7. *Biosens. Bioelectron.* **2013**, *49*, 492–498. [CrossRef]
84. Maalouf, R.; Fournier-Wirth, C.; Coste, J.; Chebib, H.; Saïkali, Y.; Vittori, O.; Errachid, A.; Cloarec, J.-P.; Martelet, C.; Jaffrezic-Renault, N. Label-free detection of bacteria by electrochemical impedance spectroscopy: comparison to surface plasmon resonance. *Anal. Chem.* **2007**, *79*, 4879–4886. [CrossRef]
85. Neufeld, T.; Schwartz-Mittelmann, A.; Biran, D.; Ron, E.Z.; Rishpon, J. Combined phage typing and amperometric detection of released enzymatic activity for the specific identification and quantification of bacteria. *Anal. Chem.* **2003**, *75*, 580–585. [CrossRef]
86. dos Santos, M.B.; Azevedo, S.; Agusil, J.P.; Prieto-Simón, B.; Sporer, C.; Torrents, E.; Juárez, A.; Teixeira, V.; Samitier, J. Label-free ITO-based immunosensor for the detection of very low concentrations of pathogenic bacteria. *Bioelectrochemistry* **2015**, *101*, 146–152. [CrossRef] [PubMed]
87. Guo, X.; Kulkarni, A.; Doepke, A.; Halsall, H.B.; Iyer, S.; Heineman, W.R. Carbohydrate-based label-free detection of *Escherichia coli* ORN 178 using electrochemical impedance spectroscopy. *Anal. Chem.* **2012**, *84*, 241–246. [CrossRef] [PubMed]
88. Sedki, M.; Chen, X.; Chen, C.; Ge, X.; Mulchandani, A. Non-lytic M13 phage-based highly sensitive impedimetric cytosensor for detection of coliforms. *Biosens. Bioelectron.* **2020**, *148*, 111794. [CrossRef] [PubMed]
89. Yao, L.; Wang, L.; Huang, F.; Cai, G.; Xi, X.; Lin, J. A microfluidic impedance biosensor based on immunomagnetic separation and urease catalysis for continuous-flow detection of *E. coli* O157:H7. *Sens. Actuators B Chem.* **2018**, *259*, 1013–1021. [CrossRef]
90. Pankratov, D.; Bendixen, M.; Shipovskov, S.; Gosewinkel, U.; Ferapontova, E. Cellulase-linked immunomagnetic microbial assay on electrodes: Specific and sensitive detection of a single bacterial cell *Anal. Chem.* **2020**, *92*, 12451–12459. [CrossRef]
91. Chan, K.Y.; Ye, W.W.; Zhang, Y.; Xiao, L.D.; Leung, P.H.; Li, Y.; Yang, M. Ultrasensitive detection of *E. coli* O157:H7 with biofunctional magnetic bead concentration via nanoporous membrane based electrochemical immunosensor. *Biosens. Bioelectron.* **2013**, *41*, 532–537. [CrossRef]
92. Hua, R.; Hao, N.; Lu, J.; Qian, J.; Liu, Q.; Li, H.; Wang, K. A sensitive Potentiometric resolved ratiometric photoelectrochemical aptasensor for *Escherichia coli* detection fabricated with non-metallic nanomaterials. *Biosens. Bioelectron.* **2018**, *106*, 57–63. [CrossRef]
93. Joung, C.K.; Kim, H.N.; Lim, M.C.; Jeon, T.J.; Kim, H.Y.; Kim, Y.R. A nanoporous membrane-based impedimetric immunosensor for label-free detection of pathogenic bacteria in whole milk. *Biosens. Bioelectron.* **2013**, *44*, 210–305. [CrossRef]
94. Burrs, S.L.; Bhargava, M.; Sidhu, R.; Kiernan-Lewis, J.; Gomes, C.; Claussen, J.C.; McLamore, E.S. A paper based graphene-nanocauliflower hybrid composite for point of care biosensing. *Biosens. Bioelectron.* **2016**, *85*, 479–487. [CrossRef]
95. Laczka, O.; Maesa, J.M.; Godino, N.; del Campo, J.; Fougt-Hansen, M.; Kutter, J.P.; Snakenborg, D.; Munoz-Pascual, F.X.; Baldrich, E. Improved bacteria detection by coupling magneto-immunocapture and amperometry at flow-channel microband electrodes. *Biosens. Bioelectron.* **2011**, *26*, 3633–3640. [CrossRef]
96. Li, Y.; Fang, L.; Cheng, P.; Deng, J.; Jiang, L.; Huang, H.; Zheng, J. An electrochemical immunosensor for sensitive detection of *Escherichia coli* O157:H7 using C60 based biocompatible platform and enzyme functionalized Pt nanochains tracing tag. *Biosens. Bioelectron.* **2013**, *49*, 485–491. [CrossRef] [PubMed]
97. dos Santos, M.B.; Agusil, J.P.; Prieto-Simon, B.; Sporer, C.; Teixeira, V.; Samitier, J. Highly sensitive detection of pathogen *Escherichia coli* O157:H7 by electrochemical impedance spectroscopy. *Biosens. Bioelectron.* **2013**, *45*, 174–180. [CrossRef]
98. Altintas, Z.; Akgun, M.; Kokturk, G.; Uludag, Y. A fully automated microfluidic-based electrochemical sensor for real-time bacteria detection. *Biosens. Bioelectron.* **2018**, *100*, 541–548. [CrossRef] [PubMed]

99. Tian, F.; Lyu, J.; Shi, J.; Tan, F.; Yang, M. A polymeric microfluidic device integrated with nanoporous alumina membranes for simultaneous detection of multiple foodborne pathogens. *Sens. Actuators B Chem.* **2016**, *225*, 312–318. [CrossRef]
100. Zhang, F.; Keasling, J. Biosensors and their applications in microbial metabolic engineering. *Trends Micorbiol.* **2011**, *19*, 323–329. [CrossRef] [PubMed]
101. Wang, Y.-P.; Lei, Q.-Y. Metabolite sensing and signaling in cell metabolism. *Signal Transduct. Target. Ther.* **2018**, *3*, 30. [CrossRef]
102. Macario, A.J.L.; de MacArio, E.C. *Monoclonal Antibodies against Bacteria*; Academic Press: Orlando, FL, USA, 1986.
103. Wang, L.; Wang, R.; Wei, H.; Li, Y. Selection of aptamers against pathogenic bacteria and their diagnostics application. *World J. Microbiol. Biotechnol.* **2018**, *34*, 149. [CrossRef]
104. Banada, P.P.; Bhunia, A.K. Antibodies and Immunoassays for Detection of Bacterial Pathogens. In *Principles of Bacterial Detection: Biosensors, Recognition Receptors and Microsystems*; Zourob, M., Elwary, S., Turner, A., Eds.; Springer: New York, NY, USA, 2008; pp. 567–602.
105. Li, L.; Li, Q.; Liao, Z.; Sun, Y.; Cheng, Q.; Song, Y.; Song, E.; Tan, W. Magnetism-resolved separation and fluorescence quantification for near-simultaneous detection of multiple pathogens. *Anal. Chem.* **2018**, *90*, 9621–9628. [CrossRef]
106. Yu, M.; Wang, H.; Fu, F.; Li, L.; Li, J.; Li, G.; Song, Y.; Swihart, M.T.; Song, E. Dual-recognition Förster Resonance Energy Transfer based platform for one-step sensitive detection of pathogenic bacteria using fluorescent vancomycin–gold nanoclusters and aptamer–gold nanoparticles. *Anal. Chem.* **2017**, *89*, 4085–4090. [CrossRef]
107. Hayman, R.B. Fiber Optic Biosensors for Bacterial Detection. In *Principles of Bacterial Detection: Biosensors, Recognition Receptors and Microsystems*; Zourob, M., Elwary, S., Turner, A., Eds.; Springer: New York, NY, USA, 2008; pp. 125–137.
108. Eltzov, E.; Marks, R.S. Miniaturized flow stacked immunoassay for detecting *Escherichia coli* in a single step. *Anal. Chem.* **2016**, *88*, 6441–6449. [CrossRef]
109. Verma, J.; Saxena, S.; Babu, S.G. ELISA-Based Identification and Detection of Microbes. In *Analyzing Microbes*; Springer: Berlin/Heidelberg, Germany, 2013; pp. 169–186.
110. Anton, G.; Wilson, R.; Yu, Z.H.; Prehn, C.; Zukunft, S.; Adamski, J.; Heier, M.; Meisinger, C.; Romisch-Margl, W.; Wang-Sattler, R.; et al. Pre-analytical sample quality: Metabolite ratios as an intrinsic marker for prolonged room temperature exposure of serum samples. *PLoS ONE* **2015**, *10*, e0121495. [CrossRef] [PubMed]
111. Henning, D.; Nielsen, K. Peroxidase-labelled monoclonal antibodies for use in enzyme immunoassay. *J. Immunoass.* **1987**, *8*, 297–307. [CrossRef] [PubMed]
112. Gosling, J.P. A decade of development in immunoassay methodology. *Clin. Chem.* **1990**, *36*, 1408–1427. [CrossRef] [PubMed]
113. Porstmann, T.; Kiessig, S.T. Enzyme immunoassay techniques. An overview. *J. Immunol. Methods* **1992**, *150*, 5–21. [CrossRef]
114. Khatkhatay, M.I.; Desai, M. A comparison of performances of four enzymes used in ELISA with special reference to beta-lactamase. *J. Immunoass.* **1999**, *20*, 151–183. [CrossRef]
115. Liu, C.; Skaldin, M.; Wu, C.; Lu, Y.; Zavialov, A.V. Application of ADA1 as a new marker enzyme in sandwich ELISA to study the effect of adenosine on activated monocytes. *Sci. Rep.* **2016**, *6*, 31370. [CrossRef]
116. Qu, Z.; Xu, H.; Xu, P.; Chen, K.; Mu, R.; Fu, J.; Gu, H. Ultrasensitive ELISA using enzyme-loaded nanospherical brushes as labels. *Anal. Chem.* **2014**, *86*, 9367–9371. [CrossRef]
117. Engvall, E. The ELISA, enzyme-linked immunosorbent assay. *Clin. Chem.* **2010**, *56*, 319–320. [CrossRef]
118. Elder, B.L.; Boraker, D.K.; Fives-Taylor, P.M. Whole-bacterial cell enzyme-linked immunosorbent assay for *Streptococcus sanguis* fimbrial antigens. *J. Clin. Microbiol.* **1982**, *16*, 141–144. [CrossRef]
119. Aydin, S. A short history, principles, and types of ELISA, and our laboratory experience with peptide/protein analyses using ELISA. *Peptides* **2015**, *72*, 4–15. [CrossRef]
120. Mirhosseini, S.A.; Fooladi, A.A.I.; Amani, J.; Sedighian, H. Production of recombinant flagellin to develop ELISA-based detection of *Salmonella Enteritidis*. *Braz. J. Microbiol.* **2017**, *48*, 774–781. [CrossRef] [PubMed]
121. Hans, R.; Yadav, P.K.; Sharma, P.K.; Boopathi, M.; Thavaselvam, D. Development and validation of immunoassay for whole cell detection of *Brucella abortus* and *Brucella melitensis*. *Sci. Rep.* **2020**, *10*, 8543. [CrossRef] [PubMed]

122. Zhu, L.; He, J.; Cao, X.; Huang, K.; Luo, Y.; Xu, W. Development of a double-antibody sandwich ELISA for rapid detection of *Bacillus Cereus* in food. *Sci. Rep.* **2016**, *6*, 16092. [CrossRef] [PubMed]
123. Sharifi, S.; Vahed, S.Z.; Ahmadian, E.; Dizaj, S.M.; Eftekhari, A.; Khalilov, R.; Ahmadi, M.; Hamidi-Asl, E.; Labib, M. Detection of pathogenic bacteria via nanomaterials-modified aptasensors. *Biosens. Bioelectron.* **2020**, *150*, 111933. [CrossRef]
124. Galikowska, E.; Kunikowska, D.; Tokarska-Pietrzak, E.; Dziadziuszko, H.; Los, J.M.; Golec, P.; Wegrzyn, G.; Los, M. Specific detection of Salmonella enterica and *Escherichia coli* strains by using ELISA with bacteriophages as recognition agents. *Eur. J. Clin. Microbiol. Infect. Dis.* **2011**, *30*, 1067–1073. [CrossRef]
125. Eltzov, E.; Guttel, S.; Kei, A.L.Y.; Sinawang, P.D.; Ionescu, R.E.; Marks, R.S. Lateral flow immunoassays—From paper strip to smartphone technology. *Electroanalysis* **2015**, *27*, 2116–2130. [CrossRef]
126. Zhao, Y.; Wang, H.; Zhang, P.; Sun, C.; Wang, X.; Wang, X.; Yang, R.; Wang, C.; Zhou, L. Rapid multiplex detection of 10 foodborne pathogens with an up-converting phosphor technology-based 10-channel lateral flow assay. *Sci. Rep.* **2016**, *6*, 21342. [CrossRef]
127. Kim, H.J.; Kwon, C.; Lee, B.S.; Noh, H. One-step sensing of foodborne pathogenic bacteria using a 3D paper-based device. *Analyst* **2019**, *144*, 2248–2255. [CrossRef]
128. Jung, B.Y.; Jung, S.C.; Kweon, C.H. Development of a rapid immunochromatographic strip for detection of *Escherichia coli* O157. *J. Food Prot.* **2005**, *68*, 2140–2143. [CrossRef]
129. Xu, D.; Wu, X.; Li, B.; Li, P.; Ming, X.; Chen, T.; Wei, H.; Xu, F. Rapid detection of *Campylobacter jejuni* using fluorescent microspheres as label for immunochromatographic strip test. *Food Sci. Biotechnol.* **2013**, *2*, 585–591. [CrossRef]
130. Niu, K.; Zheng, X.; Huang, C.; Xu, K.; Zhi, Y.; Shen, H.; Jia, N. A colloidal gold nanoparticle-based immunochromatographic test strip for rapid and convenient detection of *Staphylococcus aureus*. *J. Nanosci. Nanotechnol.* **2014**, *14*, 5151–5156. [CrossRef] [PubMed]
131. Ferapontova, E.E. Electrochemical assays for microbial analysis: How far they are from solving microbiota and microbiome challenges. *Curr. Opin. Electrochem.* **2020**, *19*, 153–161. [CrossRef]
132. Kuss, S.; Amin, H.M.A.; Compton, R.G. Electrochemical detection of pathogenic bacteria—Recent strategies, advances and challenges. *Chem. Asian J.* **2018**, *13*, 2758–2769. [CrossRef]
133. Campuzano, S.; de Ávila, B.E.-F.; Yuste, J.; Pedrero, M.; García, J.L.; García, P.; García, E.; Pingarrón, J.M. Disposable amperometric magnetoimmunosensors for the specific detection of *Streptococcus pneumoniae*. *Biosens. Bioelectron.* **2010**, *26*, 1225–1230. [CrossRef]
134. Liébana, S.; Lermo, A.; Campoy, S.; Cortés, M.P.; Alegret, S.; Pividori, M.I. Rapid detection of *Salmonella* in milk by electrochemical magneto-immunosensing. *Biosens. Bioelectron.* **2009**, *25*, 510–513. [CrossRef] [PubMed]
135. Ruan, C.; Yang, L.; Li, Y. Immunobiosensor chips for detection of *Escherichia coli* O157:H7 using electrochemical impedance spectroscopy. *Anal. Chem.* **2002**, *74*, 4814–4820. [CrossRef] [PubMed]
136. Zhang, Y.; Zhou, D. Magnetic particle-based ultrasensitive biosensors for diagnostics. *Expert Rev. Mol. Diagn.* **2012**, *12*, 565–571. [CrossRef]
137. Nemr, C.R.; Smith, S.J.; Liu, W.; Mepham, A.H.; Mohamadi, R.M.; Labib, M.; Kelley, S.O. Nanoparticle-mediated capture and electrochemical detection of methicillin-resistant *Staphylococcus aureus*. *Anal. Chem.* **2019**, *91*, 2847–2853. [CrossRef]
138. Ferapontova, E.E.; Hansen, M.N.; Saunders, A.M.; Shipovskov, S.; Sutherland, D.S.; Gothelf, K.V. Electrochemical DNA sandwich assay with a lipase label for attomole detection of DNA. *Chem. Commun.* **2010**, *46*, 1836–1838. [CrossRef]
139. Fapyane, D.; Ferapontova, E.E. Electrochemical assay for a total cellulase activity with improved sensitivity. *Anal. Chem.* **2017**, *89*, 3959–3965. [CrossRef]
140. Fapyane, D.; Nielsen, J.S.; Ferapontova, E.E. Electrochemical enzyme-linked sandwich assay with a cellulase label for ultrasensitive analysis of synthetic DNA and cell-isolated RNA. *ACS Sens.* **2018**, *3*, 2104–2111. [CrossRef] [PubMed]
141. Malecka, K.; Pankratov, D.; Ferapontova, E.E. Femtomolar electroanalysis of a breast cancer biomarker HER-2/neu protein in human serum by the cellulase-linked sandwich assay on magnetic beads. *Anal. Chim. Acta* **2019**, *1077*, 140–149. [CrossRef] [PubMed]

142. Zhong, M.; Yang, L.; Yang, H.; Cheng, C.; Deng, W.; Tan, Y.; Xie, Q.; Yao, S. An electrochemical immunobiosensor for ultrasensitive detection of *Escherichia coli* O157:H7 using CdS quantum dots-encapsulated metal-organic frameworks as signal-amplifying tags. *Biosens. Bioelectron.* **2019**, *126*, 493–500. [CrossRef] [PubMed]

143. Kuss, S.; Couto, R.A.S.; Evans, R.M.; Lavender, H.; Tang, C.C.; Compton, R.G. Versatile electrochemical sensing platform for bacteria. *Anal. Chem.* **2019**, *91*, 4317–4322. [CrossRef]

144. Zhou, Y.; Marar, A.; Kner, P.; Ramasamy, R.P. Charge-directed immobilization of bacteriophage on nanostructured electrode for whole-cell electrochemical biosensors. *Anal. Chem.* **2017**, *89*, 5734–5741. [CrossRef]

145. Jafari, H.; Amiri, M.; Abdi, E.; Navid, S.L.; Bouckaert, J.; Jijie, R.; Boukherroub, R.; Szunerits, S. Entrapment of uropathogenic *E. coli* cells into ultra-thin sol-gel matrices on gold thin films: A low cost alternative for impedimetric bacteria sensing. *Biosens. Bioelectron.* **2019**, *124–125*, 161–166. [CrossRef]

146. Ranjbar, S.; Nejad, M.A.F.; Parolo, C.; Shahrokhian, S.; Merkoçi, A. Smart chip for visual detection of bacteria using the electrochromic properties of polyaniline. *Anal. Chem.* **2019**, *91*, 14960–14966. [CrossRef]

147. Cheng, M.S.; Lau, S.H.; Chow, V.T.; Toh, C.-S. Membrane-based electrochemical nanobiosensor for Escherichia coli detection and analysis of cells viability. *Environ. Sci. Technol.* **2011**, *45*, 6453–6459. [CrossRef]

148. Ahmed, A.; Rushworth, J.V.; Wright, J.D.; Millner, P.A. Novel impedimetric immunosensor for detection of pathogenic bacteria *Streptococcus pyogenes* in human saliva. *Anal. Chem.* **2013**, *85*, 12118–12125. [CrossRef]

149. Chiriacò, M.S.; Parlangeli, I.; Sirsi, F.; Poltronieri, P.; Primiceri, E. Impedance sensing platform for detection of the food pathogen *Listeria monocytogenes*. *Electronics* **2018**, *7*, 347. [CrossRef]

150. Labib, M.; Zamay, A.S.; Kolovskaya, O.S.; Reshetneva, I.T.; Zamay, G.S.; Kibbee, R.J.; Sattar, S.A.; Zamay, T.N.; Berezovski, M.V. Aptamer-based impedimetric sensor for bacterial typing. *Anal. Chem.* **2012**, *84*, 8114–8117. [CrossRef] [PubMed]

151. Li, Z.; Fu, Y.; Fang, W.; Li, Y. Electrochemical impedance immunosensor based on self-assembled monolayers for rapid detection of Escherichia coli O157:H7 with signal amplification using lectin. *Sensors* **2015**, *15*, 19212–19224. [CrossRef] [PubMed]

152. Jasim, I.; Shen, Z.; Mlaji, Z.; Yuksek, N.S.; Abdullah, A.; Liu, J.; Dastider, S.G.; El-Dweik, M.; Zhang, S.; Almasri, M. An impedance biosensor for simultaneous detection of low concentration of *Salmonella* serogroups in poultry and fresh produce samples. *Biosens. Bioelectron.* **2019**, *126*, 292–300. [CrossRef] [PubMed]

153. Brosel-Oliu, S.; Ferreira, R.; Uria, N.; Abramova, N.; Gargallo, R.; Muñoz-Pascual, F.-X.; Bratov, A. Novel impedimetric aptasensor for label-free detection of *Escherichia coli* O157:H7. *Sens. Actuators B Chem.* **2018**, *255*, 2988–2995. [CrossRef]

154. Salimian, R.; Kékedy-Nagy, L.; Ferapontova, E.E. Specific picomolar detection of a breast cancer biomarker HER-2/neu protein in serum: Electrocatalytically amplified electroanalysis by the aptamer/PEG-modified electrode. *Chem. Electro. Chem.* **2017**, *4*, 872–879. [CrossRef]

155. Jijie, R.; Kahlouche, K.; Barras, A.; Yamakawa, N.; Bouckaert, J.; Gharbi, T.; Szunerits, S.; Boukherroub, R. Reduced graphene oxide/polyethylenimine based immunosensor for the selective and sensitive electrochemical detection of uropathogenic *Escherichia coli*. *Sens. Actuators B Chem.* **2018**, *260*, 255–263. [CrossRef]

156. Qiao, Z.; Fu, Y.; Lei, C.; Li, Y. Advances in antimicrobial peptides-based biosensing methods for detection of foodborne pathogens: A review. *Food Control* **2020**, *112*, 107116. [CrossRef]

157. Mannoor, M.S.; Zhang, S.; Link, A.J.; McAlpine, M.C. Electrical detection of pathogenic bacteria via immobilized antimicrobial peptides. *Proc. Natl. Acad. Sci. USA* **2010**, *107*, 19207–19212. [CrossRef]

158. Janczuk-Richter, M.; Marinović, I.; Niedziółka-Jönsson, J.; Szot-Karpińska, K. Recent applications of bacteriophage-based electrodes: A mini-review. *Electrochem. Commun.* **2019**, *99*, 11–15. [CrossRef]

159. Moghtader, F.; Congur, G.; Zareie, H.M.; Erdem, A.; Piskin, E. Impedimetric detection of pathogenic bacteria with bacteriophages using gold nanorod deposited graphite electrodes. *RSC Adv.* **2016**, *6*, 97832–97839. [CrossRef]

160. Richter, Ł.; Bielec, K.; Leśniewski, A.; Łoś, M.; Paczesny, J.; Hołyst, R. Dense layer of bacteriophages ordered in alternating electric field and immobilized by surface chemical modification as sensing element for bacteria detection. *ACS Appl. Mater. Interfaces* **2017**, *9*, 19622–19629. [CrossRef]

161. Niyomdecha, S.; Limbut, W.; Numnuam, A.; Kanatharana, P.; Charlermroj, R.; Karoonuthaisiri, N.; Thavarungkul, P. Phage-based capacitive biosensor for Salmonella detection. *Talanta* **2018**, *188*, 658–664. [CrossRef]

162. Golabi, M.; Kuralay, F.; Jager, E.W.H.; Beni, V.; Turner, A.P.F. Electrochemical bacterial detection using poly(3-aminophenylboronic acid)-based imprinted polymer. *Biosens. Bioelectron.* **2017**, *93*, 87–93. [CrossRef] [PubMed]
163. MagaÑA, S.; Schlemmer, S.M.; Leskinen, S.D.; Kearns, E.A.; Lim, D.V. Automated Dead-End Ultrafiltration for Concentration and Recovery of Total Coliform Bacteria and Laboratory-Spiked *Escherichia coli* O157:H7 from 50-Liter Produce Washes To Enhance Detection by an Electrochemiluminescence Immunoassay. *J. Food Prot.* **2013**, *76*, 1152–1160. [CrossRef] [PubMed]
164. Spehar-Délèze, A.-M.; Julich, S.; Gransee, R.; Tomaso, H.; Dulay, S.B.; O'Sullivan, C.K. Electrochemiluminescence (ECL) immunosensor for detection of *Francisella tularensis* on screen-printed gold electrode array. *Anal. Bioanal. Chem.* **2016**, *408*, 7147–7153. [CrossRef]
165. Guo, Z.; Sha, Y.; Hu, Y.; Yu, Z.; Tao, Y.; Wu, Y.; Zeng, M.; Wang, S.; Li, X.; Zhou, J.; et al. Faraday cage-type electrochemiluminescence immunosensor for ultrasensitive detection of *Vibrio vulnificus* based on multi-functionalized graphene oxide. *Anal. Bioanal. Chem.* **2016**, *408*, 7203–7211. [CrossRef] [PubMed]
166. Chen, S.; Chen, X.; Zhang, L.; Gao, J.; Ma, Q. Electrochemiluminescence detection of *Escherichia coli* O157:H7 Based on a novel polydopamine surface imprinted polymer biosensor. *ACS Appl. Mater. Interfaces* **2017**, *9*, 5430–5436. [CrossRef] [PubMed]
167. Hao, N.; Zhang, X.; Zhou, Z.; Hua, R.; Zhang, Y.; Liu, Q.; Qian, J.; Li, H.; Wang, K. AgBr nanoparticles/3D nitrogen-doped graphene hydrogel for fabricating all-solid-state luminol-electrochemiluminescence *Escherichia coli* aptasensors. *Biosens. Bioelectron.* **2017**, *97*, 377–383. [CrossRef]
168. Yue, H.; He, Y.; Fan, E.; Wang, L.; Lu, S.; Fu, Z. Label-free electrochemiluminescent biosensor for rapid and sensitive detection of *Pseudomonas aeruginosa* using phage as highly specific recognition agent. *Biosens. Bioelectron.* **2017**, *94*, 429–432. [CrossRef]
169. Kulasingam, V.; Jung, B.P.; Blasutig, I.M.; Baradaran, S.; Chan, M.K.; Aytekin, M.; Colantonio, D.A.; Adeli, K. Pediatric reference intervals for 28 chemistries and immunoassays on the Roche cobas®6000 analyzer—A CALIPER pilot study. *Clin. Biochem.* **2010**, *43*, 1045–1050. [CrossRef]
170. Reta, N.; Michelmore, A.; Saint, C.P.; Prieto-Simon, B.; Voelcker, N.H. Label-free bacterial toxin detection in water supplies using porous silicon nanochannel sensors. *ACS Sens.* **2019**, *4*, 1515–1523. [CrossRef] [PubMed]
171. Lin, L.; Zheng, Q.; Lin, J.; Yuk, H.-G.; Guo, L. Immuno- and nucleic acid-based current technologies for *Salmonella* detection in food. *Eur. Food Res. Technol.* **2020**, *246*, 373–395. [CrossRef]
172. Ohlsson, P.; Evander, M.; Petersson, K.; Mellhammar, L.; Lehmusvuori, A.; Karhunen, U.; Soikkeli, M.; Seppä, T.; Tuunainen, E.; Spangar, A.; et al. Integrated Acoustic Separation, Enrichment, and Microchip Polymerase Chain Reaction Detection of Bacteria from Blood for Rapid Sepsis Diagnostics. *Anal. Chem.* **2016**, *88*, 9403–9411. [CrossRef] [PubMed]
173. Mettakoonpitak, J.; Boehle, K.; Nantaphol, S.; Teengam, P.; Adkins, J.A.; Srisa-Art, M.; Henry, C.S. Electrochemistry on Paper-based Analytical Devices: A Review. *Electroanalysis* **2016**, *28*, 1420–1436. [CrossRef]
174. Shen, L.-L.; Zhang, G.-R.; Etzold, B.J.M. Paper-Based Microfluidics for Electrochemical Applications. *ChemElectroChem* **2020**, *7*, 10–30. [CrossRef] [PubMed]
175. Available online: https://www.pointofcare.abbott/int/en/offerings/istat/istat-test-cartridges (accessed on 25 September 2020).

© 2020 by the authors. Licensee MDPI, Basel, Switzerland. This article is an open access article distributed under the terms and conditions of the Creative Commons Attribution (CC BY) license (http://creativecommons.org/licenses/by/4.0/).

Review

Capacitive Field-Effect EIS Chemical Sensors and Biosensors: A Status Report

Arshak Poghossian [1,*] and Michael J. Schöning [2,*]

1 MicroNanoBio, Liebigstr. 4, 40479 Düsseldorf, Germany
2 Institute of Nano- and Biotechnologies (INB), FH Aachen, Campus Jülich, Heinrich-Mußmannstr. 1, 52428 Jülich, Germany
* Correspondence: a.poghossian@gmx.de (A.P.); schoening@fh-aachen.de (M.J.S.)

Received: 31 August 2020; Accepted: 29 September 2020; Published: 2 October 2020

Abstract: Electrolyte-insulator-semiconductor (EIS) field-effect sensors belong to a new generation of electronic chips for biochemical sensing, enabling a direct electronic readout. The review gives an overview on recent advances and current trends in the research and development of chemical sensors and biosensors based on the capacitive field-effect EIS structure—the simplest field-effect device, which represents a biochemically sensitive capacitor. Fundamental concepts, physicochemical phenomena underlying the transduction mechanism and application of capacitive EIS sensors for the detection of pH, ion concentrations, and enzymatic reactions, as well as the label-free detection of charged molecules (nucleic acids, proteins, and polyelectrolytes) and nanoparticles, are presented and discussed.

Keywords: chemical sensor; biosensor; field effect; capacitive EIS sensor; pH sensor; enzyme biosensor; label-free detection; charged molecules; DNA biosensor; protein detection

---

1. Introduction

Research in the field of biochemical sensors is one of the most fascinating and multidisciplinary topics and has enormously increased over recent years. The global market for biosensor devices grows rapidly and is expected to reach $20 billion by the year 2020 [1]. Due to the small size and weight, fast response time, label-free operation, possibility of real-time and multiplexed measurements, and compatibility with micro- and nanofabrication technologies with the future prospect of a large-scale production at relatively low cost, semiconductor field-effect devices (FEDs) based on an electrolyte-insulator-semiconductor (EIS) system are one of the most exciting approaches for chemical and biological sensing. Ion-sensitive field-effect transistors (ISFET) [2–5], extended-gate ISFETs [6], capacitive EIS sensors [7–9], light-addressable potentiometric sensors [10–13], silicon nanowire FETs (SiNW-FET) [14–17], graphene-based FETs [18,19], and carbon nanotube-based FETs [18,20] constitute typical examples of transducer structures for chemically/biologically sensitive FEDs. At present, numerous FEDs modified with respective recognition elements have been developed for the detection of pH, ion concentrations, substrate–enzyme reactions, nucleic acid hybridizations, and antigen–antibody affinity reactions, just to name a few. Moreover, the possibility of an on-chip integration of FED arrays with microfluidics make them very attractive for the creation of miniaturized analytical systems, such as lab-on-a-chip or electronic tongue devices. The possible fields of the application of FED-based chemical sensors and biosensors reach from point-of-care medicine, biotechnology, and environmental monitoring over food and drug safety up to defense and homeland security purposes.

The simplest FED is the capacitive EIS structure, which represents a biochemically sensitive capacitor. In contrast to conventional ISFETs or silicon nanowire FETs, capacitive EIS sensors are simple in layout, easy, and cost-effective in fabrication (typically, without photolithographic or encapsulation process steps). The present status report overviews recent advances and current trends in the research and development of chemical sensors and biosensors based on capacitive field-effect EIS structures. The fundamental concepts, functioning principle, and application of EIS sensors for the detection of pH, ion concentrations, and substrate–enzyme reactions, as well as the label-free detection of charged molecules and nanoparticles, are presented and discussed. The paper also encompasses some key developments of former works. For biochemical sensors based on other types of FEDs, the interested reader is referred to reviews [19,21–25].

2. Functioning Principle and Measurement Modes of Capacitive EIS Sensors

Figure 1a schematically shows a typical layer structure of a capacitive EIS sensor and a simplified electrical equivalent circuit. The EIS sensor consists of a semiconductor substrate (in this case, p-type silicon) separated from the solution by a thin (10–100 nm) gate insulator layer (or stack of layers) and a rear-side contact layer (e.g., Al). The gate insulator is assumed to be ideal—that is, no current passes through the insulator. For the operation of EIS sensors, a gate voltage (V_G) is applied between the reference electrode (RE, e.g., conventional Ag/AgCl liquid-junction electrode) and the rear-side contact to regulate the capacitance and set the working point; a small alternating voltage (~10–50 mV) is superimposed to measure the capacitance of the structure. For a proper measurement, the reference electrode should provide a stable potential independent of the pH value of the solution or concentration of the dissolved species.

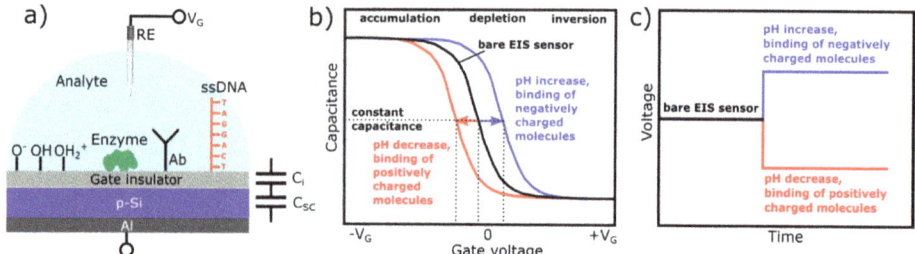

Figure 1. Layer structure of a capacitive electrolyte-insulator-semiconductor (EIS) sensor with different receptor functionalities (pH-/ion-sensing, enzyme, antibody, and DNA) and simplified electrical equivalent circuit (**a**); typical shape of high-frequency capacitance-voltage (C–V) curves (**b**); and ConCap response (**c**) of the bare and modified p-type EIS sensor. RE: reference electrode, V_G: gate voltage, Ab: antibody, DNA: deoxyribonucleic acid, C_i: gate-insulator capacitance, C_{SC}: space-charge capacitance, ssDNA: single-stranded DNA.

The electrical equivalent circuit of the EIS sensor is complex and involves components related to the semiconductor, gate insulator, electrolyte/insulator interface, bulk electrolyte, and reference electrode [26,27]. However, for the usual range of a gate insulator thickness and appropriate experimental conditions used (electrolyte solution with ionic strength of >0.1 mM and measurement frequencies of <1 kHz), the equivalent circuit of an EIS sensor can be simplified as a series connection of the gate-insulator capacitance, C_i, and the variable semiconductor space-charge capacitance, C_{sc} (V_G,φ), which is, among others, a function of the gate voltage, V_G, and the electrolyte–insulator interfacial potential, φ [27]. Hence, the expression for the total capacitance, C, of the bare EIS sensor is given in Equation (1):

$$C = \frac{C_i C_{sc}(V_G,\varphi)}{C_i + C_{sc}(V_G,\varphi)} = \frac{C_i}{1 + C_i/C_{sc}(V_G,\varphi)} \quad (1)$$

EIS sensors are basically characterized by means of the capacitance-voltage (C–V) and/or constant-capacitance (ConCap) mode [28,29]. The typical shape of a high-frequency C–V curve for a p-type EIS sensor with characteristic regions of accumulation, depletion, and inversion is exemplarily shown in Figure 1b (black curve); note, an n-type EIS sensor exhibits an identical C-V curve; however, the voltage polarity is reversed. If a negative potential ($V_G < 0$) is applied to the gate, the positively charged holes (majority carriers) will be attracted and accumulated at the semiconductor/insulator interface. In accumulation regime, $C_i << C_{sc}(V_G,\varphi)$, i.e., the overall capacitance of the EIS structure is determined by the geometrical capacitance of the gate insulator, $C = C_i$, and, thus, corresponds to its maximum capacitance.

When applying a small positive potential ($V_G > 0$) to the gate, the holes will be pushed away from the interface semiconductor/insulator. As a result, a space-charge region is formed at the semiconductor/insulator interface, which is depleted of mobile carriers (so-called depletion region). The width of the depletion layer is determined by different parameters, such as the applied voltage, doping concentration within the semiconductor, dielectric constant, and insulator thickness. Increasing the amplitude of the applied gate voltage results in an increase of the width of the depletion layer and, consequently, to a decrease of the total capacitance. If the magnitude of the positive gate potential is sufficiently high, the Fermi level bends below the intrinsic level: the concentration of electrons near the semiconductor/insulator interface exceeds the hole concentration, i.e., a thin layer of n-type silicon (so-called inversion layer) is formed, although the substrate is a p-type. By strong inversion, the width of the depletion layer reaches its maximum, and the high-frequency total capacitance of the EIS structure approaches its minimum value.

Equation (1) describes the total capacitance of the EIS sensor without defining the origin of the potential generation at the interface electrolyte/insulator. When fixing the applied gate voltage, V_G, the only variable component is the interfacial potential, φ, which is analogous to the effect of applying an additional voltage to the gate. Since FEDs are potential-/charge-sensitive devices, any kind of chemical and/or electrical change at or nearby the interface electrolyte/gate can be detected by the EIS sensor. Those changes can be induced by biochemical reactions, when the capacitive field-effect sensor is functionalized with a particular chemical and/or biological recognition element, such as a pH-sensitive layer, ionophore, enzyme, antibody, nucleic acid, etc. For (bio-)chemical sensor applications and for investigating charge effects in such capacitive EIS structures, the shift of the C–V curves along the voltage axis (ΔV_G) in the depletion region (Figure 1b) is more important. The direction of these potential shifts depends on the charge sign of the adsorbed chemical and/or biological species. For example, in case of a p-type EIS structure, an increase of the analyte's pH value or binding of the negatively charged species to the gate surface will decrease the width of the depletion layer, yielding an increase of the depletion capacitance in the Si. By this, the total capacitance of the sensor will increase, and the C–V curve will shift to the direction of more positive (or less negative) gate voltages (Figure 1b, blue curve). Conversely, a pH decrease or the electrostatic adsorption or binding of positively charged species to the gate surface will lead to an increase of the width of the depletion layer; the space-charge capacitance will decrease. As a consequence, the total capacitance of the EIS sensor will also decrease, resulting in a shift of the C–V curve towards more negative (or less positive) gate voltages (Figure 1b, red curve).

The amplitude, as well as the direction of potential shifts, can directly be determined from dynamic ConCap-mode measurements (see Figure 1c). In addition, the ConCap mode enables real-time monitoring of the sensor signal and investigation of the response time, drift, and hysteresis of the EIS sensor. In the ConCap mode, the total capacitance of the EIS sensor at the working point is kept constant by using a feedback circuit, which applies an instantly sign-inverted voltage to the EIS sensor. Usually, the working point for the ConCap mode is set within the linear range of the depletion region of the C–V curve.

In the following sections, an origin of different mechanisms of interfacial potential generation (e.g., pH and ion-concentration changes, enzymatic reactions, adsorption, and binding of charged molecules and nanoparticles) is described, which enables EIS devices to be sensitive to numerous chemical and biological species, as well as to discuss the physicochemical phenomena underlying the transduction mechanism of EIS-based chemical sensors and biosensors.

3. Chemical Sensors and Biosensors Based on Capacitive EIS Structures

3.1. EIS pH Sensor

Field-effect pH sensors based on an EIS system detect potential (charge) changes at the electrolyte/gate-insulator interface, resulting from the changes in the local or bulk pH. It is known that the gate-insulator material in the first ISFET was SiO_2, which is not the best pH-sensitive material, having a low sensitivity, a narrow linear pH range, a relatively high drift, and a large hysteresis (see, e.g., [9,30,31]). Therefore, other oxides, like Al_2O_3 [32–34], Ta_2O_5 [29,35,36], ZrO_2 [37], HfO_2 [38–41], CeO_2 [42], Gd_2O_3 [43,44], Ti-doped Gd_2O_3 [45], Lu_2O_3 [46], Nd_2O_3 [47], Yb_2O_3 [48], Dy_2TiO_5 [49], Er_2TiO_5 [50], $PbTiO_3$ [51], YTi_xO_y [52], $Tm_2Ti_2O_7$ [53], and barium strontium titanate (BST) [54–56], as well as Si_3N_4 [32,57] and nanocrystalline diamond (NCD) [58], have been proven as pH-sensitive gate insulators for EIS sensors. Some of the recent results, including the pH-sensitive material used, deposition technique, pH sensitivity, pH range, drift, and hysteresis, are summarized in Table 1.

In most cases, these pH-sensitive materials are deposited on top of a SiO_2 layer by means of different deposition techniques (e.g., thermal oxidation, chemical vapor deposition, electron-beam evaporation, sputtering, pulsed laser deposition, atomic layer deposition, and sol-gel technique), thus forming a stacked gate insulator (e.g., SiO_2–Si_3N_4, SiO_2–Al_2O_3, or SiO_2–Ta_2O_5). The upper layer of the double-insulator structure typically serves as the pH-sensitive material, whereas the SiO_2 layer provides a stable Si-SiO_2 interface with a low density of states.

The main parameters, which determine the analytical characteristics of the EIS pH sensors, are sensitivity, selectivity, stability (or drift), linear pH range, hysteresis, and response time. Other important parameters include the temperature stability, light insensitivity, reproducibility, and lifetime. These characteristics are most thoroughly studied for Si_3N_4, Al_2O_3, and Ta_2O_5 layers, which belong to the best pH-sensitive materials. At present, Si_3N_4, Al_2O_3, and Ta_2O_5 serve as pH-sensitive gate insulator materials in commercial pH-ISFETs available from many companies producing electrochemical sensors. Other more exotic pH-sensitive materials described in Table 1 sometimes show nearly Nernstian sensitivity but have been only rarely studied.

At present, the generally accepted and successfully applied model describing the functional mechanism of pH-sensitive FEDs with inorganic gate insulators (e.g., oxides and nitrides) is the so-called site-binding model, which was originally developed to describe the charging mechanism of oxide surfaces immersed in solution [59]. Due to the hydration, the surface of oxides contains neutral amphoteric hydroxyl groups in the particular cases of Ta_2O_5 and TaOH groups. Dependent on the pH value of the solution, these surface sites are either able to bind or release a proton (H^+), resulting in protonated ($TaOH_2^+$) or deprotonated (TaO^-) groups according to the following reactions in Equations (2) and (3):

$$TaOH \rightleftharpoons TaO^- + H^+ \qquad (2)$$

$$TaOH_2^+ \rightleftharpoons TaOH + H^+ \qquad (3)$$

At a pH range of pH > pH_{pzc}, the oxide surface of the capacitive field-effect sensor will be negatively charged, whereas at a pH regime with pH < pH_{pzc}, it is positively charged; the value pH_{pzc} is defined as the pH value at the point of zero charge. Consequently, the pH-dependent surface charge of the gate insulator will modulate the space-charge capacitance in the Si and, finally, the overall capacitance of the capacitive field-effect structure. Note that, in contrast to oxides, the pH sensitivity of Si_3N_4 can be explained by utilizing the modified site-binding theory. This theory considers the presence of two different types of surface sites: SiOH (silanol) and SiNH (amine) groups [60].

Table 1. Characteristics of electrolyte-insulator-semiconductor (EIS) pH sensors with different pH-sensitive materials and deposition techniques.

pH-Sensitive Material	Deposition Method	pH Sensitivity, mV/pH	pH Range	Drift, mV/h	Hysteresis, mV	Reference
SiO_2	LPCVD	41.5	2–10	19.6	19.4	[31]
SiO_2 structured	LPCVD	52	2–10	1	11	[31]
SiO_2 textured with SiO_2 particles	TO of Si	43–54	4–10	16–40	5–6	[9]
SiO_2	TO of Si	35–38	3–9	-	-	[30]
Si_3N_4	LPCVD	50	2–12	6	-	[57]
Si_3N_4	LPCVD	50	3–12	4	21	[32]
Al_2O_3	ALD	55	3–12	5.5	-	[33]
Al_2O_3	ALD	54.5	3–12	2	14	[32]
Al_2O_3	PLD	56	2–12	<1	3	[34]
Ta_2O_5	TO of Ta	57 ± 1.5	3–10	0.5	4	[29]
Ta_2O_5	TO of Ta	56	1–10	-	5	[35]
ZrO_2	TO of Zr	50.6	2–10	-	-	[37]
HfO_2	ALD	59.6	2–12	1	4.3	[41]
HfO_2	RFS	51	2–10	1	25	[40]
HfO_2	RFS	58.3	2–12	0.65	1.7	[39]
CeO_2	RFS	58.8	2–12	1	6	[42]
Gd_2O_3	TO of Gd	53	2–10	5.4	-	[43]
Gd_2O_3	RFS	55	2–10	1.2	-	[44]
Ti-doped Gd_2O_3	RFS	55	2–12	1.4	3.6	[45]
Lu_2O_3	RFS	56	2–12	1.3	2.2	[46]
Nd_2O_3	RFS	56	2–12	1.3	4.7	[47]
Yb_2O_3	RFS	55.5	2–12	1.5	3.8	[48]
BST	sputtering	48–56	2–10	-	-	[54]
BST	PLD	57.4	3–11	-	2	[56]
NCD	MPECVD	54–57	4–11	-	-	[58]
Dy_2TiO_5	co-sputtering Dy/Ti	57.6	2–12	0.4	0.2	[49]
Er_2TiO_5	co-sputtering Er/Ti	58.4	2–12	1.2	4.6	[50]
$PbTiO_3$	sol-gel	56–59	2–12			[51]
YTi_xO_y	sol-gel	58.5	2–12	0.1	2.6	[52]
$Tm_2Ti_2O_7$	co-sputtering Tm/Ti	59.4	2–12	2.4	0.6	[53]

BST: barium strontium titanate, NCD: nanocrystalline diamond, LPCVD: low-pressure chemical vapor deposition, RFS: radio frequency sputtering, TO: thermal oxidation, ALD: atomic layer deposition, PLD: Pulsed laser deposition, and MPECVD: microwave plasma-enhanced chemical vapor deposition.

Commonly, the pH sensitivity of capacitive field-effect sensors is determined as the potential change (φ) at the interface electrolyte/gate-insulator, referred to as the change in the bulk pH; see Equations (4) and (5) [61]:

$$\frac{\delta\varphi}{\delta pH} = -2.3\frac{kT}{q}\alpha, \text{ with} \tag{4}$$

$$\alpha = \frac{1}{\left(2.3\, kTC_{diff}/q^2\beta_{int}\right)+1} \tag{5}$$

In the equations, α represents a dimensionless sensitivity parameter that varies between 0 and 1, β_{int} is the surface intrinsic buffer capacity, which characterizes the ability of the oxide surface to release or bind protons, C_{dif} is the differential double-layer capacitance (depending on the ion concentration of the solution), k is the Boltzmann's constant, T is the temperature, and q is the elementary charge (1.6×10^{-19} C).

From Equations (4) and (5), the maximum Nernstian sensitivity (59.3 mV/pH at 25 °C) can be obtained only if $\alpha \approx 1$, i.e., in the case of a large value of the surface-buffer capacity (high density of surface-active sites) and a low value of the double-layer capacitance (low electrolyte concentration). For materials with $\alpha < 1$, a sub-Nernstian response can be expected. Thus, oxides with a high density of the surface sites (e.g., Ta_2O_5 with ~10^{15} sites/cm^2 or Al_2O_3 with 8×10^{14} sites/cm^2 [61]) possess a high pH sensitivity, whereas for SiO_2 with less surface sites (5×10^{14} sites/cm^2 [61]), a low pH sensitivity could be expected that, in fact, was observed in experiments.

Determination of the pH value is one of the most important measurements in many fields, including clinical diagnostics, biotechnology, food, pharmaceutical and cosmetic industries, agriculture, environmental monitoring, water purification, etc. Currently, the most often used electrode is a traditional pH glass membrane electrode. However, glass electrodes have two main problems, namely the fragility of the glass membrane and easy fouling in aggressive media.

In comparison to pH glass electrodes, FEDs are often advertised due to their resistance to breakage (unbreakable pH sensor). Therefore, breakable pH glass electrodes are gradually replaced by nonglass, unbreakable pH-ISFETs in many in-line process-monitoring systems [62]. Besides being unbreakable, often, additionally, those sensors must be CIP-(cleaning-in-place) or SIP-(sterilization-in-place) feasible. According to [62], commercially available pH-ISFETs have a limited lifetime by CIP procedures, which use highly caustic media and high temperatures to clean process vessels. Therefore, Schöning et al. studied the CIP suitability of pH-sensitive Ta_2O_5 EIS sensors, where the Ta_2O_5 films were fabricated by thermal oxidation of a Ta layer [29]. For these sensors, even after running 30 CIP cycles, a nearly-Nernstian sensitivity of 57 ± 1.5 mV/pH was recorded. Note, each CIP cycle included cleaning procedures in a 4% NaOH solution at 80 °C during 15 min and, subsequently, in 0.65% HNO_3 solution at 80 °C during 5 min. No visible degradation of the Ta_2O_5 films of the EIS sensors was observed. These experiments demonstrated that, in addition to the high pH-sensitive behavior, Ta_2O_5 films have also high corrosion-resistant properties. Such sensors can be placed in direct contact with food for pH measurements without the risk of broken glass fragments. In further experiments, the authors also demonstrated the suitability of Ta_2O_5-gate EIS structures for SIP procedures [63]. These sensors were integrated into a lab-scale bioreactor and successfully tested for continuous pH monitoring during cell-culture fermentation processes [64]. In addition, Ta_2O_5-gate EIS pH sensors were applied for measuring the extracellular acidification rate of *Escherichia coli* upon glucose pulses [65]. Further examples are EIS sensors with BST films as pH-sensitive material for the pH control in a biogas digestate [56] or an estimation of the acid content in rancid butter samples with a Si_3N_4-gate EIS pH sensor [66]. Finally, EIS pH sensors with Ta_2O_5 films prepared by radio frequency magnetron sputtering were utilized for the real-time quantitative detection of DNA (deoxyribonucleic acid) amplification via loop-mediated isothermal amplification [36]. Here, during the elongation reaction, protons are produced (proportional to the number of nucleotides incorporated), thus resulting in a pH shift of the surrounding solution that is detected by the EIS pH sensor.

3.2. Ion-Sensitive EIS Sensors

EIS sensors selective towards ions other than protons can be obtained by modification of the original gate insulator—for instance, with additional organic or inorganic ion-sensitive membranes, suitable recognition molecules, or ion implantation. The easiest and most convenient method is deposition (e.g., via spin-, dip-, or drop-coating techniques) of an ion-selective polymeric membrane containing a respective ionophore atop the gate-insulator surface. Since the ion-sensitive membrane is permeable for ions, generally, the ion-exchange process is responsible for the potential generation at the electrolyte/membrane interface (given by the well-known Nernst or modified Nernst-Nikolsky equation), which is detected by the EIS sensor. Principally, any membrane composition implemented in conventional ion-selective electrodes can also be used in EIS sensors. Poly(vinyl chloride) is one of the most commonly applied matrices for ionophore-containing membranes. For example, a K^+-sensitive EIS sensor consisting of a p-Si-SiO_2-Si_3N_4 structure covered with a valinomycin (ionophore)-containing poly(vinyl chloride) membrane was realized in [67]. Before membrane deposition, the Si_3N_4 surface was silanized in order to suppress the intrinsic pH sensitivity of the Si_3N_4 layer and to reduce the influence of the pH of the solution on the K^+-sensitive response, as well as to improve the membrane adhesion. Sensors exhibited a K^+-sensitivity of 53 ± 2 mV/pK in a linear concentration range of 10^{-5}–10^{-1} M, a lower detection limit of 5×10^{-6} M, a small hysteresis of ~2 mV, and a fast response time of 5–12 s. The developed sensors were capable for K^+-ion concentration measurements for at least six months. In a further work, the authors studied the impact of the membrane impedance on the characteristics of the K^+-sensitive EIS sensor [68]. It was shown that a high series resistance of the ion-sensitive membrane can lead to a frequency-dependent distortion of the C–V curves, especially by measurements at high frequencies. In [69], an EIS calcium sensor was developed for the determination of the risk of urinary stone formation. In this application, an Al–p-Si–SiO_2–Ta_2O_5 structure was utilized that was modified with an ion-selective poly(vinyl chloride) membrane with the Ca^{2+} ionophore ETH 1001. The EIS calcium sensor had a Ca^{2+} sensitivity of 27 ± 2 mV/pCa in the concentration range of 0.1–10 mM and a lower detection limit of 0.01-mM Ca^{2+}; its response time was about ~30 s. Measurements in real test samples have been done for the determination of the Ca^{2+} concentration in native urine. Another membrane (siloprene) containing trioctylphosphine oxide as the ionophore deposited on a Si–SiO_2–Si_3N_4 structure was utilized to realize an EIS sensor sensitive towards hexavalent chromium (CrVI) [70]. A good selectivity, a quasi-Nernstian sensitivity of 27.6 mV/pCr in the concentration range of 10^{-4}–10^{-1} M Cr(VI), and a detection limit of 10^{-5} M Cr(VI) was reported. More recently, a perchlorate anion (ClO_4^-)-sensitive EIS sensor was developed by modification of the HfO_2 gate surface with a Co(II) phthalocyanine acrylate polymer (Co(II)Pc-AP) [71].

The main problems of ion-sensitive EIS sensors based on polymeric membranes are the poor adhesion of the membrane to the gate surface and possible ionophore leakage, thus limiting the sensor lifetime. These problems can be overcome by using ion-sensitive inorganic materials, which can be deposited onto the gate surface by thin-film deposition techniques, being compatible with semiconductor technology and, therefore, with the EIS fabrication. For example, a fluoride (F^-)-selective EIS sensor was developed by the thermal deposition of polycrystalline lanthanum fluoride (LaF_3) films on the SiO_2 gate surface [72]. The sensor exhibited a high sensitivity of 52.3 mV/pF in the F^--ion concentration range of 10^{-2}–10^{-6} M, a relatively low hysteresis (5.1 mV), and a small drift (0.67 mV/h).

Due to the low solubility in water and the possibility of deposition and patterning of thin-film membranes with various compositions by microelectronic techniques, chalcogenide-glass films are very promising for the development of EIS sensors sensitive towards various heavy metal ions. For example, EIS sensors with thin-film chalcogenide-glass membranes of $CdSAgIAs_2S_3$ and $PbSAgIAs_2S_3$ were developed for the detection of Pb^{2+}- and Cd^{2+} ions, respectively [73], which belong to the most toxic species of superficial and ground waters. The chalcogenide-glass films were prepared onto the Ta_2O_5 gate surface by means of a pulsed laser deposition technique that enables the stoichiometric transfer of these multicomponent materials from the original target to the sensor surface. The sensitivity to Pb^{2+}- and Cd^{2+} ions was 24 mV/pPb and 23 mV/pCd, respectively, in a linear concentration range of about 5×10^{-6}–10^{-2} M. A detection limit of ~3×10^{-6} M and response time of 1 min was reported for both sensors.

A completely other approach to make the gate surface ion-sensitive is based on its modification via ion implantation—a technique also compatible with silicon technology. Moreover, multiple membranes sensitive to various ions can be fabricated on the wafer level using multiple implants and different photolithographic masks. The ion-implantation technique was applied to develop Na^+- and K^+-sensitive EIS sensors by implanting Na^+ or K^+ ions into the oxidized silicon nitride through an Al buffer layer [74,75]. Although, these sensors demonstrate good sensitivity (52 mV/pNa and 49 mV/pK [75]), the response to interfering ions and pH was non-negligible, evidencing a poor selectivity of implanted ion-sensitive layers. Finally, several types of EIS sensors sensitive towards Ni^{2+} [76], Cu^{2+} [77], and Hg^{2+} [78] were realized via the deposition of ion-recognition molecules (macrocyclic compounds such as calixarenes [79]) directly onto the gate surface.

3.3. Enzyme-Modified EIS Biosensors

Due to their catalytic activity, enzymes are frequently used as bioreceptors. Enzyme-modified capacitive EIS (EnEIS) biosensors are typically constructed via the immobilization of enzymes onto the gate surface. In general, the operation principle of an EnEIS biosensor can be explained as follows: During the enzymatic reaction of the enzyme with its substrate, either reactants are consumed or products are generated; this concentration change is detected by the EIS sensor. A vast majority of EnEIS biosensors are built up of pH-sensitive EIS structures, which detect hydrogen ions produced or consumed during the enzymatic reaction. Equations (6) and (7) present typical examples of enzymatic reactions generating hydrogen ions (pH decrease) and consuming hydrogen ions (pH increase) by using penicillin/penicillinase and urea/urease as the model substrate/enzyme system:

$$\text{penicillin} + H_2O \xrightarrow{\text{penicillinase}} \text{penicilloic acid} + H^+ \qquad (6)$$

$$NH_2\text{-}CO\text{-}NH_2 \text{ (urea)} + 2H_2O + H^+ \xrightarrow{\text{urease}} 2NH_4^+ + HCO_3^- \qquad (7)$$

For example, in the case of a penicillin-sensitive EnEIS biosensor, the enzyme penicillinase catalyzes the hydrolysis of penicillin to penicilloic acid, yielding an increase of the H^+-ion concentration near the gate region of the EIS sensor. A resulting local pH change near the surface of the pH-sensitive layer will alter the gate-surface charge (similar to the pH-sensitive EIS sensors described in Section 3.1), which, in turn, will modulate the space-charge capacitance in the Si and, consequently, the total capacitance of the EnEIS biosensor. Hence, the amplitude of the output signal of the biosensor will be determined by the concentration of penicillin in the sample solution. Owing to the described working principle, a large sensor signal and high analyte sensitivity can be expected for EnEIS structures with gate-insulator materials exhibiting a high pH sensitivity, as well as by measurements in low buffer capacity solutions.

The choice of the appropriate enzyme immobilization strategy is essential for the development of enzyme biosensors with good performance (high sensitivity, selectivity, operational and storage stability, and fast response time) and represents one of the most critical points for the construction of EnEIS biosensors. Therefore, a broad spectrum of enzyme immobilization methods, such as physical adsorption, the layer-by-layer (LbL) technique, covalent binding, crosslinking, affinity coupling, entrapment within a polymeric membrane or hydrogel beads, etc., has been utilized. Each of these immobilization strategies has its own pros and cons. Common drawbacks are often a low or unreproducible enzyme-surface density, as well as the necessity of complicated surface functionalization or modification procedures.

At present, a large group of EnEIS biosensors by applying various gate materials, enzyme membrane compositions, or immobilization methods have been developed for the detection of different analytes such as acetoin [80], creatinine [81–83], cyanide [84], formaldehyde [85], glucose [53,81,82,86–95], pesticides of paraoxon [96] and atrazine [97], antibiotics of ampicillin, amoxicillin [98], and penicillin [58,98–103], triglycerides [66], and urea [82,89–92,95,104–111]. Some recent developments are summarized in Table 2.

Table 2. Typical characteristics of enzyme-modified capacitive EIS (EnEIS) biosensors.

Analyte/Enzyme	pH Layer	Immobilization	Sensitivity	Detection Range, mM	LDL, µM	Ref.
acetoin/AR	Ta_2O_5	crosslinking	65 mV/dec	0.01–0.1	-	[80]
creatinine/creatinine deaminase	Dy_2TiO_5	covalent on magnetic bead	22–29 mV/dec	0.01–10	-	[81]
creatinine/creatinine deaminase	Dy_2TiO_5	entrapment in alginate bead	105 mV/dec	0.01–10	1	[82]
creatinine/creatinine deaminase	$Tm_2Ti_2O_7$	entrapment in alginate gel	82 mV/dec	0.01–15	-	[83]
cyanide/cyanidase	Ta_2O_5	covalent	4 mV/dec	0.001–10	-	[84]
formaldehyde/FDH	Si_3N_4	entrapment	31 mV/dec	0.01–20	10	[85]
glucose/GOx	Dy_2TiO_5	entrapment in alginate bead	12 mV/mM	2–8	62	[82]
glucose/GOx	Ta_2O_5	LbL, PAH/ZnO/CNT/GOx	12 mV/dec	0.5–20	-	[92]
glucose/GOx	ZnO	crosslinking	3.1 V/mM	2–7	-	[89]
glucose/GOx	$Tm_2Ti_2O_7$	encapsulation within hydrogel	14.7 mV/mM	2–8	-	[53]
glucose/GOx	Mg/ZnO	crosslinking	10.7 mV/mM	2–7	-	[95]
paraoxon/OPH	Ta_2O_5	crosslinking	~1 mV/µM	0.002–0.05	2	[96]
penicillin/PEN	Ta_2O_5	LbL, PAMAM/CNT/PEN	100 mV/dec	0.025–25	25	[94]
penicillin/PEN	NCD	adsorptive	85 mV/dec	0.005–2.5	5	[58]
penicillin/PEN	SiO_2	LbL, PAH/PENe	100 mV//dec	0.025–10	20	[101]
Penicillin/PEN	Ta_2O_5	adsorptive	46 mV/dec	0.05–10	50	[102]
penicillin/PEN	Ta_2O_5	TMV nanocarrier	92 mV/dec	0.1–10	50	[103]
urea/urease	SiO_2	LbL, Fe_3O_4-NP/PE/urease	32 mV/dec	0.1–100	100	[104]
urea/urease	Mg/ZnO	crosslinking	8.4 mV/mM	2–32	-	[95]
urea/urease	Dy_2TiO_5	entrapment in alginate gel	118 mV/dec	1–32	-	[105]
urea/urease	Ta_2O_5	LbL PAMAM/CNT/ urease/CNT	33 mV/dec	0.1–100	-	[106]
urea/urease	HfO_2	crosslinking	117 mV/dec	0.1–10	-	[107]

NCD: nanocrystalline diamond, dec: decade, LbL: layer-by-layer, PAH: poly(allylamine hydrochloride), PAMAM: polyamidoamine dendrimer, CNT: carbon nanotube, GOx: glucose oxidase, AR: acetoin reductase, OPH: organophosphorus hydrolase, FDH: formaldehyde dehydrogenase, PEN: penicillinase, NP: nanoparticle, PE: polyelectrolyte, TMV: tobacco mosaic virus, and LDL: lower detection limit.

During the last years, several new pH-sensitive gate materials and enzyme immobilization strategies have been proposed to improve the working parameters of EnEIS biosensors. Examples are the adsorption of enzymes onto or within a LbL-deposited dendrimer/carbon nanotube (CNT) multilayer [99,106], the functionalization of EIS sensors with gold [86,112] or magnetic nanoparticles [81,88,104] covered with immobilized enzymes, and the encapsulation of enzymes within alginate beads or a gel layer [82,93,105,108]. In addition, a novel strategy for enhanced field-effect biosensing utilizing capacitive EIS devices modified with a pH-responsive weak polyelectrolyte (PE)/enzyme multilayer was proposed in [101]. The EnEIS biosensor responds to both the local pH change near the gate surface induced via the enzymatic reaction and the pH-dependent charge changes of weak PE macromolecules, resulting in a large sensor signal and higher analyte sensitivity (see Figure 2). Moreover, by the incorporation of enzymes within a multilayer, a larger amount of immobilized enzymes per active sensor area, reduced enzyme-leaching effects, and an enhanced biosensor lifetime can be expected.

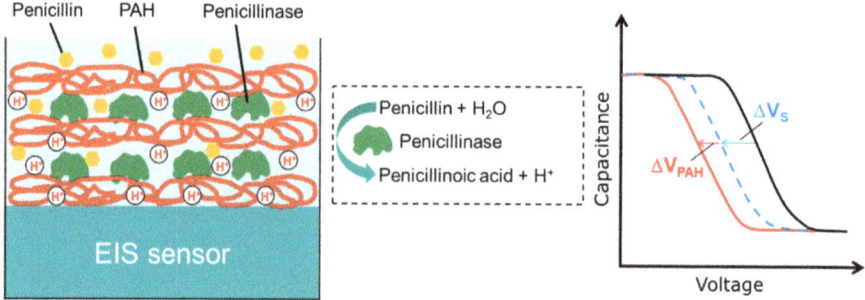

Figure 2. Functioning principle of a penicillin-sensitive EIS biosensor modified with a pH-responsive weak polyelectrolyte (PE)/enzyme multilayer: schematic structure (left), enzymatic reaction of the catalyzed hydrolysis of penicillin by the enzyme penicillinase (middle), and expected shift of the C–V curves of the EIS sensor (right). ΔV_S: shift of the C–V curve due to local pH change near the gate surface induced via the enzymatic reaction and ΔV_{PAH}: additional shift of the C–V curve induced by charge changes of weak PE macromolecules. Reproduced from Ref. [113] with permission from Springer International Publishing.

This generic concept was demonstrated by developing a penicillin-sensitive EnEIS biosensor based on a capacitive p–Si–SiO$_2$ structure functionalized with a LbL-prepared poly(allylamine hydrochloride) (PAH)/penicillinase multilayer. The developed penicillin biosensor possesses a high sensitivity of 100 mV/dec in a linear range of 25 µM–10 mM and a low detection limit of 20 µM [101]. The loss of penicillin sensitivity after two months was about 10%. A similar strategy was used in [106] for the development of a urea biosensor based on a Ta$_2$O$_5$-gate EIS structure modified with a dendrimer/CNT/urease/CNT LbL multilayer. The sensor arrangement with the enzyme urease sandwiched between two CNT layers showed an approximately two-fold higher urea sensitivity in comparison to an arrangement with the enzyme urease immobilized atop of the dendrimer/CNT multilayer. To increase the dynamic range of urea detection (0.1–100 mM), the EIS sensor surface was covered with ferric oxide (Fe$_3$O$_4$) magnetic nanoparticles modified with a LbL multilayer of PAH/PSS (poly(sodium 4-styrenesulfonate))/PAH/urease [104]. The immobilization of enzymes on magnetic beads was also applied for the development of glucose- and creatinine-sensitive EnEIS biosensors [81]. The enzymes glucose oxidase or creatinine deaminase were covalently immobilized on the magnetic beads, and then, the beads were positioned (via an external magnetic force) on the surface of a Dy$_2$TiO$_5$-gate EIS sensor integrated within a microfluidic chip. In further works, the alginate microbead-containing magnetic particles were used as enzyme carriers for the creation of EIS-based glucose, creatinine, and urea biosensors (see Figure 3) [82,93].

The alginate microbeads with embedded magnetic particles and the enzymes glucose oxidase or creatinine deaminase or urease were positioned onto the pH-sensitive Dy$_2$TiO$_5$ gate via an external magnet. The main advantages of this approach are the capability for detection of multiple analytes using the same sensor chip and possibility of replacing alginate microbeads by new beads once the encapsulated enzyme is consumed.

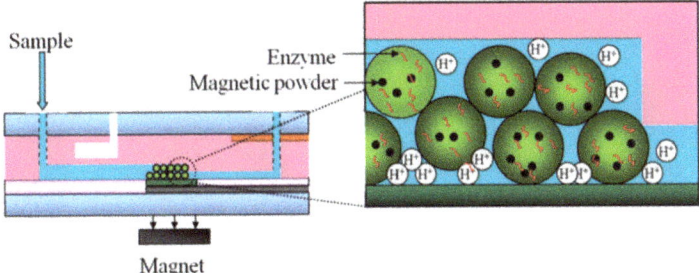

Figure 3. Operating principle of a microfluidic chip. Enzyme-carrying alginate microbeads are immobilized on the EIS-sensor surface by means of an external magnetic field. The sample is injected into the microchannel and reacts with the enzyme contained in alginate beads. The change in potential on the sensor surface induced by the release of hydrogen ions during the reaction process is measured. Reproduced from Ref. [82] with permission of Elsevier.

Another concept for a reusable EnEIS glucose biosensor using a disposable hydrogel/enzyme layer was proposed in [53]. Here, the $Tm_2Ti_2O_7$-gate surface was covered by a thermosensitive poly(N-isopropylacrylamide) hydrogel layer containing the enzyme glucose oxidase. Owing to the phase-tunable characteristics of the hydrogel, the enzyme/hydrogel film was easily loaded onto the surface of the EIS transducer at an increased temperature and then, after measurements, completely removed from the surface by decreasing its temperature, while the EIS sensor underneath was preserved.

More recently, a novel promising approach for the development of EnEIS biosensors was described in [103,114], where a highly sensitive penicillin biosensor with a superior lifetime was realized by means of modification of a Ta_2O_5-gate EIS structure with *tobacco mosaic virus* (TMV) particles as scaffolds for the dense immobilization of enzymes. The TMV has a nanotube-like structure with an average length of 300 nm, an outer diameter of 18 nm, and an internal channel of 4 nm in diameter [115,116]. The TMV surface holds thousands of sites capable for the coupling of various biological receptors, including enzymes. The enzyme penicillinase was immobilized onto the biotinylated TMV surface via a bio-affinity binding of commercially available streptavidin–penicillinase conjugates to biotin. Figure 4 shows the schematic structure of the EnEIS biosensor modified with TMV particles as enzyme nanocarriers (a), a scanning electron microscopy image of TMV particles on the Ta_2O_5 surface (b), and the sensor signal in buffer solution with different penicillin concentrations (c).

The biosensor had a high penicillin sensitivity of ~92 mV/mM in the linear range of 0.1–10 mM, a low detection limit of 50 µM, and an exceptional long-term stability of at least one year. Most likely, this novel approach may be adapted to other enzymes. The results obtained in [103] demonstrate a great potential for the integration of plant virus/receptor nanohybrids with electronic chips, thereby opening new opportunities in advanced biosensing technologies.

The existing enzyme-layer deposition methods often apply manual techniques (e.g., dip- or drop-coating), which are simple but poorly reproducible and relatively time-consuming. Advanced multisensor array and biochip technologies require the controlled and spatially resolved immobilization of a defined amount of biomolecules on the particular transducer surface. Therefore, a nano-spotter, a device for noncontact ultra-low volume dispensing, has been examined for spatially resolved deposition of the enzyme penicillinase onto the Ta_2O_5-gate surface of an EIS structure [102,117]. The nano-spotted penicillin biosensor exhibited identical sensing characteristics (in terms of sensitivity, linear range, and lower and upper detection limit) as the drop-coated EIS sensor counterpart. However, the advantage of nano-spotting is its capability for creating an array of patterned micro-spots immobilized with various enzymes.

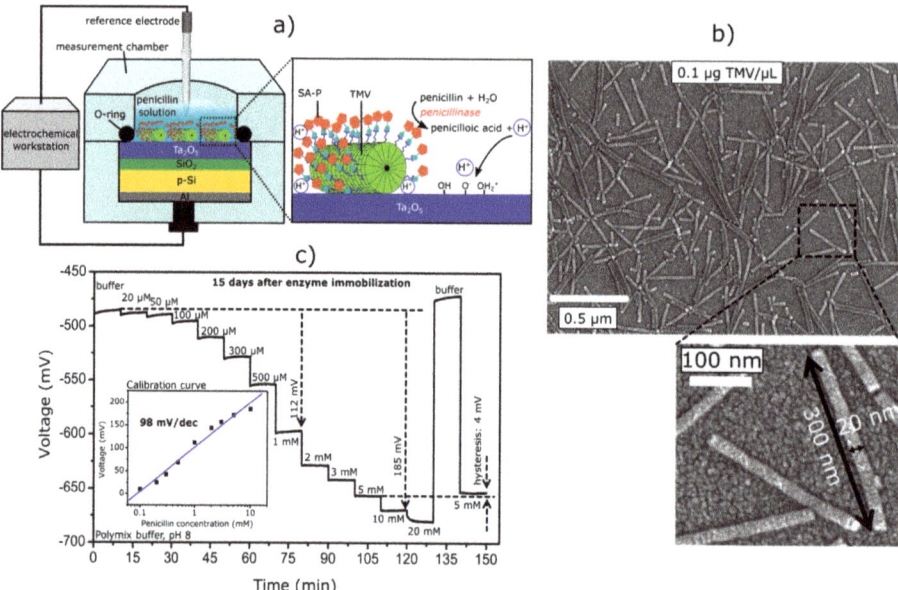

Figure 4. Schematic structure of an enzyme-modified capacitive EIS (EnEIS) biosensor modified with *tobacco mosaic virus* (TMV) particles as enzyme nanocarriers (**a**), scanning electron microscopy image of TMV particles on the Ta_2O_5 surface (**b**), and the sensor signal in buffer solution with different penicillin concentrations (**c**). Adapted from Ref. [103] with permission of Elsevier.

An evaluation of the results on EnEIS biosensors reported in the literature and partially included in Table 2 reveals that a direct comparison of their basic working characteristics is difficult because of different gate insulators, enzyme-immobilization methods, enzyme activity, or buffer capacity used. Typical problems of EnEIS biosensors are similar to that of enzyme-modified FETs and include, for instance, the rather narrow linear measurement range, the detection limit, the relatively slow response, and the dependence of the sensor signal on the enzyme-immobilization method, buffer capacity, and pH value of the test sample. There are only a few papers, where the hysteresis effect, operational and storage stability, reproducibility, and lifetime of EnEIS biosensors have been discussed. Although, during the last years, a number of technological solutions have been proposed and tested to solve these problems and to improve the working parameters of EnEIS biosensors, their transfer from research laboratories to real-life applications remains rather slow. Some examples for practical applications include EnEIS biosensors for the detection of glucose [81,82,87], creatinine [81,82], urea [82], and total triglyceride level [66] in human serum, glucose level in whole blood [88], acetoin in diluted white wine samples [80], and penicillin in bovine milk [103].

Another application field of EnEIS biosensors is their use in enzyme-based logic gates that mimic the working principle of electronic logic gates—basic elements of conventional computing. An integration of biomolecular—in particular, enzyme logic principles—with electronic transducers could facilitate novel digital biosensors with a logic output signal in YES/NO format, logically triggered actuators and drug-release devices, and even intelligent closed-loop sense/act/treat systems with enormous potential in advanced point-of-care diagnostics, personalized medicine, and theranostics [118–124]. The possibility of interfacing of enzyme logic gates with FEDs was first demonstrated in [125], where a capacitive field-effect EIS sensor consisting of an Al–p–Si–SiO_2 structure modified with pH-responsive gold nanoparticles (AuNP) was applied for designing single **AND** and **OR** logic gates. The operation of EIS-based enzyme logic gates developed in [125] was based on bulk pH changes induced by biochemical reactions activated by different combinations of chemical input signals (substrates). The enzymatic part of the system is responsible for sensing of the chemical signals and their logic treatment. As a result of bulk pH changes, the EIS sensor generates an electronic signal corresponding to the logic output produced by the enzymes.

The first example demonstrating the successful transfer of biomolecular logic principles from the bulk solution to the surface of FEDs was reported in [126], where **AND-Reset** and **OR-Reset** logic gates were realized by immobilizing multi-enzymes onto Ta_2O_5-gate EIS structures via entrapment within a polymeric membrane. In contrast to [125], the operation of these enzyme logic gates is based on local pH changes induced by an enzymatic reaction (or cascade of reactions), while the pH value of the bulk solution remains practically unchanged. Thereby, multiple enzyme logic gates, or even logic systems working in the same solution, as well as individual addressing and switching of the respective logic gates, is possible. In further works, other enzyme logic gates such as **AND-Reset** and **OR-Reset** gates with an integrated **Reset** function, **Controlled NOT (CNOT)**, and **XOR** were developed by immobilization or physical adsorption of multi-enzymes onto Ta_2O_5-gate EIS structures [127–129], demonstrating the successful interfacing of enzyme logic principles with semiconductor FEDs.

3.4. Label-Free Detection of Charged Molecules

The detection of adsorption and binding of charged molecules is of great interest for numerous application fields, ranging from clinical diagnostics, environmental monitoring, genetics and the drug industry over biosensors, DNA-chips, and protein-microarray technology up to the fundamental studies of molecular interactions at the solid/liquid interface. Since, in the majority of cases, biological molecules are difficult to detect via their intrinsic physical properties, biosensors often require the labeling of target analytes with different markers or reagents (e.g., enzymatic, redox, or fluorescent) to facilitate the signal readout. In spite of their high sensitivity, label-based techniques suffer from the fact of being time-consuming and labor- and cost-intensive. For the development of fast, simple, and inexpensive biosensors, label-free technologies, which utilize intrinsic physical properties of the analyte molecule to be detected (e.g., charge, electrical impedance, molecular weight, dielectric permittivity, or refractive index), are more favorable.

Since EIS sensors represent charge-sensitive devices (see discussion in Section 2), they can also detect any kind of charged molecules adsorbed or bound onto their gate surface. This way, the coupling of charged molecules, nanoparticles, and even inorganic/organic nanohybrids onto capacitive field-effect sensors is a very promising strategy to actively tune their electrochemical properties, especially with regards to label-free biosensing. Recent examples towards the label-free, direct electrical detection with the help of capacitive EIS sensors consider various kinds of charged molecules [7,27,130–132] and charged nanoobjects (nanoparticles and nanotubes) [112,113,133,134]. In this section, key developments of label-free EIS biosensors will be introduced, which mainly focus on electrostatic DNA detection, the detection of proteins, and oppositely charged PE macromolecules; all of them are monitored by their intrinsic molecular charge. In addition, nanoparticle-modified EIS sensors will be presented.

3.4.1. Detection of DNA Molecules

In recent years, DNA biosensors have been increasingly recognized as powerful tools in many fields of application, including molecular diagnostics, pathogen identification, drug screening, food safety, forensic and parental testing, or detecting biowarfare agents. The vast majority of DNA-modified EIS (DNA-EIS) biosensors reported in the literature is based on detecting a DNA-hybridization reaction [7,27,135–138], although the detection of single-stranded DNA (ssDNA) [8,132] and double-stranded DNA (dsDNA) [132,139–141], as well as other DNA-recognition events, like single-base mismatch [130], the by-product (protons) of the nucleotide base incorporation reaction [36], and DNA amplification by polymerase chain reaction (PCR) [139,140,142–144], have been demonstrated as well.

To our knowledge, the first successful experiment on DNA-hybridization detection with an EIS structure using synthetic homo-oligomers as a model system was reported in [145]. During the DNA-hybridization event, the probe ssDNA molecules with known sequences identified their complementary target DNA molecules (cDNA), and a dsDNA helix structure with two complementary strands was formed. The hybridization reaction was highly efficient and specific, even in the presence of noncomplementary nucleic acids. Usually, capacitive DNA-EIS biosensors detect the hybridization event on-chip: First, probe ssDNA molecules (of known sequences) are immobilized onto the gate surface; this is done by, e.g., adsorption [130,132,137] or covalent attachment [27,146]. In the next step, the target cDNA molecules are detected by in-situ real-time monitoring or ex situ. For in-situ real-time monitoring, the sensor signal is directly recorded during the hybridization process [130], whereas, for ex-situ detection, the response of the EIS sensor is compared before and after hybridization [7,132,135,137]. Capacitive EIS sensors detect DNA molecules electrostatically by their intrinsic molecular charge. During the hybridization process, the negatively charged (due to the phosphate-sugar backbone) target cDNA molecules will effectively alter the charge applied to the gate surface of the EIS sensor, which, in turn, will modulate the space-charge distribution in the semiconductor and capacitance of the EIS structure.

One obstacle of all kinds of FEDs—also including DNA-EIS biosensors for electrostatic DNA detection—is the screening of the DNA charge by mobile counter ions in the surrounding solution. Capacitive field-effect sensors are able to detect charge/potential changes occurring directly at the gate surface or within the order of the Debye length from the surface; note, the Debye length is inversely proportional to the ionic strength of the analyte (e.g., in physiological solutions (~150 mM), it amounts to ~0.8 nm [22]). In the case of DNA molecules tethered to the gate surface, the DNA charge is not confined directly to the interface, but it is distributed through some distance away from the surface, which depends on the DNA length. As it was discussed in [147], the effectivity of electrostatic coupling between the DNA charge and gate surface and, thus, the generated DNA-hybridization sensor signal will strongly drop with increased distance between the DNA charge and gate surface. The use of additional long linker molecules for ssDNA immobilization could also result in a smaller hybridization signal. In contrast, if DNA molecules preferentially lie flat on the gate surface of the capacitive EIS sensor, a higher hybridization signal can be expected. Therefore, in addition to the ionic strength of the solution, the orientation of DNA molecules to the sensor surface has a strong impact on the expected DNA-hybridization signal [147–149]; the method of immobilization of probe ssDNA must be tailored correspondingly.

To achieve a high hybridization efficiency, the DNA hybridization is typically performed in a high-ionic strength solution, while the changes in the sensor signal induced by the DNA immobilization or hybridization are often read out in a low-ionic strength solution in order to reduce the Debye screening effect and, thus, to enhance the sensor performance [135,137,146]. A post-hybridization binding of intercalators or DNA binders to dsDNA molecules may be an effective way to distinguish the hybridization signal from an undesirable background noise caused by the nonspecific adsorption of target cDNA molecules. This way, a more accurate and reliable detection of the hybridization event with EIS structures can be achieved [146]. Since DNA binders react specifically with dsDNA, the changes in the sensor signal due to the binders could serve as an indicator to verify the successful hybridization process.

Besides the measurement of the DNA-hybridization signal in a low-ionic strength solution, reducing the distance between the DNA charge and sensor surface via the immobilization of DNA molecules flat to the EIS surface with a molecular charge lying within the Debye length from the gate surface is an essential factor to enhance the sensitivity of the DNA sensor. A direct adsorption of DNA molecules onto the EIS surface is, in general, hindered because of electrostatic repulsion forces between the negatively charged phosphate groups of the DNA and the negatively charged surface of the gate insulators typically used in EIS structures (e.g., SiO_2 and Ta_2O_5). Therefore, an electrostatic adsorption of probe ssDNA molecules onto the EIS surface modified with a LBL-prepared positively charged PE layer and subsequent hybridization with cDNA molecules becomes more popular for designing DNA-EIS biosensors [130,132,137,141]. It is suggested that electrostatically adsorbed probe ssDNA molecules will be preferentially flat-oriented to the gate surface, with negatively charged phosphate groups directed to the cationic PE macromolecules, while the nucleobases will be exposed to the surrounding solution, allowing hybridization with the target cDNA molecules. In addition, due to the presence of a cationic PE layer, both the Debye screening effect and the electrostatic repulsion between probe ssDNA and target cDNA molecules will be less effective, resulting in an acceleration of the hybridization event, as well as a higher hybridization signal. Moreover, in contrast to frequently used, time-consuming, and cost-intensive covalent immobilization techniques, the LbL electrostatic adsorption method is easy, fast, and does not require complicated procedures for the functionalization of the gate surface and/or probe ssDNA molecules. For example, poly-L-lysine (PLL)-modified SiO_2-gate EIS sensors were utilized for the detection of ssDNA immobilization and the DNA-hybridization process [130]. Although the sensor was able to detect low concentrations (2 nM) of target cDNA oligonucleotides (12-mer), the hybridization signal was small (several mVs). In further experiments, these EIS sensors were applied for monitoring PCR-amplified dsDNA [139,142]. Recently, in our group, the feasibility for the label-free electrical detection of DNA with capacitive SiO_2-gate EIS sensors, which were modified with a positively charged weak PE of PAH, was demonstrated [132,137,141,144]. Figure 5 schematically shows the EIS-sensor surface before and after modification with PAH, probe ssDNA, and after cDNA hybridization; the measurement setup; and the expected ConCap response. High hybridization signals of 34 mV and 43 mV were recorded in low-ionic strength solutions of 10 mM and 1 mM, respectively [137]. In contrast, a small response of 4 mV was registered in the case of unspecific adsorption of fully mismatched DNA. These experiments demonstrated the specificity of the developed EIS sensor capable of distinguishing the complementary cDNA from fully mismatched DNA.

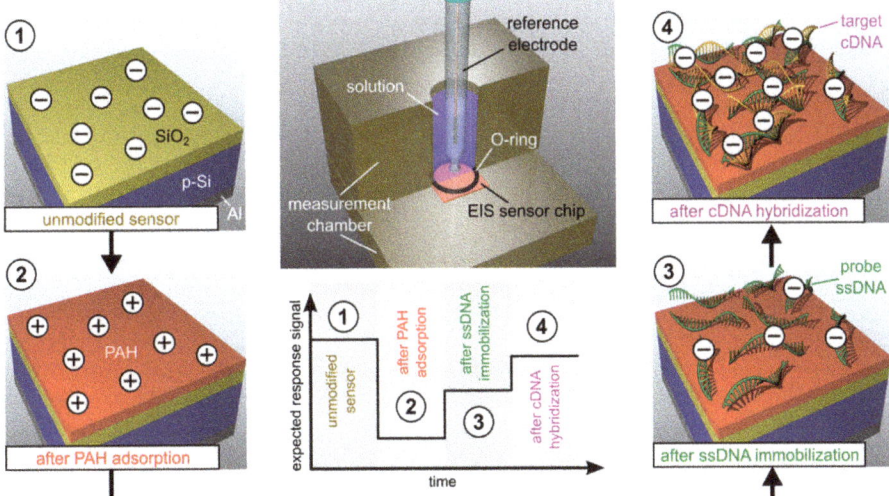

Figure 5. Modification steps of the EIS sensor, measurement setup (middle column, top), and expected ConCap response (middle column, bottom). (**1**) Unmodified sensor, (**2**) after poly(allylamine hydrochloride) (PAH) adsorption, (**3**) after single-stranded DNA (ssDNA) immobilization, and (**4**) after complementary target DNA (cDNA) hybridization. + and − symbols indicate the respective surface charges. Reproduced from Ref. [144] with permission of the American Chemical Society.

In a further work, PAH-modified EIS sensors were applied for the direct label-free electrical detection of dsDNA formed after a hybridization reaction occurred in the solution (so-called in-solution hybridization) for the first time [141]. Direct dsDNA detection could significantly simplify the surface modification procedure (because no probe ssDNA has to be immobilized onto the sensor surface) and, thus, may reduce the detection time and costs. Finally, the ability of PAH-modified EIS chips for the detection of PCR-amplified tuberculosis DNA fragments was demonstrated in [144]. The sensitivity of the sensor in artificial PCR solutions with different target cDNA (72-mer) concentrations from 1 nM to 5 µM was 7.2 mV/decade, with an estimated lower detection limit of ~0.3-nM cDNA. Such chips could serve as a sensing device for a quick verification of successful/unsuccessful DNA amplification by means of PCR.

Label-free field-effect DNA biosensors are typically disposable devices for a single-use measurement. To make DNA biosensors reusable, the complex surface/interface architecture should be regenerated, which is, in general, a complicated and time-consuming procedure [150]. On the other hand, it was reported that the surface of PE-modified DNA-EIS chips can be easily regenerated, making them suitable for multiple DNA immobilization and hybridization experiments [130,132,139]. As one example, the reusability of capacitive EIS sensors modified with PAH was examined for the detection of a ssDNA on-chip DNA hybridization event, as well as in-solution hybridized dsDNA molecules [132]. It has been demonstrated that the same biosensor can be reused for at least five DNA-detection measurements. For the experimental procedure, the simple regeneration of the gate surface of the EIS chip (covered with PAH/ssDNA or PAH/dsDNA layers) was realized by the electrostatic adsorption of a new positively charged PAH layer onto the negatively charged DNA layer. The performed experiment with the PAH-modified EIS sensors also allowed to investigate the impact of the Debye screening effect on the DNA immobilization and hybridization signal: the sensor response of the capacitive EIS sensor induced by the immobilization of ssDNA and dsDNA, as well as after the on-chip hybridization of cDNA, were recorded in solutions with different ionic strengths of 1 mM, 5 mM, 10 mM, and 20 mM. The evaluated Debye lengths amounted to approximately 9.6 nm, 4.3 nm, 3 nm, and 2.2 nm, respectively. The results of these experiments verified the assumption that, due to the more efficient screening of the DNA charge by counter ions, the amplitude of the DNA-immobilization and hybridization signal will decrease when increasing the ionic strength of the solution. For example, the on-chip cDNA-hybridization signal was reduced from 52 mV to 33 mV by increasing the ionic strength of the measuring solution from 1 mM to 20 mM [132].

The modeling of DNA-modified FEDs (DNA-FED), including DNA-EIS biosensors, is beneficial for understanding the mechanism of label-free DNA detection and for the optimization of device characteristics. Due to the high complexity of the DNA-modified gate insulator/electrolyte interface, to date, there are no exact theoretical models describing the functioning of DNA-FEDs taking into account all interfering factors. In addition to the counter-ion screening effect, other factors, like the orientation, length, and surface density of DNA molecules, charge distribution within the intermolecular spaces, distance between the DNA charge and the gate-insulator surface, gate surface-charge, etc., play a crucial role in converting the hybridization event to an electrical signal. Hence, several simplified theoretical models for DNA-FEDs were suggested and discussed. For example, a charge-plane model that takes into consideration both the Debye-screening length and the distance between the DNA charge and the gate surface was proposed in [135]. In another approach, in order to simulate the sensitive behavior of a DNA-FED, the DNA layer was modeled as an ion-permeable membrane [151]. The interplay between pH-, ion-, and charge sensitivity of FEDs modified with charged molecules was discussed in [152,153]. Moreover, the DNA hybridization-induced modulation of the ion-concentration distribution within the intermolecular spaces and ion sensitivity of the gate surface as a possible operation principle for DNA-FEDs was proposed in [154]. Recent simulations on the impact of the DNA position and orientation on the hybridization signal [149] show that the largest hybridization signal can be expected when the DNAs are parallel to the biosensor surface and distributed at equal intervals. Finally, the relation between the screening effect and the distance of the charged target molecule or particle from the electrolyte/insulator interface has been studied by using Monte Carlo simulation [155].

Summarizing this subsection, it should be emphasized that the majority of the reported label-free DNA-EIS biosensors utilize relatively short synthetic oligonucleotides as model targets and rather ideal experimental conditions. Problems may arise when dealing with real samples containing very large target DNA molecules (a thousand to several hundred thousands of base pairs) or other charged molecules (possible nonspecific adsorption).

3.4.2. Detection of Biomarkers and Other Charged Molecules

Among the biomolecular interactions, the high specificity of molecular recognition can be best typified by an antibody–antigen interaction, which is the basis of immunosensors. Immunosensors as analytical devices have been recognized as very promising tools with enormous potential applications, including, e.g., monitoring contaminants in the environment, food safety, clinical diagnostics for monitoring the functioning of the immune system, etc. One of the key challenges of immunosensors is the detection of disease biomarkers that enable the early identification of diseases and effective treatments.

Label-free immunosensors allow the direct monitoring of immunoreactions by measuring physicochemical changes induced by the antigen–antibody complex formation. Immuno-sensitive FEDs (ImmunoFEDs) for label-free protein detection via their intrinsic molecular charge have attracted considerable interest due to their excellent sensitivity, fast response time, small size, cost-efficiency, and possibility of real-time and multiplexed measurements in a small sample volume and, therefore, are considered as promising alternatives to conventional immunoassays. ImmunoFEDs are often constructed by modification of the gate surface with antibodies as recognition elements for specific biomarkers (antigens). Since antibodies and antigens are generally electrically charged in aqueous solutions, it is suggested that the formation of an antibody–antigen complex on the gate surface will modulate the surface charge, inducing a biomarker concentration-dependent response of the FED. Most ImmunoFEDs reported in the literature are based on transistor structures (different types of ISFETs or SiNWs) [22,24], while capacitive EIS-based immunosensors (ImmunoEIS) have rarely been investigated. We only found a few studies related to the label-free direct detection of protein biomarkers and other molecules by their intrinsic charge with ImmunoEIS biosensors. Some examples of successful developments are described below.

A SiO_2-gate EIS structure with APTES (3-aminopropyltriethoxysilane)-functionalized vertically aligned ZnO nanorods was utilized for the label-free detection of the prostate-specific antigen (PSA)—a biomarker strictly associated with prostate cancer [156]. The anti-PSA antibodies were covalently immobilized on the ZnO nanorods. Upon the binding of PSA of a concentration of 1 ng/mL in a 100-µM solution, a ~23-mV shift of the C–V curve was observed. In contrast, no significant voltage shift in the C–V curve was detected by measurements in a solution with an ionic strength of 10 mM, which was attributed to the counter-ion screening effect of the PSA charge. In a further work, PSA antigens were immobilized onto a polyethyleneimine-modified SiO_2-gate surface, resulting in enhanced sensitive properties [131]. The sensitivity toward PSA molecules in a low-ionic strength solution was 28.2 mV/dec and 4.7 mV/dec in the PSA concentration range of 1–10 ng/mL and 10 pg/mL–1 ng/mL, respectively. The YbY_xO_y-gate EIS device was investigated for the detection of the rheumatoid factor (RF)—a diagnostic biomarker for rheumatoid arthritis [157]. RF antibodies functionalized with N-hydroxysuccinimide were covalently immobilized on the APTES-modified $YbTi_xO_y$ surface. The sensitivity of the biosensor to serum RF antigen was ~41 mV/dec in the concentration range of 0.1 µM–1 mM. Chand et al. used specific aptamers (instead of large antibodies) immobilized onto AuNPs deposited on the surface of a SiO_2-gate EIS sensor for the label-free detection of protein kinase A (PKA) [158,159]. Wang et al. demonstrated the possibility of detecting bovine serum albumin (BSA) with HfO_2-gate EIS sensors [107]. To avoid the complicated silanization process often used for antibody immobilization, the anti-BSA antibodies were immobilized on the HfO_2 surface post-treated with NH_3 plasma. The observed shift of the C–V curve along the voltage axis after the binding of BSA to anti-BSA was about 20 mV. An EIS sensor for direct monitoring of the binding of heparin molecules via detecting their intrinsic negative charge was proposed in [160]. Heparin is well-known as an important clinical anticoagulant, where low-molecular weight heparin is used for the prophylaxis of deep venous thromboembolisms. To prevent thrombosis and avoid bleeding risks during and after surgery, monitoring and control of the heparin level in a patient's blood is important. The clinical heparin antagonist protamine or the physiological partner antithrombin III were used as heparin-specific surface probes. For the sensor, a detection limit of 0.001 U/mL was described [160], calculated from the dose-response curves; the achieved detection limit is in orders of magnitude lower than the clinically relevant concentrations. Ultimately, APTES-silanized Si_3N_4-gate capacitive EIS structures modified with magnetic nanoparticles were applied for ochratoxin A detection [161]. Ochratoxin A is one of the predominant contaminating mycotoxins in many products (e.g., dried fruits, coffee beans, beer, wine, etc.). The anti-ochratoxin A antibodies were immobilized on magnetic nanoparticles by amide bonding. The biosensor was highly sensitive (10 mV/pM in the linear range of 2.5–50 pM), with a detection limit of 4.57 pM and specific for ochratoxin A antigens, when compared to other interferences, such as ochratoxin B and aflatoxin G1.

In spite of the above-described successful experiments with ImmunoEIS sensors, detecting proteins (including disease biomarkers) and other charged molecules in real biological samples (e.g., whole blood, serum, or urine) remains a big challenge. The reasons are limitations in conjunction with the electrostatic detection of molecular charges with FEDs. The major limitation of the label-free electrostatic detection of proteins with ImmunoFEDs is the counter-ion screening of the molecular charge, already discussed in Section 3.4.1 for DNA-EIS sensors. The dimensions of some proteins (e.g., antibodies) are much larger (ca. 10–12 nm [162]) than the Debye length (~0.8 nm) in a solution (e.g., in whole blood), with an ionic strength of ~0.15 M. If immobilized antibodies are oriented such that the Fc (fragment crystallizable region) is substrate-facing (so-called end-on orientation), the distance between the binding sites and the sensor surface will be substantially greater than the Debye length. Thus, the target molecule/receptor binding will occur beyond the Debye length, making the electrostatic detection of the biomolecular charge in real samples difficult or even impossible. As a consequence, a useful measurable effect with an ImmunoEIS sensor can be expected only in low-ionic strength solutions (<10 mM). On the other hand, randomly adsorbed antibodies may have head-on, side-on, or lying-on orientations relative to the surface [163,164]. As a result, some of the bound target analyte charges can be expected to be held within the Debye length and, therefore, could be detectable by the ImmunoEIS sensor. Several strategies (e.g., measurements in desalted/filtered samples or in a low-ionic strength solution and the use of short receptors (aptamers and antigen-binding fragments)) have been proposed to reduce the influence of the counter-ion screening effect and, thus, to enhance the sensitivity of ImmunoFEDs (see reviews [22,23,153,165]).

Another issue is the nonspecific adsorption of proteins. Biological samples represent extremely complex media containing thousands of proteins and other charged chemical species (covering a very wide concentration range from pg/mL to mg/mL). They are able to nonspecifically adsorb on the gate surface of the ImmunoFED and generate false-positive signals or mask the usable signal from the target analyte of interest. This significantly hampers the sensitivity, specificity, and reliability of ImmunoFEDs and provokes false diagnostic results. To reduce/eliminate the nonspecific adsorption of proteins onto the surface of ImmunoFEDs, various strategies such as the use of blocking agents, prefiltering/purifying the biological liquids or on-chip filtering, separation, desalting, and preconcentration platforms, have been discussed for ImmunoFEDs [5,22,166].

3.4.3. Detection of the Consecutive Adsorption of Oppositely Charged PE Macromolecules

Besides the detection of charged biomolecules such as DNA or proteins discussed above, EIS sensors have also been applied for the label-free detection of LbL sequential electrostatic adsorption of cationic and anionic PE molecules and the monitoring of a PE multilayer build-up. The LbL deposition of PE multilayers provides a simple and cost-effective method for the preparation of ultra-thin films (even organic/inorganic hybrid multilayers) with a desired composition, functionality, and nanoscale control of the thickness [167]. Such PE multilayers are very attractive as coatings with functional and controllable properties, stimuli-responsive materials for actuators, microcontainer-controlled drug release systems, and biosensing applications. For the optimization and practical implementation of PE multilayer-based devices, investigation of the impact of process parameters (e.g., PE concentration, ionic strength, pH value of the solution, and surface charge of the substrate) on the formation and characteristics of the PE multilayer is essential.

Recent studies on the detection of PE macromolecules using EIS structures with different gate materials (SiO_2, SiO_2–Ta_2O_5, SiO_2–NCD, and SiO_2–AuNP) and various PE model systems (PLL-ssDNA, PAH-PSS, PAH-ssDNA, and PAH-dsDNA) have demonstrated the potential of these EIS sensors for real-time, in-situ monitoring of the PE multilayer formation with direct electrical readout [7,8,30,130,132,168–172]. For example, the effect of the semiconductor doping type on the electrical characteristics of a PAH-ssDNA-modified SiO_2-gate EIS structure was studied in [8]. The pH and ion sensitivity of a SiO_2-gate EIS sensor covered with a PE multilayer of PAH-PSS was investigated in [170]. An array of nanoplate capacitive EIS structures prepared from a silicon-on-insulator wafer was applied for the electrical monitoring of the PE multilayer formation using differential-mode dynamic ConCap measurements [7]. The feasibility of an AuNP-modified capacitive EIS sensor for the label-free detection of the consecutive adsorption of cationic weak PE PAH and anionic strong PE PSS was demonstrated in [30]. More recently, the formation of a PAH-ssDNA and PAH-dsDNA multilayer onto SiO_2-gate EIS sensors was studied in [132]. Finally, in the work of Poghossian et al., a PE multilayer stack (18 layers) of PAH-PSS onto a SiO_2-gate EIS sensor was investigated in detail to understand the effect of ionic strength of the solution, PE concentration, and number and polarity of PE layers on the sensor signal [172]. Consecutive adsorption of oppositely charged PE layers leads to alternating shifts of the sensor signal of the capacitive EIS sensor, whereas the direction of these shifts correlates with the charge sign of the terminating PE layer (PAH or PSS). To interpret it in more detail: adsorption of a positively charged PAH layer shifts the signal of the capacitive EIS sensor in the direction for an additional positive charging of the gate surface. This corresponds to a more negative sensor output signal in the ConCap mode due to the feedback-control circuit (see Figure 6a). In contrast, PSS layer adsorption shifts the potential towards the direction that results from a more negatively charged gate surface. Subsequent PAH and PSS layers adsorption show a zigzag-like behavior in potential shifts (Figure 6b) that can be explained by the charge sign of the outermost layer. Similar effects were also found for PAH-PSS and PAH-DNA multilayers elsewhere [7,8,130,132,168–171]. In addition, it has been observed that the amplitude of the signal changes has a tendency to decrease with increasing the PE layer number and ionic strength (Figure 6b). To explain the experimentally observed signal behavior of an EIS sensor modified with a PE multilayer, a simplified electrostatic model was proposed, which is based on the assumption of a reduced ionic strength and, therefore, reduced screening of PE charges by mobile ions inside the multilayer [172]. According to this model, with increasing the multilayer thickness, the electrostatic coupling between the charge of the outermost PE layer and the gate surface will drop. As a result, the potential changes generated on the gate surface by the adsorption of the outermost PE layer will gradually decrease with an increasing number of PE layers that, in fact, was observed in many experiments. However, the development of exact theoretical models quantitatively describing the influence of all interfering factors on the signal behavior of PE-modified EIS sensors still requires further experimental input.

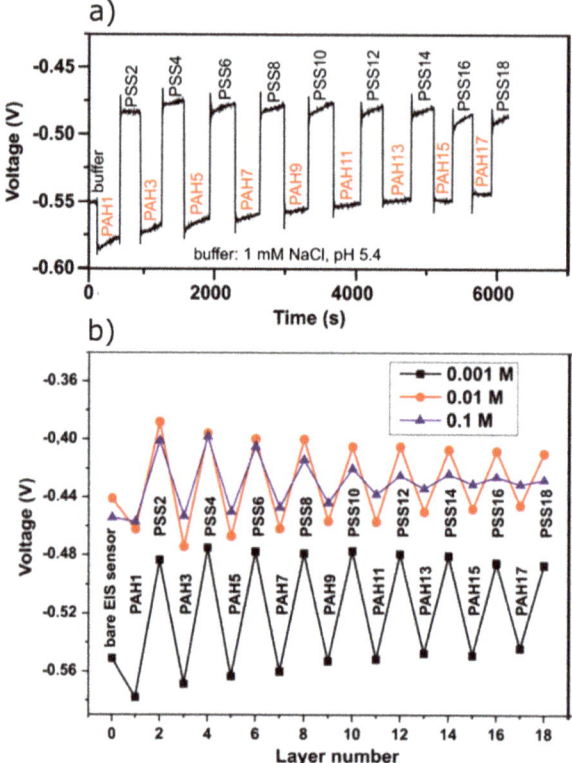

Figure 6. Electrical monitoring of the PE multilayer build-up with a p-Si–SiO$_2$ EIS sensor. (**a**) ConCap response and (**b**) potential shifts as a function of the PE layer number and ion concentration. The PAH and poly(sodium 4-styrenesulfonate) (PSS) layers were deposited from a 50 µM PE solution adjusted with different NaCl concentrations of 100, 10, and 1 mM (pH 5.4). Reproduced from Ref. [172] with permission from Springer.

3.4.4. Label-Free Biosensing with AuNP-Modified EIS Structures

Due to their unique physicochemical features, easy surface modifications with various shell molecules capable for coupling of different biochemical recognition elements, high surface-to-volume ratios, and the possibility of integration with macroscopic transducers, AuNPs are considered as highly attractive chemically and electrically tunable nanomaterials for designing electrochemical biosensors. In comparison with planar surfaces, the immobilization of receptors on nanoparticles typically provides a higher density of receptor molecules with a favorable orientation for interactions with target molecules to be detected, enhancing the transport of target molecules to the nanoparticle surface, all improving the biosensor performance. In this context, the coupling of AuNPs with FEDs represents a very promising strategy with new opportunities for label-free biosensing with direct electrical readout [30]. Since the vast majority of biomolecules are positively/negatively charged in solutions, AuNP-modified FEDs can provide a generic approach for detecting numerous charged biomolecules. Nevertheless, in spite of the popularity of AuNPs in electrochemical sensing, there are only a few articles devoted to the label-free detection of charged molecules with AuNP-modified capacitive EIS sensors [7,30,113,135,158,159].

As one example, a capacitive EIS sensor with SiO$_2$ gate was applied to detect charge changes of ligand-stabilized and bare-supported AuNPs, which were induced by oxygen plasma treatment or by exposure to aqueous oxidation and reduction solutions, respectively [133]. Another, more recent experiment described capacitive EIS sensors functionalized with negatively charged, citrate-capped AuNPs for the label-free electrostatic detection of positively charged molecules by their intrinsic molecular charge [30,113]: charge changes can be detected in those AuNP/molecule inorganic/organic nanohybrids that result from molecular adsorption or binding events. Here, the ligand-stabilized AuNPs play a dual role: (i) The AuNPs offer a simple way to couple a large variety of charged molecules on their surfaces (e.g., negatively charged citrate-capped AuNPs provide a convenient scaffold to attach positively charged molecules), and (ii) the AuNPs can serve as additional distributed, quasi-spherical nanometer-sized, local metal gates [30].

The expected modulation of the depletion layer in Si within local regions under surface areas covered with AuNPs is depicted in Figure 7: the adsorption or binding of charged molecules onto the AuNP surface will locally alter the width of the depletion layer and, therefore, the depletion capacitance. The overall capacitance of the field-effect EIS sensor will change, shifting the C–V curve along the voltage axis. The effect of coupling of charged molecules to the AuNPs is the same as if one would apply an additional voltage to the local gates. In which direction the voltage is shifting depends on the sign of the charge of the attached molecules, while the amplitude is determined by the surface density of AuNPs, the number of attached molecules per AuNP, and their intrinsic charge. Summarizing, a high sensor signal can be expected by a high AuNP surface coverage, a large number of highly charged, attached molecules per AuNP, and when performing measurements, in low-ionic strength solutions.

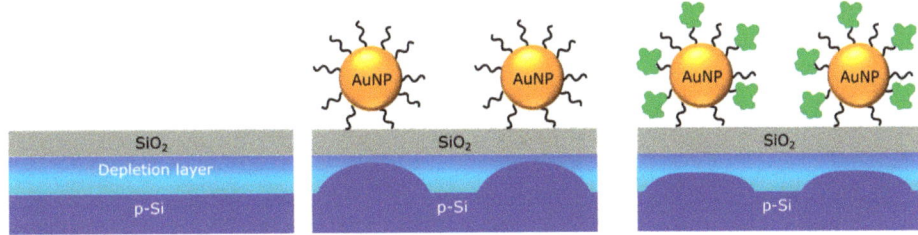

Figure 7. Schematic of a AuNP-modified capacitive EIS sensor. Expected modulation of the depletion layer after the deposition of negatively charged citrate-capped AuNPs and after the binding of positively charged molecules on AuNPs are exemplarily shown for a p-type EIS structure.

The feasibility of the proposed detection scheme has been exemplarily demonstrated by developing SiO$_2$-gate EIS sensors modified with negatively charged citrate-capped AuNPs (with an average size of ~18 nm) for the detection of typical model examples of positively charged small proteins (cytochrome c, which is a key component of the electron transport chain in the mitochondria) and macromolecules (poly-D-lysine), as well as for monitoring the consecutive adsorption of oppositely charged PE molecules [30].

The possibility of detecting protein kinase A (PKA) with an AuNP-modified EIS structure was demonstrated in [158]. PKA has several functions in the cell, including the regulation of glycogen-, sugar-, and lipid metabolisms. AuNPs (~16 nm) functionalized with a thiolated PKA-specific aptamer were deposited onto the silanized SiO$_2$-gate surface. The quantitative detection of PKA was performed by analyzing the C–V curve after the aptamer–PKA interaction. The EIS device showed a detection limit of 1-U/mL PKA. In a further work, the developed EIS sensor with an Ag/AgCl quasi-reference electrode was integrated with a polymeric microchip and tested for the label-free detection of PKA in a spiked human cell sample [159].

Finally, an array of nanoplate EIS sensors functionalized with AuNPs was applied for the label-free detection of consecutive DNA hybridization, denaturation, and rehybridization in a differential-mode setup [7,135]. Figure 8 shows a schematic of the bare (a) and AuNP-modified (5–8 nm) (b) EIS chips, combining four individually addressable EIS sensors prepared using a silicon-on-insulator wafer and a differential-mode ConCap response after consecutive DNA hybridization, denaturation, and rehybridization (c). Sensor 4 was functionalized with the probe ssDNA (20-mer) molecules perfectly matched to the complementary target cDNA sequence, while sensors 2 and 3 were immobilized with the fully mismatched ssDNA. Sensor 1 was utilized for the pH control. High differential signals of about 120 mV, 90 mV, and 80 mV were observed between sensor 4 and sensor 2 after the DNA hybridization, denaturation, and rehybridization events, respectively. The observed hybridization signal was three to four times higher than those previously reported for SiO_2-gate EIS sensors without AuNPs (24–33 mV) in [27,169].

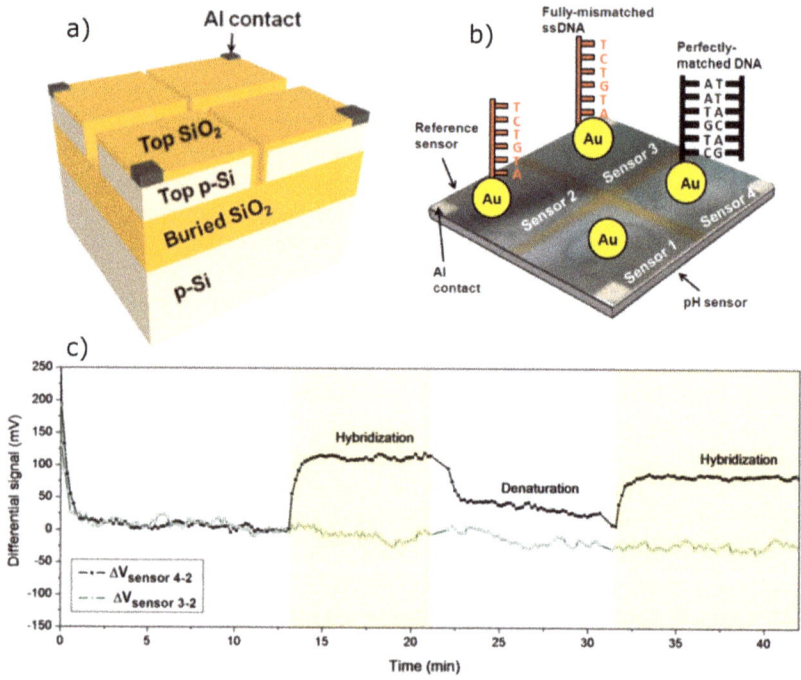

Figure 8. Schematic of the bare (a) and AuNP-modified (5–8 nm) (b) EIS chips combining four individually addressable EIS sensors prepared using a silicon-on-insulator wafer and a differential-mode ConCap response after consecutive DNA hybridization, denaturation, and rehybridization (c). $\Delta V_{4\text{-}2}$ and $\Delta V_{3\text{-}2}$: net differential signals between sensor 4 and sensor 2 and sensor 3 and sensor 2, respectively. Adapted from Ref. [135] with permission from Wiley-VCH.

4. Concluding Remarks

In spite of remarkable progress in the research and development of EIS sensors and the implementation of new strategies and ideas to improve the sensitivity characteristics in the last few years, it should be noted that there are still some issues that must be overcome before the commercialization of EIS biochemical sensors and their transfer from scientific labs to real life and widespread applications will appear. Some challenges related to ion-sensitive EIS, EnEIS, DNA-EIS, and ImmunoEIS sensors, and possible strategies proposed for their solutions, are discussed in the respective sections. As for other kinds of FEDs, due to the Debye screening effect and nonspecific adsorption, the label-free detection of charged biomolecules with EIS biosensors in untreated, real biological samples (e.g., whole blood, serum, plasma, saliva, urine, cerebrospinal fluid, and nasopharyngeal swab) with high sensitivity and specificity is still demanding.

From the application point of view, the development of an array of capacitive EIS sensors for multiplexed detection, as well as the realization of a stable, reliable, miniaturized integrated reference electrode, are two further tasks to be solved. For a correct functioning of field-effect EIS sensors, the reference electrode should provide a stable potential during measurements independent of the pH value of the solution or concentration of the dissolved species. This is usually achieved by applying the liquid-junction reference electrode (e.g., Ag/AgCl electrode), which is still bulky and often fragile and, therefore, limits seriously the large-scale application of EIS sensors. In order to achieve the full advantage of EIS sensors, a comparable small reference electrode must be realized. In this context, the development of miniaturized integrated solid-state reference electrodes compatible with Si technology is of great interest. Due to the difficulties in the miniaturization and integration of liquid-junction reference electrodes, some groups use so-called pseudo- or quasi-reference solid-state electrodes made from, for example, Pt, Au, or Ag/AgCl films. Such electrodes are often unsuitable for reliable biosensing because of non-negligible potential instabilities and the need of a long time to achieve a relatively stable potential after changing the electrolyte solution. Therefore, the results of experiments performed using quasi-reference electrodes should be carefully evaluated.

An ability of sensors for multiplexed detection—that is, to simultaneously assay for multiple chemical or biological species—could reduce both the analytical time (and cost) and sample volume. However, the integration of capacitive EIS sensors in an array format for multiparameter detection must circumvent technological difficulties in fabricating separate, individual, electrically isolated, field-effect capacitors onto the same Si chip [7]. Several capacitive EIS sensors fabricated on the same Si chip will stay interconnected through their common Si substrate. As a consequence, this can result in undesired cross-talk between the different EIS sensors on the same chip. Individually addressable on-chip fabricated EIS capacitors for multiparameter detection still remains a challenge, for which only a few studies have presented encouraging results [7,8]. The price to be paid is the loss of important advantages of capacitive EIS sensors—namely, the simple structure and easy preparation.

One widely neglected subject in most reported EIS sensors is the possible leakage current between the reference electrode and the Si substrate (ideally, no leakage current should flow). An existence of a leakage current might lead to serious experimental artifacts and even electrolysis, depending on the electrolyte composition and potentials applied to the system [173]. In spite of this fact, there are only a few works where data on leakage current levels in EIS sensors were given.

In summary, despite the above discussed issues, the research of capacitive field-effect EIS sensors is a rapidly advancing field in which novel device designs and modification methodologies are consistently being developed. This fact provides the reason for great optimism that capacitive EIS structures will play a significant role in the commercialization of FED-based chemical sensors and biosensors in the near future.

Author Contributions: The authors wrote the review article together. All authors have read and agreed to the published version of the manuscript.

Funding: This research received no external funding.

Acknowledgments: The authors would like to thank M. Jablonski and H. Iken for technical assistance in writing the article (adaption of figures and reference list).

Conflicts of Interest: The authors declare no conflict of interest.

References

1. Mu, L.; Chang, Y.; Sawtelle, S.D.; Wipf, M.; Duan, X.; Reed, M.A. Silicon nanowire field-effect transistors—A versatile class of potentiometric nanobiosensors. *IEEE Access* **2015**, *3*, 287–302. [CrossRef]
2. Poghossian, A.; Schultze, J.W.; Schöning, M.J. Multi-parameter detection of (bio-)chemical and physical quantities using an identical transducer principle. *Sens. Actuators B* **2003**, *91*, 83–91. [CrossRef]
3. Lee, C.-S.; Kim, S.K.; Kim, M. Ion-sensitive field-effect transistor for biological sensing. *Sensors* **2009**, *9*, 7111–7131. [CrossRef]

4. Jeho, P.; Hoang, H.N.; Abdela, W.; Moonil, K. Applications of field-effect transistor (FET)-type biosensors. *Appl. Sci. Converg. Technol.* **2014**, *23*, 61–71.
5. Zhou, W.; Dai, X.; Lieber, C.M. Advances in nanowire bioelectronics. *Rep. Prog. Phys.* **2017**, *80*, 016701. [CrossRef] [PubMed]
6. Pullano, S.A.; Critello, C.D.; Mahbub, I.; Tasneem, N.T.; Shamsir, S.; Islam, S.K.; Greco, M.; Fiorillo, A.S. EGFET-based sensors for bioanalytical applications: A review. *Sensors* **2018**, *18*, 4042. [CrossRef] [PubMed]
7. Abouzar, M.H.; Poghossian, A.; Pedraza, A.M.; Gandhi, D.; Ingebrandt, S.; Moritz, W.; Schöning, M.J. An array of field-effect nanoplate SOI capacitors for (bio-)chemical sensing. *Biosens. Bioelectron.* **2011**, *26*, 3023–3028. [CrossRef]
8. Garyfallou, G.Z.; de Smet, L.C.P.M.; Sudhölter, E.J.R. The effect of the type of doping on the electrical characteristics of electrolyte-oxide-silicon sensors: pH sensing and polyelectrolyte adsorption. *Sens. Actuators B* **2012**, *168*, 207–213. [CrossRef]
9. Dastidar, S.; Agarwal, A.; Kumar, N.; Bal, V.; Panda, S. Sensitivity enhancement of electrolyte-insulator-semiconductor sensors using mesotextured and nanotextured dielectric surfaces. *IEEE Sens. J.* **2015**, *15*, 2039–2045. [CrossRef]
10. Yoshinobu, T.; Ecken, H.; Poghossian, A.; Lüth, H.; Iwasaki, H.; Schöning, M.J. Alternative sensor materials for light-addressable potentiometric sensors. *Sens. Actuators B* **2001**, *76*, 388–392. [CrossRef]
11. Wu, C.; Bronder, T.; Poghossian, A.; Werner, C.F.; Schöning, M.J. Label-free detection of DNA using a light-addressable potentiometric sensor modified with a positively charged polyelectrolyte layer. *Nanoscale* **2015**, *7*, 6143–6150. [CrossRef] [PubMed]
12. Wu, C.; Poghossian, A.; Bronder, T.S.; Schöning, M.J. Sensing of double-stranded DNA molecules by their intrinsic molecular charge using the light-addressable potentiometric sensor. *Sens. Actuators B* **2016**, *229*, 506–512. [CrossRef]
13. Yoshinobu, T.; Miyamoto, K.; Werner, C.F.; Poghossian, A.; Wagner, T.; Schöning, M.J. Light-addressable potentiometric sensors for quantitative spatial imaging of chemical species. *Annu. Rev. Anal. Chem.* **2017**, *10*, 225–246. [CrossRef] [PubMed]
14. Noor, M.O.; Krull, U.J. Silicon nanowires as field-effect transducers for biosensor development: A review. *Anal. Chim. Acta* **2014**, *825*, 1–25. [CrossRef] [PubMed]
15. Wang, Z.; Lee, S.; Koo, K.-I.; Kim, K. Nanowire-based sensors for biological and medical applications. *IEEE Trans. Nanobiosci.* **2016**, *15*, 186–199. [CrossRef] [PubMed]
16. Ambhorkar, P.; Wang, Z.; Ko, H.; Lee, S.; Koo, K.-I.; Kim, K.; Cho, D.D. Nanowire-based biosensors: From growth to applications. *Micromachines* **2018**, *9*, 679. [CrossRef] [PubMed]
17. Gao, A.; Chen, S.; Wang, Y.; Li, T. Silicon nanowire field-effect-transistor-based biosensor for biomedical applications. *Sens. Mater.* **2018**, *30*, 1619–1628. [CrossRef]
18. Choi, J.; Seong, T.W.; Jeun, M.; Lee, K.H. Field-effect biosensors for on-site detection: Recent advances and promising targets. *Adv. Healthc. Mater.* **2017**, *6*, 1700796. [CrossRef]
19. Syu, Y.-C.; Hsu, W.-E.; Lin, C.-T. Review—Field-effect transistor biosensing: Devices and clinical applications. *ECS J. Solid State Sci. Technol.* **2018**, *7*, Q3196–Q3207. [CrossRef]
20. Alabsi, S.S.; Ahmed, A.Y.; Dennis, J.O.; Khir, M.H.M.; Algamili, A.S. A review of carbon nanotube field effect-based biosensors. *IEEE Access* **2020**, *8*, 69509–69521. [CrossRef]
21. Schöning, M.J.; Poghossian, A. Bio FEDs (field-effect devices): State-of-the-art and new directions. *Electroanalysis* **2006**, *18*, 1893–1900. [CrossRef]
22. Poghossian, A.; Schöning, M.J. Label-free sensing of biomolecules with field-effect devices for clinical applications. *Electroanalysis* **2014**, *26*, 1197–1213. [CrossRef]
23. Huang, W.; Diallo, A.K.; Dailey, J.L.; Besar, K.; Katz, H.E. Electrochemical processes and mechanistic aspects of field-effect sensors for biomolecules. *J. Mater. Chem. C* **2015**, *3*, 6445–6470. [CrossRef] [PubMed]
24. De Moraes, A.C.M.; Kubota, L.T. Recent trends in field-effect transistors-based immunosensors. *Chemosensors* **2016**, *4*, 20. [CrossRef]
25. Syedmoradi, L.; Ahmadi, A.; Norton, M.L.; Omidfar, K. A review on nanomaterial-based field effect transistor technology for biomarker detection. *Microchim. Acta* **2019**, *186*, 739. [CrossRef]
26. Fabry, P.; Laurent-Yvonnou, L. The *C-V* method for characterizing ISFET or EOS device with ion-sensitive membranes. *J. Electroanal. Chem.* **1990**, *286*, 23–40. [CrossRef]

27. Poghossian, A.; Ingebrandt, S.; Abouzar, M.H.; Schöning, M.J. Label-free detection of charged macromolecules by using a field-effect-based sensor platform: Experiments and possible mechanisms of signal generation. *Appl. Phys. A* **2007**, *87*, 517–524. [CrossRef]
28. Klein, M. Characterisation of ion-sensitive layer systems with a C (V) measurement method operating at constant capacitance. *Sens. Actuators B* **1990**, *1*, 354–356. [CrossRef]
29. Schöning, M.J.; Brinkmann, D.; Rolka, D.; Demuth, C.; Poghossian, A. CIP (cleaning-in-place) suitable "non-glass" pH sensor based on a Ta_2O_5-gate EIS structure. *Sens. Actuators B* **2005**, *111–112*, 423–429. [CrossRef]
30. Poghossian, A.; Bäcker, M.; Mayer, D.; Schöning, M.J. Gating capacitive field-effect sensors by the charge of nanoparticle/molecule hybrids. *Nanoscale* **2015**, *7*, 1023–1031. [CrossRef]
31. Lee, T.N.; Chen, H.J.H.; Huang, Y.-C.; Hsieh, K.-C. Electrolyte-insulator-semiconductor pH sensors with arrayed patterns manufactured by nano imprint technology. *J. Electrochem. Soc.* **2018**, *165*, B767–B772. [CrossRef]
32. Jang, H.-J.; Kim, M.-S.; Cho, W.-J. Development of engineered sensing membranes for field-effect ion-sensitive devices based on stacked high-κ dielectric layers. *IEEE Electron. Device Lett.* **2011**, *32*, 973–975. [CrossRef]
33. Oh, J.Y.; Jang, H.-J.; Cho, W.-J.; Islam, M.S. Highly sensitive electrolyte-insulator-semiconductor pH sensors enabled by silicon nanowires with Al_2O_3/SiO_2 sensing membrane. *Sens. Actuators B* **2012**, *171–172*, 238–243. [CrossRef]
34. Schöning, M.J.; Tsarouchas, D.; Beckers, L.; Schubert, J.; Zander, W.; Kordos, P.; Lüth, H. A highly long-term stable silicon-based pH sensor fabricated by pulsed laser deposition technique. *Sens. Actuators B* **1996**, *35*, 228–233. [CrossRef]
35. Chen, M.; Jin, Y.; Qu, X.; Jin, Q.; Zhao, J. Electrochemical impedance spectroscopy study of Ta_2O_5 based EIOS pH sensors in acid environment. *Sens. Actuators B* **2014**, *192*, 399–405. [CrossRef]
36. Veigas, B.; Branquinho, R.; Pinto, J.V.; Wojcik, P.J.; Martins, R.; Fortunato, E.; Baptista, P.V. Ion sensing (EIS) real-time quantitative monitorization of isothermal DNA amplification. *Biosens. Bioelectron.* **2014**, *52*, 50–55. [CrossRef]
37. Chang, L.-B.; Lee, Y.-L.; Lai, C.-S.; Hsieh, L.-Z.; Ko, H.-H. The pH-sensitive characteristics of thermal stacked ZrO_2/SiO_2 sensors. *ECS Trans.* **2007**, *2*, 1–9. [CrossRef]
38. Lai, C.-S.; Yang, C.-M.; Lu, T.-F. pH sensitivity improvement on 8 nm thick hafnium oxide by post deposition annealing. *Electrochem. Solid-State Lett.* **2006**, *9*, G90–G92. [CrossRef]
39. Yang, C.-M.; Lai, C.-S.; Lu, T.-F.; Wang, T.-C.; Pijanowska, D.G. Drift and hysteresis effects improved by RTA treatment on hafnium oxide in pH-sensitive applications. *J. Electrochem. Soc.* **2008**, *155*, J326–J330. [CrossRef]
40. Lu, T.-F.; Wang, J.-C.; Yang, C.-M.; Chang, C.-P.; Ho, K.-I.; Ai, C.-F.; Lai, C.-S. Non-ideal effects improvement of SF_6 plasma treated hafnium oxide film based on electrolyte-insulator-semiconductor structure for pH-sensor application. *Microelectron. Reliab.* **2010**, *50*, 742–746. [CrossRef]
41. Lu, T.-F.; Chuang, H.-C.; Wang, J.-C.; Yang, C.-M.; Kuo, P.-C.; Lai, C.-S. Effects of thickness effect and rapid thermal annealing on pH sensing characteristics of thin HfO_2 films formed by atomic layer deposition. *Jpn. J. Appl. Phys.* **2011**, *50*, 10PG03. [CrossRef]
42. Kao, C.H.; Chen, H.; Lee, M.L.; Liu, C.C.; Ueng, H.-Y.; Chu, Y.C.; Chen, C.B.; Chang, K.M. Effects of N_2 and O_2 annealing on the multianalyte biosensing characteristics of CeO_2-based electrolyte-insulator-semiconductor structures. *Sens. Actuators B* **2014**, *194*, 503–510. [CrossRef]
43. Chang, L.-B.; Ko, H.-H.; Lee, Y.-L.; Lai, C.-S.; Wang, C.-Y. The electrical and pH-sensitive characteristics of thermal Gd_2O_3/SiO_2-stacked oxide capacitors. *J. Electrochem. Soc.* **2006**, *153*, G330–G332. [CrossRef]
44. Yang, C.-M.; Wang, C.-Y.; Lai, C.-S. Characterization on pH sensing performance and structural properties of gadolinium oxide post-treated by nitrogen rapid thermal annealing. *J. Vac. Sci. Technol. B* **2014**, *32*, 03D113. [CrossRef]
45. Kao, C.H.; Wang, J.C.; Lai, C.S.; Huang, C.Y.; Ou, J.C.; Wang, H.Y. Ti-doped Gd_2O_3 sensing membrane for electrolyte-insulator-semiconductor pH sensor. *Thin Solid Films* **2012**, *520*, 3760–3763. [CrossRef]
46. Pan, T.-M.; Lin, C.-W. Structural and sensing properties of high-k Lu_2O_3 electrolyte-insulator-semiconductor pH sensors. *J. Electrochem. Soc.* **2011**, *158*, J96–J99. [CrossRef]
47. Pan, T.-M.; Lin, C.-W.; Lin, J.-C.; Wu, M.-H. Structural properties and sensing characteristics of thin Nd_2O_3 sensing films for pH detection. *Electrochem. Solid-State Lett.* **2009**, *12*, J96–J99. [CrossRef]

48. Pan, T.-M.; Cheng, C.-H.; Lee, C.-D. Yb$_2$O$_3$ thin films as a sensing membrane for pH-ISFET application. *J. Electrochem. Soc.* **2009**, *156*, J108–J111. [CrossRef]
49. Pan, T.-M.; Lin, C.-W. Structural and sensing characteristics of Dy$_2$O$_3$ and Dy$_2$TiO$_5$ electrolyte-insulator-semiconductor pH sensors. *J. Phys. Chem. C* **2010**, *114*, 17914–17919. [CrossRef]
50. Pan, T.-M.; Lin, C.-W.; Hsu, B.-K. Post-deposition anneal on structural and sensing characteristics of high-κ Er$_2$TiO$_5$ electrolyte-insulator-semiconductor pH sensors. *IEEE Electron. Device Lett.* **2012**, *33*, 116–118. [CrossRef]
51. Jan, S.-S.; Chen, Y.-C.; Chou, J.-C.; Cheng, C.-C.; Lu, C.-T. Preparation and properties of lead titanate gate ion-sensitive field-effect transistors by the sol-gel method. *Jpn. J. Appl. Phys.* **2002**, *41*, 942–948. [CrossRef]
52. Singh, K.; Lou, B.-S.; Her, J.-L.; Pan, T.-M. Influence of annealing temperature on structural compositions and pH sensing properties of sol-gel derived YTi$_x$O$_y$ electroceramic sensing membranes. *J. Electrochem. Soc.* **2019**, *166*, B187–B192. [CrossRef]
53. Wu, M.-H.; Lee, Y.-F.; Lin, C.-W.; Tsai, S.-W.; Wang, H.-Y.; Pan, T.-M. Development of high-k Tm$_2$Ti$_2$O$_7$ sensing membrane-based electrolyte-insulator-semiconductor for pH detection and its application for glucose biosensing using poly(N-isopropylacrylamide) as an enzyme encapsulation material. *J. Mater. Chem.* **2011**, *21*, 539–547. [CrossRef]
54. Chen, C.-Y.; Chou, J.-C.; Chou, H.-T. pH Sensing of Ba$_{0.7}$Sr$_{0.3}$TiO$_3$/SiO$_2$ film for metal-oxide-semiconductor and ion-sensitive field-effect transistor devices. *J. Electrochem. Soc.* **2009**, *156*, G59–G64. [CrossRef]
55. Buniatyan, V.V.; Abouzar, M.H.; Martirosyan, N.W.; Schubert, J.; Gevorgian, S.; Schöning, M.J.; Poghossian, A. pH-sensitive properties of barium strontium titanate (BST) thin films prepared by pulsed laser deposition technique. *Phys. Status Solidi A* **2010**, *207*, 824–830. [CrossRef]
56. Huck, C.; Poghossian, A.; Kerroumi, I.; Schusser, S.; Bäcker, M.; Zander, W.; Schubert, J.; Buniatyan, V.V.; Martirosyan, N.W.; Wagner, P.; et al. Multiparameter sensor chip with barium strontium titanate as multipurpose material. *Electroanalysis* **2014**, *26*, 980–987. [CrossRef]
57. Lai, C.-S.; Lue, C.-E.; Yang, C.-M.; Pijanowska, D.G. Fluorine incorporation and thermal treatment on single and stacked Si$_3$N$_4$ membranes for ISFET/REFET application. *J. Electrochem. Soc.* **2010**, *157*, J8–J12. [CrossRef]
58. Poghossian, A.; Abouzar, M.H.; Razavi, A.; Bäcker, M.; Bijnens, N.; Williams, O.A.; Haenen, K.; Moritz, W.; Wagner, P.; Schöning, M.J. Nanocrystalline-diamond thin films with high pH and penicillin sensitivity prepared on a capacitive Si-SiO$_2$ structure. *Electrochim. Acta* **2009**, *54*, 5981–5985. [CrossRef]
59. Yates, D.E.; Levine, S.; Healy, T.W. Site-binding model of the electrical double layer at the oxide/water interface. *J. Chem. Soc. Faraday Trans. I* **1974**, *70*, 1807–1818. [CrossRef]
60. Niu, M.-N.; Ding, X.-F.; Tong, Q.-Y. Effect of two types of surface sites on the characteristics of Si$_3$N$_4$-gate pH-ISFET. *Sens. Actuators B* **1996**, *37*, 13–17. [CrossRef]
61. Van Hal, R.E.G.; Eijkel, J.C.T.; Bergveld, P. A novel description of ISFET sensitivity with the buffer capacity and double-layer capacitance as key parameters. *Sens. Actuators B* **1995**, *24–25*, 201–205. [CrossRef]
62. Oelßner, W.; Zosel, J.; Guth, U.; Pechstein, T.; Babel, W.; Connery, J.G.; Demuth, C.; Grote Gansey, M.; Verburg, J.B. Encapsulation of ISFET sensor chips. *Sens. Actuators B* **2005**, *105*, 104–117. [CrossRef]
63. Bäcker, M.; Beging, S.; Biselli, M.; Poghossian, A.; Wang, J.; Zang, W.; Wagner, P.; Schöning, M.J. Concept for a solid-state multi-parameter sensor system for cell-culture monitoring. *Electrochim. Acta* **2009**, *54*, 6107–6112.
64. Bäcker, M.; Pouyeshman, S.; Schnitzler, T.; Poghossian, A.; Wagner, P.; Biselli, M.; Schöning, M.J. A silicon-based multi-sensor chip for monitoring of fermentation processes. *Phys. Status Solidi A* **2011**, *208*, 1364–1369. [CrossRef]
65. Huck, C.; Schiffels, J.; Herrera, C.N.; Schelden, M.; Selmer, T.; Poghossian, A.; Baumann, M.E.M.; Wagner, P.; Schöning, M.J. Metabolic responses of *Escherichia coli* upon glucose pulses captured by a capacitive field-effect sensor. *Phys. Status Solidi A* **2013**, *210*, 926–931. [CrossRef]
66. Mathew, A.; Pandian, G.; Bhattacharya, E.; Chadha, A. Novel applications of silicon and porous silicon based EISCAP biosensors. *Phys. Status Solidi A* **2009**, *206*, 1369–1373. [CrossRef]
67. Mourzina, Y.; Mai, T.; Poghossian, A.; Ermolenko, Y.; Yoshinobu, T.; Vlasov, Y.; Iwasaki, H.; Schöning, M.J. K$^+$-selective field-effect sensors as transducers for bioelectronics applications. *Electrochim. Acta* **2003**, *48*, 3333–3339. [CrossRef]
68. Poghossian, A.; Mai, D.-T.; Mourzina, Y.; Schöning, M.J. Impedance effect of an ion-sensitive membrane: Characterisation of an EMIS sensor by impedance spectroscopy, capacitance-voltage and constant-capacitance method. *Sens. Actuators B* **2004**, *103*, 423–428. [CrossRef]

69. Beging, S.; Mlynek, D.; Hataihimakul, S.; Poghossian, A.; Baldsiefen, G.; Busch, H.; Laube, N.; Kleinen, L.; Schöning, M.J. Field-effect calcium sensor for the determination of the risk of urinary stone formation. *Sens. Actuators B* **2010**, *144*, 374–379. [CrossRef]
70. Zazoua, A.; Morakchi, K.; Kherrat, R.; Samar, M.H.; Errachid, A.; Jaffrezic-Renault, N.; Boubellout, R. Electrochemical characterization of an EIS sensor functionalized with a TOPO doped polymeric layer for Cr(VI) detection. *ITBM-RBM* **2008**, *29*, 187–191. [CrossRef]
71. Braik, M.; Dridi, C.; Ali, M.B.; Ali, M.; Abbas, M.; Zabala, M.; Bausells, J.; Zine, N.; Jaffrezic-Renault, N.; Errachid, A. Development of a capacitive chemical sensor based on Co(II)-phthalocyanine acrylate-polymer/HfO$_2$/SiO$_2$/Si for detection of perchlorate. *J. Sens. Sens. Syst.* **2015**, *4*, 17–23. [CrossRef]
72. Cho, H.; Kim, K.; Meyyappan, M.; Baek, C.-K. LaF$_3$ electrolyte-insulator-semiconductor sensor for detecting fluoride ions. *Sens. Actuators B* **2019**, *279*, 183–188. [CrossRef]
73. Spelthahn, H.; Schaffrath, S.; Coppe, T.; Rufi, F.; Schöning, M.J. Development of an electrolyte-insulator-semiconductor (EIS) based capacitive heavy metal sensor for the detection of Pb^{2+} and Cd^{2+} ions. *Phys. Status Solidi A* **2010**, *207*, 930–934. [CrossRef]
74. Barhoumi, H.; Haddad, R.; Maaref, A.; Bausells, J.; Bessueille, F.; Léonard, D.; Jaffrezic-Renault, N.; Martelet, C.; Zine, N.; Errachid, A. Na$^+$-implanted membrane for a capacitive sodium electrolyte-insulator-semiconductor microsensors. *Sens. Lett.* **2008**, *6*, 204–208. [CrossRef]
75. Nouira, W.; Haddad, R.; Barhoumi, H.; Maaref, A.; Bausells, J.; Bessueille, F.; Léonard, D.; Jaffrezic-Renault, N.; Errachid, A. Na$^+$ and K$^+$ implanted membranes for micro-sensors development. *Sens. Lett.* **2009**, *7*, 689–693. [CrossRef]
76. Sakly, H.; Mlika, R.; Chaabane, H.; Beji, L.; Ouada, H.B. Anodically oxidized porous silicon as a substrate for EIS sensors. *Mater. Sci. Eng. C* **2006**, *26*, 232–235. [CrossRef]
77. Chermiti, J.; Ben Ali, M.; Jaffrezic-Renault, N. Numerical modelling of electrical behaviour of semiconductor-based micro-sensors for copper ions detection. *Sens. Lett.* **2011**, *9*, 2339–2342. [CrossRef]
78. Bergaoui, Y.; Ben Ali, M.; Abdelghani, A.; Vocanson, F.; Jaffrezic-Renault, N. Electrochemical study of Si/SiO$_2$/Si$_3$N$_4$ structures and gold electrodes functionalized with amide thiacalix[4]arene for the detection of Hg^{2+} ions in solution. *Sens. Lett.* **2009**, *7*, 694–701. [CrossRef]
79. Ijeri, V.S.; Nair, J.R.; Zanarini, S.; Gerbaldi, C. Capacitive and impedimetric sensors based on macrocyclic compounds. *Int. J. Anal. Chem.* **2011**, *2*, 60–71.
80. Molinnus, D.; Muschallik, L.; Gonzalez, L.O.; Bongaerts, J.; Wagner, T.; Selmer, T.; Siegert, P.; Keusgen, M.; Schöning, M.J. Development and characterization of a field-effect biosensor for the detection of acetoin. *Biosens. Bioelectron.* **2018**, *115*, 1–6. [CrossRef]
81. Lin, Y.-H.; Chiang, C.-H.; Wu, M.-H.; Pan, T.-M.; Luo, J.-D.; Chiou, C.-C. Solid-state sensor incorporated in microfluidic chip and magnetic-bead enzyme immobilization approach for creatinine and glucose detection in serum. *Appl. Phys. Lett.* **2011**, *99*, 253704. [CrossRef]
82. Lin, Y.-H.; Wang, S.-H.; Wu, M.-H.; Pan, T.-M.; Lai, C.-S.; Luo, J.-D.; Chiou, C.-C. Integrating solid-state sensor and microfluidic devices for glucose, urea and creatinine detection based on enzyme-carrying alginate microbeads. *Biosens. Bioelectron.* **2013**, *43*, 328–335. [CrossRef] [PubMed]
83. Pan, T.-M.; Lin, C.-W.; Lin, W.-Y.; Wu, M.-H. High-k Tm$_2$Ti$_2$O$_7$ electrolyte-insulator semiconductor creatinine biosensor. *IEEE Sens. J.* **2011**, *11*, 2388–2394. [CrossRef]
84. Turek, M.; Ketterer, L.; Claßen, M.; Berndt, H.K.; Elbers, G.; Krüger, P.; Keusgen, M.; Schöning, M.J. Development and electrochemical investigations of an EIS-(electrolyte-insulator-semiconductor) based biosensor for cyanide detection. *Sensors* **2007**, *7*, 1415–1426. [CrossRef]
85. Ben Ali, M.; Gonchar, M.; Gayda, G.; Paryzhak, S.; Maaref, M.A.; Jaffrezic-Renault, N.; Korpan, Y. Formaldehyde-sensitive sensor based on recombinant formaldehyde dehydrogenase using capacitance versus voltage measurements. *Biosens. Bioelectron.* **2007**, *22*, 2790–2795. [CrossRef]
86. Gun, J.; Schöning, M.J.; Abouzar, M.H.; Poghossian, A.; Katz, E. Field-effect nanoparticle-based glucose sensor on a chip: Amplification effect of coimmobilized redox species. *Electroanalysis* **2008**, *20*, 1748–1753. [CrossRef]
87. Her, J.-L.; Wu, M.-H.; Peng, Y.-B.; Pan, T.-M.; Weng, W.-H.; Pang, S.-T.; Chi, L. High performance GdTi$_x$O$_y$ electrolyte-insulator-semiconductor pH sensor and biosensor. *Int. J. Electrochem. Sci.* **2013**, *8*, 606–620.

88. Wu, M.-H.; Yang, H.-W.; Hua, M.-Y.; Peng, Y.-B.; Pan, T.-M. High-κ $GdTi_xO_y$ sensing membrane-based electrolyte-insulator-semiconductor with magnetic nanoparticles as enzyme carriers for protein contamination-free glucose biosensing. *Biosens. Bioelectron.* **2013**, *47*, 99–105. [CrossRef]
89. Kao, C.H.; Chen, H.; Lee, M.L.; Liu, C.C.; Ueng, H.-Y.; Chu, Y.C.; Chen, Y.J.; Chang, K.M. Multianalyte biosensor based on pH-sensitive ZnO electrolyte-insulator-semiconductor structures. *J. Appl. Phys.* **2014**, *115*, 184701.
90. Kao, C.H.; Chen, H.; Liu, C.C.; Chen, C.Y.; Chen, Y.T.; Chu, Y.C. Electrical, material and multianalyte-sensitive characteristics of thermal CeO_2/SiO_2-stacked oxide capacitors. *Thin Solid Films* **2014**, *570*, 552–557. [CrossRef]
91. Kao, C.H.; Chen, H.; Hou, F.Y.S.; Chang, S.W.; Chang, C.W.; Lai, C.S.; Chen, C.P.; He, Y.Y.; Lin, S.-R.; Hsieh, K.M.; et al. Fabrication of multianalyte CeO_2 nanograin electrolyte-insulator-semiconductor biosensors by using CF_4 plasma treatment. *Sens. Biosensing Res.* **2015**, *5*, 71–77. [CrossRef]
92. Morais, P.V.; Gomes, V.F., Jr.; Silva, A.C.A.; Dantas, N.O.; Schöning, M.J.; Siqueira, J.R., Jr. Nanofilm of ZnO nanocrystals/carbon nanotubes as biocompatible layer for enzymatic biosensors in capacitive field-effect devices. *J. Mater. Sci.* **2017**, *52*, 12314–12325. [CrossRef]
93. Lin, Y.-H.; Das, A.; Wu, M.-H.; Pan, T.-M.; Lai, C.-S. Microfluidic chip integrated with an electrolyte-insulator-semiconductor sensor for pH and glucose level measurement. *Int. J. Electrochem. Sci.* **2013**, *8*, 5886–5901.
94. Lin, C.F.; Kao, C.H.; Lin, C.Y.; Liu, Y.W.; Wang, C.H. The electrical and physical characteristics of Mg-doped ZnO sensing membrane in EIS (electrolyte-insulator-semiconductor) for glucose sensing applications. *Results Phys.* **2020**, *16*, 102976. [CrossRef]
95. Lin, C.F.; Kao, C.H.; Lin, C.Y.; Chen, K.L.; Lin, Y.H. NH_3 plasma-treated magnesium doped zinc oxide in biomedical sensors with electrolyte-insulator-semiconductor (EIS) structure for urea and glucose applications. *Nanomaterials* **2020**, *10*, 583. [CrossRef]
96. Schöning, M.J.; Arzdorf, M.; Mulchandani, P.; Chen, W.; Mulchandani, A. A capacitive field-effect sensor for the direct determination of organophosphorus pesticides. *Sens. Actuators B* **2003**, *91*, 92–97. [CrossRef]
97. Braham, Y.; Barhoumi, H.; Maaref, A. Urease capacitive biosensors using functionalized magnetic nanoparticles for atrazine pesticide detection in environmental samples. *Anal. Methods* **2013**, *5*, 4898–4904. [CrossRef]
98. Poghossian, A.; Thust, M.; Schöning, M.J.; Müller-Veggian, M.; Kordos, P.; Lüth, H. Cross-sensitivity of a capacitive penicillin sensor combined with a diffusion barrier. *Sens. Actuators B* **2000**, *68*, 260–265. [CrossRef]
99. Siqueira, J.R., Jr.; Abouzar, M.H.; Poghossian, A.; Zucolotto, V.; Oliveira, O.N., Jr.; Schoning, M.J. Penicillin biosensor based on a capacitive field-effect structure functionalized with a dendrimer/carbon nanotube multilayer. *Biosens. Bioelectron.* **2009**, *25*, 497–501. [CrossRef]
100. Abouzar, M.H.; Poghossian, A.; Razavi, A.; Besmehn, A.; Bijnens, N.; Williams, O.A.; Haenen, K.; Wagner, P.; Schöning, M.J. Penicillin detection with nanocrystalline-diamond field-effect sensor. *Phys. Status Solidi A* **2008**, *205*, 2141–2145. [CrossRef]
101. Abouzar, M.H.; Poghossian, A.; Siqueira, J.R., Jr.; Oliveira, O.N., Jr.; Moritz, W.; Schöning, M.J. Capacitive electrolyte-insulator-semiconductor structures functionalized with a polyelectrolyte/enzyme multilayer: New strategy for enhanced field-effect biosensing. *Phys. Status Solidi A* **2010**, *207*, 884–890. [CrossRef]
102. Beging, S.; Leinhos, M.; Jablonski, M.; Poghossian, A.; Schöning, M.J. Studying the spatially resolved immobilization of enzymes on a capacitive field-effect structure by means of nano-spotting. *Phys. Status Solidi A* **2015**, *212*, 1353–1358. [CrossRef]
103. Poghossian, A.; Jablonski, M.; Koch, C.; Bronder, T.S.; Rolka, D.; Wege, C.; Schöning, M.J. Field-effect biosensor using virus particles as scaffolds for enzyme immobilization. *Biosens. Bioelectron.* **2018**, *110*, 168–174. [CrossRef] [PubMed]
104. Nouira, W.; Barhoumi, H.; Maaref, A.; Jaffrézic Renault, N.; Siadat, M. Tailoring of analytical performances of urea biosensors using nanomaterials. *J. Phys. Conf. Ser.* **2013**, *416*, 012010. [CrossRef]
105. Pan, T.-M.; Lin, C.-W. High-k Dy_2TiO_5 electrolyte-insulator-semiconductor urea biosensors. *J. Electrochem. Soc.* **2011**, *158*, J100–J105. [CrossRef]
106. Siqueira, J.R., Jr.; Molinnus, D.; Beging, S.; Schöning, M.J. Incorporating a hybrid urease-carbon nanotubes sensitive nanofilm on capacitive field-effect sensors for urea detection. *Anal. Chem.* **2014**, *86*, 5370–5375. [CrossRef] [PubMed]

107. Wang, I.-S.; Lin, Y.-T.; Huang, C.-H.; Lu, T.-F.; Lue, C.-E.; Yang, P.; Pijanowska, D.G.; Yang, C.-M.; Wang, J.-C.; Yu, J.-S.; et al. Immobilization of enzyme and antibody on ALD-HfO$_2$-EIS structure by NH$_3$ plasma treatment. *Nanoscale Res. Lett.* **2012**, *7*, 179. [CrossRef] [PubMed]
108. Wu, M.-H.; Cheng, C.-H.; Lai, C.-S.; Pan, T.-M. Structural properties and sensing performance of high-k Sm$_2$O$_3$ membrane-based electrolyte-insulator-semiconductor for pH and urea detection. *Sens. Actuators B* **2009**, *138*, 221–227. [CrossRef]
109. Sahraoui, Y.; Barhoumi, H.; Maaref, A.; Jaffrezic-Renault, N. A novel capacitive biosensor for urea assay based on modified magnetic nanobeads. *Sens. Lett.* **2011**, *9*, 2141–2146. [CrossRef]
110. Pan, T.-M.; Lin, J.-C. A TiO$_2$/Er$_2$O$_3$ stacked electrolyte/insulator/semiconductor film pH-sensor for the detection of urea. *Sens. Actuators B* **2009**, *138*, 474–479. [CrossRef]
111. Pan, T.-M.; Lin, J.-C.; Wu, M.-H.; Lai, C.-S. Structural properties and sensing performance of high-k Nd$_2$TiO$_5$ thin layer-based electrolyte-insulator-semiconductor for pH detection and urea biosensing. *Biosens. Bioelectron.* **2009**, *24*, 2864–2870. [CrossRef] [PubMed]
112. Gun, J.; Rizkov, D.; Lev, O.; Abouzar, M.H.; Poghossian, A.; Schöning, M.J. Oxygen plasma-treated gold nanoparticle-based field-effect devices as transducer structures for bio-chemical sensing. *Microchim. Acta* **2009**, *164*, 395–404. [CrossRef]
113. Poghossian, A.; Schöning, M.J. Nanomaterial-modified capacitive field-effect biosensors. In *Label-Free Biosensing: Advanced Materials, Devices and Applications*, 1st ed.; Schöning, M.J., Poghossian, A., Eds.; Springer International Publishing AG: Cham, Switzerland, 2018; pp. 1–26.
114. Koch, C.; Poghossian, A.; Schöning, M.J.; Wege, C. Penicillin detection by *tobacco mosaic virus*-assisted colorimetric biosensors. *Nanotheranostics* **2018**, *2*, 184–196. [CrossRef] [PubMed]
115. Calò, A.; Eiben, S.; Okuda, M.; Bittner, A.M. Nanoscale device architectures derived from biological assemblies: The case of *tobacco mosaic virus* and (apo)ferritin. *Jpn. J. Appl. Phys.* **2016**, *55*, 03DA01. [CrossRef]
116. Bäcker, M.; Koch, C.; Eiben, S.; Geiger, F.; Eber, F.; Gliemann, H.; Poghossian, A.; Wege, C.; Schöning, M.J. *tobacco mosaic virus* as enzyme nanocarrier for electrochemical biosensors. *Sens. Actuators B* **2017**, *238*, 716–722. [CrossRef]
117. Molinnus, D.; Beging, S.; Lowis, C.; Schöning, M.J. Towards a multi-enzyme capacitive field-effect biosensor by comparative study of drop-coating and nano-spotting technique. *Sensors* **2020**, *20*, 4924. [CrossRef]
118. Wang, J.; Katz, E. Digital biosensors with built-in logic for biomedical applications—Biosensors based on a biocomputing concept. *Anal. Bioanal. Chem.* **2010**, *398*, 1591–1603. [CrossRef]
119. Lai, Y.H.; Sun, S.C.; Chuang, M.C. Biosensors with built-in biomolecular logic gates for practical applications. *Biosensors* **2014**, *4*, 273–300. [CrossRef]
120. Katz, E.; Minko, S. Enzyme-based logic systems interfaced with signal-responsive materials and electrodes. *Chem. Commun.* **2015**, *51*, 3493–3500. [CrossRef]
121. Katz, E.; Poghossian, A.; Schöning, M.J. Enzyme-based logic gates and circuits—Analytical applications and interfacing with electronics. *Anal. Bioanal. Chem.* **2017**, *409*, 81–94. [CrossRef]
122. Koushanpour, A.; Gamella, M.; Guo, Z.; Honarvarfard, E.; Poghossian, A.; Schöning, M.J.; Alexandrov, K.; Katz, E. Ca^{2+}-switchable glucose dehydrogenase associated with electrochemical/electronic interfaces: Applications to signal-controlled power production and biomolecular release. *J. Phys. Chem. B* **2017**, *121*, 11465–11471. [CrossRef] [PubMed]
123. Molinnus, D.; Bäcker, M.; Iken, H.; Poghossian, A.; Keusgen, M.; Schöning, M.J. Concept for a biomolecular logic chip with an integrated sensor and actuator function. *Phys. Status Solidi A* **2015**, *212*, 1382–1388. [CrossRef]
124. Molinnus, D.; Poghossian, A.; Keusgen, M.; Katz, E.; Schöning, M.J. Coupling of biomolecular logic gates with electronic transducers: From single enzyme logic gates to sense/act/treat chips. *Electroanalysis* **2017**, *29*, 1840–1849. [CrossRef]
125. Krämer, M.; Pita, M.; Zhou, J.; Ornatska, M.; Poghossian, A.; Schöning, M.J.; Katz, E. Coupling of biocomputing systems with electronic chips: Electronic interface for transduction of biochemical information. *J. Phys. Chem. C* **2009**, *113*, 2573–2579. [CrossRef]
126. Poghossian, A.; Malzahn, K.; Abouzar, M.H.; Mehndiratta, P.; Katz, E.; Schöning, M.J. Integration of biomolecular logic gates with field-effect transducers. *Electrochim. Acta* **2011**, *56*, 9661–9665. [CrossRef]

127. Poghossian, A.; Katz, E.; Schöning, M.J. Enzyme logic AND-Reset and OR-Reset gates based on a field-effect electronic transducer modified with multi-enzyme membrane. *Chem. Commun.* **2015**, *51*, 6564–6567. [CrossRef]
128. Honarvarfard, E.; Gamella, M.; Poghossian, A.; Schöning, M.J.; Katz, E. An enzyme-based reversible controlled NOT (CNOT) logic gate operating on a semiconductor transducer. *Appl. Mater. Today* **2017**, *9*, 266–270. [CrossRef]
129. Jablonski, M.; Poghossian, A.; Molinnus, D.; Keusgen, M.; Katz, E.; Schöning, M.J. Enzyme-based XOR logic gate with electronic transduction of the output signal. *Int. J. Unconv. Comput.* **2019**, *14*, 375–383.
130. Fritz, J.; Cooper, E.; Gaudet, S.; Sorger, P.; Manalis, S. Electronic detection of DNA by its intrinsic molecular charge. *Proc. Natl. Acad. Sci. USA* **2002**, *99*, 14142–14146. [CrossRef]
131. Kumar, N.; Kumar, S.; Kumar, J.; Panda, S. Investigation of mechanisms involved in the enhanced label free detection of prostate cancer biomarkers using field effect devices. *J. Electrochem. Soc.* **2017**, *164*, B409–B416. [CrossRef]
132. Bronder, T.S.; Poghossian, A.; Jessing, M.P.; Keusgen, M.; Schöning, M.J. Surface regeneration and reusability of label-free DNA biosensors based on weak polyelectrolyte-modified capacitive field-effect structures. *Biosens. Bioelectron.* **2019**, *126*, 510–517. [CrossRef] [PubMed]
133. Gun, J.; Gutkin, V.; Lev, O.; Boyen, H.-G.; Saitner, M.; Wagner, P.; D'Olieslaeger, M.; Abouzar, M.H.; Poghossian, A.; Schöning, M.J. Tracing gold nanoparticle charge by electrolyte-insulator-semiconductor devices. *J. Phys. Chem. C* **2011**, *115*, 4439–4445. [CrossRef]
134. Siqueira, J.R., Jr.; Abouzar, M.H.; Bäcker, M.; Zucolotto, V.; Poghossian, A.; Oliveira, O.N., Jr.; Schöning, M.J. Carbon nanotubes in nanostructured films: Potential application as amperometric and potentiometric field-effect (bio-)chemical sensors. *Phys. Status Solidi A* **2009**, *206*, 462–467. [CrossRef]
135. Abouzar, M.H.; Poghossian, A.; Cherstvy, A.G.; Pedraza, A.M.; Ingebrandt, S.; Schöning, M.J. Label-free electrical detection of DNA by means of field-effect nanoplate capacitors: Experiments and modeling. *Phys. Status Solidi A* **2012**, *209*, 925–934. [CrossRef]
136. Pan, T.-M.; Chang, K.-Y.; Lin, C.-W.; Tsai, S.-W.; Wu, M.H. Label-free detection of DNA using high-k $Lu_2Ti_2O_7$ electrolyte-insulator-semiconductors. *J. Mater. Chem.* **2012**, *22*, 1358–1363. [CrossRef]
137. Bronder, T.S.; Poghossian, A.; Scheja, S.; Wu, C.; Keusgen, M.; Mewes, D.; Schöning, M.J. DNA immobilization and hybridization detection by the intrinsic molecular charge using capacitive field-effect sensors modified with a charged weak polyelectrolyte layer. *ACS Appl. Mater. Interfaces* **2015**, *7*, 20068–20075. [CrossRef]
138. Lin, Y.-T.; Purwidyantri, A.; Luo, J.-D.; Chiou, C.-C.; Yang, C.-M.; Lo, C.-H.; Hwang, T.-L.; Yen, T.-H.; Lai, C.-S. Programming a nonvolatile memory-like sensor for KRAS gene sensing and signal enhancement. *Biosens. Bioelectron.* **2016**, *79*, 63–70. [CrossRef]
139. Hou, C.-S.J.; Milovic, N.; Godin, M.; Russo, P.R.; Chakrabarti, R.; Manalis, S.R. Label-free microelectronic PCR quantification. *Anal. Chem.* **2006**, *78*, 2526–2531. [CrossRef]
140. Branquinho, R.; Veigas, B.; Pinto, J.V.; Martins, R.; Fortunato, E.; Baptista, P.V. Real-time monitoring of PCR amplification of proto-oncogene c-MYC using a Ta_2O_5 electrolyte-insulator-semiconductor sensor. *Biosens. Bioelectron.* **2011**, *28*, 44–49. [CrossRef]
141. Bronder, T.S.; Poghossian, A.; Scheja, S.; Wu, C.; Keusgen, M.; Schöning, M.J. Label-free detection of double-stranded DNA molecules with polyelectrolyte-modified capacitive field-effect sensors. *TM. Tech. Mess.* **2017**, *84*, 628–634. [CrossRef]
142. Hou, C.S.J.; Godin, M.; Payer, K.; Chakrabarti, R.; Manalis, S.R. Integrated microelectronic device for label-free nucleic acid amplification and detection. *Lab Chip* **2007**, *7*, 347–354. [PubMed]
143. Lin, Y.-T.; Luo, J.-D.; Chiou, C.-C.; Yang, C.-M.; Wang, C.-Y.; Chou, C.; Lai, C.-S. Detection of KRAS mutation by combination of polymerase chain reaction (PCR) and EIS sensor with new amino group functionalization. *Sens. Actuators B* **2013**, *186*, 374–379. [CrossRef]
144. Bronder, T.S.; Jessing, M.P.; Poghossian, A.; Keusgen, M.; Schöning, M.J. Detection of PCR-amplified tuberculosis DNA fragments with polyelectrolyte-modified field-effect Sensors. *Anal. Chem.* **2018**, *90*, 7747–7753. [CrossRef] [PubMed]
145. Souteyrand, E.; Cloarec, J.P.; Martin, J.R.; Wilson, C.; Lawrence, I.; Mikkelsen, S.; Lawrence, M.F. Direct detection of the hybridization of synthetic homo-oligomer DNA sequences by field effect. *J. Phys. Chem. B* **1997**, *101*, 2980–2985. [CrossRef]

146. Sakata, T.; Miyahara, Y. Detection of DNA recognition events using multi-well field effect devices. *Biosens. Bioelectron.* **2005**, *21*, 827–832. [CrossRef]
147. Zhang, G.-J.; Zhang, G.; Chua, J.H.; Chee, R.-E.; Wong, E.H.; Agarwal, A.; Buddharaju, K.D.; Singh, N.; Gao, Z.; Balasubramanian, N. DNA sensing by silicon nanowire: Charge layer distance dependence. *Nano Lett.* **2008**, *8*, 1066–1070. [CrossRef]
148. Liu, Y.; Dutton, R.W. Effects of charge screening and surface properties on signal transduction in field effect nanowire biosensors. *J. Appl. Phys.* **2009**, *106*, 014701. [CrossRef]
149. Nishio, Y.; Uno, S.; Nakazato, K. Three-dimensional simulation of DNA sensing by ion-sensitive field-effect transistor: Optimization of DNA position and orientation. *Jpn. J. Appl. Phys.* **2013**, *52*, 04CL01. [CrossRef]
150. Goode, J.A.; Rushworth, J.V.H.; Millner, P.A. Biosensor regeneration: A review of common techniques and outcomes. *Langmuir* **2015**, *31*, 6267–6276. [CrossRef]
151. Landheer, D.; McKinnon, W.R.; Aers, G.; Jiang, W.; Deen, M.J.; Shinwari, M.W. Calculation of the response of field-effect transistors to charged biological molecules. *IEEE Sens. J.* **2007**, *7*, 1233–1242. [CrossRef]
152. Wunderlich, B.K.; Neff, P.A.; Bausch, A.R. Mechanism and sensitivity of the intrinsic charge detection of biomolecular interactions by field effect devices. *Appl. Phys. Lett.* **2007**, *91*, 083904. [CrossRef]
153. Lowe, B.M.; Sun, K.; Zeimpekis, I.; Skylaris, C.-K.; Green, N.G. Field-effect sensors—From pH sensing to biosensing: Sensitivity enhancement using streptavidin-biotin as a model system. *Analyst* **2017**, *142*, 4173–4200. [CrossRef]
154. Poghossian, A.; Cherstvy, A.; Ingebrandt, S.; Offenhäusser, A.; Schöning, M.J. Possibilities and limitations of label-free detection of DNA hybridization with field-effect-based devices. *Sens. Actuators B* **2005**, *111–112*, 470–480. [CrossRef]
155. Chung, I.-Y.; Seo, M.; Park, C.H. Particle simulation of ionic screening effects in electrolyte-insulator-semiconductor field-effect transistors. *J. Semicond. Technol. Sci.* **2019**, *19*, 311–320. [CrossRef]
156. Kumar, N.; Senapat, S.; Kumar, S.; Kumar, J.; Panda, S. Functionalized vertically aligned ZnO nanorods for application in electrolyte-insulator-semiconductor based pH sensors and label-free immuno-sensors. *J. Phys. Conf. Ser.* **2016**, *704*, 012013. [CrossRef]
157. Pan, T.-M.; Lin, T.-W.; Chen, C.-Y. Label-free detection of rheumatoid factor using YbY$_x$O$_y$ electrolyte-insulator-semiconductor devices. *Anal. Chim. Acta* **2015**, *891*, 304–311. [CrossRef] [PubMed]
158. Chand, R.; Han, D.; Kim, Y.-S. Rapid detection of protein kinase on capacitive sensing platforms. *IEEE Trans. Nanobiosci.* **2016**, *15*, 843–848. [CrossRef]
159. Chand, R.; Han, D.; Neethirajan, S.; Kim, Y.-S. Detection of protein kinase using an aptamer on a microchip integrated electrolyte-insulator-semiconductor sensor. *Sens. Actuators B* **2017**, *248*, 973–979. [CrossRef]
160. Milovic, N.M.; Behr, J.R.; Godin, M.; Hou, C.-S.J.; Payer, K.R.; Chandrasekaran, A.; Russo, P.R.; Sasisekharan, R.; Manalis, S.R. Monitoring of heparin and its low-molecular-weight analogs by silicon field effect. *Proc. Natl. Acad. Sci. USA* **2006**, *103*, 13374–13379. [CrossRef]
161. Bougrini, M.; Baraket, A.; Jamshaid, T.; El Aissari, A.; Bausells, J.; Zabala, M.; El Bari, N.; Bouchikhi, B.; Jaffrezic-Renault, N.; Abdelhamid, E.; et al. Development of a novel capacitance electrochemical biosensor based on silicon nitride for ochratoxin A detection. *Sens. Actuators B* **2016**, *234*, 446–452. [CrossRef]
162. Casal, P.; Wen, X.; Gupta, S.; Nicholson III, T.; Wang, Y.; Theiss, A.; Bhushan, B.; Brillson, L.; Lu, W.; Lee, S.C. ImmunoFET feasibility in physiological salt environments. *Philos. Trans. R. Soc. A* **2012**, *370*, 2474–2488. [CrossRef] [PubMed]
163. Welch, N.G.; Scoble, J.A.; Muir, B.W.; Pigram, P.J. Orientation and characterization of immobilized antibodies for improved immunoassays (Review). *Biointerphases* **2017**, *12*, 02D301. [CrossRef] [PubMed]
164. Shen, M.; Rusling, J.; Dixit, C.K. Site-selective orientated immobilization of antibodies and conjugates for immunodiagnostics development. *Methods* **2017**, *116*, 95–111. [CrossRef] [PubMed]
165. Vacic, A.; Reed, M.A. Quantitative nanoscale field effect sensors. *J. Exp. Nanosci.* **2014**, *9*, 41–50. [CrossRef]
166. Luo, X.; Davis, J.J. Electrical biosensors and the label free detection of protein disease biomarkers. *Chem. Soc. Rev.* **2013**, *42*, 5944–5962. [CrossRef]
167. Schönhoff, M. Self-assembled polyelectrolyte multilayers. *Curr. Opin. Colloid Interface Sci.* **2003**, *8*, 86–95. [CrossRef]
168. Poghossian, A.; Abouzar, M.H.; Sakkari, M.; Kassab, T.; Han, Y.; Ingebrandt, S.; Offenhäusser, A.; Schöning, M.J. Field-effect sensors for monitoring the layer-by-layer adsorption of charged macromolecules. *Sens. Actuators B* **2006**, *118*, 163–170. [CrossRef]

169. Poghossian, A.; Abouzar, M.H.; Amberger, F.; Mayer, D.; Han, Y.; Ingebrandt, S.; Offenhäusser, A.; Schöning, M.J. Field-effect sensors with charged macromolecules: Characterisation by capacitance-voltage, constant-capacitance, impedance spectroscopy and atomic-force microscopy methods. *Biosens. Bioelectron.* **2007**, *22*, 2100–2107. [CrossRef]
170. Schöning, M.J.; Abouzar, M.H.; Poghossian, A. pH and ion sensitivity of a field-effect EIS (electrolyte-insulator-semiconductor) sensor covered with polyelectrolyte multilayers. *J. Solid State Electrochem.* **2009**, *13*, 115–122. [CrossRef]
171. Abouzar, M.H.; Poghossian, A.; Razavi, A.; Williams, O.A.; Bijnens, N.; Wagner, P.; Schöning, M.J. Characterisation of capacitive field-effect sensors with a nanocrystalline-diamond film as transducer material for multi-parameter sensing. *Biosens. Bioelectron.* **2009**, *24*, 1298–1304. [CrossRef]
172. Poghossian, A.; Weil, M.; Cherstvy, A.G.; Schöning, M.J. Electrical monitoring of polyelectrolyte multilayer formation by means of capacitive field-effect devices. *Anal. Bioanal. Chem.* **2013**, *405*, 6425–6436. [CrossRef] [PubMed]
173. Janata, J. Graphene bio-field-effect transistor myth. *ECS Solid State Lett.* **2012**, *1*, M29–M31. [CrossRef]

© 2020 by the authors. Licensee MDPI, Basel, Switzerland. This article is an open access article distributed under the terms and conditions of the Creative Commons Attribution (CC BY) license (http://creativecommons.org/licenses/by/4.0/).

Review

Non-Carbon 2D Materials-Based Field-Effect Transistor Biosensors: Recent Advances, Challenges, and Future Perspectives

Mohammed Sedki [1,†], Ying Chen [2,†] and Ashok Mulchandani [2,*]

1. Department of Materials Science and Engineering, University of California, Riverside, CA 92521, USA; mabue002@ucr.edu
2. Department of Chemical and Environmental Engineering, University of California, Riverside, CA 92521, USA; ychen751@ucr.edu
* Correspondence: adani@engr.ucr.edu
† Equal contributions.

Received: 12 July 2020; Accepted: 24 August 2020; Published: 26 August 2020

Abstract: In recent years, field-effect transistors (FETs) have been very promising for biosensor applications due to their high sensitivity, real-time applicability, scalability, and prospect of integrating measurement system on a chip. Non-carbon 2D materials, such as transition metal dichalcogenides (TMDCs), hexagonal boron nitride (h-BN), black phosphorus (BP), and metal oxides, are a group of new materials that have a huge potential in FET biosensor applications. In this work, we review the recent advances and remarkable studies of non-carbon 2D materials, in terms of their structures, preparations, properties and FET biosensor applications. We will also discuss the challenges facing non-carbon 2D materials-FET biosensors and their future perspectives.

Keywords: 2D-materials; field-effect transistor; transition metal dichalcogenides; black phosphorus; phosphorene; hexagonal boron nitride; transition metal oxides; biosensors

---

1. Introduction

Field-effect transistors (FETs) are highly promising for biosensor applications due to their high sensitivity, real-time applicability, scalability, and prospect of integrating measurement system on a chip. A conventional FET system is composed of two electrodes, source and drain, connected by a semiconducting channel material. The FET sensor responds based on conductance change of the semiconductor channel material due to a gating effect of the captured analyte molecules. This gating effect modulates the electrical characteristics of the FET, such as source-to-drain current. This change in FET characteristics is transduced as a detectable signal change [1]. Bulk materials, such as gas-sensitive metal oxides and polymer membranes, were used as the first channel materials in FET chemical sensors. However, the unfavored electronic properties and limited interaction between target molecules and bulky materials limited their use, especially as they sometimes require specific operating conditions, such as high temperature for gas sensing [2–4].

One-dimensional (1D) semiconducting nanomaterials, such as carbon nanotubes (CNTs), conducting polymer nanowires (CPNWs) and silicon nanowires (SiNWs) have shown great success as channel materials in FET sensors. The high sensitivity of the 1D-FET sensors is attributed to their high surface area and high switching characteristics (current on/off ratio) of CNTs, CPNWs and SiNWs [5–7]. Despite the application of SiNWs-FET in many biosensor applications, it is challenging to scale up or commercialize them due to the low carrier mobility of SiNWs and the high device-to-device variation [8]. Even with the superior physicochemical properties of CNTs, such as excellent thermal and chemical stability [9], exceptional conductivity [10,11], and feasibility to easily immobilize bioprobes, their applications to FET sensors are limited. This can be explained by the difficulty to obtain pure semiconducting (s) or metallic (m) CNTs instead of producing a mixture of them, which destroys the electrical performance and increases the device-to-device variation [8,12].

Two-dimensional (2D) semiconducting nanomaterials, on the other hand, provide more conformal and stronger contact with electrodes. They are also easier to implement because of their relatively larger lateral sizes, which enables better control of the FET channel structure. Moreover, 2D nanosheets can be prepared in the desired shape, size, and thickness, and can be precisely transferred to the designated area of the sensor substrate [1]. Graphene, the most widely used 2D material, has a very high surface area of approximately 2630 m^2/g [13] and an exceptional mechanical strength [14] (130 GPa tensile strength and 1000 GPa modulus). Moreover, graphene has an ultra-high ideal charge carrier mobility of 200,000 cm^2 V^{-1} s^{-1}, entitling it to have excellent electrical properties of fast electron transfer [15]. This increased attention to graphene as an excellent material for FET applications [16–18]. However, the lack of intrinsic band gap in graphene, resulting in a small current on/off ratio in its FETs, limits its sensitivity and applicability in FET sensors [8].

Acknowledging all this, other families of non-carbon 2D materials have been inaugurated and are growing rapidly at present. Analogous with graphene, monolayer and few-layer transition metal dichalcogenides (TMDCs) (e.g., MoS_2, WS_2, $MoTe_2$, $MoSe_2$, WSe_2), hexagonal boron nitride (h-BN), black phosphorus (BP), transition metal oxides ($LaMnO_3$, $LaVO_3$), transition metal chalcogenides ($NbSe_3$, $TaSe_3$), and layered complex oxides have been reported [19–23]. Moreover, other 2D materials, such as silicene and germanene, have been introduced and studied [24]. The huge pool of non-carbon 2D materials covers a large number of materials with a huge variation in properties, from insulators to conductors. More details on the growing library of the 2D materials can be found in these references [25,26]. Therefore, researchers have taken those materials into many applications, including FET-based biosensors. In this article, we review the recent non-carbon 2D materials, in terms of their structures, preparations, properties and their FET biosensors. We will also discuss the challenges facing non-carbon 2D materials-FET biosensors and their future perspectives.

2. FET Platform: General Features

A biosensor is an analytical device consisting of a transducer and biological receptor as basic components that convert a biochemical response into an electronic signal. FET devices are popularly used in the electrical biosensing field due to their unique function for weak-signal and high impedance applications [27,28]. In FET-based biosensors, the gate terminal and/or the dielectric layer are modified with specific bioreceptors (antibodies, oligonucleotides, peptides, receptors, cells, enzymes, aptamers, etc.) to capture desired bio/chemical molecules [28–30]. When target bio/chemical molecules bind with bioreceptors, surface charges result in modulation of the electrical characteristics of FET devices.

The most facile and common way to build a 2D-FET platform is to use a wafer-based back-gated configuration as shown in Figure 1a. In this configuration, the deposition of electrodes (source and drain) is needed, while the bulk wafer could directly act as a back gate [31]. Two-dimensional materials could be grown or transferred via typical methods on a wafer of dielectric material (such as SiO_2) deposited on conducting substrate (such as Si) followed by deposition of metal source and drain electrodes via microfabrication process. Another widely used configuration in exploring 2D-FET biosensors is liquid-ion gating. Different from the back-gated configuration, in which bulk wafer functions as the gate, in this arrangement (as shown in Figure 1b), the ionic liquid is the gate. Hence, a double-layer is generated at the liquid-channel interface that is also the dielectric for the screening of the field [32].

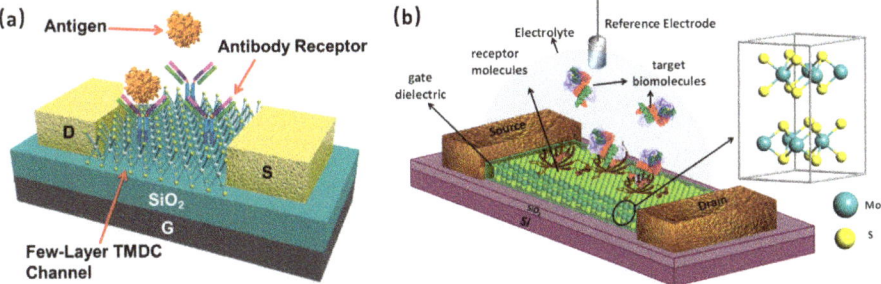

Figure 1. (**a**) Schematic illustration of the back-gated field-effect transistors (FET) biosensor with a few-layer transition metal dichalcogenide (TMDC) sensing channel [33]. Reprinted with permission from ref [33]. Copyright 2015 American Vacuum Society. (**b**) Schematic diagram of liquid-gating MoS$_2$-based FET biosensor [34]. Reprinted with permission from ref [34]. Copyright 2014 American Chemical Society. MoS$_2$ can be exchanged by other non-carbon 2D materials such as TMDCs, black phosphorus (BP) or metal oxides while antibodies can be swapped with other bioreceptors such as oligonucleotide probes, receptors, enzymes, cells or aptamers.

To detect biomolecules with specificity via 2D-FET platforms, the semiconductor channel layer has to be functionalized with biorecognition molecules specific for the target using compatible physicochemical methods [35,36]. Biological interactions of enzyme–substrate, antibody–antigen, complementary nucleic acid strands, etc., are utilized in the FET biosensors to detect target biomolecule with exquisite specificity [37]. When target molecules interact with the bioreceptor molecules, the biological interactions may cause changes in the surrounding chemical environment or material chemical structures and compositions [38]. These changes make an immediate impact to the accumulated charge carriers at the surface of the gate to transduce the biochemical interaction into an electrical signal as a measurable source-drain current. This is the principle of typical FET biosensors.

3. Non-Carbon 2D Materials

Table 1 provides an overview of the different non-carbon 2D materials, their electronic/device properties, and their FET-based biosensors.

Table 1. Overview of literature reports on non-carbon 2D material-based FET biosensors.

2D Material	2D Thickness [nm]	Mobilities [cm^2 V^{-1} s^{-1}]	I_{on}/I_{off} Ratio	Target Molecule	Detection Limit/Range	Response Time	Reference
MoS$_2$	-	1.98×10^3	7.12×10^2	miRNA-155	0.03 fM	40 min	[39]
WSe$_2$	-	-	$>10^5$	Glucose	1.0–10 mM	-	[40]
MoS$_2$	0.7	-	3.6–3.8	Kanamycin	1.06–0.66 nM	20 min	[41]
BP	10–60	-	-	IgG	10 to 500 ng/mL	on the order of seconds	[42]
Phosphorene	-	-	-	Alpha-fetoprotein	0.1 ppb–1 ppm	-	[43]
BP	30–50	468	1200	-	-	-	[44]
MoO$_3$	1.4–2.8	1100	-	Bovine serum albumin	250 μg/mL–25 mg/mL	<10 s	[45]
In$_2$O$_3$	4	19	-	Glucose	10^{-11}–10^{-5} M	-	[46]
In$_2$O$_3$	3.5	20	$>10^7$	Glucose	0.1–0.6 mM	-	[47]

473

3.1. 2D TMDCs Materials

3.1.1. Structure, Preparation and Properties

TMDCs are a group of layered materials with the general formula MX_2, where M is a transition metal from groups IV, V, or VI (Ti, Zr, Hf; V, Nb, Ta; Ct, Mo, or W) and X is a chalcogen atom (S, Se, or Te). Each layer of TMDC is composed of three planes: chalcogen, transition metal, and chalcogen. TMDCs exist in different coordination, where each transition metal atom is coordinated to six chalcogen atoms either in an octahedron or triangular prism [19,48]. The type of metal coordination preferred by these materials is highly influenced by the nature of the bond formed between the metal and chalcogen atoms. Essentially, octahedral coordination is favored by group IV transition metals, as they form strong ionic compounds, which have Coulomb repulsive forces between layers. On the other hand, group VI transition elements form more covalent bonds and coordinate in triangular prism [49,50]. Yet, group V transition elements can stabilize in octahedron and triangular prism structures due to their moderate ionicity [48].

TMDCs can be found in one of three stackings or polytypes; 1T-type stacking that dominates in bulk crystals in octahedral coordination, 2H- and 3R-type which are found with the triangular prismatic coordination. Moreover, the stable phase of MX_2 material at ambient pressure and temperature is the 2H phase with six atoms per unit cell; two metal atoms and four chalcogenides, though 1T phase can be prepared by electron beam irradiation or Li-intercalation [51,52]. It is also worth mentioning that TMDCs may undergo dimerization of metal atoms, which induces the movement of chalcogen atoms to an out-of-plane direction and results in the distortion of the 1T phase to the 1T' structure. This can be described also by means of symmetry transformation from 3- to 2-fold, with a change in the space group from (P3m1) in 1T to (P2$_1$/m) in 1T' [53].

TMDCs can be prepared in different methods that can be divided into two main approaches; top-down, in which the bulky crystals are exfoliated into mono-/few layered-TMDCs, and bottom-up, in which a thin layer of the material is built from atoms of precursors [54]. Based on these two approaches, there are several techniques/methods introduced to prepare high-quality thin layers TMDCs, including liquid phase exfoliation [55], mechanical exfoliation [56], chemical exfoliation [57,58], electrochemical deposition, and chemical vapor deposition (CVD) [25,59]. Due to its ability to prepare high quality large TMDCs layers, with a controllable number of layers and domain size, CVD is very promising among all aforementioned methods. There are many details and ongoing progress in these methods, especially in CVD-based methods, and for this, we recommend reading the work of You et al. [60] and Zhang et al. [61].

TMDC materials exhibit a wide range of electrical properties, based on the type of phase and the number of d electrons, such as metallic (e.g., NbS_2, VSe_2) [62,63], semi metallics (e.g., WTe_2, $TiSe_2$) [64,65], semiconductors (e.g., MoS_2, $MoSe_2$, WS_2, WSe_2) [66–68], and insulators (e.g., HfS_2) [69]. The first semiconducting to attract attention among the TMDCs is MoS_2, which shows a high on/off current ratio [70] that entitles it to be a good candidate for field-effect transistor applications [71]. Except for a few cases of GaSe and ReS_2, most of the TMDCs such as MoS_2 (1.8 eV), WS_2 (2.1 eV) and WSe_2 (1.7 eV) exhibit an indirect band gap with smaller energies in bulk form and a higher direct band gap in monolayer [72,73]. Nevertheless, most TMDCs, such as MoS_2 and WSe_2, are free of dangling bonds and hence make more ideal Schottky junctions than bulk semiconductors. This in turn inhibits charge transfer at the interface with bulk metals by producing Fermi energy pinning and recombination centers [74]. However, some of them exhibit high mobility, depending on metal contacts, selection of the appropriate substrate, grain boundaries, etc. MoS_2 provides a mobility of 33–151 cm^2 V^{-1} s^{-1} on BN/Si substrate at room temperature and 700 cm^2 V^{-1} s^{-1} on SiO_2/Si substrate with scandium contact [75,76]. In addition, the chemically prepared 1T MoS_2 phase is 10^7 times more conductive than its semiconducting 2H phase. On the other hand, the dichalcogenides of Ti, Ni, V, Cr, Zn, Nb essentially exhibit metallic behavior [77]. For FET, the channel material is required to be a semiconductor, as discussed in Section 2, so the semiconducting TMDCs are good candidates for FET.

On the other hand, the semimetallic and metallic TMDCs are not good channel materials for FETs, and they are better candidates for electrochemical sensors.

Superiority of TMDCs over graphene. Graphene has interesting properties that have attracted huge attention since its discovery in 2004. However, graphene does not have an intrinsic band gap, which limits its uses in the electronics industry. On the other hand, TMDCs show a tunable band gap that controls the current flow with a high on/off ratio and hence they serve as good materials for transistor applications. For example, MoS_2 shows direct bandgap (≈ 1.8 eV), large optical absorption in monolayer ($\approx 10^7$ m^{-1} in the visible range), and high current on/off ratio of $\approx 10^7$–10^8. Accordingly, it has been applied extensively in electronics and optoelectronics [78,79].

3.1.2. TMDCs-FET Biosensors

FET-based sensors are electrical systems that depend on the changes in the electrical conductivity of the semiconducting channel materials upon stimulation by target molecules. Therefore, the semiconducting TMDCs, such as MoS_2, $MoSe_2$, WS_2, and WSe_2, are the target materials among all the other TMDCs for FET sensors. Semiconducting TMDC-based FET sensors, especially MoS_2-FETs have several advantages over other materials, such as low leakage current, low power consumption, and high current on/off ratio enabling high sensitivity [80,81]. Moreover, due to their excellent abovementioned electronic properties, and mechanical flexibility, as well as their ultrathin structure, MoS_2-FET sensors are promising for the economic and low energy portable and wearable electronics [62,82]. We will discuss some of the reported TMDC-FET biosensors for the detection of different targets, such as DNA, glucose, protein, and antibiotics.

Mei et al. [83] reported the detection of DNA via hybridization with phosphorodiamidate morpholino oligos (PMO), as an ultrasensitive label-free MoS_2-FET biosensor. As shown in Figure 2i, sensor fabrication was conducted by drawing gold electrodes using photolithography and e-beam evaporation, followed by treatment with 3-Aminopropyltriethoxysilane (APTES) to cover the SiO_2/Si surface with positive charges. The negatively charged MoS_2 nanosheets were drop-casted to the positively charged channel surface and bound to it via electrostatic attraction. Then, the MoS_2 surface was modified with the DNA analogue, PMO, using 1-Pyrenebutanoic acid succinimidyl ester (PASE). The prepared PMO-MoS_2-FET biosensor showed a low limit of detection (LOD) of DNA of 6 fM, which is lower than other formerly reported DNA–DNA hybridization-based MoS_2 FET DNA biosensor. This can be attributed to the high sensitivity of the MoS_2-FET sensor and the successful and selective hybridization with PMO. Moreover, this sensor system showed applicability in the detection of DNA in serum. The signal change was recorded from the change in device current due to stimulation by the target DNA, as shown in the FET characteristic curve in Figure 2ii, and the calibration curve in Figure 2iii. Nevertheless, this system still needs more work to control the reproducibility, and the authors of this work plan to conduct it in the future. Other MoS_2-FET sensors were introduced for DNA detection, such as Lee et al.'s work in which a LOD of 10 fM was achieved [84]. Furthermore, an earlier work reported by Loan et al. used the heterostructure of MoS_2/graphene for DNA hybridization detection on FET biosensor, and they were able to achieve a very low LOD in the attomolar range [85].

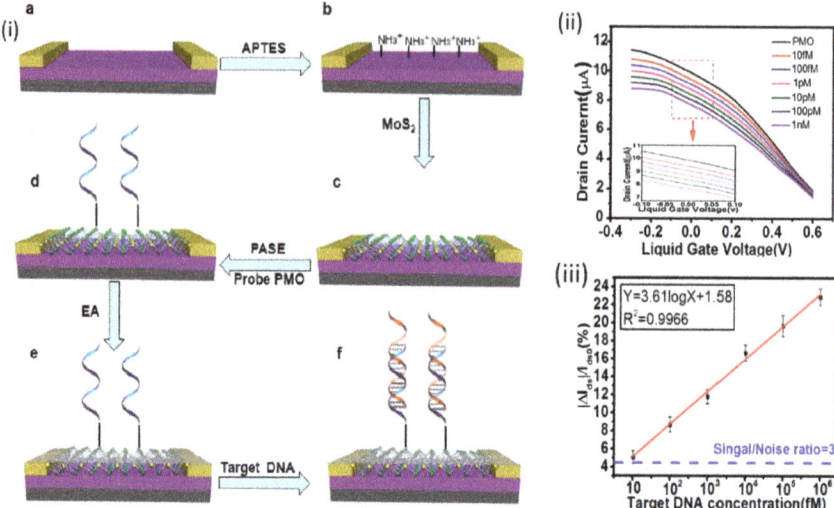

Figure 2. (i) A Schematic of the preparation of the MoS$_2$ FET biosensor for the detection of DNA; SiO$_2$/Si substrate with metal contacts (**a**), APTES functionalization of the substrate (**b**), MoS$_2$ loading (**c**), functionalization of PMO on MoS$_2$ surface using PASE linker (**d**) exposed surface passivation/blocking using EA (**e**), and target DNA capturing using the sensor (**f**). (ii) FET transfer characteristics of the complementary DNA-hybridized phosphorodiamidate morpholino oligos (PMO)-functionalized MoS$_2$ FET device, at a series of concentrations. (iii) Calibration/working curve of the MoS$_2$ FET at different concentrations of DNA. Reprinted from ref [83], Copyright (2018), with permission from Elsevier.

Majd et al. [39] developed a MoS$_2$-FET biosensor for the label-free detection of a breast cancer biomarker, miRNA-155, in cell lines and human serum. The MoS$_2$ flakes, as the channel sensing material used in this work, were prepared using the sequential solvent exchange method, and drop-casted onto the FET surface. The detection is based on direct hybridization between the immobilized probe miRNA-155 and the target miRNA-155. The prepared device showed a very high carrier mobility of 1.98×10^3 cm^2 V^{-1} s^{-1} (this number is much higher than expected mobilities of MoS$_2$, but this is what the authors claimed), and a fairly low subthreshold swing of 48.10 mV/decade. The I_{on}/I_{off} ratio (7.12×10^2) reported in this work is small compared to other reports (mostly 10^5–10^7). In terms of miRNA detection, the prepared device achieved a LOD of 0.03 fM, in a dynamic range of 0.1 fM to 10 nM. As a selectivity test, the sensor system did not show any significant response to miRNA with one base mismatch. Lastly, this sensor system was proven to be successful in the determination of miRNA-155 human breast cancer biomarker in serum samples, which enhances its clinical applicability.

Shan et al. [86] reported bilayer MoS$_2$-FET as a glucose biosensor, with the advantages of high stability, high sensitivity and rapid response. The electrical characteristic outputs of the introduced device were recorded in the absence of glucose. The effect of gate potential (V_g) (from −40 to 40 V, with a step of 5 V) on the device's source–drain current (I_{sd}) in the source–drain voltage (V_{sd}) range of −0.5 to 0.5 V, as shown in Figure 3a. The I_{sd} increased with the increase of positive gate potential. Furthermore, the (I_{sd}-V_g) FET characteristic curve exhibited the n-type behavior of the device with I_{on}/I_{off} was found 10^6, and the carrier mobility was found as 33.5 cm^2 V^{-1} s^{-1}, as illustrated in Figure 3b, which explains the high sensitivity of this sensor system. The presented system showed a LOD of 300 nM, and a sensitivity of 260.75 mA/mM. The current (I_{sd}) was directly proportional to the glucose concentration, at constant V_g and V_{sd} (Figure 3c,d). This increase in current may be attributed to the n-doping of the n-type semiconductor MoS$_2$ by electrons resulting from glucose oxidation. Moreover, the determination of unknown glucose concentration was achieved by a calibration curve plotted between the I_{sd} and glucose concentration. There is another interesting work presented by Lee et al. [40] about glucose biosensor using tungsten diselenide (WSe$_2$) field-effect transistor biosensor (WSe$_2$ BioFET) using the same concept as discussed above.

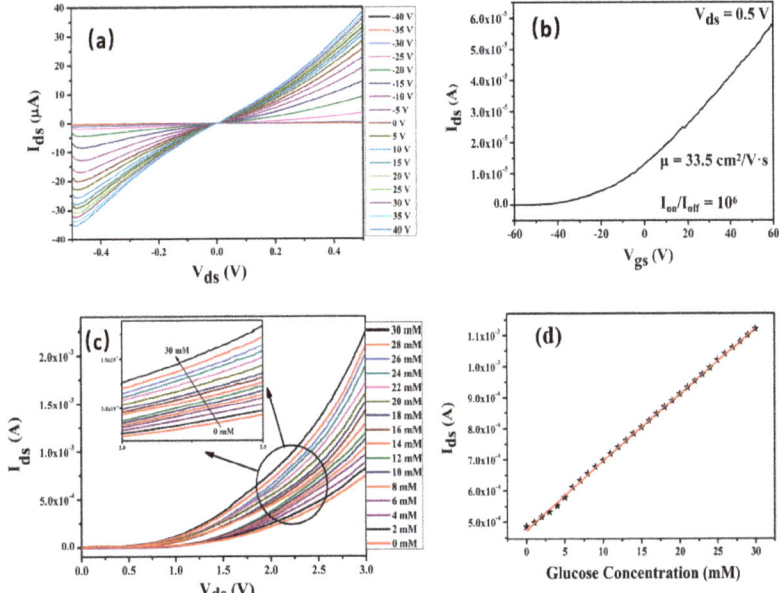

Figure 3. Basic electrical characterization graphs the device. (**a**) The effect of gate potential (V$_g$) (from −40 to 40 V, with a step of 5 V) on device's source-drain current (I$_{sd}$) in the source-drain voltage (V$_{sd}$) range of −05 to 0.5 V. (**b**) The (I$_{sd}$-V$_g$) FET characteristic curve of MoS$_2$-FET showing the n-type behavior of the device. (**c**,**d**) Increase in I$_{sd}$ with an increase in the concentration of target molecules. Reprinted with some changes from ref [86].

As an example of TMDC-FET biosensors for the detection of antibiotics, Chen et al. [41] developed an aptamer-MoS$_2$-FET biosensor for the detection of Kanamycin (KAN). Aptamer (APT) application as a selective biorecognition element for antibiotics is promising, however, its selectivity for antibiotics is still challenging due to the wide folding of APTs and the structural similarities among antibiotics. The authors implemented MoS$_2$ as the sensing material, and APT and a complementary strand DNA (CS) as recognition elements of KAN. This structure (CS-APT) of recognition elements in the proposed CS-APT-MoS$_2$-FET biosensor improved the selectivity and reliability of this sensor system and reduced the device-to-device variations. Gold nanoparticles (AuNPs) were used as a linker of DNA. The device showed an Ohmic contact and p-type behavior, as shown in Figure 4a,b. MoS$_2$ is expected to be n-type; however, in this work and others, it is p-type due to oxygen incorporation in the synthesis and fabrication processes [87]. The sensing mechanism is based on a replacement reaction, in which KAN binds to the APT and displaces the CS from the CS/APT/MoS$_2$-FET system (Figure 4c). Two control experiments of MoS$_2$-FET, and APT/MoS$_2$-FET, were used in this study for a better understanding of the sensing mechanism. In the case of MoS$_2$-FET, there was no change in device current with the addition of KAN. In the case of APT/MoS$_2$-FET, the addition of KAN resulted in a direct increase in current due to the formation of more centered structure with higher concentration of negative charges that stimulate the positive charges of the p-type MoS$_2$ and increasing current. In the main experiment, using CS/APT/MoS$_2$-FET, the addition of KAN resulted in a slow current decrease, due to the time needed for KAN to displace CS, and the mechanism is hard to determine, but the authors attribute it to the decrease of negative charges in general. Figure 4d presents the results of KAN detection. The LOD, which was relatively time-dependent, was 1.06–0.66 nM, with high selectivity of KAN (selectivity coefficient of 12.8) over the other antibiotics such as amoxicillin, tobramycin, streptomycin, and chloramphenicol.

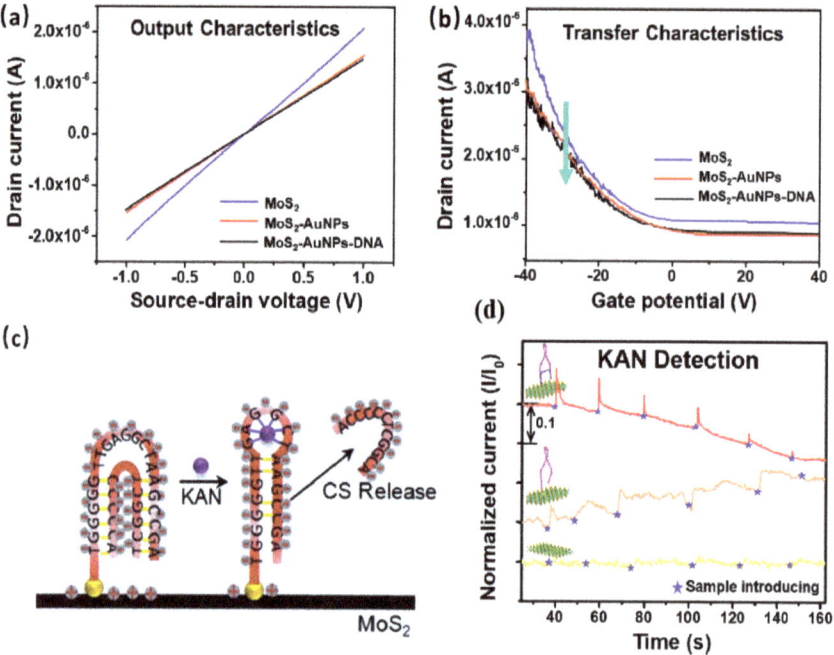

Figure 4. (**a**) I_{sd}-V_{sd} curve showing the Ohmic contact of the device. (**b**) FET characteristic curves of the device. (**c**) The proposed mechanism of Kanamycin (KAN) replacing CS. (**d**) The sensor system, with control experiments, response to KAN at different concentrations. Reprinted from ref [41], Copyright (2019), with permission from Elsevier.

3.2. Black Phosphorus/Phosphorene

3.2.1. Structure, Synthesis and Properties

Black phosphorus is a layered 2D Van der Waals material. It is the most stable allotropic substance among the phosphorus family. Its isolated single layer is popularly known as phosphorene and has attracted tremendous attention. The structure of phosphorene is an orthorhombic lattice and phosphorus atoms are covalently bonded to form a puckered honeycomb structure (as shown in Figure 5a) [88]. The bandgap, which ranges from 0.3 eV (for bulk black phosphorus) to 2.0 eV (for mono-layer phosphorene) is direct and thickness-dependent. The charge carrier mobility is large enough with the high hole mobility up to 1000 cm^2 V^{-1} s^{-1}, which is also thickness-dependent (reported in thickness less than 10 nm) [89,90].

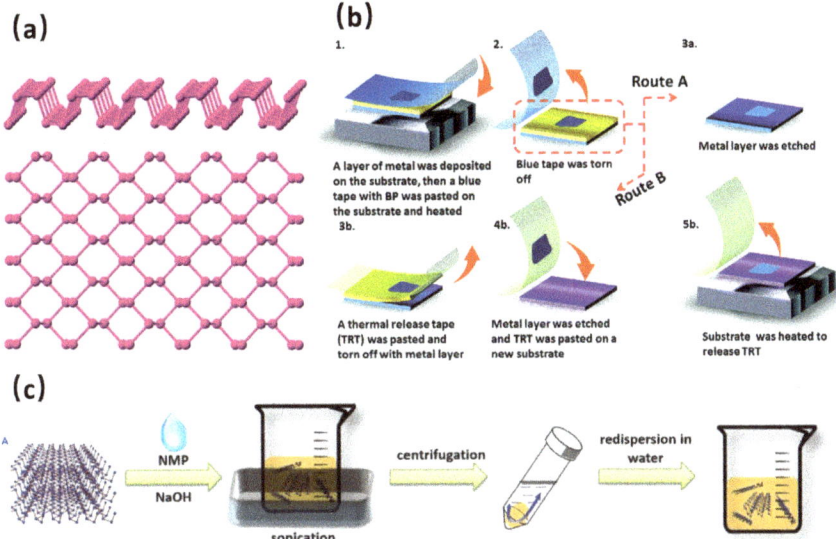

Figure 5. (a) Structure of phosphorene (side and top view) [91]. (b) Schematic diagram of the metal-assisted exfoliation process for few-layer black phosphorus [92]. Reprinted with permission from ref [92]. Copyright 2018, The Royal Society of Chemistry. (c) Schematic diagram of liquid-phase exfoliation process (basic-N-methyl-2-pyrrolidone(NMP)-exfoliated) phosphorene [93]. Reprinted with permission from ref [93]. Copyright 2015, WILEY-VCH Verlag GmbH and Co. KGaA, Weinheim.

Mechanical and liquid-phase exfoliations are the two most common methods to exfoliate layered phosphorene from bulk phosphorus [94,95]. Mechanical exfoliation with scotch tape to peel nanoflakes off from bulk crystals is easily operated and could yield high-quality black phosphorus/phosphorene flakes with low cost, which makes it perfect for fundamental research. On the other hand, its negative aspects are also apparent, such as the size is too small for most of exfoliated phosphorene, the process is labor-intensive and time-consuming, and the productivity is extremely low [95]. A further disadvantage is that the mechanically exfoliated phosphorene experiences significant irreversible deformations under ambient conditions and inconvenient for long-term storage. With these problems in mind, researchers are struggling to find ways to improve traditional mechanical exfoliation. Guan et al. [92] introduced a metal-assisted exfoliation method to obtain large-sized phosphene. A 10 nm gold layer (or silver layer) was deposited on the substrate at first and followed with normal mechanical exfoliation (Figure 5b). Then, the metal layer was etched with a solution. Few-layer phosphorene was produced with a 50 μm lateral size. The FET electronic properties proved the high quality of the as-fabricated phosphorene, which showed a hole mobility of 68.6 cm^2 V^{-1} s^{-1} and the I_{on}/I_{off} ratio of 200,000. Moreover, poly(dimethylsiloxane) (PDMS)-based substrate and semi-spherical PDMS stamp to help rapid peeling and transfer of phosphorene nanosheets were reported [96]. In addition, Ar$^+$ plasma thinning processes after regular mechanical exfoliation to obtain controllable and homogeneous monolayer phosphorene [97] were also reported accordingly.

Liquid-phase exfoliation is another common method for preparing phosphorene. Dimethylformamide (DMF), dimethyl sulfoxide (DMSO), isopropanol (IPA), N-methyl-2-pyrrolidone (NMP), and ethanol are common solvents for black phosphorus exfoliation [98,99]. With the participation of the solvent, phosphorene was prevented from air degradation and its stability for exfoliation was improved. The phosphorene produced in the liquid phase could be stored long-term and separated by centrifugation to achieve adjustable sizes. Figure 5c illustrates a basic liquid-phase exfoliation process in NMP solvent. Bulk black phosphorus was put into the solvent, followed by a four-hour ultrasonic treatment which destructed the weak interaction between the stacked sheets [93,95]. After ultrasonic treatment, the phosphorene in NMP was separated by centrifugation. Other methods, such as electrochemical exfoliation [100], chemical transport reaction [94], solvothermal method [101], are applicable for exfoliation as well. Although a lot of literature reported successfully exfoliated phosphorene, there is still a lot of work to do before the mass-production of high-quality phosphorene can be achieved.

3.2.2. Black Phosphorus/Phosphorene-FET Biosensors

Black phosphorus and phosphene-based studies have been carried out related to biological applications, such as biomedicine and biosensing [87]. Compared with exhaustive literature on biomedicine, the work on phosphorene biosensors is far less abundant. Chen et al. [42] reported the FET biosensor with few-layer BP nanosheet serving as channel materials passivated with Al_2O_3 layer to detect human immunoglobulin G (HIgG). Gold nanoparticles were deposited on the surface to immobilize anti-HIgG biorecognition molecules (Figure 6). The device's basic electrical properties showed a p-type natural of black phosphorus device. In order to test the as-fabricated sensor's dynamic response, different concentrations of HIgG antigens from 10 ng/mL to 500 ng/mL were tested for characterization. When HIgG molecules adsorbed onto biosensor surface, the source–drain current increased with the negative gating effect added. A fast response on the order of seconds with an LOD of 10 ng/mL was reported. The sensor showed good selectivity for the target antigen compared to non-specific protein avidin (Figure 6). Kim et al. [43] successfully fabricated a few-layer black phosphorus-based biosensor to detect alpha-fetoprotein (AFP) which was called "the most reliable tumor marker for diagnosis hepatocellular carcinoma". The surface functionalization process was conducted with poly-L-lysine linker to immobilize AFP antibodies. With the specific binding of AFP antigens and antibody, the detection of different concentrations of AFP antigen (1 ppm to 0.1 ppb) results showed a linear relationship between current and concentration with high sensitivity. Taking advantage of the fact that the phosphorus is the second most predominant mineral in the human body (1% of body weight) and the biocompatibility of biodegradation products of BP, Song et al. [44] developed a BP-FET that maintained its high mobility and on–off current ratio for ~36 h in body fluid before dissolving completely. Such a FET device has the potential to open a new way for transient biocompatible biosensors.

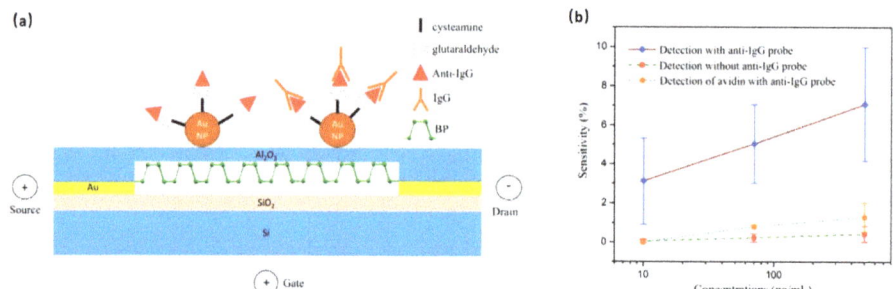

Figure 6. (a) Schematic of black phosphorus biosensor for HIgG, and (b) plot of sensitivity as a function of target and non-target antigen concentration [42]. Reprinted with permission from ref [42]. Copyright 2016, Elsevier B.V.

3.3. Metal Oxides

3.3.1. Preparations and Properties

Metal oxides are among the most varied classes of solids. They can be classified as layered and non-layered. Examples of former include MoO_3, WO_3, Ga_2O_3 and TaO_3, while ZnO, SnO_2, In_2O_3 and

CuO are examples of the latter [102]. Due to their specific structures and electrical characteristics, metal oxides have been considered as candidates to extend the library of 2D materials for transistors with a large band gap energy range (2.3–4.9 eV) and high electron mobilities (>10 cm^2 V^{-1} s^{-1}) which guarantee high sensitivity and signal-to-noise ratio in biosensing. Moreover, FETs of these materials can be processed at moderate temperatures from solutions facilitating deposition on a large scale economically and the conductivity tuned by varying crystal size, morphology, dopant, contact geometry and temperature of operation [46,102,103]. Metal oxides, to date, have been applied as electrochemical and photoelectrochemical transducers for bio/chemical sensing [30,103–106] and FET transducer for sensing of gases [107,108]. As mentioned previously, FET gas sensors of metal oxides typically operate at high temperatures, leading to higher energy needs and reliability and safety issues. On the other hand, because of oxygen atom termination of the basal surfaces, these materials are more stable in air and water [102]. As with other 2D nanomaterials, bottom-up and top-down approaches are used to synthesize many 2D metal oxide materials used in FET biosensors. The majority of the metal oxides are synthesized by hydrothermal or solvothermal procedures because these methods are facile, scalable, low temperature and low cost [109]. Typically, specific metal oxide precursors, such as metal nitrates, chlorides, and sulfates, are dissolved in water or organic solvent and reacted anywhere from 3 to 12 h or even a few days at 75–200 °C [110]. Metal oxides produced by these methods show varieties of architectures and morphologies, such as nanowires, nanowalls, nanoforests, nanoflakes, flower-like structures, and tree-like structures. [111]. Two-dimensional flakes or films may not only include layers of nanoflakes, but also aggregated nanoflakes without order. Various forms of exfoliation methods have also been implemented to achieve 2D layers of metal oxides for FET biosensors applications [45,112]. Exfoliation methods are limited to layered metal oxides.

3.3.2. Metal Oxide-FET Biosensors

Two-dimensional metal oxides have an extensive application as optical, electronic, and sensing semiconductors. Compared with a large number of one-dimensional metal oxide-based FET biosensors reported [113,114], 2D metal oxide-based FET biosensors still have a large space in FET biosensors applications. Herein, we briefly introduce several metal oxide-based FET biosensors. When considering various metal oxides, In$_2$O$_3$ has yielded many FET biosensors with good performance. Chen et al. [46] presented a 2D In$_2$O$_3$-based FET biosensor which achieved specific detection of glucose with an extremely low limit of detection (<7 fM) and showed high sensitivity. Boronic acid and glucose, respectively, acted as the receptor and target molecules (Figure 7a). The transfer curves in 0.1 M buffer solution showed clear ohmic behavior at low bias voltage and all I-V curves showed a turn-on voltage of −0.316 V with I_{on}/I_{off} ratio of 10^4. The biosensor response was linearly related to glucose concentration over a broad dynamic range from 10^{-11} to 10^{-5} M, as shown in Figure 7a. The device performances, both with respect to dynamic range and limit of detection, of this sensor was superior to other non-enzymatic FET glucose sensors using boronic acid as recognition molecule in conjunction with carbon nanotubes, graphene or reduced graphene oxide as semiconductor channel.

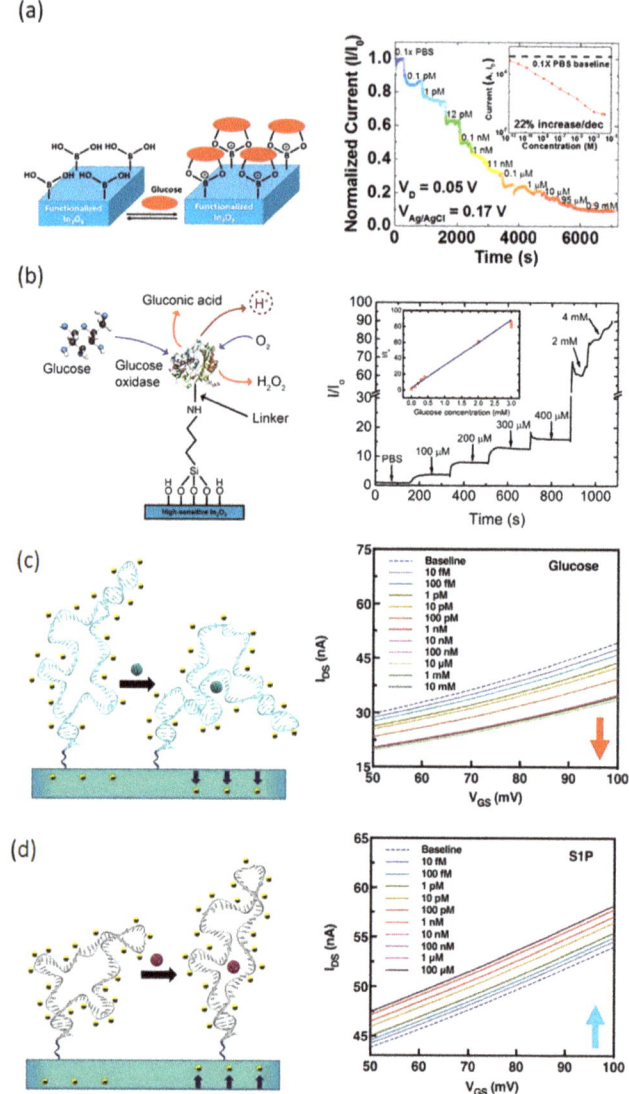

Figure 7. Schematic of In$_2$O$_3$ FET biosensors. (**a**) Illustration of the principle of glucose sensing on boronic acid-functionalized surface and responses to various concentrations of glucose [46]. Reprinted with permission from ref [46]. Copyright 2017, American Chemical Society. (**b**) Illustration of D-glucose sensing via glucose oxidase to produce gluconic acid and hydrogen peroxide (left), and responses to physiologically relevant D-glucose concentrations (right) (inset shows data from five devices) [47]. Reprinted with permission from ref [47]. Copyright 2015, American Chemical Society. Mechanism of aptamer target-induced reorientations within or near the Debye length of semiconductor channels. (**c**) Aptamers reorient closer (e.g., dopamine, glucose) to decrease transconductance (left). Transfer curves of glucose aptamer–FETs showed reductions in source-drain currents (right), (**d**) Aptamers reorient away from semiconductor channels (e.g., serotonin, S1P) to increase transconductance (left). Transfer curves of S1P aptamer–FET transfer curves increased in response to target concentrations (right). Reprinted with some changes from ref [115]. Reprinted with permission from ref [115]. Copyright 2018, Science.

The Tseng group from the University of California, Los Angeles (UCLA) [47] demonstrated an In$_2$O$_3$-based FET biosensor with ultrathin polyimide (PI) film as a substrate for conformal bioelectronics. The enzymatic oxidation of D-glucose with glucose oxidase was applied to specifically detect only D-glucose not L-glucose with a low driving voltage. The mobility and and I_{on}/I_{off} ratio of the In$_2$O$_3$-based FET biosensor were 20 cm^2 V^{-1} s^{-1} and larger than 10^7, respectively. The sensor detected physiologically relevant D-glucose concentrations (Figure 7b). The mechanism of glucose sensing was based on the protonation of In$_2$O$_3$ surface by gluconic acid produced during glucose oxidase catalyzed oxidation of D-glucose. The reported biosensor has the potential for applications in wearable, non-invasive health-monitoring technologies such as glucose levels in tears.

Highly sensitive In$_2$O$_3$-based FET biosensors with DNA aptamers for the detection of small electroneutral molecules, such as dopamine, glucose, serotonin, and sphingosine-1-phosphate (S1P) in undiluted physiological fluids of high-ionic strength was reported by the Weiss group from UCLA [115]. The DNA aptamers for the target compounds were selected by solution-phase systematic evolution of ligands by exponential enrichment (SELEX) and immobilized on the semiconductor In$_2$O$_3$ sensing channel. The sensing mechanism was the modulation in the gate conductance of the device as a result of a target-induced change in the confirmation of negatively charged phosphodiester backbones of immobilized aptamers. For example, when the glucose aptamer-FET was exposed to glucose, charged backbones of aptamers moved closer to the semiconductor channel causing an increase in electrostatic repulsion and a decrease of device transconductance (Figure 7c). On the other hand, S1P aptamers moved away from the channel surface when target S1P was captured by the aptamers, resulting in an increase of the device transconductance (Figure 7d). The integration of highly sensitive In$_2$O$_3$-based FET and specific stem-loop receptor overcame the limitations of traditional FET biosensors in detecting molecules/targets in high ionic strength solutions because of shielding created by the electrical double layer, i.e., Debye length and/or small molecules with no or few charges.

Balendhran et al. [45] reported a liquid exfoliation method to obtain 2D α-MoO$_3$ nanoflakes with lateral dimensions in the range 50–150 nm. Then, a MoO$_3$ electron conduction channel was drop-casted on a rough alumina substrate to obtain a FET biosensor for bovine serum albumin detections. The platform showed a fast response time of less than 10 s and LOD of 250 µg/mL. As more attention has been paid in 2D metal oxide materials, they will provide great opportunities in the detection of biomolecule applications.

3.4. h-BN

Hexagonal boron nitride has a similar chemical structure to graphene and has a stoichiometry of 1:1 of B and N. With the help of the covalent B-N bond, the localized electronic state provides h-BN with a high band gap from 5 eV to 6 eV and thereby excellent electrical insulation properties [116]. Like most 2D materials, single- or few-layer h-BN can be produced by either exfoliating from bulk crystals of boron nitride or by the CVD method [117]. The excellent electrical insulation property makes 2D h-BN a perfect alternative dielectric for layered 2D material based heterostructures and has recently exhibited considerable improvements on channel mobility [118,119]. Dean et al. [120] fabricated and characterized graphene on h-BN substrates devices. Nanosheets h-BN were produced from bulk single-crystal h-BN by exfoliation. Its one-atom structure contributed to the smooth surface and reduced roughness compared to SiO$_2$. Then, graphene was transferred by a typical polymethylmethacrylate (PMMA)-based method to build graphene/h-BN heterostructures. Electronic transport measurement results showed super excellent hall mobility of monolayer graphene device (140,000 cm^2 V^{-1} s^{-1}) which could compare with SiO$_2$ supported devices. In another report by Joo et al. [121], the heterostructure device of 2D MoS$_2$ on h-BN substrate was found to have much higher n-doping effect compared to MoS$_2$ directly on SiO$_2$/Si substrate. This was attributed to the h-BN layer helping reduce the oxygen p-doping with SiO$_2$, thereby lowering Schottky barrier height in MoS$_2$/hBN heterostructure. Similarly, h-BN/MoS$_2$/h-BN heterostructure was reported by Saito et al. [122] as well. When the conductance characterization was conducted within the linear region, hysteresis was small enough to be negligible. Meanwhile, in the nonlinear, hysteresis behaviors could be detected under dark condition. They also reported the illumination effects on the hysteric behaviors. When conductance characteristics operated under illumination, hysteric behaviors in the nonlinear region disappeared. The above results, while they do not present specific examples of applications of h-BN-based biosensors, provide illustrations of potentials of heterostructures of h-BN with other 2D materials in construction of biosensors with ultrahigh sensitivity.

4. Summary, Challenges and Future Perspectives

FET biosensors are very promising for clinical applications as early diagnostic tools, due to their high sensitivity, rapid response, low power operation, label-free working environment, and feasibility for commercialization. The semiconducting channel of nanomaterials plays a crucial rule in the sensing process, and alongside the recognition element, they determine the sensing characteristics of a FET biosensor. Graphene has shown to be successful as a channel material due to its 2D layered structure, high surface area, and very high charge carrier mobility. However, this success is limited, as graphene has no intrinsic band gap, i.e., semimetallic, which lowers its current switching ratio (I_{on}/I_{off} ratio is 1 to 2), and, in role, lowers the sensitivity of its corresponding graphene-FET biosensors. Other materials such as SiNWs and CNTs were reported as promising FET sensing channel materials in a huge number of studies, however, they suffer from difficulty in achieving high reproducibility, as discussed in the introduction of this article.

Keeping all that in mind, non-carbon 2D materials, which include TMDCs, BP, 2D metal oxides, h-BN, and others, represent the new candidates. Figure 8 is a schematic diagram summarizing the different types of non-carbon 2D materials used in FET biosensors and highlighting their advantages, disadvantages and future perspectives. Semiconducting TMDCs are the most promising among them, due to their 2D layered structure, relatively high stability, high surface area, and considerably high current switching ratio of 10^3–10^7 that increases the corresponding TMDC-FET sensor sensitivity and enables much lower limit of detection. Mono-to-few layer BP is another new promising material for FET biosensors, as it has a higher charge carrier mobility compared to TMDCs, and a higher conductivity with a lower band gap. BP has a lower Schottky barrier due to its better wavefunction matching with metal contacts. Non-carbon-FET biosensors for the detection of nucleic acids, proteins, cancer biomarkers, glucose, and others, have been successfully conducted on a lab scale. However, much effort is still required to reach clinical applications. The main challenges are in achieving the stability and reproducibility of these 2D-FETs devices, as well as the materials and devices fabrication methods. CVD is a promising method for the synthesis of TMDCs, but it lacks the high reproducibility from a batch to another and results in device-to-device variation. The variation can be due to changes in grain size, level of defects, film continuity, and more. Moreover, transferring the CVD-synthesized materials to the desired substrate, to build the FET sensor, introduces many defects that negatively affect the device performance. Chemical and liquid phase exfoliation methods are promising in terms of scaling up, but the drop-casting of the materials on the surface of FET is not that highly reproducible either. BP has another problem, which is its very low stability, as it degrades in air and moisture even in a few hours. In addition, biosensor applications make materials more susceptible to degradation, as they involve liquids in contact with these materials for a long time. Two-dimensional metal oxides-based biosensors are very promising due to their high chemical stability, and universal surface complexation with various receptors. In addition, metal oxides are easy to synthesize, process, and load on substrates. However, more efforts are still needed to increase carrier mobility and improve organic/inorganic interface compatibility. Monolayer h-BN is an excellent alternative for dielectric layers (e.g., SiO_2) in FET, where it helps reduce the oxygen p-doping of SiO_2, thereby lowering Schottky barrier height, and reduces surface roughness scattering and hence it improves the carrier mobility of the top sensing material such as graphene or TMDCs.

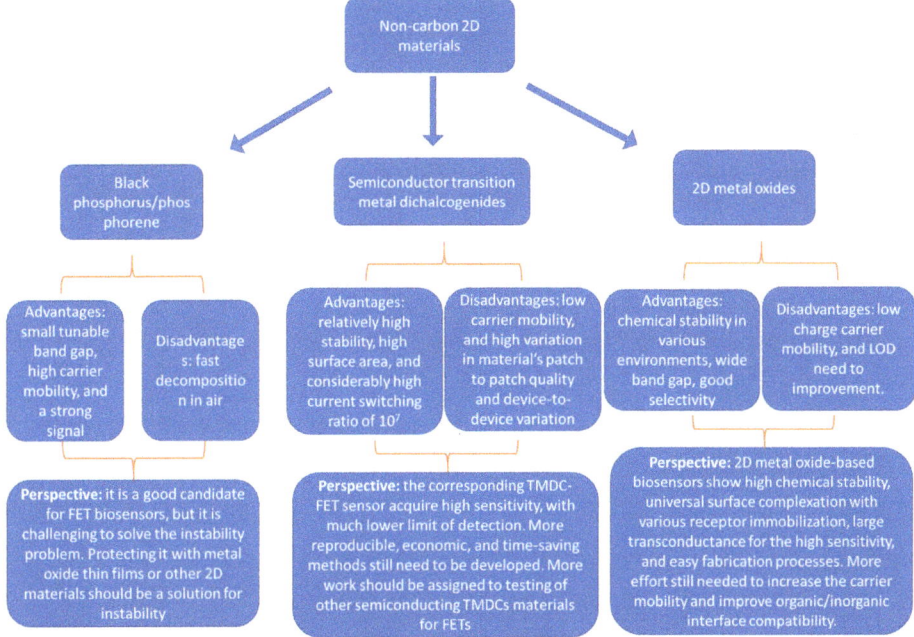

Figure 8. A schematic diagram summarizing the non-carbon 2D materials, their advantages and disadvantages as channel materials for FET biosensor applications, and the authors perspectives about them.

Another problem with non-carbon 2D materials can be that these materials are still new, and their fabrications routes are not robust and clear enough to a large portion of the research community. It is worth mentioning that the number of studies conducted on TMDCs-, BP-, or 2D metals oxides-FET biosensors is significantly limited over the last five years. Even though, the presented limited number of studies is mainly focused on MoS_2-FETs, while the other semiconducting TMDCs, such as $MoSe_2$, WS_2, and WSe_2, were just not getting their chances yet. To overcome these problems and open the way for their FET biosensors, more reproducible, economic, and time-saving methods still need to be developed. A complete library of the non-carbon 2D materials, based on matching their electronic and semiconducting properties and tabulating them for an easier choice for FETs and commercialization, is still needed. More work should be assigned to the testing of other semiconducting TMDC materials for FETs.

Author Contributions: M.S., Y.C. and A.M. wrote and edited the manuscript. All authors have read and agreed to the published version of the manuscript.

Funding: This work was supported by grants from the National Science Foundation (1842718), Department of Energy (under award number FE0030456) and UC Riverside and Korea Institute of Materials Science (Research Program (POC2930)) through UC-KIMS Center for Innovation Materials for Energy and Environment. A.M. recognizes the W. Ruel Johnson Chair in Environmental Engineering.

Conflicts of Interest: The authors declare no conflict of interest.

References

1. Sarkar, D. 2D Materials for Field-Effect Transistor–Based Biosensors. In *Fundamentals and Sensing Applications of 2D Materials*; Rout, C.S., Late, D., Morgan, H., Eds.; Elsevier: Cambridge, UK, 2019; pp. 329–377.

2. Barsan, N.; Weimar, U. Understanding the fundamental principles of metal oxide based gas sensors; the example of CO sensing with SnO2 sensors in the presence of humidity. *J. Phys. Condens. Matter* **2003**, *15*, R813.
3. Mcbride, P.T.; Janata, J.; Comte, P.A.; Moss, S.D.; Johnson, C.C. Ion-selective field effect transistors with polymeric membranes. *Anal. Chim. Acta* **1978**, *101*, 239–245. [CrossRef]
4. Pham, T.; Li, G.; Bekyarova, E.; Itkis, M.; Mulchandani, A. MoS2-based optoelectronic gas sensor with sub-part-per-billion limit of NO_2 gas detection. *ACS Nano* **2019**, *13*, 3196–3205. [CrossRef] [PubMed]
5. Wanekaya, A.K.; Chen, W.; Myung, N.V.; Mulchandani, A. Nanowire-based electrochemical biosensors. *Electroanalysis* **2006**, *18*, 533–550. [CrossRef]
6. Ramnani, P.; Gao, Y.; Ozsoz, M.; Mulchandani, A. Electronic detection of miRNA at attomolar level with high specificity. *Anal. Chem.* **2013**, *85*, 8061–8064. [CrossRef]
7. Tran, T.-T.; Mulchandani, A. Carbon nanotubes and graphene nano field-effect transistor-based biosensors. *TrAC Trends Anal. Chem.* **2016**, *79*, 222–232. [CrossRef]
8. Zhang, A.; Lieber, C.M. Nano-bioelectronics. *Chem. Rev.* **2016**, *116*, 215–257. [CrossRef]
9. Ajayan, P.M. Nanotubes from carbon. *Chem. Rev.* **1999**, *99*, 1787–1800. [CrossRef]
10. Heller, I.; Kong, J.; Heering, H.A.; Williams, K.A.; Lemay, S.G.; Dekker, C. Individual single-walled carbon nanotubes as nanoelectrodes for electrochemistry. *Nano Lett.* **2005**, *5*, 137–142. [CrossRef]
11. Gooding, J.J.; Chou, A.; Liu, J.; Losic, D.; Shapter, J.G.; Hibbert, D.B. The effects of the lengths and orientations of single-walled carbon nanotubes on the electrochemistry of nanotube-modified electrodes. *Electrochem. Commun.* **2007**, *9*, 1677–1683. [CrossRef]
12. Zhang, Y.; Zheng, L. Towards chirality-pure carbon nanotubes. *Nanoscale* **2010**, *2*, 1919–1929. [CrossRef] [PubMed]
13. Chandran, G.T.; Li, X.; Ogata, A.; Penner, R.M. Electrically transduced sensors based on nanomaterials (2012–2016). *Anal. Chem.* **2017**, *89*, 249–275. [CrossRef] [PubMed]
14. Lee, C.; Wei, X.; Kysar, J.W.; Hone, J. Measurement of the elastic properties and intrinsic strength of monolayer graphene. *Science* **2008**, *321*, 385–388. [CrossRef] [PubMed]
15. Bolotin, K.I.; Sikes, K.J.; Jiang, Z.; Klima, M.; Fudenberg, G.; Hone, J.; Kim, P.; Stormer, H.L. Ultrahigh electron mobility in suspended graphene. *Solid State Commun.* **2008**, *146*, 351–355. [CrossRef]
16. Tsang, D.K.H.; Lieberthal, T.J.; Watts, C.; Dunlop, I.E.; Ramadan, S.; Armando, E.; Klein, N. Chemically functionalised graphene FET biosensor for the label-free sensing of exosomes. *Sci. Rep.* **2019**, *9*, 1–10.
17. Tu, J.; Gan, Y.; Liang, T.; Hu, Q.; Wang, Q.; Ren, T.; Sun, Q.; Wan, H.; Wang, P. Graphene FET array biosensor based on ssDNA aptamer for ultrasensitive Hg2+ detection in environmental pollutants. *Front. Chem.* **2018**, *6*, 333. [CrossRef]
18. Colombo, L.; Venugopal, A. Graphene FET with Graphitic Interface Layer at Contacts. U.S. Patent No. 9,882,008, 30 January 2018.
19. Khan, K.; Tareen, A.K.; Aslam, M.; Wang, R.; Zhang, Y.; Mahmood, A.; Ouyang, Z.; Zhang, H.; Guo, Z. Recent developments in emerging two-dimensional materials and their applications. *J. Mater. Chem. C* **2020**, *8*, 387–440. [CrossRef]
20. Bao, X.; Ou, Q.; Xu, Z.; Zhang, Y.; Bao, Q.; Zhang, H. Band structure engineering in 2D materials for optoelectronic applications. *Adv. Mater. Technol.* **2018**, *3*, 1800072. [CrossRef]
21. Li, J.; Luo, H.; Zhai, B.; Lu, R.; Guo, Z.; Zhang, H.; Liu, Y. Black phosphorus: A two-dimension saturable absorption material for mid-infrared Q-switched and mode-locked fiber lasers. *Sci. Rep.* **2016**, *6*, 30361. [CrossRef]
22. Torrisi, F.; Coleman, J.N. Electrifying inks with 2D materials. *Nat. Nanotechnol.* **2014**, *9*, 738–739. [CrossRef]
23. Zhu, C.; Du, D.; Lin, Y. Graphene and graphene-like 2D materials for optical biosensing and bioimaging: A review. *2D Mater.* **2015**, *2*, 32004. [CrossRef]
24. Butler, S.Z.; Hollen, S.M.; Cao, L.; Cui, Y.; Gupta, J.A.; Gutiérrez, H.R.; Heinz, T.F.; Hong, S.S.; Huang, J.; Ismach, A.F. Progress, challenges, and opportunities in two-dimensional materials beyond graphene. *ACS Nano* **2013**, *7*, 2898–2926. [CrossRef]
25. Zhou, J.; Lin, J.; Huang, X.; Zhou, Y.; Chen, Y.; Xia, J.; Wang, H.; Xie, Y.; Yu, H.; Lei, J. A library of atomically thin metal chalcogenides. *Nature* **2018**, *556*, 355–359. [CrossRef] [PubMed]
26. Sankar, I.V.; Jeon, J.; Jang, S.K.; Cho, J.H.; Hwang, E.; Lee, S. Heterogeneous Integration of 2D Materials: Recent Advances in Fabrication and Functional Device Applications. *Nano* **2019**, *14*, 1930009. [CrossRef]

27. Vu, C.A.; Chen, W.Y. Field-effect transistor biosensors for biomedical applications: Recent advances and future prospects. *Sensors* **2019**, *19*, 4214. [CrossRef] [PubMed]
28. Nehra, A.; Pal Singh, K. Current trends in nanomaterial embedded field effect transistor-based biosensor. *Biosens. Bioelectron* **2015**, *74*, 731–743. [CrossRef]
29. Sakata, T. Biologically Coupled Gate Field-Effect Transistors Meet in Vitro Diagnostics. *ACS Omega* **2019**, *4*, 11852–11862. [CrossRef]
30. Syedmoradi, L.; Ahmadi, A.; Norton, M.L.; Omidfar, K. A review on nanomaterial-based field effect transistor technology for biomarker detection. *Microchim. Acta* **2019**, *186*, 739. [CrossRef] [PubMed]
31. Pham, T.; Ramnani, P.; Villarreal, C.C.; Lopez, J.; Das, P.; Lee, I.; Neupane, M.R.; Rheem, Y.; Mulchandani, A. MoS2-graphene heterostructures as efficient organic compounds sensing 2D materials. *Carbon* **2019**, *142*, 504–512. [CrossRef]
32. Jing, X.; Illarionov, Y.; Yalon, E.; Zhou, P.; Grasser, T.; Shi, Y.; Lanza, M. Engineering Field Effect Transistors with 2D Semiconducting Channels: Status and Prospects. *Adv. Funct. Mater.* **2020**, *30*, 1901971. [CrossRef]
33. Nam, H.; Oh, B.-R.; Chen, M.; Wi, S.; Li, D.; Kurabayashi, K.; Liang, X. Fabrication and comparison of MoS 2 and WSe 2 field-effect transistor biosensors. *J. Vac. Sci. Technol. B Nanotechnol. Microelectron. Mater. Process. Meas. Phenom.* **2015**, *33*, 06FG01. [CrossRef]
34. Sarkar, D.; Liu, W.; Xie, X.; Anselmo, A.C.; Mitragotri, S.; Banerjee, K. MoS$_2$ field-effect transistor for next-generation label-free biosensors. *ACS Nano* **2014**, *8*, 3992–4003. [CrossRef] [PubMed]
35. Zheng, C.; Huang, L.; Zhang, H.; Sun, Z.; Zhang, Z.; Zhang, G.J. Fabrication of Ultrasensitive Field-Effect Transistor DNA Biosensors by a Directional Transfer Technique Based on CVD-Grown Graphene. *ACS Appl. Mater. Interfaces* **2015**, *7*, 16953–16959. [CrossRef] [PubMed]
36. Cheng, S.; Hideshima, S.; Kuroiwa, S.; Nakanishi, T.; Osaka, T. Label-free detection of tumor markers using field effect transistor (FET)-based biosensors for lung cancer diagnosis. *Sens. Actuators B Chem.* **2015**, *212*, 329–334. [CrossRef]
37. Syu, Y.-C.; Hsu, W.-E.; Lin, C.-T. Review—Field-Effect Transistor Biosensing: Devices and Clinical Applications. *ECS J. Solid State Sci. Technol.* **2018**, *7*, Q3196–Q3207. [CrossRef]
38. Kaisti, M. Detection principles of biological and chemical FET sensors. *Biosens. Bioelectron.* **2017**, *98*, 437–448. [CrossRef] [PubMed]
39. Majd, S.M.; Salimi, A.; Ghasemi, F. An ultrasensitive detection of miRNA-155 in breast cancer via direct hybridization assay using two-dimensional molybdenum disulfide field-effect transistor biosensor. *Biosens. Bioelectron.* **2018**, *105*, 6–13. [CrossRef]
40. Lee, H.W.; Kang, D.-H.; Cho, J.H.; Lee, S.; Jun, D.-H.; Park, J.-H. Highly Sensitive and Reusable Membraneless Field-Effect Transistor (FET)-Type Tungsten Diselenide (WSe$_2$) Biosensors. *ACS Appl. Mater. Interfaces* **2018**, *10*, 17639–17645. [CrossRef]
41. Chen, X.; Hao, S.; Zong, B.; Liu, C.; Mao, S. Ultraselective antibiotic sensing with complementary strand DNA assisted aptamer/MoS2 field-effect transistors. *Biosens. Bioelectron.* **2019**, *145*, 111711. [CrossRef]
42. Chen, Y.; Ren, R.; Pu, H.; Chang, J.; Mao, S.; Chen, J. Field-effect transistor biosensors with two-dimensional black phosphorus nanosheets. *Biosens. Bioelectron.* **2017**, *89*, 505–510. [CrossRef]
43. Kim, J.; Sando, S.; Cui, T. Biosensor Based on Layer by Layer Deposited Phosphorene Nanoparticles for Liver Cancer Detection. In Proceedings of the ASME 2017 International Mechanical Engineering Congress and Exposition, Tampa, FL, USA, 3–9 November 2017; p. V002T02A072.
44. Song, M.K.; Namgung, S.D.; Sung, T.; Cho, A.J.; Lee, J.; Ju, M.; Nam, K.T.; Lee, Y.S.; Kwon, J.Y. Physically Transient Field-Effect Transistors Based on Black Phosphorus. *ACS Appl. Mater. Interfaces* **2018**, *10*, 42630–42636. [CrossRef] [PubMed]
45. Balendhran, S.; Walia, S.; Alsaif, M.; Nguyen, E.P.; Ou, J.Z.; Zhuiykov, S.; Sriram, S.; Bhaskaran, M.; Kalantar-zadeh, K. Field Effect Biosensing Platform Based on 2D α-MoO3. *ACS Nano* **2013**, *7*, 9753–9760. [CrossRef] [PubMed]
46. Chen, H.; Rim, Y.S.; Wang, I.C.; Li, C.; Zhu, B.; Sun, M.; Goorsky, M.S.; He, X.; Yang, Y. Quasi-Two-Dimensional Metal Oxide Semiconductors Based Ultrasensitive Potentiometric Biosensors. *ACS Nano* **2017**, *11*, 4710–4718. [CrossRef] [PubMed]
47. Rim, Y.S.; Bae, S.H.; Chen, H.; Yang, J.L.; Kim, J.; Andrews, A.M.; Weiss, P.S.; Yang, Y.; Tseng, H.R. Printable Ultrathin Metal Oxide Semiconductor-Based Conformal Biosensors. *ACS Nano* **2015**, *9*, 12174–12181. [CrossRef] [PubMed]

48. Zhang, Y.J.; Yoshida, M.; Suzuki, R.; Iwasa, Y. 2D crystals of transition metal dichalcogenide and their iontronic functionalities. *2D Mater.* **2015**, *2*, 44004. [CrossRef]
49. Zong, X.; Yan, H.; Wu, G.; Ma, G.; Wen, F.; Wang, L.; Li, C. Enhancement of photocatalytic H2 evolution on CdS by loading MoS2 as cocatalyst under visible light irradiation. *J. Am. Chem. Soc.* **2008**, *130*, 7176–7177. [CrossRef]
50. Wilson, J.A.; Yoffe, A.D. The transition metal dichalcogenides discussion and interpretation of the observed optical, electrical and structural properties. *Adv. Phys.* **1969**, *18*, 193–335. [CrossRef]
51. Lin, Y.-C.; Dumcenco, D.O.; Huang, Y.-S.; Suenaga, K. Atomic mechanism of the semiconducting-to-metallic phase transition in single-layered MoS 2. *Nat. Nanotechnol.* **2014**, *9*, 391. [CrossRef]
52. Kappera, R.; Voiry, D.; Yalcin, S.E.; Branch, B.; Gupta, G.; Mohite, A.D.; Chhowalla, M. Phase-engineered low-resistance contacts for ultrathin MoS 2 transistors. *Nat. Mater.* **2014**, *13*, 1128. [CrossRef]
53. Han, G.H.; Duong, D.L.; Keum, D.H.; Yun, S.J.; Lee, Y.H. Van der Waals metallic transition metal dichalcogenides. *Chem. Rev.* **2018**, *118*, 6297–6336. [CrossRef]
54. Zhang, H. Ultrathin two-dimensional nanomaterials. *ACS Nano* **2015**, *9*, 9451–9469. [CrossRef] [PubMed]
55. Coleman, J.N.; Lotya, M.; O'Neill, A.; Bergin, S.D.; King, P.J.; Khan, U.; Young, K.; Gaucher, A.; De, S.; Smith, R.J. Two-dimensional nanosheets produced by liquid exfoliation of layered materials. *Science* **2011**, *331*, 568–571. [CrossRef]
56. Lee, C.; Li, Q.; Kalb, W.; Liu, X.-Z.; Berger, H.; Carpick, R.W.; Hone, J. Frictional characteristics of atomically thin sheets. *Science* **2010**, *328*, 76–80. [CrossRef] [PubMed]
57. Jang, J.; Jeong, S.; Seo, J.; Kim, M.-C.; Sim, E.; Oh, Y.; Nam, S.; Park, B.; Cheon, J. Ultrathin zirconium disulfide nanodiscs. *J. Am. Chem. Soc.* **2011**, *133*, 7636–7639. [CrossRef] [PubMed]
58. Jeong, S.; Yoo, D.; Jang, J.; Kim, M.; Cheon, J. Well-defined colloidal 2-D layered transition-metal chalcogenide nanocrystals via generalized synthetic protocols. *J. Am. Chem. Soc.* **2012**, *134*, 18233–18236. [CrossRef]
59. Zheng, B.; Chen, Y. Controllable growth of monolayer MoS_2 and $MoSe_2$ crystals using three-temperature-zone furnace. In Proceedings of the IOP Conference Series: Materials Science and Engineering, Changsha, China, 28–29 October 2017; Volume 274, p. 12085.
60. You, J.; Hossain, M.D.; Luo, Z. Synthesis of 2D transition metal dichalcogenides by chemical vapor deposition with controlled layer number and morphology. *Nano Converg.* **2018**, *5*, 26. [CrossRef] [PubMed]
61. Zhang, Y.; Yao, Y.; Sendeku, M.G.; Yin, L.; Zhan, X.; Wang, F.; Wang, Z.; He, J. Recent progress in CVD growth of 2D transition metal dichalcogenides and related heterostructures. *Adv. Mater.* **2019**, *31*, 1901694. [CrossRef]
62. Zhao, S.; Hotta, T.; Koretsune, T.; Watanabe, K.; Taniguchi, T.; Sugawara, K.; Takahashi, T.; Shinohara, H.; Kitaura, R. Two-dimensional metallic NbS2: Growth, optical identification and transport properties. *2D Mater.* **2016**, *3*, 25027. [CrossRef]
63. Wang, C.; Wu, X.; Ma, Y.; Mu, G.; Li, Y.; Luo, C.; Xu, H.; Zhang, Y.; Yang, J.; Tang, X. Metallic few-layered VSe 2 nanosheets: High two-dimensional conductivity for flexible in-plane solid-state supercapacitors. *J. Mater. Chem. A* **2018**, *6*, 8299–8306. [CrossRef]
64. Li, P.; Wen, Y.; He, X.; Zhang, Q.; Xia, C.; Yu, Z.-M.; Yang, S.A.; Zhu, Z.; Alshareef, H.N.; Zhang, X.-X. Evidence for topological type-II Weyl semimetal WTe 2. *Nat. Commun.* **2017**, *8*, 1–8. [CrossRef]
65. Rasch, J.C.E.; Stemmler, T.; Müller, B.; Dudy, L.; Manzke, R. 1 T–$TiSe_2$: Semimetal or Semiconductor? *Phys. Rev. Lett.* **2008**, *101*, 237602. [CrossRef] [PubMed]
66. Ovchinnikov, D.; Allain, A.; Huang, Y.-S.; Dumcenco, D.; Kis, A. Electrical transport properties of single-layer WS2. *ACS Nano* **2014**, *8*, 8174–8181. [CrossRef] [PubMed]
67. Ganatra, R.; Zhang, Q. Few-layer MoS2: A promising layered semiconductor. *ACS Nano* **2014**, *8*, 4074–4099. [CrossRef] [PubMed]
68. Cong, C.; Shang, J.; Wang, Y.; Yu, T. Optical properties of 2D semiconductor WS2. *Adv. Opt. Mater.* **2018**, *6*, 1700767. [CrossRef]
69. Kanazawa, T.; Amemiya, T.; Ishikawa, A.; Upadhyaya, V.; Tsuruta, K.; Tanaka, T.; Miyamoto, Y. Few-layer HfS 2 transistors. *Sci. Rep.* **2016**, *6*, 22277. [CrossRef]
70. Radisavljevic, B.; Whitwick, M.B.; Kis, A. Integrated circuits and logic operations based on single-layer MoS2. *ACS Nano* **2011**, *5*, 9934–9938. [CrossRef]
71. Pu, J.; Yomogida, Y.; Liu, K.-K.; Li, L.-J.; Iwasa, Y.; Takenobu, T. Highly flexible MoS2 thin-film transistors with ion gel dielectrics. *Nano Lett.* **2012**, *12*, 4013–4017. [CrossRef]

72. Tongay, S.; Sahin, H.; Ko, C.; Luce, A.; Fan, W.; Liu, K.; Zhou, J.; Huang, Y.-S.; Ho, C.-H.; Yan, J. Monolayer behaviour in bulk ReS 2 due to electronic and vibrational decoupling. *Nat. Commun.* **2014**, *5*, 3252. [CrossRef]
73. Del Pozo-Zamudio, O.; Schwarz, S.; Sich, M.; Akimov, I.A.; Bayer, M.; Schofield, R.C.; Chekhovich, E.A.; Robinson, B.J.; Kay, N.D.; Kolosov, O. V Photoluminescence of two-dimensional GaTe and GaSe films. *2D Mater.* **2015**, *2*, 35010. [CrossRef]
74. Li, Z.; Ezhilarasu, G.; Chatzakis, I.; Dhall, R.; Chen, C.-C.; Cronin, S.B. Indirect band gap emission by hot electron injection in metal/MoS2 and metal/WSe2 heterojunctions. *Nano Lett.* **2015**, *15*, 3977–3982. [CrossRef]
75. Das, S.; Chen, H.-Y.; Penumatcha, A.V.; Appenzeller, J. High performance multilayer MoS2 transistors with scandium contacts. *Nano Lett.* **2012**, *13*, 100–105. [CrossRef] [PubMed]
76. Lee, G.-H.; Cui, X.; Kim, Y.D.; Arefe, G.; Zhang, X.; Lee, C.-H.; Ye, F.; Watanabe, K.; Taniguchi, T.; Kim, P. Highly stable, dual-gated MoS2 transistors encapsulated by hexagonal boron nitride with gate-controllable contact, resistance, and threshold voltage. *ACS Nano* **2015**, *9*, 7019–7026. [CrossRef] [PubMed]
77. Wang, Q.H.; Kalantar-Zadeh, K.; Kis, A.; Coleman, J.N.; Strano, M.S. Electronics and optoelectronics of two-dimensional transition metal dichalcogenides. *Nat. Nanotechnol.* **2012**, *7*, 699. [CrossRef]
78. Fuhrer, M.S.; Hone, J. Measurement of mobility in dual-gated MoS 2 transistors. *Nat. Nanotechnol.* **2013**, *8*, 146. [CrossRef] [PubMed]
79. Choi, W.; Choudhary, N.; Han, G.H.; Park, J.; Akinwande, D.; Lee, Y.H. Recent development of two-dimensional transition metal dichalcogenides and their applications. *Mater. Today* **2017**, *20*, 116–130. [CrossRef]
80. Kim, S.; Konar, A.; Hwang, W.-S.; Lee, J.H.; Lee, J.; Yang, J.; Jung, C.; Kim, H.; Yoo, J.-B.; Choi, J.-Y. High-mobility and low-power thin-film transistors based on multilayer MoS 2 crystals. *Nat. Commun.* **2012**, *3*, 1–7. [CrossRef]
81. Desai, S.B.; Madhvapathy, S.R.; Sachid, A.B.; Llinas, J.P.; Wang, Q.; Ahn, G.H.; Pitner, G.; Kim, M.J.; Bokor, J.; Hu, C. MoS$_2$ transistors with 1-nanometer gate lengths. *Science* **2016**, *354*, 99–102. [CrossRef]
82. Zhu, J.; Xu, H.; Zou, G.; Zhang, W.; Chai, R.; Choi, J.; Wu, J.; Liu, H.; Shen, G.; Fan, H. MoS$_2$–OH bilayer-mediated growth of inch-sized monolayer MoS2 on arbitrary substrates. *J. Am. Chem. Soc.* **2019**, *141*, 5392–5401. [CrossRef]
83. Mei, J.; Li, Y.-T.; Zhang, H.; Xiao, M.-M.; Ning, Y.; Zhang, Z.-Y.; Zhang, G.-J. Molybdenum disulfide field-effect transistor biosensor for ultrasensitive detection of DNA by employing morpholino as probe. *Biosens. Bioelectron.* **2018**, *110*, 71–77. [CrossRef]
84. Lee, D.-W.; Lee, J.; Sohn, I.Y.; Kim, B.-Y.; Son, Y.M.; Bark, H.; Jung, J.; Choi, M.; Kim, T.H.; Lee, C. Field-effect transistor with a chemically synthesized MoS 2 sensing channel for label-free and highly sensitive electrical detection of DNA hybridization. *Nano Res.* **2015**, *8*, 2340–2350. [CrossRef]
85. Loan, P.T.K.; Zhang, W.; Lin, C.; Wei, K.; Li, L.; Chen, C. Graphene/MoS2 heterostructures for ultrasensitive detection of DNA hybridisation. *Adv. Mater.* **2014**, *26*, 4838–4844. [CrossRef] [PubMed]
86. Shan, J.; Li, J.; Chu, X.; Xu, M.; Jin, F.; Wang, X.; Ma, L.; Fang, X.; Wei, Z.; Wang, X. High sensitivity glucose detection at extremely low concentrations using a MoS 2-based field-effect transistor. *RSC Adv.* **2018**, *8*, 7942–7948. [CrossRef]
87. Dolui, K.; Rungger, I.; Sanvito, S. Origin of the n-type and p-type conductivity of MoS 2 monolayers on a SiO 2 substrate. *Phys. Rev. B* **2013**, *87*, 165402. [CrossRef]
88. Carvalho, A.; Wang, M.; Zhu, X.; Rodin, A.S.; Su, H.; Castro Neto, A.H. Phosphorene: From theory to applications. *Nat. Rev. Mater.* **2016**, *1*, 1–16. [CrossRef]
89. Li, L.; Yu, Y.; Ye, G.J.; Ge, Q.; Ou, X.; Wu, H.; Feng, D.; Chen, X.H.; Zhang, Y. Black phosphorus field-effect transistors. *Nat. Nanotechnol.* **2014**, *9*, 372–377. [CrossRef]
90. Chen, P.; Li, N.; Chen, X.; Ong, W.-J.; Zhao, X. The rising star of 2D black phosphorus beyond graphene: Synthesis, properties and electronic applications. *2D Mater.* **2017**, *5*, 014002. [CrossRef]
91. Akhtar, M.; Anderson, G.; Zhao, R.; Alruqi, A.; Mroczkowska, J.E.; Sumanasekera, G.; Jasinski, J.B. Recent advances in synthesis, properties, and applications of phosphorene. *Npj 2D Mater. Appl.* **2017**, *1*, 5. [CrossRef]
92. Guan, L.; Xing, B.; Niu, X.; Wang, D.; Yu, Y.; Zhang, S.; Yan, X.; Wang, Y.; Sha, J. Metal-assisted exfoliation of few-layer black phosphorus with high yield. *Chem. Commun.* **2018**, *54*, 595–598. [CrossRef]
93. Guo, Z.; Zhang, H.; Lu, S.; Wang, Z.; Tang, S.; Shao, J.; Sun, Z.; Xie, H.; Wang, H.; Yu, X.-F.; et al. From Black Phosphorus to Phosphorene: Basic Solvent Exfoliation, Evolution of Raman Scattering, and Applications to Ultrafast Photonics. *Adv. Funct. Mater.* **2015**, *25*, 6996–7002. [CrossRef]

94. Kitada, S.; Shimizu, N.; Hossain, M.Z. Safe and Fast Synthesis of Black Phosphorus and Its Purification. *ACS Omega* **2020**, *5*, 11389–11393. [CrossRef]
95. Ge, X.; Xia, Z.; Guo, S. Recent Advances on Black Phosphorus for Biomedicine and Biosensing. *Adv. Funct. Mater.* **2019**, *29*, 1900318. [CrossRef]
96. Favron, A.; Gaufrès, E.; Fossard, F.; Phaneuf-L'Heureux, A.-L.; Tang, N.Y.W.; Lévesque, P.L.; Loiseau, A.; Leonelli, R.; Francoeur, S.; Martel, R. Photooxidation and quantum confinement effects in exfoliated black phosphorus. *Nat. Mater.* **2015**, *14*, 826–832. [CrossRef] [PubMed]
97. Lu, W.; Nan, H.; Hong, J.; Chen, Y.; Zhu, C.; Liang, Z.; Ma, X.; Ni, Z.; Jin, C.; Zhang, Z. Plasma-assisted fabrication of monolayer phosphorene and its Raman characterization. *Nano Res.* **2014**, *7*, 853–859. [CrossRef]
98. Sresht, V.; Pádua, A.A.H.; Blankschtein, D. Liquid-Phase Exfoliation of Phosphorene: Design Rules from Molecular Dynamics Simulations. *ACS Nano* **2015**, *9*, 8255–8268. [CrossRef]
99. Dhanabalan, S.C.; Ponraj, J.S.; Guo, Z.; Li, S.; Bao, Q.; Zhang, H. Emerging Trends in Phosphorene Fabrication towards Next Generation Devices. *Adv. Sci.* **2017**, *4*, 1600305. [CrossRef]
100. Xiao, H.; Zhao, M.; Zhang, J.; Ma, X.; Zhang, J.; Hu, T.; Tang, T.; Jia, J.; Wu, H. Electrochemical cathode exfoliation of bulky black phosphorus into few-layer phosphorene nanosheets. *Electrochem. Commun.* **2018**, *89*, 10–13. [CrossRef]
101. Wang, M.; Liang, Y.; Liu, Y.; Ren, G.; Zhang, Z.; Wu, S.; Shen, J. Ultrasmall black phosphorus quantum dots: Synthesis, characterization, and application in cancer treatment. *Analyst* **2018**, *143*, 5822–5833. [CrossRef]
102. Meng, Z.; Stolz, R.M.; Mendecki, L.; Mirica, K.A. Electrically-transduced chemical sensors based on two-dimensional nanomaterials. *Chem. Rev.* **2019**, *119*, 478–598. [CrossRef]
103. Șerban, I.; Enesca, A. Metal Oxides-Based Semiconductors for Biosensors Applications. *Front. Chem.* **2020**, *8*, 354. [CrossRef]
104. Rajendran, S.; Manoj, D.; Raju, K.; Dionysiou, D.D.; Naushad, M.; Gracia, F.; Cornejo, L.; Gracia-Pinilla, M.A.; Ahamad, T. Influence of mesoporous defect induced mixed-valent NiO (Ni2+/Ni3+)-TiO2 nanocomposite for non-enzymatic glucose biosensors. *Sens. Actuators B Chem.* **2018**, *264*, 27–37. [CrossRef]
105. Liu, P.; Huo, X.; Tang, Y.; Xu, J.; Liu, X.; Wong, D.K.Y. A TiO2 nanosheet-g-C3N4 composite photoelectrochemical enzyme biosensor excitable by visible irradiation. *Anal. Chim. Acta* **2017**, *984*, 86–95. [CrossRef]
106. Tripathy, N.; Kim, D.H. Metal oxide modified ZnO nanomaterials for biosensor applications. *Nano Converg.* **2018**, *5*, 1–10. [CrossRef] [PubMed]
107. Soneja, S.; Dwivedi, P.; Dhanekar, S.; Das, S.; Kumar, V. Temperature Dependent Electrical Characteristics of Nanostructured WO3 Based Ambipolar Bottom Gate FET. *IEEE Trans. Nanotechnol.* **2018**, *17*, 1288–1294. [CrossRef]
108. Yuliarto, B.; Gumilar, G.; Septiani, N.L.W. SnO2 nanostructure as pollutant gas sensors: Synthesis, sensing performances, and mechanism. *Adv. Mater. Sci. Eng.* **2015**, *2015*, 1–14. [CrossRef]
109. Rim, Y.S.; Chen, H.; Zhu, B.; Bae, S.H.; Zhu, S.; Li, P.J.; Wang, I.C.; Yang, Y. Interface Engineering of Metal Oxide Semiconductors for Biosensing Applications. *Adv. Mater. Interfaces* **2017**, *4*, 1700020. [CrossRef]
110. Rim, Y.S. Review of metal oxide semiconductors-based thin-film transistors for point-of-care sensor applications. *J. Inf. Disp.* **2020**. [CrossRef]
111. Dral, A.P.; Johan, E. 2D metal oxide nanoflakes for sensing applications: Review and perspective. *Sens. Actuators B Chem.* **2018**, *272*, 369–392. [CrossRef]
112. Alsaif, M.M.Y.A.; Field, M.R.; Daeneke, T.; Chrimes, A.F.; Zhang, W.; Carey, B.J.; Berean, K.J.; Walia, S.; Van Embden, J.; Zhang, B.; et al. Exfoliation solvent dependent plasmon resonances in two-dimensional sub-stoichiometric molybdenum oxide nanoflakes. *ACS Appl. Mater. Interfaces* **2016**, *8*, 3482–3493. [CrossRef]
113. Woong-Ki, H.; Jongwon, Y.; Takhee, L. Hydrogen plasma-mediated modification of the electrical transport properties of ZnO nanowire field effect transistors. *Nanotechnology* **2015**, *26*, 125202.
114. Su, M.; Yang, Z.; Liao, L.; Zou, X.; Ho, J.C.; Wang, J.; Wang, J.; Hu, W.; Xiao, X.; Jiang, C.; et al. Side-Gated In2O3 Nanowire Ferroelectric FETs for High-Performance Nonvolatile Memory Applications. *Adv. Sci.* **2016**, *3*, 1–7. [CrossRef]
115. Nakatsuka, N.; Yang, K.A.; Abendroth, J.M.; Cheung, K.M.; Xu, X.; Yang, H.; Zhao, C.; Zhu, B.; Rim, Y.S.; Yang, Y.; et al. Aptamer-field-effect transistors overcome Debye length limitations for small-molecule sensing. *Science* **2018**, *362*, 319–324. [CrossRef] [PubMed]

116. Wen, W.; Song, Y.; Yan, X.; Zhu, C.; Du, D.; Wang, S.; Asiri, A.M.; Lin, Y. Recent advances in emerging 2D nanomaterials for biosensing and bioimaging applications. *Mater. Today* **2018**, *21*, 164–177. [CrossRef]
117. Kim, K.K.; Lee, H.S.; Lee, Y.H. Synthesis of hexagonal boron nitride heterostructures for 2D van der Waals electronics. *Chem. Soc. Rev.* **2018**, *47*, 6342–6369. [CrossRef] [PubMed]
118. Laturia, A.; Van de Put, M.L.; Vandenberghe, W.G. Dielectric properties of hexagonal boron nitride and transition metal dichalcogenides: From monolayer to bulk. *Npj 2D Mater. Appl.* **2018**, *2*, 1–7. [CrossRef]
119. Sediri, H.; Pierucci, D.; Hajlaoui, M.; Henck, H.; Patriarche, G.; Dappe, Y.J.; Yuan, S.; Toury, B.; Belkhou, R.; Silly, M.G.; et al. Atomically Sharp Interface in an h-BN-epitaxial graphene van der Waals Heterostructure. *Sci. Rep.* **2015**, *5*, 1–10. [CrossRef]
120. Dean, C.R.; Young, A.F.; Meric, I.; Lee, C.; Wang, L.; Sorgenfrei, S.; Watanabe, K.; Taniguchi, T.; Kim, P.; Shepard, K.L.; et al. Boron nitride substrates for high-quality graphene electronics. *Nat. Nanotechnol.* **2010**, *5*, 722–726. [CrossRef]
121. Joo, M.-K.; Moon, B.H.; Ji, H.; Han, G.H.; Kim, H.; Lee, G.; Lim, S.C.; Suh, D.; Lee, Y.H. Electron Excess Doping and Effective Schottky Barrier Reduction on the MoS 2/h -BN Heterostructure. *Nano Lett.* **2016**, *16*, 6383–6389. [CrossRef]
122. Saito, A.; Ayano, T.; Nomura, S. Photoresponse in h-BN/MoS2/h-BN thin-film transistor. *Jpn. J. Appl. Phys.* **2018**, *57*, 045201. [CrossRef]

© 2020 by the authors. Licensee MDPI, Basel, Switzerland. This article is an open access article distributed under the terms and conditions of the Creative Commons Attribution (CC BY) license (http://creativecommons.org/licenses/by/4.0/).

MDPI
St. Alban-Anlage 66
4052 Basel
Switzerland
Tel. +41 61 683 77 34
Fax +41 61 302 89 18
www.mdpi.com

Sensors Editorial Office
E-mail: sensors@mdpi.com
www.mdpi.com/journal/sensors

www.ingramcontent.com/pod-product-compliance
Lightning Source LLC
LaVergne TN
LVHW070128100526
838202LV00016B/2246